International Telemetering Conference 2017

Las Vegas, Nevada, USA
23 - 26 October 2017

ISBN: 978-1-5108-5318-8

Printed from e-media with permission by:

Curran Associates, Inc.
57 Morehouse Lane
Red Hook, NY 12571

Some format issues inherent in the e-media version may also appear in this print version.

Printed by Curran Associates, Inc. (2017)

For permission requests, please contact International Telemetering Conference at the address below.

International Telemetering Conference
c/o John Jensen
5825 North Jornada Road
Las Cruces NM 88012-7247

telemetry@comcast.net
www.telemetry.org

Additional copies of this publication are available from:

Curran Associates, Inc.
57 Morehouse Lane
Red Hook, NY 12571 USA
Phone: 845-758-0400
Fax: 845-758-2633
Email: curran@proceedings.com
Web: www.proceedings.com

TABLE OF CONTENTS

SESSION 1: ANTENNAS AND LINK MARGINS

SESSION 2: RANGE INFRASTRUCTURE I

SESSION 3: SPACE SYSTEMS

SESSION 4: MODULATION AND CODING I

SESSION 5: NETWORKING AND RECORDERS I

SESSION 6: SENSORS AND DATA ACQUISITION I

SESSION 7: RF SYSTEMS I

SESSION 8: SYSTEM TOOLS AND TECHNIQUES I

SESSION 9: SENSORS AND DATA ACQUISITION II

SESSION 10: RF SPECTRUM

SESSION 11: NETWORKING AND RECORDERS II

SESSION 12: RF SYSTEMS II

SESSION 13: RANGE NETWORKING CONCEPTS

SESSION 14: SYSTEM IMPLEMENTATIONS

SESSION 15: MODULATION AND CODING II

SESSION 16: RANGE INFRASTRUCTURE II

SESSION 17: SYSTEM TOOLS AND TECHNIQUES II

INTERNATIONAL CONSORTIUM
FOR TELEMETRY SPECTRUM

2017 ICTS Annual Report

Submitted by the ICTS Officers:

Mr. Scott A. Hoschar (USA), Chairman
Mr. Guy Williams (USA), Vice-Chairman
Mr. Tim Chalfant (USA), Secretary

Regional Coordinators:

Dr. Gerhard Mayer (Germany), Region I Coordinator
Mr. Scott A. Hoschar (USA), Region II Coordinator
Mr. Tim Chalfant, (USA), Acting Region III Coordinator

ICTS Meetings & Sessions

The ICTS held Business Meetings and General Sessions in November 2017 (Glendale, Arizona, USA) and in June 2017 (Toulouse, France). In Arizona, as part of the 2016 International Telemetering Conference (ITC 2016), ICTS Chairman Mr. Mikel Ryan prepared and conducted the ICTS General Session. After his welcome and introduction the *"Regional Reports of International Telecommunication Union (ITU)-Region 1-Europe/Africa* (Mr. E. Prescott), *ITU-Region 2/The Americas* (Mr. M. Ryan) *and ITU-Region 3-Asia/Pacific* (Mr. T. Chalfant)*"* were presented, followed by briefings on "*A Concise History of the ICTS" (* M. Ryan), *"Encroachment Threats to Aeronautical Telemetry in the USA: Update #13*" (prepared by Mr. S. Hoschar and Mr. M. Ryan, both of the USA, and presented by M. Ryan (also of the USA), "*World Radiocommunication Conference 2015: The U.S.A Flight Test Perspective"(* K. Keane, USA).

Due to continuing USA Government budget cuts, and ICTS members retiring, only one of the USA's ICTS Members (G. Williams) was able to contribute at the June 2017 Toulouse, France Sessions. Still, the session was a success, with a distinguished list series of presentations (some prepared by USA ICTS Members) and an involved audience. ICTS Vice-Chairman Mr. Mikel Ryan assisted with the preparation, and ICTS member J-C Ghnassia (France) conducted the ICTS General Session as part of the European Telemetry & Test Conference (ETTC 2017). THANK YOU Jean Claude! After his welcome and introductions, the ITU Regional Reports of *"ITU Region 1-Europe/Africa"* (Dr. G. Mayer) and *"ITU Region 2-The Americas"* (Dr. G. Mayer) and *"ITU Region III – Asia/Pacific"* (Dr. G. Mayer) were delivered, followed by briefings on *"The Coded Orthogonal Frequency Division Multiplexing (COFDM) Roadmap at EMBRAER"* (Mr. S. Penna, Brazil), *"Encroachment Threats to Aeronautical Mobile Telemetry in the USA: Update #14"* (Mr. Guy W. Williams, USA, presented by J-C Ghnassia), and concluding with "*Action Item 1.16 (WRC-19): "Study to enable outdoor WAS/RLAN Operations", Threat to AMT-C-Band Operations and Necessary Protection Criteria"* (Mr. Luc Falga, France).

ICTS Membership

- Chairman/Region 2 Coordinator – Mr. Scott A. Hoschar (USA/Patuxent River MD) is the Head of the Middle Atlantic Area Frequency Coordination Office (MIDLANTAC) and the Installation Spectrum Manager of the Naval Air Warfare Center Aircraft Division (NAWCAD).
- Vice Chairman – Mr. Guy Williams (USA/Edwards AFB) Air Force Spectrum Relocation Fund (SRF) Program Manager, ENS, of the Air Force Test Center (AFTC).
- Secretary – Mr. Tim Chalfant (USA), COLSA Consulting to the Test Resource Management Center (TRMC).
- Region 1 Coordinator – Dr. Gerhard Mayer (Germany) of the University of Salzburg.
- Region 3 Coordinator – Acting - Mr. Tim Chalfant (USA), COLSA Consulting to the Test Resource Management Center (TRMC). (NOTE: position is vacant due to retirement of Mr. Darrell Ernst)
- Mr. Jean Claude Ghnassia (France) Head of Instrumentation for Airbus (retired).
- Mr. Steve Lyons (United Kingdom), QinetiQ Group.
- Mr. Luiz Fernando de Souza (Brazil) of Embraer Sociedade Anônima.
- Mr. Sergio Penna (Brazil) of Embraer Sociedade Anônima.
- Mr. Ken Keane (USA/Washington DC) of Duane-Morris LLP.

- Mr. Eric Prescott, Engineering Manager for Flight Test Instrumentation at BAE Systems Inc. (retired), currently a Telemetry Consultant for Nprime.
- Mr. Ray Faulstich (USA/Patuxent River MD) of Pacific Architects and Engineers Incorporated.

We have currently good but declining membership numbers from ITU Regions 1 and 2. ICTS members are getting the message out to the international community through the IFT, the International Test and Evaluation Association (ITEA), and the Society of Flight Test Engineers. Besides the before-mentioned biannual Sessions, ICTS Members regularly offer papers at related Working Groups, technical conferences, trade journals and newsletters that bring the ICTS and its mission forward.

'Blue Ribbon' Spectrum Panel at the ITEA 2017 Plenary Session

The ICTS Chairman (Mr. M. Ryan) coordinated and hosted a "**Telemetry Spectrum Encroachment**" Session at the 9-11 May 2017 21st ITEA Test Instrumentation Workshop in Las Vegas with the help of another ICTS member, Mr. Ken Keane who acted as panel moderator.

> *"Our telemetry electromagnetic assets are under continual encroachment pressure by powerful non-Government interests. This Workshop will provide concise histories, with both National and Field-Level perspectives, of this perpetual threat, and how we are regulating, defending and, indeed, augmenting our crucial Aeronautical Mobile Telemetry frequency bands via worthy organizations such as the ICTS, AFTRCC, S4G, RSRWG and NASCTN."*

To launch the ITEA Program at the Opening Plenary Session (begins Wednesday, 10 May 0800) a Blue Ribbon Panel (all members of the Senior Executive Service) held a two-hour session of individual "**Telemetry Spectrum Encroachment**" presentations and Q&A, moderated by Mr. Ken Keane (of Duane Morris LLP, legal counsel for the TRMC & the AFTRCC) and ICTS member:

DISTINGUISHED BLUE RIBBON PANEL MEMBERS

* Mr. Frederick D. Moorefield Jr. currently serves as the Director of Spectrum Policy & International Engagements for the DoD Chief Information Officer.
* Mr. Derrick Hinton currently serves as the Director of the TRMC.
* Mr. Paul Mann currently serves as the Deputy Director of the TRMC.
* Mr. Thomas C. Dowd currently serves as the Director, Range Dept. and is responsible for all of the Naval Air Systems Commands open air ranges and Air Vehicle Modification and Instrumentation facilities in the United States.
* Dr. Elisabetta L. "Betta" Jerome currently serves as the Technical Advisor, Armament and Weapons T&E, Eglin Air Force Base, Florida.

This Blue Ribbon Panel was followed by four more relevant presentations from 1:30 to 3:30PM that afternoon. In a related vein, Mr. Guy Williams (ICTS Vice-Chairman and AFTC, Spectrum Relocation Fund Program Manager) chaired an **"MRTFB Spectrum**

Relocation" Session:

> *With the President's Broadband Initiative to make available 500 MHz of spectrum to the commercial market, spectrum relocation and sharing has become a fact of life for our Major Range and Test Facility Bases (MRTFB). The most recent AWS-3 auction, which impacted the 1755-1780 MHz band for our AMT operations, resulted in $41.3B net revenue to the Treasury and provided $3.1B to the US Department of Defence (DoD). A portion of the DoD funds are being used to implement transition plans for relocation out of the 1755-1780 MHz AMT bands. This Workshop will provide an overview of what the Air Force, Navy, and Army MRTFBs have already started and will continue to do for the next 10 years in order to comply with the relocation transition dates, while becoming more efficient and effective in the remaining spectrum available to them.*

While the Air Force, Navy, and Army are forced to compress into upper L band and exploit more of C band for AMT, commercial entities such as the National Test Pilot School (NTPS) have implemented a C-Band airborne mobile telemetry datalink system in order to continue to fulfil their mission to educate and train military and civilian flight test personnel due to the limited availability of L-band spectrum

WRC-19 Threats to AMT and Proposed Response: The USA Perspective

For 2019 World Radiocommunications Conference (WRC-19), several potential 'AMT impact' topics are being proposed. Issues being monitored include a proposal for a new Recommendation/Report dealing with IMT emissions in the 2170-2200 MHz band, and implications for the DISH agreement (out-of-band-emissions concerns with proximity to the 2200-2290 MHz AMT band), additional allocations for Radio Local Area Networks in 5 GHz (possible negative effects for Recommendation ITU-R M. 1459), Non-Geostationary Satellite Orbit operations in 4/6 GHz, and BSS service in the 1452-1492 MHz band for ITU Regions 1 and 3.

The ICTS needs to maintain good communication in the international telemetering communities to protect IFT members who rely on these bands. The USA, Australia, New Zealand, Canada and the United Kingdom are attempting to establish a Flight Test Working Group (FTWG) and a Spectrum Working group under the MTEP/5i structure. This effort stalled in 2016 but will be discussed at the October 2017 ICTS business session. Mr. Chalfant is advising the Air Force Test Center's representative to the FTWG to cue up AMT Spectrum Encroachment. Mr. Chalfant prepared a briefing deck (with Mr. Keane's help) for a FTWG organizing meeting in September 2016 highlighting the ICTS, the WRC process and threats to AMT. At this time we have no further information on the SWG. The TRMC is the POC for Multinational Test and Evaluation Program (MTEP) involvement from the Ranges. 5i preparation will be critical to protect telemetering at the working group meeting in preparation for WRC-19. The ICTS needs to seek similar forums that include the other international telemetering organizations and corporations.

CEPT on the Way to WRC-19

The European Conference on Postal and Telecommunications Administrations (CEPT, with 48 European counties cooperating to regulate posts, radio spectrum and communications networks) and its activities are the main source of information to provide an independent assessment of ITU-Region 1 issues & positions that could impact AMT capabilities, in preparation of the WRC-19. Further sources (from meetings & reports) are monitored from the African Telecommunication Union (ATU) and the Arab Spectrum Management Group (ASMG).

ECC PT1, the Project Team 1 of the European Communications Committee (ECC) takes care of all "IMT Matters". It is responsible for implementing the *"Wireless Access Policy for Electronic Communication Services"* (WAPECS) concept (the new European flexible approach based on technology and service neutral regulation) for the MFCN and was tasked to initiate also studies for the introduction of MFCN in the L-Bands to investigate the suitability of the current ECC framework for future usage of LTE-based M2M communication.

As a follow up to the decision taken at WRC-15 where the frequency bands 1427-1452 and 1492-1518 MHz were identified for IMT, the ECC agreed to initiate studies for the introduction of mobile/fixed communication networks (MFCN) in these frequency bands and for assessing how to maintain flexibility. The objective was to develop relevant ECC harmonization measures in due time. 1427-1452 MHz is not considered to be allocated for IMT, within the CEPT.

Due to the enormous demand of spectrum for IMT in various applications the whole range of frequency spectrum above 6 GHz up to the 60 GHz limit is at discussion for studies, on a licensed (primary of co primary) basis and under the terms "Licensed Shared Access" and "Licensed Assisted Access" in the AMT S-and C-band range, as well.

ECC PT D is the second important group under ICTS observation, with a strong impact by the WRC-15 Action Item 1.16 *related to the WAS/RLAN in the frequency band 5150 – 5250 MHz).*

Allocations for AMT in 'Europe

The residual **AMT L-Band** operations in Europe remain in existence, protected by the RR Article 5 Footnotes, in

Russian Federation and allied countries	1429 – 1535 MHz
Switzerland	1435 – 1445 MHz
United Kingdom	1427 – 1452 MHz

All other European countries have completely changed their AMT operations to the **S- and C-Bands**

S - band Core	2300 – 2330 MHz
S - band Extension	2330 – 2400 MHz
C – band Global	5091 - 5150 MHz
C- band Region 1	5150 – 5250 MHz

C – band is already **used** in Austria, France, Spain, Germany, The Netherlands, Norway, Sweden, Switzerland. UK is in planning status for 2018 onward.

Threats to AMT (European View)

S - Band 2300 – 2400 MHz

The **Licensed Shared Access** (LSA) is seen as enabler to release additional spectrum for Mobile Broadband Services, sharing with incumbents on a secondary basis assessing protection of existing services. The Project Teams FM 53 worked on the regulatory conditions and FM 52 on the implementation of LSA for the shared use of the band 2.3–2.4 GHz on a secondary basis (CEPT Report 56).

Extensive LSA testing was done in Finland, France, Italy, The Netherlands and Spain (with their industrial partners), supported by the Joint Research Centre of the European Commission. The objective was to share experience on LSA implementation and testing (e.g., measurement results) between the national spectrum regulating agencies.

With that the CEPT has finalized its LSA- framework. The LSA implementation may now be considered by the national administrations.
Of special interest for the European AMT community will be the pending report of ITU(R) Working Party 5A on the "*studies of the necessary mitigation techniques*".

C-Band 5150 – 5250 MHz

Licensed Assisted Access (LAA) idea is that LTE cells operating in other (licensed) bands synchronize cells in the in the (unlicensed) part of C-band, 5150 – 5250 – (5350) MHz, presently allocable to indoor WLAN operations on a power level of +23 dBm, only !
With the LTE Advanced Pro Standard in band 5150 – 5250 MHz 5 channels with 20 MHz bandwidth each would be implemented, with a power level of +36 dBm and a power flux density of 17 dBm/MHz.

WRC-19 Action Item 1.16 *"to consider issues related to the WAS/RLAN in the frequency band 5150 – 5250 MHz ... in accordance with COM 6/22(WRC-15)"* does address studies in the frequency band 5150 – 5925 MHz related to Wireless Access Systems (WAS) inclusive Radio Local Area Networks (RLAN) applications with multimedia capability. The above resolution invites commercial users *"to perform sharing and compatibility studies between WAS/RLAN applications and incumbent services in the frequency band 5 150-5 350 MHz et.al. with the possibility of enabling outdoor WAS/RLAN operations including possible associated conditions."*
The supporters believe, that a fair sharing & coexistence with the incumbent primary services (e.g. AMT) is possible, by controlling their transmitted power and by a dynamic frequency selection (***"listen before talking")*** (!!!!).

Industry 4.0, Smart Manufacturing

Global Wireless Industrial Applications (WIA) in the process industries are still at the beginning of their growing phase (ref. CEPT M2M Workshop, March 2016):

WIA connected devices in year	2005	2012	2018 (predicted)
in millions	0.2	12	200

Their needs have been addressed in a CEPT Workshop (2 – 4 March, 2017 in Mainz / Germany) on 5G of Wireless Communications, setting up the roadmap for their spectrum requirements. It was stated that the 5G standard wireless systems cannot fulfil all needs, especially in **the process-oriented industries** (latency, longevity, reliability).

Existing wireless base technologies are not specifically designed for these industries. Distances up to 5 km have to be covered in a high-quality service level: the presently used unlicensed bands at 2.4 & 5.8 GHz are insufficient, and European productivity is suffering from the constraints. The requested band(s) in the range 1 – 6 GHz shall be specific for industrial applications with a bandwidth of (minimum) 80 MHz (or 2x40 MHz) and be globally available.

Candidate for studies are also **the 2340 – 2400 & 5150 – 5250 MHz bands.**
A lot of machine-to-machine communications is presently still done in the unlicensed 2,4 GHz band that is awfully crowded. An access to the spectrum (2340 -2400 MHz) would be the ideal option without a need to change (S-band) technology.

The lobby of the **"One M2M Partnership Project"** attempted to by-pass the regular preparation process for the WRC-19 and succeeded. **Their spectrum demand** is now covered by **WRC-19 AI 9.1.8:** *"Narrowband- and Broadband Machine Type Communication Infrastructures"* in the ITU(R) *"list of urgent requirements in preparation of the WRC-19, Resolution 958 [Com 6/1] (WRC-15).* ITU(R) Working Party 5D is invited to support and help to bring their issue to a decision level at the WRC-19.

The global **telemetry community is alerted by that rapidly-developing issue!** Beside tracking and (if needed) trying to influence the preparation process for the WRC-19 a clear strategy to determine what conditions of coexistence with those intruding services could be tolerable is urgently required. Even with plausible mitigation techniques we have to evaluate how it could work in practice. Interference and noise levels in band, out-of-band emissions and spurious domains that are generated by the signals of those WIA services have to be recorded and analysed. In short, depending as always on the bands at issue (like 2340-2400 MHz), the role of ICTS remains as important as ever.

The Potential for AMT in the Ku-, K- & Ka-Bands

A possible tasking for the ICTS involves investigating the feasibility of augmenting our current international AMT bands. We have no current spectrum allocations for AMT in the Ku-,K-and Ka-Band range (20–36 GHz), but we should actively, expeditiously determine whether we may need frequency spectrum in that range in the future, year 2020 and beyond.

Dr. Michael Rice's (USA, Brigham Young University) notable *"Aeronautical Telemetry Above 15 GHz: A Feasibility Study"* demonstrates that the market-available components and systems for the Ku-, K- and Ka-Bands are capable of accommodating some of our AMT needs. The biggest challenge to our community is the propagation issue, but methods to compensate for any loss of the link margin were identified in Dr. Rice's study. Also, the considerable regulatory challenge in securing an allocation in these bands is addressed by Dr. Rice: *"All of spectrum in these bands has been assigned by the USA's Federal Communication Commission (FCC) and the National Telecommunications and Information Administration (NTIA), and the incumbents would try to enthusiastically protect their assignments."*

In case of a need for additional AMT bands (to augment the currently allocated L-, S- and C-Band portions), the K- and Ka-Bands were already identified as promising candidates (15G-Study Group, in 2008).

The International Mobile Telecommunications (IMT) 5th generation mobile/wireless networks (5G) lobby intends to study their possible use of bands in the range 6-100 GHz, with related AIs proposed for the WRC-19. In particular, the METIS 2020 Research Project is being partially funded by the European Commission with an eye toward enabling its vision of the 5G future. Nearly one year after its November 2012 creation, METIS has laid out wireless scenarios that it hopes to address and specified the capabilities it expects of 5G networks. Note that portions of K-and Ka-Band are conspicuously targeted so the Telemetry Community needs to quickly decide if this is an applicable, profitable assignment for the ICTS. If so, we need to first decide:

> -Is there a need for additional spectrum for AMT, time horizon 2020 and beyond, or not?
> -If so, what kind of test missions needing point-to-point or networked telemetry, could live with the physical limits of the K- and Ka-Bands?
> - Apart from these two questions, whether it would be prudent to launch an initiative for additional spectrum when the telemetry community in the United States, at least, is struggling to secure domestic approval for access to bands allocated at WRC-07 ?

If the answers to these questions (especially the third) are satisfactory we should seek possible parts of those bands where a co-primary base allocation for AMT is possible, preferably in all three ITU Regions. A cursory examination of the ITU Radio Regulations (Edition 2012, Article 5) reveals that bands from 25-29.5 GHz and 31-31.3 GHz are listed for allocation to the Mobile Service (MS), on a non-interference base to other incumbent services.

Adding urgency to this issue is **WRC-19 Action Item 1.13** *"..to consider identification of frequency bands for future development of IMT... invites to conduct sharing and compatibility studies for the band 24,25 -27,5 GHz, 31,5 – 33,4 GHz et. al."*

The **Ka-Band is a favourite candidate for future AMT extension requirements within the time horizon 2020 to 2025**, a demand was already predicted by the "15G-Study Group" charted by the USA's Department of Defence's Test Resource Management Center (DOD/TRMC). Studies to investigate future AMT spectrum demand are under regional work now. The intentions of the IMT world to acquire additional spectrum in the K-and Ka band range should alert the telemetering community to quickly ponder about the requirements for additional spectrum, within a time horizon 2020 and beyond. At a **later time point** it will be **much more difficult** (or worst case with almost no chance) to **initiate that process**.

An Uncertain Future for the ICTS

As detailed in previous Annual ICTS Reports, the lack of financial support and erosion of the ICTS membership via retirement (see the **ICTS Membership** Listing in page 2) is crippling the ICTS. Although incessant international threats to our current AMT bands abound and WRC-19 coming up, due to a continuing prohibition on "unnecessary" international travel even national travel, our diminishing ranks cannot attend the national WRC preparation meetings, and we are losing the personal contacts and the ability to advise the national WRC representatives. A commendable exception is TRMC funding for Mr. Keane's and Dr. Dan Jablonski's WRC-involved activities. AMT spectrum is continually threatened with band reallocation and reductions in bandwidth, and we believe the mandate at a minimum, to defend our current electromagnetic assets means that there is as much a need for an organization with the objectives of the ICTS as there has ever been.

But how do we recover, presuming recovery is possible? Recent, candid suggestions by the ICTS membership include and discussions will continue at our fall 2017 business meeting:

1. Perhaps convert the ICTS to a standing IFT committee like the Telemetering Standards Coordination Committee (TSCC), with no real outreach efforts. The meetings would occur in a similar manner.
2. Perhaps eliminate the ICTS's Regional Coordinators, assuming there is a need or better way to collect information from around the globe.
3. Perhaps change the ICTS to a collection of National "CTSs" and work domestic issues with the active members.
4. Establish a Corporate Outreach effort (discussed below).

At the upcoming 2017 International Telemetering Conference in Las Vegas Nevada, the ICTS membership will meet, as feasible, to take stock of our dilemma.

Corporate Outreach

ICTS would like to continue its presence in the International Telecommunications Union and membership on the various working groups in the world radio-communication council. This presence is critical to the ICTS ability to advise and inform its membership and the IFT on threats to telemetry spectrum and the effect on the telemetering sciences. To continue this effort we can no longer rely on the individual ability of its members to secure separate support and travel funding. That support, previously provided by ICTS members individually or via their corporate ties, is waning.

In an effort to shore-up our financial support and membership erosion the ICTS is pursuing a stronger relationship with its corporate membership. Last fall the ICTS conducted a preliminary out-reach to corporations with significant investments in telemetry and telemetry products. The ICTS is planning to mail 80+ corporations from the 2017 ITC mailing list (60 US, 20 International) soliciting their connection to the ICTS and marketing our ability to inform them of telemetry market trends in spectrum threats, encroachment, and issues they may desire to work within their national administrations.

After the letters are mailed, the ICTS plans to follow up on those contacts and expand the outreach to additional corporate members with subscriptions to our Telemetry Spectrum

Newsletter. Decisions as to revising the letter, the distribution list, and follow-up contacts will be topics of discussion at out ICTS Meeting in October.

With an expanded corporate membership we plan to provide a periodic newsletter and news flashes to ICTS members on international spectrum issues that will provide significant value to ICTS members (and the IFT). Our goal is to be self-sufficient and not solely reliant on the IFT or individual member personal funds to support our out-reach, representation of the telemetering interests in the ITU, and news-gathering channels.

Conclusions

The main objective of the ICTS is to provide a forum for information exchange on telemetry spectrum issues, such as early warning on threats to our existing spectrum allocations. In addition this forum will be used to share information on the use and research and development of new technologies for improving spectral efficiency.

Ideally, the ICTS would continue to be the *"eyes and ears"* for the IFT/AMT community with regards to international spectrum threats and issues. The significant impact and success of our past work should be evident from this and past ICTS annual rreports and the ICTS general sessions of the ETTC and ITC -- consider the great success at WRC-07. However, our work is handicapped by a lack of travel funding and the loss of experienced, dedicated, articulate, internationally recognized and respected spokesmen. The "I" in ICTS, IFT & ITU stands for "International" so the Conferences, the negotiations, and the voting that will decide the fate of our AMT bands are conducted in necessarily widespread locales.

The importance of having on-call, immediately accessible AMT experts to amplify and clarify complex telemetry spectrum issues for delegates have been confirmed numerous times. It would be advantageous to the AMT Community to attend the relevant study group ITU Working Parties (WP) 4 & 5 for the preparation of WRC-19. These meetings are where the critical agenda points with potential to threaten AMT spectrum are under discussion. After all, WRC-07, WRC-12 and WRC-15 conclusively demonstrated that voting delegates to ITU Study Groups can be made sensitive to our encroachment concerns by ICTS Members.

HIGH-PRECISION MOTION ESTIMATION SYSTEMS FOR UUV NAVIGATION

Hua Lee and Vincent Radzicki
Department of Electrical and Computer Engineering
University of California, Santa Barbara

ABSTRACT

This paper is the summary of a sequence of research tasks in the area of 3D bearing-angle estimation for UUV homing and docking exercises. The main focus is to simplify the concept as well as computation efficiency of the homing and docking tasks, by elevating the estimation modality from the conventional twin-receiver configuration to the 2D circular arrays. The objective is to utilize the multi-element receiver array for the entire navigation procedure, including bearing-angle estimation, optimal path planning, and high-precision docking.

INTRODUCTION

The homing-docking procedures in UUV navigation involve three key components [1-4]. The first is the estimation of the bearing angle in the far field for the homing process. The second component is dynamic path planning for the optimal approach. As the UUV reaches the near field, the docking procedure is then activated, as the third and final task.

The early prototype for homing-docking operations was a twin-transmitter single-receiver system, designed for the estimation of the bearing angle in the far field. This new system extends the capability to full-scale multi-dimensional motion estimation for precision homing and docking.

The receiver arrays of the new system are in the form of multi-element circular arrays. This array configuration enables high-precision *3D* bearing-angle estimation, which leads to the accuracy improvement of dynamic path planning.

The mathematical analysis in the paper is formulated based on the concept of single-frequency coherent waveform for simplicity. It should be point out that the accuracy of the estimation process improves significantly with waveforms with increased frequency bandwidth.

In addition, to maintain the focus of the paper, the configuration of the systems is limited to the single-transmitter mode. Extension to multiple-element transmitter arrays for improved accuracy can be achieved with minor modifications.

BEARING-ANGLE ESTIMATION WITH CIRCULAR ARRAYS

One of the most well-known applications with circular arrays is the bearing-angle estimation of infrasound propagation. Figure (1) shows the physical implementation of infrasound arrays.

Figure (1): Circular infrasound arrays.

For applications in this category, the propagation of wavefront from a distant source can be modeled as the far-field case. Here we first denote the unknown bearing angle as θ_o. For the two-dimensional case, then the directional bearing-angle vector can be written in the complex form $exp(j\theta_o)$, or in the vector form,

$$[\ \alpha, \beta\ ] = [\ cos(\theta_o),\ \ sin(\theta_o)]$$

The phase offset due to the time delay corresponding to a coherent source in the far field at the incident angle $\theta = \theta_o$ can be written in the form

$$\begin{aligned} exp(j\omega\tau) &= exp(j2\pi f(R/v)\ cos(\theta - \theta_o)) \\ &= exp(j2\pi(R/\lambda)\ cos(\theta - \theta_o)) \end{aligned}$$

where τ is the time delay, R is the radius of the circular array, λ is the wavelength corresponding to the coherent frequency f, and v is the propagation speed. If the radius of the array is smaller than the wavelength, the phase profile can be fully determined without phase unwrapping as

$$\Phi(\theta) = 2\pi(R/\lambda)\, cos(\theta - \theta_o)$$

$$= \pi(R/\lambda)\, exp(-j\theta_o)\, exp(j\theta) + \pi(R/\lambda)\, exp(j\theta_o)\, exp(-j\theta)$$

As it can be seen, the phase profile $\Phi(\theta)$ is a single-mode periodic function with magnitude $2\pi(R/\lambda)$ and period 2π. This periodic function consists of a pair of complex-conjugate components. The Fourier coefficient of the first term is $\pi(R/\lambda)exp(-j\theta_o)$, of which the phase can fully define the bearing angle numerically.

This formulation is based on the single-frequency model. As we superimpose the estimates from all frequencies within the available frequency bandwidth, the estimation of the bearing angle becomes more accurate. The implementation for wideband waveforms can be further improved with linear frequency weighting.

This approach consists of several important features. The first property is that this algorithm is capable and applicable to cases with bearing angles corresponding to multiple sources. The second feature is that the accuracy of the algorithm does not introduce degradation when implemented with *FFT* procedures. In addition, the computation efficiency of this technique is high due to the simplicity of the model.

FOURIER ANALYSIS

The second approach to bearing-angle estimation is the direct Fourier transform method. The formulation of this method is based on the property that the spatial-frequency spectrum of a plane-wave component consists of one single peak along a circle. The radius of the circle is $1/\lambda$, and the angular position of the peak matches the bearing angle of the incident plane wave exactly.

If we apply the two-dimensional Fourier transform to the wavefield pattern over the circular array, the kernel of the transformation is in the form

$$exp(-j2\pi < \boldsymbol{f}, \boldsymbol{x} > \; = \; exp(-j2\pi(f_x x + f_y y))$$

where $< \boldsymbol{f}, \boldsymbol{x} >$ denotes the inner product of the frequency vector

$$\boldsymbol{f} = [f_x, f_y]$$

and the spatial-position vector

$$\boldsymbol{x} = [x, y]$$

Because of the use of circular array, the position vector is in the form

$$\mathbf{x} = [\,x, y\,] = [\,R\cos(\theta),\ R\sin(\theta)\,]$$

And, because the wavefield pattern is coherent plane wave, for the two-dimensional case, the spatial-frequency spectrum is located along a circle of radius $1/\lambda$,

$$\mathbf{f} = [\,f_x\,,\ f_y\,] = 1/\lambda\ [\,\cos(\varphi),\ \ \sin(\varphi)\,]$$

As a result, the transform kernel becomes

$$\begin{aligned} exp(-j2\pi(f_x\,x + f_y\,y)) &= exp(-j2\pi(R/\lambda)\ (\cos(\theta)\cos(\varphi) + \sin(\theta)\sin(\varphi)) \\ &= exp(-j2\pi(R/\lambda)\ (\cos(\theta - \varphi)) \end{aligned}$$

Then the Fourier transform of the phase variation of wavefield pattern translates into the form

$$\int_0^{2\pi} exp(j2\pi(R/\lambda)\cos(\theta - \theta_o))\ exp(-j2\pi(R/\lambda)(\cos(\theta - \varphi))\ d\theta = Q(\varphi - \theta_o)$$

where $Q(\varphi)$ is the autocorrelation of the periodic function $exp(j2\pi(R/\lambda)\ \cos(\theta))$. Thus the Fourier spectrum of the coherent wavefield pattern detected over the circular aperture has one single peak at $\varphi = \theta_o$. Figure (2) shows the autocorrelation of one sample coherent component. The superposition of the full collection of the spectra will further improve the accuracy of the estimate.

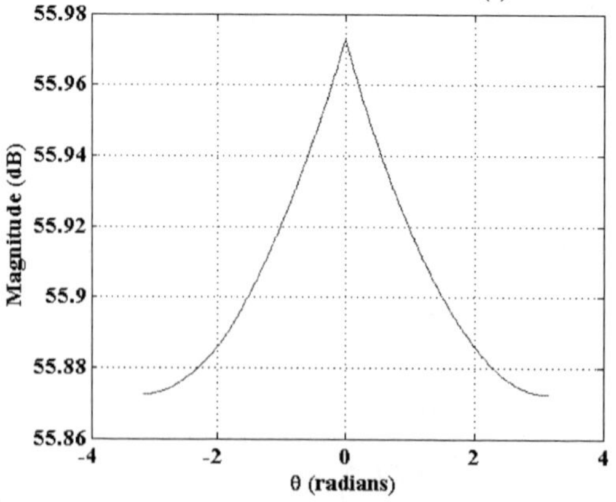

Figure (2): Autocorrelation of one coherent component

THREE-DIMENSIONAL CASE

Three-dimensional bearing-angle estimation is involved in a wide range of applications. More recently, it has become one of the critical components of homing-docking procedures for UUV's in underwater operations.

Consider a planar data-acquisition configuration, where the source plane is located at $z = 0$, and the planar receiving aperture is at $z = z_0$.

$$\begin{aligned} s(x, y, z_0) &= s(x, y, 0) * h(x, y, z_0) \\ &= s(x, y, 0) * \frac{1}{j\lambda r} \; exp(j2\pi r/\lambda) \\ &= \iint s(x', y', 0)\, h(x - x', y - y', z_0)\, dx'dy' \end{aligned}$$

For a centered transmitter at $(x, y, z) = (0, 0, 0)$, the wavefield pattern over the aperture is

$$h(x, y, z_0) = \frac{1}{j\lambda r} \; exp(j2\pi r/\lambda)$$

where r is

$$r = [x^2 + y^2 + z_0^2]^{1/2}$$

For a small translational spatial offset of $(\Delta x, \Delta y)$ on the x-y plane, the received wavefield changes slightly because

$$r = [(x - \Delta x)^2 + (y - \Delta y)^2 + (z_0)^2]^{1/2}$$

$$= z_0 \, [1 + (\frac{x - \Delta x}{z_0})^2 + (\frac{y - \Delta y}{z_0})^2]^{1/2}$$

If the propagation distance z_0 is relatively large, we can approximate the term by taking the low-order terms of the Taylor's series

$$r \approx z_0 \, [1 + \frac{1}{2} (\frac{x - \Delta x}{z_0})^2 + \frac{1}{2} (\frac{y - \Delta y}{z_0})^2]$$

$$= z_0 + \frac{1}{2z_0} (x^2 - 2\Delta x \, x + \Delta x^2) + \frac{1}{2z_0} (y^2 - 2\Delta y \, y + \Delta y^2)$$

To simply the procedure, we regroup them into four terms,

$$R = z_0 + \frac{1}{2z_0}(x^2 + y^2) + \frac{1}{2z_0}(\Delta x^2 + \Delta y^2) - \frac{1}{z_0}(\Delta x\, x + \Delta y\, y)$$

Then we place the terms of parameter r back into the convolution kernel. Subsequently, it can be partitioned into five terms

$$s(x, y, z_0) = h(x - \Delta x, y - \Delta y, z_0) = \frac{1}{j\lambda r} exp(j2\pi r/\lambda)$$

$$\approx \frac{1}{j\lambda z_0} exp(j2\pi z_0/\lambda)\, exp(j\pi(x^2+y^2)/\lambda z_0)$$

$$exp(j\pi(\Delta x^2+\Delta y^2)/\lambda z_0)\, exp(-j2\pi(\Delta x\, x+\Delta y\, y)/\lambda z_0)$$

We should note that the complex amplitude of the first term $1/j\lambda r$ is approximated as $1/j\lambda z_o$. This approximation is feasible, because amplitude of the integration kernel is less sensitive to approximation errors.

Removing the term through demodulation,

$$A(x, y) = \frac{1}{j\lambda z_0} exp(j2\pi z_0/\lambda)\, exp(j\pi(x^2+y^2)/\lambda z_0),$$

the modified wavefield at the aperture becomes $s'(x, y, z_0)$

$$s'(x, y, z_0) = exp(j\pi(\Delta x^2+\Delta y^2)/\lambda z_0)\, exp(-j2\pi(\Delta x\, x+\Delta y\, y)/\lambda z_0)$$

Now we denote

$$cos\theta_o = \Delta x/(\Delta x^2 + \Delta y^2)^{1/2}$$

$$sin\theta_o = \Delta y/(\Delta x^2 + \Delta y^2)^{1/2}$$

and

$$sin\varphi_o = (\Delta x^2 + \Delta y^2)^{1/2}/z_0$$

For a circular aperture of radius R, the receiver position can be rewritten in the form

$$x = R\, cos\theta \text{ and } y = R\, sin\theta$$

Thus, the wavefield is now in the form

$$s'(x, y, z_0) = exp(j\pi(\Delta x^2+\Delta y^2)/\lambda z_0)\ exp(-j2\pi(\Delta x\, x+\Delta y\, y)/\lambda z_0)$$

$$= S\ exp(-j2\pi(\Delta x\, x+\Delta y\, y)/\lambda z_0)$$

$$= S\ exp(-j2\pi(R/\lambda)\ sin\varphi_o\ (cos\theta_o\ cos\theta\ +\ sin\theta_o\ sin\theta))$$

$$= S\ exp(-j2\pi(R/\lambda)\ sin\varphi_o\ cos(\theta - \theta_o\ ))$$

where S denotes the complex amplitude

$$S = exp(j\pi(\Delta x^2+\Delta y^2)/\lambda z_0)$$

Upon close examination, the phase term is similar to that of the two-dimensional case

$$\Omega(\theta) = 2\pi(R/\lambda)\ sin\varphi_o\ cos(\theta - \theta_o)$$

with the additional scaling factor $sin\varphi_o$. Physically, the *3D* bearing-angle vector can be written as

$$[\ \alpha, \beta, \gamma\ ] = [\ sin(\varphi_o)cos(\theta_o),\ \ sin(\varphi_o)sin(\theta_o),\ \ cos(\varphi)\ ]$$

During the homing-docking process, the objective is to incrementally adjust the approaching angle for the convergence of the *3D* bearing-angle vector toward

$$[\ \alpha, \beta, \gamma\ ] = [\ 0, 0, 1\ ]$$

The most unique and important feature of this approach is to formulate that the relationship between the bearing-angle estimation and translational displacement in the docking process. This enables the dynamic path planning for high-precision homing-docking procedures. The mathematical analysis in this section is limited to one coherent frequency for simplicity. The accuracy of the estimation improves when substantial frequency bandwidth is available.

CONCLUSION

This paper is the summary of a sequence of research tasks in the area of 3D bearing-angle estimation for UUV homing and docking exercises. The main focus is to elevate the estimation modality from the conventional twin-receiver configuration to the 2D circular arrays. The objective is to utilize the multi-element receiver array for the complete task, including bearing-angle estimation, optimal path planning, and high-precision docking.

The utilization of circular arrays as the format of the signal processing techniques in the paper is because of the direct application to UUV navigation. In many other applications, the configuration of the 2D arrays can be rectangular arrays or, for better data-acquisition efficiency, hexagonal arrays [5-6]. Because the mathematical formula is relatively independent of the array configurations, the use of non-uniform arrays is also feasible.

REFERENCES

1. Hua Lee, "Signal Processing Algorithms for UUV Navigation and Collision Avoidance," *Journal of the Acoustical Society of America,* 125(4), p. 2537, April 2009.

2. Hua Lee, "Underwater Acoustical Imaging and Sensing Systems for Homing and Docking, Navigation, and Collision Avoidance," *Acoustical Imaging*, Vol. 30, pp. 295-304, 2011.

3. Michael Lee, Daniel Doonan, Michael Liebling, and Hua Lee, "Resolution Analysis and System Integration of a Dynamically Reconfigurable FMCW Medical Ultrasound Imaging System," *Proceedings of International Telemetering Conference*, 2012.

4. Melissa Johnson and Hua Lee, "Estimation of Angle of Arrival with Two-Dimensional Circular and Non-Uniform Arrays," *Proceedings of International Telemetering Conference*, pp. 457-466, 2015.

5. Hua Lee, "Estimation of Angle of Arrival with Non-Uniform Reconfigurable Receiver Arrays," *Journal of Acoustical Society of America*, 139(4), 2016.

6. Hua Lee, "Optimization of Computation Efficiency in Underwater Acoustic Navigation System," *Journal of Acoustical Society of America*, 139(4), 2016.

NTPS AERONAUTCAL MOBILE TELEMETRY C BAND CONVERSION

Stephanie Luongo
National Test Pilot School
Mojave, CA 93501
sluongo@ntps.edu

Faculty Advisor:
Marco Lotterio

ABSTRACT

Due to the limited availability of L Band spectrum for non-government use, the National Test Pilot School (NTPS) has implemented a C Band airborne mobile telemetry (AMT) datalink system in order to continue to fulfill its mission to educate and train military and civilian flight test personnel. Instrumented aircraft were upgraded to support a PCM stream with higher resolution data and higher data rates required during performance & flying qualities and loads & flutter testing. This technology upgrade included a ground system that was both L and C Band capable and used a Smartronix Telemetry Receiver Processor to convert RF-to-IRIG Chapter 10 UDP packets for a direct input to an IADS® server in the NTPS control room. To achieve the similar performance to the legacy L Band datalink, a high gain, C Band antenna and low noise amplifier were installed as well as a new antenna rotator that allowed for higher precision movements and 360-degree continuous rotation in the pan axis. This paper will discuss the design analyses for and flight testing of the NTPS C Band datalink.

INTRODUCTION

Legacy NTPS Operations' use of aeronautical mobile telemetry in the L Band required de-confliction with users at Edwards Air Force Base. Although NTPS had been granted use of this spectrum on a non-interference basis, the process of requesting use of the L Band frequency was difficult and sorties were often rescheduled because another user had priority. As part of an NTPS technology upgrade, a new TM system was architected to not only break away from the existing frequency, but also to take advantage of present-day off-the-shelf technology.

Ground and flight evaluations of the C Band Telemetry system were performed using an AermacchiMB-326M Impala aircraft as the test platform. A build-up approach was taken to test and evaluate the C Band datalink that began with ground testing the individual components and subsystems. Functional checkouts, EMI/EMC testing, instrumentation calibrations and IADS® displays verification were performed; however, this paper will discuss only the system characterization flight testing performed at the NTPS in Mojave, California, during November of 2016.

SYSTEM DESCRIPTION

The C Band Telemetry system comprised of both airborne and ground components. The operating frequency was fixed at 5.1 GHz. The modulation type was PCM/FM with a bit rate of 250 Kbps. The total RF transmit power was 20 Watts (+43 dBm).

Airborne Components

The airborne components included a Nginuity signal conditioner and Data Acquisition System (DAS), a Quasonix C Band transmitter, a 50/50 RF splitter, and 2 omni-directional blade antennas, as shown in Figure 1. The DAS digitized instrumentation sensor outputs and sent PCM telemetry frames to the transmitter over a 4-wire (data and clock) RS-422 serial interface. The transmitter randomized the PCM data and modulated it using a carrier frequency of 5.1 GHz. The RF output of the transmitter was then divided by a 50/50 splitter and fed to two aircraft antennas, one located on the bottom fuselage and one located on the top side of the fuselage.

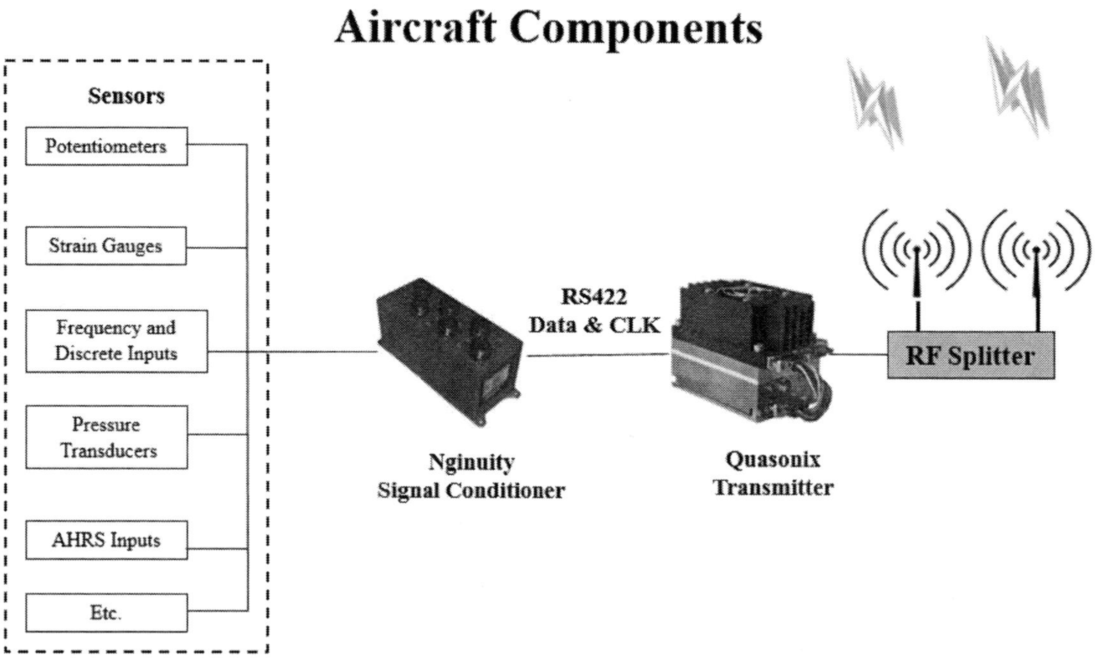

Figure 1 Aircraft Components.

Ground Components

The ground components included a high-gain parabolic dish antenna, a Low Noise Amplifier (LNA), a 36-inch coxial cable, a Moog antenna rotator, a 50-ft low-loss coaxial cable, and a two-channel, dual-band Smartronix receiver and telemetry processor. The downlinked signal was received by the rooftop antenna and passed through the rotary joint of the antenna rotator and coaxial cabling into the building and eventually to the receiver. The receiver performed the RF-to-IF down-conversion and demodulated the signal to TTL-level data and clock inputs. The TTL signals were fed to a PCM module within the receiver for processing into IRIG Chapter 10 UDP packets. These packets were broadcasted onto the NTPS network in which an IADS®™ server was configured to decommutate and pass IADS® data to client computers in the TM Control Room. Figure 2 shows the ground components setup.

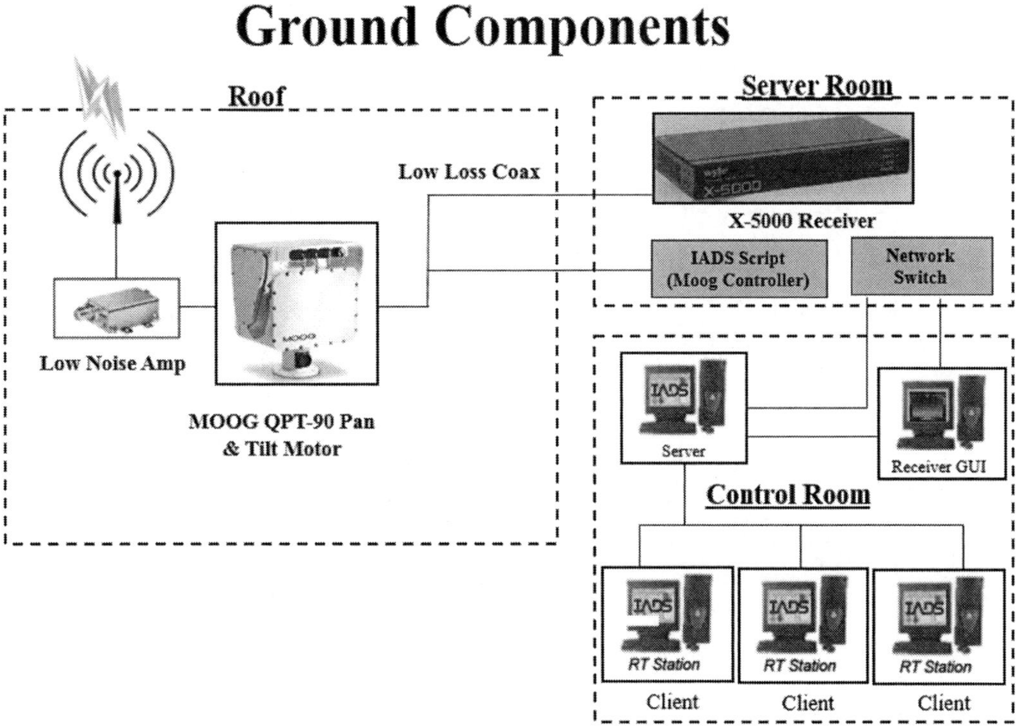

Figure 2 Ground Components.

LINK MARGIN AND NOISE FIGURE ANALYSES

Noise figure and link margin analyses were completed using guidance from [1] and are presented in Table 1 and Table 2, respectively. Due to a limitation in the RF test equipment available for measurements, values used in the analyses were obtained from component datasheets. An operating range of 40 Nmi (a limitation of the experimental aeronautical mobile telemetry license) was used in the maximum range analysis and a margin of 10.9 dB was calculated. An operating range of 150 ft (the shortest distance expected between the aircraft and the ground station) was used in the minimum range analysis and a received signal strength at the receiver of -15 dBm was calculated. The purpose of minimum range calculations was to ensure avoidance of receiver saturation and/or damage.

Table 1 System Noise Figure Analysis.

	LNA	Short Coax	Rotator	Long Coax	Receiver
Component Gain (dB)	24.0	-1.0	-1.8	-10.0	variable
Component Noise Figure (dB)	2.0	1.0	1.8	10.0	6.0
Noise Factor contribution by stage (absolute)	1.6	0.0	0.0	0.1	0.2
				System Noise Factor (absolute)	**1.9**
				System Noise Figure (dB)	**2.7**

Table 2 Link Margin Analysis.

	TRANSMIT PATH		
1	Frequency	5.1	GHz
2	Power TX	43.5	dBm
3	RF Splitter	3.4	dB
4	Cable Loss	0.5	dB
5	TX Antenna Gain	0.0	dBi
6	EIRP Max.	39.6	dBm
	RECEIVE PATH		
7	Range	40.000	Nmi
8	Atmostpheric Atten	144.0	dB
9	Scattering/Reflections Loss	10.0	dB
10	RX Antenna Gain	26.8	dBi
11	Circular Polarization Loss	3.0	dB
12	Low Noise Amplifier	24.0	dB
13	Cable 1 Loss	1.0	dB
14	Rotator Pass Thru Loss	1.8	dB
15	Cable 2 Loss	10.0	dB
16	TOTAL LOSS	119.0	dBm
17	Power at RCVR (Line 6 - Line 16)	-79.4	dBm
	SYSTEM TOTALS		
18	RCVR Sensitivity @ 1 Mbps***	-93.0	dBm
19	System Noise Figure	2.7	dB
20	**Link Margin (Line 18+Line19-Line17)**	**10.9**	**dB**
	***Sensitivity value of the RCVR was provided by vendor. ***Actual bit rate implemented in system was 250 Kbps.		

SYSTEM CHARACTERIZATION FLIGHT TEST

General

The objectives of the system characterization flight were to verify a maximum range of greater than 40 Nmi and to verify a robust telemetry datalink at different aircraft aspects relative to the ground station and aircraft attitudes. Telemetered data processed by IADS® as well as receiver RF signal strength and spectrum data were displayed in real-time and were also recorded for post-flight analysis.

Maximum Range Testing

A maximum range test was performed to determine if the datalink could operate out to a maximum range of 40 Nmi (as granted by the FCC AMT License). The aircraft was flown in the north-northeast direction from Mojave Air and Space Port (KMHV) at 12,500 FT MSL with a tail-on aspect. (A total loss of link was defined as the receiver losing lock for at least a 3-second duration.) No intermittent dropouts were observed during the entire leg outbound. The aircraft was vectored back to the southwest once beyond maximum range. A screenshot of the IADS® display and spectrum at the receiver is shown in Figure 3. The greatest range achieved before turning inbound was 43.21 Nmi and no loss of signal was observed. The signal strength measured at this range was -82 dBm, which agreed to within 2 dBm of the link analysis. The maximum range requirement of 40 Nmi was met therefore allowing school sorties to be flown within the NTPS operating working areas with reliable telemetry.

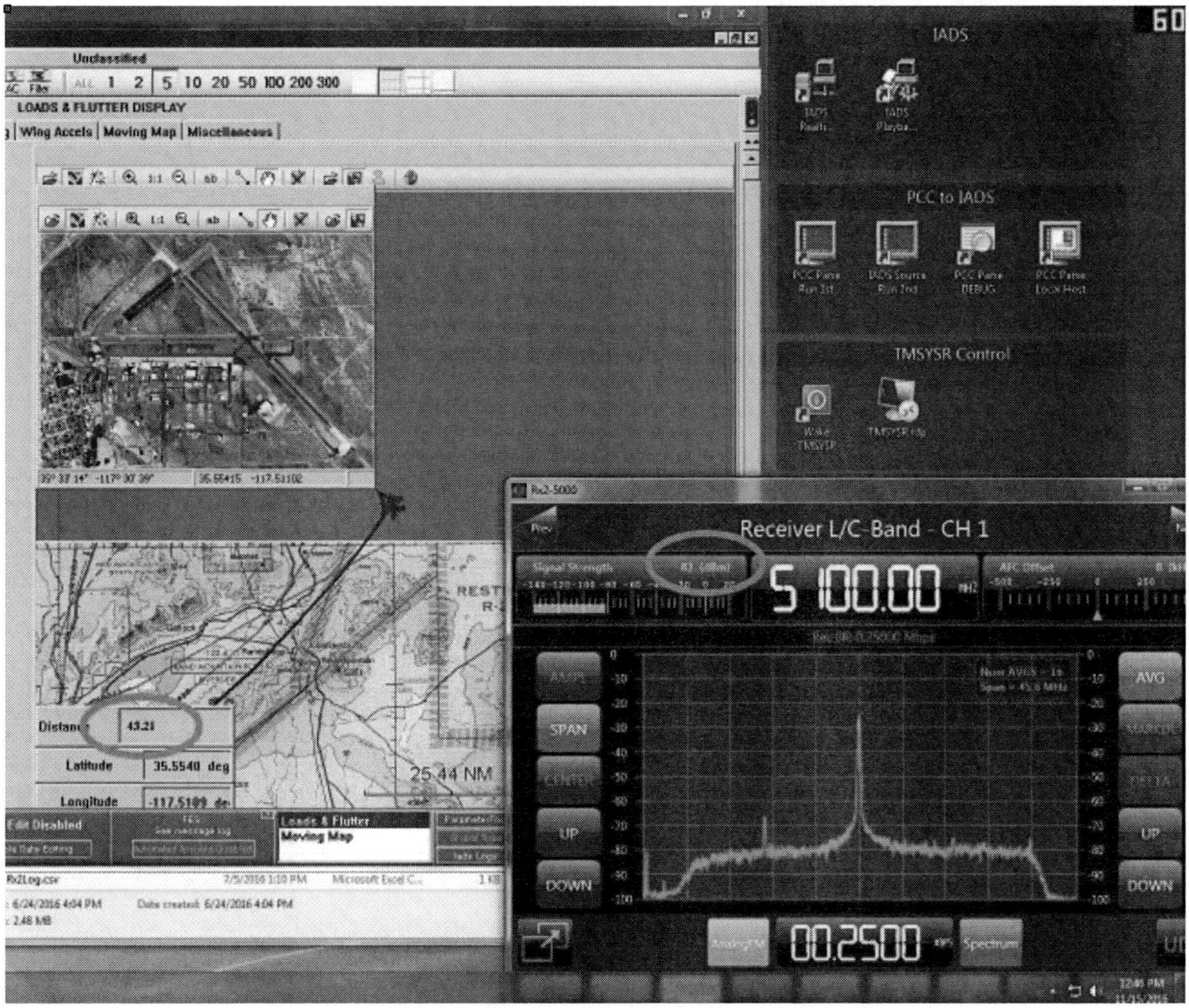

Figure 3 Maximum Range Test Screen Capture.

Aspect Testing

RF performance of the datalink at different aircraft aspects was verified. One clockwise and one counter clockwise square ground track pattern were flown at 16,500 ft MSL at a range of approximately 15 Nmi. The counter clockwise track is shown in Figure 4. The range of signal strengths measured and the percentage of time during the leg that telemetry dropouts occurred are summarized in Table 3.

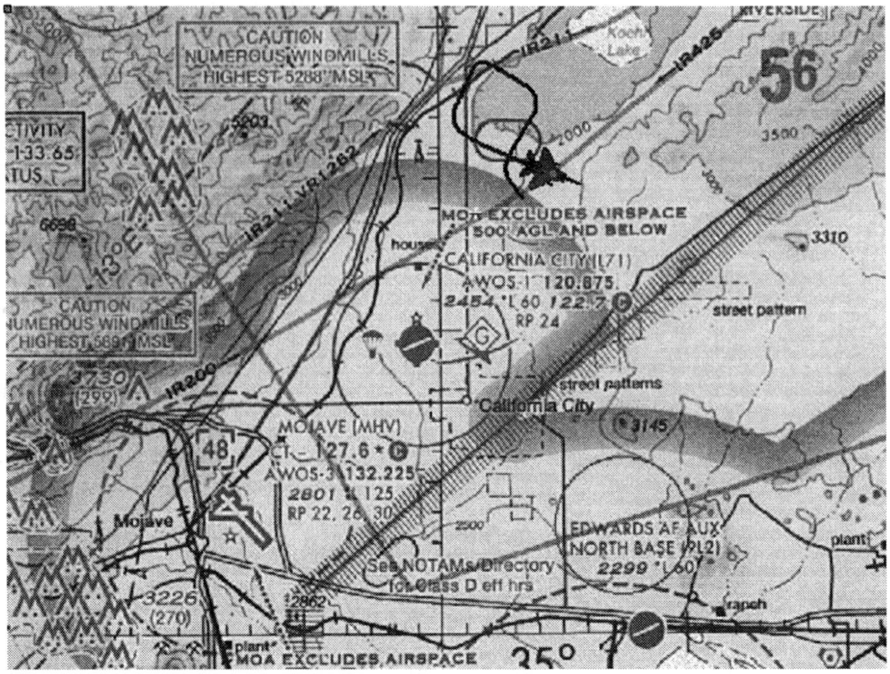

Figure 4 Ground Track Pattern during Aspect Testing.

Table 3 RF Performance during Aspect Testing.

Direction	Aspect	RF Signal Strength	% of Time w/ Dropouts
Clockwise	Tail On	-79 dBm to -70 dBm	No Dropouts
	Right Wing	-93 dBm to -71 dBm	No Dropouts
	Nose On	-87 dBm to -80 dBm	No Dropouts
	Left Wing	-83 dBm to -74 dBm	No Dropouts
Counter-clockwise	Tail On	-83 dBm to -68 dBm	No Dropouts
	Right Wing	-89 dBm to -79 dBm	No Dropouts
	Nose On	-96 dBm to -80 dBm	No Dropouts
	Left Wing	-98 dBm to -83 dBm	No Dropouts

For both directions the tail-on aspect showed the best performance and the use of 2 antennas of equal power output operating simultaneously did not appear to have any adverse effects, e.g. destructive interference. Though the signal strength varied, the receiver never exhibited a dropout or loss of lock. With the new system, a sortie which involves flying the aircraft at different aspects relative to the ground antenna will experience high telemetry reliability and dropouts will be infrequent.

Roll Attitude Performance

The RF performance at various varying roll attitudes was verified. While inbound on a nose-on aspect and at an altitude of 12,500 ft MSL and range of approximately 30 Nmi, wing-rocks (bank variations) were performed. Bank angles during the maneuvering varied between $\pm 60^0$ and were flown to a roll rate of less than 10 ^{0}/sec (or approximately ½ lateral stick deflection). Small variations in heading and yaw were permitted. The ground track and spectrum at the receiver are shown in Figure 5. No RF dropouts or loss of lock were observed while maneuvering.

Figure 5 Roll Attitude Performance Testing.

Pitch Attitude Performance

Similar to roll attitude testing, the RF performance at varying pitch attitudes was verified. Pitch attitude evaluation maneuvers were performed by continuously varying the pitch attitude ($\pm 45^0$) of the aircraft in a sinusoidal/roller coaster pattern for 3 cycles. The wings remained level during the pull-ups and pushovers, and the pitch rate did not exceed 3 ^{0}/sec. No RF dropouts were observed while maneuvering.

CONCLUSIONS

The system characterization flight test results presented above showed evidence of a robust datalink. The maximum range was greater than 40 Nmi, a limit hindered only by the FCC experimental mobile telemetry license. The system performed reliably with no dropouts as aircraft aspect, attitude, and range from the ground station was varied. The NTPS AMT C Band conversion will allow for a continued telemetry capability in order to support the school's mission of flight test training and education.

REFERENCES

[1] D. M. Pozar, *Microwave and RF Design of Wireless Systems*, John Wiley & Sons, Hoboken N.J., 2001.

AN UNBALANCED FEED DESIGN FOR WIDEBAND PHASED ARRAYS

David W. Landgren, Daniel J. P. Dykes, and Kenneth W. Allen
Advanced Concepts Laboratory, Georgia Tech Research Institute,
Georgia Institute of Technology, 30332 Atlanta, GA, USA
Corresponding author: kenneth.allen@gtri.gatech.edu

ABSTRACT

In this work, a planar phased array antenna was engineered with ultra-wideband (UWB) performance that covers portions of the L-, S-, and C-bands. The unit cell design contains a driven layer that is directly connected to a single coaxial feed and a parasitic layer located above the driven layer separated by free space. This design does not require a balun or any vias between the antenna ground plane and the driven layer, resulting in a simple antenna stack-up consisting of only planar layers, aside from the feed structure. As a consequence, the complexity, and potentially sensitivity to mechanical tolerances, is reduced. The simulated results of this unbalanced phased array are discussed and experimentally validated.

INTRODUCTION

Recently, the government auctioned off sections of the electromagnetic spectrum to industry and other entities. This has resulted in a transition of certain government systems from L-Band to higher frequencies such as C-Band. In this work, an UWB antenna system was designed to support fire-control radar testing that will instantaneously cover both the legacy L-Band frequencies and the new C-Band frequencies. Additionally, the planar array has a reduced form factor (thickness $< \lambda/10$ at the highest frequency of operation) compared to existing, more traditional solutions, e.g. parabolic dish antenna. The key to minimizing antenna thickness is designing the array without baluns or commercial-off-the-shelf (COTS) combiner networks; this proceeding will focus on a technique to design arrays without baluns.

Traditionally, wideband planar arrays are designed with a differential feeding scheme. That is, each element (or unit cell) of the phased array is fed with a pair of transmission lines that are ideally supplied with signals that are equal in amplitude and have a 180-degree phase difference. Assuming the aperture geometry is also symmetric, this feeding scheme results in the flow of electrical current in a preferred direction across the unit cell, a common technique for producing a linear polarization state. Furthermore, this feature of the electrical current supports UWBs from planar arrays as described by the well-known current sheet concept developed by Wheeler [1].

Phased arrays with unbalanced feeding schemes simplify the front-end antenna system since no baluns are required; however, these antenna arrays have resulted in more complex architectures for the unit cell. In some cases, this has required vias to be positioned in very close spatial proximity to the feed structure [2-7]. The spatial proximity leads to tight electromagnetic coupling between the vias and feed structure; consequently, the antenna performance of such designs can be sensitive to mechanical deviations from the nominal design values. In this work, a more stable solution was engineered by removing the vias altogether. It will be shown that this can be done without a significant reduction in performance, as the realized gain of a fully driven array is within 1.5 dB of the area gain limit for arrays larger than 8x8.

The remainder of this work is organized as follows. The first section will provide details of the unit cell geometry. The second section will present two designs: (*i*) a single-layer design and (*ii*) a two-layer design, comprised of a driven layer and a parasitic director layer. The third section will provide a discussion of how performance is affected by array size and options for implementing a power combiner. Finally, a conclusion will summarize this work and project how this technology can be utilized in the future.

PHASED ARRAY UNIT CELL GEOMETRY

The antenna unit cell consists of multiple layers of low-permittivity (ε_r) materials, a standard 0.085" coaxial feed, and multiple patterned copper layers (more specifically fragmented aperture layers). The specific materials are a low-density General Plastics foam FR-4540 ($\varepsilon_r = 2.0$) and Rogers 5880LZ ($\varepsilon_r = 1.96$). The low-density foam provides mechanical support and separation between the aperture and the ground plane with the desirable properties of light weight, machinability, and minimum electromagnetic interaction due to the low-permittivity. The two Rogers 5880LZ boards have 0.5 oz copper that is patterned for the radiating layer and continuous for the ground plane. An example stack-up is shown in Figure 1, illustrating a single feed and no additional conductors or vias. This is an indication that no baluns are required.

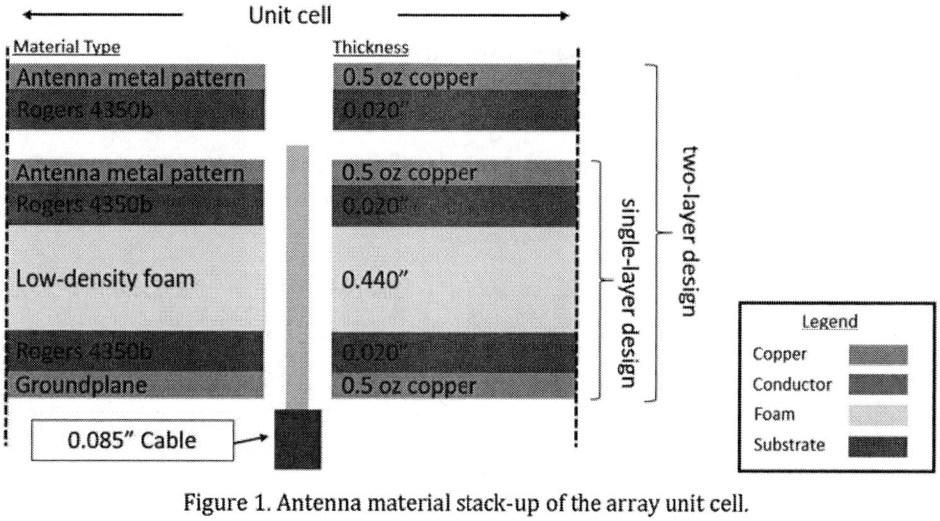

Figure 1. Antenna material stack-up of the array unit cell.

The exact details of the fragmented layer are determined by a process developed by the Georgia Tech Research Institute (GTRI) over the last decade and a half [8]. This process leverages an evolutionary algorithm to design antennas; however, this process can be extended to evaluate, characterize, and generate other electromagnetic structures [9-17]. The general form factor is chosen based on physical principles (e.g., area gain limit, avoiding grating lobes, ground plane spacing, etc.) and application-specific features (e.g., feed structure, restricted footprint, etc.). The aperture is integrated within this form factor, Figure 1, and is a fragmented layer consisting of discrete areas, Figure 2, that can be toggled between conducting and dielectric by a genetic algorithm to optimize the performance of the antenna. A full 3D finite-difference time-domain (FDTD) electromagnetic simulation is conducted for each candidate antenna solution, and the performance is compared against some predefined goal (e.g., gain vs. frequency). In the case of this work, the goal was to maximize realized gain vs. frequency from L-band to C-band.

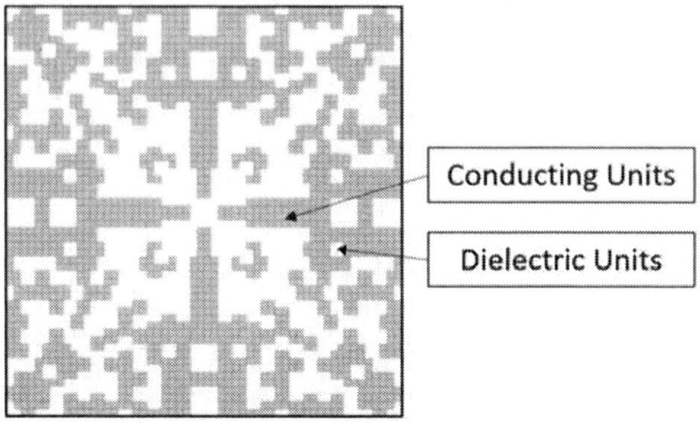

Figure 2. Example fragmented layer.

OPTIMIZED DESIGNS OF THE PHASED ARRAY UNIT CELL

A unit cell was designed with a single fragmented radiating layer (i.e., driven layer). This layer is directly excited by the center conductor of a coaxial feed. The simulated (using an in-house FDTD code) realized gain, reflection coefficient, and complex input impedance of this design is shown Figure 3. The realized gain has been normalized to 0 dBiL, corresponding to 100% efficiency. This design has two minima in the gain across the optimized band, at approximately 2.5 and 5.25 GHz. Inspecting the input impedance at these two frequencies reveals that the antenna is more capacitive than the rest of the operating band. This insight led to the idea of adding an additional layer for tuning to help improve the complex input impedance, where the imaginary component can be dominated by capacitance or inductance, to provide a flattened resistance and reactance profile over the frequency bands of interest.

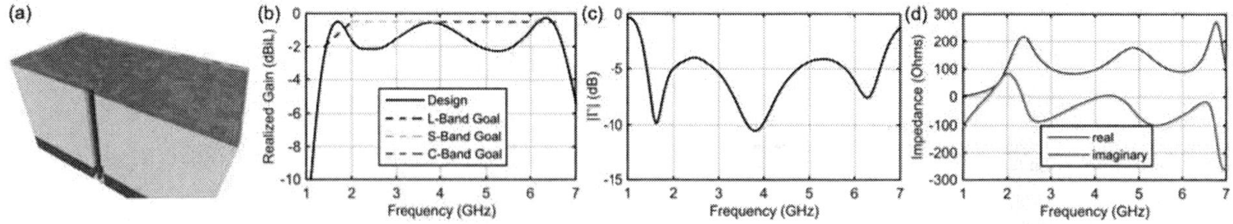

Figure 3. Single-layer antenna design. (a) An image showing half of the unit cell, (b) simulated results of the infinite array gain, (c) reflection coefficient, and (d) complex input impedance as a function of frequency.

In order to address the degradation in gain of the single-layer design, a two-layer antenna was designed. The second layer acts as a parasitic tuning mechanism that resides at some distance (i.e., 7.5 mm) away from the driven layer, determined by the optimizer. During this optimization, the fragmented pattern of the driven layer was held fixed to match the single-layer design, Figure 3(a). The predicted performance is shown in Figure 4. The additional layer has helped increase the infinite array gain by approximately 0.5 dBiL, an appreciable improvement over the previous design. Given that all of the materials for this design are low-loss dielectrics ($tan\delta < 0.01$), the majority of the improvement is associated with the impedance match. The impedance plot in Figure 4(d) confirms this, given that both components of the complex impedance are flatter and have reduced extrema across the operating band.

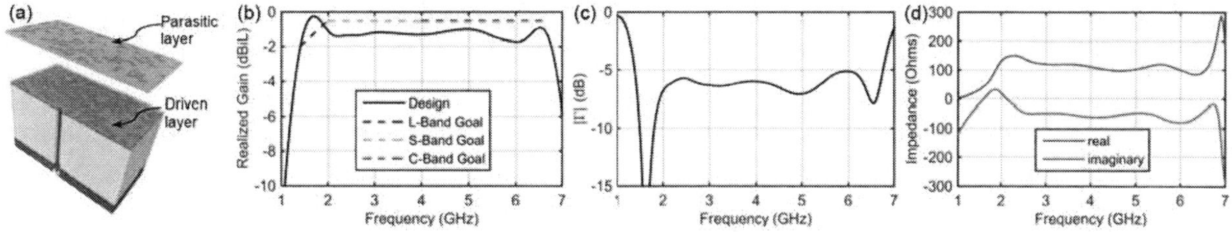

Figure 4. Two-layer antenna design. (a) An image showing half of the unit cell, (b) simulated results of the infinite array realized gain, (b) reflection coefficient, and (d) input impedance as a function of frequency.

EXPERIMENTAL RESULTS

A small 4x4 array prototype was built and measured based on the single-layer design. An image of this antenna array, along with an inset of the unit cell, is shown in Figure 5(a). The array is fully cabled, i.e. each element is directly connected to a cable, so that each element can be characterized individually. In order to mechanically support the 16 cables and the antenna, a fixture was built with plastic posts and a metal plate, as shown in Figure 5(a). It should be noted that the fixture is electromagnetically shielded by the antenna ground plane and should have minimal impact on the antenna performance. The total gain is calculated by performing a complex sum of all 16 ports of the 4x4 array using uniform amplitude weighting. This mimics a perfect power combiner and therefore is ideal. The experimental results of the measured realized gain agree with the simulation, as shown in Figure 5(b). For telemetry applications, a combiner network would be integrated with the antenna array, resulting in a single RF-connector output. However, the development of the combiner network goes beyond the scope of this proceeding.

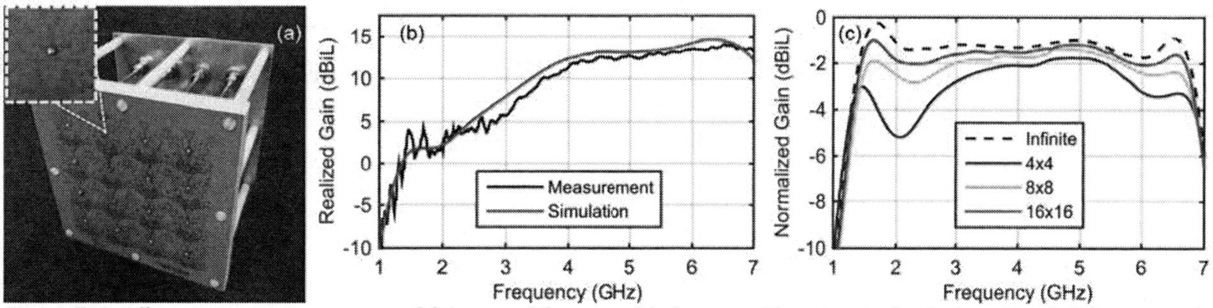

Figure 5. Small prototype antenna array. (a) image of the assembled array with an inset of a single unit cell, (b) measured broadside gain compared with the simulated result, and (c) simulated results for varying area sizes compared with an infinite array in terms of normalized gain.

DISCUSSIONS

Ultimately, a larger array (tentatively 16x16) will be fabricated in order to support the gain requirements for specific test ranges. Smaller arrays can significantly deviate from performance of the infinite case due to array truncation effects. These deleterious effects are especially present at the lower frequencies, where the antenna is less than a wavelength in some cases. Generally, as the array becomes larger, the performance converges towards the infinite array result. This is illustrated in Figure 5(c), which shows the trend in performance for various sized arrays: 4x4, 8x8, and 16x16. The 16x16 is nearly converged to the results from the infinite array result, i.e. periodic boundary conditions applied in the plane of the unit cell. This helps build confidence that the array designed in the infinite array setting will perform as anticipated, if tiled to 16x16 or greater. It should also be noted that the experimental results presented are for the single-layer design without the tuning layer. Prototyping of the two-layer design will be the subject of future work.

COTS products are available to support a 16:1 power combiner network. However, in-house combiners are also being designed, leveraging the same design procedure used for the fragmented antenna aperture. These fragmented combiners are part of an initiative to explore the more general space of fragmented circuitry. Fragmented combiner networks could potentially provide superior performance as the combiner network could be optimized simultaneously with the antenna aperture, tuning for deviations from the assumed 50 Ω load.

CONCLUSIONS

In this work, a planar phased array antenna was demonstrated with UWB performance that covers portions of the L-, S-, and C-bands with FDTD simulations, validated by experimental measurements. The unit cell design is simplified in comparison to the planar unbalanced modular arrays (PUMA) in refs. [2-7] and offers a new approach to this problem. The unit cell includes a DC-connected driven layer with a parasitic layer above to tune the input impedance of the array. This simple stack-up does not require a balun or vias from the driven layer to the ground plane. Consequently, this simple antenna stack-up consists of only planar layers, aside from the feed structure. Therefore, the structural complexity is reduced which could lead to a more robust solution in terms of sensitivity to mechanical deviations. The concepts developed in this work are flexible and a suitable solution for scaling to higher frequencies, seeing as the structures are mainly dependent on the spatial resolution of the planar features. Regarding intended telemetry applications, the antenna will be integrated with an appropriate combiner network resulting in a single RF feed for interfacing with test range systems.

ACKNOWLEGEMENTS

This project is managed by the Test Resource Management Center (TRMC) and funded through the Spectrum Access R&D program via Picatinny Arsenal under Contract No. WL5QKN-15-9-1004. The Executing Agent and Program Manager work out of the AFTC. Any opinions, findings, and conclusions or recommendations expressed in this material are those of the authors and do not necessarily reflect the views of TRMC, the SAR&D Program and/or the Picatinny Arsenal.

REFERENCES

[1] Wheeler, Harold. "Simple relations derived from a phased-array antenna made of an infinite current sheet." *IEEE Transactions on Antennas and Propagation* 13.4 (1965): 506-514.

[2] Holland, Steven S., and Marinos N. Vouvakis. "Design and fabrication of low-cost PUMA arrays." *Antennas and Propagation (APSURSI), 2011 IEEE International Symposium on*. IEEE, 2011.

[3] Logan, John T., and Marinos N. Vouvakis. "Planar ultrawideband modular antenna (PUMA) arrays scalable to mm-waves." *Antennas and Propagation Society International Symposium (APSURSI), 2013 IEEE*. IEEE, 2013.

[4] Lee, Michael Y., et al. "Simplified design of 6:1 PUMA arrays." *Antennas and Propagation & USNC/URSI National Radio Science Meeting, 2015 IEEE International Symposium on*. IEEE, 2015.

[5] Holland, Steven S., Daniel H. Schaubert, and Marinos N. Vouvakis. "A 7–21 GHz dual-polarized planar ultrawideband modular antenna (PUMA) array." *IEEE Transactions on Antennas and Propagation* 60.10 (2012): 4589-4600.

[6] Holland, Steven S., and Marinos N. Vouvakis. "The planar ultrawideband modular antenna (PUMA) array." *IEEE Transactions on Antennas and Propagation* 60.1 (2012): 130-140.

[7] Logan, John T., et al. "A review of planar ultrawideband modular antenna (PUMA) arrays." *Electromagnetic Theory (EMTS), Proceedings of 2013 URSI International Symposium on*. IEEE, 2013.

[8] Maloney, James Geoffrey, et al. "Fragmented aperture antennas and broadband antenna ground planes." U.S. Patent No. 6,323,809. 27 Nov. 2001.

[9] Allen, Kenneth W., et al. "An X-band waveguide measurement technique for the accurate characterization of materials with low dielectric loss permittivity." *Review of Scientific Instruments* 87.5 (2016): 054703.

[10] Scott, Mark M., et al. "Permittivity and permeability determination for high index specimens using partially filled shorted rectangular waveguides." *Microwave and Optical Technology Letters* 58.6 (2016): 1298-1301.

[11] Allen, Kenneth W., et al. "Metasurfaces guided by evolution: Ultra-wideband Ku-bandpass frequency selectivity is fit for survival," Submitted (2017).

[12] Westafer, Ryan S., et al. "Micropatterned W-band Antenna Tiles," *IMS 2017 - IEEE MTT-S INTERNATIONAL MICROWAVE SYMPOSIUM,* June 4-7 (2017).

[13] Rahmat-Samii, Yahya, and Eric Michielssen. "Electromagnetic optimization by genetic algorithms." *Microwave Journal* 42.11 (1999): 232-232.

[14] Weile, Daniel S., and Eric Michielssen. "Genetic algorithm optimization applied to electromagnetics: A review." *IEEE Transactions on Antennas and Propagation* 45.3 (1997): 343-353.

[15] Johnson, J. Michael, and V. Rahmat-Samii. "Genetic algorithms in engineering electromagnetics." *IEEE Antennas and Propagation Magazine* 39.4 (1997): 7-21.

[16] Haupt, Randy L., and Douglas H. Werner. *Genetic algorithms in electromagnetics*. John Wiley & Sons, 2007.

[17] Nagar, Jogender, and Douglas H. Werner. "A Comparison of Three Uniquely Different State of the Art and Two Classical Multiobjective Optimization Algorithms as Applied to Electromagnetics." *IEEE Transactions on Antennas and Propagation* 65.3 (2017): 1267-1280.

Modeling and Simulation with Numbers!

Juan Guadiana, Telemetronics Inc.
James Baird, Telemetronics Inc.
Curtiss Tackitt, Blue Origin LLC.
Ana Dubois, Mannatek Solutions.
Dorse Dubois, Mannatek Solutions.

October 19, 2017

ABSTRACT

Moore's Law predicts trends in our technology well and we are witnessing the relentless march of software solutions deep into what was the hardware domain. Link models are normally the realm of scientific software packages like Mathematica, MatLab or Satellite Tool Kit. Here we use a spreadsheet application, Numbers (Excel works too) to apply Frii's Transmission equation to perform a link model.

Modeling a single link is easy and a staring antenna Array is modeled as many links. Creating the model does require planning just as creating any software application would, but the "coding" is fairly straight forward. Results are stunning graphical plots. A cool simulation is created with the spreadsheet depicting the array's Graphics User Interface (GUI). Very low cost, an excellent way for students to learn to model and simulate their systems. The work serves as a prototype to experiment with before investing in expensive software or software development. Spreadsheets do break easily so plan to back up frequently.

Introduction

The Model was created using Apple *Numbers* and does export to Microsoft *Excel. S*ome features like steppers or sliders break as do a few plots. Surprisingly capable this software has potential for even more ambitious work. There are three basic models used in the system model. A classical Bessel function models each antenna in each subarray and transmitting antenna of the vehicle under test. The trajectory model has three profiles. Third, post detection combining is modeled. Major components of the system model reside in their own spreadsheets (tabs) and are linked together to keep the complexity manageable. Although challenging, the whole thing is managed by the System Monitor (first sheet). There is a separate sheet for each subarray, trajectory model, receiving/transmitting antenna models, Array Summary (Post detection combining model) and finally the Simulation.

Distributed Array Remote Telemetry Sensor (DARTS)

The Distributed Array Remote Telemetry Sensor system is modeled as might be deployed at White Sands Missile Range. The DARTS System concept employs 10 subarrays (each with 6 elements) distributed about the test range. Arrays can provide more uniform coverage plus redundancy is naturally exploited by post detection combining for better results. Understanding DARTS at least casually will help make better sense of the entire exercise. The system modeled is based on a 6 element (Horn) Subarray. A twelve element DARTS subarray variant (with two rows of six horns) is shown below in Figure 1. The Simulation is also based on this variant. For the purposes of this model we use the lower row.

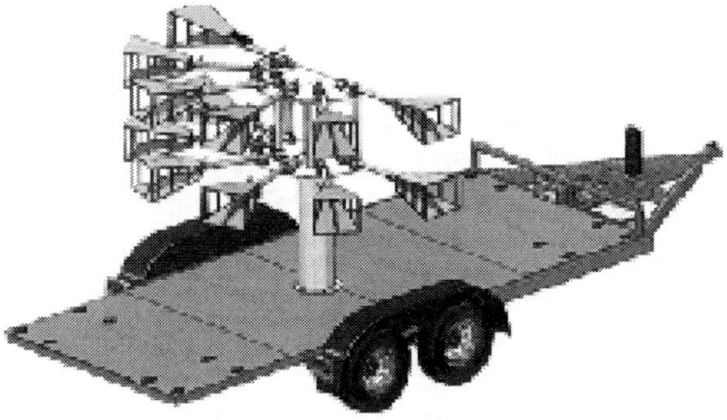

Figure 1. 12 Element Array (1 of 10)

Receive Antenna Model.

Let's take a look at the base antenna model first before diving into the system model. There are 10 subarrays each with 6 elements (horns) for a total of 60 elements. The transmitting antenna makes it 61 elements to be modeled. One way to do this is with a Bessel function demonstrated in Figure 2 below. This receive antenna model resides in the "Scan to RX Antenna" tab.

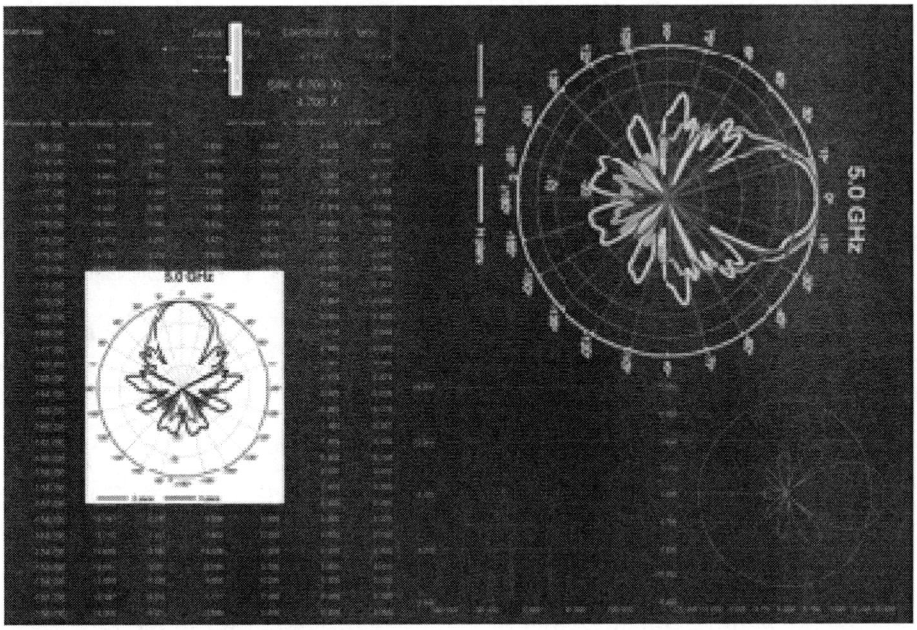

Figure 2 Receive Antenna Model

The model begins with the scan of an antenna pattern taken from the manufacturers literature (inset). The scan is rotated and placed as the lowest layer at the upper right hand graphic. The calculated antenna pattern is superimposed directly over and in red. By manipulating the slider control shown or entering directly, the coefficient of the Bessel function may be manipulated until the antenna pattern falls between the E and H plane curves. This coefficient is used by the system model for all 60 Receive Antenna Patterns. The Transmit Antenna Model Bessel coefficient is currently static.

SubArray Model

As mentioned, six horns are deployed at one site and make up a subarray. The position convention goes like this: array azimuth of 0 degrees describes the center of the composite of antenna patterns; six horns (30° Beam-width) spaced 30 degrees apart; oriented at -75,-45,-15,15,45 and 75 degrees. So when you want to look directly east you enter an Array Offset of 90 degrees at the System Monitor Table (discussed later).

The elevation control let's you select whether the an antenna uses an optimal, median or designated elevation offset, by entering 1,2 or 3 respectively. Designated Elevation uses the angle designated at the System Monitor table, median is the median elevation the target exhibits through out the entire trajectory, and optimal is the median elevation exhibited through that individual antenna sector. Median doesn't always give the best answer. Changes can always be made directly in the Antenna Block to override the automated calculation. Let's look at DARTS 1's sheet shown in Figure 3 below.

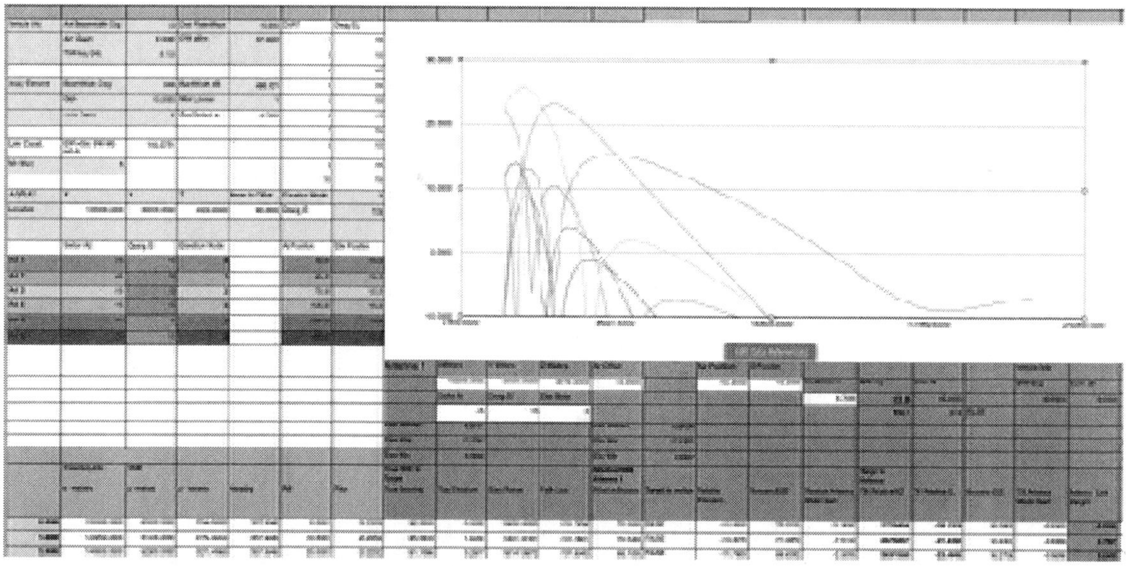

Figure 3 DARTS 1 Subarray Sheet

DARTS 1's Array Block (upper left) is populated from the System Monitor Table. The subarray sheet calculates the Link models. Link geometry is based on the trajectory data (lower left). The blue block is Antenna 1 specific and there are five additional blocks for that subarray. There are nine additional subarray sheets just like this one, let's take a look at the big picture.

SYSTEM MONITOR

The System Monitor (sheet) is the system's Graphical User Interface. Shown below in Figure 4, the Map display details the location of each subarray and the trajectory of the subject vehicle. Remaining plots detail composite performance of each array and score the post detection combining performance. The table displays the settings that control the model. Array coordinates and orientation are listed and maybe edited. Vehicle transmitter characteristics and the Trajectory Control are shown. It is easy to place an array anywhere on the range as it is to command the type of trajectory simulation desired.

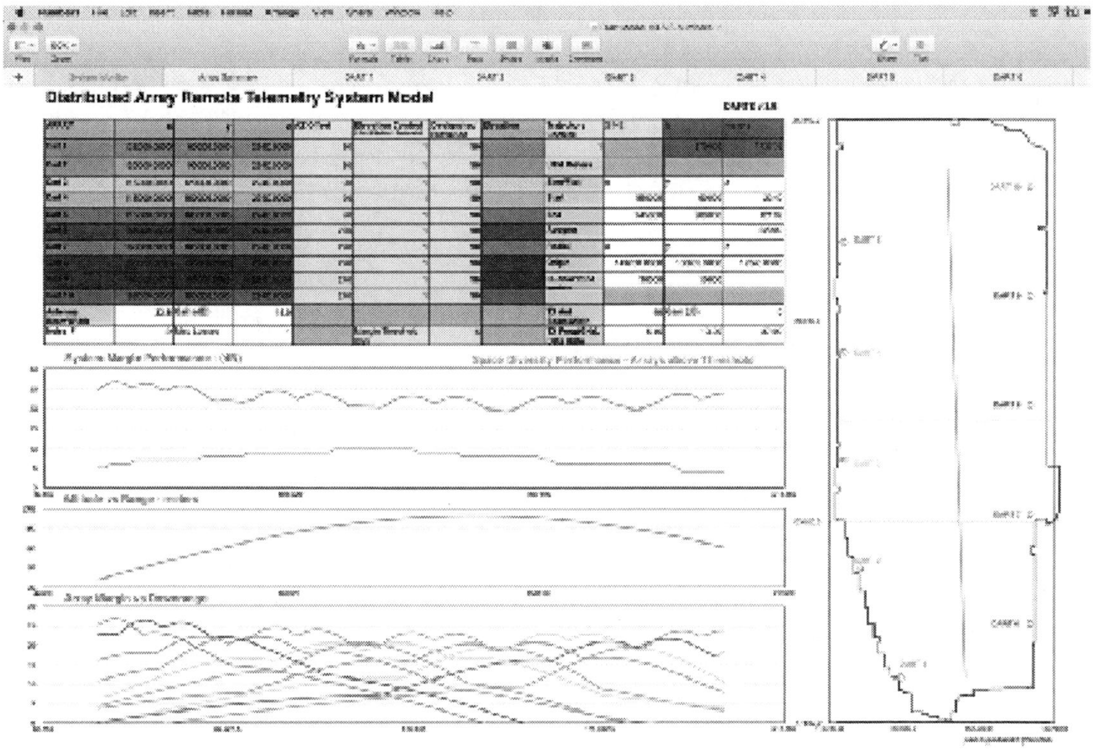

Figure 4 System Monitor Sheet

The Trajectory tab has three profiles that are the stimuli for evaluating how a staring array system might perform in a nominal range environment. Each profile contains 100 trajectory points and link calculations mentioned are made with all 60 elements using the geometry from each element (60,000 link calculations). If you have a 6 DOF simulation you want to use, just paste right into the sheet. Universal Transverse Mercator (UTM) coordinate system is used here, any cartesian system will work. You may edit coordinates to drive an array anywhere on the range and designate offsets or elevation modes. More on this table later.

The Array Summary sheet aggregates data from each subarray to build the System Analyses. Three plots on the System Monitor Page are actually copied from this page. These are the system Link Margin composites for each of the ten arrays, a classical “Altitude vs Downrange” trajectory plot and the System Link Margin Composite. Finally spatial diversity performance prediction of the number of arrays able to maintain acquisition at any point in the trajectory. At White Sands Missile Range three sites must be acquired to capitalize on the Post Detection Combining using NetAcquire Correlating Source Selectors that realize equal or better performance than any single tracking asset on the range. Subarray performance is calculated on the sheet bearing it’s name (i.e. DART 1). Simply select an array in the sheet tabs to view it’s data. A composite plot is created for each subarray and viewed just below table 1 in any of the “DARTS” sheets.

For Excel Users

The Excel export of this sheet sometimes requires a little help. The (right side) of the System Monitor Page is made up of two plots and a jpeg graphic, as shown in Figure 4. When exported to excel the layers may not position properly.

When calibrating the map it is necessary to overlay the plots to achieve the map shown at far right in Fig 5. When selecting the plot or graphic keep in mind you will have difficulties selecting the graphic unless it is in "Front". To achieve that, go to Excel's "Arrange" feature and select what is on top and send to the "back". This may need to be done twice to get this right. Overlay the two plots (Array Locations and Trajectory Plot), (see Figure 5) resize one plot so

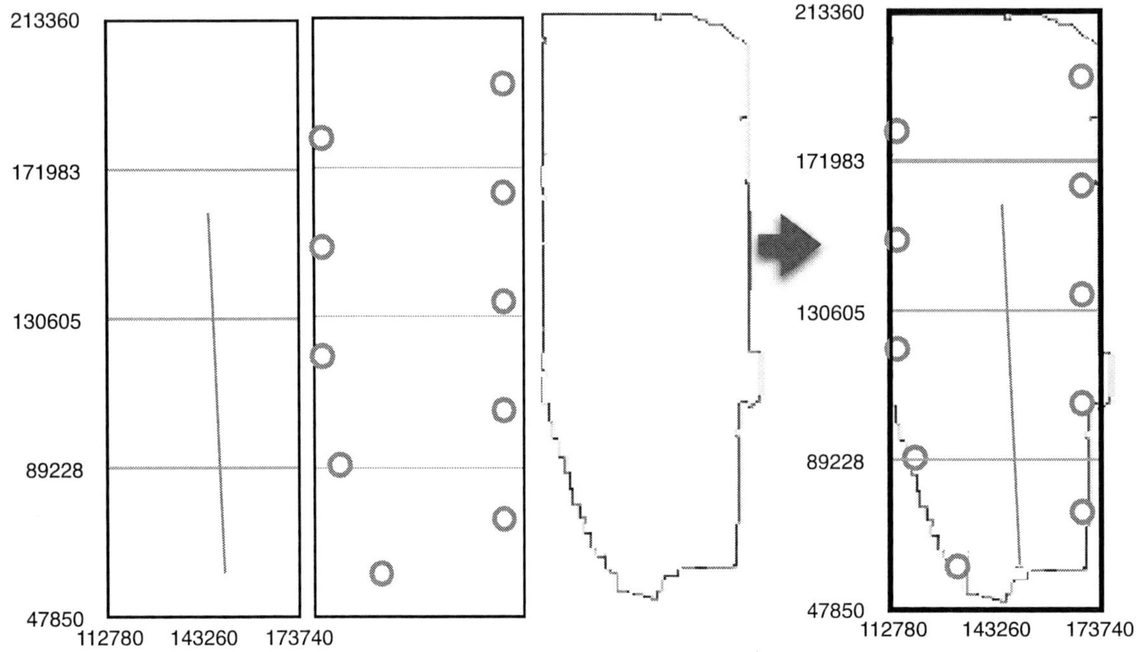

Figure 5 Range Map Components

that the borders over lay each other. Move them to the back and then resize the jpeg Range Boundary graphic so it fits the plot borders. Send the jpeg Range Boundary graphic to the back so that the other two plots maybe be viewed. Using range boundary coordinates would eliminate this step. The Excel export keeps improving and so this step may not be necessary before long. Once you've done this, the program is set up and ready to go. The Numbers version since native is ready to go out of the box.

Using the Model

System Monitor Table changes propagate through out the model. Subarray coordinates used in this application were arbitrarily chosen. Each array's bore-sight azimuth and elevation mode is commanded here. The Trajectory simulation is also controlled here. Sine, Flat and Radial trajectories are selected by entering 1,2 or 3 below the cell labeled Trajectory Control.

Normally, we would just play with the subarray coordinates shown on the System Monitor table. Blue cells pertain to subarray DART 1, DART 2 is green. and so on. Punch in different coordinates, drive the subarray anywhere on the Range and observe position update on the map.

The first simulation is a Sine profile where the start, end and apogee points of the trajectory are assigned (see Figure 4). The Flat trajectory uses the same start and end points but maintains the start altitude throughout. The radial trajectory is defined by its origin, altitude, start radius and end radius, as shown in Figure 6.

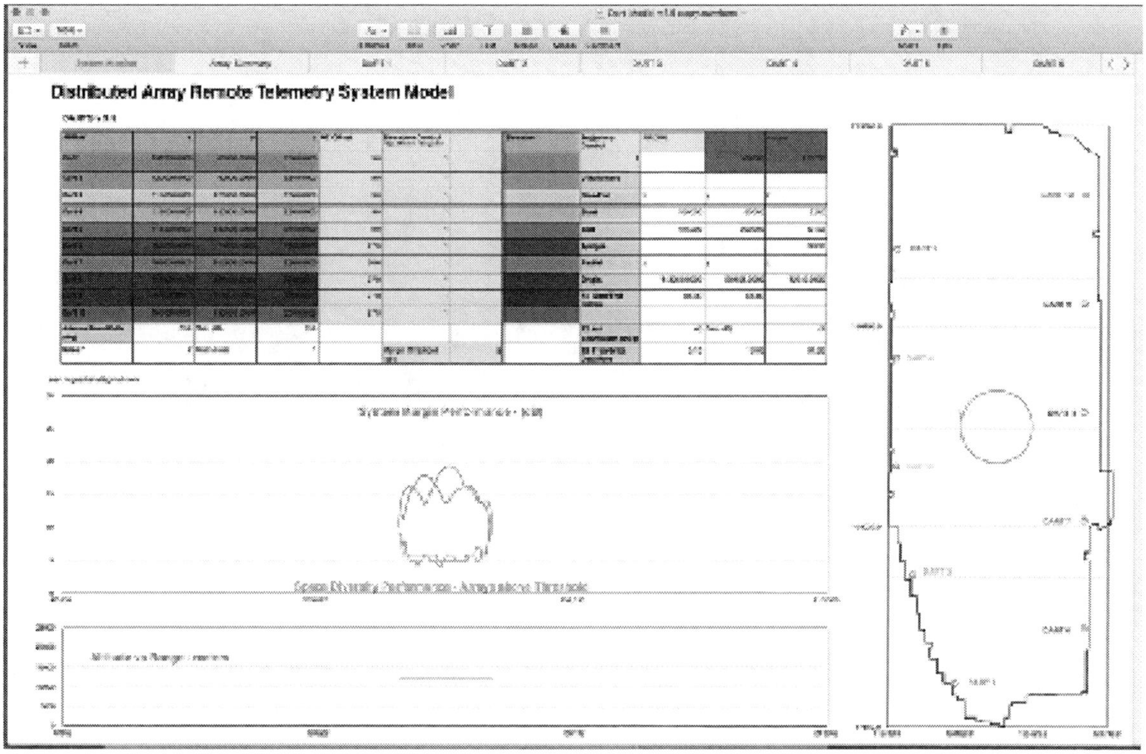

Figure 6 Radial Trajectory

The DARTS1 sheet takes each data sample from the Trajectory Sheet and does the link calculation for each receiving antenna in the subarray. There are 100 samples in this case, but can be as many as desired. Each Antenna Block calculates the displacement and relative orientation to estimate effective receive and transmit antenna gain and calculate space loss.

At each element (antenna block) the offsets and parameters may be inspected. Make all Global changes on the System Monitor Table, changes that affect a subarray should be made in the array block (upper left most) part of the appropriate DARTS sheet. To change the characteristics of a particular antenna use the specific antenna block. Above each respective block are two plots that report the performance of that antenna. Upper plot is Margin vs Downrange, the lower plot is Margin vs Time. When the trajectory is set to radial the upper plot looks strange, but the lower plot is normal. Below the table there is a composite plot for the subarray.

Modeling Experiments

Transmit Antenna Pattern Study

Simple investigations are done simply by playing with the trajectory or with array coordinates. DART sheets let you study the array's individual antenna pattern and in this case the transmit antenna pattern.

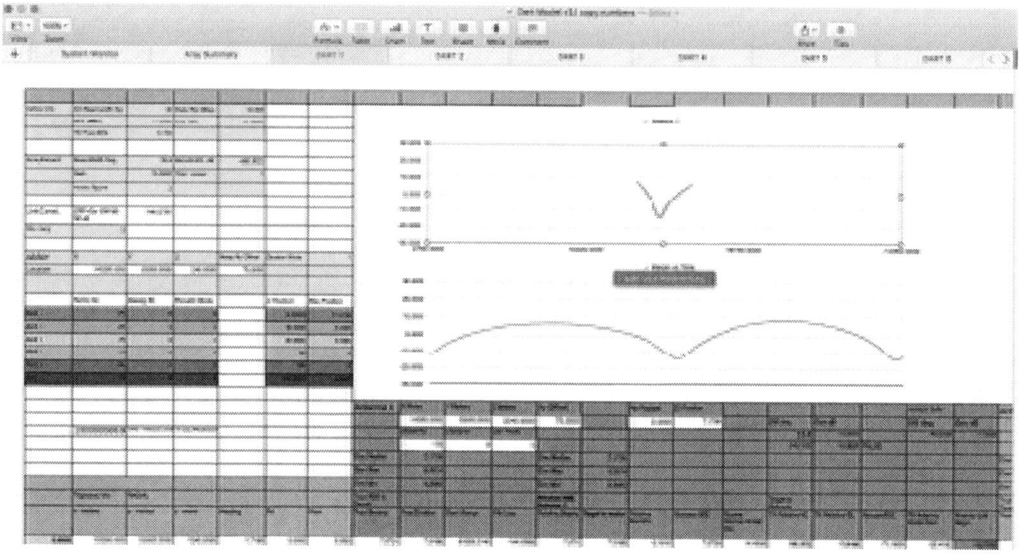

Figure 7 Transmit Antenna Pattern

Commanding the DARTS 1 array azimuth offset to 75 degrees, allows the radial trajectory volume to be contained in DART 1 - Antenna 1's Beam. We can take a look at antenna transmit power as the target revolves around the radial. The classical axial null is evident in both plots. It is helpful that the results are plotted both vs downrange and vs time.

Array Antenna Pattern Study (Receive)

Set DARTS1's coordinates to the Radial Trajectory origin (change X and Y, leave Z unchanged). The System Monitor reflects DART 1's location as shown in Figure 8. The Transmit antenna gain column (as viewed in the DART 1 sheet) shows transmit antenna gain actually is pretty stable and very close to zero dB. This is interesting and quite similar to data captured at an antenna range or anechoic chamber. The plot is affected by the receive pattern, but normalized our test range becomes an antenna range.

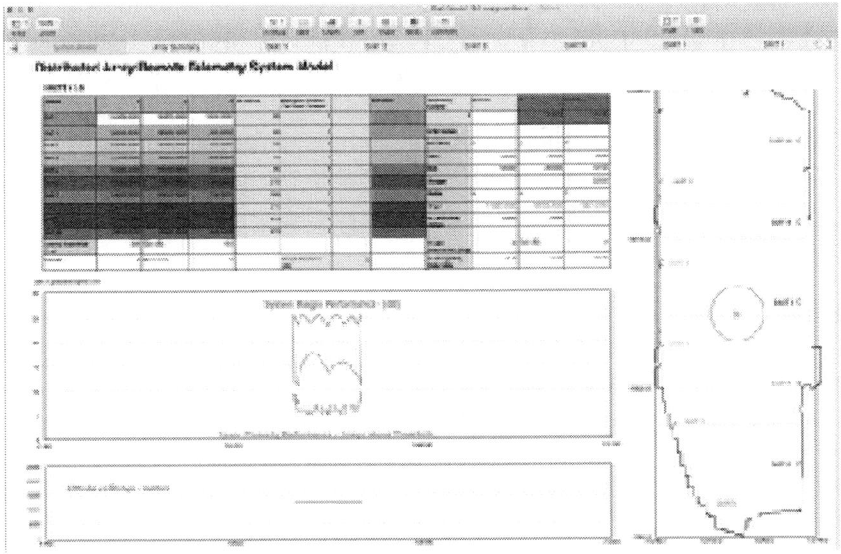

Figure 8 Radial Origin and DARTS1 Array co-located

Simulation

To create the simulation, a mock up of the Graphic User Interface is created on the spreadsheet for the DART 3 subsystem shown in Figure 9.

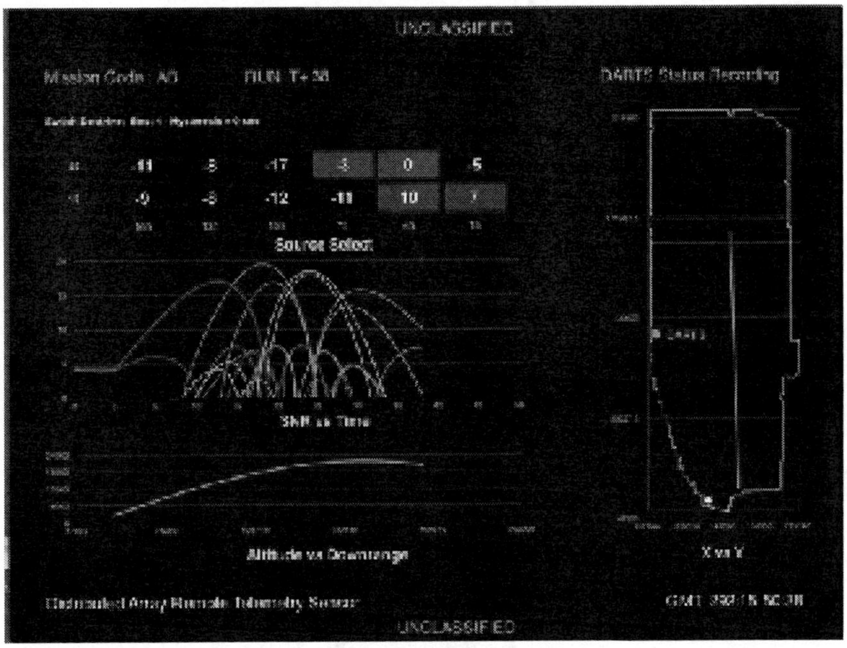

Figure 9 DARTS Simulation

Again a 12 element DART subsystem is simulated here. The Range Map appearing on the right hand side is the map from the model. Several pieces of the model are reused in the simulation such as altitude vs range plot and antenna signal power plots. New is an indicator for each of the 12 elements. The logic in this model identifies four elements with the greatest received signal power and accentuates them in blue. The DART sequencer switches antenna signals to the receivers with the intent of processing only the best four signals. The four signals are subjected to post detection combining to achieve spatial diversity and improve data quality. The stepper feature advances the simulation and Quicktime's screen recording feature captures the GUI and narration as it is stepped through 100 trajectory samples.

Conclusion

The Model and the Simulation are pleasing to the eye. As can be seen by inspecting the model, the number of DARTS subsystems can be reduced without impacting the data quality significantly, demonstrating its use as an analyses tool. The model serves as a mission planning tool facilitating the evaluation of candidate locations of each DARTS without having to physically move an array, provide power, network access then test. Configurations can be evaluated prior to changing components such as receiving antennas or LNA components.

Staring Arrays are compelling because they do not miss-track, nor do they need to be manned. The number of vehicles that can be acquired with a classical tracker is one. This constraint is relieved with staring array systems. The model's most significant contribution is demonstrating staring arrays will work.

Related papers

A Brave New C Band Architecture   J. Guadiana   ITC Proceedings 2015

Authors contact information

Juan Guadiana	JGuadiana@Telemetronics, Inc.
James Baird	JBaird@Telemetronics, Inc.
Curtiss Tackitt	CTackitt@BlueOrigin.com
Anna Dubois	anna@MannatekSolutions.com
Dorse Dubois	Dorse@MannatekSolutions.com

Acknowledgements

Brecken Uhl   Scientific Value LLC

Our friends, the professionals at Pax River and White Sands Missile Range.

Steered Conformal Array for the Rotating Airframe Missile

Kevin Cook, Kathy Bowland
Advanced Concepts Laboratory
Georgia Tech Research Institute
Atlanta, GA, 30318
[Kevin.Cook, Kathy.Bowland]@gtri.gatech.edu

Theresa Brunasso
D&S Microwave
Snellville, GA, 30078
Theresa.Brunasso@dsmicrowave.com

ABSTRACT

Recent reallocation of L-Band spectrum that was heavily used to support Aeronautical Mobile Telemetry has led to a significant spectrum access shortfall. The DoD was authorized to use portions of C-Band to address this shortfall. One system directly affected by this is the Rotating Airframe Missile (RAM).

An antenna system for the RAM is being developed that will switch between sections of a conformal antenna array, activating small subarrays at any given moment. This approach minimizes the variation in gain for any particular aspect angle as the RAM rotates, reducing the required transmit power. The antenna aperture is designed using GTRI's fragmented aperture technology to provide a very efficient radiator in the space provided. Small subarrays of this conformal array are passively combined directly behind the aperture. Inside the RAM body there are two actively controlled switches that alternatingly switch which subarrays are fed, always ensuring that the two adjacent subarrays pointing in the desired beam direction are active. These switches are controlled by an on-board algorithm using data captured from the RAM's Inertial Measurement Unit (IMU) and rate sensors to determine the attitude of the missile at all times and dictate which subarrays should be active.

INTRODUCTION

In January 2015, the FCC closed bidding on the AWS-3 auction. This auction sold off portions of L band that were previously used for telemetry needs, reallocating that spectrum for commercial use [1]. The DoD was authorized to use portions of the C-band spectrum to address the resulting shortfall. The existing telemetry system on the RAM uses a portion of the sold spectrum, so a redesign is required to move this to the newly assigned portion of C-band. As part of this auction, $500M was set aside to fund R&D efforts necessary to facilitate transitions like the one described here. This effort is funded from this money under the Spectrum Access R&D Program.

The previous RAM telemetry antenna was an omnidirectional antenna, but this design approach is ineffectual at C-Band for two reasons: first, the amplifiers are not as efficient at this band, which would require higher power consumption and generate more heat; second, the higher directivity of the antennas would result in more variation in signal strength received at the base station as the platform rolls. This paper describes the proposed antenna array that GTRI is designing to allow the RAM's telemetry system to move to the C-Band, first giving an overview of the technical approach, then detailing the system requirements, the radiating elements, and the beamformer, and wrapping up with an overview of how the system will be fabricated.

TECHNICAL APPROACH

GTRI is designing a conformal fragmented aperture antenna array that will wrap around the telemetry section of the RAM. This array will be divided into subarrays, each covering a discrete angular sector. Onboard rate sensors and accelerometers will be used to determine the attitude of the missile and a pair of switching beamformers to select which subarrays are excited at any given time. Figure 1 shows a conceptual rendering of this system. Switching between antenna elements to steer a beam has been successfully demonstrated previously [2], [3]. The two switching beamformers will feed passive splitters sending the RF signal to two subarrays simultaneously. Figure 2 shows the rotational element positions for a single cross-section of the array. The solid circle encloses the elements of a driven pair of subarrays and the two dashed circles show which elements would be driven for the two adjacent beams. Using this dual-beamformer approach allows for higher gain beams than a single beamformer driving the same size subarray because of the increased active aperture size of adding a second driven subarray. The angular coverage is kept to the fraction of the cylinder covered by a single subarray, resulting in better cross-over points than if the larger aperture size were fed with a single beamformer. Figure 3 shows the array factors for a driven beam and its adjacent beams calculated at either end of the band of interest.

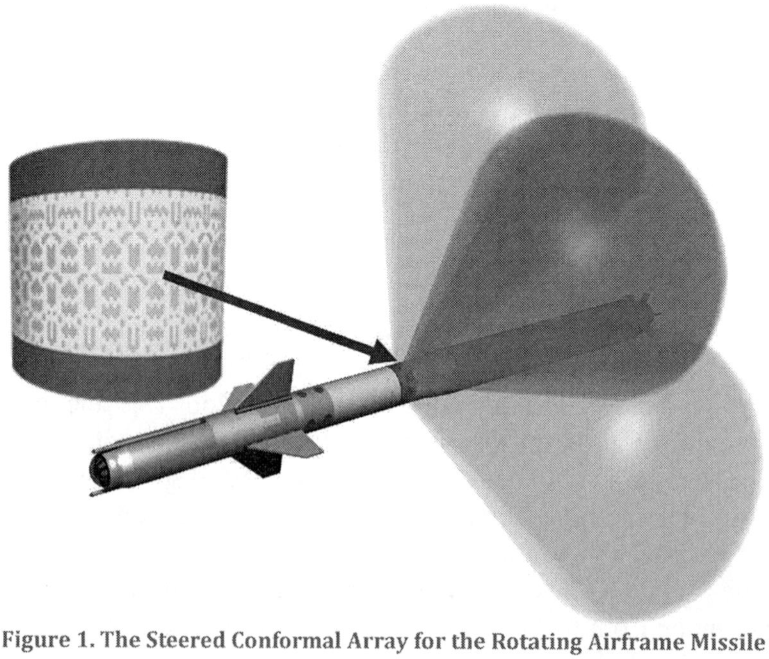

Figure 1. The Steered Conformal Array for the Rotating Airframe Missile

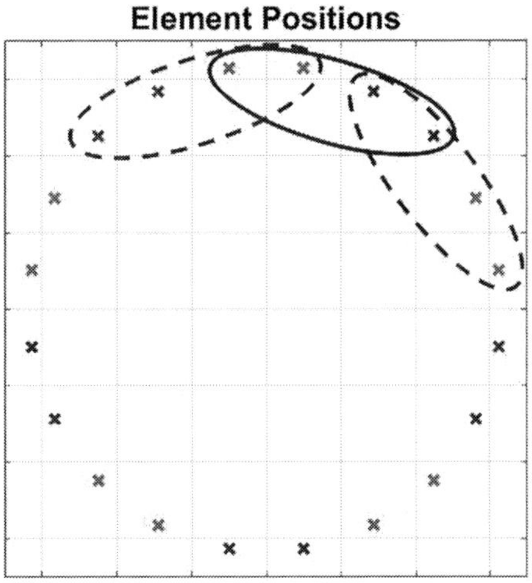

Figure 2. Driven Elements for a Single Beam Shown with Driven Elements for the Adjacent Beams.

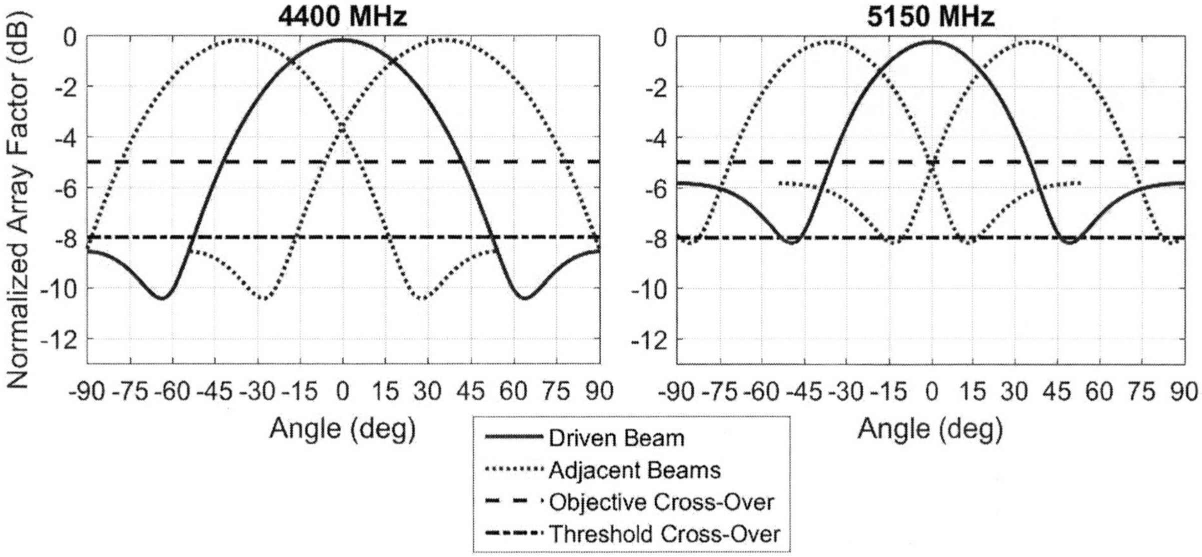

Figure 3. Calculated Array Factors at Both Ends of the Band of Interest

REQUIREMENTS

The requirements for this effort are detailed in Table 1. These requirements are intended to guide the technical design choices made for this effort to allow the system to best meet the telemetry needs of the RAM. The Axial direction is defined as the direction along the axis of the missile. Axial beamwidth is the width of the beam in the plane that is parallel to this axis. The antenna must achieve the -7 dBiL gain in order to exceed the gain of the previous omni-directional antenna, otherwise the stated goal of reducing the required transmit power wouldn't be achieved.

Table 1. System Requirements

Metric	Requirement
Frequency of Operation	4400 MHz – 5150 MHz
Polarization	Single Linear
Gain	Objective: As high as possible without requiring an on-board amplifier. Threshold: -7 dBiL
RAM Rotation Rate	Objective: 30 Hz Threshold: 10 Hz
Acceptable Variation in Gain with Airframe Rotation	Objective: 5 dB Threshold: 8 dB
Antenna Dimensions:	Thickness: 0.150" Axial Length: 2.938" Outside Diameter: 5"
Axial Beamwidth:	Threshold: 90° Objective: 170°

RADIATING ELEMENTS

The radiating elements designed for this effort employ fragmented aperture technology as described in [4]. A fragmented aperture is a patterned metal surface that relies heavily on computational electromagnetics and computer optimization. The antennas resulting from this design method are not restricted to canonical structures, allowing for highly efficient antennas that routinely approach the theoretical limit of antenna performance based on a uniformly illuminated aperture. These antennas can be tailored to particular requirements or installations. Some examples where this technology could be leveraged include wide bandwidths [5], conformal applications [6], and custom radiation patterns [7].

Careful antenna optimization setup is necessary to achieve the desired antenna characteristics. In the rotational dimension, the antenna pattern is dominated by the array factor described in the technical approach section. To achieve the desired beamwidth in the axial direction, the available length is divided into four antenna elements and amplitude array beamspoiling is used to exchange broadside gain for a wide beamwidth axial pattern. To mitigate performance issues due to edge effects from the elements on the top and bottom of this axial length, each edge is allowed to optimize independently. GTRI has shown this approach to result in significant reduction of edge effects and improved efficiency for small arrays [8]. Due to the amplitude beamspoiling, the optimal edge-element feeds have minimal effect on the antenna performance so they have been replaced by "dummy" elements that are just patterned copper and have no feed. This significantly simplifies the antenna passive beamformer design and the antenna construction. Additionally, letting the edges optimize independently frees up the design to add features to these elements that are not in the driven elements, like the mounting screws shown in Figure 4.

Figure 4. Optimized Array

This antenna design process can result in a very large solution space. For the application described in this paper, each radiating element is defined by 42 bits, resulting in 4.4 x 10^{12} different possible element designs. Freeing the top and bottom elements to correct for edge effects, as described in the previous paragraph increases the number of unknowns to 126 and the number of possible solutions to 8.5 x 10^{37}. Evaluating all possible solutions is not feasible, so a genetic algorithm is employed to efficiently find one of the best cases. This optimization technique will find the optimal solution in most cases while evaluating a very small fraction of all possible solutions [9].

Figure 5 shows the simulated principal plane pattern cuts for a pair of driven subarrays, the axial pattern on the left and the rotational pattern on the right. Asymmetry in the antenna pattern is introduced by the independently optimized edge elements, which are crucial to achieving the beamwidth requirements. Figure 6 shows the broadside antenna gain, which is well above the threshold value of -7 dBiL across the band.

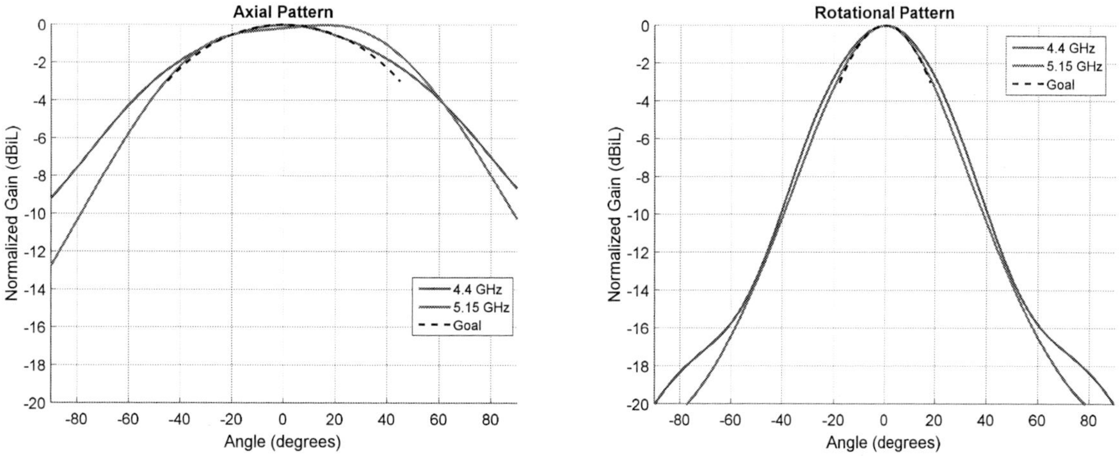

Figure 5. Principal Plane Patterns for a Single Driven Subarray

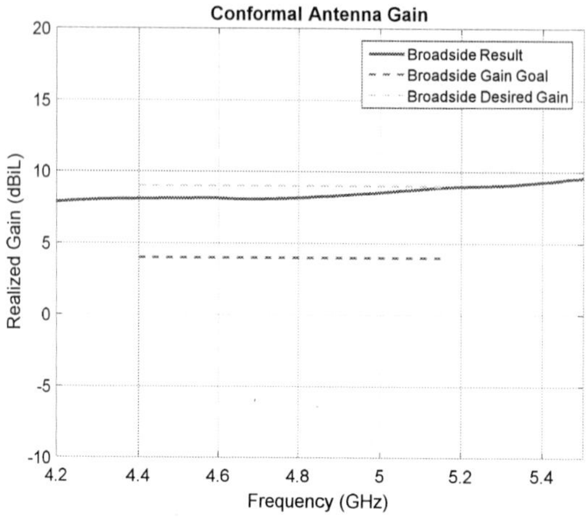

Figure 6: Broadside Antenna Gain

BEAMFORMER

As shown in Figure 7, the switched beamformer consists of a power divider and two single throw, five pole switches (SP5Ts). As the RAM rotates, the two SP5Ts will alternate switching through their five positions, so that two adjacent ports are selected. This results in the ten positions listed in the table included in Figure 7.

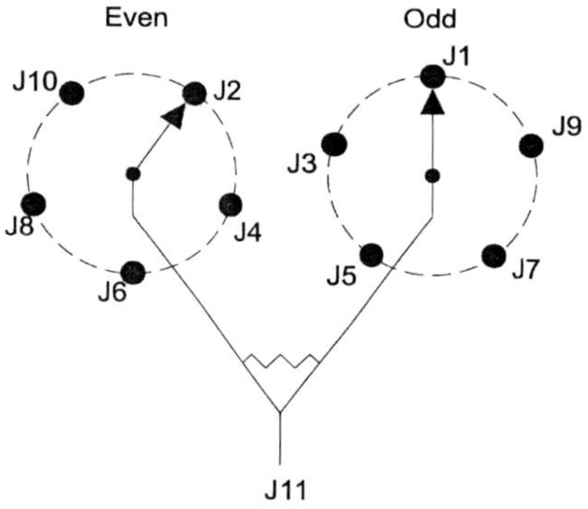

	Selected Port	
Position	Odd	Even
1	J1	J2
2	J3	J2
3	J3	J4
4	J5	J4
5	J5	J6
6	J7	J6
7	J7	J8
8	J9	J8
9	J9	J10
10	J1	J10

Figure 7. Switched Beamformer RF Schematic and Positions

To fit in the RAM, the design approach was to keep the switch boards as small and lightweight as possible. This goal has the added advantage of keeping the RF traces short which minimizes losses. A MACOM PIN diode with integrated bias network was selected for the SP5Ts. For the objective rotation rate of 30Hz, each switch will have to switch every 6.7ms, well within the part's 80ns capability [10]. To keep the RF traces narrow and for consistent operation over temperature, the SP5Ts are mounted on Rogers TMM10. The 1.5" diameter switch board with its SP5T and drivers is shown in Figure 8. Short RF cables will connect the switch board to the passive beamformers. The input port is offset slightly so that when the two switch boards face each other with the input ports lined up, the output ports going to the antenna subarrays will be staggered.

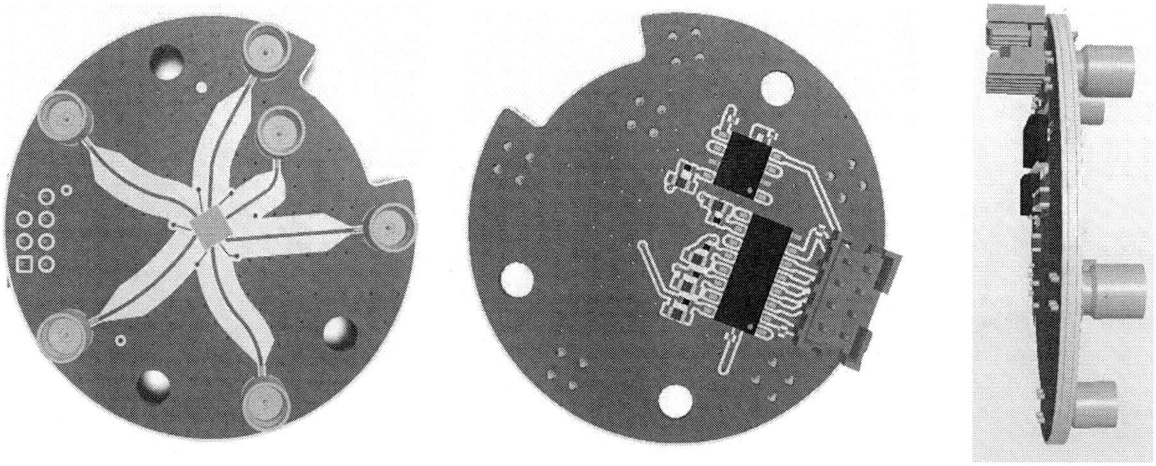

Figure 8. Switch Board

A separate board connects the two switch boards. Like the switch boards, the divider board is designed to be compact and lightweight. A Mini-Circuits divider was selected based on is compact size and excellent performance over the frequency band. The divider board is shown in Figure 9.

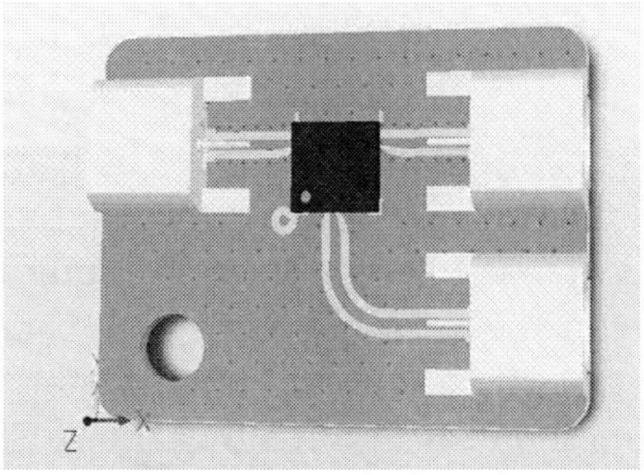

Figure 9. Divider board

FABRICATION

The antenna and passive beamformer assembly will be conformal to the exterior surface of the RAM. The antenna artwork will be etched on Taconic TLY-5 substrate. The passive beamformer assembly will be a stack-up combining TLY-5 and FR-EZ22p prepreg that will be fabricated using traditional Printed Circuit Board techniques. The space between the two PCB components will be filled with Cumming Microwave C-Stock AK-500 plastic, which is rigid and must be machined to shape. The antenna, plastic, and passive beamformer will be glued together under vacuum into a single flush stack-up. Figure 10 shows an early test that was performed using leftover PCB boards from a previous program. Pins will be soldered between the passive beamformer outputs through the dielectric foam to the antenna feeds. The full conformal antenna assembly will be mounted to the RAM body with 40 metal screws on the edge of the antenna, allowing for the removal of the antenna assembly if necessary.

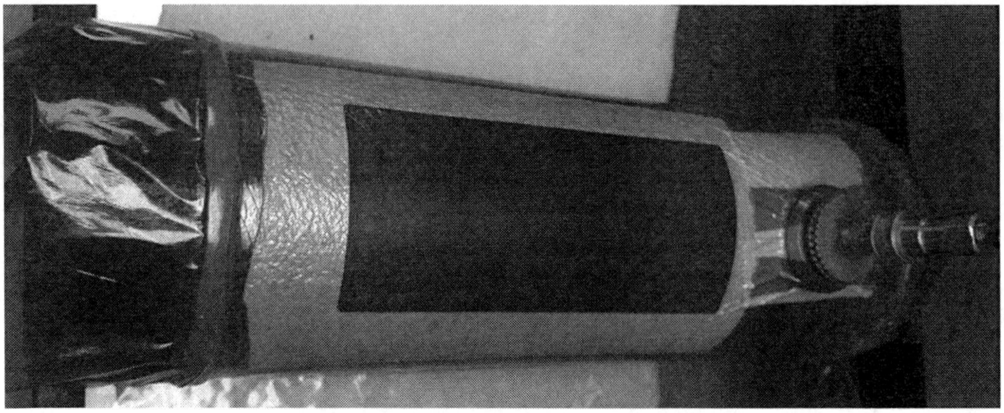
Figure 10. Testing the Assembly Bonding Process

The PCBs described in the beamformer section will be fabricated using traditional PCB techniques. The press-fit SMP connectors on the active switch boards will attach to small sections of semi-rigid coaxial cable that will connect to the subarray feeds that will pass through the shell of the RAM to connect to the antenna assembly. Also inside the cylinder will be a small control board that will connect to the switch boards by ribbon cables.

CONCLUSIONS

This system will allow telemetry data from the RAM to be transmitted to the ground stations during a test in a very efficient manner. Driving a pair of subarrays in the alternating fashion described increases the gain vs using a single subarray, but allows the angular coverage to remain narrow. This results in an aperture gain that is well above the threshold requirement while also achieving cross-over points between beams that will keep the variation in gain well below the objective. By spoiling the beam along the axis of the RAM, the beamwidth is widened enough to meet the threshold.

ACKNOWLEDGEMENTS

This project is managed by the Test Resource Management Center (TRMC) and funded through Spectrum Access R&D Program via Picatinny Arsenal under Contract No. W15QKN-15-9-1004. The Executing Agent and Program Manager work out of the AFTC. This material has been approved for public release; distribution is unlimited; 412TW-PA-17388.

Any opinions, findings and conclusions or recommendations expressed in this material are those of the author(s) and do not necessarily reflect the views of the TRMC, the SAR&D Program and/or the Picatinny Arsenal.

The authors would like to thank Kathy Gilbreath for generating the rendered system image shown in Figure 1, John Morse for the work he did laying out the printed circuit boards and generating the renderings shown in Figure 8 and Figure 9, and Jeramy Marsh for the test assembly shown in Figure 10.

REFERENCES

[1] AWS-3 Transition, National Telecommunications & Information Administration (NTIA), https://www.ntia.doc.gov/category/aws-3-transition

[2] W. L. Williams et al, "Lightweight agile beam antennas for UAVs using injection molded materials," IEEE AP-S International Symposium, Honolulu, HI, June 2007.

[3] H. Wang, Z. Zhang, Y. Li, M. F. Iskander, "A Switched Beam Antenna With Shaped Radiation Pattern and Interleaving Array Architecture," IEEE Trans. on Antennas and Propagation, vol. 63, no. 7, pp. 2914-2921, July 2015.

[4] J.G. Maloney, M.P. Kesler, P.H. Harms, T.L. Fountain, and G.S. Smith, "The Fragmented Aperture Antenna: FDTD Analysis and Measurement," Millennium Conference on Antennas and Propagation (AP 2000), Davos, Switzerland, 4 pages, April 2000.

[5] US Patent #6323809, "Fragmented aperture antennas and broadband antenna ground planes," J. G. Maloney, M. P. Kesler, P. H. Harms, and G. S. Smith, Nov 27, 2001.

[6] R. T. Lee, J. G. Maloney, B. N. Baker, and D. W. Landgren, "FDTD in Curvilinear Coordinates using a Rectangular FDTD Formulation," in 2011 IEEE Antennas and Propagation Society International Symposium and USNC/URSI National Radio Science Meeting, 3-8 July 2011, Piscataway, NJ, USA, 2011, pp. 2326-9.

[7] J. G. Maloney, B. N. Baker, R. T. Lee, G. N. Kiesel, and J. J. Acree, "Wide scan, integrated printed circuit board, fragmented aperture array antennas," in 2011 IEEE Antennas and Propagation Society International Symposium and USNC/URSI National Radio Science Meeting, 3-8 July 2011, Piscataway, NJ, USA, 2011, pp. 1965-8.

[8] K. R. Cook, B. N. Baker, J. Perez, D. J. P. Dykes, D. W. Landgren, "Independent Edge-Element Optimization For Efficiency of Planar Arrays," 2015 Allerton Antenna Applications Symposium, September 22-24 2015, Monticello, IL, USA.

[9] J. G. Maloney, R. T. Lee, D. W. Landgren, "Genetic algorithms for fragmented aperture antennas: A complete evaluation of a 24-bit design," in 2013 IEEE Antennas and Propagation Society International Symposium and USNC/URSI National Radio Science Meeting, 7-13 July 2013, Lake Buena Visa, FL, USA, 2013, pp. 115.

[10] MACOM TDS for MA4SW510B-1. Rev. V4. M/A-COM Technology Solutions Inc. (MACOM). https://cdn.macom.com/datasheets/MA4SW510B-1.pdf

Spectrum Access R&D (SARD) Program: An Update on the Conformal C-Band/Multi-band Antenna Project

Scott Kujiraoka, Russell Fielder, Maxim Apalboym and Michael Chavez
NAVAIR-Point Mugu and China Lake, California

ABSTRACT

In September 2016, work was initiated on the subprojects which comprise the Conformal C-Band/Multiband Antenna project: SARD #1: Broadband Conformal C-Band Missile Wraparound Antennas; SARD #2: Beam Switching Array Antennas; SARD #3: Multiband Conformal Antennas for Aircraft Applications; SARD #4: High Altitude Coronal Efforts on Antenna Performance; and SARD #5: Small, Medium Gain Multiband Receive Antennas. A brief status of each of them will be discussed and detail the technology areas being developed by each.

KEY WORDS

C-Band Telemetry, Conformal Wraparound Antennas, Spectrum Access, Spectrum Selloff.

BACKGROUND

The Spectrum Access Research and Development (SARD) Program is a $500 million Program created by the Federal Government selloff of frequency spectrum to commercial wireless companies such as Verizon, AT&T, Sprint and T-Mobile. The funding is used to support the augmentation of telemetry operations at Government Test Ranges to operate in the C-Band frequency range. Since this project is mainly related to Telemetry Operations, it is being managed out of the Test Resource Management Center (TRMC).

BROADBAND CONFORMAL C-BAND MISSILE WRAPAROUND ANTENNAS

The initial increment of funding was received at NAWC-WD, however it is was so small that it provided for only minimal support. Additional funding has now been received so all efforts can fully commence. Below are the efforts that will be developed under this project.

Further studies will be conducted with the assets developed under the previous C-Band Telemetry (TM) S&T effort. These studies would include conducting additional test flights over the water to further characterize the effects of multipath on C-Band TM reception. Contrary to the original theory that the multipath effects operating at C-Band would be less disruptive than at S-Band; initial flight data while taxiing on the flight line prior to takeoff showed that the multipath interference was greater at C-Band to the point of not receiving C-Band TM data until the test asset was airborne. Using data retrieved from these additional test flights flying close to the surface of the water, studies will be conducted to mitigate the effects of multipath on TM data reception.

Once again taking assets from the original C-Band TM S&T effort, continue the development on the C-Band TM antenna to cover the entire frequency range (4400-5150 MHz) as well as stabilize the antenna gain over this frequency range. The original design will be optimized to simplify the fabrication of the antenna and to increase the operating frequency range of the antenna. Some design tradeoffs may have to be done such as stabilizing the antenna gain at the expense of decreasing the operational bandwidth of the antenna.

BEAM SWITCHING ARRAY ANTENNAS

In a rolling missile, in order to maximize gain and directivity, it is desirable to be able to steer the beam to always be pointing in the direction the TM receiving stations. By accomplishing this feat, the benefit would be the need for lower transmitter power and antenna gain. This would partially alleviate the overheating issue within the TM unit as well as simply the TM antenna design. In addition, the developed system needs to maintain the ability to track a spinning five inch diameter missile with a Roll Rate in the range of 10-30 Hz.

This subproject was awarded to Georgia Tech Research Institute (GTRI). For their initial design, a fractal antenna was used to achieve the broadband antenna performance. Multiple elements were combined to cover 60 degrees around the circumference of the missile. This design included a microprocessor design to control a switch matrix to turn on and off antenna elements. This involves the use of an Inertial Measurement Unit (IMU) to determine the location of each antenna element as it was spinning. Currently there are issues locating a reasonably inexpensive IMU which can track Roll Rates as high as 30 Hz. A more detailed discussion of this topic is covered in another paper presented at this Conference entitled "Steered Conformal Array for the Rotating Airframe Missile".

MULTI-BAND CONFORMAL ANTENNAS FOR AIRCRAFT APPLICATIONS

In order to support the augmentation of C-Band TM operations in Navy jets, the existing upper (Door 4) antenna panel has to replace the existing S-Band TM antenna with a tri-band (L, S and C-Band) one. This subproject has been awarded to Toyon Corporation (Goleta, CA). They are currently awaiting the finalization of the requirements document in order to be fully engaged in the development effort. Originally Space Time Coding (as defined in RCC-106-15) was to be used to address the issues of antenna nulling caused by the close proximity of the two panels. However it has since been dropped from the Requirements Document.

HIGH ALTITUDE ANTENNA PERFORMANCE UNCERTAINTY

Like all of the other subprojects, this Technology Gap called out in theTri-Service C-Band Roadmap Study (TSCRS see Reference [1]) was submitted by the US Army at the White Sands Missile Range (WSMR). Due to increased atmospheric losses of operating in C-Band and other effects, these TM operations will require an increase in the power requirements. As a result, TM operations in C-Band will lead to unquantifiable coronal ionization discharges of RF/EMI prior to the RF energy radiating from antenna elements. In order to characterize this issue, high altitude chamber tests are planned to investigate the interaction of C-Band frequencies coupled to RF transmission components in the presence of low density gasses injected into a vacuum. It should be noted there were no bidders in the original Call for Proposals. The personnel at WSMR have re-edited the solicitation announcement so that the Request for Proposals (RFPs) will be resubmitted during the next phase of the SARD Program.

SMALL, MEDIUM-GAIN MULTI-BAND ANTENNAS

This Technical Gap from the TSCS Report (Reference [1]) was submitted by Aberdeen Proving Grounds (APG). They had required a small (2'-4' in diameter) ground station antennas for surface vehicle weapons systems test. This requirement came from the need to have high slew rate tracking antennas for close in surface test flights. Vendors claim comparable performance to single band tracking antenna systems, however actual testing of multi-band antennas have shown degraded performance in the lower L-Band. APG's lack of commercially available small, medium-gain (15-25 dBi) multi-band tracking antennas has driven the requirement for this effort. This subproject was also awarded to Georgia Tech Research Institute (GTRI). They are also using a Fractal Broadband Antenna to cover the three frequency bands (L, S & C Band). In addition, they are using an in-house design to support the 8:1 Combiner Design which will be used in the Digital Beamforming Network.

FOLLOW ON THOUGHTS

The original business model of SARD was to send all $500M to Picatinny Arsenal for contract award. Unfortunately, they did not anticipate that use of 'in-house' government resources to perform some of the development work. As a result it has been extremely difficult having funding sent directly to Government Field Activities. Hopefully in the Follow-on $500M Program, many of these funding issues can be straighten out.

REFERENCES

[1] Tri-Service C-Band Roadmap Study (TSCRS): Gap Analysis and Technology Shortfalls Report (dated 24 September 2014).

CHALLENGES IN THE DESIGN AND IMPLEMENTATION OF AIRBORNE TELEMETRY PROCESSING SYSTEMS

John Otranto, Bill Eckman, Dana Irvin, Felix Tao, Kirill Lokshin, Amit Puri
Ingenicomm, Inc.
14120 Parke Long Court #210, Chantilly, VA 20151
jotranto@ingenicomm.com

ABSTRACT

While typical telemetry processing systems are fixed, ground-based assets, certain mission profiles or telemetry acquisition models may involve telemetry processing systems which reside on other platforms, such as ships, mobile vehicles, or airplanes. The design and implementation of telemetry processing systems for these platforms poses unique challenges, which may include requirements for unusual mechanical packaging, heightened electromagnetic sensitivity, or specialized electrical interfaces. This paper presents some of the key challenges involved in the design and implementation of an airborne telemetry processing system and discusses how lessons learned from solving these challenges may be applied to future telemetry processing system designs.

KEY WORDS

Ruggedized hardware; airborne system; ATR chassis; MIL-STD-810.

INTRODUCTION

The conventional modern architecture for telemetry processing systems began to emerge in the mid-1980s with the development of systems like NASA's TCDS (Telemetry and Communications Data System) and NCPS (Network Command Processing System). With the evolution of commercial x86-based computer systems in the early 1990s, telemetry systems converged towards the 19-inch rackmount computer server as a common form factor, with the average system footprint usually being around 3U to 4U. Over time, this form factor has in many cases shrunk to a 1U to 2U system, largely as a consequence of a reduced need for custom hardware interfaces, but has otherwise remained the default for a typical ground-based telemetry system.

Despite being ubiquitous within ground segments around the world, conventional rack-mounted telemetry capture and processing systems are not suitable for every mission profile; many missions, particularly within the defense sector, require mobile telemetry processing capability that can be transported on a variety of platforms, including aircraft, ships, or ground vehicles.

While attempts have been made to utilize conventional telemetry processing systems in these roles, it is typically difficult for such systems to operate within the extreme environments often posed by mobile platforms. Consequently, providing reliable telemetry processing capability for these missions can require the design and implementation of a completely new system architecture, which is designed from the ground up to meet the unique environmental requirements of a particular target platform. Such designs pose a number of unique challenges which are not present in the design of conventional, ground-based telemetry processing systems.

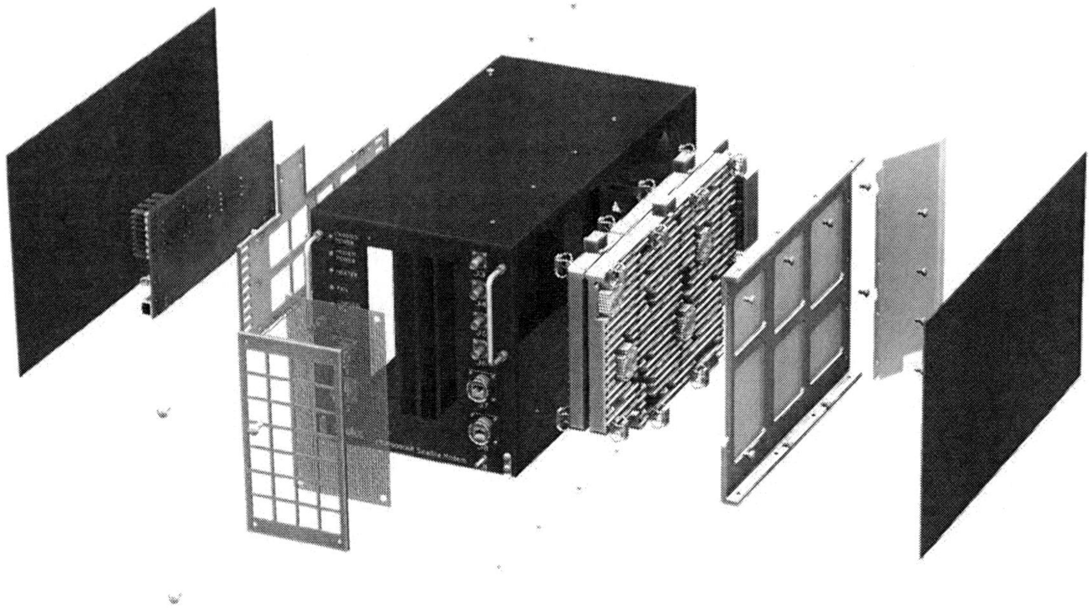

Figure 1: Airborne Modem – Exploded View

Figure 1 shows an exploded view of a representative airborne modem / telemetry processor, which will be referenced as a case study in this paper. The example airborne modem includes many of the same commercial off-the-shelf (COTS) components that might be included in a conventional, ground-based unit; however, these components are integrated into a specialized Aircraft Transport Rack (ATR) chassis, and are coupled with specialized environmental control hardware that manages the environment inside the chassis. The integrated unit is designed to meet the MIL-STD-810 [1] and DO-160C [2] standards for deployment and operation aboard a typical military aircraft.

DESIGN AND IMPLEMENTATION CHALLENGES

The ruggedization of COTS components for use in an airborne system presents many design challenges; when equipment that is designed for use in a commercial environment is used in airborne applications, care must be taken to ensure that the commercial equipment is not exposed to environmental extremes it was not designed to handle. Typical environmental and operating challenges for airborne deployment include:

- Temperature: COTS electronics are typically designed to work in a temperature range from 0°C to 70°C, while the operating temperature range for avionics is typically from -45°C to 120°C. Further, while conventional electronics are typically deployed in a temperature-controlled environment, such as an enclosed server room, this is not the case for airborne electronics, which often experience rapid changes in temperature as an aircraft maneuvers.

- Airflow: Unlike a conventional server room, which typically features extensively filtered airflow, the operating environment on board an aircraft can contain extreme amounts of dust, sand, and other particles. In addition, airflow on board the aircraft can vary wildly depending on the specific flight profile and operating altitude.

- Vibration: For typical commercial electronics, vibration is not a concern during operation. Avionics are regularly exposed to vibration extremes while an aircraft is in flight; the vibration parameters specified in the DO-160C and MIL-STD-810 specifications are well outside the operating range of most COTS equipment.

- Pressure: COTS ground electronics, which remain at a fixed altitude, rarely need to operate with any pressure change at all, much less the variable pressure environment specified in DO-160C.

- Power: +28V DC is a common power source used in avionics, and commercial equipment used in avionics must be adapted to this power source. Additionally, commercial electronics must be designed to comply with EMC, EMI, and static electricity requirements specific to the environment on board an operating aircraft.

- Humidity: Commercial electronics may require modifications to handle the humidity requirements that avionics are commonly subjected to while an aircraft is in flight.

The following sections discuss several of these key challenges, and present some of the unique solutions necessary to address them.

KEY CHALLENGE: TEMPERATURE

Commercial electronic components and assemblies are typically designed to be operated at a significantly narrower range of temperature extremes that is typically found in avionics applications, and use of such components on board an aircraft requires an enclosure that keeps the commercial equipment within a safe temperature range. Doing so requires active monitoring and control of temperature, as well as fail-safe power control to ensure that the electronic components are not operated when the temperature is outside of the acceptable range.

In designing the representative airborne modem, a number of discrete measures were adopted to provide resilient temperature regulation for the commercial electronic components, which included a set of particularly temperature-sensitive demodulator boards.

As a first step, the components were enclosed within a custom-machined solid aluminum heatsink superstructure. Adding heat dissipation fins to the surface of the superstructure enabled the boards to effectively conduct heat into the aluminum, which then transfers this heat to an air stream forced across the reverse side of the superstructure. The components would normally be cooled by direct convection; this convection process was thus transferred to the aluminum superstructure. A thermal analysis of this cooling mechanism is shown in Figure 2.

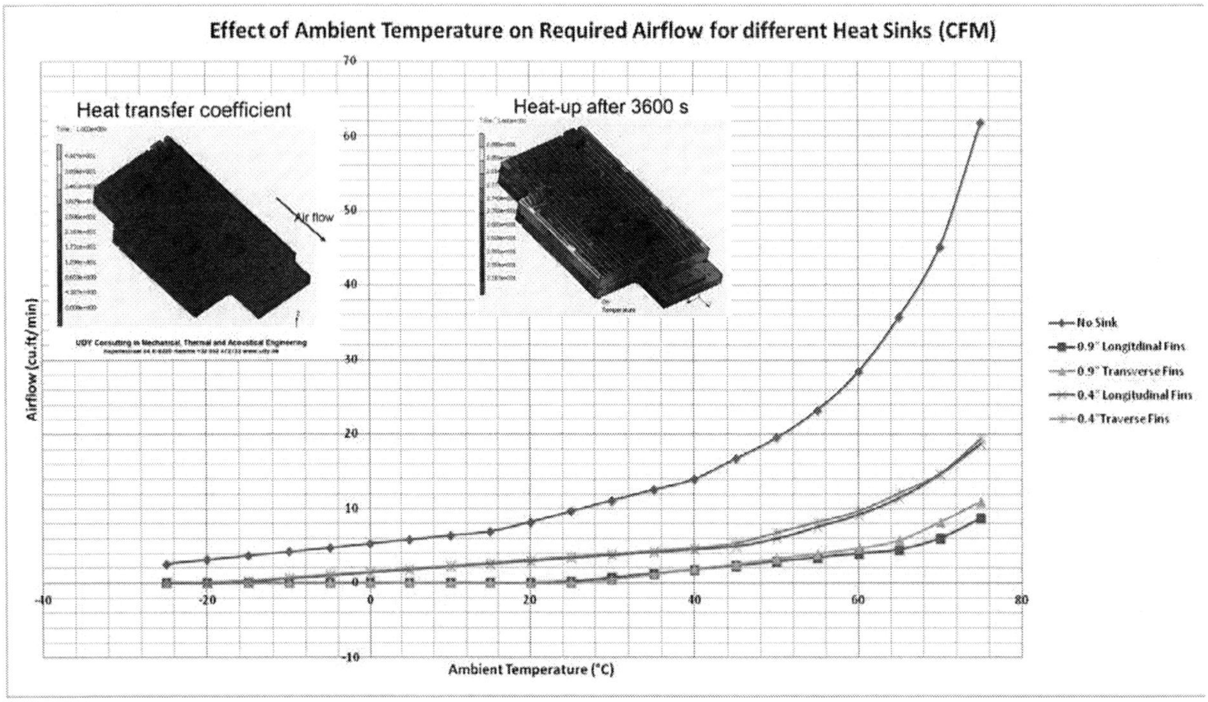

Figure 2: Thermal Analysis

In order to ensure sufficient heat transfer between the heat-generating board circuits and the aluminum structure, gap padding and pyrolytic graphite sheeting were added to the demodulator boards, simultaneously electrically isolating the board components from the superstructure and providing high thermal conductivity across the connection gap.

However, purely passive temperature control is not sufficient to meet the environmental requirements for airborne operation, particularly on the lower end of the aircraft's ambient temperature range. These controls were therefore augmented with a dynamic active temperature regulation mechanism. Inside the representative modem, an embedded controller (shown in Figure 3) monitors, in real time, the variations in its own temperature, the temperature of the demodulator boards and the aluminum superstructure, and the system intake and exhaust air temperatures.

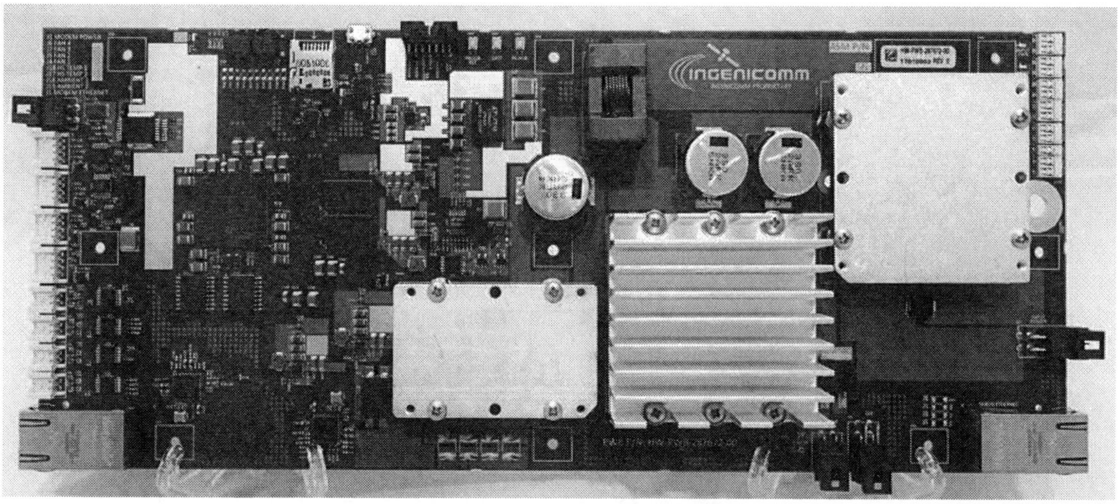

Figure 3: Thermal Controller

Using this data, the controller autonomously controls power relays that feed current to the modem components, integrated heating elements, and system ventilation fans, providing dynamic temperature adjustment and preventing the temperature-sensitive components from being powered on if the temperature is outside their safe operating range.

KEY CHALLENGE: AIRFLOW

To ensure sufficient airflow through the modem chassis, and particularly across the superstructure, a number of changes were required to the internal air distribution channels within the chassis; the changes, which are shown in Figure 4, ensured that airflow was biased to the superstructure and more equally directed across both sides of it.

To accommodate the extreme airflow through the system, a custom filter was designed to provide sufficient air inlets while preventing dust and debris from penetrating the unit. The filter was designed as a large, rectangular orifice in the middle of the front of the chassis, and consisted of a filter frame with thumbscrews at each of the four corners, which was designed to secure filter material behind the frame. The filter material chosen was corrosion-resistant 316 stainless steel wire cloth, woven in a 200 x 200 mesh, with a 0.0016" wire diameter.

While providing excellent dust filtration, this lightweight filter fabric proved insufficiently rigid to remain properly positioned inside the filter frame. As a result, a supporting structure was designed to permanently encase the filter fabric; the filter material is sandwiched between two layers of aluminum sheeting with large, square openings that mimic the openings in the filter frame.

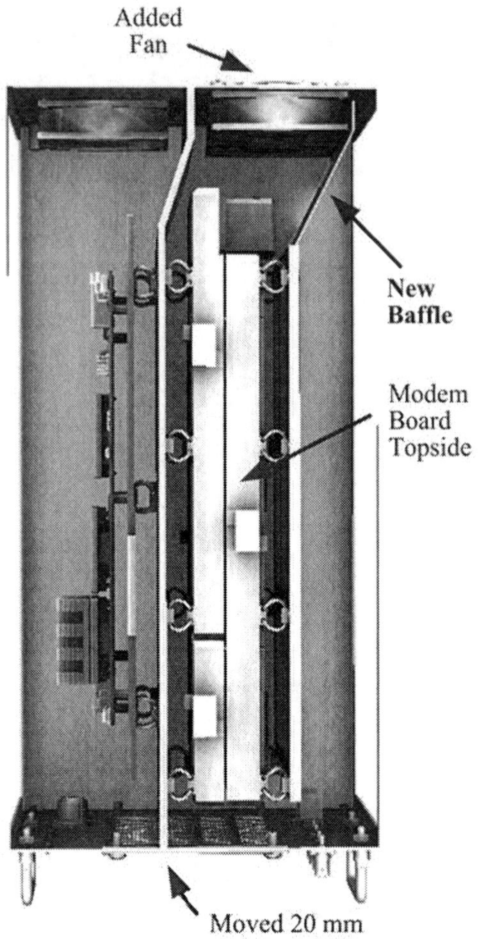

Figure 4: Internal Airflow Channeling

KEY CHALLENGE: VIBRATION

A number of the commercial components utilized in the representative modem design are highly sensitive to vibration. In particular, the modem includes a two-part demodulator board assembly, with a mechanically vulnerable connector joining the two circuit boards. In order to protect this connector, the aluminum superstructure, which encases the boards, provides structural support on both sides of the circuit boards and eliminates stress on the board set connector, which cannot endure any flex. Further, the entire superstructure assembly is mounted within the system chassis using wire rope isolators, preventing direct transfer of shock and vibration to the assembly.

Vibration isolation became a significant challenge during the final assembly of the modem chassis. Cable routing required careful planning to avoid excessive wiring inside the chassis but to allow sufficient slack for components mounted on isolators to move without stressing cable attachments to components that were not vibration-isolated, such as the fans and front panel connectors. Further, cables required secure routing mechanisms to prevent gradual movement over time. This objective was met by adding tie-down points for cables that were integrated with isolated components, eliminating the possibility of relaxed cables striking isolated components during operation.

CONCLUSIONS

The design of the representative airborne telemetry system described in this paper showcases some of the key design challenges and principles involved with airborne equipment, including unique design elements to account for environmental constraints and extremes with respect to temperature, airflow, and vibration. The demanding operating environment on a typical aircraft imposes requirements that are not commonly encountered in conventional telemetry system design, and forces system implementers to consider unique solutions to address them.

At the same time, the representative design demonstrates that a fully aircraft-qualified system can be developed using many of the same COTS components that would be present in a conventional, ground-based telemetry system. Such component reuse reduces the cost and risk of developing airborne systems by allowing significant system elements – including, in many cases, entire boards and assemblies – to be adapted from existing, operationally qualified telemetry systems.

REFERENCES

[1] MIL-STD-810G Working Group, (USA), *Department of Defense Test Method Standard, Environmental Engineering Considerations and Laboratory Tests*" MIL-STD-810G, Oct 2008.

[2] SC-135 RTCA, Inc., *Environmental Conditions and Test Procedures for Airborne Equipment*, RTCA DO-160C, Dec 1989.

Data Security Architecture Considerations for Telemetry Post Processing Environments

Jeff Kalibjian

DXC Technology Company

ABSTRACT

Telemetry data has great value, as setting up a framework to collect and gather it involve significant costs. Further, the data itself has product diagnostic significance and may also have strategic national security importance if the product is defense or intelligence related. This potentially makes telemetry data a target for acquisition by hostile third parties. To mitigate this threat, data security principles should be employed by the organization to protect telemetry data. Data security is in an important element of a layered security strategy for the enterprise. The value proposition centers on the argument that if organization perimeter/internal defenses (e.g. firewall, IDS, etc.) fail enabling hostile entities to be able to access data found on internal company networks; they will be unable to read the data because it will be encrypted.

After reviewing important encryption background including accepted practices, standards, and architectural considerations regarding disk, file, database and application data protection encryption strategies; specific data security options applicable to telemetry post processing environments will be discussed providing tangible approaches to better protect organization telemetry data.

KEYWORDS

encryption, key management, data security, disk encryption, file encryption, database encryption, application encryption, Hardware Security Module, FIPS 140-2, Common Criteria, tokenization, Format Preserving Encryption (FPE)

INTRODUCTION

Although encryption technology has been available to organizations for many years, it really wasn't until rather recently (2011-2012 period according to the Ponemon Institute 2016 Global Encryption Trends Study sponsored by Thales e-security [1]) that the number of organizations that actually had an encryption strategy outnumbered the organizations that did not. Today in the year 2017 an encryption strategy is a necessary requisite for all organizations.

Organizations deploy encryption technologies to protect classes of data in their organization that have great value or data that if compromised will result in penalties/fines (as required by industry or government governance mandates) or that will result in negative publicity. Some examples of these data classes include:

- Employee Personal Identifiable Information (PII)
- Intellectual property
- Customer Personal Identifiable Information (PII)
- Business partner information
- Organization financial information
- Customer credit card or bank account numbers
- Organizational records

When deploying encryption technologies organizations are trying to mitigate data loss risk. That is, if their external cyber defenses fail (e.g. firewall, IPS, IDS, etc.) adversaries will indeed be able to have access to the organization internal networks; but if sensitive data is encrypted; the adversary will only be able to retrieve encrypted data that will be useless to them. Organizations may also be trying to address insider threats. In other words, if sensitive data is encrypted with proper access controls on the cryptographic keys used to encrypt the data, even organization employees will be restricted in accessing data they should not be able to review.

ENCRYPTION CHALLENGES

The phrase in the last sentence, "with proper access controls on the cryptographic keys" is extremely important. If care is not taken in managing cryptographic keys then even though sensitive data may be encrypted it is still be very vulnerable to compromise because adversaries (whether internal or external to the organization) will be able to gain access to those keys and hence access to the sensitive data. Thus, in implementing an encryption solution particular attention must be paid to validating the proposed key management strategy.

Deploying encryption technologies in an organization can also be challenging because it will necessitate some amount of change to business process. The goal is to try to find a solution which will satisfy organization data security requirements while absolutely minimizing business process change. In the ideal world the recommended security model would be for the security to travel with the data and be persistently encrypted on or in use by any platform and application. However, implementing that level of security into an already functioning business can be difficult. Other challenges include: determining what data should be encrypted, initial deployment of the encryption solution, and training organization personnel to use the encryption solution effectively.

IMPORTANT SECURITY CERIFICATION STANDARDS

Certifications are important in security products as they imply that a neutral third party has evaluated how well a product adheres to important accepted standards. When evaluating encryption products it is very desirable that they have been engineered to meet accepted security standards. Two security standards are very important in the industry: FIPS 140-2 and Common Criteria.

The NIST Federal Information Processing Standard (FIPS) 140-2 [2] is a hardware specification that is used to evaluate hardware that protects cryptographic keys and performs sensitive cryptographic operations on data at rest and in motion. This hardware is often called a Hardware Security Module (HSM). There are four levels of merit in the specification, with each increasing level implementing more robust security safeguarding measures.

• **Level 1**—No specific hardware physical security is required. Instead, it primarily revolves around the use of approved NIST cryptographic algorithms, for example, Advanced Encryption Standard (AES) and Triple DES.

• **Level 2**—In addition to Level 1 specifications, requirements involve hardware fixtures on the computing platform that can show evidence of tampering. Examples of such fixtures would be pick-resistant locks or covers, coatings, or seals.

• **Level 3**—In addition to Levels 1 and 2 measures, Level 3 includes tamper detection and response circuitry to prevent any cryptographic keying material from being lost to an adversary.

• **Level 4**—In addition to Levels 1, 2, and 3 measures, Level 4 includes protection against a key compromise from environmental attacks, for example, temperature and voltage changes. Additionally, the system must be completely, formally modelled.

Systems that have achieved Level 3 or Level 4 evaluations are very robust; due to the difficulty in achieving those evaluations, Level 2 is more commonly attempted. In the financial services industry, only Level 3 or above evaluated devices are used for managing money. Certain National Security Agency (NSA) devices are most likely between Levels 3 and 4 and have enhanced protections against Electronic Counter Measures (ECM) oriented attacks.

Common Criteria (CC) is an International Organization for Standardization/International Electrotechnical Commission (ISO/IEC) standard [3] and is recognized internationally by 25 countries. These countries have signed a memorandum stating that a Common Criteria evaluation done in one country to a particular Evaluation Assurance Levels (EAL) will be recognized as if that evaluation was done in their country. It's important to note that the memorandum is signed with respect to a certain EAL level—specifically, EAL-4. For levels above this, some amount of re-validation (or total re-evaluation) by a laboratory in the specific country will be required.

Common Criteria evaluations are usually undertaken on pure software systems, although CC evaluations on devices—for example, copy machines or printers—are also done. Common Criteria is really a methodology for evaluating security claims about a product. It's not looking at the entire product, only its security features. It's an established methodology for not only creating a specification based on a those claims, but also a process for evaluating if they are valid.

The CC includes specifying a Security Target (ST) document that defines the security functionality and assurance methodologies used to verify that those functionalities are actually achieved. The desired EAL level being pursued dictates the rigor of how those assurances are verified. The ST document also defines the Target of Evaluation (TOE)—which is the product or subset of product for which the methodology will be performed.

A Protection Profile (PP) is a ST that is written in a general way for a particular product type—for example, firewall products or intrusion prevention products. This provides consumers with a framework to make an "apples-to-apples" comparison of products in the same category that have a common criteria evaluation.

The Common Criteria seven Evaluation Assurance Levels are:

- **EAL1—functionally tested**—Evaluation of TOE includes testing against the specification with respect to the documentation.
- **EAL2—structurally tested**—Design and test information is evaluated.
- **EAL3—methodically tested and checked**—There is a thorough evaluation of the TOE.
- **EAL4 —methodically designed/tested/reviewed**—This is the highest level a "retrofit" can achieve on an existing product line.
- **EAL5—semi-formally designed and tested**—Rigorous commercial development techniques, planned at design time, are used.
- **EAL6—semi-formally verified design and tested**—Development techniques are consistent with a TOE that is in a high-threat environment.
- **EAL7—formally verified design and tested**—Rigorous formal evaluation techniques require TOE to have very tight and well-defined security behaviors.

Observe, for CC, that the higher evaluation level only implies better security. It's assumed if a more rigorous evaluation—higher EAL level—was undertaken and confirmed, the more likely it is that the secure features were indeed implemented correctly. That being said, EAL-5 and above products are very robust and implies a formal security design process was carried out on the product. However, with an EAL-4 or below rating, the methodology could have just been "bolted on" without true security focus at the inception of design.

When evaluating a vendor Hardware Security Module or key management product, A FIPS 140-2 evaluation at Level 2 or higher is an absolute must. A Common Criteria evaluation of EAL 3 or above is also desirable, but not absolutely necessary. However a product with both FIPS 140-2 evaluation and a Common Criteria certification is more preferred. While enterprise encryption applications will utilize a HSM and/or hardware based key management as a back-end to provide key management and cryptographic functionality, the encryption application agent itself will likely be comprised of all software and run outside the HSM and/or key manager. In this case, it is very desirable that the application itself be Common Criteria evaluated to demonstrate that it has been architected with care to prevent compromise. These concepts are summarized in Table 1.

Table 1. Encryption Element Certification Desirability

Product	FIPS 140-2	Common Criteria
HSM	Must: Level 2 and above	Desirable: EAL 3 or above
Key Manager	Must: Level 2 and above	Desirable: EAL 3 or above
Software Application/Agent	Must: Level 1	Desirable: EAL 3 or above

BEST ENCRYPTION PRACTICES

There are basically four encryption approaches used to protect enterprise server and database data. In addition to the server/database encryption strategy, the enterprise must also be concerned with encryption strategies for data in transit, data when backed up, end-point data on employee devices and e-mail. However these are beyond the scope of this paper. An organization may not opt to solely pursue just one encryption approach. Instead, they may choose a hybrid approach in which two or more approaches are pursued in subsets of their IT infrastructure.

Disk Encryption. Encryption employed by the disk itself provides the least amount of data security to the organization. This is because when the disk spins up and the proper authentication/password information is presented, the entire content of the disk is available for decrypted use by the host operating system (OS) and any higher layer application with no further protections on the data except what the operating system might offer. Thus, the only real benefit of encryption at this level is if the disk is physically stolen, or inadequately erased when it is decommissioned. When powered down, even if the actual data on the disk is physically read it will be indecipherable. On power-up the system will not be able to decrypt unless the appropriate authentication/password information is presented.

File Encryption. Encryption used on files provides a fair amount of data security to the organization employing it. Typically in this mode of operation, there are two types of encryption technology: agents and agentless. If an agent is used, it would be deployed at the OS level to intercept all file read/write operations, evaluate established policy and determine if applications and/or users attempting to access data in files (flat files, database files, etc.) have a right to do so. If agentless technology is used, an encryption solution which operates as a storage proxy running as a 'bump-in-the-wire' is leveraged. It provides transparent encryption and decryption of contents stored in enterprise Network Attached Storage (NAS), Storage Area Network (SAN), Direct Attached Storage (DAS), tape libraries, Virtual Tape Libraries (VTL), Content Addressable Storage (CAS), cloud storage and various RESTful (Representational State Transfer) object stores for authorized hosts and applications. This technology moves the encryption processing to a FIPS 140-2 validated appliance and minimizes the workload on the storage server.

The truly enormous advantage of encryption at the file level is no modifications needs to be made to existing applications that use data in files or databases. However, the downside is that for the period of time the files are decrypted (this may occur when an application or even database administrator needs to access the data), all the data in the files are theoretically vulnerable to compromise. Thus, if files being decrypted for an application happened to be database files, then for the period of time those files were decrypted, all the data in the database would be potentially vulnerable. However, another advantage of this approach is that key management policies and data access policies can be centrally orchestrated with dedicated (and redundant) management applications controlling all the file encryption/decryption agents.

Database Encryption. Database encryption can almost be considered a subset of application encryption. Basically in database encryption very specific subsets of the data in the database can be encrypted utilizing the native encryption capabilities of the database. This implies that even when the DBA is in the database that data can still be protected from compromise by the DBA. The downside of this approach is that if the organization has many different vendor databases; each database will have its own key management and policy applications, making it somewhat tedious to centrally manage all the policies. These solutions also have to decrypt and re-encrypt when a key rollover occurs. This a time-consuming process which in many cases takes the database off line during the process.

Application Encryption. At the application level, very specific subsets of data can be encrypted. This granular control affords data in the enterprise to have the most protection as data is essentially encrypted to the very last moment until it is required by the application. Contrast this with file level encryption; in which, for example, an entire set of database files might be pre-emptively decrypted because an application is going to access one particular column in that database. While application level encryption affords the organization the best security; it also involves some complexity in that the applications will need to be specifically modified to leverage the granular encryption services. It is at the application level that both tokenization and Format Preserving Encryption (FPE) approaches can also be introduced. In tokenization a non-sensitive entity is substituted for a sensitive data item where the substitution method is very difficult to reverse if the tokenization system is not present. In FPE, the output of an encryption process is encrypted but in the same format as the input. For instance, if the input was in a credit card number format, the output would be in the same format. This allows the FPE encrypted items to be directly placed in the same database schema intended for use with the original non-encrypted data. Compounding the complexity is the need to migrate the existing data sets to the tokenized or format preserved encryption form. However, this is a one-time operation, and while that roll-over may necessitate a one-time interruption in the production environment; once it is done it never needs to be done again, because any new data coming in is immediately tokenized or FPE"ed" before coming into the production environment.

Overall application performance can be improved as the data is protected once and then is used in its protected state throughout the network. There is no need for encrypt/decrypt cycles at every server in the network. Security policy is based on the data element no matter where it resides; it is not specific to a platform.

ENCRYPTION IMPLEMENTAION IN TELEMETRY POST PROCESSING ENVIRONMENTS

The most preferred data security option to implement in telemetry post processing environments is application level encryption. The reason being the simple fact that with application encryption strategies (as discussed in Best Encryption Practices section above) one can absolutely minimize the amount of time any data element will need to remain unencrypted and thus potentially vulnerable to compromise. That being said the reality of the situation is that organizations will leverage one or more third party data analysis applications to analyze telemetry data. Those applications also need to manage the data they consume and may also utilize some combination of databases or flat files to store and manage the data. When a third party data analysis application is utilized, the organization is constrained by the data security features the vendor has implemented in that application.

Table 2 Preferred Options for Implementing Data Security in Telemetry Post Processing Environments

Encryption Preference	Strategy Description
Preference 1	Seek out vendors that have implemented application level encryption in their analysis application that has been FIPS 140-2 and Common Criteria evaluated
Preference 1A	Seek out vendors that have implemented database encryption in their analysis application that has been FIPS 140-2 and Common Criteria evaluated.
Preference 2A	If the organization has already purchased telemetry analytic tools that do not implement encryption use transparent file and database encryption features of a selected vendor encryption product that has been FIPS 140-2 and Common Criteria evaluated secure database and flat files the analytic application uses

Preference 2B	If the organization is implementing their own telemetry analysis application select an application encryption vendor whose products have been FIPS 140-2 and Common Criteria evaluated. Or consider leveraging database encryption.
Preference 3	Implement file level security on flat files and databases that contain telemetry data. Select an encryption vendor whose product is FIPS 140-2 and Common Criteria evaluated.
Preference 4	Deploy disks that implement disk level encryption. Favor disk vendors which have FIPS 140-2 and Common Criteria evaluation on their encrypted disk products

If solutions cannot be identified implementing application or database encryption with the analysis feature set desired, or if the organization has already procured applications not implementing application level security or database security; then the next best option is file level security. While not the most optimal, this strategy can still be quite effective as the file level encryption tool can transparently protect both databases and flat files without need to modify the application. The downside being, that when just one data element needs to be accessed that is in a large file with many data elements or a large database, the entire flat file or database must be decrypted! When this is done, the data element being accessed not only is vulnerable to compromise but all the rest of the data in the file or database that was also decrypted.

To mitigate risks in this circumstance, one can use the audit features of the file encryption solution to monitor access of the entirely decrypted file or database to ensure only authorized entities are accessing the data. While leveraging the audit feature provides the organization with another check, it is still possible other attack vectors might be used to access the decrypted data that the audit system may not be able to detect and record. And this is one of the reasons why the file based encryption strategy is only a second choice. However often times organizations must resort to this second option because their analytics application vendor implemented no data security at all.

File level security is still superior to disk level encryption because discrete sets of data can still be secured from general operating system level access (in separate files) even when the disk has been spun up and decryption password already entered. This is not true in disk encryption, once the disk spins up and the password is input to decrypt the disk, all the data on that disk is only protected from access by the host operating system access control features. The file level encryption vendors integrate their access control agents with the commonly used operating systems to provide a separate, robust, access control layer beyond the operating system.

If application encryption can't be used and for some other reasons file level encryption also is not an option; then disk level encryption should be deployed. It at least protects the organization against data compromise if disks are physically stolen. The encryption options are summarized in Table 2. Finally no matter what type of encryption solution is selected, it is imperative the solution has at least undergone a successful FIPS 140-2 evaluation at Level 2 or higher and preferable that it has also achieved Common Criteria evaluation at EAL Level 3 or higher.

SUMMARY

Application level encryption is the best way to protect a telemetry data product. Unfortunately, organizations may not be able to implement an application encryption strategy if their commercially procured data analysis tools do not support application encryption. In such circumstances the next most optimal approach would be to proceed deploying a file based encryption solution. Unfortunately, a third encryption option, self-encrypting disks, provide little practical data security once the disk has spun up and the password has been entered to decrypt the disk. This is because after the disk has been decrypted, only the host operating system access control protections are available to mediate access to the data on the disk which has already been decrypted. All encryption solutions (whether application, file or disk based) should be minimally FIPS 140-2 Level 2 evaluated and preferably also have undergone Common Criteria evaluation at a level of EAL of 3 or higher.

References

[1] Ponemon Institute LLC, ***2016 Cost of Data Breach Study: Global Analysis***, Ponemon Institute LLC, 2016, https://www.thales-esecurity.com/cpn/2016-global-encryption-trends-study, (Thales Inc. sponsored research)

[2] National Institute of Standard and Technology (NIST), ***Security Requirements for Cryptographic Modules***, National Institute of Standard and Technology, May 2001, http://nvlpubs.nist.gov/nistpubs/FIPS/NIST.FIPS.140-2.pdf

[3] International Organization for Standardization/International Electrotechnical Commission (ISO/IEC), ***Common Criteria for Information Technology Security Evaluation, Version 3.1, Revision 5***, International Organization for Standardization/International Electrotechnical Commission (ISO/IEC), April 2017, http://www.commoncriteriaportal.org/files/ccfiles/CCPART1V3.1R5.pdf

DEVELOPMENT OF A PORTABLE TM GROUND STATION FOR KSLV-II WITH A RS422-TO-ETHERNET TM PCM DECODER AND MAGALI TM SOFTWARE

Wonju Yoon and Seongmin Noh
Launcher Electronics Team, Korea Aerospace Research Institute
{wj.yoon, smnoh}@kari.re.kr

ABSTRACT

The onboard telemetry (TM) system of Korea Space Launch Vehicle-II (KSLV-II) transmits PCM frame data to the TM ground station via RS422 before flight or via RF link during flight. This paper describes the development of a portable TM ground station for KSLV-II that can decommutate the PCM frame data received via RS422. The developed portable TM ground station consists of two major components: a RS422-to-Ethernet TM PCM decoder and a laptop running MAGALI commercial TM software. The TM PCM decoder transforms a serial bit stream of the PCM data received via RS422 into multicast UDP packets. The MAGALI TM software supports an Ethernet acquisition module for the reception of the PCM frame data, which enables decommutation of the UDP packet-based PCM frame data received from the TM PCM decoder. Furthermore, the use of multicast UDP enables simultaneous decommutation of the PCM frame data at multiple laptops running MAGALI.

INTRODUCTION

Korea Aerospace Research Institute (KARI) has been developing the Korea Space Launch Vehicle-II (KSLV-II), a 3-stage launch vehicle, to put a 1.5 ton satellite into low orbit at 600$\sim$800 km altitude. Each stage of the KSLV-II has an independent onboard telemetry (TM) system whose architecture is based on the traditional RS-422 serial communication interconnection between one Master Data acquisition Unit (MDU) and multiple Remote Data acquisition Units (RDUs). The MDU gathers data from various sensors and other onboard electronic units, which are acquired by multiple RDUs and the MDU itself, and then generates PCM frame data according to IRIG Standard 106, Chapter 4. The generated PCM frame data are transmitted using randomized non-return-to-zero-level (RNRZ-L) code, which is defined in IRIG Standard 106, Appendix D, via two RS-422 interfaces: 1) one for a S-band TM transmitter that transmits the PCM frame data via RF link during flight, and 2) another for a TM ground station (TGS) that decommutates the PCM frame data to check the sensor data and the state of health of onboard electronic units before flight.

During the whole period of the onboard TM system development and the KSLV-II system integration, many diverse tests are conducted at multiple sites. Although the TM PCM frame data should be checked in real-time for each test, it is difficult to build TGSs on all the test sites. A portable TGS is thus required. We first used a commercial off-the-shelf (COTS) portable TGS, MAG_200/TGS-M from NEXEYA France. However, it could not provide enough portability for our needs, and furthermore it was difficult to adopt multiple portable TGSs because of their high cost. We therefore decided to develop our own portable TGS.

This paper describes the development of our own portable TGS to satisfy our needs for high portability and low cost. The developed portable TGS consists of two major components: 1) a RS422-to-Ethernet TM PCM decoder that is developed by KARI and 2) a laptop running MAGALI that is COTS TM software developed by NEXEYA France. The TM PCM decoder transforms a serial bit stream of the PCM frame data received via RS422 into multicast UDP packets and then sends them to Ethernet network. MAGALI running on a laptop decommutates the UDP packet-based PCM frame data received via Ethernet network. The details are described in the following sections.

MOTIVATION

As described earlier, we first adopted a COTS portable TGS, MAG_200/TGS-M. Although there are several COTS portable TGSs in the TM market, we should choose one based on MAGALI because of its compatibility with the Naro Space Center. The TGS in the Naro Space Center was built with MAGALI-based equipment for launching KSLV-I (Naroho), and it will be reused with a system upgrade for the launch of KSLV-II. MAGALI is COTS TM software developed by NEXEYA France, which provides an IRIG 106 Class II PCM decommutation module.

MAG_200/TGS-M has slightly heavy weight of about 20 kg, and although it can be carried by hand it is inconvenient. Frequent moving sometimes results in damage to the TGS hardware. Furthermore, its high cost makes it difficult to adopt multiple portable TGSs. These problems caused considerable inconvenience for the use of MAG_200/TGS-M and thus we decided to develop our own portable TGS satisfying the following requirements.

- Light weight for high portability
- Low cost for budget savings
- Easy replacement in the event of hardware failure for easy maintenance

Because MAGALI must be used for decommutation of the PCM frame data, the key aspect of the development of our portable TGS was the method of inputing the PCM frame data generated by the MDU to MAGALI. MAGALI supports some acquisition modules for the reception of the PCM frame data, and among them, we focused on the Ethernet acquisition module. It was because the Ethernet is a very widely used communication interface and all the laptops have at least one more Ethernet interface. If the PCM frame data are formatted to the Ethernet packets that are inputted to MAGALI with the Ethernet acquisition module, a general-purpose laptop can be used for MAGALI to decommutate the PCM frame data. The use of a general-purpose laptop can provide much more flexibility to build a portable TGS and many advantages of lighter weight, lower cost, and easier hardware replacement.

DEVELOPMENT OF A RS422-TO-ETHERNET TM PCM DECODER

The PCM frame data from the MDU of the KSLV-II has a bit rate of up to 10 Mbps and are transmitted via the RS422 serial interface. To utilize MAGALI with the Ethernet acquisition module on a general-purpose laptop for decommutation of the PCM frame data, we developed a RS422-to-Ethernet TM PCM decoder that transforms a serial bit stream of the PCM frame data received via RS422 into Ethernet packets to be inputted to the Ethernet acquisition module of MAGALI. Our TM PCM decoder was developed to meet the following functional requirements that are listed in the order in which they are processed by the developed TM PCM decoder.

1. Reception of the PCM frame data of up to 10 Mbps via RS422 interface
2. Bit-level decoding of the PCM frame data with bit synchronization and de-randomization
3. Extraction of each minor frame data by detecting sync word
4. Addition of IRIG-B timestamp at the detection of sync word to each minor frame data
5. Transformation of each minor frame data embedding IRIG-B timestamp into a multicast UDP packet and transmission of the packet

Figure 1 shows a functional block diagram of the developed TM PCM decoder. An Altera FPGA, an Ethernet controller, RS422 transceivers, and a flash memory were used for the development of the TM PCM decoder. Most of the required functions were implemented in the FPGA using Verilog HDL. There are four processing modules (PCM frame decoding module, IRIG-B time decoding module, Ethernet control module, and Nios II processor module) and two dual-port memories (data memory and configuration memory) inside the FPGA.

The Nios II processor is a 32-bit soft core processor for the Altera FPGA. The Nios II processor module is basically in charge of the access of the external flash memory where the configuration data are stored. The configuration data include the bit rate and the size of the PCM frame data, network setting for the TM PCM decoder (MAC address, IP address, multicast address, and so on), and feature options. When the power is turned on, the processor module loads the stored configuration data from the external flash memory and passes the data to the PCM frame decoding module and the Ethernet control module so that they can be initialized. The stored configuration data can be updated via Ethernet using developed user software, and the details of this are described later.

The PCM frame decoding module is the main processing module of the TM PCM decoder. It decodes the serial bit stream of the PCM frame data received via RS422 serial interface by performing bit synchronization and de-randomization. For the bit synchronization, the TM PCM decoder uses the bit rate of the PCM frame data that is pre-configured and stored in the external flash memory. In addition, the PCM frame data are encoded with RNRZ-L code and thus de-randomization of

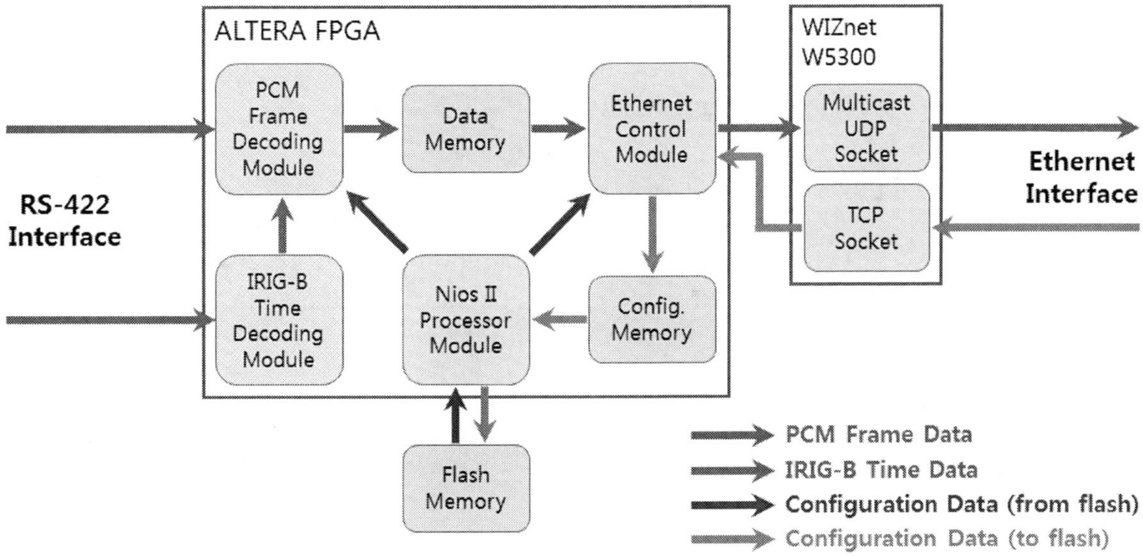

Figure 1: Functional Block Diagram of the RS422-to-Ethernet TM PCM Decoder

the data has to be accomplished together. The decoded bits are stored in the registers within the FPGA.

The next processing step of the PCM frame decoding module is searching for the sync word within the decoded bits. Every minor frame that constitutes the PCM frame data has the sync word at the beginning of the minor frame data. When the sync word is detected, minor frame data of a preconfigured size are extracted and written to the data memory. At this time, the timestamp at the detection of the sync word is appended to the minor frame data and written to the data memory together. The IRIG-B time decoding module provides IRIG-B time information in real-time to the PCM frame decoding module by decoding the received IRIG-B time signal. The provided IRIG-B time information is 8 bytes and has a resolution of 1 us. This decoding process repeats every minor frame.

The Ethernet controller IC used in the TM PCM decoder is W5300 from WIZnet. The W5300 is a hardwired TCP/IP embedded Ethernet controller that enables easier internet connection for embedded systems. Its internal resister can be accessed via 8/16 bit bus interface and internet connection using TCP or UDP sockets can be easily made by resister read/write. The W5300 can use 8 independent hardware sockets simultaneously.

The Ethernet control module within the FPGA controls the W5300 to interface with Ethernet network. Two sockets within the W5300 are used: one is the multicast UDP socket for transmitting the PCM frame data to MAGALI, and the other is the TCP socket for receiving the updated configuration data from the developed user software. When the power is turned on, the Ethernet control module first performs initialization of the W5300 for network configuration using the configuration data loaded from the external flash memory by the processor module, and then waits for the minor frame data to be filled in the data memory. If the entire data of one minor frame are filled in the data memory, the Ethernet control module reads the data from the memory and writes them to the tx registers for multicast UDP socket of the W5300, which generates a multicast UDP packet for the data and transmits it to the Ethernet network.

The Ethernet control module also processes the reception of the updated configuration data. The developed user software generates the configuration data updated by user and sends the data to the TCP socket of the W5300 via Ethernet. If the configuration data are received at the TCP socket of the W5300, an rx interrupt is issued to the Ethernet control module, which reads the data from the rx registers for the TCP socket of the W5300, writes them to the configuration memory, and generates an interrupt to the Nios II processor module to inform of the reception of the updated configuration data. Finally, the configuration data in the configuration memory are moved to the external flash memory by the Nios II processor module and the update process of the configuration data is completed.

The developed TM PCM decoder also has a simulation mode for test. When the simulation mode is enabled, the PCM frame decoding module is disabled and the Nios II processor pushes the dummy PCM frame data, which are generated every minor frame period, to the data memory. With this simulation mode, it is possible to perform the interface test with MAGALI without the MDU and its real PCM frame data.

The basic functional verification of the developed TM PCM decoder was performed using Wireshark, widely used network protocol analyzer software. Wireshark showed that the developed TM PCM decoder converts a serial bit stream of the PCM frame data into multicast UDP packets embedding the IRIG-B timestamp successfully. Figure 2 shows a picture of the developed RS422-to-Ethernet TM PCM decoder. The size is 13 cm X 7.3 cm X 2.6 cm and the weight is about 220 grams. The TM PCM decoder is powered via a USB so that it can be easily used with laptops. In addition, the manufacturing cost is very low because all used ICs including the FPGA are inexpensive.

BUILDING THE PORTABLE TGS WITH THE TM PCM DECODER AND THE MAGALI

When the development of the RS422-to-Etheret TM PCM decoder was completed, the next step in building our portable TGS was to interface the TM PCM decoder with the MAGALI. Because MAGALI already has an Ethernet acquisition module, the initial interfacing of the TM PCM decoder with MAGALI could be successfully completed after some trials and errors. However, the original MAGALI software did not have a feature to use the timestamp embedded in each UDP packet-based minor frame data when decommutating the PCM frame data. The original MAGALI generated timestamps at the reception of each UDP packet using the system time of the computer on which it was running, and this was used for the decommutation. Since packet transmission jitter exists in the Ethernet network, although minor frames are actually generated with a constant time interval of the defined minor frame period, the timestamp interval between the received UDP packets is not constant. This problem makes the analysis of the PCM frame data very difficult.

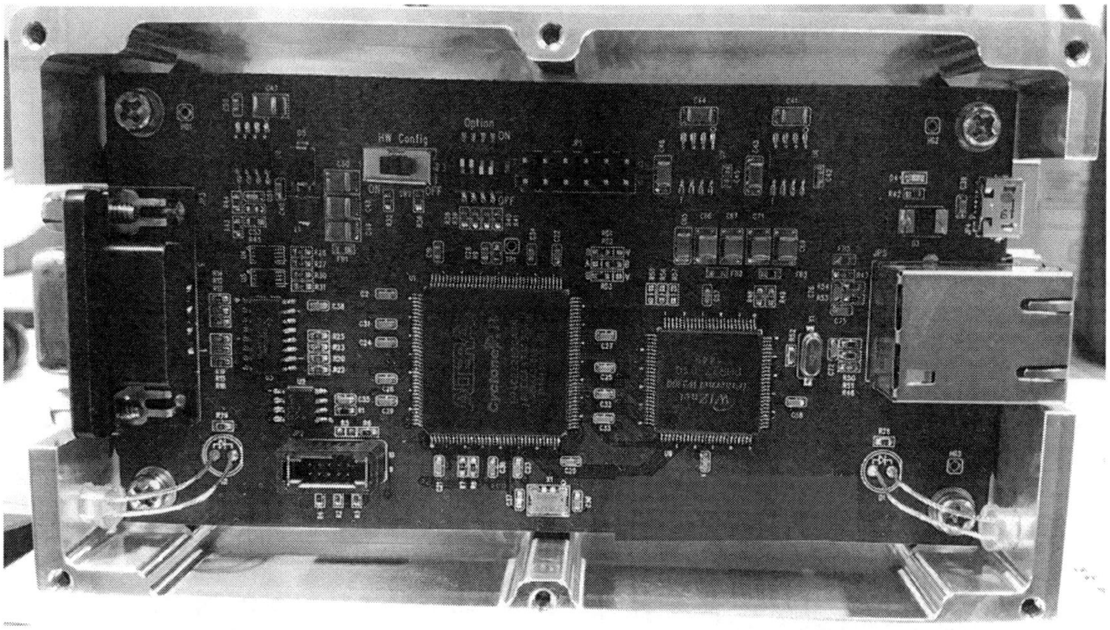

Figure 2: Picture of the developed RS422-to-Ethernet TM PCM decoder

To solve this problem, we required NEXEYA France, a developer of MAGALI, to add a feature to use the timestamp embedded on each minor frame for decommutation. After adding the embedded timestamp feature, the jitter problem could be clearly resolved and interfacing the TM PCM decoder with MAGALI could be completed successfully.

To build the portable TGS, any general-purpose laptop can be used if it has an Ethernet interface to receive multicast UDP packets from the TM PCM decoder and a USB port to supply 5 V power to the TM PCM decoder. Figure 3 shows a picture of one of our developed portable TGSs. Because an ultra-slim laptop whose weight is less than 1 kg is chosen, the total weight of our portable TGS, including a laptop with a power adaptor, a TM PCM decoder, and cables, can be less than 2 kg.

As described earlier, we had three main requirements for the portable TGS: light weight for high portability, low cost for budget savings, and easy hardware replacement for easy maintenance. As shown in the example presented in Figure 3, our portable TGS has very low total weight, which can provide extremely high portability. Because the manufacturing cost of the TM PCM decoder is low and a general-purpose laptop can be used for the portable TGS, most of the building cost of the portable TGS is occupied by the cost of the MAGALI software. Furthermore, when a hardware failure occurs, the low-cost TM PCM decoder can be replaced and the use of the general-purpose laptop also provides easy repair or replacement in the event of failure of the laptop. Consequently, our portable TGS satisfies the three main requirements.

Our portable TGS using an Ethernet interface and multicast UDP connections between the TM PCM decoder and the laptop running MAGALI offers some advantages. The multicast UDP packets sent from one TM PCM decoder can be delivered to multiple laptops, making it possible for multiple laptops running MAGALI to simultaneously decommutate the PCM frame data. In addi-

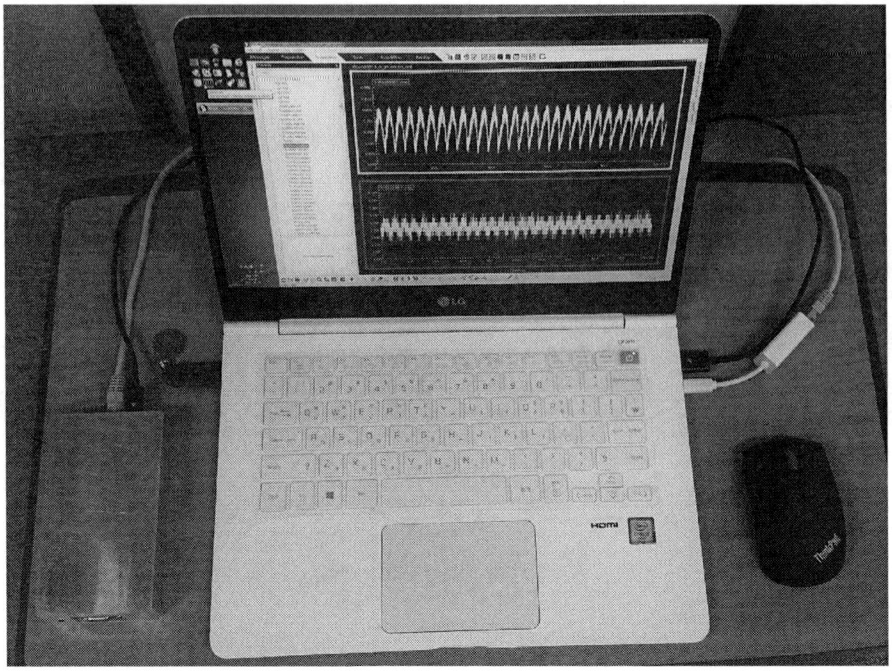

Figure 3: Picture of one of the developed portable TGSs

tion, the use of Ethernet allows the laptop to be remotely operated far from the TM PCM decoder. These features provide flexibility of the test configuration using our portable TGS according to variable test environments.

CONCLUSIONS

This paper described a method to develop and build our own portable TGS for KSLV-II. Our portable TGS consists of two major components: a RS422-to-Ethernet TM PCM decoder that is developed by KARI and a laptop running the COTS TM software MAGALI. The TM PCM decoder transforms a serial bit stream of the PCM frame data via RS422 from the MDU into multicast UDP packets in units of minor frame data, and MAGALI running on a general-purpose laptop receives the multicast UDP packet-based PCM frame data to decommutate them. The developed portable TGS satisfies our requirements of high portability, low cost, and ease of replacement in the event of hardware failure.

We have already been using multiple portable TGSs successfully for diverse tests at several sites from the functional verification test of the onboard TM system to the system integration test of KSLV-II. Our portable TGS has already played an important role in the development of an onboard TM system as well as in the program success of the KSLV-II and will continue to do so in the future.

THE VIVALDI PORTAL – AN APPLICATION FOR METADATA VALIDATION AND VERIFICATION

Timothy Darr, Dan Corlette, Atul Jain, Ryan Bollweg
Knowledge-Based Systems, Inc.
College Station, TX, 77840
tdarr@kbsi.com

Jon Morgan
JT3, LLC
Edwards Air Force Base, CA
jon.morgan.2.ctr@us.af.mil

ABSTRACT

This paper describes the VIVALDI (Verification and Validation of Metadata for Test & Evaluation) portal, a web-based application that gives test and instrumentation engineers and T&E stakeholders the ability to define rules and apply these rules to support the verification and validation (V&V) of metadata for multi-vendor test and instrumentation systems.

INTRODUCTION

There currently exist generic standards for representing rules for verification and validation (V&V). However, rules written in these standards are not accessible to the typical instrumentation engineer-the rules are often described in a computer programming language or a language meant to be processed by a machine. Few authors and users in the T&E community have the time, inclination or interest in learning these languages. To address this, the VIVALDI (Verification and Validation of Metadata for Test & Evaluation) portal captures rules in an intuitive, domain specific, unambiguous controlled natural language (CNL). A CNL allows those that know the rules most intimately-the engineers-to define the rules without having to learn or even know about the underlying rule standard.

The VIVALDI portal is an implementation of a CNL-based method in which users capture and evaluate V&V rules within and across metadata instance documents. The method uses a CNL syntax for the T&E metadata V&V rule set in order to abstract the highly technical rule languages to a more natural T&E form. As a result, the users can easily specify, validate and manage the specification and validation of the rules themselves. Our approach is very flexible in that under the hood, the method automatically translates rules to a host of target rule languages. The rules are captured in natural language, and used to perform V&V within a single metadata instance document and across multiple metadata instance documents.

Our paper outlines provides an overview of the VIVALDI portal. Key capabilities include: an extensible web-based architecture for capturing and executing rules; catalogs of TMATS [1], MDL [2] and IHAL [3] [4] rules captured in CNL; translators from CNL to target rule languages; and rule execution on open source rule engines.

BRIEF TOUR OF THE PORTAL

This section provides a brief tour of the VIVALDI portal.

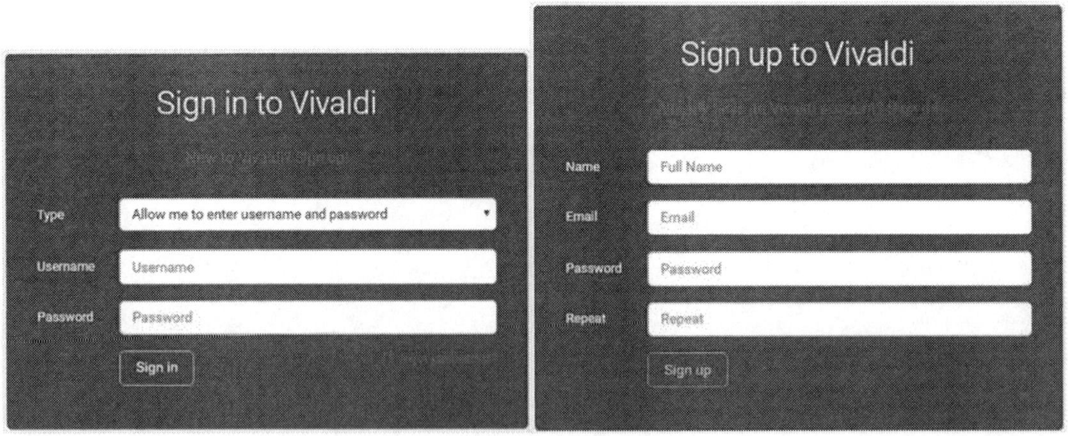

Figure 1 – Authentication

User Authentication

Users of the VIVALDI portal will need to have a login. Figure 1 shows the screens for authentication. At the left of the figure is the login screen where the user enters an email and password. At the right of the figure is the registration screen for users that do not have a login. To get authentication credentials, the user must provide their name, email and a password.

Dashboard

The VIVALDI portal dashboard is the landing page for the VIVALDI portal. Figure 2 shows the dashboard. After login, the user will be redirected to this page. The central area of the dashboard will include a summary of activity in the VIVALDI portal, to include: number of rules, number of instance documents, active users, jobs processed, supported standards, etc.

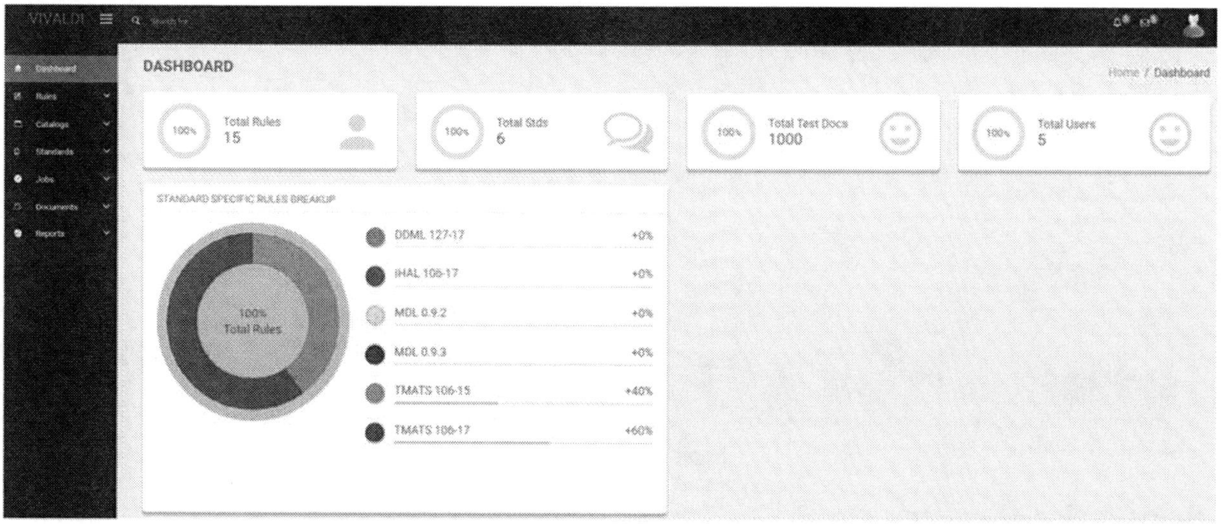

Figure 2 – VIVALDI Portal Dashboard

The left side of the dashboard is where the user accesses the VIVALDI portal capabilities.

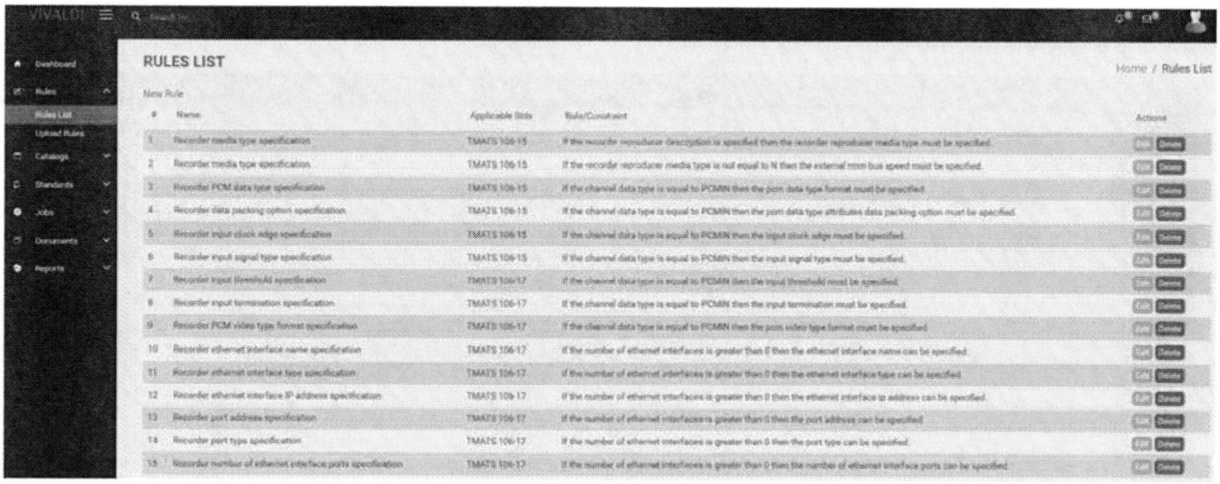

Figure 3 – Registered Rules

Rule Management

At the heart of the VIVALDI portal capability is the management (create, read, update, delete) of rules. The notion of a "registered rule" is a rule that can be used by any authenticated user of the VIVALDI portal system. The VIVALDI portal will allow the user to view all the registered rules. Figure 3 shows a screen of the registered rules. Each row includes metadata about the rule, including a name, description, the standard(s) that the rule applies to and the machine-understandable representation of the rule (this can be hidden based on user profile). Users can edit or delete a rule by selecting the appropriate button.

Figure 4 shows a screen for creating a rule. The rule author provides a name, and the standard(s) to which the rule applies. The user authors the rule using the VIVALDI portal predictive editor. This editor ensures that the rule can be unambiguously understood by both a human and machine. The editor suggests possible rule completions.

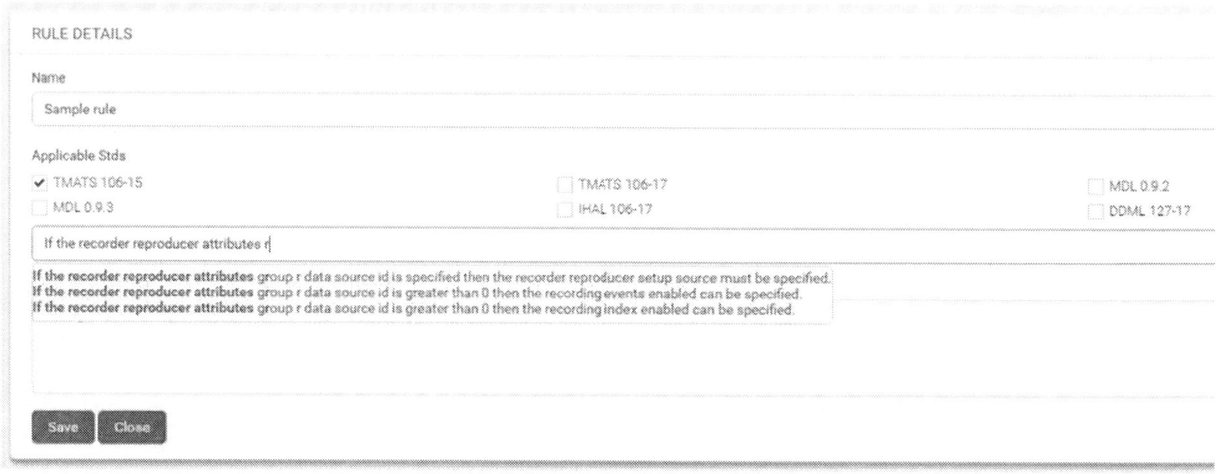

Figure 4 – Rule Creation

Rule Catalog Management

Rules are organized into catalogs so that they can be sent to a validation engine to validate an instance document. Figure 5 shows a screen that lists the rule catalogs. Each row includes metadata about the rule catalog, including a name and description. Users can edit or delete a rule catalog by selecting the appropriate button.

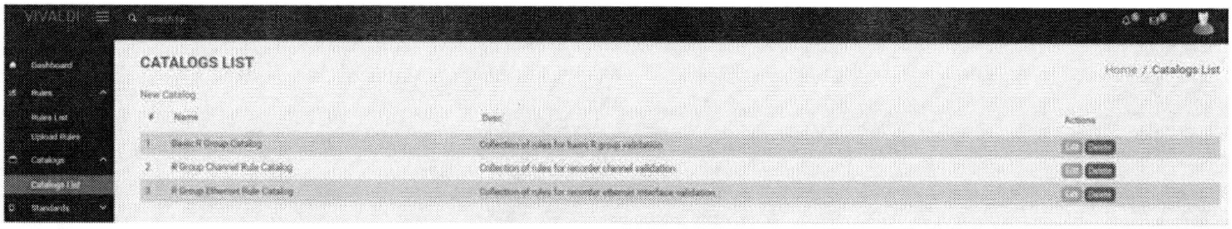

Figure 5 – Rule Catalogs

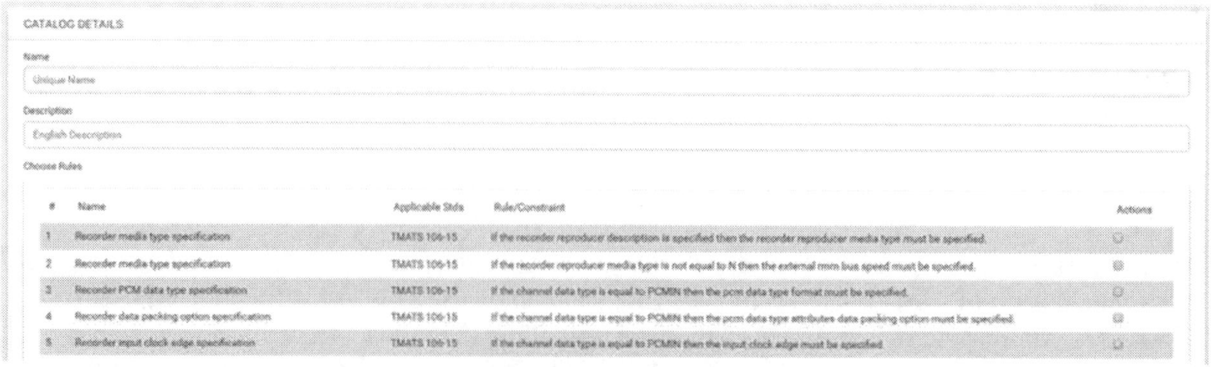

Figure 6 – Create Rule Catalog

Figure 6 shows the screen to create a rule catalog. The rule catalog author provides a name and description (creator, creation / modification dates are automatically assigned) and selects the rules to be included in the catalog.

VIVALDI PORTAL ARCHITECTURE

Figure 7 shows the current VIVALDI portal architecture. The VIVALDI portal is implemented using the Node.js® framework [5]. Basic authentication is provided using a datastore managed by the VIVALDI portal application. If necessary in the future, we will implement domain authentication using the appropriate framework. The connection from the user to the Node.js® server and the authentication databases will be encrypted. The VIVALDI portal on the Node.js® server connects to the supporting capabilities via REST APIs over encrypted channels. The REST APIs will be standardized as described later in this paper.

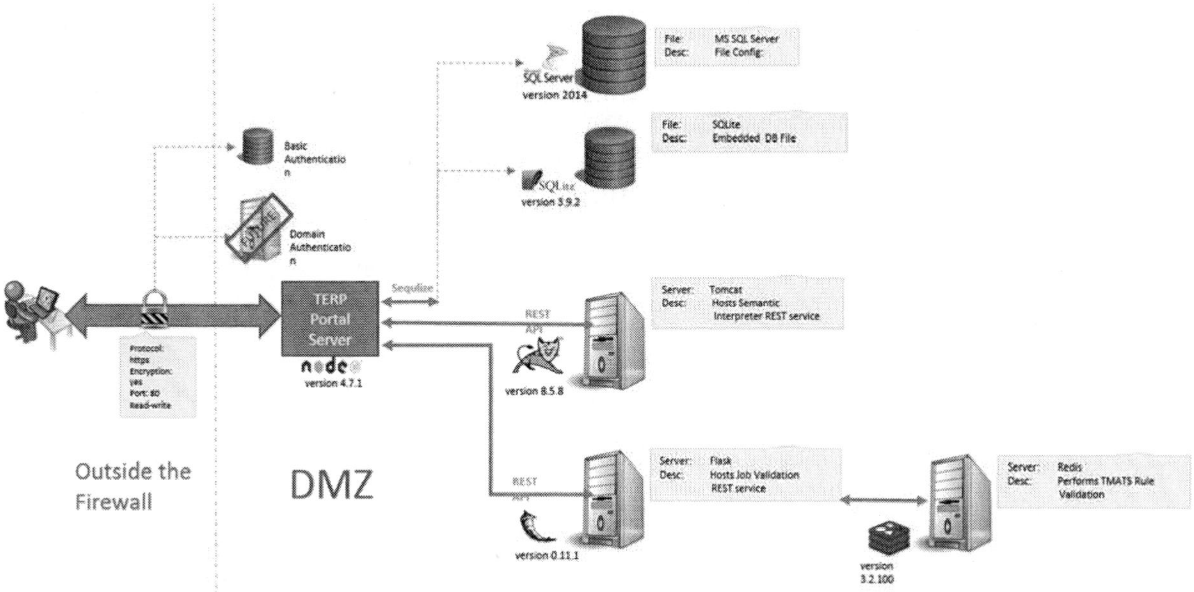

Figure 7 – VIVALDI Portal Architecture

The backend capabilities include a SQL database for storing rules, rule catalogs, jobs, etc., a Tomcat server [6] that hosts the language processing capabilities, a Flask server for processing jobs and a Redis server [7] for performing TMATS validation.

VIVALDI PORTAL LANGUAGE TECHNOLOGY

In this section, we outline the major components for the VIVALDI portal rule processing.

Controlled natural languages (CNLs) are subsets of natural languages that are obtained by restricting the grammar and vocabulary in order to reduce or eliminate ambiguity and complexity. Since the language has a restricted vocabulary and defined grammar, it would very difficult to write a statement without auto-completion or predictive editor support, making the rule authoring process unobtrusive and effortless. Use of error messages, domain term highlighting, predictive feedback and conceptual authoring were some of the techniques that were found to support the CNL writing process.

One of the most obvious ways to support writing restricted languages like CNL is error highlighting. While they author rules, the interface parses the text and validates the text against the CNL rules. If the validation fails, the interface will try to identify the cause of the error and provide suggestions for fixing the error.

VIVALDI Portal Grammar: The VIVALDI Portal Grammar is a controlled natural language (CNL) grammar that is an extension of the ACE "Attempto Control English" controlled natural language grammar [8]. These extensions construct a new language specific to the problem domain that ensures that the underlying grammar does not contain syntactic ambiguities. The greatest challenge in processing natural language with computing systems is the ambiguity that exists in both the grammar and semantics. The grammar makes the resulting system scalable in the sense that as new rules are added that conform to the CNL they do not collide with existing statements.

CNL Rules: These rules represent the input statements the system must support and are representable within the CNL grammar.

Predictive Editor: The predictive editor is the GUI-based input device that is used for users to author rules and store them in the system. The predictive editor recognizes the approved structures of the CNL and is able to look-ahead with in the defined grammar. The author gets immediate feedback while the text is being written and helps prevent the author from entering sentences that are not allowed in the CNL grammar. While writing, the author can request rule suggestions. The suggestions provided to the user are context aware: i.e., the editor provides suggestions based on the cursor positions and the applicable grammar rule(s).

Free-Text Input: The free-text input capability takes a rule typed in by a user and maps that statement to the best controlled natural language statement based on how the system was configured. This allows for the system to take in ambiguous statements and map them to the "best" known rule that has been vetted by the system designers. The user has the ability to accept or reject the mapping. Once accepted, the user has confidence that the accepted sentence is a CNL sentence that can be processed by the system.

Semantic Interpreter: The semantic interpreter takes a CNL rule as input and creates a semantic representation with the help of a small CNL ontology. The semantic interpretation process incorporates a small ontology that seeks to make minimal commitments to external ontologies. The goal of the semantic representation is to provide a representation that is well suited to augmentation with external ontologies while at the same time carrying enough meaning to be immediately useful. Users of the system can take the semantic interpretation directly and drive other computing systems or components.

CNL Ontology: The CNL ontology is a light-weight ontology that contains some basic semantics that supports pulling semantic information directly from a CNL syntactic structure. The goal is to not tie the representation to external ontologies so that the final semantic interpretation is flexible and easily mapped or used to drive richer semantic representations.

Semantic Interpretation: This is the final output of the VIVALDI portal language components. The current form of this interpretation is a set of triples in an SVO (subject-verb-object) format. The intent behind the semantic interpretation is to provide a representation that contains as much semantics as possible without the use of external ontologies.

The following statements are characteristic of rules that VIVALDI portal can support:

- The measurement identified 0x0100 has a data rate of 480 pps.
- The measurement identified by 0x0200 has a sample rate of 48 Hz.
- The measurement identified by 0x0300 has data length of 2 bytes.
- The measurement identified by 0x0600 has resolution of 5 percent.

Figure 8 shows some controlled natural language representations and semantic representations for the rules above. The web interface supports entities; which are selectable from a dropdown or other control. The semantic interpretation is a representation of the rule in a triple format.

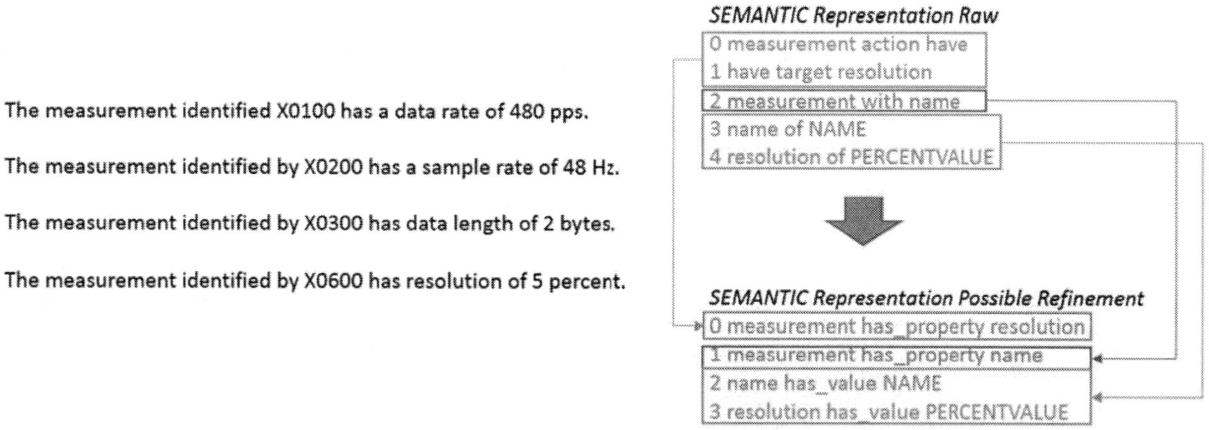

Figure 8 - Semantic Representations for Statements

STANDARDIZATION

KBSI proposes to support the standardization of VIVALDI portal capabilities with RCC standardization committees so that the T&E community end users can take advantage of its features. Figure 9 overviews the expected VIVALDI portal standards (shown in green). The VIVALDI portal and services are shown as well as a vendor configuration service.

We expect at least four standards:

- The rule catalog contains a collection of rules that are useful to the T&E community and can be applied across T&E missions.
- The CNL standard as the CNL grammar to support rule capture.
- The API as an interface standard for utilizing the portal capabilities within a broader computer program infrastructure.
- A method for capturing rules in CNL, translating the rules into a target language, and executing the rules against XML instance documents.

Initial thoughts of an API would include the following:

- The validate() method takes a metadata instance document and a collection of CNL rules as input and returns XML validation report as output; the validation report documents the rules that are satisfied and those that are violated.
- The translate() method takes a CNL rule set and target rule language as input and returns the target rule instances translated from the CNL as output.

- The optimize() method takes a metadata instance document and a CNL rule set as input and returns an optimized metadata instance document as output; the CNL grammar would be extended to include optimization criteria to support this method.
- The deconflictRules() method takes a CNL rule set as input and returns a de-conflicted[1] CNL rule set as output.
- The deleteRule() method takes a CNL rule and a rule catalog as input and removes the rule from the catalog.
- The addRule() method takes a CNL rule and a rule catalog as input and adds the rule to the catalog.
- The updateRule() method takes a CNL rule and a rule catalog as input and updates the rule in the catalog.

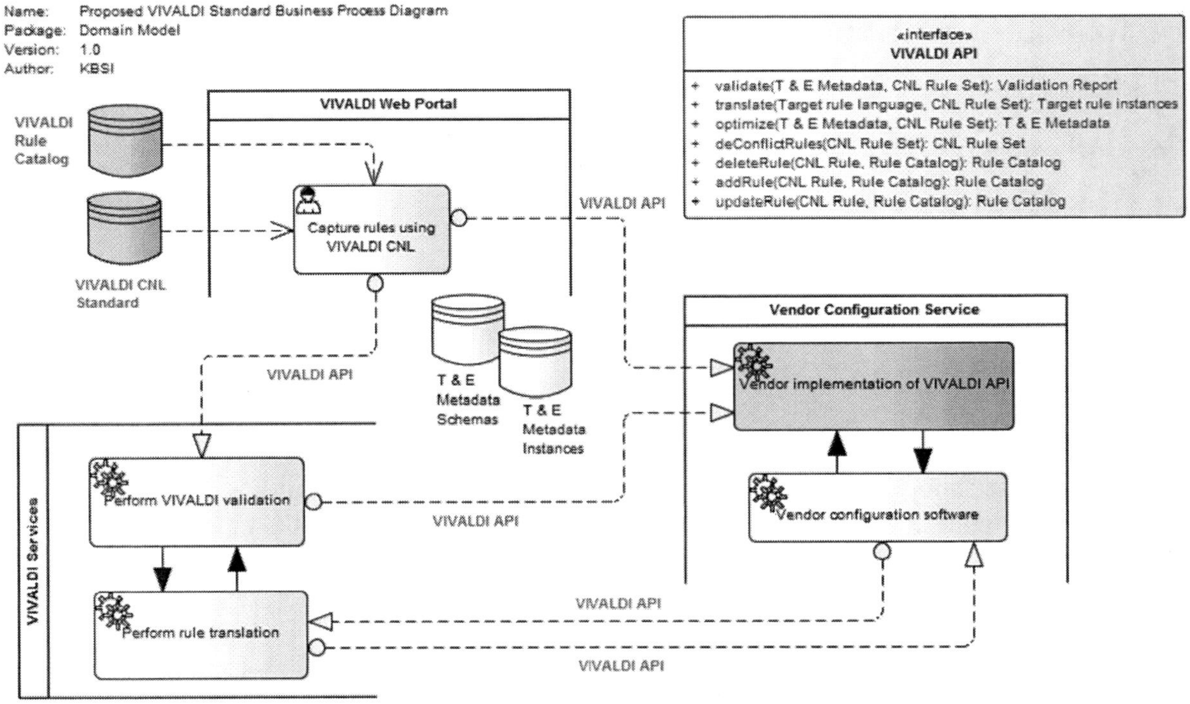

Figure 9 – Overview of the Expected Standardization

As shown in Figure 9, the standards are used as follows:

- During the rules capture step, the user selects rules from the standard rule catalog or authors new rules using the standard CNL language.
- Once the CNL rules have been captured, there are two possible scenarios:
  - The user validates a metadata instance document against the rules using the VIVALDI portal services.
  - The user validates a metadata instance document against the rules using a third-party vendor implementation of the API. In this scenario, the vendor software can

[1] Two rules conflict if there is no way that both rules can be satisfied.

access VIVALDI portal capability for CNL translation, or validation using the API.

CONCLUSIONS

Verification and validation (V&V) of metadata and systems design are critical functions in systems engineering because they help ensure that requirements are realizable, that designs are correct and consistent, that designs satisfy all requirements, and help reduce costs by optimizing over design variables. In short, V&V reduces lifecycle costs by ensuring correctness in design phase. The VIVALDI portal includes: (i) a natural language based syntax for the T&E metadata V&V rule set in order to abstract the "computer science"-heavy rule languages to a domain-specific CNL-based syntax, (ii) CNL rules can be transformed into any target rule language, (iii) there is no translation from XML to another format for rule execution, and (iv) the approach is based on open sources and established standards.

These benefits will apply directly to instrumentation design for T&E ranges and to the entire DoD T&E community by standardization of various technology components. The proposed standards enable interoperability among T&E users and vendor configuration software at the level of English-language statements of the intended rule. Unambiguous statements of the rule semantics mean that the users can clearly communicate rules in nothing more than a text editor. Translators from the CNL to a target rule language enable V&V of the rules at the machine level using the right target language and rule engine.

REFERENCES

[1] Range Commander's Council Telemetry Group, "Chapter 9 Telemetry Attributes Transfer Standard," in *Telemetry Standards, IRIG Standard 106-13 (Part 1)*, US Army White Sands Missile Range, NM, Secretariat, Range Commander's Council, 2013.

[2] M. Moore, "Metadata Description Language: The iNET Metadata Standard Language," in *Proc. International Telemetering Conf.*, Las Vegas, NV, 2009.

[3] J. Hamilton, R. Fernandes, P. Koola and C. H. Jones, "An Instrumentation Hardware Abstraction Language," in *42nd International Telemetering Conference*, San Diego, CA, 2006.

[4] J. Hamilton, R. Fernandes, M. Graul and C. Jones, "Extensions to the Instrumentation Hardware Abstraction Language (IHAL)," in *44th International Telemetering Conference*, San Diego, CA, 2008.

[5] "Node JS," Node.js Foundation, [Online]. Available: https://nodejs.org/en/. [Accessed 12 May 2017].

[6] "Apache Tomcat," The Apache Software Foundation, [Online]. Available: http://tomcat.apache.org/. [Accessed 12 May 2017].

[7] "redis," redislabs, [Online]. Available: https://redis.io/. [Accessed 12 May 2017].

[8] N. E. Fuchs and R. Schwitter, "Attempto Controlled English (ACE)," Katholieke Universiteit Leuven, 1996.

INTEGRATING STL INTO LOW-SWAP ASSURED POSITION, NAVIGATION, AND TIMING SOLUTIONS FOR MILITARY INFORMATION SYSTEMS

ITC/USA 2017
23-26 October 2017
Bally's Las Vegas
Nevada

Jon Sinden
Product Manager
Spectracom
Phone: 585-321-5840
Supporting Author: Lisa Perdue

Abstract

Today's military communications and information dominance systems rely on available and accurate GNSS PNT signals to provide leaders the information required to make timely and effective decisions. The proliferation of GNSS-degrading and denying devices across state and non-state actors puts this critical information overmatch capability in jeopardy. Traditional PNT devices utilize various alternative internal components to maintain holdover of the PNT data during loss of GNSS input but are susceptible to drift over time. This presentation discusses how the Satellite Time and Location (STL) service integrated into low Size, Weight, and Power (SWAP) assured PNT devices can provide critical continuity of operations when GPS is unavailable. Comparison timing testing between a GPS-disciplined oscillator in holdover and STL is presented. Future considerations for STL as an alternative position and navigation signal of opportunity are also discussed.

| 10/18/2017

orolia spectracom

Introduction

- TODAYS CRITICAL INFRASTRUCTURE RELIES ON GPS FOR POSITION, NAVIGATION, AND TIMING (PNT)REFERENCE
- MANY MILITARY SYSTEMS ALSO RELY ON THIS CRITICAL CAPABILITY
- HOW CAN ALTERNATIVE PNT SIGNALS SUPPORT CRITICAL CAPABILITY IF GPS IS LOST OR CORRUPTED?

| 10/18/2017

orolia @spectracom

OUR CRITICAL INFRASTRUCTURE – RELIANT ON **TRUSTED** GPS TIME

- POWER GRIDS
- FINANCIAL MARKETS
- EMERGENCY SERVICES
- INDUSTRIAL CONTROL
- TELECOM

ACCURACIES REQUIRED:

<1 MICROSECOND TO UTC

Notes: A wide array of todays critical civil infrastructure require synchronization on common position, navigation, and timing references to function. Financial markets, communications systems, and public utilities all trust on GPS to provide this information.[1]

4 | 10/18/2017

orolia @spectracom

DEFENSE SYSTEMS – RELIANT ON GPS FOR POSITION, NAVIGATION, AND TIMING

- THREATS TO GPS AFFECT DEFENSE SYSTEMS AS WELL AS CIVIL CRITICAL INFRASTRUCTURE
- LOSS OF GPS ACCESSIBILITY AND INTEGRITY = DEGRADED OPERATIONS
- WHAT TECHNOLOGIES CAN SUPPLEMENT GPS?

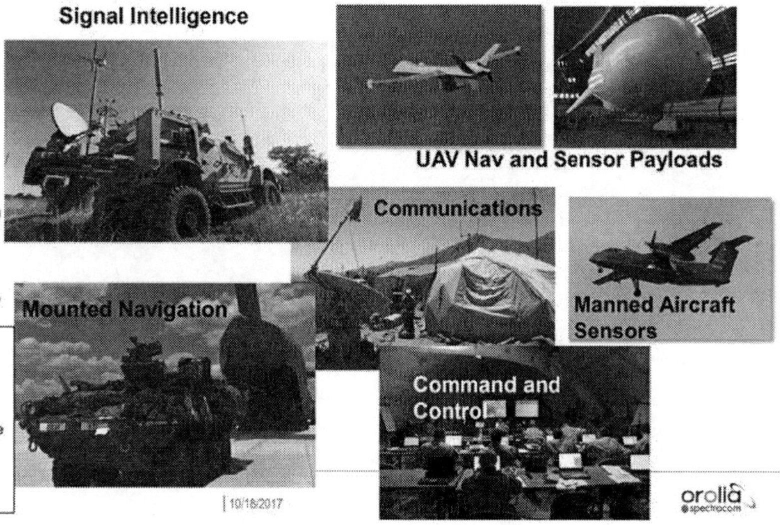

Notes: Like civil infrastructure, many of our digital military systems are reliant on GPS to provide the trusted PNT data that allows our forces to maneuver in, communicate across, understand, and make decisions over the modern battlefield. Loss of access or trust of the PNT information removes a critical force multiplier our modern armed forces require to dominate adversaries given the current force structure.[2]

RESILIENCY, RELIABILITY AND PROTECTION WITH <u>**INTEGRITY AND TRUST**</u>

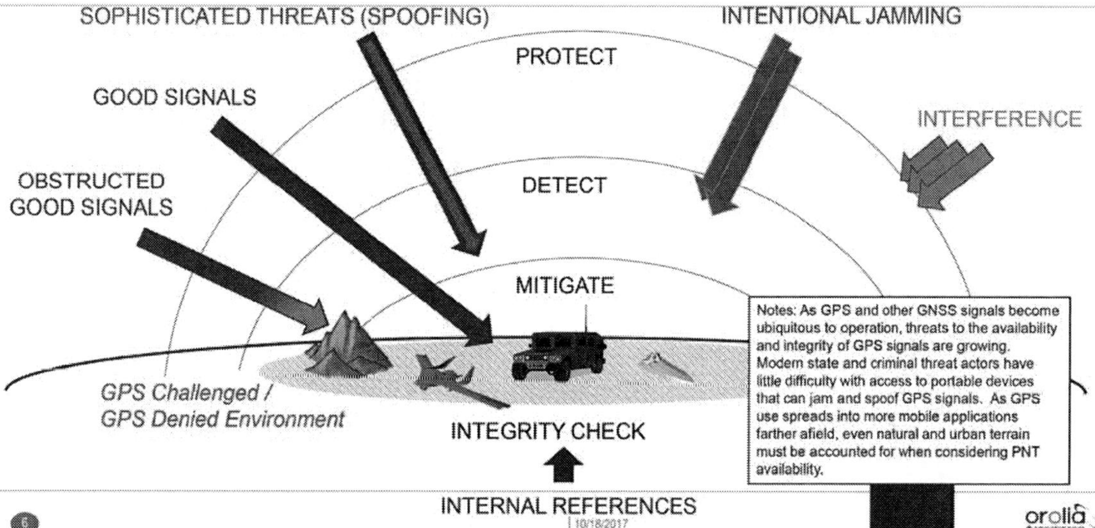

Notes: As GPS and other GNSS signals become ubiquitous to operation, threats to the availability and integrity of GPS signals are growing. Modern state and criminal threat actors have little difficulty with access to portable devices that can jam and spoof GPS signals. As GPS use spreads into more mobile applications farther afield, even natural and urban terrain must be accounted for when considering PNT availability.

SIGNALS OF OPPORTUNITY FOR POSITION, NAVIGATION, AND TIMING

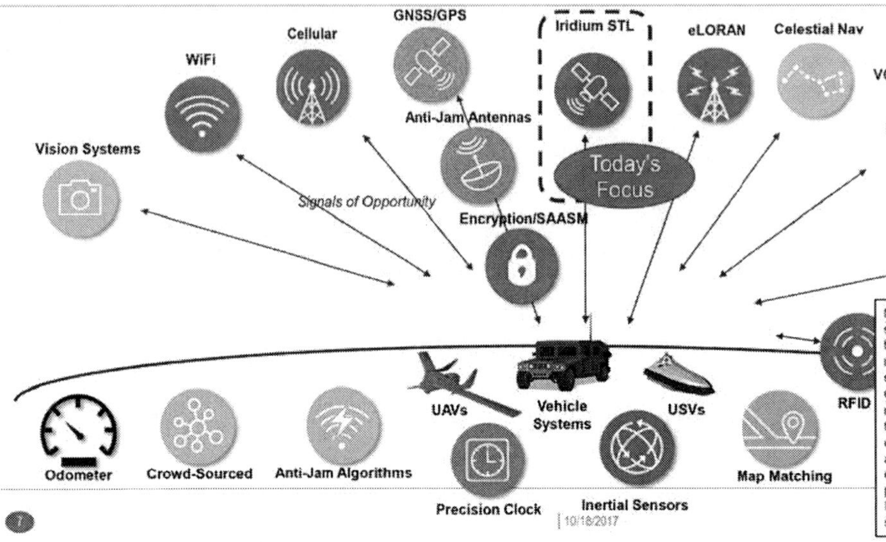

Notes: To combat threats to GPS signal availability and trust, other techniques and signals of opportunity must be considered as alternate sources of PNT information when GPS is not available. While GPS will remain the most accurate system for the time being, correctly integrating combinations of other signals can help a system survive operations for the duration of GPS signal loss. In this presentation, we will look at the Iridium STL system as an alternative source of PNT.

7 | 10/18/2017

Augmenting GPS with STL

- STL PRESENTS AN ALTERNATIVE PNT SIGNAL TO GPS
- HOW DOES STL COMPARE TO A TIMING REFERENCE HOLDOVER OSCILLATOR WHEN GPS IS LOST?

| 10/18/2017

STL – AN ALTERNATIVE PNT SOURCE

Notes: Satelles' Satellite Time & Location (STL) service provides an alternate space-based PNT source the GPS constellation. It uses the Iridium system's pager channel to provide a PNT signal accessible by terrestrial receivers.[3]

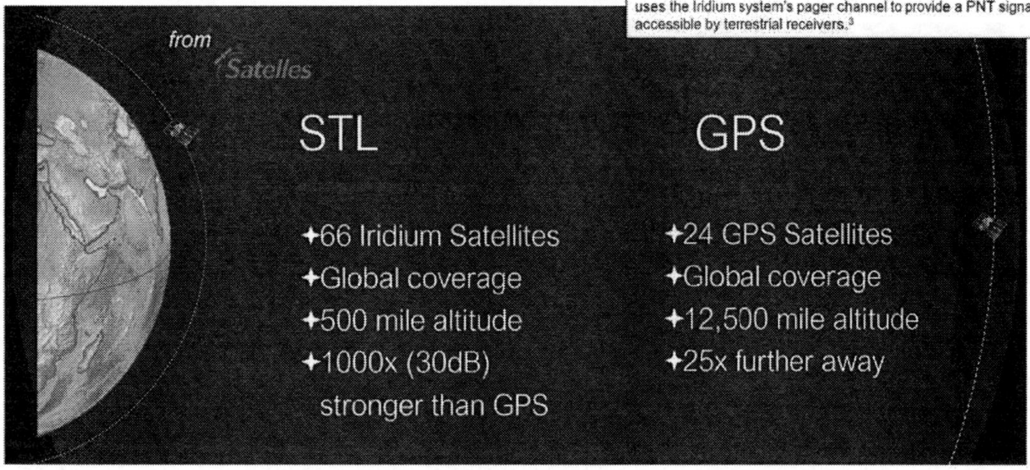

SIGNAL COMPARISON TO GPS

Notes: While STL is not as accurate as GPS, it is 30dB stronger, allowing it to be received inside structures where GPS is blocked. Depending on the application, STL is potentially a viable backup when GPS access or trust is lost for a period of time.

	GPS L1	STL
Timing accuracy to UTC	~20 ns	~200 ns
Positioning accuracy	~3 meters	30-50 meters
Time To First Fix -- Timing	~100 seconds	~few seconds
Time To First Fix -- Positioning	~100 seconds	~10 minutes
Anti-spoofing	Military Only	Yes, encrypted signal (all)
Coverage	Global	Global
Availability Outdoors	With view of horizon to get accurate position with low DOP	Limited view of sky lengthens convergence time
Availability Indoors	No	Yes: 30 – 40 dB stronger

TIMING SIGNAL - HOW IT WORKS

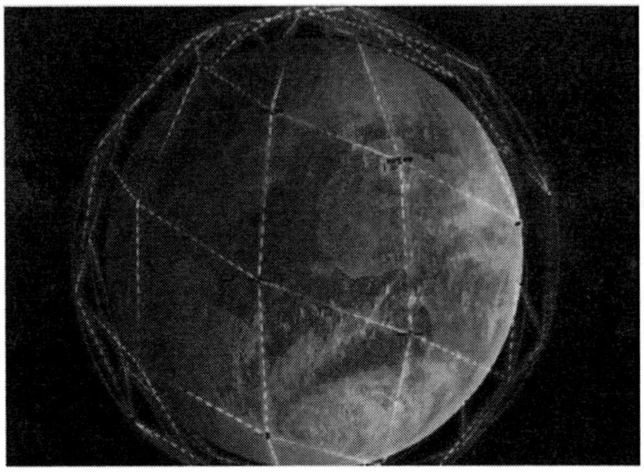

RE-PURPOSED PAGING CHANNELS

1620 MHZ BAND, 25 KHZ CHANNEL, QPSK

SPREAD SPECTRUM CODED SIGNAL, 90 MSEC FRAME, ~1.4 SEC BURST ON AVERAGE

UTC SYNCHRONIZED TIMING
- +/- 500 nsec spec
- 100-200 nsec typical

TIMING UPDATES TO A PRECISION LOCAL OSCILLATOR FOR CONTINUOUS TIME AND FREQUENCY SOURCING

LOW EARTH ORBIT (LEO) => STRONG SIGNAL WITH SPOT BEAMS

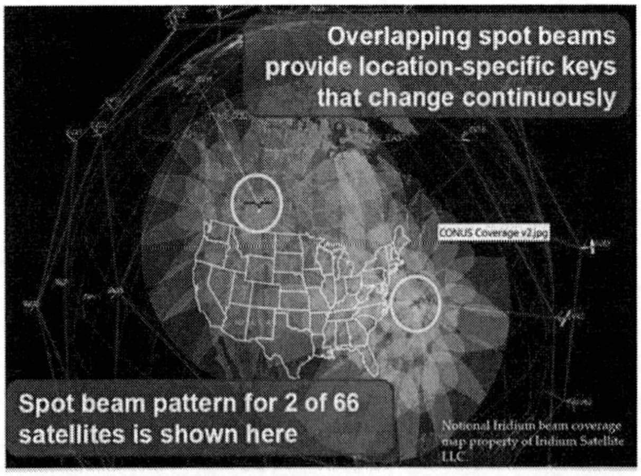

IRIDIUM: ~100 MINUTE ORBIT

GPS: 12 HOUR ORBIT

ENCRYPTED SIGNAL WITH SUBSCRIBER KEYS

POSITIONING DETERMINATION
- Spot beam location (gross)
- Range and Doppler updates
- Convergence over minutes

GEO-LOCATION SECURITY

STL PERFORMANCE

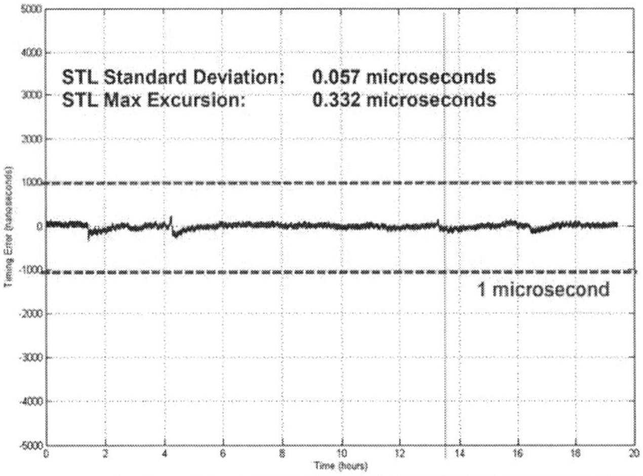

GPS / STL FUNCTIONAL TIMING TEST

Notes: These are the test conditions used for the STL timing performance experiment

Oscillator Disciplining
- No disciplining algorithms were changed for the different references

After 900 seconds (15 minutes) capture time CW jamming signal is introduced
- Unit was disciplined by GPS for at least 12 hours prior to test start

GPS is lost and configured backup reference is taken into use
- If none then unit enters holdover

Measurements taken from time and frequency server 1PPS
- Not from receivers directly

HOLDOVER SETUP

Notes: A Spectracom SecureSync unit provided GPS, oscillator holdover, and STL timing information.

- TIME AND FREQUENCY SERVER SYNCHRONIZED TO GPS
- INTERFERENCE SIGNAL IS ADDED
- WITH NO SECONDARY REFERENCE SPECIFIED, UNIT USES HOLDOVER OSCILLATOR
    - OCXO used in test
- STL PERFORMANCE IS COMPARED TO HOLDOVER PERFORMANCE

15 | 10/18/2017 orolia spectracom

GPS AND OSCILLATOR HOLDOVER TEST RESULTS -24HR

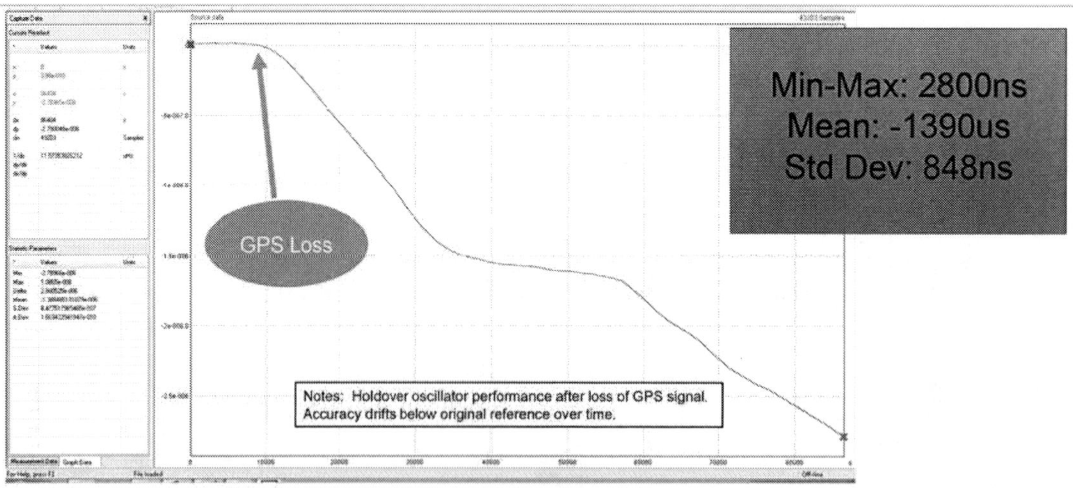

Notes: Holdover oscillator performance after loss of GPS signal. Accuracy drifts below original reference over time.

16 | 10/18/2017 orolia spectracom

STL TEST RESULTS -24HR

Notes: STL timing performance after loss of GPS signal. The STL signal drifts above and below the original reference, but maintains better spread, mean, and standard deviation than just relying on the holdover oscillator.

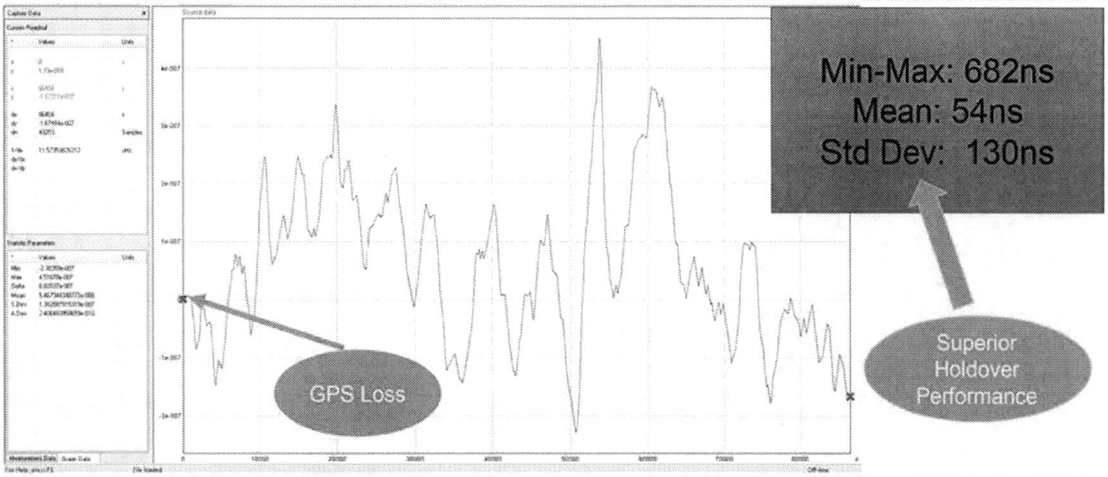

Next Steps

POSITIONING DETERMINATION FROM STL

Notes: While the STL signal provides a useful timing alternative to GPS, its positioning error is 10x – 50x greater than the standard GPS FOM. Stationary and slow moving STL-enabled applications show the best accuracy performance. This is not precise enough for many use cases, even as a backup to GPS.[3]

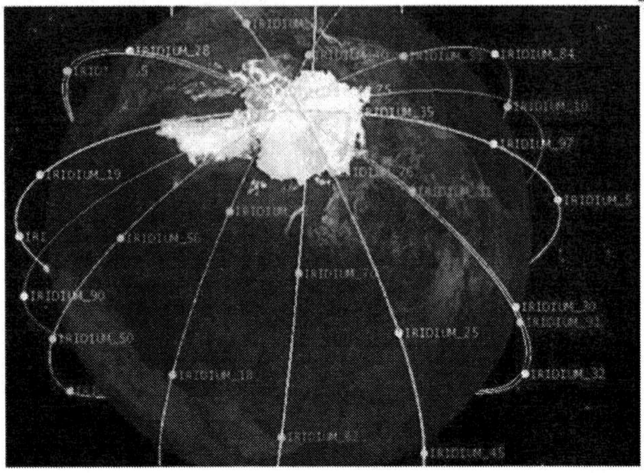

TRACK RANGE AND DOPPLER FROM A COUPLE OF SATELLITES

AS SATELLITES MOVE RAPIDLY THROUGH THE SKY, GEOMETRY IMPROVES

~30 minutes horizon-horizon

Polar orbits – North-South

WORKS WELL FOR STATIONARY OR SLOW MOVING VEHICLES

Requires IMU for complete nav solution

Ranging/Doppler measurement updates reduce IMU drift

POSITIONING CONSIDERATIONS

Notes: STL position and navigation is generally less precise due to the mechanics of a lower, faster moving constellation. Our next steps will look at ways STL can be utilized to augment GPS position and navigation operations in a degraded environment while taking into account these STL limitations.[3]

- *Timing requires only one satellite*
- *Positioning requires at least 4 (x, y, z, t)*
- *Needs at least one satellites in each direction*

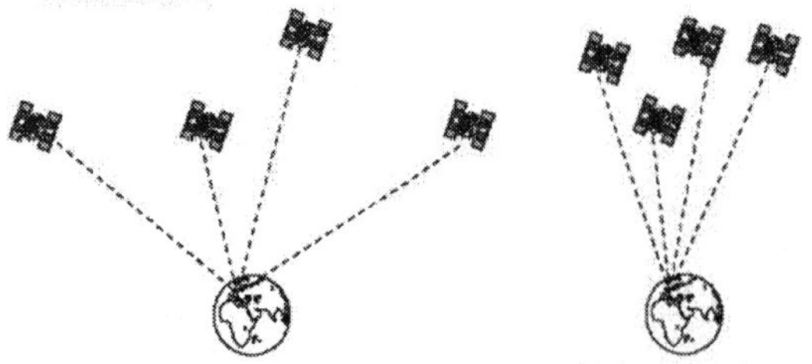

STL OPERATES DIFFERENTLY THAN GPS

Only see 1-3 satellites at a time vs. GPS with 8-12

INSTANTANEOUS TTFF

~500km on first packet because of spot beams

DILUTION OF PRECISION

Too few satellites for triangulation, wrong geometry

Position determined over many minutes as satellites move across the sky

Not a standalone GPS replacement, but how can it augment a GPS+INS system?

PNT + STL IN A GNSS DEGRADED/DENIED ENVIRONMENT

- NORMAL OPERATION

GNSS+IMU -> Positioning Solution
GNSS+Oscillator -> Timing Solution

- GNSS DEGRADED/DENIED

IMU provides positioning resiliency
- Useable for **seconds to minutes**

Oscillator provides accurate timing resiliency
- Useable for **hours to days**

- STL AUGMENTATION

Add STL to the GNSS+IMU+Oscillator PNT System
- Reduce the Position error curve

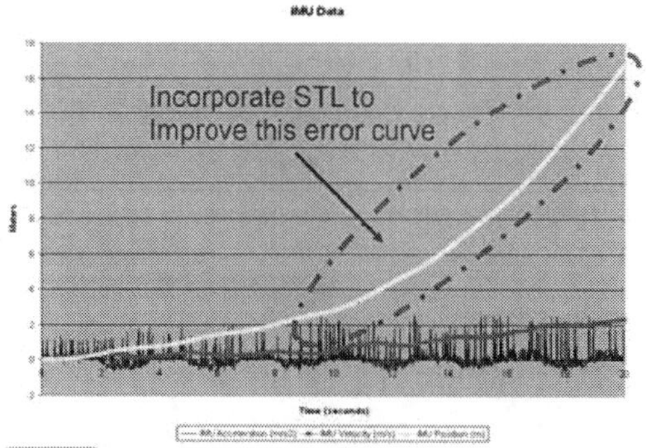

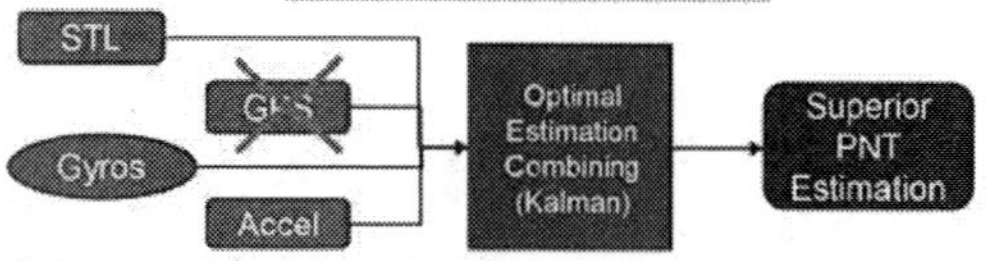

While STL may not be sufficient as a standalone positioning and navigation alternative to GPS, how can it be integrated into a INS-enabled PNT solution? We will be testing the effect of coupling the STL signal with a GPS, IMU, and timing oscillator through Kalman filtering. As an INS drifts over time after loss of GPS, will the addition of STL information to the filter improve the error curve and maintain greater position accuracy for a longer period than just a standard GPS/IMU solution?[4]

BETTER PNT ASSURANCE WITH ADDITIONAL SIGNALS AND TECHNIQUES

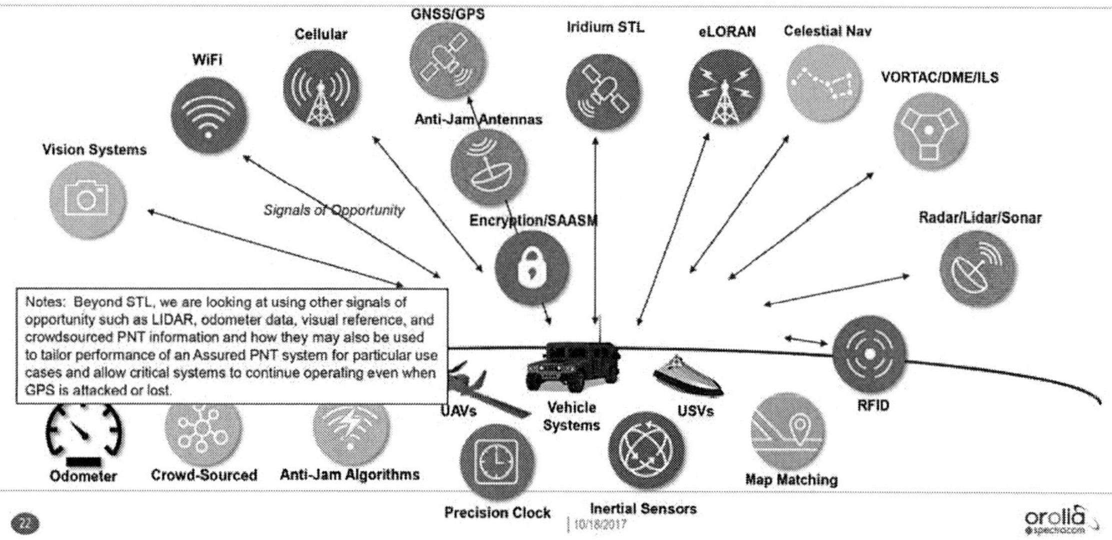

Notes: Beyond STL, we are looking at using other signals of opportunity such as LIDAR, odometer data, visual reference, and crowdsourced PNT information and how they may also be used to tailor performance of an Assured PNT system for particular use cases and allow critical systems to continue operating even when GPS is attacked or lost.

Conclusion

- GPS PROVIDES ACCURATE PNT DATA TO CRITICAL INFRASTRUCTURES, INCLUDING MILITARY SYSTEMS
- THREATS OF JAMMING AND SPOOFING GPS SIGNALS
- STL CAN PROVIDE EFFECTIVE BACKUP TIMING REFERENCES IN THE EVENT OF GPS LOSS
- THE NEXT STEP IS DETERMINING THE VALUE OF STL IN AN COUPLED PNT DEVICE

REFERENCES

1. CARROLL, J., (2003), VULNERABILITY ASSESSMENT OF THE U.S. TRANSPORTATION INFRASTRUCTURE THAT RELIES ON THE GLOBAL POSITIONING SYSTEM, *JOURNAL OF NAVIGATION, 56*(2), 185-193.
2. KELLER, J, (2016), "GPS JAMMING IS A GROWING THREAT TO SATELLITE NAVIGATION, POSITIONING, AND PRECISION TIMING," RETRIEVED FROM HTTP://WWW.MILITARYAEROSPACE.COM/ARTICLES/2016/06/GPS-JAMMING-SATELLITE-NAVIGATION.HTML.
3. SATELLES, (2016), SATELLES TIME AND LOCATION, RETRIEVED FROM HTTP://WWW.SATELLESINC.COM/WP-CONTENT/UPLOADS/2016/06/SATELLES-WHITE-PAPER-FINAL.PDF.
4. MAGNUSSON, N., ODENMAN, T., (2012), "IMPROVING ABSOLUTE POSITION ESTIMATES OF AN AUTOMOTIVE VEHIVLE USING GPS IN SENSOR FUSION," 2012:EX048, CHALMERS UNIVERSITY OF TECHNOLOGY, GOTEBORG, SWEDEN 2012.

METER-LEVEL TDOA MULTILATERATION SYSTEM BASED ON HIGH PRECISION PHASE SYNCHRONIZATION OVER FIBER OPTICAL LINKS

Jie Tian[1,*], Dong Hou[2], Nan Xie[1], Yongsheng Cheng[1]
[1]Institute of Electronic Engineering, China Academy of Engineering Physics, Mianyang, Sichuan 621900, China
[2]School of Automation Engineering, University of Electronic Science and Technology of China, Chengdu, Sichuan 611731, China
*Email: tianliudou@gmail.com

ABSTRACT

The accuracy of TDOA estimation between sites affects the final positioning calculating the performance of the target. The problem needs to be solved in the multi-station localization systems, requires highly accurate phase synchronization for consistency errors cancellation of receivers. In this paper, the impact of phase error on TDOA localization is analyzed briefly, and we proposed a TDOA multi-station positioning system based on a phase synchronization technique via a long round-trip FOL. By comparing with an actual GPS timing TDOA based method, we demonstrated that the performance of the proposed technique can approximately triple the positioning precision. The simulation results show that the FOL synchronization technique can be used as a potential synchronization method of TDOA multilateation system due to its effective phase error cancellation.

INTRODUCTION

The time difference of arrival(TDOA) measurement based multilateration system, which possesses relatively high (positioning) accuracy, networking and anti-strike capability, is regarded as one important technique of passive loca-lization methods[1]. It would be the accuracy of TDOA estimation between sites that affects the final positioning ca-lculating performance of the target. However, one problem is still need to be solved in such multilateration systems, which requires high accurate phase synchronization for consistency errors cancellation of receivers. Current phase s-ynchronization methods for multi-station application include GPS, TWSTFT, communication links, microwave sig-nal etc[2]. However, there are scenarios that cannot take advantages of GPS due to jamming or interference, and the environmental affection over long distance may cause severe problems for communication links based methods. Therefore, Fiber Optic Link(FOL) could be one of the potential direct synchronization techniques due to its low attenuation and high reliability[3-5]. Some previous works have experimentallyrevealed the feasibility of phase synchronization by using FOL in radar systems [6]–[8]

In this paper, we proposed a TDOA multi-station positioning system based on a phase synchronization techniq-ue via a long round-trip FOL. By comparing with an actual GPS timing TDOA based method, we demonstrated that the performance of the proposed technique can approximately triple the positioning precision.

PRINCIPLE AND THE RESULTS OF THE PROPOSED SCHEME

Usually, a Multi-Station positioning System realizes the 3-Dimension localization by making use of TDOA method. In our scheme, the target emits spread spectrum signal to improve the anti-jamming ability and phase measurement reliability. As shown in Figure 1, 1 Master Station(MS) and 4 Slave stations(SS) are deployed in the estimated area, after acquiring the pseudo-code phase information of the wireless signal emitted by the target, the phase data recei-ved by each station was sent to the Data Processing Unit (DPU), where the phase/time differences and estimated po-sition could be calculated. To improve the clock system stability and estimation performance, an atomic clock is int-roduced in MS as time reference source. In the interest of measuring simultaneous pseudo-code phases, strict sync-hronization among all stations is required, for this purpose, GPS timing synchronization method is usually used.

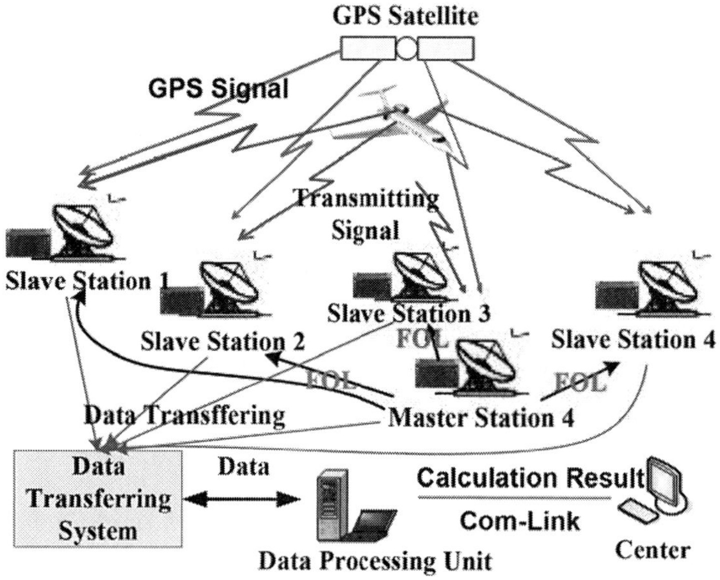

Figure 1 Scheme of acutal multi-station positioning system, GPS timing was used to be compared with our FOL based phase synchronization method;

Time delay relationship between the output baseband signals could be derived as (i = 1, MS), $x_{i,BB}(t) = x_{1,BB}(t - \Delta\tau)e^{j(2\pi\Delta f_{mix}(t-\Delta\tau)+\Delta\varphi_{mix})}, i = 2, \ldots, N$, where $x_{i,BB}(t), i = 1, \ldots, N$ is the baseband signal of the i th st-ation, $\Delta\tau$ represents the relative delay between two signals, as well as the TDOA between two stations. $\Delta\varphi_{mix} = \varphi_{1,analog} + \varphi_{1,digit} - \varphi_{2,analog} - \varphi_{2,digit}$, which shows that the phase consistency error of the mixers can jointly affect the time delay relationship. The frequency errors between sites is negligeable due to the same LO sources.

According to the comparasion between the direct link and the phase-correction link phase synchronization methods in our previous work, we know that a long direct FOL will introduce the unacceptable phase noise and that may result in failure to synchronization among the sites. Quite the contrary, the introduced phase compensator can correct the phase error to reduce the fluctuation and realize the high precise phase synchronization performance. So we extend the application of Bi-station Phase-correction link synchronization to multi-station scenario as below.

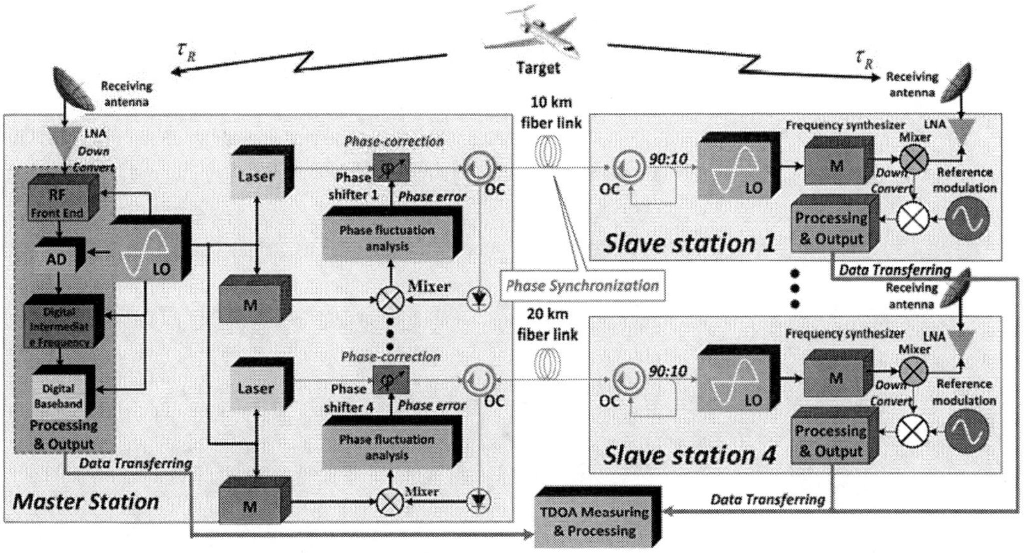

Figure 2 Scheme of the Star network of multi-station phase synchronization, 4 phase-correction components are integrated into the MS.

Our proposed method is shown in Figure 2, we used mode-locked laser for the deployment of phase-correction via FOL between Master Sation and each Slave Station, and formed a star network. As shown in Figure 2, the phase-correction link synchronization (solid line in the figure) was operated by using the phase shifter. In the transmitter, a phase shifter was incorporated between the laser and the fiber link. The phase shifted laser beam was transferred over the long distance practical round-trip FOL. In the receiver, the beam was split into two beams. One beam was converted into the electronic signal which was used as the receivers LO. The other returned to the transmitter site, and was mixed with the transmitters reference LO to calculate out the phase error. Then, the error signal was applied to the phase shifter to perform phase correction.

As the center of the star type network, Master Sation has a phase-correction component for each Slave Station, 1 Master Sation and 4 Slave Station share the same LO source (atomic clock). As shown in Figure 3 (b～f), we have the time jitter measurements of each SS during the 30 minutes experiment, and the distance between MS and SS1～SS4 is 20/19/10/25km respectively, matches the site deployment shown in Figure 3 (a).

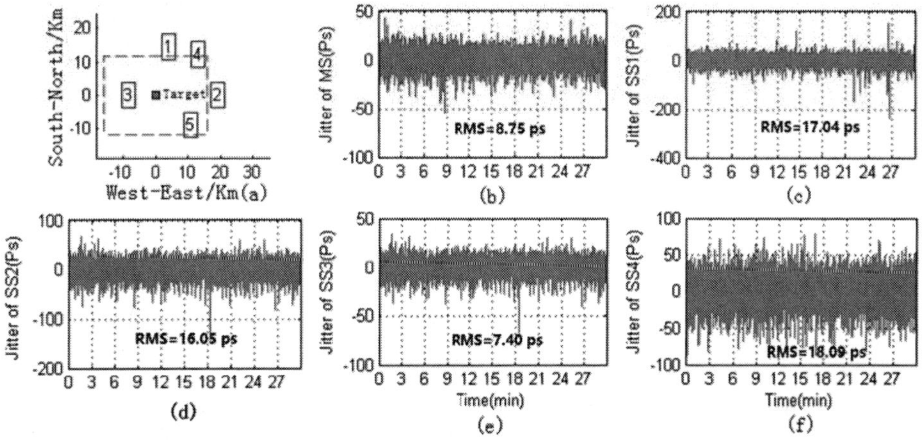

Figure 3 (a) Station deployment; (b) Timing jitter of Master Station; (c) Timing jitter of Slave Station 1; (d) Timing jitter of SS2; (e) Timing jitter of SS3; (f) Timing jitter of SS4;

THE EXPERIMENT DESIGN AND THE NUMERICAL RESULTS

In this experiment, we use the previous timing jitter measurements to calculate the positioning precision for the FOL phase-synchronization based multilateraion system. And the hardware we use for the semi-physical simulation is as follows: Figure 4 (a) is the erbium-doped fiber mode-locked laser we used as a laser source, Figure 4 (b) is the long distance Fiber Optic Link, Figure 4 (c) is the Phase Compensator for the phase correction as introduced in Section II.

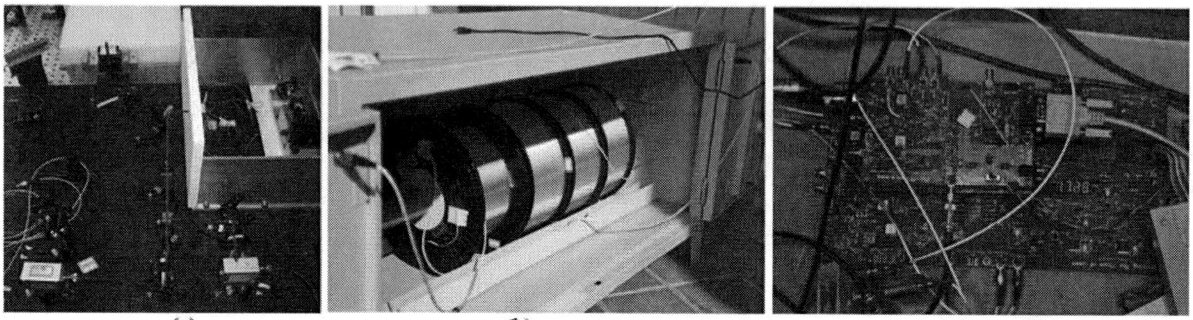

(a) (b) (c)

Figure 4 (a) Laser source; (b) 50Km Fiber Link; (c) Phase Compensator

Without loss of generality, we set the carrier frequency as 5GHz, The bandwidth of the spreading spectrum waveform was 10MHz, data rate was 4Mchip/s, 200km from basel-ine to the landing point, attenuation factor was 0.8. For multilateration system, the source of positioning errors inclu-de signal transmitting error $\varepsilon_c \leq 3ns$, timing error between sites $\varepsilon_{s,FOL} \leq 50ps$(as comparison, $\varepsilon_{s,GPS} \leq 10ns$), Do-ppler effect error $\varepsilon_r \leq$ 10ns, LO errors is negligeable(frequency stability 0.1ppb), and channel errors (group delay) $\varepsilon_d \approx n \times 0.35/B_W + \varepsilon_{d,filter} \leq 5ns$, then the systematic overall timing DOP error could be calculated directly by $\Delta_{FOL,max} \leq \sqrt{2\varepsilon_c^2 + \varepsilon_s^2 + 2\varepsilon_v^2 + \varepsilon_d^2 + \varepsilon_r^2}$ ≈6.6ns, is 48% lower than GPS Timing based method($\Delta_{GPS,max} \leq$ 14.1ns), and we can infer from Figure 5 (a) (b), the DOP error distribution of FOL Phase-Correction based method is much better than that of GPS timing based method. And with consideration of positioning error of GPS(≤8m), we obtain the positioning precision (8.2m) which is one third of the GPS Timing based method (19.6m).

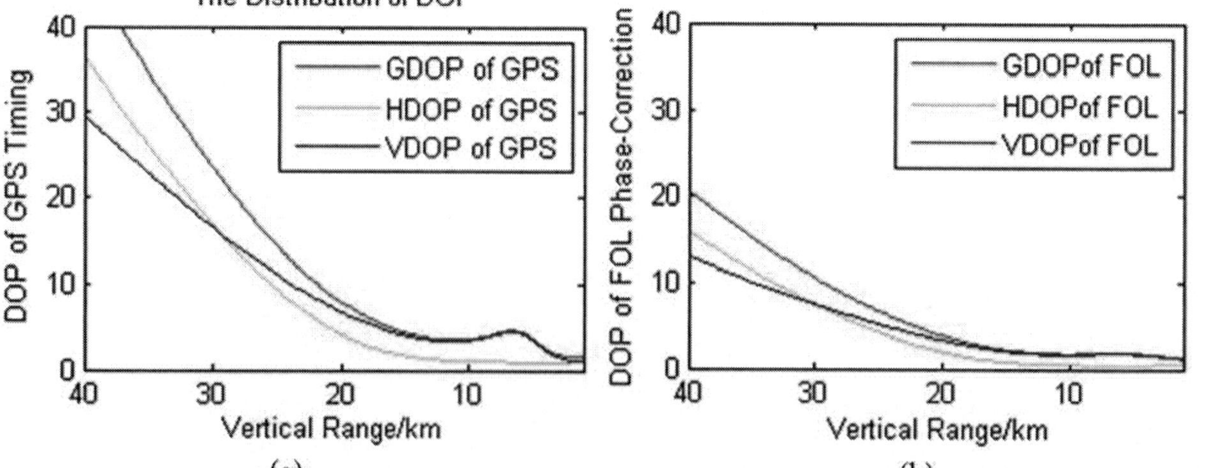

(a) (b)

Figure 5 (a) The GPS timing based DOP error distribution; (b) The FOL Phase-Correction based DOP error distribution;

CONCLUSIONS

Literatures have described the synchronization technique for Bistatic ISAR systems and other advanced applications [9-10]. However, the synchronization of ground-based Multilateration systems without GPS has been little studied. In this paper, an actual multi-station positioning system based on highly precise phase synchronization via a long round-trip FOL is demonstrated. Based on the calculation with the actual phase errors, the agreement of simulation and real measurement of localization results proves that the proposed technique has better performance than GPS timing based method. The simulation multi-station localization results with the FOL synchronization show that the phase-correction synchronization scheme is an effective approach for improving the accuracy of TDOA positioning method. For future work, we will make more experimental validation for the FOL synchronization of real multilateration systems.

This work was supported by the the National Natural Science Foundation of China (Grant No. 61601084).

REFERENCES

[1] Zhong, S.; Xia, W.; He, Z. Multipath Time Delay Estimation Based on Gibbs Sampling under Incoherent Reception Environment. *IEICE Trans. Fundam.* 2015, E98A, 1300–1304.

[2] N. Okello, F. Fletcher, D. Musicki, et al. Comparison of recursive algorithms for emitter localisation using TDOA measurements from a Pair of UAVs[J]. *IEEE Trans. Aerosp. Electron. Syst.*, 2011, 47(3):1723–1732.

[3] S. M. Foreman, K. W. Holman, D. D. Hudson, D. J. Jones, and J. Ye, "Remote transfer of ultrastable frequency references via fiber network-s, " *Review of Scientific Instruments*, vol. 78, no. 2, 2007.

[4] B. Wang, X. Zhu, C. Gao, Y. Bai, J. W. Dong, and L. J. Wang, "Square kilometer array telescope - precision reference frequency synchronis-ation via 1f-2f dissemination," *Sci. Rep.* 5, 13851(2015).

[5] J. Tian, Y. Cheng, N. Xie and D. Hou, "Bistatic ISAR imaging based on phase synchronization with fiber optic link," *2016 IEEE Radar Conf-erence (RadarConf)*, Philadelphia, PA, 2016, pp. 1-5

[6] T. Lane and J. Scheer, "Fiber optic line phase noise and drift effects on bistatic imaging radar performance," in Aerospace and Electronics Conference, 1996. NAECON 1996., *Proceedings of the IEEE 1996 National*, vol. 2, pp. 648–654 vol.2, May 1996.

[7] J. Roman, L. Nichols, K. Wiliams, R. Esman, G. Tavik, M. Livingston, and M. Parent, "Fiber-optic remoting of an ultrahigh dynamic range radar," *Microwave Theory and Techniques, IEEE Transactions on*, vol. 46, pp. 2317–2323, Dec 1998.

[8] J. Palmer, I. Longstaff, M. Martorella, and B. Littleton, "Isar imaging using an emulated multistatic radar system," *Aerospace and Electronic Systems, IEEE Transactions on*, vol. 41, pp. 1464–1472, Oct 2005.

[9] M. Martorella, "Analysis of the robustness of bistatic inverse synthetic aperture radar in the presence of phase synchronisation errors," *Aerospace and Electronic Systems, IEEE Transactions on*, vol. 47, pp. 2673–2689, OCTOBER 2011.

[10] W. Wang, C. Ding, and X. Liang, "Time and phase synchronisation via direct-path signal for bistatic synthetic aperture radar systems," *Radar, Sonar Navigation, IET*, vol. 2, pp. 1–11, February 2008.

TELEMETRY SYSTEM FOR INTERCOLLEGIATE ROCKET ENGINEERING COMPETITION VEHICLE

Seth Kitchen and Daniel Klinger(Students), Kurt Kosbar (Adviser)
Telemetry Learning Center
Department of Electrical and Computer Engineering
Missouri University of Science and Technology
Rolla, MO, 65409

ABSTRACT

This paper describes a telemetry system for a high-powered rocket entered in the Intercollegiate Rocket Engineering Competition hosted by the Experimental Sounding Rocket Association. On-board the rocket GPS coordinates, acceleration, magnetic field and lux readings are collected, along with other data. The data is sent between internal systems using commercial Internet-of-Things boards that utilize IEEE 802.11 wireless protocols. The aggregated data is transmitted to a ground station through a monopole transmitting antenna and custom designed helical receiving antenna such that in the event of a crash, data is not lost. The ground station data recovery is performed using a commercial XBee transceiver, before being displayed in real time for tracking and safety purposes, and stored for future data analysis. The target apogee is 9 km, so real time GPS data will be useful for both tracking and vehicle recovery operations.

INTRODUCTION

Rocket Design Team (RDT) constructs a high-powered rocket for the Intercollegiate Rocket Engineering Competition (now Spaceport America Cup) each year. The main objective of the competition is to launch a four kilogram payload as close as possible to a target altitude of 9.2 km and recover the rocket in a state where it can be launched again without repair.

RDT's initial design for 2017 used a motor-fed staging system. It differs from a normal two-stage rocket because there is no separation of the bottom half of the rocket. The first motor simply slides out with a recovery system. The distinction is illustrated in Figure 1. The electronics were designed and finished for this system, but design, manufacturing, and propulsion could not finish in time. A new single stage system was designed and the electronics were altered to work in nearly the same way. Both uses are included in this paper. The Spaceport America test launch was scheduled for June 24, 2017. We plan to describe the results of the competition in a future paper.

Figure 1: Differences Between Stagings

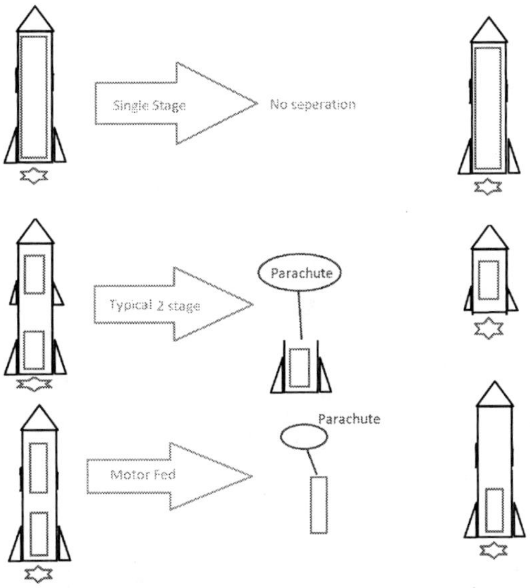

In previous years, if the rockets crashed the SD card holding all of its telemetry data would be crushed. Even when the rocket did not crash, very little data was written (usually only max altitude) because no effort had been made to make a telemetry system. Most importantly when the rockets reached high altitudes, they drifted into the desert and it was challenging to locate and recover them. For the 2017 competition the RDT decided to fix the problem by transmitting data to the ground through an on-board antenna, measuring four different sensors worth of data (lux, acceleration, magnetic field strength, gyro), and adding a GPS module for finding the rocket afterward.

Four electronics bays were designed for the motor fed system. To log which parts of the rocket were working, a client-server relationship was set up between the bays. Because of the complexity of the system and need for Wi-Fi connectivity, the Raspberry Pi 3 was selected for 3 of the bays. The Raspberry Pi 3 is able to run an entire Internet of Things (IoT) operating system, support 4 USB devices, support up to 40 General-Purpose Input/Output (GPIO) connected modules, and send audio to speakers through an audio jack. To have a back-up board tested in case of Raspberry Pi 3 failure, the fourth bay used an ESP8266 micro-controller board[6] which has Wi-fi capability, but is unable to run a full IoT operating system.

IOT COMMUNICATION FOR MOTOR-FED DESIGN

A. Bay 1: Determining Motor 1 Burnout

The proper time to eject the first motor is right after all of its propellant has been fired. Before the first motor is ejected and the second motor is slid down, electronics must detect a decrease in acceleration. RDT decided to accomplish this by connecting an accelerometer to the Raspberry Pi 3 also in Bay 1. The team selected an accelerometer module with multiple accelerometers attached, so that if some did not work well, others could be used. The selected board quadrupled as an MPU6000, MPU6050, MPU9250, and MPU9255. All of these models also work as magnetometers and gyros.

The team selected the Python programming language because open source libraries were plentiful for sensors. The code used for this project targeted the MPU9255. Features for the MPU9255 are listed in Figure 2.

Figure 2: MPU9255 Technical Specs

- Digital-Output 6-Axis MotionFusion Data from DMP; 9-Axis Fused Data from Motion Processing Library
- Tri-Axis Angular Rate Sensor (Gyro) with Full-Scale Range of ±250, ±500, ±1000, and ±2000 dps
- Tri-Axis Accelerometer with a Programmable Full-Scale Range of ±2 g, ±4 g, ±8 g and ±16 g
- Tri-Axis Compass with a Full-Scale Range of ±4800 µT
- Reduced settling effects and sensor drift by elimination of board-level cross-axis alignment errors between accelerometer, gyroscope, and compass
- Digital Motion Processing™ (DMP™) engine offloads complex MotionFusion, sensor timing synchronization, and gesture detection
- MotionApps™ Platform support for Android, Windows 8 and Embedded Applications
- Embedded algorithms for run-time bias and compass calibration. No user intervention required.
- Digital-Output Temperature Sensor
- Digital Input on FSYNC Pin to Support Video Electronic Image Stabilization and GPS
- Programmable interrupt supports gesture recognition, panning, zooming, scrolling, tap detection, and shake detection
- VDD Supply Voltage Range of 2.4 V–3.46 V; VDDIO of 1.7 V to VDD
- Gyro Operating Current: 3.2 mA (Full Power, Gyro 1 kHz ODR)
- Gyro + Accelerometer Operating Current: 3.4 mA (Full Power, Gyro at 1 kHz ODR, Accelerometer at 4 kHz ODR)
- Accelerometer Low Power Mode Operating Current: 6.4 µA at 0.98 Hz
- 400 kHz Fast-Mode I²C or Up to 1 MHz SPI-Serial Host Interfaces
- On-Chip Timing Generator with ±1% Frequency Variation Over Full Temperature Range
- User Self Test
- 10,000 *g* Shock Tolerant
- Smallest and Thinnest Package for Portable Devices (3 mm x 3 mm x 1 mm)
- RoHS and Green Compliant

The python code for the MPU9255 first imports external python libraries which provide the tools for communicating over I2C, use math functions like square root, use timers, use separate threads, communicate over serial, communicate over networks, and play sound. The code then enables I2C, stores the accelerometer/gyro address, and stores the magnetometer address. Note the accelerometer and gyro I2C addresses are the same, but the magnetometer address is different. The power management mode, user control mode, and interrupt pin configuration are set up so the sensors are given power and can be read from at default speed.

The data from the accelerometer, gyro, and magnetometer comes in two byte chunks. The team is not concerned with the byte value of the data, but what the decimal value is. So the code reads in the two bytes and converts them to decimal.

Since the MPU9255 gives the acceleration, gyro position, and magnetic field strength in a coordinate direction, it is useful to take the magnitude of the combined directions. RDT used the 3-dimensional distance function to do so.

The read functions are called to get raw data. The accelerometer data is then scaled as an understandable value in g's (where 1 g = 9.8 m/s^2) in each coordinate direction. The distance function combines all of the coordinate directions into a single magnitude in g's. The same procedure can be done to read gyro and magnetometer data.

When the magnitude of the acceleration realized by the sensor exceeds 5 g, a boolean value checking if the rocket has launched turns to true. When the acceleration drops under 5 g's the rocket is losing acceleration and therefore the first motor is burning out. After a short time delay, the client board triggers a relay to ignite a black powder charge to push the second motor down and eject the first motor.

The relay used was modeled after a commercially available product[10], but was redesigned to fit in a smaller volume and uses a 3.3 Volt relay. When the control pin is turned on, the relay opens.

The opening of the relay closes the ignition circuit which lights an electronic match connected on another header. RDT took out the IoT board header on the relay board and just soldered wires to the through-hole resistors to save space. The final relay board design is shown in Figure 3. The printed and soldered board is shown in Figure 4.

Figure 3: Relay Board Design

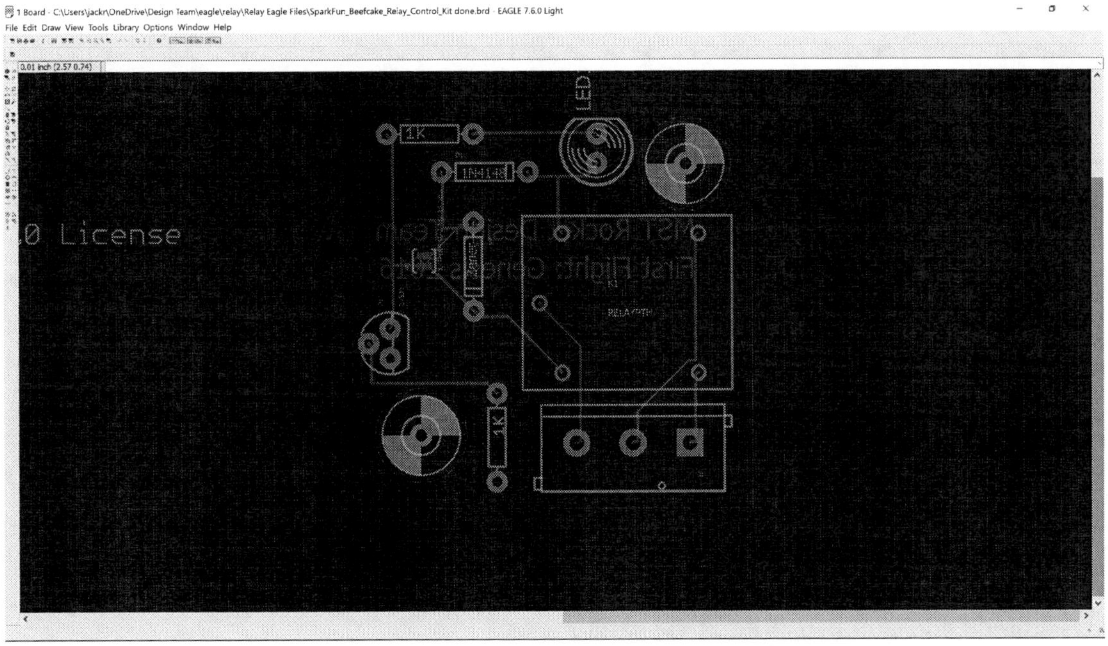

Figure 4: Printed Relay Board Connected

B. Bay 2: Recovery of Motor 1

To protect from parachutes being deployed while the first motor is still in the rocket, it is critical to have independent measurements that the motor has separated from the rocket. RDT decided to check the amount of light and the altitude.

When the first motor is inside the rocket, there is less light. When the first motor is outside the rocket, there is more light. We selected a light sensor[7] that can give lux values between .1 and 40,000, runs on 2.7-3.6 Volts, and can withstand temperatures between -30 and 80 degrees C. Communication is done via I2C.

The light sensor was connected to the micro-processor in bay 4, and programmed using a variant of the C++ programming language. Libraries for the ESP8266 micro-controller need to be installed for the code to compile.

Again the team imports the necessary libraries for communicating over Wi-fi, using the light sensor, and communicate over I2C. Then the team's code initializes global objects which represent the Wi-fi functionalities of the microcontroller and the features of the light sensor. The gain and delay (in milliseconds) is also initialized globally. The gain is dependent on the application, but for the rocket's purposes 16 is a good value. There is no need for a delay in this application, so by default it is set as 0.

Next the code sets up a debug serial session at 115200 baud rate. To make sure the code is working as expected *Serial.print* can be placed at points in the code to see if the code is reaching that particular line of code. The last bit of configuration is looking for the light sensor and identify it if found. If the light sensor is not attached, the code will continue after printing the error.

In the main loop variables are set up to store the raw data received from the light sensor. The loop waits for data from the lux sensor and when it receives the data it stores it. If no data can be pulled, the error is recorded and printed to the serial connection. Again the code converts the raw data from the sensor to an actual lux value. If the transition from raw data to lux data cannot be done, the boolean good will be set to false and "BAD" will be printed via serial. Otherwise the lux value is stored and "Good" is printed via serial.

If the lux is less than 100, the sensor is in the rocket. If the lux is greater than 100, the sensor is outside of the rocket.

An altitude check is also important because if the parachute is deployed at too high of an altitude, the first motor will float far away. There is also no use deploying the parachute at too low of an altitude because the parachute will shred and be wasted. A previously tested altimeter[8] was selected for this check. Figure 5 shows the features of the altimeter.

Figure 5: Entacore Altimeter Features

- Small form factor: 70mmx25mmx15mm (L x W x H).
- Pressure sensor with 30 cm (1 foot) resolution.
- Integrated temperature sensor.
- Battery input from 3.7V to 12V
- Can run off a single Lithium polymer battery cell (3.7 V).
- Lower current draw (14 mA)
- Maximum MSL value of 11,770 meters (38,615 feet).
- Completely self powered when connecting to a PC or laptop.
- Two high current (6 amp) lines for firing ejection charges.
- Component free underside - simpler to install.
- Data
  - output status
  - continuity
  - battery voltage
  - temperature
- Ejection settings
  - Apogee
  - Time
  - Accent altitude
  - Decent altitude
  - Peak velocity
- Peak velocity ejection option (great for staging!).
- Time delay options on both lines (for redundancy and other special requirements).
- Adjustable line firing time for both ejection lines (you can now control an electric motor etc. for a given time).
- Minimum battery voltage alarm settings. Now you can set your altimeter up for any battery type!
- Completely upgradable firmware, making sure you always have the latest!
- Least squares apogee detection.
- Over 30 minutes of total flight time!
- Connect to a PC or laptop to download flight profiles and modify settings.
- Data stored at 0.1 second intervals in non-volatile flash memory.
- Continually records output status, continuity, battery voltage and temperature.
- Imperial or metric beep-out of altitude after launch.
- Free custom designed software for downloading data and chaning settings.
- No data kit required - connect directly to the altimeter with a mini-USB cable (not included).

There is no coding necessary to use the altimeter. When it is plugged into a laptop or desktop, the altitude range desired can be input.

C. Bay 3: Motor 2 Ignition Bay

Bay 3 is responsible for checking that motor 2 is in place and then igniting motor 2. To check that motor 2 is in place Rocket Design Team decided to put a magnet near the bottom of the rocket. When motor 2 slides down, it comes closer to the magnet and so the magnetic field increases. Using the MPU9255 in the same way mentioned in Bay 1 allows the detection to occur. After a small time delay, the MPU9255 opens a relay which ignites the second motor.

D. Bay 4: Nosecone Master Bay

All client bays would send their logs and data to master Raspberry Pi 3 in the nosecone. The master Raspberry Pi 3 is running a TCP server. In addition, for redundancy in telemetry information, an IMU[9] is attached to the Pi. The IMU gathers acceleration, pressure, gyro, and magnetic field on a single chip. A monopole antenna hooked up to an XBee transmitter plugs into the Pi via USB. The data can be written to the transmitter via serial UART. A helical antenna on the ground has an Xbee receiver and compiles the data.

Actual Implementation in Single Stage Rocket

Since the checks and ignitions are no longer needed in a single stage rocket, the telemetry system described above was converted to one where the data is just logged. Two bays were used instead of four. In a coupler in the middle of the rocket, the MPU9255 is hooked up to a Raspberry Pi 3 and the light sensor is hooked up to the ESP8266 microcontroller board. Both boards send their data to the nosecone bay. The nosecone bay has the Raspberry Pi 3 running the TCP server, collection from the IMU, and sending the data to the ground station via the monopole antenna.

ANTENNA

The antenna system consists of two parts: A small monopole antenna on a low-power transmitter in the nosecone of the rocket, and a large high-gain helical antenna at a ground station. Operating on the 900 MHz ISM band, and partially surrounded by structural steel, the system is not exactly ideal for transmitting long distances, but should be powerful enough to transmit data to a ground station from over 30,000 feet away according to RDT's link budget.

The antenna system in the nose is quite simple. It consists of an XBee 900MHz radio module[1]2 connected to a small rubber duck antenna via RP-SMA connector. The radio module transmits at a maximum power of 50 mW, which is low power but gives us enough power to work with. This antenna[1]1 is vertically polarized with approximately 2 dBi gain. Unfortunately, as shown in Figure 6 below, the antenna is placed near several plates of $\frac{1}{4}$ in. steel in the nosecone. This design limits the antenna's ability to transmit in the direction of the plate, although it does radiate in all other directions.

Figure 6: Monopole Antenna in Rocket Bay

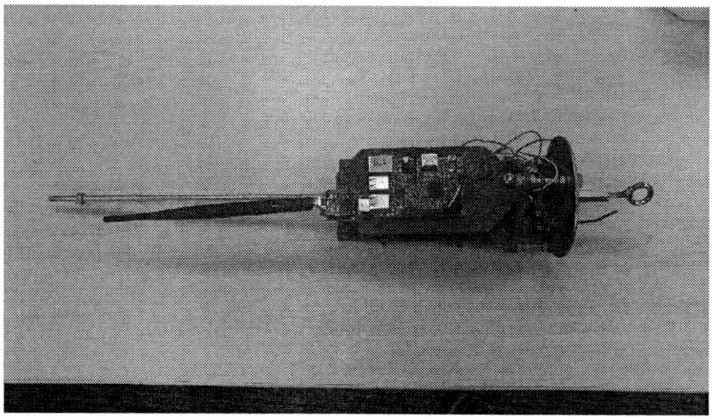

On the ground, there is a slightly more complicated antenna system to receive the data. The receiving radio module is the same as the one in the nosecone. It has -100 dBm sensitivity and communicates with a receiving computer via serial connection. From the radio module, there is a 10 ft. pigtail cable[1]3 of LMR-200 with RP-SMA connected to the radio, and an N-Type connector that goes to a bulkhead on the antenna. This cable claims to have a 2.9 dB insertion loss and 1.0 dB distance loss at 900 MHz. The antenna itself, shown in Figure 7, is a custom-built 10-turn helical antenna that should give approximately 15 dBi of directivity. The antenna consists of a large length of insulated 12 gauge SWG wire wrapped around a treated cellulose tube. At the base, the wire is connected to the N-Type bulkhead in the center of the ground plane, the ground plane being a $\frac{3}{4}$ wavelength diameter copper plate. To provide an optimal helical antenna geometry[1], the antenna windings have a circumference of 1.1 wavelengths and a winding separation of one quarter wavelength.

Figure 7: Helical Antenna

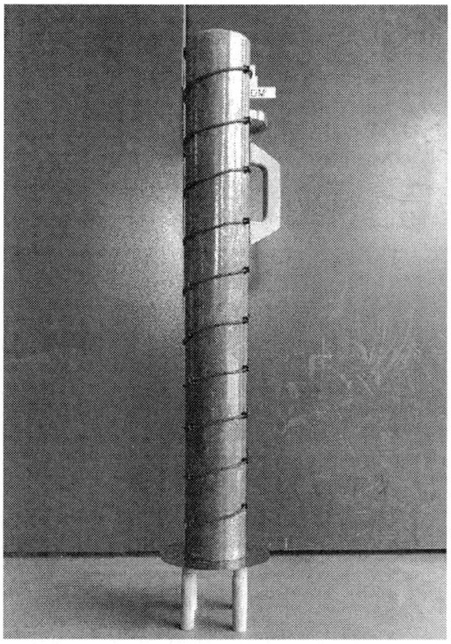

A helical antenna was chosen for several desirable qualities. First, once the rocket reaches apogee and deploys its parachute, the nosecone and the transmitting antenna will be tumbling in an unknown orientation. If the antenna settles into an undesirable orientation, cross-polarization losses could exceed 20 dB, an unacceptable amount. With the helical antenna, the transmission will have a steady 3 dB of loss from a linearly polarized transmitter, regardless of transmitter orientation[4]. Second, the helical antenna is simple to construct. The most difficult part of construction was finding a properly sized shape to wrap the wire around, everything else was comparatively easy. Third, high gain helical antennas are fairly tolerant to error in construction and setup. To help compensate for any imperfections in antenna construction, an antenna with a relatively high half-power beamwidth and bandwidth was deemed important, and helical antennas fit that bill[5].

However, there are also several drawbacks to using a helical antenna. Most importantly, the characteristic impedance of a helical antenna is 140 Ohm, and the coaxial cable we obtained is 50 Ohm. Many solutions were considered, such as baluns, pi networks, and commercial impedance-matching solutions. We chose an approach[2] that involves adding a wide conductive strip at the base of the wire a certain distance from the ground plane to satisfactorily match the impedances. However, a spiraling conductive strip proved difficult to mount mechanically to the wire near the reflector. Spectrum analysis of the completed antenna showed a VSWR of around 2:1 at 900 MHz. This VSWR may prove problematic for receiving very weak signals, but hopefully the link budget will account for this extra loss.

Early on in research for antennas, a link budget was made to determine the feasibility of receiving a 50 mW signal from almost 6 miles away. Using a modified logarithmic form of the Friis Energy Equation (Eq. 1), and a conservative estimate for receiver sensitivity (which was not known at the time), it was determined that an antenna with moderately high gain would be able to receive the signal.

$$P_{RX} = P_{TX} + G_{TX} - L_{TX} - L_{FS} - L_M + G_{RX} - L_{RX} \quad (1)$$

where P_{RX} is received power (dBm), P_{TX} is transmitter output power (dBm), G_{TX} is transmitter antenna gain (dBi), L_{TX} is transmitter losses (coax, connectors...) (dB), L_{FS} is path loss, usually free space loss (dB), L_M is miscellaneous losses (fading margin, body loss, polarization mismatch, other losses...) (dB), G_{RX} is receiver antenna gain (dBi), and L_{RX} is receiver losses (coax, connectors...) (dB). The path loss can be approximated with (Eq. 2).

$$L_{FS}(\text{dB}) = 20\log_{10}\left(4\pi\frac{\text{distance}}{\text{wavelength}}\right) \quad (2)$$

where distance and wavelength are in the same units.[3] With this equation, the following link budget was constructed, following project design and construction.

Positive Factors

1. Transmitter Power: 17 dBm (50 mW)
2. Transmitter Directivity: 2 dBi
3. Receiver Directivity: 15 dBi

Negative Factors

1. Path Losses: 112 dB (30,000 feet at 900 MHz)
2. Feedline Losses: 4 dB
3. Polarization Mismatch: 3 dB

This ideally gives us around -85 dBm of power, which is within the receiver's -100 dBm sensitivity. That allows us about 15 dB of link budget to account for antenna inefficiencies, VSWR mismatch, fast data transfer, and any other losses we might encounter. However, the XBee modules use a networking protocol without error correction mechanisms, so packets of data must be received in full or no data gets received. This means that connections very near the limit of receiver sensitivity will not be received. Anecdotal evidence suggests that the effective sensitivity may be closer to -80 dBm, giving us a link budget of only 5 dB to account for other losses.

CONCLUSIONS

The design, construction and testing of the telemetry system has been completed. The team looks forward to testing the design at IREC 2017 (June 24, 2017). A later edition of the paper will supply and analyze the results.

References

1. Weeratumanoon, Eakasit. "Survey of Helical Antennas." *Virginia Tech*. Date Accessed: 17 June, 2017. <https://theses.lib.vt.edu/theses/available/etd-02102000-19330046/unrestricted/07chapter2.PDF>.

2. Kraus, J. "A 50-ohm input impedance for helical beam antennas." IEEE. Volume: 25 Issue: 6. Date Accessed: 17 June, 2017. <http://ieeexplore.ieee.org/document/1141687/>.

3. "Link Budget." Wikipedia. *Wikimedia Foundation*. Date Accessed: 17 June, 2017. <https://en.wikipedia.org/wiki/Link_budget#Equation>.

4. Reisert, Joseph H. "Antenna Polarization." *Astron Wireless*. Date Accessed: 17 June, 2017. <http://www.astronwireless.com/topic-archives-antennas-polarization.asp>.

5.Slade, Bill. "The Basics of Quadrifilar Helix Antennas." *Orban Microwave*. Date Accessed: 17 June, 2017. <http://orbanmicrowave.com/wp-content/uploads/2015/03/TheBasicsofQuadrifilarHelixAntennas.pdf>.

6. Sparkfun Thing. <https://www.sparkfun.com/products/13231>.

7. TSL2561 Light Sensor. <https://www.adafruit.com/product/439>.

8. Entacore Altimeter. <http://entacore.com/electronics/aimusb>.

9. BerryIMU. Ozzmaker. <http://ozzmaker.com/product/berryimu-accelerometer-gyroscope-magnetometer-barometricaltitude-sensor/>.

10. Sparkfun Beefcake Relay. <https://www.sparkfun.com/products/13815>.

11. GSM-09 Antenna. Chang Hong Information Co. Ltd. Sparkfun. <https://www.sparkfun.com/datasheets/Wireless/Antenna/GSM-09.pdf>.

12. Xbee 900. Sparkfun. <https://www.sparkfun.com/datasheets/Wireless/Zigbee/XBee-900-Datasheet.pdf>.

13. Pigtail Cable. TP-Link. Amazon. <https://images-na.ssl-images-amazon.com/images/I/91v2TywSufL.pdf>.

AN EVALUATION OF TERRESTRIAL WIRELESS NETWORK MODELING APPROACHES FOR THE SPACE MOBILE NETWORK

Todd A. Newton[1], Christopher J. Roberts[2], Gregory G. Fletcher[1], Daniel S. Rossiter[1]

[1]Southwest Research Institute® San Antonio, Texas

[2]NASA Goddard Space Flight Center Greenbelt, Maryland

todd.newton@swri.org, christopher.j.roberts@nasa.gov, gregory.fletcher@swri.org, dan.rossiter@swri.org

ABSTRACT

The Space Mobile Network (SMN) is NASA's next generation architecture concept for communications services between ground and space-based assets. The SMN calls for a paradigm shift in space communications. The transition will move from an approach based on static, preplanned communications over point-to-point channels to a dynamic, event-driven, and network-based approach that facilitates service-oriented communications. In doing so, the SMN is able to leverage some concepts and technologies present in today's terrestrial wireless networks, while others must be extended or adapted to the space communications domain. This paper provides background on key SMN architectural concepts and an evaluation of the suitability of terrestrial wireless network modeling tools to be used and applied for proving out SMN concepts.

KEYWORDS

Space Mobile Network, IP, Network Simulation

INTRODUCTION

The Space Mobile Network (SMN) is NASA's next generation architecture concept for space communications services. The SMN calls for a paradigm shift in space communications. The transition will move from an approach based on static, preplanned communications over point-to-point channels to a dynamic, event-driven, and software defined network-based approach that facilitates service-oriented communications. The goal of the SMN is to provide mission users of space communications services with an experience analogous to a terrestrial mobile wireless user. A terrestrial mobile wireless user can access services and communicate with other users without a need for detailed knowledge about the underlying network infrastructure, protocols, and processes that enable those communications. SMN seeks to allow its users to connect to the SMN architecture in the same way, eliminating human-supported prescheduling and providing reliable access for data to reach the end user.[1] SMN features, including optical communications, delay tolerant networking (DTN) services, user initiated services (UIS), and position, navigation and timing (PNT) services, would enable SMN to provide this type of user experience.

The SMN would connect users via a continuous, low-rate data link for low-latency resource control and management, and it would allow user platforms to automatically request additional services as needed. Figure 1 illustrates how a low gain omnidirectional antenna can enable this capability.

Figure 1 - Notional Concepts of the Space Mobile Network

The SMN would provide a simpler, faster, and automated system for space-based communications and would enable providers to support higher data rates and greater volumes of data. This would enable new or improved mission concepts and use cases, including variable data rates and collaborative platform operations. For instance, multiple operationally independent and spatially diverse user mission platforms monitoring the sun could each downlink data using lower rates to perform day-to-day functions and monitor for significant events. When a user platform detects an unexpected science event or an event that will generate an unknown quantity of data, the SMN dynamically orchestrates the necessary resources: the user platform automatically sends a service request or notification (for higher data rate links from space or ground assets, to alert other subscribed user mission platforms), using the low-rate link and associated control protocol. The SMN processes and determines the parameters of the request, determines how and when the communications need can be met, and provides a response. SMN PNT services enable precise autonomous orbit determination, allowing for satellite formation flying and improving spatial-temporal science data correlation across platforms.

NASA's long-term vision for the SMN is to see their spacecraft communicate and interoperate with commercial spacecraft. A rendering of this vision can be seen in Figure 2 which highlights this interoperability of the spacecraft, various data link speeds, a user-driven approach to network services, and a sense of always being connected at all times.

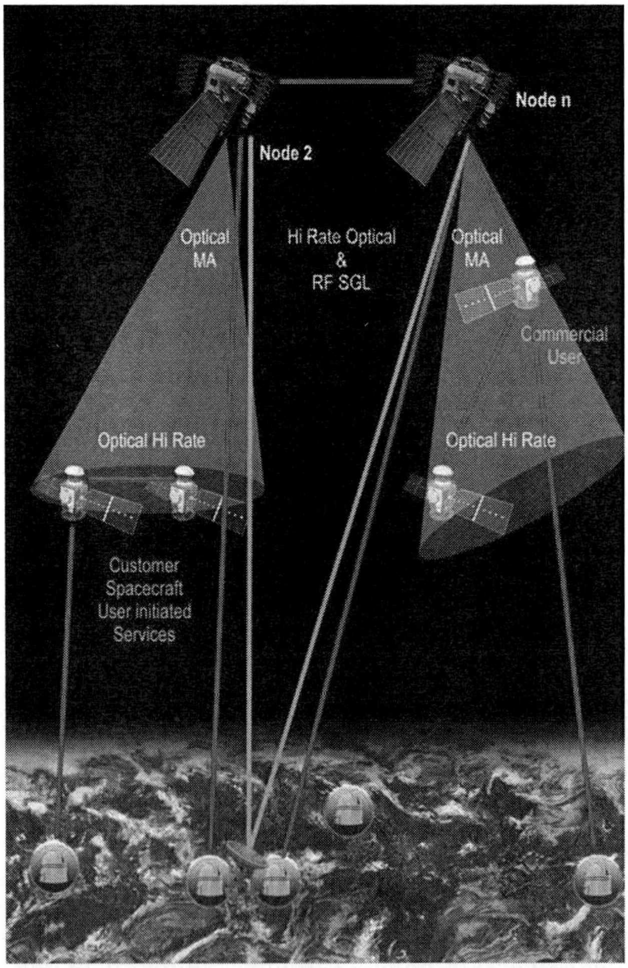

Figure 2 - Long-Term Vision of the Space Mobile Network

The vast majority of network research is targeting terrestrial networks. There are several existing commercial and open source software tools available for network simulation, performance monitoring, and troubleshooting. Though these tools are almost explicitly utilized for terrestrial applications, the networking concepts that they deal with are directly applicable to the SMN concepts. Current research in network scheduling policies and the associated performance metrics apply in both realms. As such, the SMN community stands to gain valuable insight of SMN performance through the lessons learned from terrestrial network scheduling algorithms and the extension of current network simulation tools for the SMN environment. Prominent network simulation environments such as OPNET or ns-3 can be extended to provide simulations of the SMN. This paper provides some background on key SMN architectural concepts and an evaluation of the suitability of terrestrial wireless network modeling tools to be used and applied for proving out SMN concepts.

ATTRIBUTES OF THE SPACE MOBILE NETWORK

Delay Tolerant Networking Services

One of the major distinguishing characteristics of SMN users from terrestrial wireless network users is the intermittent availability of high data rate space links due to spatial geometry and a corresponding tolerance for delays in data receipt for most high-volume space mission users. In general, missions have two types of data latency requirements: either the user needs the data to reach its final destination quickly, possibly requiring near real-time delivery, or the user needs to offload the data from the instruments' recorder (the store and forward method) due to data storage constraints, only requiring a link from the spacecraft to the communications asset. Delay Tolerant Network (DTN) protocols can support store-and-forward operations while opening up the availability of high-rate links when near real-time delivery is necessary.[2] DTN also bundles data in smaller increments, increasing the efficiency of the network by allowing for greater availability of assets to service different mission data priorities. These features would allow SMN to create a standardized and scalable network.

Independent from DTN, SMN would require hardware that enables flight systems to move and process data at higher speeds; however, near-term DTN protocol-specific implementations could include hardware acceleration of the Bundle Protocol (BP), Licklider Transmission Protocol (LTP), or Bundle Protocol Security (BPSec). In the future, as the operational network grows, DTN will require advancements in DTN routing and network management, and implementation of DTN functionality will be realized within ground stations, space relay nodes, and user ground systems.[1]

Optical Communications Services

Direct-to-Earth (DTE) communications links from low Earth orbit (LEO) spacecraft can be hundreds of times shorter than links from LEO to geostationary relays, making it possible for DTE links to deliver immensely greater volumes of data. Currently, service providers cannot take full advantage of this significantly shorter link length due to the lack of available spectra necessary to transmit the data.[3] Optical links would enable providers to offer such high data rates without being constrained by spectra availability. Past demonstrations have shown initial capabilities of optical communications; for instance, in the Lunar Laser Communications Demonstration, DTE optical downlinks performed at rates higher than 600 Mbps from the moon.[4] Demonstrations and tests involving optical links continue. For instance, the Laser Communications Relay Demonstration (LCRD), which will launch in 2019, will spend two years conducting experiments to test optical links from a geosynchronous terminal, measuring the effects of various settings and conditions such as weather on the links. With such new technologies being developed and tested, even greater data rate capability is expected. Furthermore, due to reduced size, weight, and power requirements for optical communications systems, missions could use multiple access systems with continuously available, low-rate optical links at lower costs and with less operational complexity. Producing a reduced user burden than RF links allow, optical multiple access systems would be easier for missions to assume and utilize. [1]

User Initiated Services

Today, spacecraft communication network users rely on flight operations teams to determine many aspects of the network scheduling process, which can result in long lead times when scheduling a connection. SMN would maintain support of today's pre-planned scheduled services while engaging user initiated services (UIS) to support an increasing number of spacecraft communication networks users, allowing for a more responsive and efficient network. UIS would enable platform-triggered requests and disposition of services through continuously available low rate links, allowing any user to autonomously request services at any time. The SMN would, in turn, automatically de-conflict such requests based on state, priority, and constraint information about the network's nodes and user capabilities. The SMN would respond by providing the service as requested, by finding different yet compatible parameters to satisfy the request, or by declining to provide the service as requested. Along with developing a protocol for spacecraft communication network users to negotiate a service request, UIS would also rely on a scheduling system that could field requests dynamically; the system would compare requests against available resources, schedules, and priorities, and it would dispatch the scheduled service control details to the user systems and provider elements. In addition, this concept may be federated through peering agreements, enabling a combination of various, possibly dynamically entering and exiting, providers and scheduling systems to provide the services.[5] UIS would reduce operations costs and complexity for mission designers with this fundamentally unique operations concept.

Position, Navigation, and Timing Services

In the vein of automated services, SMN would transition beyond the use of ground-based processes to calculate spacecraft orbits, instead moving to adopt onboard navigation (aboard user spacecraft) and establish a global beacon service to create positioning and navigation autonomy. Onboard navigation would use GPS to allow the spacecraft to determine its own position and location, thus enabling the spacecraft, for instance, to orient itself and point accurately to offload data or assist the UIS system in coordinating communications services. However, some spacecraft cannot support GPS SWaP or their orbits are outside the reach of GPS. The Next Generation Broadcast Service (NGBS) beacons would broadcast from at least three geosynchronous nodes, such as relay satellites, providing user spacecraft with the location of the node, thus allowing the user spacecraft to determine its own position and course. In this way, the beacons would serve as a global means to enhance user auto-navigation capabilities.[6]

NETWORK MODELING FOR TERRESTRIAL WIRELESS NETWORKS

Network simulation has often been the first step in designing and validating robust terrestrial networks and protocols. A network simulation environment reduces deployment risk and can save money in the long run by discovering limitations and problems and addressing them prior to full scale deployment rather than having to re-deploy physical network infrastructure. A simulation environment also allows complex networks with hundreds or thousands of nodes to be simulated, which is likely beyond the scope of financial viability. Simulations can be utilized alongside real components of hardware such that hardware testing can be performed against the simulated nodes and networks for verifying proper behavior of the real hardware. When considering deployments in space, the deployment costs alone are great; being able to validate design and proper protocol implementations prior to deployment are critical.

OPNET and ns-3 are the industry leading network simulation tools in the commercial and open source realms, respectively. Both offer a large set of network and component models, including fully-simulated networks and simulation-in-the-loop where simulation is combined with some physical hardware to create a hybrid network between physical and simulated components. Models for various layers of the Open Systems Interconnection (OSI) model are available, allowing different hardware technologies such as copper interfaces or WiFi interfaces to be modeled with their particular network characteristics at the physical layer and different models to be built on top of those physical layer models. Both simulation tools support wireless network simulation including WiFi and Long-Term Evolution (LTE) networks, allowing various simulation scenarios to be constructed and measured for performance based on the associated models being simulated.

Before we can discuss how a simulation would be devised for a given scenario, some of the relevant network simulation key terms are defined:

- **Node:** A node in a simulation represents the component being simulated. The component contains one or more network interfaces for connecting to the network as well as any applications that may run on the simulated node, such as a specific type of traffic generator or receiver. It can be represented as a single computer in the network that is interconnected with other nodes to form a network.
- **Net Device:** An abstraction representing a network interface card (NIC) plus the associated driver software. This would be attached to a node to support connection to the network via a specific interface.
- **Channel:** A channel provides the interconnection between nodes. A given node may have multiple incoming/outgoing channels corresponding to the various net devices

attached such as ones supporting Ethernet, WiFi, or Bluetooth. Each channel may have its own set of performance characteristics, such as speeds, loss, etc.

- **Application:** In the simulation context, an application is a piece of code that generates or consumes network traffic on the node it is installed on. An application running on one node in the simulation may be the data generator for an application running on another node in the simulation.

When approaching a simulation of a network, the process can either be relatively simple or quite complex depending on what is being simulated. A simple network with a few nodes all using well-known net devices, channels, and applications can be pieced together in a matter of minutes by an experienced individual. This is because well-known net devices, channels, and applications have software modules for simulation readily available, meaning that the only software to be written is to instantiate and populate the simulated nodes with the corresponding Net Device(s) and Application(s). If, however, custom components are required, then the required effort is increased. Modifications to existing modules or creation of new modules must occur to simulate these new concepts. This is the case with simulation of the Space Mobile Network, where there are custom net devices, channel characteristics, and applications that must be simulated to accurately represent the proposed network. Though some effort is required here, it is encouraging to see that the networking tools available today for modeling and simulating terrestrial networks is extensible to being able to model the SMN.

MODELING OF NETWORK SCHEDULING POLICY OPTIMIZATION

As part of an internal research project, Southwest Research Institute is evaluating methods for optimizing user-perceived network performance through use of a constraints-dependent and requirements-aware packet scheduler. The proposed scheduler would make use of network performance parameters collected in near-real time along with domain knowledge of what data is being scheduled in order to dynamically select an appropriate scheduling policy for the current network condition. As the network condition changes, so too will the policy selected by the algorithm and implemented by the scheduler. In this way, the algorithm will continually update the scheduling policy to keep up with dynamic network changes through time and space. This multi-policy scheduler is most applicable for highly-nomadic clients connected to wireless networks in both the terrestrial and space networking domains, where the network topology is all but certain to change relative to the connected client. In these situations, a single generalized scheduling policy cannot provide an appropriate network performance for all scenarios, but instead must adapt by selecting the policy that performs the best across all scenarios. The research is aimed at identifying the conditions in which certain policies perform better than others as well as identifying those logical transition points when a policy change would be beneficial to the network. Figure 3 provides a high-level view of the network scheduling policy selector concept.

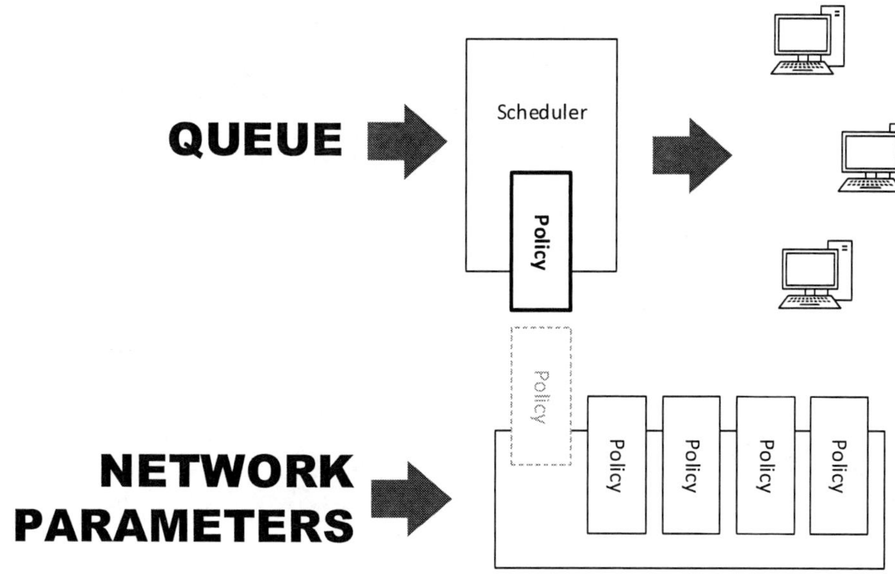

Figure 3 – Network Scheduling Policy Selector

To measure the efficacy of this scheduler and the various policy selections in various network topologies, a network testbed is being used. This testbed is composed of both physical and simulated components. This hybridized simulation approach will provide significant flexibility in testing a wide array of network environments from both earth and space. Testing will focus on accurately representing the target environments (Space Mobile Network and terrestrial mobile networks) and determining which scheduling policies provide the most significant performance improvements compared to standard scheduling techniques.

MODELING THE SPACE MOBILE NETWORK

There are several commonalities between modeling the SMN and terrestrial mobile device networks. In both cases, the goal is to allow the user to have no knowledge of the specific assets used or path traversed for the data to reach its destination (see Figure 4). This is the case for terrestrial networks; however, traditional communication to and from spacecraft or space borne assets requires very specific knowledge of other space borne components (i.e. TDRSS) and/or ground stations. The SMN is intended to enable user initiated services similar to the terrestrial mobile user experience. The user determines a need for communication at a time determined by the needs of the user asset and initiates a data transfer; the network is then responsible for delivering the data. Terrestrial network models generally accommodate a user moving between fixed communication assets. The SMN model will need to track ephemeris not only for the user asset, but also for network assets, which will be in a variety of orbits and available ground stations in relation to their positions.

Figure 4 depicts a simplified comparison of how the user experience for terrestrial networks and the SMN don't require the user to have detailed knowledge of all the assets in the network.

Figure 4 – Comparison of Terrestrial Wireless Network and Space Mobile Network Communication Paths

DTN models currently exist and are relatively simple, requiring pre-planning of the spacecraft overpass of a particular ground station. The duration of the communication event depends on orbital altitude, inclination, transmission frequency, and data volume. The SMN model takes into account users' needs to offload their data at whatever rate their communications link allows while optimizing flexibility on the provider side to implement, allocate, and schedule all the nodes and links along the rest of the end-to-end path. In the near-earth environment, some scenarios can be supported using IP for the network layer services.

The SMN model incorporates DTNs with the rest of the network. Efficient and effective scheduling policies must determine the priority of data packets. For example, on terrestrial networks, voice data (Voice over Internet Protocol, or VoIP) is often prioritized over other types of data, such as one-way streaming video, because end-to-end latency and jitter adversely affect communication, while streaming video can be buffered.

The SMN model will have the capability to incorporate different scheduling policies, network protocols, and network assets and to determine the effect these parameters have on overall network performance. The model will measure quality using metrics like: jitter, network latency, end-to-end delay (caused by assets within the network, like distance, protocols and congestion), throughput, packet loss, and goodput. The model will use these metrics to determine which applications would be adversely affected.

Modeling the SMN in this way provides data to optimize the network given its expected use. Scenarios can be run to test multiple nodes, paths, and users, as well as to determine the effect of the dynamic addition or loss of assets within the network. Worst-case scenarios can be run to determine the network performance during heavy instantaneous network demand.

As has been stated, the SMN represents a drastic shift from previous network models, focusing more on scalability and flexibility and less on complete predictability. With this shift comes many benefits, but it also introduces some risks, such as data delivery failure during route changes within the SMN. Network modeling provides a cost-effective method for both quantifying these risks and for developing verifiable strategies to mitigate these risks before significant cost is spent deploying this next generation of space communication infrastructure.

CONCLUSION

The SMN represents a shift in spacecraft communications from the static, preplanned communications approach to a dynamic, network-based approach that is event-driven and service-oriented, similar to terrestrial wireless networks today. SwRI's current research in network scheduling optimization, particularly over wireless networks, is one of these similarities that has applicability and relevance to the SMN. Network performance is being measured in near-real time, and the network scheduling policies are being adapted when necessary in order to meet the critical demands of the wireless network. The same concepts exist for the SMN, where the availability of some links change and/or the network is congested for a period of time but the SMN still needs to deliver certain data flows within a specific latency window. Because of the similarities with terrestrial wireless networks, the SMN is able to leverage similar concepts and common network test tools for evaluating SMN concepts. Most of the networking tools available today are extensible to cover the SMN-specific protocols and characteristics.

REFERENCES

[1] Israel, David J., et al. "Enabling Communication and Navigation Technologies for Future Near Earth Science Missions." 14th International Conference on Space Operations. 2016.

[2] Davis, Faith A., Jane K. Marquart, and Greg Menke. "Benefits of Delay Tolerant Networking for Earth Science Missions." In *Aerospace Conference, 2012 IEEE*, pp. 1-11. IEEE, 2012.

[3] Boroson, Don M., et al. "A New Optical Communication Architecture for Delivering Extremely Large Volumes of Data from Space to Ground." AIAA SPACE 2015 Conference and Exposition, 2015.

[4] "Earth Regimes Network Evolution Study," NASA. 1 May 2015.

[5] Israel, David J., Gregory W. Heckler, and Robert J. Menrad. "Space mobile network: a near earth communications and navigation architecture." *Aerospace Conference, 2016 IEEE.* IEEE, 2016

[6] Valdez, Jennifer E., Benjamin Ashman, Cheryl Gramling, Gregory W. Heckler, and Russell Carpenter. "Navigation Architecture for a Space Mobile Network." AAS GN&C Conference, Breckenridge, CO, 2016.

3D LOCALIZATION FOR LAUNCH VEHICLE USING COMBINED TOA AND AOA

Soonho Kwon[1, 2], Donghyun Kim[1], Jeongwoo Han[1], Chunwon Kim[1], Dae-Oh Kim[1], Intae Hwang[2]

[1]Flight Tracking Technology Team, Naro Space Center
Korea Aerospace Research Institute, Daejeon, Republic of Korea
[2]Department of Electronics and Computer Engineering
Chonnam National University, Gwangju, Republic of Korea

ABSTRACT

Generally, a ground telemetry station for launch vehicle (LV) has tracking function only; therefore, position measurements depend on radar. Time of arrival (TOA) and angle of arrival (AOA) are typical location techniques for emitting targets. In this paper, we propose a Combined TOA and AOA localization method for LV using two ground stations. When transmitter (Tx) time is not known, it is necessary to make virtual onboard timer for TOA estimation. The virtual onboard timer generates time stamps of streaming frame according to data rate. First station which is located in space center has no tracking function. But it can generate the virtual onboard timer. Second station has tracking function, so it generates AOA information. By solving sphere equation(s) of TOA from at least one station and a line equation of AOA, target position in three-dimensions (3D) can be obtained. We confirm the localization performance by means of comparison with an on-board GPS of a real launch mission.

Keyword: AOA, Ground Telemetry Station, Launch Vehicle, Localization, TOA.

1. INTRODUCTION

Various techniques have been proposed for location estimation, depending on the measurements collected by the wireless networking infrastructure. These techniques can be broadly classified as time-based and angle-based localization. Time of arrival (TOA) is a form of time-based localization, in which a circle centered on the base station (BS) is produced and the target position is determined by the intersection of at least three circles. Angle of arrival (AOA) uses angle-based localization, in which defines a line of bearing from the BS to the target is defined, and the position calculated from the intersection of a minimum of two bearing lines. Due to noisy measurements, the position is typically determined by statistical estimation, in which a set of nonlinear equations constructed from the TOA and/or AOA statistics is processed using knowledge of the BS geometry[1][2].

Generally, in a launch mission for a launch vehicle (LV), a ground telemetry station contains only tracking function, which uses an antenna control unit (ACU) to obtain the AOA from the

LV on the azimuth and elevation axes. There is no function for measuring the LV's range (or distance). Therefore, the 3D LV position measurements depend on radar or received on-board LV telemetry data. Each ground telemetry station is synchronized with GPS time, and its location has already been measured or synchronized with GPS position. Furthermore, since there is no time synchronization between the LV and ground stations, TOA measurement is not available. The on-board telemetry transmitter sends streaming frames, which include a frame counter.

Under the previous described conditions, we suggest a Combined TOA and AOA 3D localization method for LV, using two ground stations. First station which is located in space center has no tracking function. But it can generate virtual onboard timer to estimate TOA. Second station has tracking function, so it obtain AOA. We make nonlinear equations with the TOA and AOA, then solve these to find 3D position of LV.

The remainder of this paper is organized as follows. Section 2 describes typical localization methods, while section 3 provides details of the proposed proposed TOA and AOA method. In section 4, the performances of the proposed methods are evaluated by means of comparison with measured results from an on-board GPS during an LV flight test.

2. TYPICAL LOCALIZATION

2.1 TOA METHOD

If only time delay estimates are available, at least three or four stations are required for 2D or 3D localization, respectively. TOA algorithms are based on the signal delay from the target to station, in order to estimate the distance. Each distance estimate can be represented as a circle's radius, the center of which is the station's location. Therefore, we obtain a number of circles depending on the number of stations, and the point of intersection of all the circles is the target's position [3][4].

Let (x, y) be the coordinates of the unknown target and (x_i, y_i) those of ith station, with $i = 1, 2, 3 \dots N$ and N being the number of stations; then, the distance between the target and station is

$$\widehat{d_\iota} = \sqrt{(x - x_i)^2 + (y - y_i)^2} + n_i, \tag{1}$$

where n_i is zero mean Gaussian; that is, $n_i \sim \left(\mathcal{N}\left(0, \sigma_{n_i}^2\right)\right)$.

Clearly, (1) is non-linear and can be solved by using iterative techniques; however, a linear least-squares method can also be used by first linearizing (1). Let $\hat{d}_i^2$ be the 2D noisy distance estimates:

$$\hat{d}_i^2 \approx (x - x_i)^2 + (y - y_i)^2 \tag{2}$$

Then, a reference station is selected and its distance equation is subtracted from (2) for $i = 1, 2, 3 \dots N \ (i \neq r)$. Let d_r represent this reference station's reference distance. Hence, we obtain:

$$(x_i - x_r)x + (y_i - y_r)y = 0.5[(x_i^2 + y_i^2) - (x_r^2 + y_r^2) + \hat{d}_r^2 - \hat{d}_i^2\ ] \tag{3}$$

In matrix form, we have:

$$\mathrm{A}_t u_t = 0.5\widehat{b}_t, \tag{4}$$

$$\text{where } \mathrm{A}_t = \begin{bmatrix} x_1 - x_r & y_1 - y_r \\ x_2 - x_r & y_2 - y_r \\ \vdots & \vdots \\ x_N - x_r & y_N - y_r \end{bmatrix} \in \mathbb{R}^{N\times 2}, \quad u_t = \begin{bmatrix} x \\ y \end{bmatrix} \in \mathbb{R}^{2\times 1},$$

$$\hat{b}_t = \frac{1}{2}\begin{bmatrix} (x_1^2 + y_1^2) - (x_r^2 + y_r^2) + \hat{d}_r^2 - \hat{d}_1^2 \\ (x_2^2 + y_2^2) - (x_r^2 + y_r^2) + \hat{d}_r^2 - \hat{d}_2^2 \\ \vdots \\ (x_N^2 + y_N^2) - (x_r^2 + y_r^2) + \hat{d}_r^2 - \hat{d}_N^2 \end{bmatrix} \in \mathbb{R}^{N\times 1}.$$

The Moore-Penrose pseudo inverse is taken on both sides to obtain the location estimates:

$$\hat{u}_t = 0.5(\mathrm{A}_t^T \mathrm{A}_t)^{-1} \mathrm{A}_t^T \hat{b}_t \tag{5}$$

2.2 AOA METHOD

If only angle estimates are available, only two or three stations are required for 2D or 3D localization, respectively. Each station forms a line on a 2D plane, on which the station and target are situated; hence, we obtain a certain number of lines depending upon the number of stations. The point of intersection of these lines is the target's estimated position. The AOA system generally exhibits effective results; however, the estimation error increases significantly as the target moves away from the stations [3][5].

Using the same notation as for TOA, we have

$$\hat{\theta}_i \approx \arctan\left[\frac{(y - y_i)}{(x - x_i)}\right] + m_i, \tag{6}$$

where m_i represents the zero mean Gaussian noise in the estimate of the ith angle; that is, $m_i \sim \left(\mathcal{N}\left(0, \sigma_{m_i}^2\right)\right)$.

Equation (6) can be written in matrix form as

$$\mathrm{A}_a u_a = \hat{b}_a, \tag{7}$$

$$\text{where } \mathrm{A}_a = \begin{bmatrix} \tan\hat{\theta}_1 & -1 \\ \tan\hat{\theta}_2 & -1 \\ \vdots & \vdots \\ \tan\hat{\theta}_N & -1 \end{bmatrix} \in \mathbb{R}^{N\times 2}, \quad u_a = \begin{bmatrix} x \\ y \end{bmatrix} \in \mathbb{R}^{2\times 1},$$

$$\hat{b}_a = \begin{bmatrix} x_1 \tan\hat{\theta}_1 - y_1 \\ x_2 \tan\hat{\theta}_2 - y_2 \\ \vdots \\ x_N \tan\hat{\theta}_N - y_N \end{bmatrix} \in \mathbb{R}^{N\times 1}.$$

For the standard linear least-squares estimator, the solution is given by

$$\hat{u}_a = (\mathrm{A}_a^T \mathrm{A}_a)^{-1} \mathrm{A}_a^T \hat{b}_a \tag{8}$$

3. PROPOSED LOCALIZATION

3.1 Combined TOA and AOA

In this study, the TOA spheres of two station and AOA line are used for estimating a 3D point. Figure 1 shows that TOA spheres of two stations are intersected and it generate the intersected plane and circle.

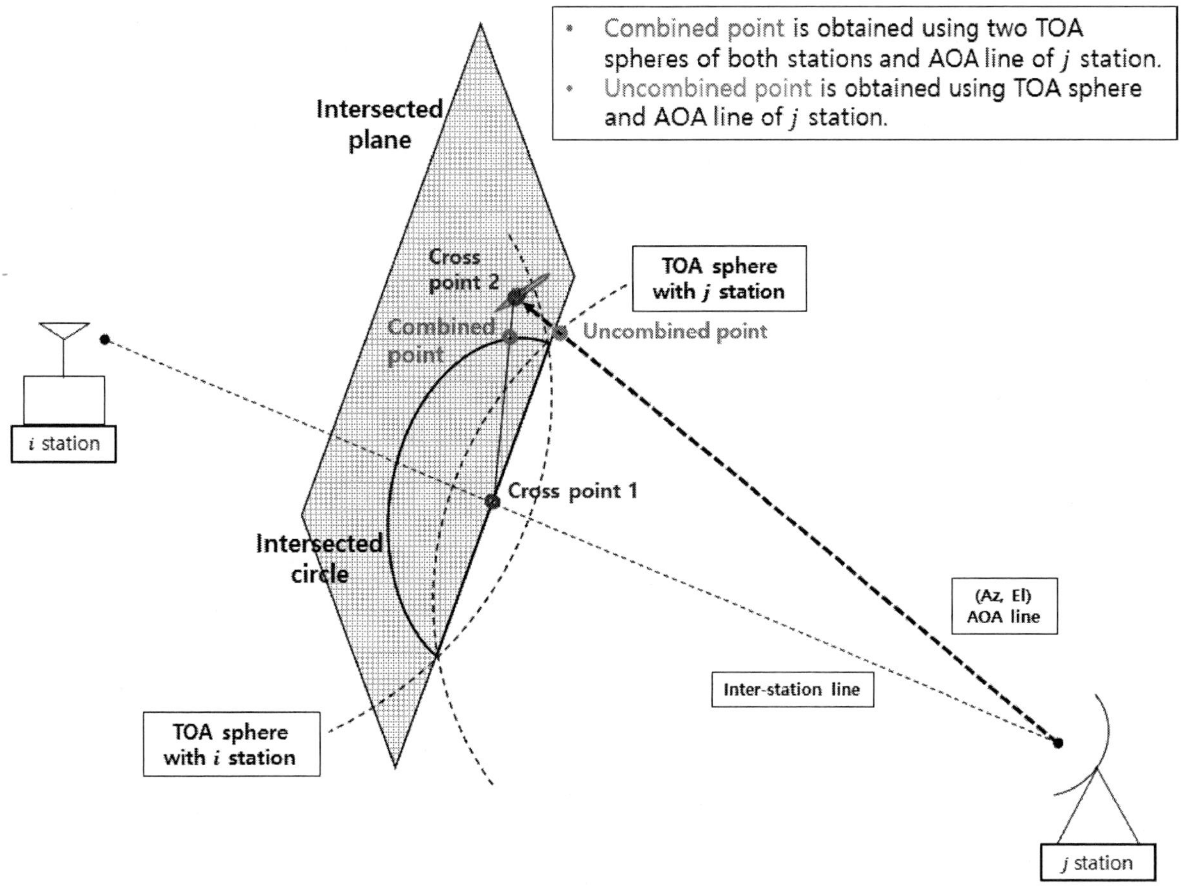

Figure 1. Combined TOA and AOA localization

In order to combine the TOA and AOA methods, we create sphere equations (9) and (10) with the estimated TOA range as radius, a connection line (11) between *i* and *j* station, and AOA line (12) at a reference station which is *j* station.

TOA sphere from *i* station:

$$r_i = (Tx - Rx_i) \times c$$
$$(x - x_i)^2 + (y - y_i)^2 + (z - z_i)^2 = r_i^2, \quad (9)$$

TOA sphere from *j* station:

$$r_j = (Tx - Rx_j) \times c$$
$$(x - x_j)^2 + (y - y_j)^2 + (z - z_j)^2 = r_j^2, \quad (10)$$

Inter-station line between i and j station:

$$\frac{x-x_j}{s_{xj}}=\frac{y-y_j}{s_{yj}}=\frac{z-z_j}{s_{zj}}=k$$

$$x=ks_{xj}+x_j,\ y=ks_{yj}+y_j,\ z=ks_{zj}+z_j, \tag{11}$$

AOA line of j station:

$$\frac{x-x_j}{a_{xj}}=\frac{y-y_j}{a_{yj}}=\frac{z-z_j}{a_{zj}}=t$$

$$x=ta_{xj}+x_j,\ y=ta_{yj}+y_j,\ z=ta_{zj}+z_j, \tag{12}$$

where i,j: station index,

$P_i(x_i,y_i,z_i)$, $P_j(x_j,y_j,z_j)$: point of each station,

r_i,r_j: TOA range,

c: speed of light; 299792458 (m/s),

$\vec{a_j}=(a_{xj},a_{yj},a_{zj})$: AOA unit vector, it is coordinate converted of $\overrightarrow{AOA_j}\ (range_j=1, El_j, Az_j)$, and

$$\vec{s_j}=\frac{P_i(x_i,y_i,z_i)-P_j(x_j,y_j,z_j)}{\left\|P_i(x_i,y_i,z_i)-P_j(x_j,y_j,z_j)\right\|_2}.$$

Intersection plane of the spheres of i and j station is obtained in (13) using equations (9) and (10).

$$x(x_i-x_j)+y(y_i-y_j)+z(z_i-z_j)=\frac{r_i^2-r_j^2-x_i^2+x_j^2-y_i^2+y_j^2-z_i^2+z_j^2}{-2} \tag{13}$$

Cross point 1 is obtained in (14); cross point between the intersection plane (13) and inter-station line (11).

$$XP_1(x_1,y_1,z_1)=P_j(x_j,y_j,z_j)+k\cdot\vec{s_j} \tag{14}$$

(15) is radius of the circle on the intersection plane of i and j station

$$r=\sqrt{r_j^2-k^2} \tag{15}$$

Cross point 2 is obtained in (16); cross point between the intersection plane (13) and AOA line (12).

$$XP_2(x_2,y_2,z_2)=P_j(x_j,y_j,z_j)+t\cdot\vec{a_j} \tag{16}$$

Combined point can be obtained in (17) using two TOA sphere and a AOA line equations.

$$P(x,y,z)=XP_1(x_1,y_1,z_1)+r\cdot\overrightarrow{xp}, \tag{17}$$

where $\overrightarrow{xp}$ is unit vector between cross point 1 and 2.

Finally, combined range of the LV is obtained in (18)

$$r_{combined} = \overrightarrow{xp}_j \cdot \vec{a}_j, \tag{18}$$

where $\overrightarrow{xp}_j$ is line between j station and Combined point $P(x, y, z)$.

In this paper, we only estimate range information. There is no adjustment on the AOA information of j station. The combined range (18) is inner product between vector $\overrightarrow{xp}_j$ and AOA unit vector $\vec{a}_j$ of j station. This is estimated range information using two TOA sphere and an AOA line.

3.2 Test Configuration

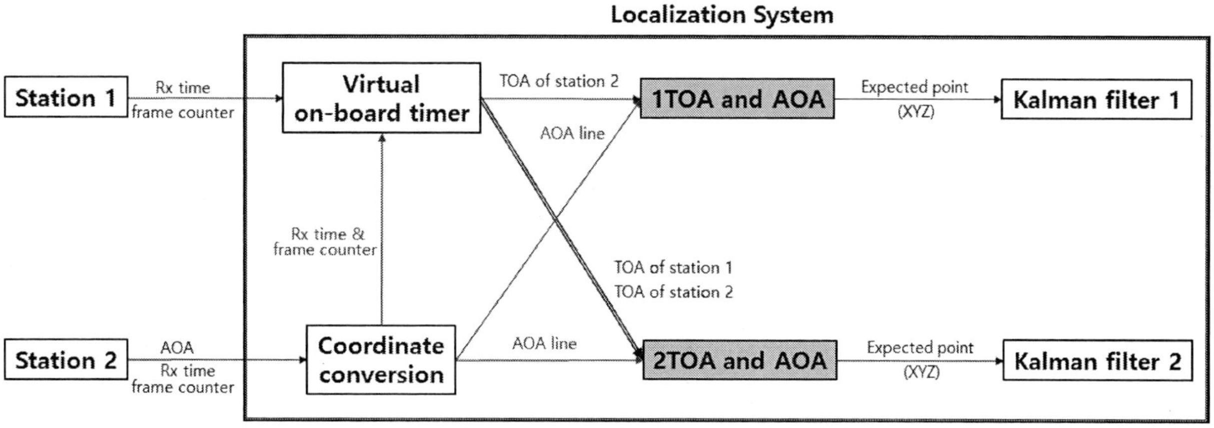

Figure 2. Proposed localization system.

In this study, we configure two ground telemetry stations. Station 1 is at the space center where the launch pad is located. It has no tracking function, but contains a wide beam to receive telemetry signals. Station 2 is located approximately 150 km from the space center. It contains a monopulse tracking beam of $\pm 0.4^{\circ}$ and exhibits a tracking accuracy of 0.05°. At Station 1, there is no AOA information, but a virtual on-board timer can be created. At Station 2, both TOA and AOA are available.

Figure 2 displays the proposed localization system. For the virtual on-board timer, the input parameters are received frame counters and Rx time stamps of each station. The TOA sphere of each station can be obtained using this on-board timer. In this study, we selected Station 2 as the reference station for the two localization methods. For the 1TOA and AOA method, the input parameters are the virtual TOA sphere and AOA line for Station 2. For the 2TOA and AOA method, the input parameters are the virtual TOA spheres for both stations and AOA line for Station 2.

4. PERFORMANCE RESULTS

Figure 3 illustrate the 3D GPS trajectory and Combined TOA and AOA results for the LV. At Station 1, TOA was available during CT 0 ~ 300s. At Station 2, TOA and AOA were available during CT 10 ~ 521s. Therefore, Combined TOA and AOA method use only one TOA of Station 2 when TOA of Station 1 is not available.

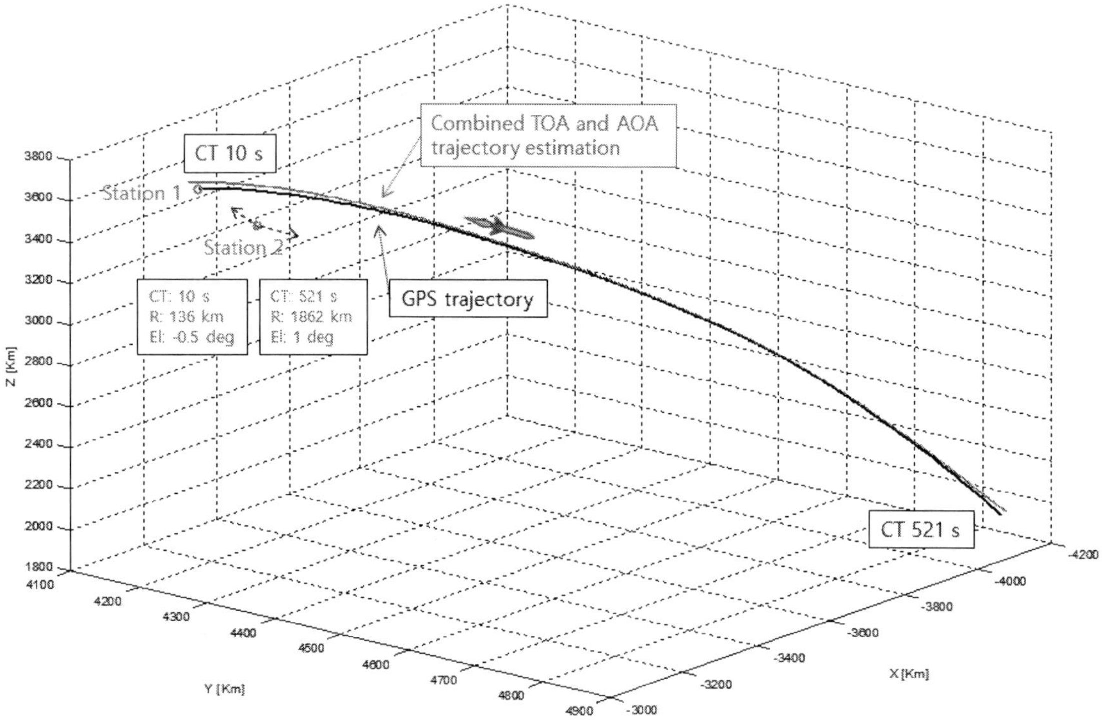

Figure 3. 3D Trajectory comparison.

Figures 4 show the slant range comparison with GPS for the each localization methods; 1TOA and AOA, and 2TOA and AOA methods. As mentioned above, 2TOA and AOA method uses only TOA of Station 2 around at CT 100 s when TOA of Station 1 is not available. It shows that 2TOA and AOA method is more closed to GPS than 1TOA and AOA; refer to 1TOA_AOA_R and 2TOA_AOA_R. Figure 5 shows the slant range errors of each localization method and the KALMAN filtering results compared to GPS; refer to 1TOA_AOA_R_error, 1TOA_AOA_Kalman_R_error , 2TOA_AOA_R_error, and 2TOA_AOA_ Kalman_R_error. It is clear that there is a smoothing effect with the KALMAN filters for each method.

The 1TOA and AOA method shows range error of +26 km at CT 50 s and +7 km at CT 300 s. On the other hand, 2TOA and AOA method shows range error of +24 km and +5 km at the time. We can reduce range errors of 2 km by combining two TOAs information. As described section in 3.2, TOA information is estimated using the virtual on-board timer. Therefore, there is estimation errors; which is corresponding range errors in Figure 5.

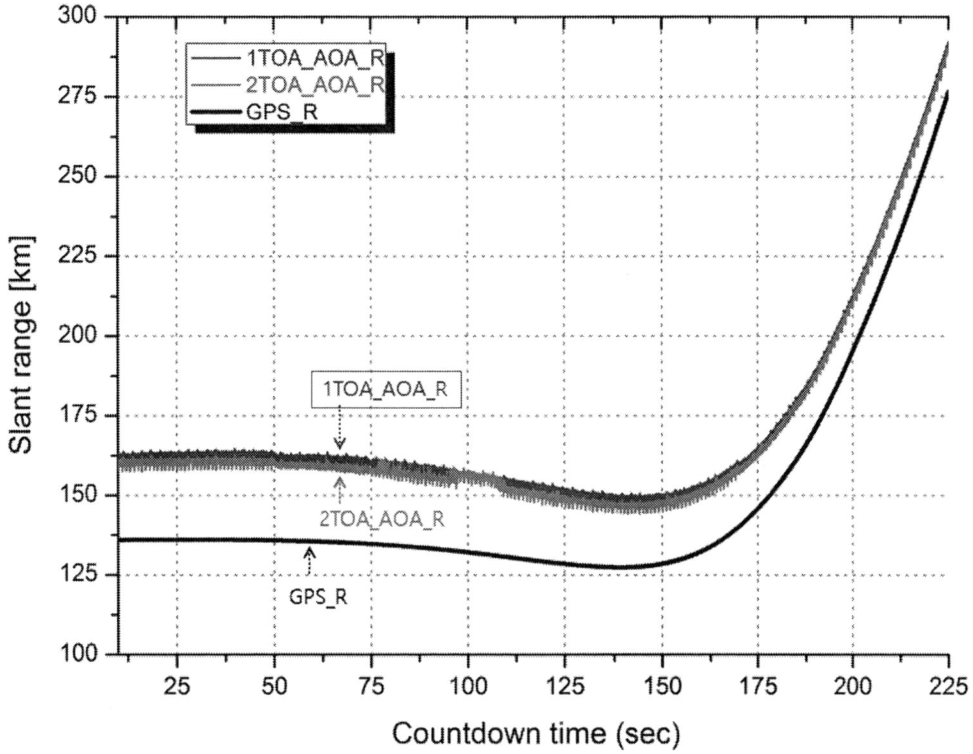

Figure 4. Slant range comparison for each method.

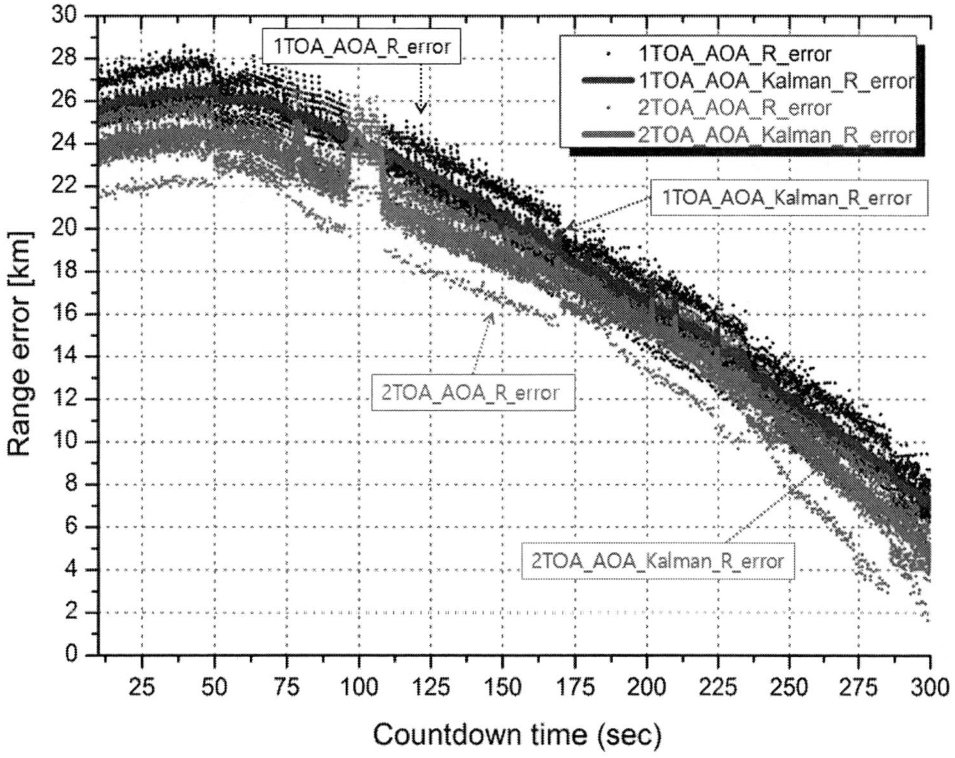

Figure 5. Slant range error with KALMAN filtering.

5. CONCLUSIONS

In this paper, we have proposed a 3D localization system for an LV. For the proposed system, Combined TOA and AOA method were applied using two ground telemetry stations; no-tracking station at the space center was used for the virtual on-board timer generation, while a monopulse tracking station was used for the AOA information. The performance of proposed system was compared to that of an on-board GPS, which was obtained during the LV flight test. The analysis results indicate that if two TOAs are used, the range errors are reduced about 2 km. And the results demonstrate a similar performance to GPS and illustrate that the obtained 3D position can be applied to other systems, such as radar.

REFERENCES

[1] Yiu-Tong Chan, H. Yau Chin Hang, Pak-chung Ching, “Exact and approximate maximum likelihood localization algorithms,” IEEE Transactions on Vehicular Technology, vol. 55, no. 1, pp.10–16, 2006.

[2] Mohamed Zhaounia, Mohamed Adnan Landolsi, Ridha Bouallegue, “Hybrid TOA/AOA approximate maximum likelihood mobile localization,” Journal of Electrical and Computer Engineering, vol. 2010, 2010.

[3] M. W. Khan, N. Salman, A. H. Kemp, “Enhanced hybrid positioning in wireless networks I: AoA-ToA,” 2014 International Conference on Telecommunications and Multimedia (TEMU), IEEE, pp. 86-91, 2014.

[4] Ismail Guvenc, Sinan Gezici, Fujio Watanabe, Hiroshi Inamura, “Enhancements to Linear Least Squares Localization Through Reference Selection and ML Estimation,” in Proc. IEEE Wireless Commun. Networking Conf. (WCNC), pp. 284-289, 2008.

[5] Rong Peng, Mihail L. Sichitiu, “Angle of Arrival Localization for Wireless Sensor Networks”, Sensor and Ad Hoc Communications and Networks 2006, IEEE, Sept 2006.

[6] Stephen Boyd, Lieven Vandenberghe, *Convex Optimization*, Cambridge University Press, Cambridge, 2009.

AUTOMATING VERIFICATION FOR LEGACY SYSTEMS: A CASE STUDY OF TECHNOLOGY SUSTAINMENT WITHIN THE NASA SPACE NETWORK

Dana Irvin, John Otranto, Kirill Lokshin, Amit Puri
Ingenicomm, Inc.
14120 Parke Long Court #210, Chantilly, VA 20151
dirvin@ingenicomm.com

ABSTRACT

The NASA Space Network (SN), which consists of the geosynchronous Tracking and Data Relay Satellite (TDRS) constellation and its associated ground elements, is a critical national space asset that provides near-continuous, high-bandwidth telemetry, command, and communications services for numerous spacecraft and launch vehicles.

The Space Network includes several key ground system elements, one of which is the White Sands Complex Data Interface Service Capability (WDISC). The WDISC has undergone multiple cycles of modification and technology refresh over its lifetime, making test automation an attractive option for reducing system verification and validation cost.

This paper considers the implementation of automated testing for the WDISC as a case study in technology sustainment, discusses the principal benefits and challenges of implementing test automation for a legacy system, and presents findings that demonstrate the effectiveness of such automation models.

KEY WORDS

Test automation; TDRSS; WDISC; legacy system sustainment.

INTRODUCTION

The NASA Space Network (SN) was established in the early 1980s to replace NASA's worldwide network of ground tracking stations, with increased footprint and availability for space missions. It consists primarily of the geosynchronous Tracking and Data Relay Satellite (TDRS) constellation and its associated ground elements, and provides near-continuous, high-bandwidth telemetry, command, and communications services for numerous missions, spacecraft and launch vehicles. It operates as a bent-pipe relay system between various customer platforms and their ground facilities.

One of the key elements of the TDRS ground system is the White Sands Complex Data Interface Service Capability (WDISC). WDISC implements a service interface which provides uplink and downlink connectivity to the various Mission Operations Centers (MOCs) which utilize TDRS services. Originally fielded in early 2003, the WDISC consists of several commercial off-the-shelf (COTS) Programmable Telemetry Processor (PTP) systems. These systems are used to support multiple missions, including Landsat-7, Solar Dynamics Observatory (SDO), GLAST/Fermi, Swift, THEMIS, WISE, and others. [1]

In order to ensure the continued operation of the WDISC, NASA commissioned the WDISC Sustainment (WDISC-S) project, which encompassed upgrades and technology refreshes of the WDISC PTP systems, including a refresh of the COTS PTP software, which had undergone multiple generational upgrades since the initial deployment of the WDISC units, as well as improvement and modernization of WDISC hardware components.

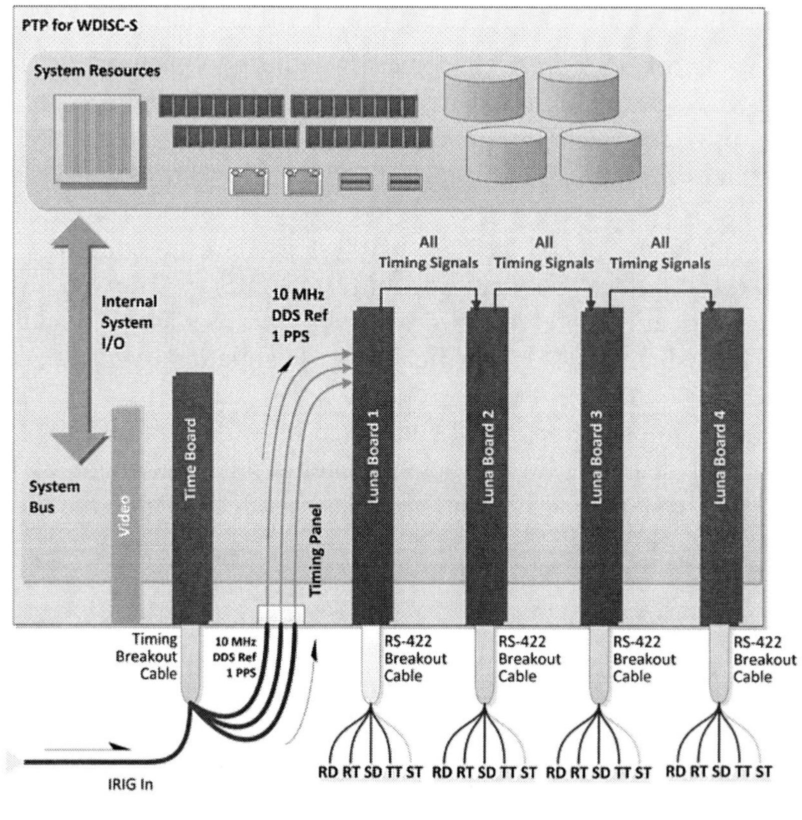

Figure 1: WDISC-S PTP Conceptual Design

A depiction of the upgraded WDISC-S system architecture is provided in
Figure *1*. Each WDISC-S system provides four processing software instances and four dual-channel serial interfaces, allowing for a matrix of simultaneous forward and return operations executions.

To ensure that the upgraded systems were capable of functioning as drop-in replacements for the existing WDISC PTP units, it was necessary to develop a comprehensive regression test program, which would be capable of testing the various interfaces currently supported by WDISC, as well as providing non-intrusive automated testing functionality to automate regression testing for future software upgrades and expanded mission capabilities without placing an excessively onerous burden on test engineers in the White Sands Complex (WSC).

INTERACTIVE AND AUTOMATED TESTING

The COTS PTP digital data processing and simulation software package, which provides the key functions of the upgraded WDISC-S system, includes capabilities for interactive test configuration and execution. Each data interface supported by WDISC may be independently verified by a test engineer operating the local system or accessing it via a remote console. However, performing all the regression tests for a single system refresh could take hours or days, depending on the number of tests, the length of each test execution, and the desired number of test iterations. For example, if 400 test scenarios were to be interactively executed by a test engineer, with each test execution requiring 15 minutes, it would take one test engineer 100 hours to execute those tests – once – on a single WDISC unit.

WDISC-S additionally implements test automation capability through the icAutomata software, which was previously utilized by NASA to support automated testing for the Space Network Ground Segment Sustainment (SGSS) program. The icAutomata software uses the Remote Interface Library (RIL) API provided with the PTP data processing and simulation software to address each PTP server instance, configure individual software parameters, start and stop data flows, and monitor and evaluate the status of the software and its component elements.

The icAutomata software is responsible for translating the parameters and steps defined in each regression test scenario into configuration values and operations. Each test scenario defines a set of verification points, which are collected during the execution of a scenario. The verification points may be simple numerical statistics, such as the number of bits transmitted or the number of pattern errors detected, or more complex Boolean criteria, such as the results of a bit-level comparison between input and output files. Once a test scenario has been configured and data flow begins, icAutomata monitors the various status indicators and statistics generated by the PTP data processing software, and collects the verification points as defined in the test scenario.

The verification points collected by icAutomata are used to generate reports for the test program, shown in Figure 2, which not only inform the test engineer whether a test passed or failed, but can also provide a detailed test record, including the verification statistics for each executed scenario. The icAutomata software also supports requirements tagging, where certain steps having success or failure criteria can be associated with specific requirement identifiers. When such tagging is employed, the detailed test report requirement verification summary that can be directly cross-referenced to the program requirements traceability and verification matrix, further enhancing the test granularity and directly tracing test steps to requirements.

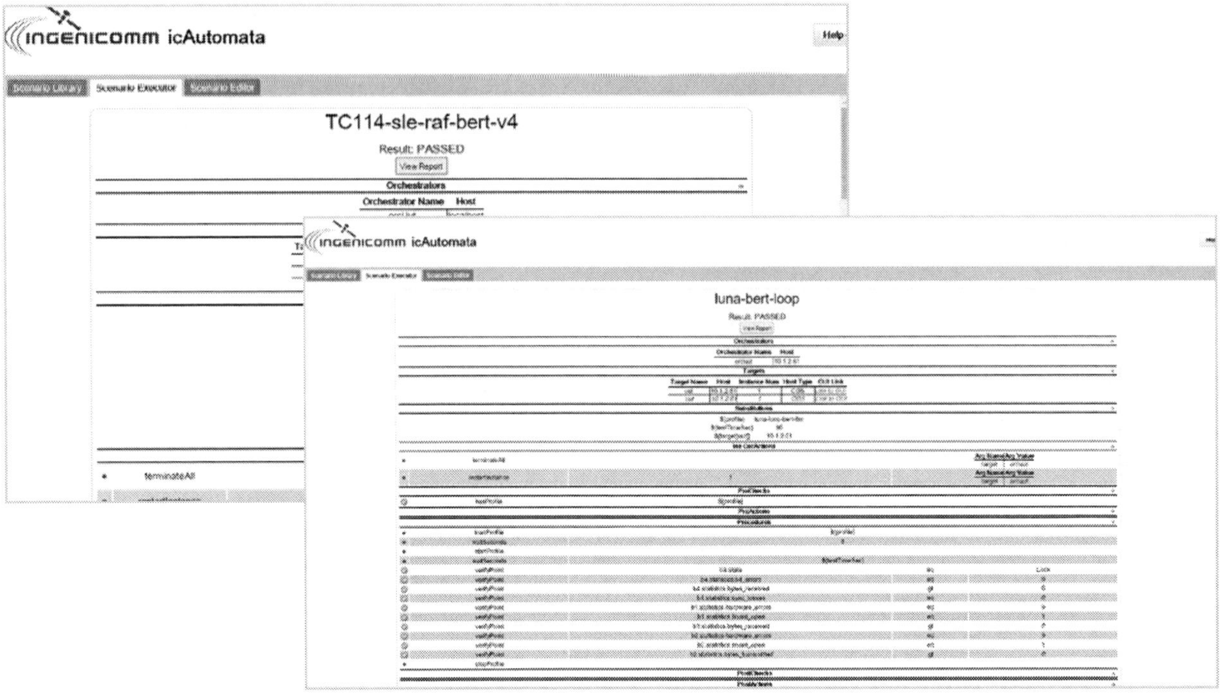

Figure 2: Example icAutomata Test Execution Reports

WDISC-S TEST PREPARATION

As the WDISC-S program encompassed a replacement of the existing WDISC systems, the test program focused on verifying that the refreshed systems function identically to the legacy systems. Further, the test program was intended to demonstrate that the legacy system configuration files could be reused without alteration on the replacement systems, even though the PTP software that utilizes those configurations had changed significantly since the original delivery of the WDISC systems in 2003.

The operational WDISC systems utilize more than 400 discrete mission configuration files, called "desktops", which define the specific PTP software modules used in each operational configuration, the parameters and properties set within each modules, and the routing of data between modules. In order to utilize these configuration files with the icAutomata software, each module's status and configuration properties and parameters were added to the icAutomata test tool as a target manifest, enabling the test scripts to add or remove the modules from a desktop configuration, set configuration values in each module, and monitor and compare the status values generated by each module.

In order to execute the WDISC-S test program, it was additionally necessary to developing a WDISC testbed, depicted in Figure 3), including specific system configurations, network resources, and simulators to emulate the TDRSS environment at NASA WSC.

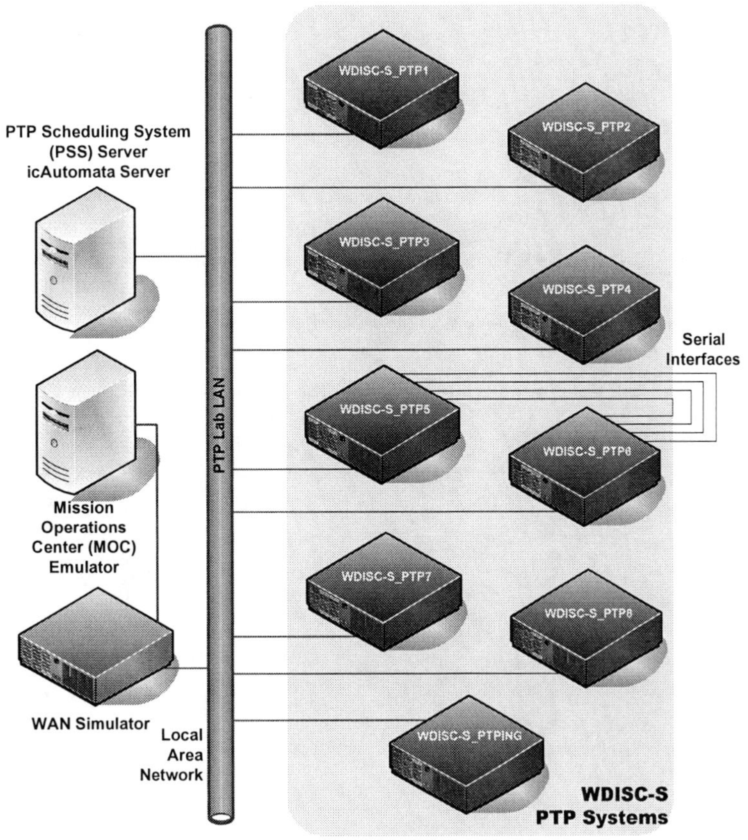

Figure 3: WDISC-S Testbed

The WDISC-S testbed included, in addition to eight production PTP units and one test PTP unit, a server that hosted both the NASA's PTP Schedule System (PSS) and the icAutomata software, an uplink/downlink emulator for serial communications I/O, a WAN simulator for long-distance network simulation, and a MOC emulator for transmitting/receiving data to and from the PTP via TCP/IP socket connections. The MOC emulator is connected through the WAN emulator to simulate remote command and telemetry data transfers to the WDISC-S systems.

WDISC-S TEST METHODOLOGY

As part the WDISC-S test program, each of the more than 400 WDISC desktop configuration files had to be demonstrated to operate on each of the eight production WDISC-S units. Because many of the desktops had similar configurations, it was possible to use a limited set of standard testbed configurations to demonstrate all desktop's operation. The testbed configuration for most automated tests, shown in Figure 4, utilized a serial I/O system as an uplink/downlink emulator, a single WDISC-S production system, and the MOC emulator to send and receive IP-based data to and from the WDISC-S system.

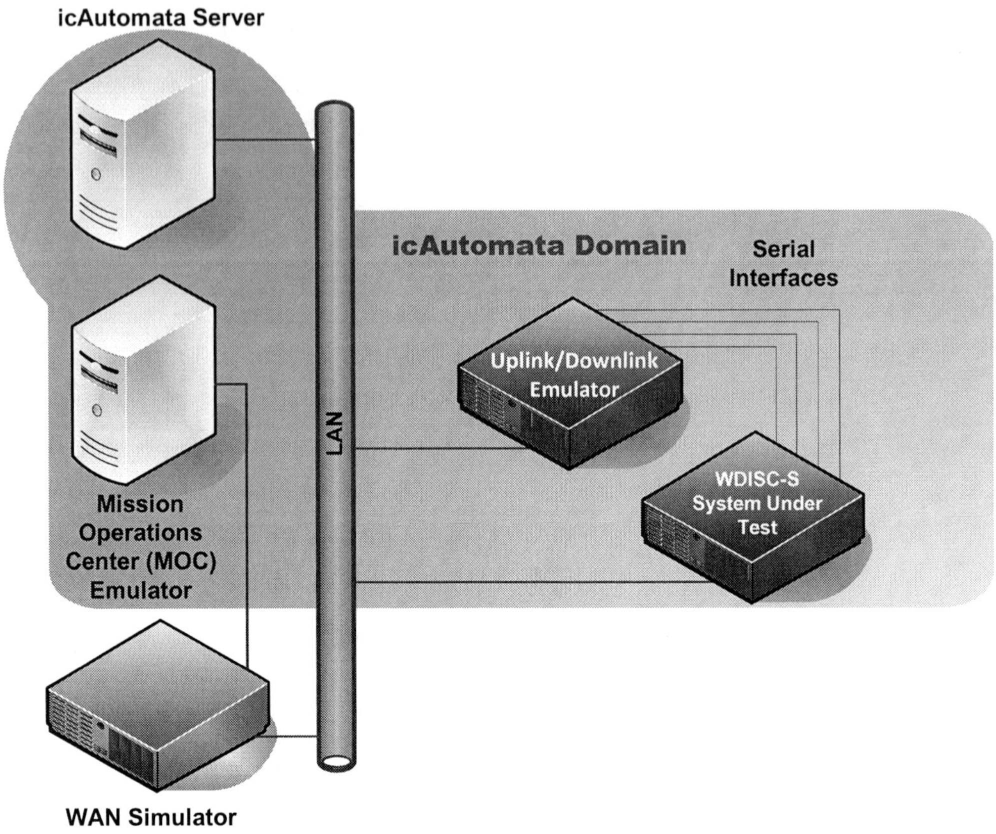

Figure 4: icAutomata Controls Scenario Execution for Multiple Testbed Systems

In order to utilize icAutomata, automation scenarios had to be created for each of the WDISC-S desktop configurations. This was, again, aided by the similarity of many desktops; once a single automation scenario was configured, it was possible to copy the scenario, make a few key changes, and save it as a new scenario. Because the automation scenarios are defined in XML, a text editor could be used to modify scenarios when needed. Further, it was possible to generate scripts that would copy, modify, and save scenarios, which was more accurate than manual updates and eliminated the possibility of errors due to operator error or fatigue associated with making repetitive changes.

The icAutomata tool supports the definition of hierarchical scenarios that allow for an automation scenario to execute another automation scenario, or multiple automation scenarios in sequence. This feature was utilized to run the final set of WDISC-S tests, where all 400+ scenarios are exercised. The master scenario would execute a separate scenario for each WDISC desktop configuration. The desktop-specific automation scenario would load the WDISC desktop configuration, enable the uplink/downlink emulator, enable the MOC emulator, and finally enable the system under test. The scenario would run for a specific set period of time, and then disable all elements in the test. Once completed, control would return to the master automation scenario, which would proceed to the next instance to test the next desktop configuration.

The master scenario allows test execution to continue through the full set of tests, or to halt when an error condition is encountered. Using this approach, it was possible to kick off the testing of a WDISC-S system and allow the testing to run automatically through to completion. A detailed test report was automatically generated, and could be examined to determine the results of the testing.

CONCLUSIONS

The introduction of automation into system testing has proven beneficial from both cost-saving and accuracy improvement perspectives. Testing is performed more readily, executing tests is easier, tests are more repeatable, and reporting is more complete and more uniform. It is possible to expand the amount of testing conducted, so as to better ensure the system is fully functional both during initial deployment, and during subsequent modifications and upgrades. However, the use of automated tools requires some dedication at the outset, and the tools need to serve the needs of the customer with respect to clarity of purpose, utility of implementation, and production of artifacts that assure documentation, configuration management, and repeatability.

In the case of WDISC-S, NASA deployed an automation testbed similar to the one described here at WSC to support automated testing following the completion of the upgrade. Currently, NASA engineers are developing site-specific automated test scenarios and automating multi-level pre-deployment regression test programs. While the initial effort to implement automation in a legacy environment may have been greater than merely conducting a test by hand, the automation can now significantly reduce the effort required to re-test the system whenever systems are updated for security patching, new software is deployed to add features or address bugs, or new mission configurations are added. The WDISC-S systems can be retested and recertified in a fraction of the time previously required, with substantially reduced effort, and with reduced potential for errors in testing.

REFERENCES

[1] National Aeronautics Space Administration; Goddard Space Flight Center, *Space Network Users' Guide (SNUG)*, Revision 10, NASA Code 452, Space Network Project, document number 450-SNUG, pp 3-6-20, Aug 2012.

IMPLEMENTATION OF DIGITAL SIGNAL PROCESSING WITH HIGH DATA RATE SENSORS FOR DATA COMPRESSION

Seongmin Noh
Korea Aerospace Research Institute, Daejeon, South Korea
smnoh@kari.re.kr,

Hoyoung Yoo
Chungnam National University, Daejeon, South Korea
hyyoo@cnu.ac.kr

ABSTRACT

The onboard telemetry system of Korea Space Launch Vehicle-II (KSLV-II) acquires acoustic, vibration, and piezoelectric pressure sensors that require high data rate over several kilo samples per second, so the compression method is needed to expand link margin of telemetry system. This paper implements third-octave and FFT signal processing algorithms to reduce sensor data with high compression ratio depends on data acquisition requirements. The developed signal processing hardware module is composed of analog signal conditioning block and digital signal processing block on FPGA, and the digital block is fully implemented with dedicated hardware using HDL. For digital hardware implementation, multistage structure with ANSI standard octave filter bank is used for third-octave processing, and pipelined architecture is used for FFT. The performance of data acquisition and signal processing is evaluated and compared to the commercial data acquisition equipment.

INTRODUCTION

The Korean launch vehicle telemetry system (hereinafter referred to as KSLV TLM) is mounted on each stage of three stages, and transmits sensor data and status data mounted on the launch vehicle to the ground in real time. Some of the sensor data, such as sound, vibration, shock, acceleration, and pressure sensors, require a high sample rate of several kilo samples per second or more, and sensor data rates need to be minimized to ensure limited RF link margins for the launch vehicle. The KSLV TLM performs compression on TLM data by performing signal processing at onboard equipment for channels requiring a high sampling rate. Such signal processing methods include third-octave signal processing for acoustic sensors load measuring acoustic load, and FFT (Fast Fourier Transform) signal processing for acceleration and vibration sensors.

An example of the application of signal processing technology to a telemetry on launch vehicle is an Indian launch vehicle [1]. By using digital signal processor (DSP), FFT processing of vibration data and third-octave data processing of acoustic signal have reduced data amount by

1/50 ~ 1/60. In addition, several hardware and software using DSP are being developed in the commercial TLM data acquisition unit manufacturer [2]. DSP has a merit that various algorithms can be flexibly developed. However, in the field of launch vehicles, since the applied signal processing algorithm is not various and generally measurement requirements are fixed, FPGA based signal processing can be used for high-speed and high-performance processing of multi-channel acquisition. In this paper, we present hardware implementation and test results of TLM signal processing device called signal processing unit (SPU) using FPGA based digital signal processing technology.

SYSTEM DESCRIPTION

The proposed signal processing unit consists of a power supply board, an encoder board, and a data acquisition board. The power supply board supplies power to the other boards in the unit. The encoder board receives the measured data from the data acquisition board, generates PCM frames, and transmits them to the ground via the RF interface. The data acquisition board measures current excitation type acoustic, vibration and pressure sensor signals and performs the signal processing and finally transmits the results to the encoder board. This unit is not designed for signal processing only, but it is designed considering versatility and extensibility by sharing the same platform as a general acquisition device that performs real-time data measurement as well as signal processing.

The data acquisition board, which plays the most important role in the unit, has 12 analog channels in total, and each channel has a circuit to supply the excitation current for the sensor, an amplifier, and an analog filter circuit. The digital signal processing is designed based on the FPGA using HDL, and it can be divided into the part performing the third-octave signal processing function and the part performing the FFT signal processing functionally.

DIGITAL SIGNAL PROCESSING

First, third-octave signal processing is mainly used as a signal processing algorithm for acoustic sensor data, and conforms to ANSI S1.11-2004 standard [3]. For SPU, the multi-stage structure using FIR filter is applied to the detailed implementation method of the algorithm, and the signal processing structure is shown in the following figure. Such a multi-stage structure not only reduces the calculation complexity but also has the advantage of reducing the instability of the digital filter.

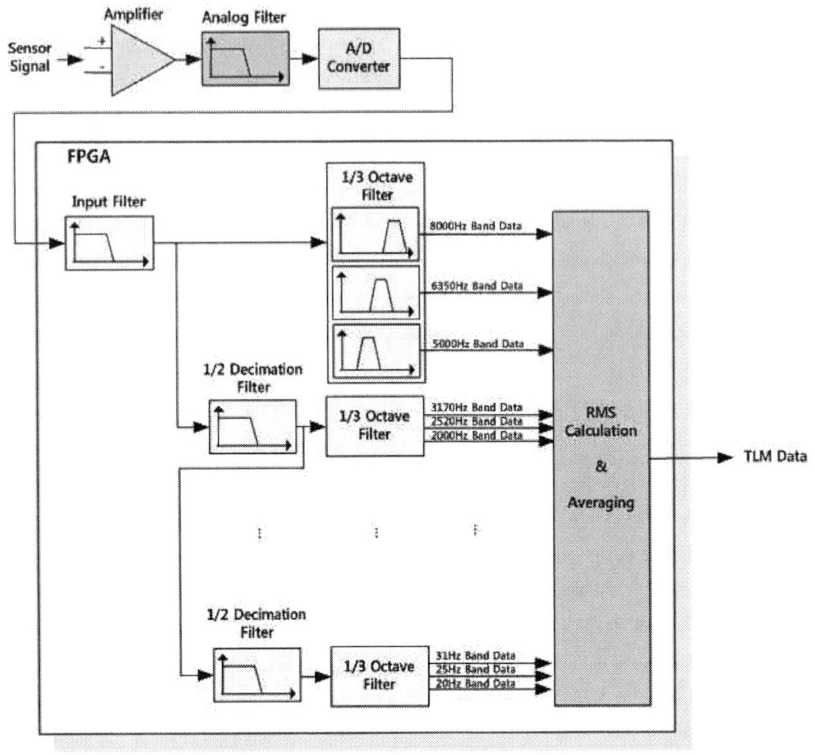

Figure 1 Structure of Third-Octave Signal Processing

The analog sensor signal is oversampled at a high sample rate over than 100k samples per second and decimated with input digital filter to minimize aliasing due to the digital conversion of the analog signal. This digital filter is a low-pass filter with decimation and attenuates aliasing within the 8kHz band by more than 80dB.

The third-octave signal processing is performed by calculating a linear average of RMS (Root Mean Square) values during 0.25 second for each of the 17 frequency bands (ANSI No. 13 to 39) having a center frequency of 20 Hz to 8 kHz. 1/2 decimation filter and 1/3 octave filter is implemented by 64th order FIR filter can be operated by sharing multiple channels. Finally, the data compression rate through processing is at least 1/150 times compare to real time acquisition with 16K samples per second.

The minimum specification of the octave filter proposed by the ANSI standard is the 3rd order Butterworth IIR filter. In this design, the 64th order FIR filter with higher attenuation rate for the cutoff band is applied. The magnitude response of the filter is shown in the following figure. Designed with 64 filter taps, the octave filter satisfies the standard class-2 specification [4], and can be implemented in a more rigorous class-0 or class-1 with increase in hardware. Also, the SPU's third-octave signal processing is designed to flexibly change channel extensions, frequency bands, octave filter specifications, and averaging methods.

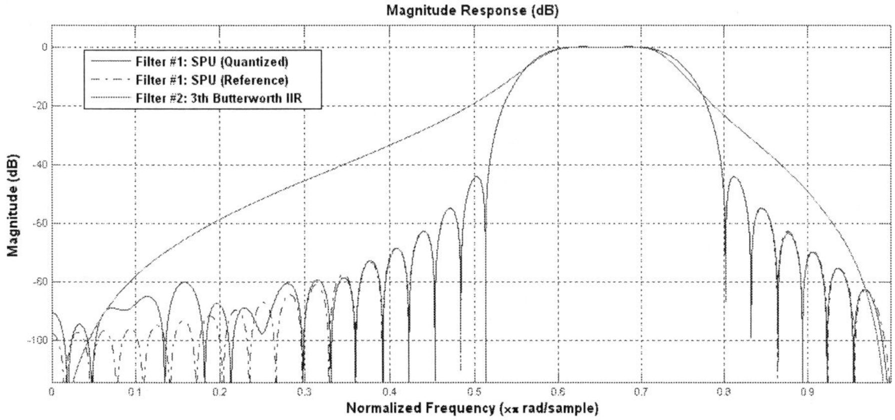

Figure 2 Magnitude Response of Third-Octave Filter

The second signal processing algorithm, FFT, is used as an algorithm for analyzing the frequency response of sensor data. The FFT signal processing structure of SPU is shown in the following figure.

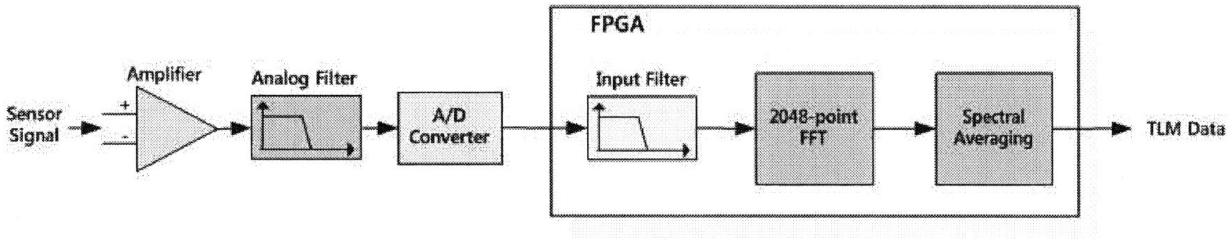

Figure 3 Structure of FFT Signal Processing

Same with case of the third-octave processing, measurement of the sensor signal is performed by oversampling at a high sampling rate and filtered using input digital filter to minimize aliasing. The FFT signal processing is performed in a manner that computes a linear average of the frequency response during one second and results are magnitude data with 5 Hz intervals for the frequency band of 0 Hz to 2 kHz. The FFT IP (Intellectual Property) of the Intel FPGA (previously Altera FPGA) is used as the FFT core, and this IP is operated in a single delay feedback structure of a radix-2^2, which is one of the well-known pipeline structures for FFT calculation. The number of points of FFT is 2048 and the one FFT core is shared by multiple sensor channels.
The data compression rate through this signal processing is 1/10 times compare to real time acquisition with 4K samples per second. FFT signal processing is designed to flexibly change the frequency band, frequency interval, and averaging method.

TEST RESULTS

The functional test of the signal processing unit was performed by inputting the same arbitrary signal to the signal processing unit and the commercial signal acquisition equipment. For accurate reference equipment, National Instruments' NI-6351 model with 16-bits ADC is selected. The first test was performed by inputting noise, and the two devices have similar results, and detailed results are shown in the following figure. The power reduction in the last 8 KHz band is due to the cutoff frequency of the input digital filter being 8 KHz. The second test uses sinusoids with three frequencies 100Hz, 300Hz, 1KHz as the input signal. The signal processing results for two devices show an accuracy within 1 dB error, which is expected to be improved in the subsequent development model design. In the case of the attenuation for the band outside the signal band, the signal processing unit has higher attenuation than the reference NI equipment, which is due to the high attenuation rate of the octave band filter applied to the SPU.

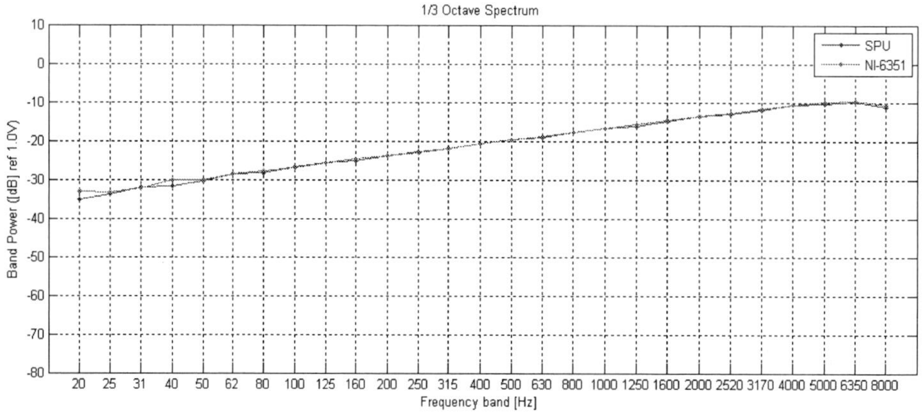

Figure 4 Test Result for Third-Octave Processing with White Noise

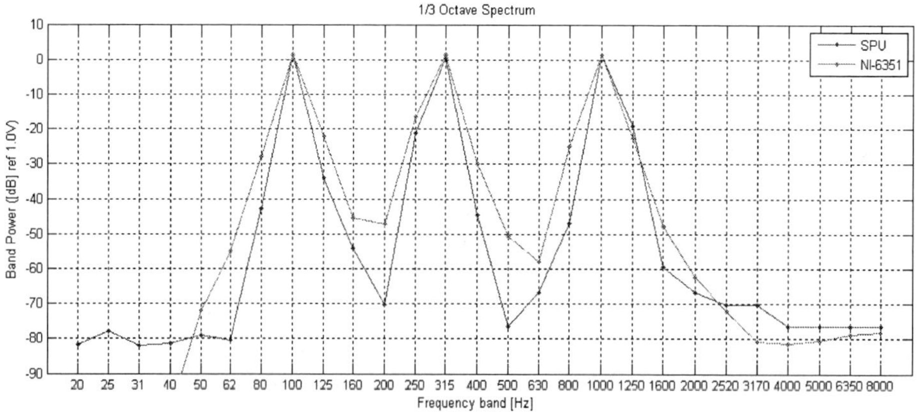

Figure 5 Test Result for Third-Octave Processing with Sinusoidal Wave

The results of the FFT function test are shown in the following figure by inputting the sinusoidal wave signal (100Hz, 300Hz, 1kHz) to the SPU and reference equipment. In the result, it is confirmed that the FFT result corresponding to the sinusoidal frequency has an accuracy within 1 dB error. However, some noise or harmonic components are appeared at SPU and not detected in the reference equipment, it is necessary to improve the acquisition performance in analog signal conditioning circuit of SPU in the future.

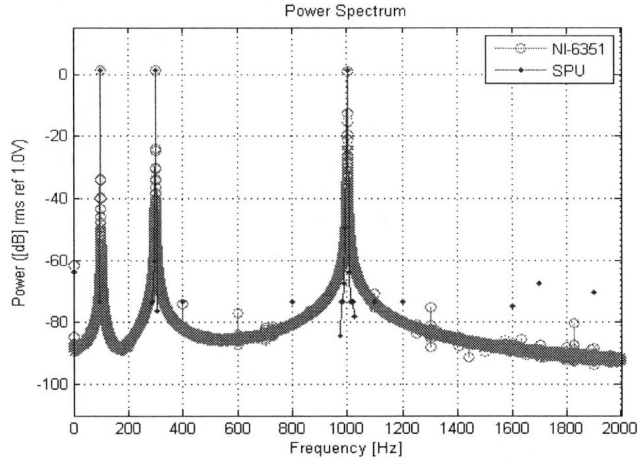

Figure 6 Test Result for FFT Processing with Sinusoidal Wave

CONCLUSION

The FPGA-based signal processing unit proposed in this paper can reduce the amount of data of sensor signals with high data rate through digital implementation of multi-stage octave signal processing and pipelined FFT signal processing. The developed unit confirmed its validity by comparing the performance with the precise reference equipment. This unit has higher performance than DSP-based signal processing, so it can perform signal processing that requires higher precision and high computation complexity with a larger number of acquisition channels. Based on this implementation, other types of digital signal processing algorithms will be developed in the future and processing unit will be developed as part of a general telemetry device, not just signal processing. Through the sensor signal processing for data compression, it will be possible to perform various space environmental measurement for flight tests of the Korean space launch vehicle.

REFERENCES

[1] T.N., Santhosh Kumar, et al, "DSP Based Signal Processing Unit for Real Time Processing of Vibration and Acoustic Signals of Satellite Launch Vehicles," in *Proc. Int. Telemetering Conf.*, (Las Vegas, NV), Nov. 1995.

[2] Dave Buckley, "Moving Data Analysis into the Acquisition Hardware", in *Proc. Int. Telemetering Conf.*, (San Diego, CA), Oct. 2014.

[3] American National Standards Institute, *Specification for Octave-band and Fractional-octave-band Analog and Digital Filters ANSI S1.11-2004 (R2009)*, 2009.

[4] Yu-Ting Kuo, et al, "Design and Implementation of Low-Power ANSI S1.11 Filter Bank for Digital Hearing Aids," *IEEE Trans. Circuit and Systems*, vol. 57, pp. 1684–1696, July. 2010.

POWER TELEMETRY ONBOARD A SEMI-AUTONOMOUS MARS ROVER

Rebecca C. Marcolina, Olugbenga O. Osibodu
Missouri University of Science and Technology, Rolla, MO, USA

Advised By: Dr. Kurt Kosbar, Dr. Melanie Mormile
Missouri University of Science and Technology, Rolla, MO, USA

ABSTRACT

This paper explores the telemetry of the power distribution system utilized onboard a semi-autonomous Mars rover. The Missouri S&T Mars Rover Design Team designs and fabricates such a rover to compete in the University Rover Challenge, a competition whose tasks simulate a future manned mission to Mars. To maximize efficiency during competition, the rover's modular power distribution system consists of three separate units: a 72 Watt-hour, Lithium-polymer battery pack; a custom Battery Management System (BMS); and a central power board. The BMS and power board measure and process electrical and environmental data autonomously, creating a self-regulating system onboard the rover. The two also form a communication chain between team teleoperators and the battery pack. This continuous stream of real-time data enables the team to quickly monitor the rover's safe operation, to make informed decisions during competition, and to apply this data to the design of future power systems.

INTRODUCTION

The Mars Rover Design Team (MRDT) at Missouri University of Science and Technology was created five years ago as an opportunity for students to develop and innovate space technology. The team's primary purpose is to design and fabricate a next-generation Mars rover over the course of a year to compete in the University Rover Challenge (URC) (Figure 1) [1]. Hosted by the Mars Society in Hanksville, Utah, the competition tasks simulate a future manned mission to Mars. Each rover must be able to assist astronauts, collect and analyze a series of samples, service various pieces of equipment, and traverse rugged terrain [1].

To complete these tasks quickly and efficiently, MRDT's 2017 competition rover, Gryphon, utilizes a custom power system whose specifications meet not only URC requirements, but also those that arise from the power needs of the rover's other sub-systems. Team members create a power budget to reference throughout the design and development cycle. However, as designs for other systems onboard the rover have the potential to change dramatically throughout the year, the power budget anticipates extra current draw to provide security in the case of unexpected power needs.

Figure 1: MRDT's 2017 Competition Rover, Gryphon

By creating an adaptable, versatile power distribution system, the team has flexibility in its design cycle without sacrificing or taxing the rover's power needs. This adaptability also presents additional challenges. A vehicle built to encounter harsh terrain, the components of a rover must be both durable and well secured. Special consideration must be given to the power distribution system - it must be able to protect itself amid a mechanical failure. Should electrical components loosen, there is a possibility that they could short or spark and harm connected devices or other sub-systems. On the other hand, the power distribution system must also remain both easily serviceable and accessible for the dynamic design and development cycles of the other sub-systems. In response, the team has created a modular power distribution system characterized by its independent component designs. This design method also provides an additional benefit through the formation of a communication chain between the rover and team operators. To function safely and properly, individual units of the power distribution system continuously relay data about their present operating conditions. The team can then monitor these conditions in real time, and adjust them as necessary while the rover is in operation. The result is an innovative, effective solution to the challenges of competition.

THE POWER DISTRIBUTION SYSTEM

The purpose of the rover's power distribution system is to create, allocate, and manage power across each of its other five major sub-systems. These include the following: Drivetrain, Ground Support Systems, Manipulators, Rovecore, and Science [2]. Gryphon's power distribution system is composed of three distinct units: a Lithium-Polymer battery pack, a custom Battery Management System (BMS), and a central power board. To streamline this design, the battery pack and BMS are contained within a single structure known as the battery box. The system's configuration is illustrated below in Figure 2.

Each aspect of Gryphon's power system was developed according to additional team-mandated requirements. First and foremost, the team wanted each element of the system to be easily accessible for testing and maintenance throughout its development. Second, as each sub-system has its own current and voltage requirements, the power distribution system must be compatible with a wide range of devices without decreasing its power output. Lastly, MRDT wanted to incorporate both manual and automatic safety features and controls into the battery pack, BMS, and power board. These additions protect the electrical components and devices onboard the rover in case of a major failure.

Figure 2: The Power Distribution System Positioned Between Gryphon's Motor Controllers

To meet these requirements, the team chose a modular design for Gryphon's power distribution system. This means that each unit of the system operates almost completely independently of one another, which allows the team greater control of the entire system. They can also be tested and developed simultaneously, affirming the system's versatile nature. Modularity represents the biggest design consideration of Gryphon's power distribution system – it ensures that each device onboard the rover is interchangeable yet able to operate according to its own unique power needs and specifications.

THE BATTERY PACK

To power Gryphon, the team chose Lithium-Polymer (LiPo) 18650 cells for their excellent energy density, as their energy to size ratio is significantly higher than other battery options [3, 4]. These batteries have also been utilized in previous MRDT rovers, and have shown to be a reliable power source during competition. The battery pack must have the potential to supply a maximum sustained peak current of 180A to the rover - a situation that only arises when all of the drive motors, telecommunications systems, camera gimbals, and the robotic arm are operating simultaneously. Additionally, after establishing the power budget and studying the performance of the team's previous rover, it was determined that a battery pack of the same type and size - but not mechanical structure - should be utilized in Gryphon's power distribution system. The structure of the previous pack arranged the cells in a flat sheet, but it proved to be flimsy and difficult to service. Gryphon's battery pack, therefore, was designed to be sturdier, and is shown below in Figure 3. Individual LiPo cells are contained in 3-D printed cell holders in groups of ten, and are then connected in parallel to create a single battery block. Gryphon's battery pack is composed of eight of these blocks connected in series.

Figure 3: The Battery Pack

The LiPo cells are arranged in a 2x5 array within each block, and each block features conductive tabbing spot-welded across its top and bottom. By attaching bus bars across one end of the battery pack and two separate contacts at the other, this design enables connection when the blocks are simply bolted together. The bus bars are composed of electrical grade aluminum, while the contact points are electrical grade aluminum washers placed into recessed ports in the battery pack's plastic endcaps. These cell holders also feature small holes through which team members can measure the voltage of individual blocks. Digital temperature sensors are also embedded within the pack to ensure the batteries' safe operation. The innovative structure of Gryphon's battery pack enables MRDT operators to remove and replace individual cell blocks easily in the case of malfunction or damage. Its design streamlines the pack's wiring and voltage measurements, both electrically and manually, while the pack's assembly maintains the proper condition of the cells and their connections. Additionally, the recessed design of the pack's contacts ensure that they can only be accessed deliberately, lowering the chance of both injury and accidental pack shortage.

THE BATTERY MANAGEMENT SYSTEM

The primary purpose of the BMS is to monitor and manage the condition of the battery pack. As LiPo batteries that operate below or beyond their rated voltage can ignite or explode, it is imperative that the team is able to continually assess their condition while operating the rover. Most of the concepts behind Gryphon's BMS were adapted from the design of the BMS onboard the team's previous rover. Its execution has been very different, however, as the former BMS was never fully operational. As a result, MRDT carefully considered the design and fabrication of

Gryphon's BMS. The device encompasses a custom printed circuit board coupled with an MSP432 microcontroller, various MOSFETs, and an array of individual sensors (Figure 4) [3].

Figure 4: The BMS

These additions enable the BMS to continuously measure and record the temperature, voltage, and current draw of the battery pack. As data is collected, the microcontroller communicates this information to team operators and, if necessary, protects the pack from danger. All readings are transmitted to the team's base station, an external communication center. It allows team operators to collect telemetry from each onboard system, display it for analysis and discussion, and send commands back to the rover. By analyzing these readings, the team can monitor Gryphon's status during competition and make informed decisions as it runs. Additional abilities of the BMS are completely autonomous. The device can assess and adjust dangerous or unfavorable temperature conditions within the battery pack by operating up to four fans positioned throughout the battery box. Furthermore, the BMS protects the entire power distribution system against overcurrent by automatically shutting down the rover should it detect current levels greater than the allotted 180A output of the batteries. Coupled with a manual emergency stop button that disconnects the rover's batteries from the power board completely, these two emergency stop functions ensure the safety of team operators should any type of problem arise concerning the rover's sub-systems. In the case of the batteries reaching dangerously low voltage, the BMS will turn off power to the rover, and then itself, by means of an electrically resettable rocker switch. It will also automatically turn itself off to save power if it is left on while the rover is off for more than an hour. The streamlined design of Gryphon's BMS offers a functional, practical approach to navigating competition tasks safely, while its communication chain of power system data enables team members to operate the rover efficiently and effectively.

THE POWER BOARD

The power board is responsible for regulating and distributing the power received from the battery box to the rest of the systems onboard the rover. It does so by relying upon a combination of hardware and software power controls. While influenced by previous designs, the layout of Gryphon's power board was reorganized to give team operators more control over the total distribution of power by separating critical and noncritical system devices. This separation helps to protect the rover and its systems in the case of an electrical short. Should there be a shortage in the rover's robotic arm, for example, this design prevents other systems from shorting as well.

As illustrated in Figure 5, power is allocated from the batteries to the other sub-systems through a collection of eleven different buses. Seven of these buses are responsible for distributing power directly from the battery pack to Gryphon's motor controllers, a group of devices that control the rover's propulsion system. The remaining four buses - communications, logic, actuation, and extra - power devices across the rest of the rover's sub-systems. As these four buses must receive power at a constant rate of 12V, pack voltage and current first pass through one of three different 12V buck converters to accommodate this restraint. This is necessary because the typical range of pack voltage is 25 - 32V.

Figure 5: The Power Board with Its Seven Motor Buses and Four Specialized Buses

The communications bus powers all of the communication devices onboard Gryphon, while the logic bus powers all of its other integrated circuits and microcontrollers. The actuation bus powers the higher current devices, such as the cameras onboard the rover or the motors that open and close the rover's drop bay. These devices may draw anywhere from 1 - 6A continuously, as opposed to most of the rover's other devices that operate within a 100 - 500mA range. The extra bus provides additional power to any systems if necessary. It was added as a security measure for cases where systems need more current than they were originally allotted in the power budget.

Each of the eleven buses have their own independent current sensors, switching MOSFETS, and fusing. The current sensors report data about each bus to the microcontroller which has software defined current limits. Should the current draw for any bus rise above the set limit, the microcontroller will automatically disable that bus. The fuse for each bus is installed in series to provide a hardware backup should the microcontroller fail. The motor buses connect directly to their specified motor controller by means of Anderson Powerpole connectors on the power board, while the remaining buses pass through an additional printed circuit board known as the patch panel before connecting to their portion of the rover. The patch panel is a separate board mounted onto the power board that provides each bus with multiple vertical Anderson connector pairs. Both the power board and the patch panel are shown below in Figure 6.

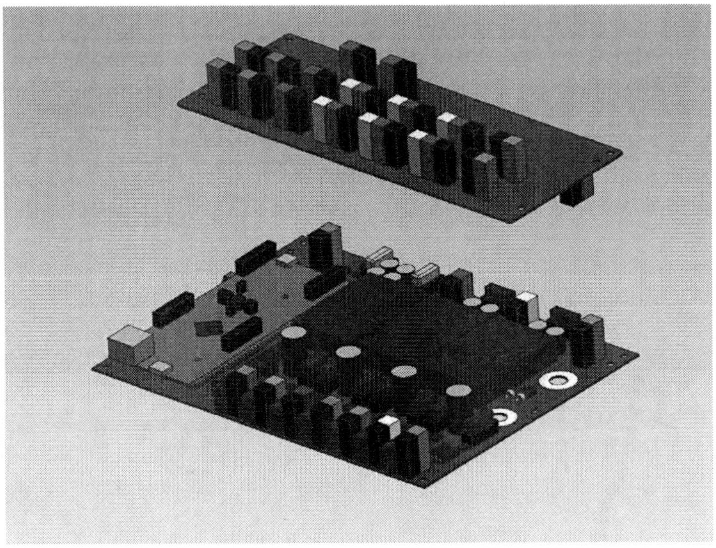

Figure 6: 3-D Model of Gryphon's Power Board (lower) and Patch Panel Attachment (upper) Generated in Siemens NX Modeling Software

Every competition task requires its own specialized devices, so it is extremely valuable for the team to be able to connect different devices to the power distribution system. The same connection is used for both the rover's science arm and its robotic arm, for example, and the team must remove and replace these devices between the science and equipment servicing tasks. Gryphon's power board is designed to offer additional protection and utility through the separation of each bus to manage situations such as these. The practice of separating the buses streamlines and manages the power output from the batteries safely and rapidly. This method also protects Gryphon from major electrical damage due to a short, as it would affect an individual system as opposed to all boards and components. The separated buses ultimately allow MRDT operators greater control of specific devices on the rover, as well as more reliability as vital devices are independently powered.

SYSTEM INTEGRATION AND TELEMETRY

Gryphon's power distribution system has two separate communication methods that are bridged within the power board. There is an internal communication system for the power board and BMS to exchange data, and an external system which is used to communicate between the rest of the rover and the base station. The internal communication is achieved with a three wire RS-232 connection, whereas the external communication utilizes MRDT's custom protocol, RoveComm. This divide between internal and external communication is necessary due to the lack of support for RoveComm on the hardware of the BMS. The low-power MSP432 microcontroller on the BMS does not support the Ethernet connectivity necessary for RoveComm, so RS-232 was chosen as a suitable alternative. By using these two communication methods in tandem, the team receives a continuous stream of reliable data that allows operators to make informed decisions during competition.

INTERNAL COMMUNICATION

Power board to BMS communication is implemented by combining the RS-232 protocol with a three-wire serial line. RS-232 uses +15V as a logical low and -15V as a logical high. This design decision was made because the standard Universal Asynchronous Receive/Transmit (UART) protocol generated by the microcontrollers on the BMS and power board only have a range of 0V - 3.3V. As the layout of the rover was not finalized when the team selected the communication protocols, the team needed to employ one that would be reliable regardless of the noise level surrounding the signal wire. For this reason, the team chose to utilize RS-232. Its higher voltage range reduces noise on the signal line caused by the large buck converters on the power board itself in addition to the noise generated by the motor controllers situated on either side of it. By implementing this communication protocol on Gryphon, the team increases the reliability of messages sent between the power board and BMS while retaining the unit's flexible design. Messages sent from the power board include commands to turn on and off the fans in the battery box, to reset the battery pack power output, and to shut down the pack output. Upon receiving these commands, the BMS acts accordingly and then responds to the power board. It primarily sends data about the current being drawn from the battery pack, the battery pack voltage, and the temperature of the battery pack to the power board. This data sent between the power board and BMS is crucial to the safe operation of the rover. Most importantly, it allows the team to monitor and adjust the condition of the battery pack. To ensure that these messages are properly delivered and received, MRDT chose this method of communication because it reduces the distortion and damage of the messages sent between the BMS and power board. This, in turn, improves the rover's performance during both development and competition.

EXTERNAL COMMUNICATION

Base station operators utilize a team generated software known as Rover Engagement Display (RED) to relay commands to the devices onboard the rover during operation. This is made possible through a custom communication protocol known as RoveComm, which connects the base station to the rover's power board. It is inherited from MRDT's 2016 competition rover and is based on the lightweight User Datagram Protocol (UDP), a protocol that provides no guarantee of message delivery or packet duplication protection. RoveComm takes advantage of UDP's lightweight messages but adds its own message verification system. This implementation makes RoveComm a reliable communication system without being as resource intensive as Transmission Control Protocol (TCP). The header of a RoveComm message includes data about the version of the protocol used to send the message, the type of data being sent, the size of the data, and a flag to specify whether an acknowledgement response is required. Its packet structure allows for a variety of message types to be sent to the different devices on the rover.

As RoveComm is a protocol used by every device on the rover, its numerous types of data are divided according to each device's purpose. Thus, a section of RoveComm is reserved specifically for messages to and from the power board. These commands are typically used to disable and enable the motor, actuation, extra, and logic buses, but the power board also continuously sends data about the current readings of each bus to the base station. This communication channel is what allows the team to direct Gryphon during operation, as the base station operators use the information conveyed by the power board to determine their next command to Gryphon. They view the status and current draw of each of the various buses on the power board via a readout generated by RED, shown in Figure 7. This information is then used to make decisions about the rover's performance during its operation. For example, the URC science task requires a team's rover to take soil samples from a chosen location. MRDT operators would typically direct Gryphon to lower its science arm and begin drilling into the soil, but the cameras onboard the rover cannot always view the drill. This means that the team does not always know if the drill is spinning. To compensate, the team studies the current draw levels indicated by RED. As the science arm is connected to the actuation bus, the current level of the bus will fluctuate as the drill is used. The changes in the actuation bus's current level are shown in RED, so the team can evaluate the situation and respond accordingly. As each component of the power board listed in RED acts in a similar manner, team operators can monitor each sub-system onboard the rover during its operation. Developed by MRDT to enhance a rover's performance, RoveComm contributes to the safe, efficient operation of Gryphon and the communication chains the rover relies upon to succeed in competition.

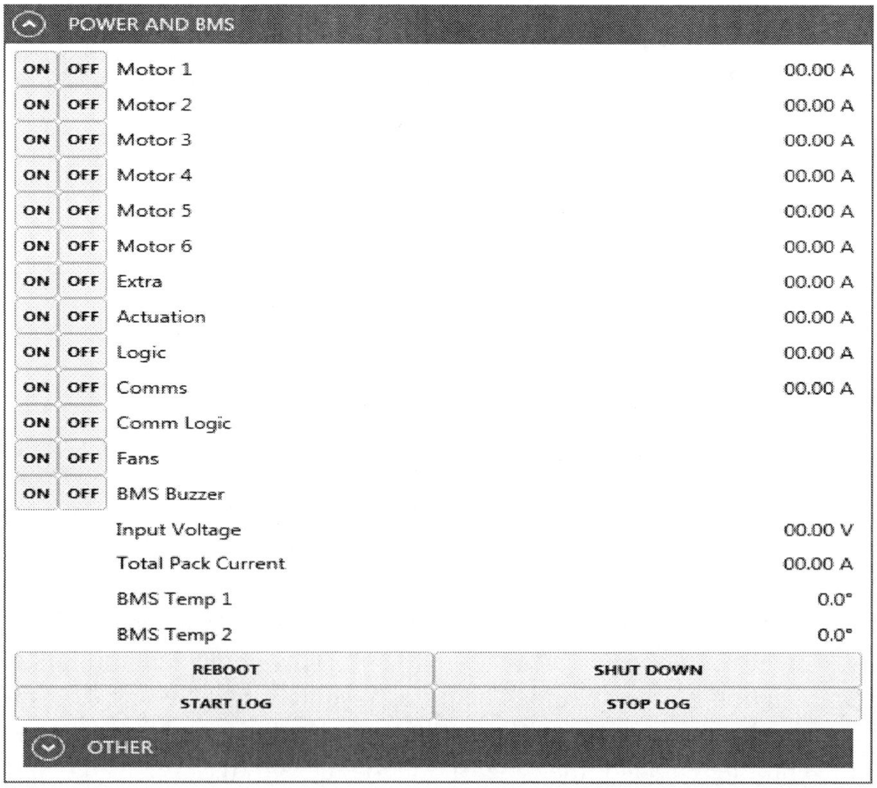

Figure 7: Example Readout of the Power System Toggles in RED

CONCLUSION

The power distribution system is an integral part of a rover's design and functionality. While Gryphon's reliable, modular structure ensures the rover's safe and effective operation, MRDT strives to further develop its designs for future competition rovers. By analyzing the performance of Gryphon's power distribution system, the team has chosen to begin by introducing active monitoring and, eventually, active balancing of individual blocks within the battery pack. Doing so would allow the team more precise control of the health and safe operating conditions of the battery pack. If the BMS could measure the voltage of individual battery blocks, it could register that a block's voltage level was approaching its lower limit and shut off the battery in response. While team members tried to incorporate active battery monitoring into Gryphon's power distribution system, they were never able to do so successfully. Active monitoring is a more accurate means of prevention than the method that the team currently uses, which is to assume that all of the battery blocks have equal voltage levels. Unfortunately, this practice requires that team members overcompensate the minimum useable voltage as a result. Should MRDT employ active monitoring, it could also implement active balancing more easily. This practice would enable the BMS to alter the current draw from each of the individual battery blocks to ensure that their voltage levels were evenly balanced - a condition that would extend the life and available use of the battery pack during the rover's operation. Yet even without these features, Gryphon's power distribution system remains an innovative solution to the challenge of powering a semi-autonomous rover. Its adaptability accommodates the needs of the rover's other sub-systems without sacrificing its own design considerations, while its constant communication of device conditions on both internal and external levels allows team operators to formulate tactical responses during competition. The implementation of a modular design is a simple way to allocate energy to all of Gryphon's sub-systems and devices rapidly and effectively, and has proven to be a strong foundation for the team's future design visions.

ACKNOWLEDGEMENTS

The authors would like to acknowledge Emily 'Ellis' Sansone, the MRDT Power Systems sub-team lead, as well as the other sub-team members for their support and assistance throughout the completion of this paper. They would also like to credit the hard work and dedication of every MRDT member, as well as their outstanding commitment to the team's shared vision, "Today. Tomorrow. Forever.".

REFERENCES

[1] Mars Society, "University Rover Challenge", Lakewood, CO, http://urc.marssociety.org/, Accessed 2017, June 1.

[2] Mars Rover Design Team, "Sub-teams", Rolla, MO, http://marsrover.mst.edu/subteams/, Accessed 2017, June 1.

[3] Mars Rover Design Team, "Power System Definition Document", Rolla, MO, 2017.

[4] BATTERY Bro, "The 18650 Battery", https://batterybro.com/, Accessed 2017, June 1.

PERFORMANCE OF OPTIMIZED VLC GSSK SYMBOL SETS IN DISPERSIVE MEDIA

Mandip Shrestha and Deva K. Borah
Klipsch School of Electrical & Computer Engineering
New Mexico State University
Las Cruces, NM 88003, USA
Email: mandip@nmsu.edu, dborah@nmsu.edu

ABSTRACT

Visible Light Communication (VLC) systems using an array of LEDs as a transmitter and a digital camera as a receiver can provide reliable communication. Such systems can employ various modulation formats that include the generalized space shift keying (GSSK), where only the indices of the LEDs are used to send information. This paper investigates the modeling of GSSK VLC systems under different atmospheric conditions as well as for underwater channels, and investigates the problem of selecting the best set of modulation symbols. The behavior of such systems is modeled using the point spread function, and the inter-symbol Euclidean distance is maximized to obtain the best symbol set. The paper adopts the recently proposed symbol set selection algorithm from the literature exploiting the clustering property of GSSK VLC systems. The results show that the symbol set design algorithm can provide several dBs of gain in the symbol error rate (SER) performance over randomly selected symbol sets in GSSK systems under various atmospheric conditions as well as for underwater channels.

INTRODUCTION

There are multiple modulation formats that can be used in VLC systems. Examples include the on-off keying (OOK), orthogonal frequency division multiplexing (OFDM), generalized spatial modulation (GSM), spatial modulation (SM) and generalized space shift keying (GSSK) modulation [1] - [3]. Our work in this paper focuses only on the GSSK VLC systems. Such systems have low implementation complexity over SM and GSM systems, where information is present in both LED indices and modulated symbols.

In GSSK systems, each group of transmit bits is mapped to a symbol that actives specific LEDs representing the symbol. Consider a transmitter with an array of L LEDs. If L_a LEDs are activated out of L transmit LEDs, then there are $M_{all} = C(L, L_a)$ possible combinations of transmit LEDs, where $C(n,k) = n!/(n-k)!k!$. For M-ary GSSK systems, we need M to be a power-of-two, and the highest possible M is $M = 2^K$, where $K = \lfloor log_2(M_{all}) \rfloor$ and $\lfloor . \rfloor$ is the floor function. As an example, consider L = 16 and L_a = 3 giving $C(16,3) = 560$ combinations of transmit LEDs. If the GSSK system is 16-ary, we have to choose the best set of 16 symbols (or combinations) from

$C(560, 16) \approx 3.60 \times 10^{30}$ sets of active LEDs. This is a challenging problem, and algorithms to obtain the best symbol set for GSSK are presented in [4], and extended to GSM systems in [5]. The performance of the algorithm given in [4] has been studied only for indoor systems with no atmospheric effects. Our work here considers outdoor systems including underwater channels. In outdoor systems, the weather may cause poor visibility that makes it hard for an imaging receiver to identify the active LEDs. Weather conditions, such as fog, rain, haze and snow, can cause link failures, and need to be studied. Similarly, in the case of underwater systems, scattering properties of the small particles determine the received signal. Since multiple scattering of light in underwater is a complex process, modeling of VLC systems in underwater channels is a challenging task.
Our study uses the point spread function (PSF) to analyze the communication channel behavior. The PSF gives the system response due to a point source and includes the effect of multiple scattering and attenuation in the medium. Once the PSF is known for a given system, the image due to any light source is obtained using the convolution of the PSF with the original image to get the received image signal under the given weather or underwater conditions.

SYSTEM MODEL

The transmitter consists of an array of L LEDs arranged in a square of size $\sqrt{L} \times \sqrt{L}$. The received signal $s_l(x, y; x_l, y_l)$ at position (x, y) in the xy-plane (i.e., the image plane) due to the l-th LED is given by [6]

$$s_l(x, y; x_l, y_l) = A \sum_{i=1}^{\nu} (\frac{\mu}{\mu_0})^2 \frac{e_i}{2\pi\sigma_i^2} \exp(-\frac{(x - x_l)^2 + (y - y_l)^2}{2\sigma_i^2}) \quad (1)$$

where (x_l, y_l) is the image center of the l-th LED, ν is the number of summation terms, $A = 1/p^2$, p is the pixel length, e_i is an LED model parameter, σ_i^2 is the variance of the i-th mixture term, $\mu = f_0/d$ is the image magnification, f_0 is the camera focal length, d is the distance between the transmitter and a receiver, μ_0 is the image magnification at a reference distance d_0.
Consider the 2-D image at the receiver by the l-th LED without any weather or underwater conditions. Let $\mathbf{S}_l'$ be the sample matrix produced by (1). The (i, j)-th element of $\mathbf{S}_l'$ is given as $s_l'(i, j; x_l, y_l) = s_l(i\delta_s, j\delta_s; x_l, y_l)$, where i and j are positive integers, and δ_s is a real number representing the spatial sampling interval along the x- and y- directions in the image plane. Let u_l define the 'on' or 'off' status of the l-th LED. In the presence of a dispersive medium with a PSF $\mathbf{P}$, the received image is modified as $\sum_l (\mathbf{S}_l' * \mathbf{P}) u_l$, where $*$ denotes 2D convolution. The model of the PSF $\mathbf{P}$ is described in the next subsections. The matrix $\mathbf{S}_l' * \mathbf{P}$ for each LED l is rearranged as a column vector by stacking its columns as in [4]. Let $\mathbf{s}_l$ denote this column vector. Define a new matrix $\mathbf{S}$ whose l-th column is $\mathbf{s}_l$. The signal vector at the receiver sensor plane is given by

$$\mathbf{r} = \mathbf{Hu} + \mathbf{w} \quad (2)$$

where $\mathbf{H} = \tilde{\mathbf{A}}\mathbf{S}$ is the overall channel matrix, and $\tilde{\mathbf{A}}$ is the pixelization matrix [4]. The vector $\mathbf{u}$ represents a symbol and is of size $L \times 1$. Each element of $\mathbf{u}$ is either a 0 or a 1. The vector $\mathbf{w}$ is a noise vector. Each element w_m of $\mathbf{w}$ is zero-mean Gaussian with a variance $\sigma_{w,m}^2 = \alpha\ (\mathbf{Hu}+\mathbf{q}_b)_m + \beta$, α and β are constants, q_b is the background illumination vector. We consider $\sigma_{w,m}^2 \approx \sigma_w^2$ for high background illumination.

POINT SPREAD FUNCTION UNDER ATMOSPHERIC EFFECTS

Light from the transmitter gets scattered in the channel before arriving at the receiver. Although multiple scattering can be neglected in clear air, it is very crucial in weather conditions such as fog or haze. Multiple scattering of light has been studied by many authors. Generally there are two ways to study the scattering of light. One is the Monte-Carlo simulation technique using ray tracing, and the other is the use of the radiative transfer equation (RTE). Monte-Carlo simulation method is too complex computationally when the number of scattering particles is large. RTE method has been extensively studied by the authors in [7]. Generally, two parameters are used to define an atmospheric condition: (1) the optical thickness, T, that determines the amount of depletion that a beam of radiation undergoes as it passes through the atmosphere. T is related to the atmospheric visibility V as $T = \sigma d$, where d is the distance between the transmitter and the receiver, and $\sigma = 3.912/V$. The parameter σ is called the extinction coefficient. (2) The parameter, q, called the forward scattering parameter, whose value is between 0 and 1. The value of q determines the weather conditions as shown in Table 1.

q	0.0 - 0.2	0.2 - 0.7	0.7 - 0.8	0.8 - 0.85	0.85 - 0.9	0.9 - 1
Weather condition	Air	Aerosols	Haze	Mist	Fog	Rain

Table 1: Forward scattering parameter and the corresponding weather conditions [8].

PSF for a given atmospheric condition mentioned in [7] is only valid for $T > 1$. The authors in [8] have approximated the PSF in [7] to make it work for every value of T using the modified generalized Gaussian distribution as

$$P_a(x, y; q, T) = \frac{\exp\left(- \frac{(x^2+y^2)^{\frac{k_1 T}{2}}}{|F(k_1 T, \frac{1-q}{q})|^{k_1 T}} \right)}{4\Gamma^2(1 + \frac{1}{k_1 T}) F(k_1 T, \frac{1-q}{q})^2} \tag{3}$$

where $F(p, \gamma) = (\frac{\gamma^2 \Gamma(\frac{1}{p})}{\Gamma(\frac{3}{p})})^{\frac{1}{2}}$ is the scale parameter, $\Gamma(.)$ is the Gamma function, i.e., $\Gamma(z) = \int_0^\infty e^{-t} t^{z-1} dt$, $z > 0$ and $k_1 = 0.5$. The 2D sample matrix produced by (3) in the image plane is $\mathbf{P}$, whose (i, j)-th element is given as $P_a(i\delta_s, j\delta_s; q, T)$, where i and j are positive integers.

Figure 1 shows that the value of T determines the shape of the peak of the APSF. As the value of T increases, the shape of the peak becomes broader. Similarly, decreasing the value of q broadens the shape of the PSF as long as the atmosphere is not too thin ($T < 0.3$). At a very thin atmosphere i.e., higher visibility ($T = 0.1$) and clean air ($q = 0.1$), the PSF is almost identical to the delta function. Note that Fig. 1(d) with $T = 0.1$ cannot be obtained using the PSF given in [7]. The cross-sections of the PSF in Fig. 1 indicate that at thick atmosphere ($T > 0.4$) and under adverse weather conditions for VLC, e.g., haze, fog or rain, the received image of any light source has a glow around it. For pure air ($q < 0.1$), the visibility is very high so the value of T is always low. For adverse weather conditions, such as haze or fog, the visibility is low so T is always high.

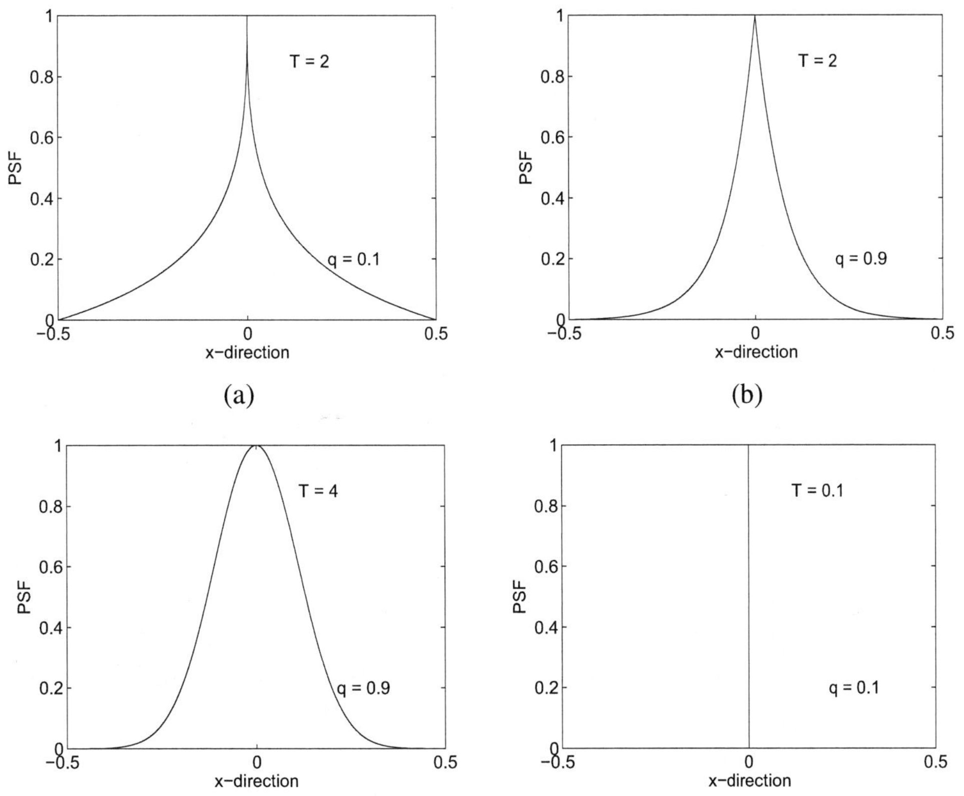

(a) (b) (c) (d)

Figure 1: Cross-sections of APSF in different weather conditions

The received image due to foggy weather is shown in Fig. 2. The parameters used are $f_0 = 4.9$ mm, p = 5.5 μm, $d = 20$ m, $\nu = 2$, $d_0 = 0.1$ m, e_1 =5.444$\times10^{-8}$, $e_2 = 1.126 \times10^{-6}$, $\sigma_1 = 2.1027 \times10^{-6}$ m, $\sigma_2 = 3.1392 \times10^{-6}$ m, $L = 16$, $T = 4$ and $q = 0.9$.

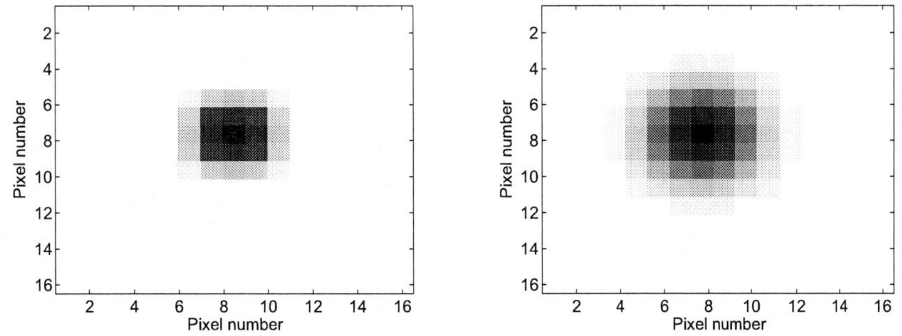

(a) Image without any dispersive media.

(b) Foggy weather image in thick atmosphere.

Figure 2: Signal acquired by the camera receiver due to 16 LEDs at d = 20 m. The darker pixels represent relatively higher intensities.

POINT SPREAD FUNCTION FOR UNDERWATER CONDITIONS

The scattering and absorption of light cause blurring of images in underwater, thus degrading the performance of VLC systems. Multiple scattering cannot be neglected in underwater communication. This phenomenon has been studied by many authors, and different models of PSF have been given. Using the model given in [9], the PSF for the underwater image is given by

$$\begin{aligned} G(\theta_b, \tau_b) &= \frac{\delta(\theta_b)}{\pi\theta_b} e^{-\tau_b} + 0.525 \frac{\tau_b}{\theta_b} e^{(-2.6\theta_b^{0.7} - \tau_b)} + \frac{\beta_2^2}{2\pi}[2 - (1+\tau_b)e^{-\tau_b}] e^{(-\beta_1(\beta_2\theta_b)^{\frac{1}{3}} - (\beta_2\theta_b)^2 + \beta_3)}, \\ \beta_1 &= \frac{6.857 - 1.5737\tau_b + 0.143\tau_b^2 - 6.027 \cdot 10^{-3} \cdot \tau_b^3 + 1.069 \cdot 10^{-4} \cdot \tau_b^4}{1 - 0.1869\tau_b + 1.97 \cdot 10^{-2} \cdot \tau_b^2 - 1.317 \cdot 10^{-3} \cdot \tau_b^3 + 4.31 \cdot 10^{-5} \cdot \tau_b^4}, \\ \beta_2 &= \frac{0.469 - 7.41 \cdot 10^{-2}\tau_b + 2.78 \cdot 10^{-3} \cdot \tau_b^2 + 9.6 \cdot 10^{-5} \cdot \tau_b^3}{1 - 9.16 \cdot 10^{-2} \cdot \tau_b - 6.07 \cdot 10^{-3} \cdot \tau_b^2 + 8.33 \cdot 10^{-4} \cdot \tau_b^3}, \\ \beta_3 &= \frac{6.27 - 0.723\tau_b + 5.82 \cdot 10^{-2} \cdot \tau_b^2}{1 - 0.072\tau_b + 6.3 \cdot 10^{-3} \cdot \tau_b^2 + 9.4 \cdot 10^{-4} \cdot \tau_b^3}, \end{aligned} \tag{4}$$

where θ_b is the scattering angle at which the light is refracted away from its original direction, $\tau_b = \tau\omega$, τ is the optical length that determines the amount of depletion that a beam of radiation undergoes as it passes through water, and ω is the single scattering albedo which is the ratio of light scattered to the total light attenuation. The optical length is given by $\tau = \sigma d$, where σ is the total attenuation coefficient and d is the distance between the transmitter and the receiver. δ is related to the direct beam and can be neglected during normalization. For the imaging receiver, we project the PSF $G(\theta_b, \tau_b)$ into the xy- plane. For small field angle, first we project PSF in θ_b to get PSF in x- plane using $x = d\theta_b$ and rotate it to get the PSF in xy- plane as $P_u(x, y; \omega, \tau)$. We use the sampled version of P_u. The (i, j)-th element of the sample matrix **P** produced by P_u is given as $P_u(i\delta_s, j\delta_s; \omega, \tau)$, where i and j are positive integers.
The received underwater image is shown in Fig. 3. The parameters used are $f_0 = 4.9\,\text{mm}$, pixel width $5.5\,\mu\text{m}$, $d = 10\,\text{m}$, $\nu = 2$, $d_0 = 0.1\,\text{m}$, $e_1 = 5.444 \times 10^{-8}$, $e_2 = 1.126 \times 10^{-6}$, $\sigma_1 = 2.1027 \times 10^{-6}\,\text{m}$, $\sigma_2 = 3.1392 \times 10^{-6}\,\text{m}$, $L = 16$, $\tau = 15$ and $\omega = 0.6$.

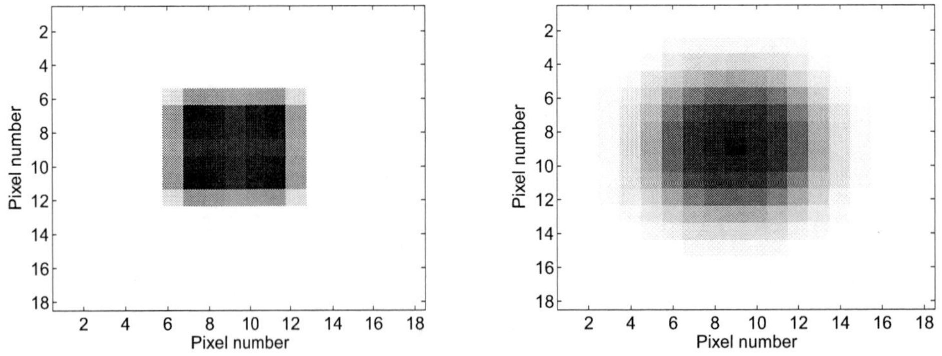

(a) Image without any dispersive media.

(b) Image through murky water.

Figure 3: Signal acquired by the camera receiver due to 16 LEDs at d = 10 m. The darker pixels represent relatively higher intensities.

To understand the shape of the PSF, Fig. 4 shows the impact of the values of τ and ω. As τ and ω increase, the shape of the PSF becomes broader. It implies that for pure water (low τ and ω) the visibility is very high as the shape of the PSF is very thin.

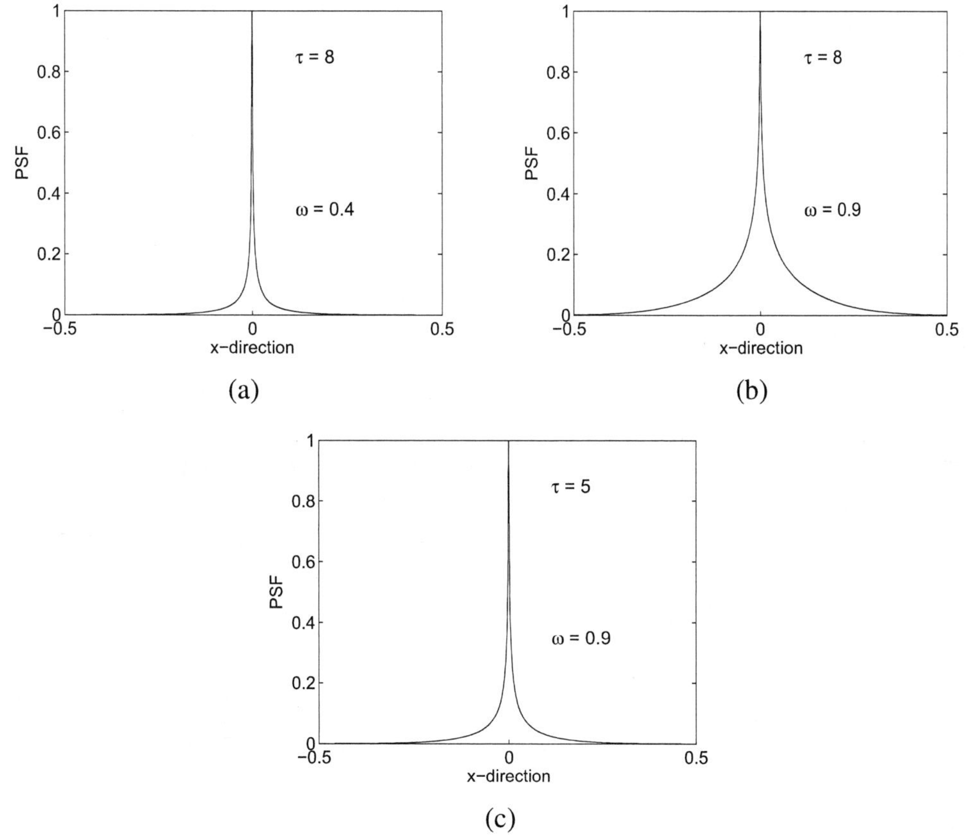

Figure 4: Cross-sections of PSF in different water conditions.

SYMBOL SET DESIGN

Maximum likelihood (ML) detection is used to detect the active LEDs as

$$\hat{\mathbf{u}} = \arg\min_{\mathbf{u}} ||\ \mathbf{r} - \mathbf{H}\mathbf{u}\ ||^2 \tag{5}$$

where $\hat{\mathbf{u}}$ is the detected transmit signal vector. The symbol error rate (SER) at the receiver can be approximated as in [4] using the nearest neighbor upper bound,

$$P_e \approx \sum_{i=0}^{i_{max}} \bar{N}_i Q(\frac{D_i}{2\sigma_w}) \tag{6}$$

where $\bar{N}_i$ is the average number of neighbors at Euclidean distance (ED) D_i, $D_i = ||\ \mathbf{H}\mathbf{u}_j - \mathbf{H}\mathbf{u}_k\ ||$ $\forall j, k, j \neq k$ and $Q(x) = 1/\sqrt{2\pi} \int_x^{\infty} e^{-y^2/2} dy$.

Designing the optimal symbol set is numerically too complex, because an exhaustive search requires searching over $C(M_{all}, M)$ combinations. For VLC systems, the authors in [4] propose to consider only those symbols that involve active LEDs that are close neighbors. We adopt the same algorithm that drastically reduces the search space in finding the best symbol set.

Step 1: Consider only those symbols that involve active LEDs located within a cluster radius of R_c from a specific LED, called the cluster center. Each LED can be a cluster center. From each cluster center choose a group of L_a LEDs including the cluster center that are within a distance of R_c from the cluster center. Remove repeated symbols. This reduces the set of total symbols from M_{all} to M_r symbols, where $M_r \ll M_{all}$ usually.

Step 2: If L is large, M_r can still be too large, so we need to reduce the set of M_r symbols to a set of M_s symbols so that $M_s < M_r$. We calculate the inter-symbol ED between each pair of symbols in the set containing the M_r symbols. Next, find the minimum value $D_0 = \min_{j,k,j \neq k} || \mathbf{Hu}_j - \mathbf{Hu}_k ||$, and identify the symbols that have the largest number of neighbors at D_0 and remove them from the set. Repeat this step until M_s symbols are left.

Step 3: This step reduces M_s symbols to M symbols. For this, generate all possible subsets of M symbols from the M_s symbols using a tree structure as given in [4]. For each such subset, compute the inter-symbol distance D_i=D_{jk} between each pair of symbols $\mathbf{u}_j$ and $\mathbf{u}_k$ in the subset and the corresponding average number of neighbors. Select the subset that has the largest D_{min} [4].

NUMERICAL RESULTS

In our numerical results, we use $f_0 = 4.9\,\text{mm}$, $p = 5.5\,\mu\text{m}$, inter-LED distance $\Delta = 1.5$ cm, $\nu = 2$, $d_0 = 0.1\,\text{m}$, $e_1 = 5.444 \times 10^{-8}$, $e_2 = 1.126 \times 10^{-6}$, $\sigma_1 = 2.1027 \times 10^{-6}\,\text{m}$, $\sigma_2 = 3.1392 \times 10^{-6}\,\text{m}$, $L = 16$, $L_a = 2$ and $M = 8$. The signal-to-noise ratio (SNR) is defined as the ratio of electrical power from the averaged intensity received per symbol and the noise variance, $(1/\sigma_w^2)((L_a/L)\sum_{i,j} h_{i,j})^2$ where $h_{i,j}$ is the (i,j)-th element of $\mathbf{H}$. Random symbol case uses averaging over 30 randomly selected symbol sets.

Figure 5 shows performance comparison between optimal and randomly selected symbol sets. The analytical and simulation results show accurate match. The optimal symbols using the tree search method provide about 12 dB of gain over randomly selected symbols for the given weather conditions of $T = 1$ and $q = 0.8$. We also show results where the optimal symbol design ignores the knowledge of the **P** matrix. We call this as partial channel information (PCI). Optimal symbol set design using PCI has about 7 dB of gain over random symbols case. Locally optimal symbols obtained using a tree search with $M_s = 20$ and $R_c = 1$ give nearly 10 dB of gain.

As the atmosphere becomes thicker, SER performance becomes worse as shown in Fig. 6(a). If the atmospheric density (T) is constant with changing weather, the SER performance is better as we move to the right in Table 1 from aerosols to rain which is shown in Fig. 6(b). Higher q value and lower T is expected to give better SER performance at given SNR as long as the atmospheric density is thick ($T > 0.4$) because of decrease in PSF width. SER performance with almost constant gain in the region between the q value of 0.2 and 0.6 in Fig. 6(b) is because it denotes the same weather aerosols in the channel. The drastic change in the SER performance from $q = 0.8$ to $q = 0.9$ is because of change in the weather from a misty to the foggy to the rainy weather.

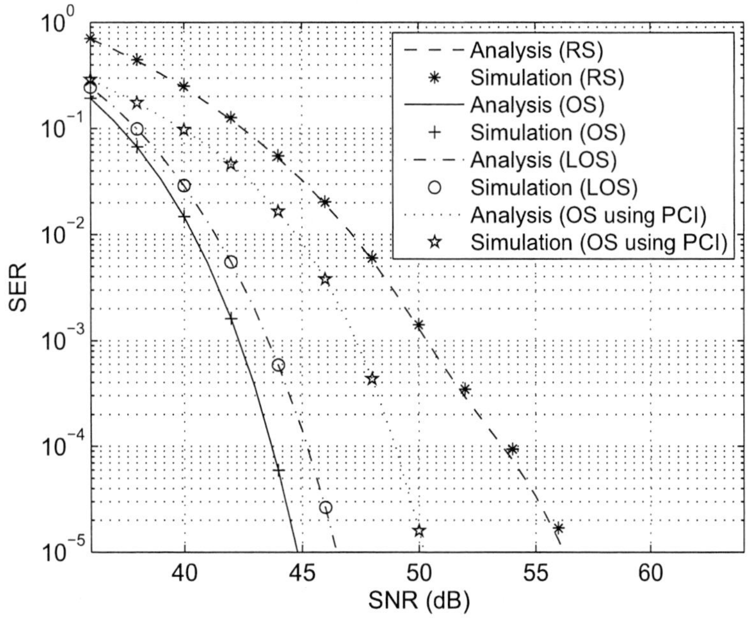

Figure 5: SER performance comparison at misty weather and thick atmosphere defined by $T = 1$ and $q = 0.8$ at a distance of $d = 20\,\text{m}$. 'RS' denotes random symbols, 'OS' denotes optimal symbols and 'LOS' denotes locally optimal symbols. LOS uses $M_s = 20$ and $R_c = 1$.

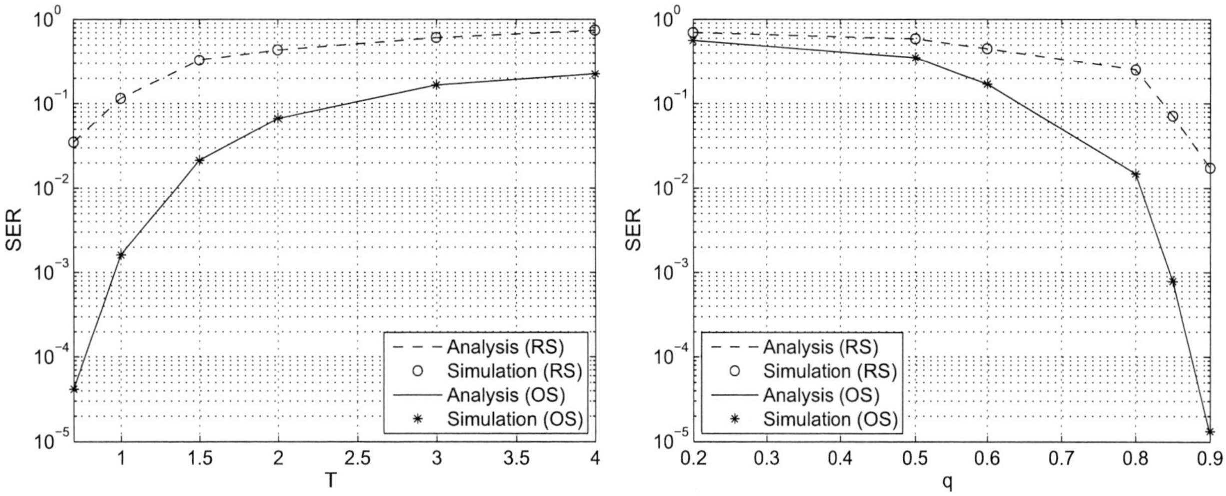

(a) SER vs T keeping $q = 0.8$ and SNR = 42 dB. (b) SER vs q keeping $T = 1$ and SNR = 40 dB.

Figure 6: SER performance comparison of optimal symbols with randomly selected symbols under different weather conditions at a distance of $d = 20\,\text{m}$. 'RS' denotes random symbols and 'OS' denotes optimal symbols.

In the case of underwater channels, Fig. 7 presents SER performance with the analysis and simulation results for a given water condition. Fig. 7 shows that the optimal symbol using tree search provides about 11 dB of gain over randomly selected symbols for given water condition of $\tau = 20$ and $\omega = 0.84$. Optimal symbol set design using PCI has about 3 dB gain over random symbol case. The locally searched symbol set with $M_s = 15$ and $R_c = 1$ has about 6 dB gain over randomly selected symbols. As the scattering of light becomes more (i.e., as ω increases), the SER performance becomes worse as shown in Fig. 8(a). Similarly if attenuation of signal is more (increasing of τ or σ) in a given water condition, the SER performance degrades as shown in Fig. 8(b). Lower attenuation of light and less scattering of light improve SER performance at given SNR.

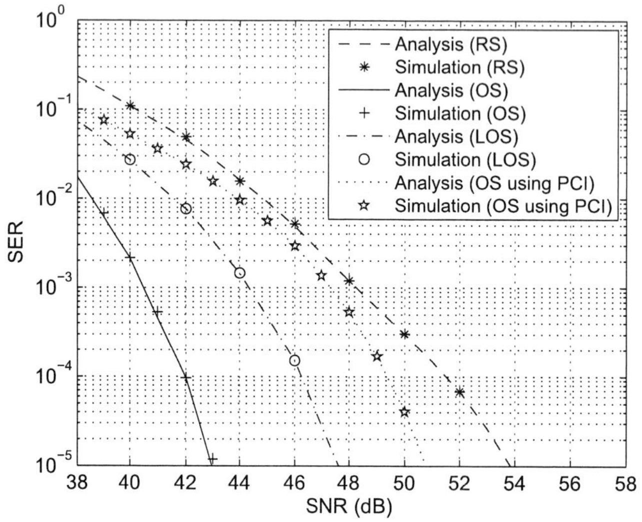

Figure 7: SER performance study for underwater VLC with water conditions given by $\tau = 20$ and $\omega = 0.84$ at a distance of $d = 10\,m$. 'RS' denotes random symbols, 'OS' denotes optimal symbols and 'LOS' denotes locally optimal symbols. LOS uses $M_s = 15$ and $R_c = 1$.

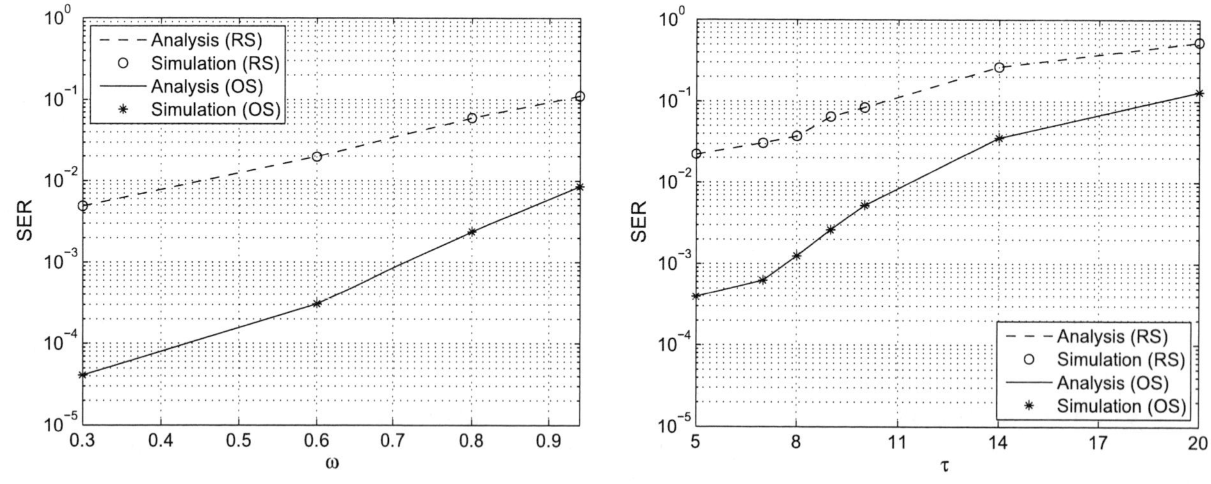

(a) SER versus ω with $\tau = 9.5$ and SNR = 35 dB. (b) SER versus τ with $\omega = 0.64$ and SNR = 33 dB.

Figure 8: SER performance comparison for different water conditions at a distance of $d = 10\,\text{m}$. 'RS' denotes random symbols and 'OS' denotes optimal symbols.

CONCLUSIONS

This paper models the GSSK VLC systems for outdoor channels. The effect of the weather and the underwater channel conditions are used to design the optimal symbol sets. For an 8-ary modulated GSSK VLC system, 12 dB SER gain is obtained using the optimal symbol set over randomly selected symbol sets for the weather parameters $T = 1$ and $q = 0.8$ at a distance of $d = 20\,\text{m}$. Similarly, a gain of 11 dB is obtained using the optimal symbol set over randomly selected symbol sets for the water condition described by $\tau = 20$ and $\omega = 0.84$ at a distance of $d = 10\,\text{m}$.

References

[1] S. P. Alaka, T. L. Narasimhan, and A. Chockalingam, "Generalized Spatial Modulation in Indoor Wireless Visible Light Communication," *Proc. 2015 IEEE Global Communications Conference (GLOBECOM)*, pp. 1–7, San Diego, CA, USA 2015.

[2] P. M. Butala, H. Elgala, and T. D. C. Little, "Performance of optical spatial modulation and spatial multiplexing with imaging receiver," *Proc. 2014 IEEE Wireless Communications and Networking Conference (WCNC)*, pp. 394–399, Istanbul 2014.

[3] W. Popoola, E. Poves, and H. Haas, "Generalised space shift keying for visible light communications," *Proc. 2012 8th International Symposium on Communication Systems, Networks and Digital Signal Processing (CSNDSP)*, pp. 1–4, Poznan 2012.

[4] Y. Sun, D. K. Borah, and E. Curry, "Optimal Symbol Set Selection in GSSK Visible Light Wireless Communication Systems," *IEEE Photonics Technology Letters*, vol. 28, pp. 303–306, Feb. 1 2016.

[5] E. Curry, D. K. Borah, and J. M. Hinojo, "Optimal Symbol Set Design for Generalized Spatial Modulations in MIMO VLC Systems," *Proc. 2016 IEEE Global Communications Conference (GLOBECOM)*, Washington D.C. 2016.

[6] J. Perez-Ramirez and D. K. Borah, "A Single-Input Multiple-Output Optical System for Mobile Communication: Modeling and Validation," *IEEE Photonics Technology Letters*, vol. 26, pp. 368–371, Feb. 15 2014.

[7] S. G. Narasimhan and S. K. Nayar, "Shedding Light on the Weather," *Proc. 2003 IEEE Computer Society Conference on Computer Vision and Pattern Recognition*, vol. 1, pp. 665–672, 2003.

[8] S. Metari and F. Deschenes, "A New Convolution Kernel for Atmospheric Point Spread Function Applied to Computer Vision," *Proc. IEEE 11th International Conference on Computer Vision*, pp. 1–8, Rio de Janeiro 2007.

[9] K. Oliver, W. Hou, and S. Wang, "Image feature detection and matching in underwater conditions," *Ocean Sensing and Monitoring - SPIE 7678*, vol. 2, 2010.

SPARSE MMSE EQUALIZER FOR GTR-STBC IN AERONAUTICAL TELEMETRY

Md. Shah Afran
Mohammad Saquib
The University of Texas at Dallas

Michael Rice
Brigham Young University

ABSTRACT

This paper investigates the performance of sparse minimum mean squared error (MMSE) equalizer for generalized time-reversed space-time block codes (GTR-STBC) in aeronautical telemetry. GTR-STBC equipped with MMSE equalizer performs the best trade-off between the signal-to-noise ratio and inter-symbol interference by allocating unequal power over aeronautical telemetry channels. However, aeronautical telemetry channels are in general consists of larger delay spreads which make the MMSE equalization of aeronautical channels with GTR-STBC computationally complex. Interestingly enough, in spite of larger delays aeronautical channels are made of few sparsely distributed multipaths and therefore their MMSE equalizers are highly compressible. In this paper, compressed sensing based greedy algorithm is used for the design of sparse MMSE equalizer and a convex curve-fitting algorithm is used to find the sub-optimum power allocation parameter at the same sparsity level for GTR-STBC. Our simulation results show that 75-90% of the non-zero equalizer taps can be reduced with a slight relaxation of the mean-squared error (or equivalently slight degradation of bit-error rate performance). It is also observed that the optimum transmitter power profile for the sparse MMSE equalizer is different than that of the non-sparse equalizer.

INTRODUCTION

Aeronautical telemetry deals with a wideband air-to-ground communication link that usually comprises of frequency selective fading channels. In addition, severe size, weight and power limitations are imposed on the airborne transmitter. Highly power-efficient transmitters, such as radio frequency (RF) power amplifiers operating in full saturation, and constant-envelope modulation schemes are employed to overcome these limitations. Since frequency selective fading leads to inter-symbol interference (ISI), techniques for mitigating ISI using the aeronautical telemetry standard IRIG 106 [1] [especially the shaped offset quaternary phase shift keying version TG

(SOQPSK-TG)] are of immediate interest. The goal therefore is one involving single-carrier, constant-envelope modulation operating in a frequency selective channel. This is clearly an equalization and/or diversity problem. Diversity reception, especially by widely separated antennas, is an obvious but expensive solution. Another feasible solution is transmit diversity when the size, weight and power constraints allow it. The most common scenario will be a system involving multiple transmit antennas with a single receive antenna, which can increase the reliability or throughput in multipath fading channels.

Generalized time-reversed space-time block codes (GTR-STBC), which is a modified version of TR-STBC and incorporates the unequal power allocation ρ, is introduced in [2]. Because TR-STBC and its generalized version involve the combination of diversity processing and equalization, a suitable minimum mean squared error (MMSE) equalization technique for SOQPSK-TG was also designed in [2]. Equipped with GTR-STBC and a suitable equalization technique, post-equalizer mean-squared error was used to identify the optimum ρ. The conceptual tool of unequal power allocation together with GTR-STBC is used as a simple transmit diversity scheme based on the partial knowledge of the channel by the transmitter. Here, the transmitter only needs to know ρ, which is easy to compute at the receiver and send back to the transmitter. This simple scheme includes transmit selection diversity ($\rho = 0$ or 1) and traditional TR-STBC ($\rho = 1/2$), and therefore, performs better than both of them.

Number of equalizer taps grows proportionally to the product of the number of input and output antennas for multiple-input multiple-output (MIMO) systems [3, 4]. Therefore, sparse equalizers will be crucial for GTR-STBC to reduce the computational complexity of equalization which is proportional to the number of non-zero equalizer taps. Sparse equalizers also offer other benefits over their non-sparse counterpart. As the number of non-zero taps is small, a sparse equalizer can adapt rapidly to changing channel conditions and can support systems with high mobility. Sparse equalizer can even perform better than non-sparse equalizer under non-ideal channel estimation. In [5], it was shown that more than 50% of the non-sparse equalizer taps has a very small contribution to the equalizer output signal-to-noise ratio (SNR). These filter taps which would be almost zero for ideal case contribute mostly noise in non-ideal channel estimation case. As a result, elimination of these taps could lead to performance improvement of the equalization.

In this paper, we explore compressed sensing based sparse equalization in the context of GTR-STBC. A joint optimization problem is designed to calculate the sparse GTR-STBC equalizer coefficients and unequal power allocation parameter ρ. The criterion for selecting the optimum ρ for sparse equalizer is to minimize the non-zero support of equalizer and also the post-equalizer mean-squared error (MSE) simultaneously given a maximum allowable loss in MSE. This behavior is illustrated in Figure 1, from which, we can also conclude that this joint optimization problem is mathematically untractable (MSE curves are in general non-convex). Therefore, we take a sub-optimum approach where we use a convex curve fitting method to calculate the sparse equalizer and also the power allocation parameter ρ for sparse GTR-STBC equalization.

Our numerical results show that a significant reduction in number of active taps can be achieved with a slight relaxation of the mean-squared error (or equivalently slight degradation of bit-error rate performance). The optimum transmitter power profile for the sparse MMSE equalizer is also observed to be different than that of the non-sparse equalizer.

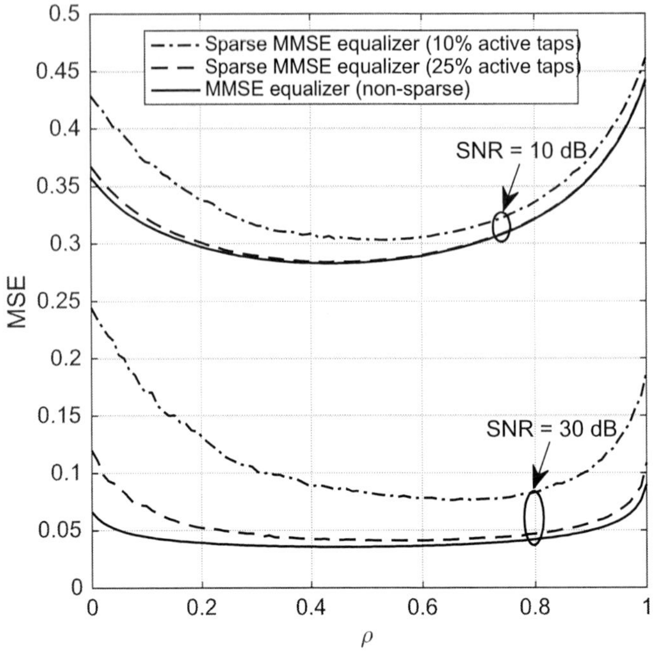

Figure 1: Plot of MSE vs power allocation parameter for GTR-STBC with both sparse and non-sparse MMSE equalizers. A pair of measured channel impulse responses from Cairns Army Airfield, Ft. Rucker, Alabama were used.

SPARSE MMSE EQUALIZATION OF SOQPSK-TG

The system considered here is summarized in Figure 2. As before, the complex-valued baseband equivalent representation [6] is used for all signals. Starting with the block diagram of Figure 2 (a), the SOQPSK-TG signal $x_c(t)$ is transmitted through a channel impulse response $h_c(t)$ whose output, accompanied by thermal noise, forms the receive signal $r_c(t)$. After the application of an anti-aliasing low-pass filter with impulse response $h_a(t)$, T-spaced samples of $r_c(t)$ are produced by an A/D converter. Assuming the anti-aliasing filter does not distort the received signal, the samples of the received signal may be expressed as

$$r(n) = x(n) * h(n) + w(n) = \sum_{k=-L_1}^{L_2} h(k)x(n-k) + w(n), \tag{1}$$

where

$$r(n) = r_c(nT), \qquad x(n) = x_c(nT), \qquad h(n) = h_c(t) * h_a(t)\big|_{t=nT}.$$

In (1) $w(n)$ is the n-th sample of a zero-mean complex-valued Gaussian random sequence with autocorrelation function

$$R_w(k) = \frac{1}{2}\mathrm{E}\left\{w(n)w^*(n-k)\right\} = \sigma_w^2\delta(k). \tag{2}$$

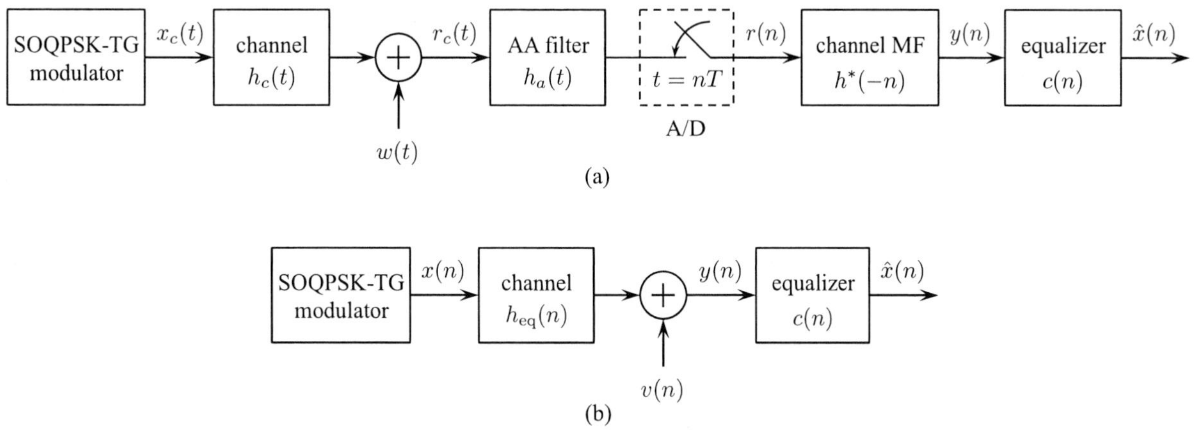

Figure 2: A block diagram of the system that applies an MMSE equalizer to samples of a received SOQPSK-TG signal: (a) the system showing the continuous-time signals, the anti-aliasing filter, and A/D converter; (b) the equivalent discrete-time system.

Note that (1) assumes the discrete-time channel has support on $-L_1 \leq n \leq L_2$. As a first step, we apply the samples $r(n)$ to a filter matched to the discrete-time channel to produce $y(n)$:

$$\begin{aligned} y(n) &= r(n) * h^*(-n) \\ &= x(n) * \underbrace{h(n) * h^*(-n)}_{h_{\text{eq}}(n)} + \underbrace{w(n) * h^*(-n)}_{v(n)} \quad (3) \\ &= \sum_{k=-L_{\text{eq}}}^{L_{\text{eq}}} h_{\text{eq}}(k) x(n-k) + v(n), \quad (4) \end{aligned}$$

where $L_{\text{eq}} = L_1 + L_2$ and $v(n)$ is a complex valued Gaussian random sequence with zero mean and autocorrelation function

$$R_v(k) = \frac{1}{2} \mathrm{E}\left\{ v(n) v^*(n-k) \right\} = \sigma_w^2 h_{\text{eq}}(k). \quad (5)$$

The samples $y(n)$ form the input to an MMSE equalizer. The MMSE equalizer is an FIR filter with coefficients $c(n)$ for $-L_c \leq n \leq L_c$ designed to minimize the mean squared error between the equalizer filter output $\hat{x}(n)$ and the sequence $x(n)$. The entire system may be represented by an equivalent discrete-time system shown in Figure 2 (b).

Following the development in [2], the relationship between $x(n)$ and the equalizer output $\hat{x}(n)$ is

$$\hat{x}(n) = c(n) * y(n) = \sum_{k=-L_c}^{L_c} c(k) y(n-k). \quad (6)$$

The vector of filter coefficients that minimizes the mean squared error

$$\mathcal{E} = \mathrm{E}\left\{ \left| x(n) - \hat{x}(n) \right|^2 \right\}, \quad (7)$$

is given by [2]

$$\mathbf{c} = \left[\mathbf{G}\mathbf{G}^\dagger + \frac{2}{E_b/N_0}\mathbf{H}_{\text{eq}}\right]^{-1} \mathbf{g}^\dagger, \tag{8}$$

where $\mathbf{c}$ is the $(2L_c+1)\times 1$ vector of filter coefficients, $\mathbf{G}$ is the $(2L_c+1)\times(2L_c+1+2L_{\text{eq}})$ matrix

$$\mathbf{G} = \begin{bmatrix} h_{\text{eq}}(L_{\text{eq}}) & \cdots & h_{\text{eq}}(-L_{\text{eq}}) & & & \\ & h_{\text{eq}}(L_{\text{eq}}) & \cdots & h_{\text{eq}}(-L_{\text{eq}}) & & \\ & & \ddots & & & \\ & & & h_{\text{eq}}(L_{\text{eq}}) & \cdots & h_{\text{eq}}(-L_{\text{eq}}) \end{bmatrix}; \tag{9}$$

$\mathbf{H}_{\text{eq}}$ is the $(2L_c+1)\times(2L_c+1)$ matrix given by

$$\mathbf{H}_{\text{eq}} = \begin{bmatrix} h_{\text{eq}}(0) & \cdots & h_{\text{eq}}(-2L_c) \\ \vdots & & \vdots \\ h_{\text{eq}}(2L_c) & \cdots & h_{\text{eq}}(0) \end{bmatrix}, \tag{10}$$

where it is understood that $h_{\text{eq}}(k) = 0$ for $|k| > L_{\text{eq}}$.

Vector $\mathbf{g}$ is the $1\times(2L_c+1)$ vector given by

$$\mathbf{g} = \left[\mathbf{0}_{1\times(L_c-L_{\text{eq}})} \;\; \mathbf{h}_{\text{eq}} \;\; \mathbf{0}_{1\times(L_c-L_{\text{eq}})}\right], \tag{11}$$

where $\mathbf{0}_{1\times(L_c-L_{\text{eq}})}$ is a row vector comprising $L_c - L_{\text{eq}}$ zeros (here we assume $L_c > L_{\text{eq}}$, i.e., the equalizer is longer than the channel), and $\mathbf{h}_{\text{eq}}$ is the $1\times(2L_{\text{eq}}+1)$ vector given by

$$\mathbf{h}_{\text{eq}} = \left[h_{\text{eq}}(L_{\text{eq}}) \;\; \cdots \;\; h_{\text{eq}}(0) \;\; \cdots \;\; h_{\text{eq}}(-L_{\text{eq}})\right]. \tag{12}$$

The corresponding mean squared error is

$$\mathcal{E}_{\text{min}} = \sigma_x^2\left(1 - \mathbf{g}\left[\mathbf{G}\mathbf{G}^\dagger + \frac{2}{E_b/N_0}\mathbf{H}_{\text{eq}}\right]^{-1}\mathbf{g}^\dagger\right). \tag{13}$$

Now for the design of sparse equalizer from (7), at first we write $\mathcal{E}$ as a functions of $\mathbf{c}_s$. Putting $\left[\mathbf{G}\mathbf{G}^\dagger + \frac{2}{E_b/N_0}\mathbf{H}_{\text{eq}}\right] = \mathbf{M}$ we can write [4]

$$\mathcal{E}(\mathbf{c}_s) = \sigma_x^2\left(\underbrace{1 - \mathbf{g}\mathbf{M}^{-1}\mathbf{g}^\dagger}_{\mathcal{E}_{\text{min}}/\sigma_x^2} + \underbrace{(\mathbf{c}_s - \mathbf{c})\mathbf{M}(\mathbf{c}_s - \mathbf{c})^\dagger}_{\mathcal{E}_{\text{ex}}}\right). \tag{14}$$

Here, $\mathbf{c} = \mathbf{M}^{-1}\mathbf{g}^\dagger$ and $\mathcal{E}_{\text{ex}}$ is the term which represents the increase in normalized mean squared error due to the sparse equalizer $\mathbf{c}_s$. This term is equal to zero when we use equalizer $\mathbf{c}$ instead, but MMSE equalizer is generally non-sparse. Therefore, we allow a tradeoff between performance and complexity by $\mathcal{E}_{\text{ex}}$. Interestingly enough, results from sparse equalization literature ([4] and references therein) show that, in general a slight relaxation of $\mathcal{E}$ through $\mathcal{E}_{ex}$ can achieve highly sparse filters. Using either Cholesky factorization or eigen decomposition of matrix $\mathbf{M}$, we can write $\mathbf{M} = \mathbf{M}_1\mathbf{M}_1^\dagger$, which formulates our sparse equalizer design problem as

$$\hat{\mathbf{c}}_s = \underset{\mathbf{c}_s \in \mathbb{C}^{2L_c+1}}{\text{argmin}} \|\mathbf{c}_s\|_0 \quad \text{subject to} \quad \|\mathbf{M}_1^\dagger(\mathbf{c}_s - \mathbf{c})\|_2^2 \leq \epsilon, \tag{15}$$

where $\|\mathbf{M}_1^\dagger(\mathbf{c}_s - \mathbf{c})\|_2^2 = \mathcal{E}_{ex}$ and $\epsilon > 0$ is a design parameter which amounts the relaxation in MSE for the design of sparse filter. Here $\|\mathbf{c}_s\|_0$ is used to denote the number of nonzero entries of $\mathbf{c}_s$ while $\|\cdot\|_2^2$ denotes square of l_2 norm of a vector. However, in general, the optimal design of sparse filter is NP-hard [7]. Therefore, we will use sub-optimum algorithms described in [4]. In this paper, we use orthogonal matching pursuit (OMP) algorithm for sparse equalizer design as it gives us better control on calculating nonzero taps [8]. We denote the OMP operation by $\hat{\mathbf{c}}_s = \text{OMP}(\mathbf{b}, \mathbf{A}, \text{stopping criterion})$, where $\mathbf{b}$ is a known data vector and $\mathbf{A}$ denotes the sensing matrix. OMP is a greedy algorithm which tries to iteratively select a set of columns from $\mathbf{A}$ that are most correlated with $\mathbf{b}$ and then it solves a restricted least-squares problem using the selected columns. The stopping criterion can be a predefined sparsity level (number of nonzero entries) of $\hat{\mathbf{c}}_s$ or the upper-bound on loss in MSE (ϵ). In our numerical study we will use the first criterion. Using either $\mathbf{M} = \mathbf{M}_1\mathbf{M}_1^\dagger$ (Cholesky factorization) or $\mathbf{M} = \mathbf{U}\mathbf{\Lambda}\mathbf{U}^\dagger$ (eigen decomposition) with $\mathbf{c} = \mathbf{M}^{-1}\mathbf{g}^\dagger$ in (15), we can make following call to the OMP algorithm for the calculation of $\mathbf{c}_s$

$$\hat{\mathbf{c}}_s = \text{OMP}\left(\mathbf{M}_1^{-1}\mathbf{g}^\dagger, \mathbf{M}_1^\dagger, \|\hat{\mathbf{c}}_s\|_0 = k\right). \tag{16}$$

SPARSE MMSE EQUALIZATION FOR GENERALIZED TR-STBC (GTR-STBC)

An abstraction (to the sample level) for a 2×1 TR-STBC system is illustrated in Figure 3. Here the system transmits $2L$ samples of an SOQPSK-TG signal sampled at N samples/bit over two transmit antennas to one receive antenna. The equivalent discrete-time channel between transmit antenna 1 and the receive antenna is represented by the impulse response $h_1(n)$ for $-L_{1,1} \leq n \leq L_{1,2}$ whereas the equivalent discrete-time channel between transmit antenna 2 and the receive antenna is represented by the impulse response $h_2(n)$ for $-L_{2,1} \leq n \leq L_{2,2}$.

The TR-STBC encoder partitions the sample sequence $x(0), \ldots, x(2L-1)$ into two sequences $x_1(n)$ and $x_2(n)$ as shown in Figure 3. The length-$2L$ packet is transmitted in two intervals[1] each spanning L sample intervals. During the first interval $x_1(0), \ldots, x_1(L-1)$ is transmitted from antenna 1 whereas $x_2(0), \ldots, x_2(L-1)$ is transmitted from antenna 2. During the second interval, $x_2^*(L-1), \ldots, x_2^*(0)$ is transmitted from antenna 1 whereas $-x_1^*(L-1), \ldots, -x_1^*(0)$ is transmitted from antenna 2.

Power division using $0 \leq \rho \leq 1$ is accomplished by modified the TR-STBC system along the lines illustrated in Figure 3. Amplitude scaling is applied to the signals entering each channel so as to divide the power between the channels. Here ρ represents the proportion of total power allocated

[1]In a practical implementation, a guard interval at least as long as the longest channel impulse response must be inserted between the two intervals. Here, such an interval is assumed, although we won't complicate the notation to make this explicit.

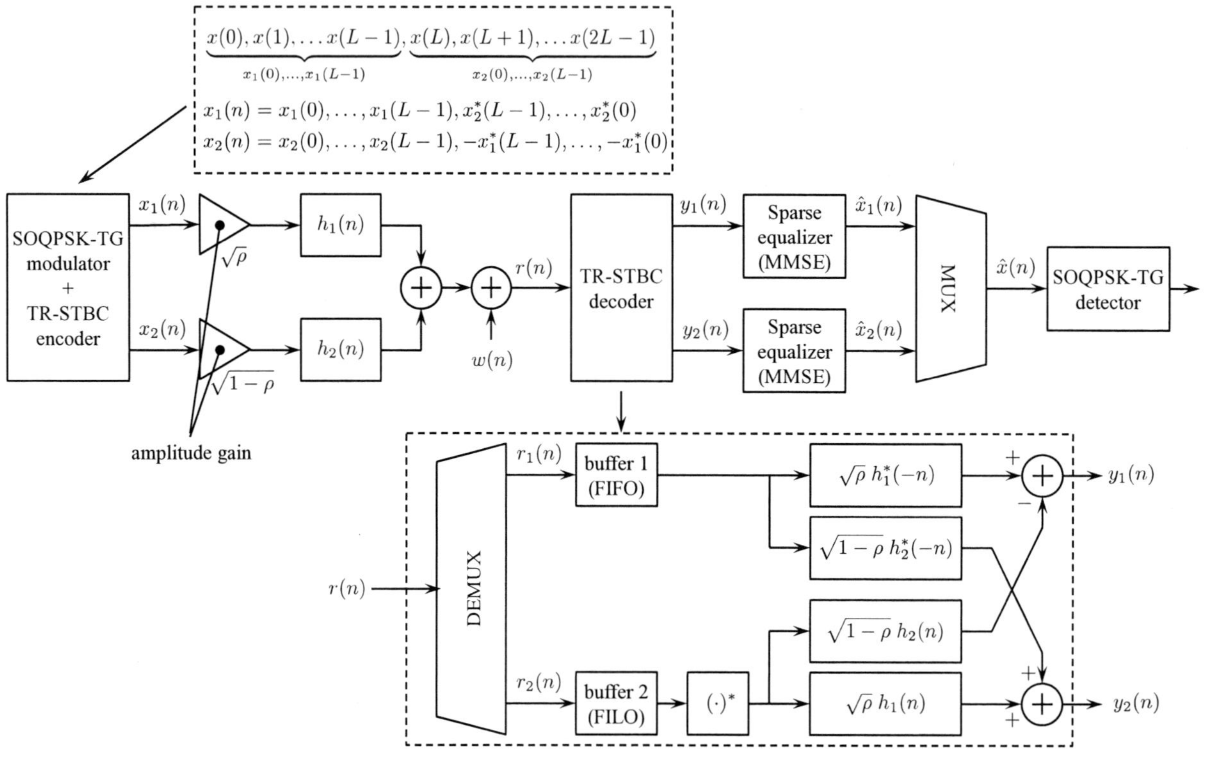

Figure 3: A block diagram of the TR-STBC system based on sparse equalizer and unequal power allocation using $0 \leq \rho \leq 1$.

to transmit antenna 1. The traditional TR-STBC system[2] is a special case for which $\rho = 1/2$. The square-root is used in Figure 3 because the amplitudes are what are being modified—the energy (or power) is the square of the amplitude.

The received signal $r(n)$ is given by

$$r(n) = \sqrt{\rho}\, x_1(n) * h_1(n) + \sqrt{1-\rho}\, x_2(n) * h_2(n) + w(n), \tag{17}$$

where $w(n)$ is a complex-valued Gaussian random sequence with zero mean and autocovariance function given by (2). The TR-STBC decoder partitions $r(n)$ into $r_1(n)$ and $r_2(n)$ as follows:

$$\begin{aligned} r_1(n) &= r(n) \text{ for } 0 \le n \le L-1 \\ r_2(n-L) &= r(n) \text{ for } L \le n \le 2L-1. \end{aligned} \tag{18}$$

These two sequences are given by

$$r_1(n) = \sqrt{\rho}\, x_1(n) * h_1(n) + \sqrt{1-\rho}\, x_2(n) * h_2(n) + w_1(n) \tag{19}$$

$$r_2(n) = \sqrt{\rho}\, x_2^*(-n) * h_1(n) - \sqrt{1-\rho}\, x_1^*(-n) * h_2(n) + w_2(n), \tag{20}$$

where $w_1(n)$ and $w_2(n)$ are related to $w(n)$ the same way $r_1(n)$ and $r_2(n)$ are related to $r(n)$. The TR-STBC decoder processes $r_1(n)$ and $r_2(n)$ using a bank of filters based on the channel impulse responses $h_1(n)$ and $h_2(n)$ as shown. The result of this processing is a pair of parallel sequences $y_1(n)$ and $y_2(n)$ which may be expressed as

$$\begin{aligned} y_1(n) &= r_1(n) * \sqrt{\rho}\, h_1^*(-n) - r_2^*(-n) * \sqrt{1-\rho}\, h_2(n) \\ &= x_1(n) * \underbrace{\left[\rho\, h_1(n) * h_1^*(-n) + (1-\rho)\, h_2(n) * h_2^*(-n)\right]}_{h_{\text{eq}}(n)} \\ &\qquad + \underbrace{w_1(n) * \sqrt{\rho}\, h_1^*(-n) + w_2^*(-n) * \sqrt{1-\rho}\, h_2(n)}_{v_1(n)}, \end{aligned} \tag{21}$$

and

$$\begin{aligned} y_2(n) &= r_1(n) * \sqrt{1-\rho}\, h_2^*(-n) + r_2^*(-n) * \sqrt{\rho}\, h_1(n) \\ &= x_2(n) * \underbrace{\left[(1-\rho)\, h_2(n) * h_2^*(-n) + \rho\, h_1^*(-n) * h_1(n)\right]}_{h_{\text{eq}}(n)} \\ &\qquad + \underbrace{\sqrt{1-\rho}\, w_1(n) * h_2^*(-n) + \sqrt{\rho}\, w_2^*(-n) * h_1(n)}_{v_2(n)}. \end{aligned} \tag{22}$$

These equations show that the equivalent composite channel for non-equal power allocation is

$$h_{\text{eq}}(n) = \rho\, \underbrace{h_1(n) * h_1^*(-n)}_{\eta_1(n)} + (1-\rho)\, \underbrace{h_2(n) * h_2^*(-n)}_{\eta_2(n)}. \tag{23}$$

[2] In the traditional TR-STBC system, $\rho = 1/2$ is included in neither the development nor the notation because the same power is assumed to be applied to each channel. Hence there is no need to account for it, other than in normalizing the noise variance.

Because the support for $h_1(n)$ is $-L_{1,1} \le n \le L_{1,2}$, the support for $\eta_1(n)$ is $-(L_{1,1}+L_{1,2}) \le n \le (L_{1,1}+L_{1,2})$. Similarly because the support for $h_2(n)$ is $-L_{2,1} \le n \le L_{2,2}$, the support for $\eta_2(n)$ is $-(L_{2,1}+L_{2,2}) \le n \le (L_{2,1}+L_{2,2})$. Consequently, the support for $h_{\text{eq}}(n)$ is $-L_{\text{eq}} \le n \le L_{\text{eq}}$ where

$$L_{\text{eq}} = \max\left\{L_{1,1}+L_{1,2}, L_{2,1}+L_{2,2}\right\}. \tag{24}$$

The noise sequences $v_1(n)$ and $v_2(n)$ are complex-valued Gaussian random sequences each with zero mean and autocorrelation and cross correlation functions

$$\frac{1}{2}\mathrm{E}\left\{v_1(n)v_1^*(n-k)\right\} = \frac{1}{2}\mathrm{E}\left\{v_2(n)v_2^*(n-k)\right\} = \sigma_w^2 h_{\text{eq}}(k). \tag{25}$$

$$\mathrm{E}\left\{v_1(n)v_2^*(n-k)\right\} = 0. \tag{26}$$

By way of summary, the TR-STBC system presents to the equalizers the sequences $y_1(n)$ and $y_2(n)$ which may be represented by

$$y_1(n) = x_1(n) * h_{\text{eq}}(n) + v_1(n), \tag{27}$$
$$y_2(n) = x_2(n) * h_{\text{eq}}(n) + v_2(n), \tag{28}$$

where $h_{\text{eq}}(n)$ is given by (23). The noise terms $v_1(n)$ and $v_2(n)$ are uncorrelated zero-mean Gaussian random sequences each with autocorrelation function (25). A pair of equalizers operate in parallel on $y_1(n)$ and $y_2(n)$. Because the the noise sequences $v_1(n)$ and $v_2(n)$ are statistically equivalent and $h_{\text{eq}}(n)$ is common to both, the pair of equalizers operating on $y_1(n)$ and $y_2(n)$ are identical as long as $a_1(n)$ and $a_2(n)$ are statistically equivalent (the usual case). Any equalizer can be applied here (linear or non-linear, with or without noise whitening) with the usual performance-complexity tradeoffs. Here, we apply the approximate MMSE equalizer introduced in the previous section because the MMSE equalizer permit a mathematically tractable analysis for the resulting mean-squared error. We leverage the analytical expression to find the value of ρ that minimizes the mean squared error.

The equalizer filters of Figure 3 are identical, and the vector of filter coefficients is given by (8) and the corresponding mean squared error is given by (13). In these equations, $\mathbf{G}$, $\mathbf{g}$, and $\mathbf{H}_{\text{eq}}$ may be expressed in terms of power allocation parameter ρ. Using (23), it is straightforward to show that

$$\mathbf{G} = \rho\mathbf{G}_1 + (1-\rho)\mathbf{G}_2, \tag{29}$$
$$\mathbf{g} = \rho\mathbf{g}_1 + (1-\rho)\mathbf{g}_2, \tag{30}$$
$$\mathbf{H}_{\text{eq}} = \rho\mathbf{H}_1 + (1-\rho)\mathbf{H}_2, \tag{31}$$

where $\mathbf{G}_1$, $\mathbf{g}_1$, and $\mathbf{H}_1$ are formed from $\eta_1(n)$ the same way $\mathbf{G}$, $\mathbf{g}$, and $\mathbf{H}_{\text{eq}}$ are formed from $h_{\text{eq}}(n)$, respectively. Similar definitions apply to $\mathbf{G}_2$, $\mathbf{g}_2$, and $\mathbf{H}_2$ with $\eta_2(n)$. Making the substitutions for $\mathbf{G}(\rho)$, $\mathbf{g}(\rho)$, and $\mathbf{H}_{\text{eq}}(\rho)$ gives

$$\mathbf{c}(\rho) = \mathbf{M}(\rho)^{-1}\mathbf{g}(\rho)^{\dagger}, \tag{32}$$

and

$$\mathcal{E}_{\text{min}}(\rho) = \sigma_x^2\left[1 - \mathbf{g}(\rho)\mathbf{M}(\rho)^{-1}\mathbf{g}(\rho)^{\dagger}\right]. \tag{33}$$

Here $\mathbf{c}(\rho)$ is non-sparse MMSE equalizer and $\mathcal{E}_{\text{min}}(\rho)$ is the MSE, where

$$\mathbf{M}(\rho) = \left[\mathbf{G}(\rho)\mathbf{G}(\rho)^{\dagger} + \frac{2}{E_b/N_0}\mathbf{H}_{\text{eq}}(\rho)\right]. \tag{34}$$

Let $\rho_{\text{opt,ns}}$ is the optimum ρ for non-sparse equalizer which minimizes (33). Therefore, $\mathcal{E}_{\text{min}}(\rho_{\text{opt,ns}})$ is the minimum attainable MSE of the system with equalizer $\mathbf{c}(\rho_{\text{opt,ns}})$. Now, following the development in the previous section, for sparse GTR-STBC equalizer $\mathbf{c}_{\text{s}}(\rho)$, we can find MSE as a function of both ρ and $\mathbf{c}_{\text{s}}(\rho)$

$$\mathcal{E}(\rho, \mathbf{c}_{\text{s}}(\rho)) = \sigma_x^2\left[1 - \mathbf{g}(\rho)\mathbf{M}(\rho)^{-1}\mathbf{g}(\rho)^{\dagger} + \left(\mathbf{c}_{\text{s}}(\rho) - \mathbf{c}(\rho)\right)\mathbf{M}(\rho)\left(\mathbf{c}_{\text{s}}(\rho) - \mathbf{c}(\rho)\right)^{\dagger}\right], \tag{35}$$

where similar to (16) $\mathbf{c}_{\text{s}}(\rho)$ can be computed by following call to the OMP algorithm

$$\hat{\mathbf{c}}_{\text{s}}(\rho) = \text{OMP}\left(\mathbf{M}_1^{-1}(\rho)\mathbf{g}(\rho)^{\dagger}, \mathbf{M}_1^{\dagger}(\rho), \|\hat{\mathbf{c}}_{\text{s}}(\rho)\|_0 = k\right). \tag{36}$$

Since $\mathcal{E}_{\text{min}}(\rho_{\text{opt,ns}})$ is the minimum attainable MSE, the term $\mathcal{E}_{\text{excess}}$ which represents the increase of MSE for GTR-STBC in dB will be calculated as

$$\mathcal{E}_{\text{excess}}(\text{dB}) = 10\log_{10}\left(\frac{\mathcal{E}(\rho_{\text{opt}}, \mathbf{c}_{\text{s}}(\rho_{\text{opt}}))}{\mathcal{E}_{\text{min}}(\rho_{\text{opt,ns}})}\right). \tag{37}$$

However, unlike the MSE for non-sparse case as in (33), MSE for sparse equalization as given in (35) is mathematically untractable due to the non-convex nature of MSE function as shown in Figure 1. Therefore, finding the optimum ρ for sparse equalization is computationally challenging. As a result, we have taken a sub-optimum approach to calculate ρ_{opt} which will try to minimize $\mathcal{E}(\rho, \mathbf{c}_{\text{s}}(\rho))$. Our algorithm fits a convex curve after calculating $\mathcal{E}(\rho, \mathbf{c}_{\text{s}}(\rho))$ at few candidate values of ρ chosen from 0 to 1 with an interval of 0.25 (along with $\rho_{\text{opt,ns}}$). Later we use this convex curve to seek for lower MSE value.

NUMERICAL RESULTS

The forgoing analysis was applied to a helicopter-to-ground radio link using 1000 channel impulse responses from Cairn Army Airfield, Ft. Rucker, Alabama, captured during the channel sounding experiment described in [9]. For the modulation, we assume 10 Mbit/s SOQPSK-TG with the equalizers operating at 2 samples/bit. In the results shown below, $h_1(n)$ is the equivalent discrete-time channel between the nose antenna and the receive antenna and $h_2(n)$ is the channel between the tail antenna and the receive antenna.

Natural normalization is applied to the channels. Let $h_{1,\text{u}}(n)$ and $h_{2,\text{u}}(n)$ be unnormalized channel impulse responses for the two equivalent discrete-time channels obtained directly from the channel sounding data, and let

$$E_1 = \sum_{n=-L_{1,1}}^{L_{1,2}} |h_{1,\text{u}}(n)|^2; \qquad E_2 = \sum_{n=-L_{2,1}}^{L_{2,2}} |h_{2,\text{u}}(n)|^2, \tag{38}$$

be the energies in two channels. The natural normalization uses

$$h_1(n) = \frac{1}{\sqrt{E}} h_{1,\mathrm{u}}(n); \qquad h_2(n) = \frac{1}{\sqrt{E}} h_{2,\mathrm{u}}(n), \tag{39}$$

where $E = \max\{E_1, E_2\}$. This normalizes the stronger of the two channels to unit energy. We call this the natural normalization because in real multi-antenna scenarios, it is often the case that one of the channels is stronger than the other.

The numerical results were produced as follows. The values of ρ that minimize (33) and (35) for $L_c = 5 \times L_{\mathrm{eq}}$ [10] were numerically computed for non-sparse and sparse equalizers, respectively. After that, (37) is used to calculate $\mathcal{E}_{\mathrm{excess}}$ (dB). The corresponding results are shown in Figures 4 and 5, for $E_b/N_0 = 10$ dB and $E_b/N_0 = 16$ dB, respectively. Figure 4 shows that, at $E_b/N_0 = 10$ dB, a reduction of 75% equalizer taps is possible with a relaxation in MSE of $<$ 0.06 dB. A relaxation in MSE of around 0.4 dB would allow us to eliminate more than 90% of the equalizer taps for the same E_b/N_0. However, for $E_b/N_0 = 16$ dB, Figure 5 indicates that more relaxation of MSE (more performance degradation) is required for similar sparsity levels. This is consistent with findings in [5], where it is shown that as E_b/N_0 increases the sparse equalizers need more active taps to maintain the similar performance in output SNR.

Figures 6 and 7 plots the optimum power profiles at $E_b/N_0 = 10$ dB and $E_b/N_0 = 16$ dB, respectively. We observe that the profile of ρ is different for sparse equalizers than the non-sparse MMSE equalizers. The profile of ρ for sparse equalizers also depend on the sparsity level (or equivalently on increase of MSE). Figure 1 explains this curious behaviour where it is observed that an increase of sparsity level changes the MSE curves along with the optimum choice of ρ.

In Figure 8, we compare the bit error rate (BER) of our proposed sparse MMSE equalizers with non-sparse MMSE equalizers for GTR-STBC. It is interesting to note from Figure 8 that at a 25% sparsity level both sparse and non-sparse equalizers have almost similar BER performances for E_b/N_0 of 10-12 dB. With a target BER of 10^{-4}, sparse equalizers with 25% active taps require only about 0.1 dB relaxation in E_b/N_0. For similar performance with a sparsity level of 10% about 1.5 dB relaxation is required.

CONCLUSIONS

In this paper, we investigated the performance of sparse MMSE equalizers for GTR-STBC using a set of measured channel impulse responses collected from helicopter-to-ground channel sounding experiments. Unlike the non-sparse case, sparse MMSE equalizers produce a non-convex MSE function which is mathematically untractable. Therefore, we proposed a convex curve-fitting based algorithm to find the ρ that sub-optimally minimizes MSE for sparse equalizers. Simulation results showed that at low to moderate SNR a huge reduction of non-zero equalizer taps can be achieved at the cost of a slight relaxation of MSE. It is also observed that depending on the system conditions the optimum value of ρ is very sensitive to the sparsity level of the equalizers.

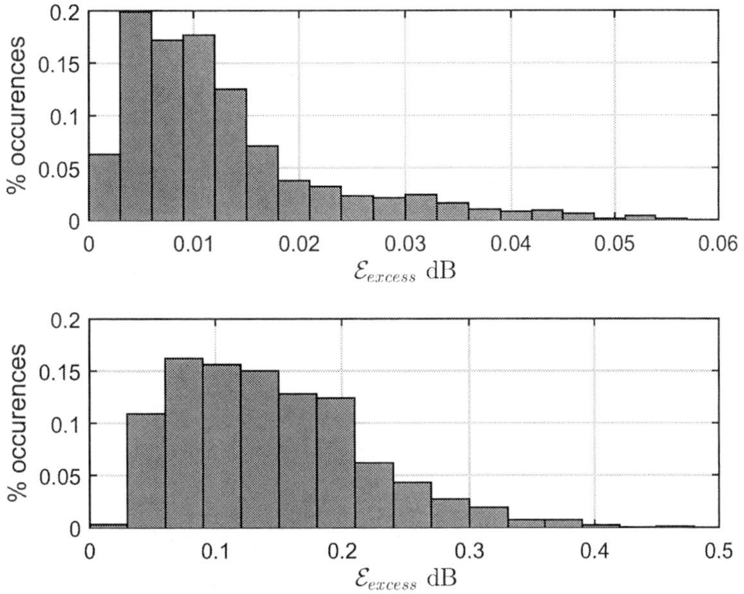

Figure 4: Increase of MSE with $E_b/N_0 = 10$ dB for 1000 pairs of measured channel impulse responses from Cairns Army Airfield, Ft. Rucker, Alabama: (top) sparse MMSE equalizer with 25% active taps; (bottom) sparse MMSE equalizer with 10% active taps.

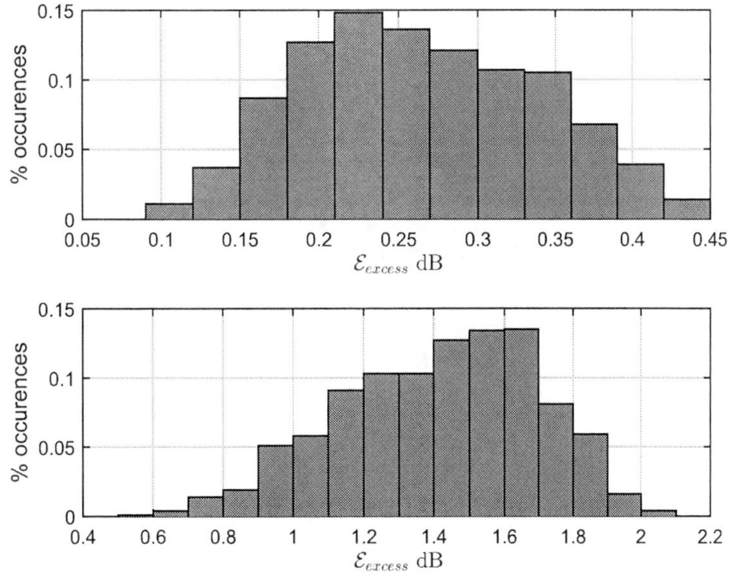

Figure 5: Increase of MSE with $E_b/N_0 = 16$ dB for 1000 pairs of measured channel impulse responses from Cairns Army Airfield, Ft. Rucker, Alabama: (top) sparse MMSE equalizer with 25% active taps; (bottom) sparse MMSE equalizer with 10% active taps.

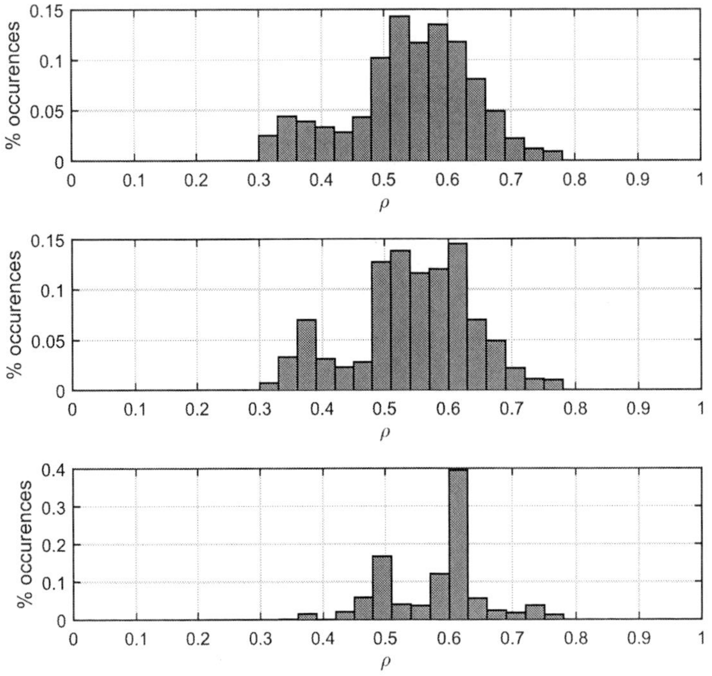

Figure 6: Optimum power allocations in the mean-squared error sense with $E_b/N_0 = 10$ dB for 1000 pairs of measured channel impulse responses from Cairns Army Airfield, Ft. Rucker, Alabama: (top) non-sparse MMSE equalizer; (middle) sparse MMSE equalizer with 25% active taps; (bottom) sparse MMSE equalizer with 10% active taps.

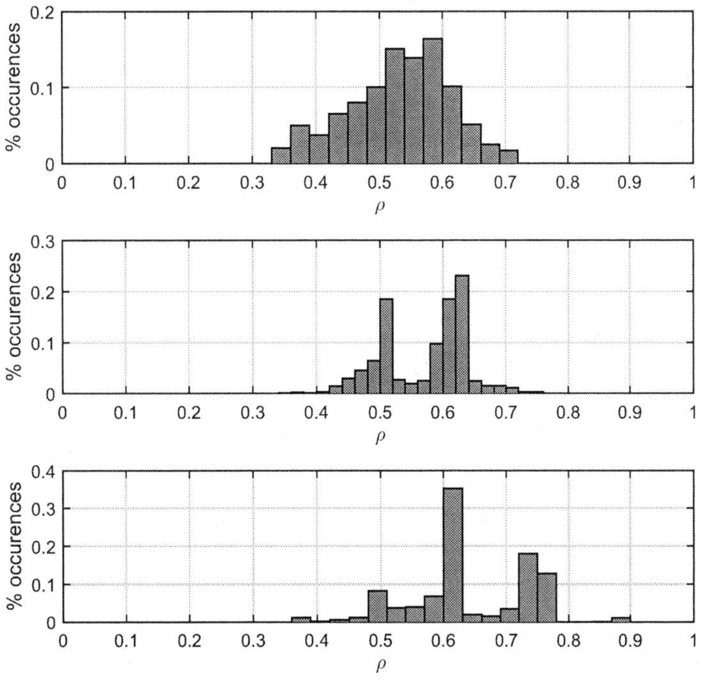

Figure 7: Optimum power allocations in the mean-squared error sense with $E_b/N_0 = 16$ dB for 1000 pairs of measured channel impulse responses from Cairns Army Airfield, Ft. Rucker, Alabama: (top) non-sparse MMSE equalizer; (middle) sparse MMSE equalizer with 25% active taps; (bottom) sparse MMSE equalizer with 10% active taps.

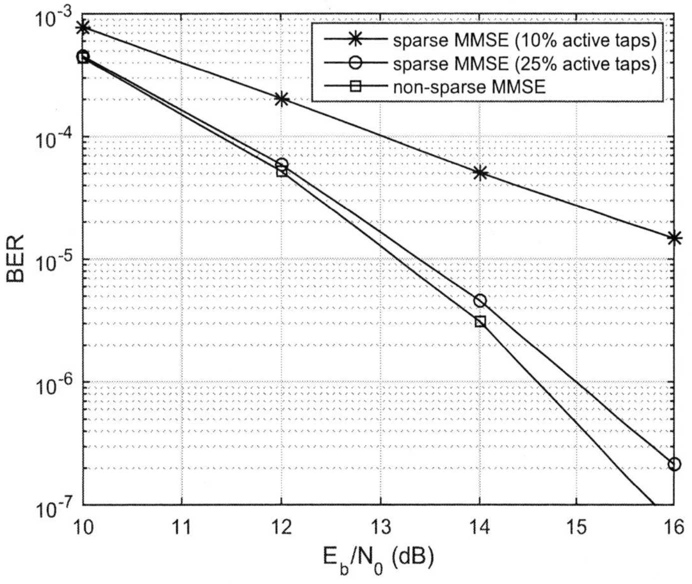

Figure 8: Comparison of average simulated BER for 1000 randomly selected channels from 39300 channel impulse responses measured at Cairns Army Airfield, Ft. Rucker, Alabama.

ACKNOWLEDGEMENTS

This project is funded by the Test Resource Management Center (TRMC) Test and Evaluation, Science & Technology (T&E/S&T) Program through the U.S. Army Program Executive Office for Simulation, Training, and Instrumentation (PEO STRI) under Contract No. W900KK-13-C-0026 (PAQ).

REFERENCES

[1] *IRIG Standard 106-04: Telemetry Standards*, Range Commanders Council Telemetry Group, Range Commanders Council, White Sands Missile Range, New Mexico, 2004, (Available on-line at http://www.wsmr.army.mil/RCCsite/Pages/Publications.aspx).

[2] M. Rice, M. Afran, and M. Saquib, "Equalization in aeronautical telemetry using multiple transmit antennas," *IEEE Transactions on Aerospace and Electronic Systems*, vol. 51, no. 5, pp. 2148 – 2165, jul 2015.

[3] D. Wei, C. Sestok, and A. Oppenheim, "Sparse filter design under a quadratic constraint: Low-complexity algorithms," *IEEE Transactions on Signal Processing*, vol. 61, no. 4, 2013.

[4] A. Gomaa and N. Al-Dhahir, "A new design framework for sparse FIR MIMO equalizers," *IEEE Transactions on Communications*, vol. 59, no. 8, 2011.

[5] G. Kutz and D. Raphaeli, "Determination of tap positions for sparse equalizers," *IEEE Transactions on Communications*, vol. 55, no. 9, 2007.

[6] J. Proakis and M. Salehi, *Digital Communications*, 5th ed. New York: McGraw-Hill, 2007.

[7] B. Natarajan, "Sparse approximate solutions to linear systems," *SIAM Journal on Computing*, vol. 24, 1995.

[8] J. Tropp and A. Gilbert, "Signal recovery from random measurements via orthogonal matching pursuit," *IEEE Transactions on Information Theory*, vol. 53, no. 12, pp. 4655–4666, 2007.

[9] M. Rice and M. Jensen, "Mulitpath propagation for helicopter-to-ground MIMO links," in *Proceedings of the IEEE Military Communications Conference*, Baltimore, MD, October 2011.

[10] J. Treichler, I. Fijalkow, and C. Johnson, "Fractionally spaced equalizers: How long should they be?" *IEEE Signal Processing Magazine*, May 1996.

ON THE DESIGN OF A SQUARE-ROOT NYQUIST PULSE SHAPING FILTER FOR AERONAUTICAL TELEMETRY

Chad Josephson
Brigham Young University
Department of Electrical and Computer Engineering
Provo, UT, 84602
chadcjosephson@gmail.com

Faculty Advisor:
Dr. Michael Rice

ABSTRACT

Bandwidth efficient linear modulations require a pulse shape with finite support in the time domain while simultaneously achieving good spectral containment in the frequency domain. The square-root Nyquist pulse achieves zero intersymbol interference (ISI) at its matched-filter output but does so with infinite support in the time domain. This paper investigates three different methods for generating an FIR approximation of a square-root Nyquist pulse.

INTRODUCTION

In an effort to improve spectral efficiency in aeronautical telemetry, linear, non-binary modulations are now being investigated. The transmitter comprises an in-phase and quadrature modulator with a pulse shape as shown in Figure 1. The look-up tables (LUTs) store the constellation points. Conceptually, the LUT outputs are thought of as a pair of weighted impulse trains with period T_s. Pulse shaping is performed by filtering the weighted impulse trains with a pair of identical filters whose impulse responses are the desired pulse shape. The result is a pair of weighted pulse trains that form the baseband inputs to a quadrature mixer as shown.

The pulse shape determines the spectrum of the signal. Pulse shapes with finite support in the time domain have infinite support in the frequency domain and are therefore not attractive from the spectral occupancy point of view. Pulse shapes with finite support in the frequency domain have infinite support in the time domain. Pulse shapes that span more than one symbol introduce the possibility of intersymbol interference (ISI). Pulse shapes with infinite support in the time domain whose corresponding matched-filter output has no ISI must satisfy the Nyquist no-ISI criterion [1].

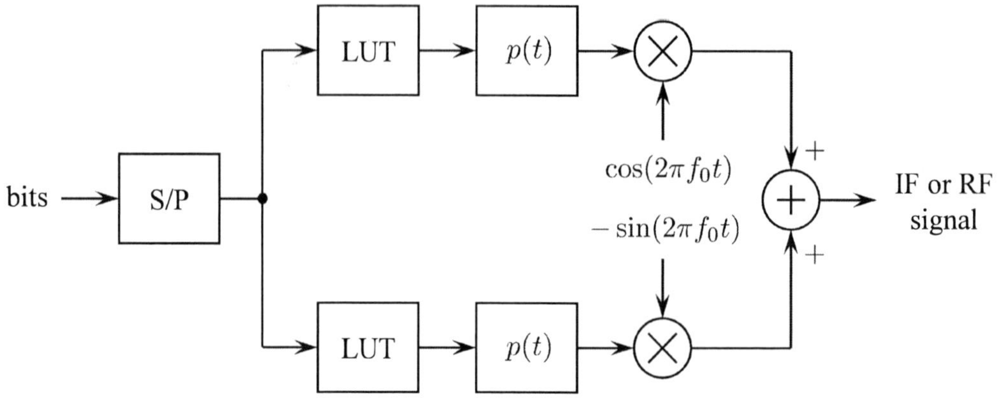

Figure 1: A block diagram of an I/Q modulator used for linear modulation. The look-up tables (LUT) store the constellation points and the filter with impulse response $p(t)$ performs the pulse shaping.

The most popular class of pulse shapes that satisfy the Nyquist no-ISI criterion is the square-root raised-cosine (SRRC) given by

$$p_{\text{ideal}}(t) = \frac{1}{\sqrt{T_s}} \frac{\sin\left(\pi(1-\alpha)\frac{t}{T_s}\right) + \frac{4\alpha t}{T_s}\cos\left(\pi(1+\alpha)\frac{t}{T_s}\right)}{\frac{\pi t}{T_s}\left[1-\left(\frac{4\alpha t}{T_s}\right)^2\right]}, \tag{1}$$

for $-\infty < t < \infty$ where T_s is the symbol time and $0 \le \alpha \le 1$ is the excess bandwidth [1] as illustrated in Figure 2 (a). The frequency-domain representation of $p_{\text{ideal}}(t)$ is given by

$$|P_{\text{ideal}}(f)|^2 = \begin{cases} T_s & 0 \le |f| \le \frac{1-\alpha}{2T_s} \\ \frac{T_s}{2}\left[1+\cos\left(\frac{\pi|f|T_s}{\alpha} - \frac{\pi(1-\alpha)}{2\alpha}\right)\right] & \frac{1-\alpha}{2T_s} \le |f| \le \frac{1+\alpha}{2T_s} \\ 0 & |f| > \frac{1+\alpha}{2T_s} \end{cases} \tag{2}$$

and is illustrated in Figure 2 (a). The baseband bandwidth of the pulse shape is

$$B_c = \frac{1+\alpha}{2T_s} \quad \text{cycles/s.} \tag{3}$$

Note that the RF bandwidth is twice the baseband bandwidth. The corresponding autocorrelation function is

$$r_{p,\text{ideal}}(\tau) = \int_{-\infty}^{\infty} p_{\text{ideal}}(t)p_{\text{ideal}}(t-\tau)dt \tag{4}$$

and describes the matched-filter output.

All modern transmitters use discrete-time processing to generate the SRRC pulse shape. The unit-

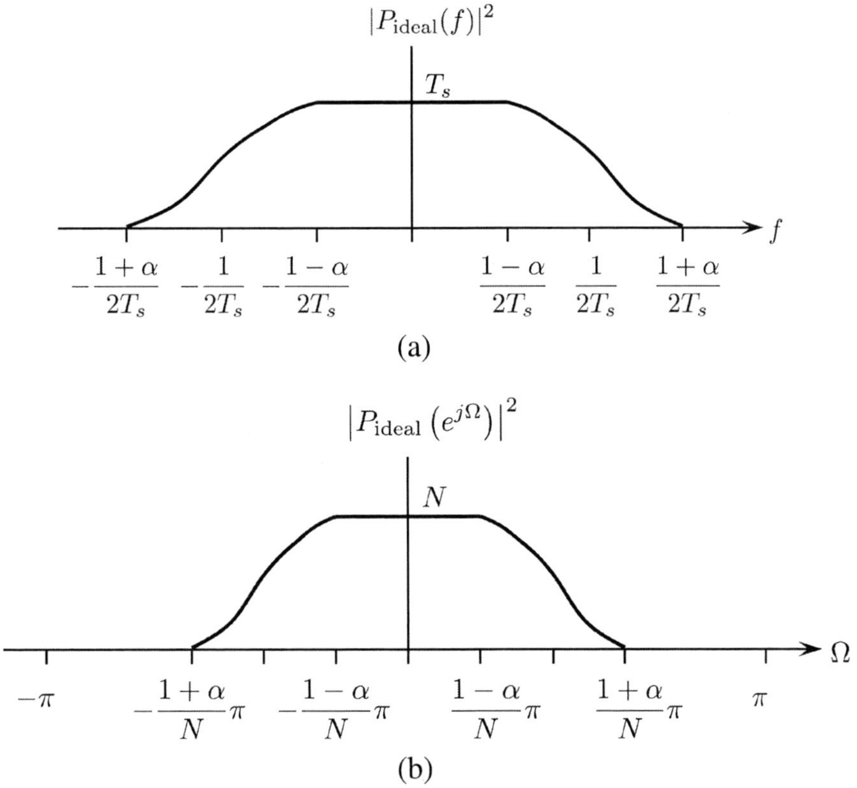

Figure 2: Frequency-domain representations of the ideal SRRC pulse shape:
(a) $|P_{\text{ideal}}(f)|^2$ given by (2); (b) $|P_{\text{ideal}}(e^{j\Omega})|^2$ given by (7).

energy preserving discrete-time version is

$$p_{\text{ideal}}(nT) = \frac{1}{\sqrt{N}} \frac{\sin\left(\pi(1-\alpha)\frac{n}{N}\right) + \frac{4\alpha n}{N}\cos\left(\pi(1+\alpha)\frac{n}{N}\right)}{\frac{\pi n}{N}\left[1 - \left(\frac{4\alpha n}{N}\right)^2\right]} \tag{5}$$

where T is the sample time, $-\infty < n < \infty$ is an integer, and N is the ratio of symbol time to sample time

$$N = \frac{T_s}{T} \quad \text{samples/symbol.} \tag{6}$$

The DTFT of the ideal discrete-time SRRC pulse shape (5) is [1]

$$|P_{\text{ideal}}(e^{j\Omega})|^2 = \begin{cases} N & 0 \le \left|\frac{\Omega}{2\pi}\right| \le \frac{1-\alpha}{2N} \\ \frac{N}{2}\left[1 + \cos\left(\frac{|\Omega|N}{2\alpha} - \frac{\pi(1-\alpha)}{2\alpha}\right)\right] & \frac{1-\alpha}{2N} \le \left|\frac{\Omega}{2\pi}\right| \le \frac{1+\alpha}{2N} \\ 0 & \left|\frac{\Omega}{2\pi}\right| > \frac{1+\alpha}{2N} \end{cases} \tag{7}$$

and is illustrated in Figure 2 (b). The baseband bandwidth is

$$W_d = \frac{1+\alpha}{N}\pi \ \text{ rads/sample} \quad \rightarrow \quad B_d = \frac{1+\alpha}{2N} \ \text{ cycles/sample.} \tag{8}$$

As before, the bandpass bandwidth is twice the baseband bandwidth.

Two properties of the discrete-time SRRC pulse shape are important to the design techniques described in the next three sections.

1. Any realistic discrete-time implementation requires a finite-length pulse shape, using a finite-length impulse response (FIR) filter of length L. L is usually parameterized by its *span* d which is the time support of the pulse shape measured in symbols. It is customary to express the time-support of the FIR approximation $p(n)$ as

$$-\frac{Nd}{2} \leq n \leq \frac{Nd}{2} \tag{9}$$

and to express the length of the pulse shape as

$$L = Nd + 1. \tag{10}$$

The corresponding autocorrelation function has support on $-Nd \leq n \leq Nd$ and has length $2Nd + 1$.

2. The autocorrelation function is used to quantify the residual ISI accompanying $p(n)$, the FIR approximation of $p_{\text{ideal}}(nT)$. For the pulse shape $p(n)$, the discrete-time autocorrelation function is

$$r_p(m) = \sum_n p(n)p(n-m), \qquad -L+1 \leq m \leq L-1. \tag{11}$$

An example of the discrete-time autocorrelation corresponding to a pulse shape sampled at $N = 10$ samples/symbol is shown in Figure 3. The filled markers indicate the samples used for detection. Because $r_{p,\text{ideal}}(kN) = 0$ for $k \neq 0$, the ideal pulse shape has no ISI. In this case, the sample at delay $= 0$ (shown by the filled-in diamond) is the only contributor to detection. When an FIR approximation of the discrete-time SRRC pulse shape is used, the samples corresponding to the filled-in circles are not exactly zero. The normalized ISI is given by

$$S = \frac{\displaystyle\sum_{\substack{n=-d \\ n\neq 0}}^{d} r_p^2(nN)}{r_p^2(0)}. \tag{12}$$

The challenge with the SRRC pulse shape is that a realistic implementation uses an FIR filter. Limiting the time support of the discrete-time pulse shape creates sidelobes in the corresponding spectrum. The peak sidelobe level, the normalized ISI, and the filter length represent the elements in the trade-off space defining this problem. In the application motivating this paper, the span

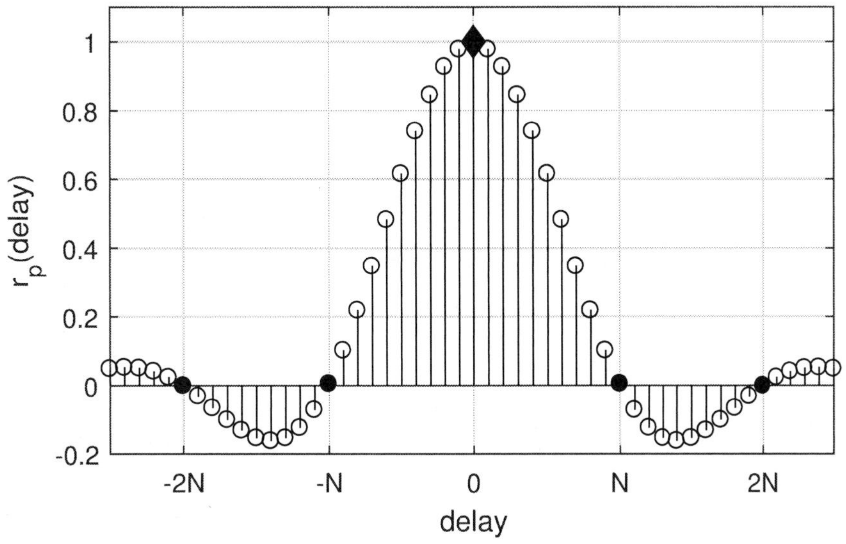

Figure 3: The autocorrelation function corresponding to the discrete-time SRRC pulse shape for $\alpha = 0.5$ and $N = 10$.

can be no greater than 32 and the peak sidelobe level must be 60 dB below the main-lobe level. The question is the following: how to create an FIR approximation of the SRRC pulse shape that meets a spectral containment requirement of -60 dB while simultaneously minimizing the ISI which inevitably accompanies an FIR approximation? Three different approaches will be taken to answer this question.

SQUARE-ROOT NYQUIST FILTERS BASED ON WINDOWED SRRC FILTERS

One method of creating an FIR approximation of the ideal SRRC is by windowing the ideal SRRC. That is

$$p(n) = w(n)p_{\text{ideal}}(nT) \tag{13}$$

where $w(n)$ is a window that is zero outside the interval $-Nd/2 \leq n \leq Nd/2$.

The simplest window is the vector of ones corresponding to the time support (9). This simply truncates the ideal SRRC pulse shape, but produces relatively high sidelobes in the corresponding spectrum. Windowing with smoother functions reduces the peak sidelobe level [2], but increases the normalized ISI. Results for the Blackman, Blackman-Harris, Chebyshev, Flattop, Gaussian, Hamming, Hann, Kaiser, Nuttall, and Taylor windows are summarized in Figure 5. Some windows, such as the Kaiser window, have parameters. These parameters were tuned to minimize the normalized ISI while simultaneously meeting the -60 dB sidelobe constraint.

An example using the Blackman window is shown in Figure 4. Dashed lines have been plotted vertically at $B = 1.5/2$ cycles/symbol and horizontally at the -60 dB point to show the constraints. Notice how the Blackman window meets the constraints, whereas the all-ones window does not.

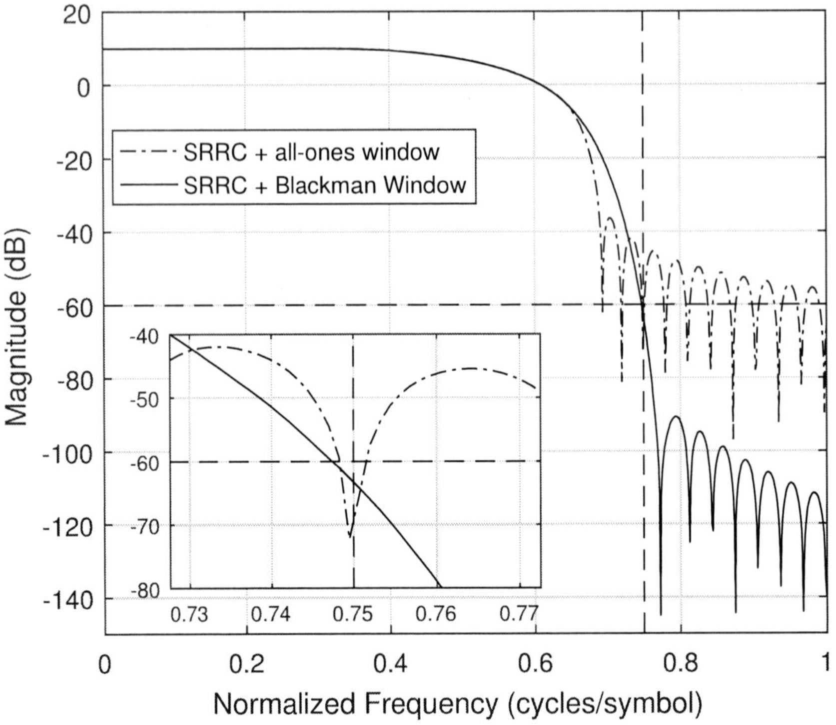

Figure 4: SRRC pulse with the all-ones window vs. an SRRC pulse with the Blackman window for $d = 32$ symbols, $\alpha = 0.5$, and $N = 10$ samples per symbol.

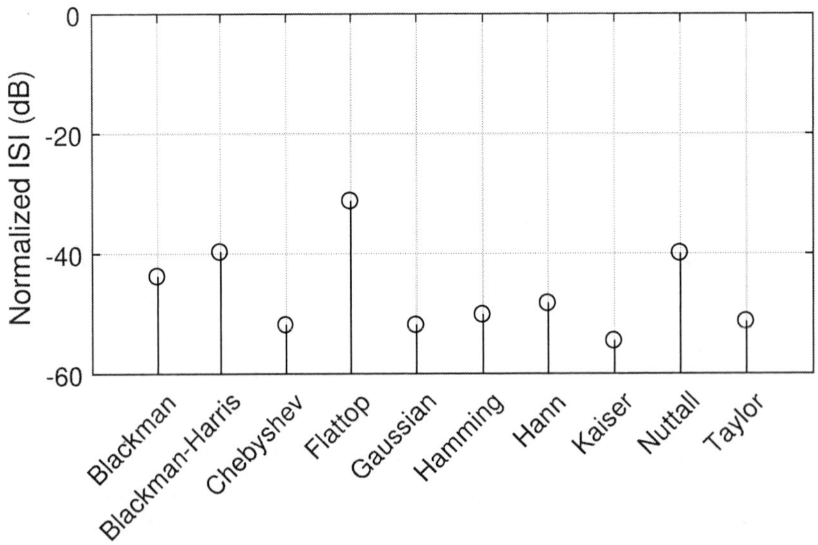

Figure 5: Normalized ISI vs. window type for $d = 32$ symbols, $\alpha = 0.5$, and $N = 10$ samples per symbol.

The Kaiser window with $\beta = 0.43$ has the best result at $S = -55$ dB. The Gaussian and Chebyshev windows also deliver competitive results at $S = -51$ dB. It is concluded that the SRRC pulse shape when windowed by the Kaiser window will deliver the least ISI within the given constraints among the windowing functions that were investigated.

SQUARE-ROOT NYQUIST FILTERS BASED ON COMMON FIR FILTER DESIGN TECHNIQUES

Another way to build $p(n)$ is to use a finite impulse response low-pass filter (FIR LPF) transformation as explained in [1]. The iterative algorithm in [1] can be based on any available FIR low-pass filter design technique. Two common techniques are the Parks-McClellan (PM) algorithm and the least squares (LS) algorithm. These two algorithms are similar in that they both attempt to calculate a finite number of filter coefficients that will match as closely as possible to a given ideal filter that has infinite or a large number of filter coefficients. The MATLAB functions firpm() and firls() derive from these two algorithms. These two MATLAB functions accept as parameters a passband frequency f_p and stopband frequency f_s. It is desired to find f_p and f_s such that the pulse created by firpm() or firls() meets all the constraints and has low ISI. The iterative algorithm in [1] is used to find an optimum f_p and f_s.

Figure 6 shows the results of the iterative algorithm using firpm() compared to a truncated SRRC pulse shape with parameters $\alpha = 0.5$, $d = 32$ symbols, and $N = 10$ samples per symbol. Note that the pulse shape generated using the PM algorithm meets the constraints indicated by the dashed lines.

The same procedure is used for the LS method by using firls() and the same iterative algorithm. Care must be taken when selecting d of the pulse created by firls(). If d is chosen to be too long, MATLAB returns an error. After trial and error, it was found that a good choice is $d = 23$ symbols. Figure 7 shows the results of the pulse shape created using firls() plotted with a truncated SRRC pulse shape with parameters $d = 23$ symbols, $\alpha = 0.5$, and $N = 10$ samples per symbol.

The normalized ISI of the pulses created using the PM and LS algorithms is computed by (12). The PM algorithm produces a pulse shape whose normalized ISI is $S = -48.89$ dB and the LS algorithm produces a pulse shape whose normalized ISI is $S = -54.14$ dB.

SQUARE-ROOT NYQUST FILTERS OPTIMIZED TO REDUCE OUT-OF-BAND EMISSIONS

Wilson and Lo [3] developed a square-root Nyquist pulse shape design technique based on minimizing the out-of-band power of the pulse shape. For a sequence $p(n)$ with DTFT $P(e^{j\Omega})$ the

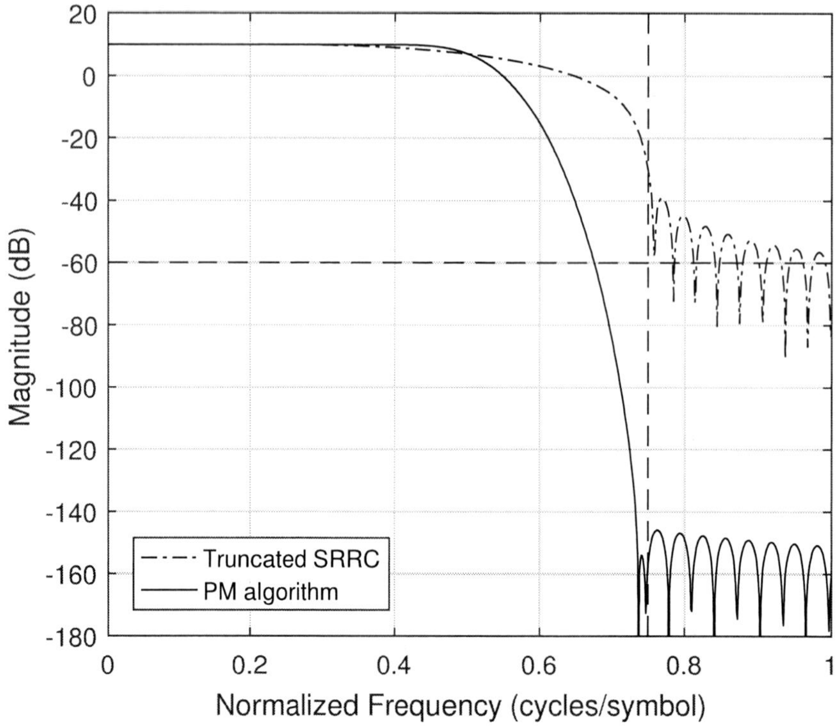

Figure 6: DTFT of the truncated (all-ones windowed) SRRC pulse shape and the pulse shape created using the PM algorithm for $d = 32$ symbols, $\alpha = 0.5$, and $N = 10$ samples per symbol.

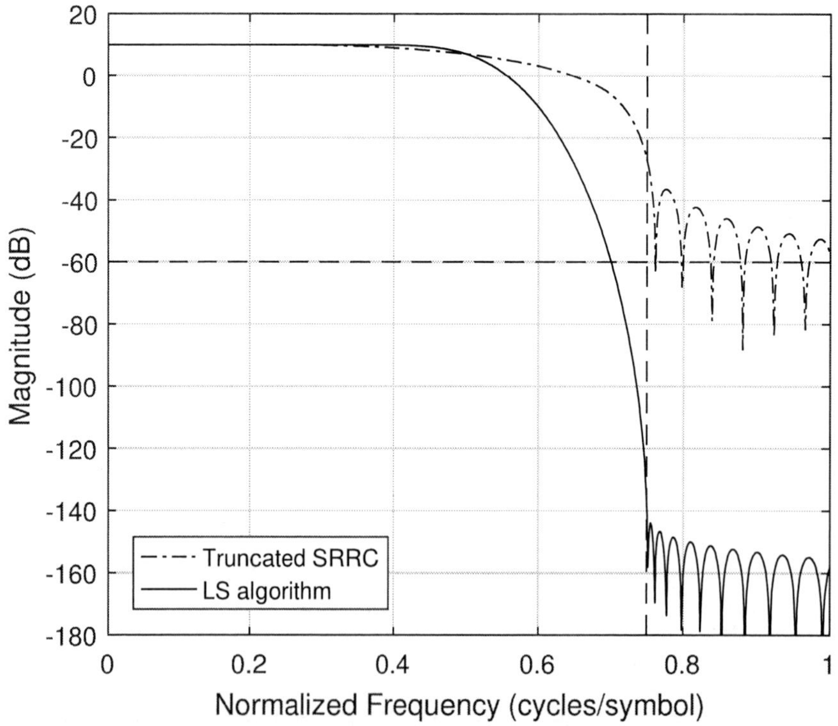

Figure 7: DTFT of the truncated (all-ones windowed) SRRC pulse shape and the pulse shape created using the LS algorithm for $d = 23$ symbols, $\alpha = 0.5$, and $N = 10$ samples per symbol.

out-of-band power is [3]

$$P_{\text{oob}} = \frac{1}{\pi}\int_{W_d}^{\pi} |P(e^{j\Omega})|^2 d\Omega = \frac{1}{\pi}\left[r_p(0)\left(\pi - W_d\right) - 2\sum_{m=1}^{L-1} r_p(m)\frac{\sin mB_d}{m}\right], \tag{14}$$

where L is the length of the sequence $p(n)$ given by (10), W_d is given by (8), and $r_p(m)$ is given by (11). The filter coefficients are [see (9)]

$$p\left(-\frac{Nd}{2}\right), \ldots, p(0), \ldots p\left(\frac{Nd}{2}\right). \tag{15}$$

Because a linear-phase pulse shape is required for linear modulations, the pulse shape must possess the even symmetry [4]: $p(n) = p(-n)$ for $n = 1, \ldots, Nd/2$. Consequently, only half of the pulse shape samples need to be computed. The Wilson-Lo design selects the filter coefficients $p(0), p(1), \ldots, p(Nd/2)$ to minimize the out-of-band power (14) subject to a constraint on the normalized ISI S given by (12). In mathematical terms, the Wilson-Lo pulse shape is the solution to

$$\min_{p(0),\ldots,p(Nd/2)} P_{\text{oob}} \qquad \text{subject to} \qquad S \leq \gamma^2 \tag{16}$$

for some small positive constant γ. The constraint on S is required to approximate the Nyquist no-ISI condition. Given the results of the previous two sections, $\gamma^2 = 10^{-5}$ is required to generate a competitive pulse shape. For the results presented below, the optimization algorithm was initialized using the truncated SRRC pulse shape.

The result for $\gamma^2 = 10^{-5}$, $N = 10$, $d = 32$, and $\alpha = 0.5$ gives a normalized ISI of $S = -80.56$ dB. A plot of the DTFT of this pulse shape and the DTFT of the truncated SRRC pulse shape is shown in Figure 8.

COMPARISON OF DIFFERENT METHODS

A comparison of all the methods investigated in this paper is shown in Figure 9. As shown, the pulse shape with the best ISI is the one created using the Wilson-Lo technique. This technique is so efficient that it far surpasses the closest competitor by 26 dB.

FPGA IMPLEMENTATION

In an FPGA implementation the filter coefficients are quantized using b-bit words. Because $b = 16$ is a common value, the impact of quantization using fixed-point 16-bit words on the Wilson-Lo pulse shape is investigated. Figure 10 shows the impact of quantization in the frequency domain. The quantized pulse shape still meets the spectral requirement. The normalized ISI of the 16-bit fixed point Wilson-Lo pulse shape is $S = -79.29$ dB meaning that the finite precision fixed point version only suffers a 1.27 dB degradation in normalized ISI.

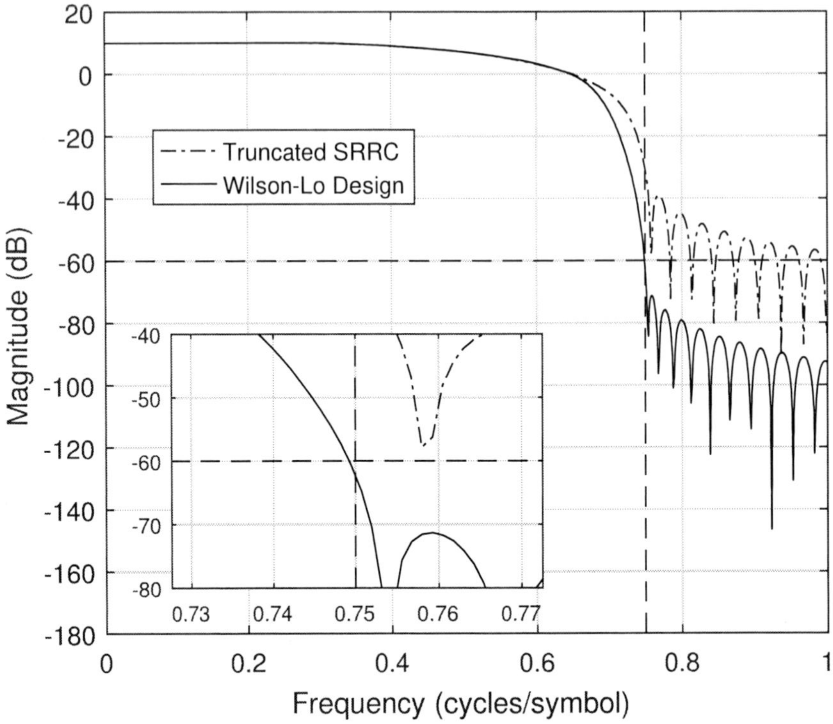

Figure 8: A truncated SRRC pulse shape and a pulse shape created using the Wilson-Lo technique [3].

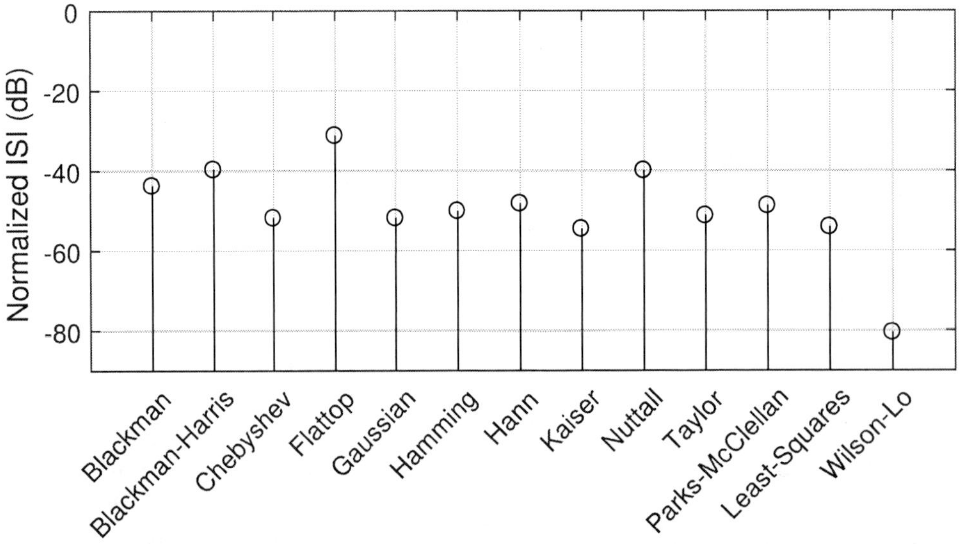

Figure 9: A comparison of all methods investigated in this paper.

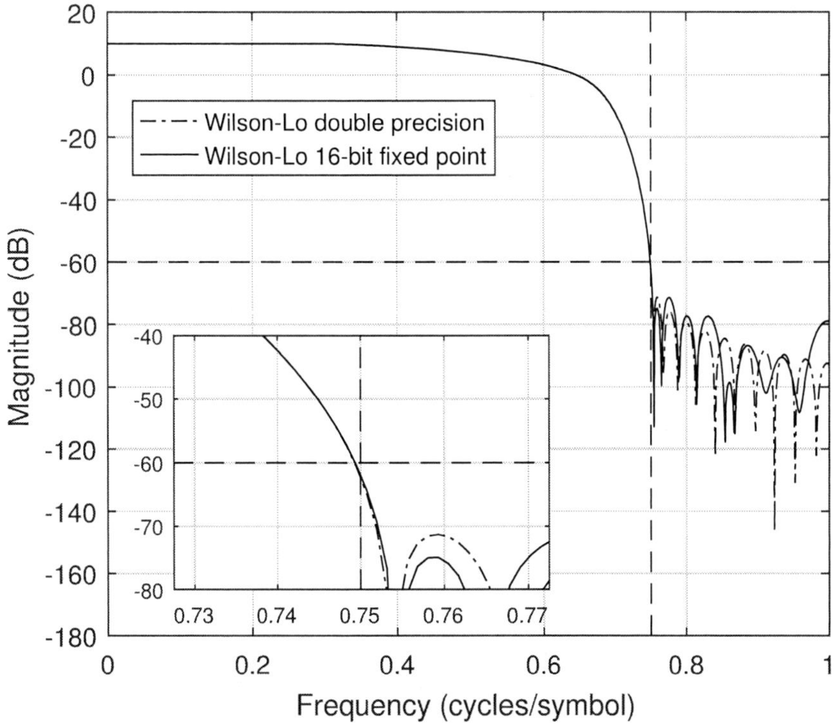

Figure 10: The DTFT of the Wilson-Lo pulse shape plotted and its quantized version.

CONCLUSIONS

Three digital filter design techniques were used to create discrete-time pulse shapes that meet the design requirements of length, peak side-lobe level, and normalized ISI. The results show that the Wilson-Lo constrained optimization produces the best results. When quantizing the coefficients using 16-bit fixed-point words, the spectral containment criterion is still met and the normalized ISI is at an acceptably low level.

ACKNOWLEDGEMENTS

This work was completed as part of the project "Coded APSK for Improved Spectral Efficiency in Aeronautical Mobile Telemetry" (NSC-06-0130) funded by the Spectrum Access Research and Development Program through the National Spectrum Consortium Base Agreement No. 2017-319 and the University of Kansas Center for Research, Inc.

REFERENCES

[1] M. Rice, *Digital Communications: A Discrete-Time Approach.* Upper Saddle River, NJ: Pearson Prentice-Hall, 2009.

[2] f. harris, "On the use of windows for harmonic analysis with the discrete Fourier transform," *Proceedings of the IEEE*, vol. 66, no. 1, pp. 51–83, 1978.

[3] S. Wilson and H. Lo, "Optimum digital pulse shaping filters," in *Proceedings of the IEEE Military Communications Conference*, (Baltimore, MD), 1–3 November 2016.

[4] A. Oppenheim and R. Schafer, *Discrete-Time Signal Processing.* Upper Saddle River, NJ: Pearson Prentice-Hall, third ed., 2010.

OPTIMUM PARAMETER COMBINATIONS FOR MULTI-H FULL RESPONSE CONTINUOUS PHASE MODULATION

Ding Xingwen Chang Hongyu Chen Ming

Beijing Research Institute of Telemetry, China

ABSTRACT

According to IRIG 106-15, the ARTM CPM waveform, a kind of multi-h partial response continuous phase modulation (CPM), has almost three times the spectral efficiency of PCM/FM and approximately the same detection efficiency of PCM/FM. But the improved spectral efficiency of ARTM CPM comes at the price of computational complexity in the receiver. This paper focuses on multi-h full response CPM, which generally has less detection complexity than ARTM CPM, but also has good spectral efficiency and detection efficiency. Taking the minimum Euclidean distance, spectral efficiency and detection complexity as judgment criterions, optimum parameter combinations for multi-h full response CPM are presented.

KEY WORDS

Multi-h Full Response Continuous Phase Modulation, Optimum Parameter Combinations, Minimum Euclidean Distance, Spectral Efficiency, Detection Complexity.

INTRODUCTION

CPM is a non-linear modulation with memory, since the consecutive symbols are correlative with each other. CPM signal with the characteristic of constant envelope has a certain tolerance against the non-linear distortion caused by non-linear power amplifiers, so as to greatly simplify the design complexity of communication system. Also, the characteristic of continuous phase can reduce the spectrum sidelobe, so as to decrease the channel interference. It is known that the performance of CPM signals can be improved by combining with multi-h phase codes [1].

In general, CPM signals can be represented as

$$s(t,\mathbf{a}) = \sqrt{\frac{2E}{T}} \cos(2\pi f_c t + \phi(t,\mathbf{a}) + \phi_0), \qquad -\infty < t < \infty \tag{1}$$

Where E denotes symbol energy, T denotes symbol duration, f_c denotes carrier frequency, $\phi(t,\mathbf{a})$ denotes the information bearing phase and ϕ_0 is the initial phase.

For multi-h full response CPM signal, its modulation index changes cyclically at the end of every symbol interval over k values $H_k = (h_0, h_1, \cdots, h_{k-1})$. In equation (1), $\phi(t,\mathbf{a})$ can be expressed as:

$$\phi(t,\mathbf{a}) = 2\pi a_n h_n q(t - nT) + \pi \sum_{i=-\infty}^{n-1} a_i h_i = \theta(t, a_n) + \theta_n, \qquad nT \le t \le (n+1)T \tag{2}$$

where $\theta(t, a_n)$ denotes the instantaneous phase, which means the changing portion of the phase in the current symbol interval; θ_n denotes the cumulative phase before the current symbol; a_i denote information symbols in the M-ary alphabet: $a_i \in \{\pm 1, \pm 3, ..., \pm M - 1\}$; h_i are modulation indices, h_i = $h_{i \bmod k}$; $q(t)$ is the phase function, which can be expressed as the integral of a certain frequency pulse $g(t)$:

$$q(t) = \int_0^t g(\tau) d\tau \tag{3}$$

Two commonly used frequency pulse shapes $g(t)$ are listed in Table 1.

Table 1 Frequency pulse shapes of multi-h full response CPM

REC	$g(t) = \begin{cases} \frac{1}{2T}, 0 \le t \le T \\ 0, otherwise \end{cases}$
RC	$g(t) = \begin{cases} \frac{1}{2T}[1 - \cos(\frac{2\pi t}{T})], 0 \le t \le T \\ 0, otherwise \end{cases}$

ANALYSIS ON ERROR PERFORMANCE

The minimum Euclidean distance between all possible pairs of the transmitted signal sequences is the most useful criterion for evaluating the error performance of CPM schemes with MLSD. The squared Euclidean distance between any two signals $s(t,\boldsymbol{\alpha})$ and $s(t,\boldsymbol{\beta})$ corresponding to the two symbol sequences $\boldsymbol{\alpha}$ and $\boldsymbol{\beta}$, which split apart at time t = 0 and remerge later, is defined as [3]

$$D^2 = \sum_{i=0}^{n-1} \int_{iT}^{(i+1)T} \left[s(t,\boldsymbol{\alpha}) - s(t,\boldsymbol{\beta}) \right]^2 dt = \frac{2E}{T} \sum_{i=0}^{n-1} \int_{iT}^{(i+1)T} \left(1 - \cos\left[\phi(t,\boldsymbol{\alpha}) - \phi(t,\boldsymbol{\beta}) \right] \right) dt \tag{4}$$

where n is observation interval length, D^2 is additive from interval to interval and depends on the phase difference $\phi(t,\boldsymbol{\alpha}) - \phi(t,\boldsymbol{\beta})$.

The minimum Euclidean distance D_{min} is the minimum overall Euclidean distances corresponding to all possible merging events. For multi-h schemes, the minimization also is performed over all cycled shifts of h_i values. When n is sufficiently large, the largest minimum Euclidean distance (i.e. the free distance) is obtained. It can be shown that any pair of signal sequences achieves the free distance within a small number of symbol intervals [4]. Generally, the minimum squared Euclidean distance d_{min}^2 is normalized with respect to the bit energy E_b as [3]

$$d_{min}^2 = \frac{D_{min}^2}{2E_b} \tag{5}$$

where

$$E_b = \frac{E}{\log_2 M} \tag{6}$$

The error rate performance of any wideband CPM signal system is usually expressed in terms of the minimum squared Euclidean distance of signals. In fact, only at high SNR, the asymptotic error probability is approximately given by [2],

$$P_e \approx Q\left(\sqrt{\frac{d_{min}^2 E_b}{N_0}} \right) \tag{7}$$

where $Q(.)$ is the standard Q function.

It is seen from equation (7) that the performance of a CPM system can be improved by increasing the minimum Euclidean distance. Multi-h phase code is one such technique, which can be used to increase the minimum squared Euclidean distance [1][2].

An upper bound on the minimum squared Euclidean distance, called d_B^2, can be obtained by calculating the distance for all pairs of phase paths in the trellis giving a first inevitable merge, a merge which occurs independently of the values of modulation indices [1]. The basic idea behind the multi-h continuous phase modulation (MHCPM) concept is to delay the first inevitable merge, thus creating a possibly larger minimum squared Euclidean distance.

ANALYSIS ON SPECTRAL PERFORMANCE

The bandwidth that a signal occupies is an important aspect of its total performance, as important as its error performance. Many methods have been developed for calculating the power spectra of CPM [1,5-10]. Most of these methods fall into three categories: the direct approach [5-6],the Markov chain approach [7-8], and the autocorrelation function approach [9-10].

In the direct approach, the power spectrum is obtained by the Fourier transform of a finite-time segment and taking the expectation or average value with respect to the random phase and data sequence. This general method has the ability to provide exact spectral density calculations, via numerical integration. The power spectrum of *M*-ary multi-h full response CPM signals is obtained by this method in [5], where a *M*-ary multi-h full response CPM is treated as an M^k-ary signal set over *KT*. Since M^k distinct MHCPM waveforms must be taken into account, computation of the spectrum is roughly proportional to M^k.

Since the MHCPM signal has a periodic phase trellis with *q* phases at each level, time-varying with a period of *k*, then the MHCPM signal can be described as a Markov chain approach, numerical integration usually is not required. The major drawback of this approach is the need to pay careful attention to the state modeling of the multi-h phase code. The computational complexity grows exponentially with *M*. Furthermore, the state model changes for different modulation index sets $H_k = (h_0, h_1, \cdots, h_{k-1})$.

In the autocorrelation function approach, the power spectrum is computed by the numerical Fourier transform, so the numerical integration is required. The power spectrum of MHCPM is calculated by the efficient method applicable for correlated data symbols [9], where the computational complexity is polynomial in *k*.

In general, the roll-off of spectral sidelobe of MHCPM is determined by the smoothness of the frequency pulse shapes $g(t)$. It is known that the spectral envelope decays as $f^{-2(m+1)}$, if the m-th order derivative of $g(t)$ is the lowest order, which is not continuous everywhere [11]. The width of the spectral main lobe increases with increasing h and M.

NUMERICAL RESULTS

Generally, multi-h full response CPM signals with *REC* frequency pulse shape can produce higher minimum squared Euclidean distances than multi-h full response CPM signals with *RC* frequency pulse shape. And for multi-h full response CPM signals, the spectral properties of *RC* frequency pulse shape almost have little superiority, compared with *REC* frequency pulse shape. Thus, we only focus on 2-ary multi-h full response CPM signals with *REC* frequency pulse shape.

Table 2 Optimum parameter combinations for 2-ary multi-h full response CPM with *REC* frequency pulse shape

Multi-h Code H_k	Minimum Squared Euclidean Distances d^2_{min}	Normalized 99% Bandwidth
10/16 11/16 12/16	4.34	1.78
7/16 8/16	3.31	1.18
4/15 5/15	1.56	1.00

In this section, the minimum squared Euclidean distances and normalized 99% bandwidths of 2-ary multi-h full response CPM signals are presented. Table 2 lists three representative optimum parameter combinations for 2-ary multi-h full response CPM with *REC* frequency pulse shape. $H_k = (10/16, 11/16, 12/16)$ is the representation of high coding gain. $H_k = (4/15, 5/15)$ is the representation of high bandwidth efficiency. $H_k = (7/16, 8/16)$ is the representation of a tradeoff between coding gain and bandwidth efficiency. As references, for CPFSK signal ($h = 0.7$, pre-modulation filter with -3 dB point at 0.7 times the data rate), the minimum squared Euclidean distances d^2_{min} is 2.43, the normalized 99% bandwidth is 1.17; for ARTM CPM signal, the minimum squared Euclidean distances d^2_{min} is 1.29, the normalized 99% bandwidth is 0.56. Furthermore, All these 2-ary multi-h full response CPM signals have much less detection complexity than ARTM CPM, but have more detection complexity than CPFSK.

CONCLUSION

This paper analyzes the error performance and spectral performance of multi-h full response CPM signals, which generally have less detection complexity than ARTM CPM, but also has good spectral efficiency and detection efficiency. And three representative optimum parameter combinations for 2-ary multi-h full response CPM are proposed.

REFERENCES

[1] J. B. Anderson, T. Aulin and C. E. Sundberg, Digital Phase Modulation, Plenum Press, New York, 1986.

[2] T. Aulin and C. E. Sundberg, "On the minimum Euclidean distance for class of signal space codes," IEEE Trans. On Inform. Theory, Vol. IT-28, January 1982, 43-55.

[3] T. Aulin and C. E. Sundberg, "Continuous Phase Modulation, Part I: Full Response Signaling," IEEE Trans. Commun., Vol. COM-29, No 3, March 1981, 196-209.

[4] J. B. Anderson and R. De Buda, "Better Phase-Modulation Error Performance Using Trellis Phase Code," Electron. Lett., Vol. 12, No 22, October 1976, 587-588.

[5] S. G. Wilson and R. Gaus, "Power Spectra of Multi-h Phase Codes," IEEE Trans. Commun., Vol. COM-29, No 3, March 1981, 250-256.

[6] R. R. Anderson and J. Satz, "Spectra of Digital FM," Bell Syst. Tech. J., Vol. 44, 1965, 1165-1189.

[7] P. K. M. Ho and P. J. McLane, "Power Spectral Density of Digital Continuous Phase Modulation with Correlated Data Symbols Part 2 : Rowe-Prabhu Method," IEE Proc., Vol. 133, Pt. F, No 1, February 1986, 106-114.

[8] P. Galko and S. Pasupathy, "The Mean Power Spectral Density of Markov Chain Driven Signals," IEEE Trans. Inform. Theory, Vol. IT-27, No 6, November 1981, 746-756.

[9] P. K. M. Ho and P. J. McLane, "Power Spectral Density of Digital Continuous Phase Modulation with Correlated Data Symbols Part 1 : Autocorrelation Function Method," IEE Proc., Vol. 133, Pt. F, No 1, February 1986, 95-105.

[10] T. Aulin and C. E. Sundberg, "Calculating Digital FM Spectra by Means of Auto Correlation," IEEE Trans. Commun., Vol. COM-30, No 5, May 1982, 1199-1208.

[11] T. J. Baker, "Asymptotic Behavior of Digital FM Spectra," IEEE Trans. Commun., Vol. COM-22, No 10, October 1974.

A PROPOSED REVISION TO IRIG 218 BASED ON REAL WORLD EXPERIENCE

Gary A. Thom
GDP Space Systems
300 Welsh Road, Horsham, PA 19044
gthom@delta-info.com

Abstract

The Range Commanders Council has been attempting to standardize Telemetry over IP (TMoIP) for many years now. While the attempt has been valiant, the outcome to date has not been very successful. As a result, many vendors have implemented their own proprietary methods for sending PCM data over IP networks resulting in a lack of interoperability. As telemetry ground stations are finally making the move toward network centric architectures, it is worth considering the lessons learned over the previous 10 years of designing, installing, troubleshooting and optimizing telemetry data distribution over IP networks. This paper describes a proposed revision to IRIG 218 based on these real life experiences. It discusses the critical decisions and architectural decisions to be made and some of the pitfalls to be avoid.

Key Words: IRIG 218, TMoIP, IP, TCP, UDP, network, PCM.

1 Introduction

The motivation for moving to TMoIP was twofold: first, to find cost effective PCM data distribution and second, to provide reliable and robust PCM data distribution regardless of the destination. The global explosion of IP networking has provided a built in infrastructure with access to the most remote destinations. A wide variety of transport mechanisms for IP traffic provides ubiquitous connectivity, whether twisted pair, fiber optic cable, microwave links, satellite links, analog modems and cell phones, IP connectivity is everywhere.

This ubiquity and global deployment has driven down the cost of networking components such as routers and switches. It provided dynamic routing and redundant paths, improving reliability and fault tolerance. The insatiable appetite for more data has resulted in ever increasing bandwidth availability.

The result is a reliable, cost effective infrastructure for PCM data distribution, whether on private IP networks or globally via the public internet.

2 Network Basics

Much of the support needed by TMoIP is provided by the various layers of the network protocol stack. This section discusses these layers and the services that they provide.

2.1 Physical Layer

The Physical Layer consists of the hardware networking interface. This layer provides for the transmission of bits over some physical media. Media types include, but are not limited to optical fiber, coax cable or twisted pair cable. The specification of the Physical Layer includes features such as connector type, electrical signal levels, modulation, frequency, data rate and line coding.

The primary functionality that the Physical Layer provides is the efficient delivery of bits over a selected medium from one point to another.

2.2 Data Link Layer

The Data Link Layer provides for the transmission of data between nodes on the same network segment. For TMoIP, the Ethernet [5] protocol provides this layer functionality. There are two important aspects of this layer. The first is addressing. The Ethernet protocol provides for the use of the Media Access Control (MAC) address. A MAC address is defined for each node on the network segment and Ethernet packets are sent from one node to another based on the MAC address.

The second is the Ethernet packet. The Ethernet packet shown in Figure 1, consists of a header, a payload and a trailer.

Preamble	Start of frame delimiter	MAC destination	MAC source	Optional VLAN Tag	Ethertype or length	Payload Data	Frame check sequence	Inter-packet gap
		Packet Header					Trailer	
		Ethernet Frame						
7 bytes	1 bytes	6 bytes	6 bytes	4 bytes	2 bytes	46–1500 bytes	4 bytes	12 bytes

Figure 1 - Ethernet Packet Format

The Ethernet Frame has a minimum size of 64 bytes. For that reason, it has a minimum payload size of 46 bytes after accounting for the mandatory header bytes. If less data is to be sent, it will be padded out with zeros to reach the 64 byte minimum. The maximum size of the Ethernet Payload is 1500 bytes. Since the Length field is 2 bytes, payload lengths greater that 1500 bytes can be generated. These are called Jumbo Frames. Jumbo Frames are not universally supported, so their use may hamper interoperability and therefore should not be used in TMoIP systems.

A key feature of the Ethernet Frame is the Frame Check Sequence. This error detection mechanism will identify errors in data transmission over the link. Typically, the Data Link Layer will drop packets that fail the FCS check and will not pass the resulting payload up to the next layer in the protocol stack.

2.3 Internet Layer

The Internet Layer provides connectivity between network segments. To do this, the Internet Protocol (IP) is used to provide a higher level address, the IP Address, which is routable between network segments. This is what allows the routing of data across the global internet.

Today there are two IP protocols: IPv4 [6] and IPv6 [7]. IPv4 was the original internet protocol. It provided for a 32-bit address. The address is divided between a network address and a subnet address. The size of these address components is determined by the subnet mask. This addressing information is carried in the header of the IPv4 packet as shown in Figure 2.

31	30	29	28	27	26	25	24	23	22	21	20	19	18	17	16	15	14	13	12	11	10	09	08	07	06	05	04	03	02	01	00	W
Version				Hdr Len				Type of Service								Total Length																1
Identification																Flags				Fragment Offset												2

Time to Live	Protocol	Header Checksum	3
Source Address			4
Destination Address			5
Payload Data (0 to 1480 bytes)			

Figure 2- IPv4 Packet Format

Besides the addresses, there are a few other useful fields in the IPv4 header. The Total Length field specifies the size of the IP header plus the payload. This is very useful when the IP packet is smaller than 46 bytes because it tells how much of the Ethernet frame payload is data and not fill. It is also useful for packets which are larger than a single Ethernet frame and used with the Fragment Offset field allow large payloads to be fragmented into multiple Ethernet frames. The Protocol field indicates the contents of the packet Payload. The Header Checksum only protects the header information. It does not protect the data. If the checksum check fails, the packet is discarded. Since the FCS is also being checked at the Ethernet level, this check will only catch the rare errors that get through that check or some implementation error at the IP layer.

IPv6 was developed primarily because the world was running out of IPv4 addresses. IPv6 expands the IP address to 128 bits. There is a similar split between network address and subnet address, the sizes of which are determined by the IP Address Prefix. The IPv6 packet is shown in Figure 3.

<table>
<tr><td>31</td><td>30</td><td>29</td><td>28</td><td>27</td><td>26</td><td>25</td><td>24</td><td>23</td><td>22</td><td>21</td><td>20</td><td>19</td><td>18</td><td>17</td><td>16</td><td>15</td><td>14</td><td>13</td><td>12</td><td>11</td><td>10</td><td>09</td><td>08</td><td>07</td><td>06</td><td>05</td><td>04</td><td>03</td><td>02</td><td>01</td><td>00</td><td>W</td></tr>
<tr><td colspan="4">Version</td><td colspan="8">Traffic Class</td><td colspan="20">Flow Label</td><td>1</td></tr>
<tr><td colspan="16">Payload Length</td><td colspan="8">Next Header</td><td colspan="8">Hop Limit</td><td>2</td></tr>
<tr><td colspan="32" rowspan="4">Source Address</td><td>3</td></tr>
<tr><td>4</td></tr>
<tr><td>5</td></tr>
<tr><td>6</td></tr>
<tr><td colspan="32" rowspan="4">Destination Address</td><td>7</td></tr>
<tr><td>8</td></tr>
<tr><td>9</td></tr>
<tr><td>10</td></tr>
<tr><td colspan="32">Payload Data (0 to 1460 bytes)</td><td></td></tr>
</table>

Figure 3- IPv6 Packet Format

In addition to the address change, IPv6 removed the unnecessary Header checksum and some other fields.

2.4 Transport Layer

The Transport Layer provides the data delivery services. There are two common data delivery services used over IP. These are User Datagram Protocol (UDP) [9] and Transport Control Protocol (TCP) [8].

UDP provides a connectionless, best-effort datagram delivery service. There is no connection established. Packets are just sent to the destination address. There is no packet loss detection, no packet acknowledgement or retransmission. One benefit of this service is low latency transmission. Buffering for acknowledgement and retransmission is not required and buffering increases latency. Another benefit is datagram delivery. That is the payload of the packet is considered an atomic message which is delivered intact. UDP has a packet format shown in Figure 4.

<table>
<tr><td></td><td>30</td><td>29</td><td>28</td><td>27</td><td>26</td><td>25</td><td>24</td><td>23</td><td>22</td><td>21</td><td>20</td><td>19</td><td>18</td><td>17</td><td>16</td><td>15</td><td>14</td><td>13</td><td>12</td><td>11</td><td>10</td><td>09</td><td>08</td><td>07</td><td>06</td><td>05</td><td>04</td><td>03</td><td>02</td><td>01</td><td>00</td><td>W</td></tr>
<tr><td colspan="16">Source Port Number</td><td colspan="16">Destination Port Number</td><td>1</td></tr>
<tr><td colspan="16">Packet Length</td><td colspan="16">Checksum</td><td>2</td></tr>
<tr><td colspan="32">Payload Data (0 to 1476 bytes)</td><td></td></tr>
</table>

Figure 4- UDP Packet Format

The UDP header provides length information for the entire UDP packet and a checksum that protects the header and the data. The checksum is optional for IPv4 and mandatory for IPv6. The port numbers are used to route the payload to a specific application at the Application Layer.

Typical UDP transmission is unicast where one source sends packets to one destination. The connectionless nature of UDP also allow for multicast transmission of packets. In this case, packets sent to special multicast IP addresses can be distributed to multiple destinations which join the multicast group. This provides not only an efficient delivery mechanism, but also provides a receiver directed addressing model where the receiving device determines what packet stream to receive.

TCP provides a connection oriented, guaranteed delivery byte-stream protocol. First a connection is established between the source and destination device using a handshaking protocol. After the connection is established a predetermined number of packets are sent before waiting for an acknowledgement. As acknowledgements are received, more packets are sent. If an acknowledgement is not received, the packet is retransmitted. The packets contain a byte stream and not atomic packets. That is, the data that is sent may not arrive in a packet of the same size as the original data. TCP fragments the packets as needed to support the acknowledgement and retransmission process. This requires the data to be parsed to find the start of a payload header which must have some unique synchronization information. The TCP Packet is shown in Figure 5.

<table>
<tr><td>31</td><td>30</td><td>29</td><td>28</td><td>27</td><td>26</td><td>25</td><td>24</td><td>23</td><td>22</td><td>21</td><td>20</td><td>19</td><td>18</td><td>17</td><td>16</td><td>15</td><td>14</td><td>13</td><td>12</td><td>11</td><td>10</td><td>09</td><td>08</td><td>07</td><td>06</td><td>05</td><td>04</td><td>03</td><td>02</td><td>01</td><td>00</td><td>W</td></tr>
<tr><td colspan="16">Source Port Number</td><td colspan="16">Destination Port Number</td><td>1</td></tr>
<tr><td colspan="32">Sequence Number</td><td>2</td></tr>
<tr><td colspan="32">Acknowledgement Number</td><td>3</td></tr>
<tr><td colspan="4">Offset</td><td colspan="12">Flags</td><td colspan="16">Window Size</td><td>4</td></tr>
<tr><td colspan="16">Checksum</td><td colspan="16">Urgent Pointer</td><td>5</td></tr>
<tr><td colspan="32">Payload Data (0 to 1464 bytes)</td><td></td></tr>
</table>

Figure 5 - TCP Packet Format

TCP uses port number as described above for UDP. Sequence Number, Acknowledgement Number, Window Size are used for acknowledgement and retransmission.

In addition to higher latency and packet parsing requirements, another drawback of TCP is that it requires bi-directional communications making it unusable for some applications that do not provide a return path.

The need for a checksum at this layer is questionable. Studies have shown that some errors get through the Ethernet FCS check and may be caught at this layer. In addition, it has been reported that errors are introduced in intermediate network devices. In any case, packets which fail the checksum check at this layer are also discarded and ultimately retransmitted.

2.5 Application Layer

TMoIP is our Application Layer. It needs to provide those functions that are not provided by the other layers of the protocol but which are required by the application.

The Range Commanders Council (RCC) has developed the IRIG 218 standard [1], the latest version having been published in 2010. This is a very efficient TMoIP header consisting of only four bytes as shown in Figure 6**Error! Reference source not found.**. It is derived from the Pseudo-Wire protocol [2][3][4] defined by the IETF.

31	30	29	28	27	26	25	24	23	22	21	20	19	18	17	16	15	14	13	12	11	10	09	08	07	06	05	04	03	02	01	00
Res				L	R	M		Res		Length						Sequence Number															
Payload Data (0 to 1472 bytes)																															

Figure 6- IRIG 218-10 Packet Format

The IRIG 218 header contains six bits that are reserved (Res) and set to zeros. It contains Four bits (L R, M) that carry alarm information. The alarm bits are a carryover from circuit switched telephony over IP defined in the Pseudo-Wire standard. These bits are remapped from forward and reverse alarm indication to a local and remote error indication. These indications are poorly defined in IRIG 218 and meaningless for TMoIP.

The Length field is used only when the payload is less than 64 bytes. If the Length field is zero, then the payload length is determined from lower level protocols. This may have been added because of the minimum payload size requirement for Ethernet, which is 64 bytes. However, the IP and UDP layers provide correct packet size information even when fill is inserted to meet the minimum Ethernet payload size requirement. This makes this field unnecessary.

So the only real useful information in the IRIG-218-10 header is the Sequence Number. This field allows the receiving device to determine if packets are lost or if they arrive out of order.

The IRIG 218-10 packet format does not contain any information that would be useful in reconstructing the PCM output timing at the receiving device. Lacking any other timing information, an Ethernet-to-PCM device will need to measure the arrival time difference between two packets and knowing the number of bits in the packet, it can calculate a data rate. However, there could be significant jitter in the packet arrival times, requiring several measurement to be averaged in order to remove the effects of the jitter. This will result in either a delayed startup while waiting for sufficient packets to average or a drifting output clock rate as the calculation is improved.

The standard makes a reference to using the Real Time Protocol (RTP) [10] to provide clock recovery support for TMoIP. The required RTP header fields are shown in Figure 7.

31	30	29	28	27	26	25	24	23	22	21	20	19	18	17	16	15	14	13	12	11	10	09	08	07	06	05	04	03	02	01	00	W
Ver		P	X	CSRC				M	Payload Type							Sequence Number																1
Timestamp																																2
SSCR Identifier																																3

IRIG 218 Packet	

Figure 7 - RTP Packet Format

While RTP does provide a 4-byte Timestamp which could be used to calculate the data rate, none of the other information in the RTP header is relevant. There is 2-byte sequence number which is redundant with the IRIG 218 header. The remaining fields: Payload Type, CSRC Count, X and P Flags and SSRC Identifier provide no additional value and are undefined in IRIG 218 for use in TMoIP.

The format and resolution of the one useful field in the RTP header, the Timestamp, is undefined in IRIG 218. This will lead to interoperability problems as different vendors use different time bases and resolutions.

The use of RTP seems to be a waste of space for the value received and an opportunity for confusion. For this reason, many vendors of TMoIP products have developed their own TMoIP packet formats which they feel better serve the application.

3 What do we want to use TMoIP for?

There are two main applications for TMoIP. The first is for PCM to PCM distribution over IP networks. In this application, the goal is to replace more traditional distribution systems such as coax cable, fiber optic cable, microwave links, matrix switches, patch panels, etc. with packet based transmission over IP networks. This is the application that the current IRIG 218-10 addresses. The application uses two TMoIP devices: one to convert PCM data to Ethernet packets and the second to convert the Ethernet packets back to PCM as shown in Figure 8.

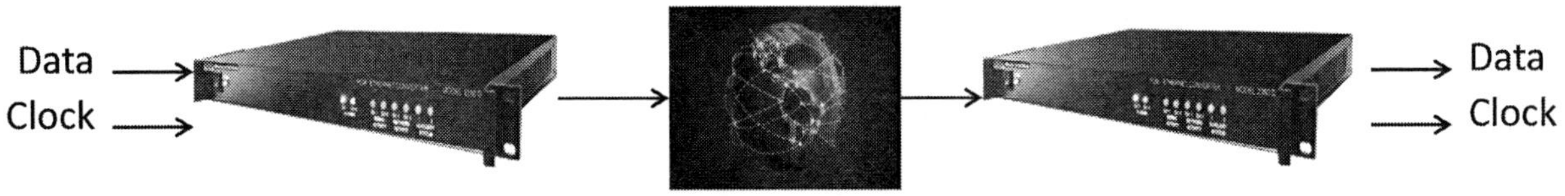

Figure 8 - PCM to PCM Data Distribution

The second application is for PCM to Computer distribution. In this application, the goal is to provide a convenient interface to get PCM data into a computer for software decommutation and/or recording. The current IRIG-218 does not address this application. In addition to converting serial data to IP packets, this application requires frame/subframe synchronization and frame alignment of synchronized data to the packet boundaries. This application only requires a single TMoIP device as shown in Figure 9.

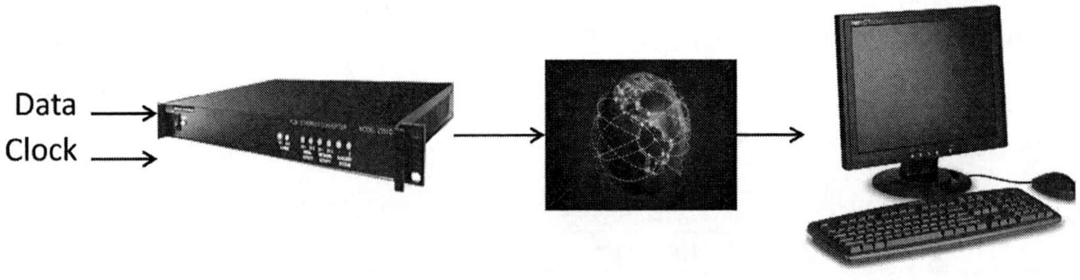

Figure 9 – PCM-to-Computer Data Distribution

4 What is needed for a TMoIP Protocol?

In order to answer this question, we need to divide the problem it several parts. For the first part, let's assume that we will be using UDP because of the connectionless, low-latency characteristics of the transmission. Later we will look at TCP.

4.1 UDP

As we have seen above, the network layers up through UDP provide addressing, routing, errored packet removal, payload size information and packet alignment.

4.1.1 Packet Data Alignment

Since UDP datagrams are atomic, the data sent by the source device in a UDP packet will be the same as the data received at the destination, assuming MTU sizes are not violated. That is, the packet size and location of data within the payload will be unchanged. To take advantage of this feature, we need to require that the first byte of the TMoIP header is the first byte of the UDP payload.

In the general case, the PCM data in the TMoIP packet is unaligned. However, as will be discussed later, there are cases where alignment may be helpful.

4.1.2 Packet Size

It is not necessary for the TMoIP Layer to explicitly indicate the packet size because this information can be derived from the UDP layer. Adjusting the packet size provides one means of controlling the latency and adapting to the data rate of the TMoIP stream. Packets from one byte to the maximum UDP payload size should be supported.

4.1.3 Packet Loss Detection

Since errored packets are discarded by the lower protocol layers and since UDP does not provide packet acknowledgement and retransmission, we must detect missing packets. This can be accomplished with a Sequence Number which increments for every packet transmitted. The 16-bit Sequence Number allows for detection of up to 65534 missing packets.

Missing packets need to be accounted for in the output PCM stream by inserting a "dummy" packet. If they are not, then there will be a dip in the output data rate or a discontinuity in the output clock. It may be desirable to fill the "dummy" packets with alternating ones and zeros in order to keep up the transition density, so as to not cause problems for downstream bit synchronizers which do not like long runs without transitions.

Sequence numbers will also allow the detection of packets that arrive out of order due to routing differences. Detecting an out-of-order packet and doing something about it are two different things. Correcting an out-of-order packet can only be done if the time to use that packet has not passed. Since the UDP layer is not providing any buffering, any buffering required to correct out-of-order packets would need to be handled by the TMoIP application. This will increase latency. In low latency applications, out-of-order packets should be discarded.

4.1.4 Data Rate Recovery

It is desirable to have a quick way to determine the PCM data rate in order to set the output PCM clock at the destination device. It can be argued that no additional information is required. You know how often you are receiving packets and you know how much data is in each packet so you can calculate the data rate. However, due to network jitter, the arrival time of packets may not be consistent, so many packets will need to be received and their arrival times averaged in order to get a good approximation of the data rate. While this is true, it will cause either a delay in the start of the PCM output or it will result in a change in the PCM output clock as the data rate calculation accuracy improves.

There are two alternate approaches. The first approach measures the data rate of the PCM input at the source device and sends that information in the TMoIP header. This method provides the fastest startup time and good stability. Fine control of the output clock must still be performed by a control loop in order to match the input and output clocks to avoid buffer overflow or underflow.

The second method timestamps the first bit of the PCM data in each packet. The difference in timestamps of consecutive packets along with the payload size can be used to calculate the data rate. While this has the downside of requiring two packets before starting the PCM output it removes any network jitter from the calculation and has some additional benefits that will be discussed later.

In either case, it is necessary to completely specify the range, accuracy and resolution of the measurement.

4.2 TCP

Like UDP, the layers up through TCP also provide addressing, routing, errored packet removal and payload size information. However, they do not provide packet alignment.

TCP does provide guaranteed delivery of packets, so we do not need to worry about lost packet detection. This eliminates the need for a Sequence Number in the header.

Since TCP packets may be segmented by the protocol stack and intermediate network devices, the received TCP packet may not contain a complete TMoIP payload. Assuming that the first byte sent over a new TCP connection is the first byte of a TMoIP header, we only need to know the TMoIP payload size in order to find the next header.

5 The Proposed TMoIP Protocol

In order to accommodate the functionality described in the previous section, the following TMoIP packet header and rules are proposed. The recommended TMoIP Header is shown in Figure 10.

<table>
<tr><td>31</td><td>30</td><td>29</td><td>28</td><td>27</td><td>26</td><td>25</td><td>24</td><td>23</td><td>22</td><td>21</td><td>20</td><td>19</td><td>18</td><td>17</td><td>16</td><td>15</td><td>14</td><td>13</td><td>12</td><td>11</td><td>10</td><td>09</td><td>08</td><td>07</td><td>06</td><td>05</td><td>04</td><td>03</td><td>02</td><td>01</td><td>00</td><td>W</td></tr>
<tr><td colspan="4">Version</td><td colspan="4">Reserved</td><td colspan="8">Flags</td><td colspan="16">Sequence Number / Payload Size</td><td>1</td></tr>
<tr><td colspan="32">Timestamp[Course]</td><td>2</td></tr>
<tr><td colspan="2">RES</td><td colspan="30">Timestamp[Fine]</td><td>3</td></tr>
<tr><td colspan="33">Payload Data (0 to</td></tr>
</table>

Figure 10 - Proposed TMoIP Header

This header contains the following fields:

- A 4-bit Version Number which indicates the format of the packet. Since the original IRIG 218 header had Reserved bits in this position which should always be set to '0000', we can use this field as a Version number. This new header format would use '0001'.
- A 4-bit field Reserved for future use. Shall be set to '0000'
- The 8-bit Flag field provides information about the header or the data.
    - Bit 16: Timestamp Present – If set to one, the Timestamp fields would be present. If set to zero, the Timestamp would be dropped making the packet very similar to the current IRIG 218-10 packet header.
    - Bit 17: Test Data – If set to one, payload contains 2^15-1 PN Pattern for automatic BER testing. If set to zero, payload contains PCM data.
    - Bit 19-18: Frame Alignment – 00 = no frame alignment, 01 = Frame aligned, first packet (Bits 23-20 contains frame/sub-frame lock status), 10 = Frame aligned, continuation packets (Bits 23-20 not used), 11 = Reserved.
    - Bits 21-20: Minor Frame Sync Status – 00 = Search, 01 = Check, 10 = Lock, 11 = Flywheel.
    - Bits 23-22: Major Frame Sync Status – 00 = Search, 01 = Check, 10 = Lock, 11 = Flywheel.
- The Sequence Number / Payload Size field is dual use. When UDP is used, this field contains a 16-bit Sequence Number which increments for every packet and is used to detect missing or out of order packets. When TCP is used, this field contains an 16-bit Payload Size which indicates the number of TMoIP data bytes following the header.
- The Timestamp field is made up of three sub-fields. The full Timestamp marks the time that the first data bit of the packet payload is received at the ingest device relative to the selected time system (i.e. IRIG B, NTP, PTP). The Timestamp consists of the following sub-fields:
    - 32-bit Course Timestamp. This represents Seconds since 12:00 AM 1/1/1970 which is the same as the Posix epoch and can easily be derived from IRIG B, GPS, NTP or PTP.
    - 30-bit Fine Timestamp. This represents fractions of the second and is usually derived by phase locking a local oscillator to the time source.
    - 2-bit Fine Timestamp Resolution indicator. This field indicates the resolution of the Fine Time field as follows:
        - 00 – Milliseconds (0 – 999)
        - 01 – Microseconds (0 – 999,999)
        - 10 – Nanoseconds (0 – 999,999,999)
        - 11 – Reserved

The following rules apply:

- The first byte of the UDP Payload shall be the first byte of the TMoIP Header.
- The UDP packet TMoIP Header shall contain a Sequence Number. The Sequence Number shall increment by one for every packet transmitted and shall roll over from all ones to all zeros.
- The first byte of the TCP Payload after the connection is established shall be the first byte of the TMoIP header.
- The TCP packet TMoIP Header shall contain a Payload Size. The Payload Size shall indicate the number of bytes in the TMoIP Payload (the number of bytes until the next TMoIP Header).
- An Ingress device shall insert a Timestamp in every packet. It shall indicate the resolution of the fine timestamp based on the resolution that the device can support.
- In a frame aligned packet, the first byte of the frame shall be the first byte of the TMoIP Payload and the Frame Alignment flags shall be set to '01'. Additional frame bytes shall continue contiguously in subsequent TMoIP payloads with the Frame Alignment flags set to '10'. The Timestamp shall not be present in these packets.

6 Conclusion

This proposed TMoIP header provides all of the information necessary to transmit PCM data over an IP network. It provides packet loss detection using the Sequence Number field. It provides data rate determination using the Timestamp. It provides data alignment for TCP using the Payload Size field. It supports both PCM-to-PCM distribution applications as well as frame aligned PCM-to-Computer applications. It is efficient adding only 3 32-bit words to the packet overhead. It provides for future expansion by making use of the Version and Reserved fields. The payload of the TMoIP packet can be any serial stream of PCM data. The proposed protocol is independent and unaware of the content.

7 References

[1]	IRIG STANDARD 218-10	Telemetry Transmission Over Internet Protocol (TMoIP) Standard.
[2]	IETF RFC 3985	Pseudo Wire Emulation Edge-to-Edge (PWE3) Architecture.
[3]	IETF RFC 3916	Requirements for Pseudo-Wire Emulation Edge-to-Edge (PWE3).
[4]	IETF RFC 4553	Structure Agnostic Time Division Multiplexing (TDM) over Packet (SAToP).
[5]	IEEE 802.3	IEEE Standard for Ethernet.
[6]	IETF RFC 791	Internet Protocol.
[7]	IETF RFC 2460	Internet Protocol, Version 6 (IPv6) Specification
[8]	IETF RFC 793	Transmission Control Protocol.
[9]	IETF RFC 768	User Datagram Protocol.
[10]	IETF RFC 3550	RTP: A Transport Protocol for Real-Time Applications

EXTENDING CHAPTER 10 RECORDING WITH TELEMETRY NETWORK STANDARDS

Myron L. Moodie, Todd A. Newton
Southwest Research Institute
San Antonio, Texas
myron.moodie swri.org, todd.newton swri.org

ABSTRACT

RCC 106 Chapter 10 has established the standard for interoperable flight test recording on the DOD ranges. The growth of network and distributed technologies in flight test instrumentation (FTI) has led to the recent adoption of Chapters 21 through 28 to provide standards for implementing interoperable telemetry networks. However, the new standards have led to confusion and concern that the investment in Chapter 10 recorders will be lost. This paper first clarifies the complementary nature of the RCC 106 chapters and proposes one possible path to extending the current capability of a Chapter 10 recorder with telemetry network capability while minimizing impact to existing recording and support systems.

KEYWORDS

Chapter 10, recording, TmNS, network

INTRODUCTION

RCC 106 Chapter 10 has established the standard for interoperable flight test recording on the DOD ranges. The success of the standard has provided interoperability between airborne and ground recorders across the distributed range infrastructure and across vendors. This commonality has allowed the development of standard data processing and analysis tools for use in ground stations. By leveraging the standard, the return on the investment by the ranges in these systems can be maximized.

The growth of network and distributed technologies in flight test instrumentation (FTI) has provided the benefits of flexibility, scalability, and compatibility with prevalent computing capabilities. In order to achieve interoperability with network FTI similar to what has been achieved with Chapter 10 recorders, the RCC adopted Chapters 21 through 28 in the 2017 IRIG 106 standard release to provide standards for implementing interoperable telemetry networks.

With thcsc standards now in place test programs and vendors have a common target to allow maximizing the return on the investment in network technologies.

However, the new network standards have led to confusion and concern that the investment in Chapter 10 recorders will be lost. This paper first clarifies the complementary nature of the RCC 106 chapters and proposes one possible path to extending the current capability of a Chapter 10 recorder with telemetry network capability while minimizing impact to existing recording and support systems.

CHAPTER 10 RECORDER OVERVIEW

The Chapter 10 recorder standard covers the wide breadth of instrumentation types while providing a uniform recording format and data download interface to allow a standard set of data processing and analysis tools. To achieve this, the standard covers the key aspects of a recorder including recording file format, packet structure, recorder control and status, and data download interface. Figure 1 provides an overview of the functionality covered by the Chapter 10 standard and the interaction between each function. The standard has evolved and expanded over multiple years and thus grown in complexity. To help manage this complexity the standard was restructured in the 2017 release by separating the data format definition, with all of the instrumentation type-specific features, into a standalone Chapter 11. Likewise, recorder control and status was consolidated into the existing Chapter 6. For simplicity, this paper will simply refer to a “Chapter 10” recorder to match the terminology used in the industry.

The Chapter 10 recorder collects acquired data and multiplexes it into a series of packets for storage. The packet types cover numerous specifc input types including PCM, analog, MIL-STD-1553, and ARINC 429. In addition to the data packets, additional time and computer generated packet types are used to contain additional information about the recorded data and allow multiplexing of multiple data streams, data search, and data protection. The stream of packets are stored on the media. While the standard does not mandate a specific storage approach on the media, it does mandate a specific file system and file structure that is derived from STANAG 4575 be presented to the user through the required data download interface. This data download interface is available over either FibreChannel or FireWire from the On-Board recorders and Ethernet for the Ground-Based recorders.

The data interoperability is achieved through a tightly defined packet, file, and directory structure. In fact, much of the standard wording for Chapters 10 and 11 are focused on these areas.

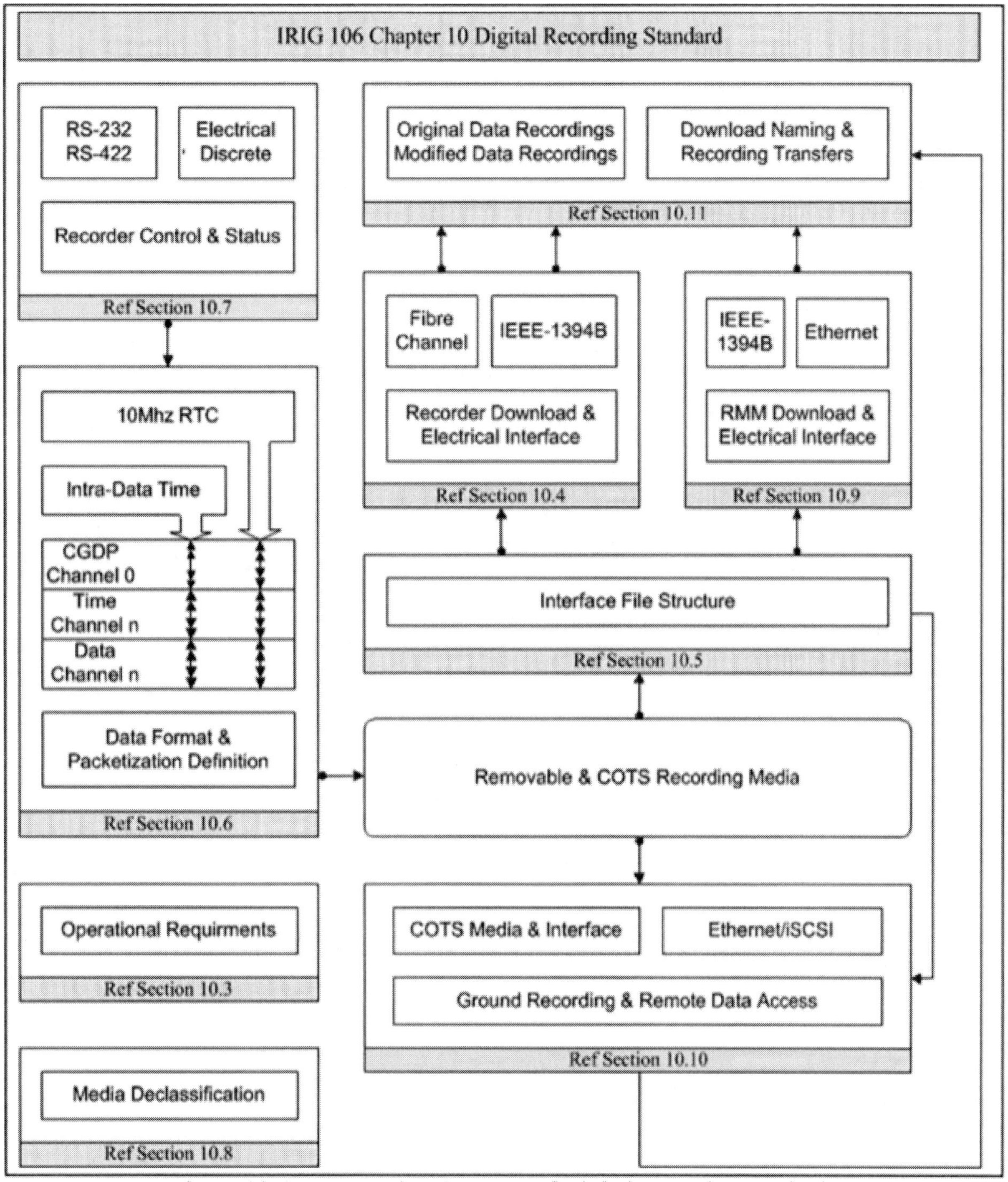

Figure 1. Chapter 10 Functional Breakdown 1

TMNS INTERFACES OF INTEREST

The network standards defined in Chapters 21 through 28 of IRIG 106 collectively are referred to as the Telemetry Network Standard (TmNS). The TmNS is focused on interoperable network interfaces between components for data transfer, timing, configuration, control, and status. The TmNS interfaces leverage the Internet Protocol (IP) at their root and so are compatible on any physical medium that supports IP transport. In addition, Chapters 27 and 28 specifically define a two-way network telemetry to provide end-to-end IP network connectivity between the test articles and the ground stations.

While a recorder is clearly an important component of a test system, the TmNS does not focus on a recorder (or any other component). Instead it defines standard interfaces that can be used on a recorder (or any other component) to achieve the required functionality of that component. In fact, besides a few references to a recorder as an example to explain the use case of a particular interface defined in the TmNS, the only other mention of recorder is in the tmnsRecorder portion of the Management Resources. Some of the TmNS interfaces of interest to a recorder include:

Inter a e	*Re rder se*
TmNSMessage [2]	Provides packet structure for transporting acquired data including IEEE 1588 timestamp
Reliability Critical (RC) Delivery Protocol [3]	Provides a data retrieval mechanism for transferring TmNSMessages
Metadata Description Language (MDL) [4]	Provides configuration of recorder resources
tmnsTmaCommon and tmnsRecorder Management Resources [5]	Provides status of recorder and control of recording state

EXTENDING CHAPTER 10 WITH TMNS

Concern has been raised by Chapter 10 recorder users that the emergence of the TmNS chapters of IRIG 106 will cause changes that risk the large investment in Chapter 10 recorders and associated processing and analysis tools. However, since much of the interoperability required for the installed base of tools relies on the packet and file formats specified in Chapter 10 (and 11) and since the TmNS chapters do not address recording or file format at all, we believe that these concerns can be alleviated. The interfaces specified in the TmNS are complementary to the requirements specified in the Chapter 10 portions of the IRIG 106 standards and thus should not conflict. Also, the interfaces in the TmNS portion of the standards are intended to be a "menu" of interfaces, not what is required of any particular component. Component specifications external of the standard are intended to specify which interfaces are required for a particular component.

An initial start to extending Chapter 10 recorders with TmNS capabilities is to add the RC Delivery Protocol. This will allow retrieval of data recorded in the usual Chapter 10 recording approach to be made available to other TmNS-based consumers through TmNSMessages. Since the recorded file format and existing Chapter 10 data download interface would remain unchanged then typical Chapter 10 processing and analysis would continue as usual.

Once the recorder is capable of producing TmNSMmessages through the RC Delivery Protocol, the natural next extension would be to support configuration through MDL. This will allow configuration of the TmNSMessage shapes produced by the recorder and other parameters needed by TmNS consumers. Adding system management support through the tmnsTmaCommon and tmnsRecorder management resources and associated interfaces in Chapter 25 will allow the Chapter 10 recorder to be consistently managed with other TmNS-based components.

Since the current Chapter 10 standard supports acquisition and recording of Ethernet inputs and since TmNSMessages are typically transported on Ethernet, the Chapter 10 recorder currently can record TmNSMessages. For further extension, the Chapter 11 data formats could be extended to have native TmNSMessage support which would allow indexing into the data based on the TmNS-specific parameters such as MDID and PDID and would also reduce storage requirements slightly by not having to record the Ethernet or UDP/IP headers.

CONCLUSION

The investment in Chapter 10 recorders has led to interoperability and tool reuse across the range infrastructure. As the new capabilities provided by the recent addition of network telemetry through the TmNS chapters of IRIG 106 are leveraged it is important to not lose the investment in Chapter 10 recorders. This paper has illustrated how the TmNS is complementary in nature and suggested ways to extend Chapter 10 recorders with these capabilities without losing the current mission of the recorders. These are just some of the possibilities and the needs of the test community will drive which combination of TmNS interfaces will become prevalent on future recorder upgrades.

REFERENCES

[1] "Digital Recording Standard", IRIG RCC 106-17 Chapter 10, August 2017.

[2] "Message Formats", IRIG RCC 106-17 Chapter 24, August 2017.

[3] "TmNSDataMessage Transfer Protocol", IRIG RCC 106-17 Chapter 26, August 2017.

[4] "Metadata Configuration", IRIG RCC 106-17 Chapter 23, August 2017.

[5] "Management Resources", IRIG RCC 106-17 Chapter 25, August 2017.

IRIG 218 UPDATE STATUS

Shawn T. Perry
NAWCAD Pax River
Atlantic Test Range RTPS
Patuxent River, MD, 20670
shawn.perry@navy.mil

ABSTRACT

The intent of this presentation is to relay the current progress of the Range Commanders Council (RCC) Timing and Telecommunications Group's (TTG) IRIG 218 TMoIP update task. Current Range utilization of TMoIP will be discussed, as well as, the vision for future use cases. The requirements gathered for this update will be evaluated. The pink sheet draft of the IRIG 218-18 standard will be presented for interactive dialogue.

INTRODUCTION

As Range transport datalinks transitioned from synchronous ATM systems to asynchronous IP networks, the Range Commanders Council (RCC) Timing and Telecommunications Group (TTG) saw a need to establish a means of transporting Serial Streaming Telemetry (SST) in the emerging environment. In 2010, the IRIG 218 standard was published leveraging other existing standards such as pseudowire. While consideration was given to the future growth of 218, as indicated by references to IPv6 and RTP, the intent of the standard was to remain as a simple, efficient transport mechanism for Range communications activities. The roll out of 218-based equipment has been a great success for Ranges, leading to a boom in distributed telemetry mission support and the dawn of fully interconnected Ranges for Live, Virtual, and Constructive (LVC) testing. Implementation of 218 was not perfect, however. Unique customer requirements, feature development by equipment manufacturers, and some ambiguity in the standard led to proprietary variations of 218 and interoperability issues.

TABLE E-1. TMoIP PACKET SUMMARY				
Field	**Description**		**Length**	**P/C/F** [1]
Ethernet Dest Addr	Identifies station(s) to receive frame		6	P
Ethernet Src Addr	Identifies station that originated frame		6	C
802.1Q Length/Type	Virtual LAN (VLAN) tag length/type		2	F = 0x8100
VLAN Tag Ctrl Info	Bit	Description	2	
	0 - 2	User Priority Field		P
	3	Canonical Format Indicator (CFI)		F = 0
	4 - 15	VLAN Identifier (VID)		P
Length/Type			2	F = 0x0800
IP Header	Byte	Description		
	0	Version + IP header length	1	F = 0x45
20 Bytes Total	1	TOS	1	P
	2 - 3	Total length of IP packet	2	C
	4 - 5	16 bit ID	2	C/F
	6 - 7	Flags + Fragment Offset	2	F
	8	TTL	1	F/P
	9	Protocol (UDP)	1	F = 0x11
	10 - 11	IP Header checksum	2	C
	12 - 15	Source IP address	4	P
	18 - 19	Destination IP address	4	P
UDP Header	Byte	Description		
	0 - 1	Source Port	2	P
8 Bytes Total	2 - 3	Destination Port	2	P
	4 - 5	UDP Length	2	C
	6 - 7	UDP Checksum	2	C
Payload	TMoIP Control Word		4 (or 12)	C
	TM Raw Packet Data		Var	C
Ethernet FCS	Ethernet Frame Check Sequence		4	C

[1] P = Programmable by user.
C = Calculated or placed in packet without user intervention.
F = Fixed.
Var = Variable.

BODY

Over time, telemetry equipment manufacturers began to voice their concerns to the RCC Telemetry Group (TG) that there were deficiencies in 218, especially the lack of a timestamp. At the same time, Ranges implementing 218 equipment were growing frustrated with interoperability issues between manufacturers. The TG brought these concerns to the attention of the TTG for resolution. The TTG began a quest to identify requirements through joint discussion with the TG and manufacturers. The progress of those discussions is the purpose of this presentation. The TTG's goal is to identify any remaining interoperability issues and finalize the updated standard draft for pink sheet review in the next few months, followed by an updated standard release, 218-18, next year.

The traditional PCM transport model that was the basis of 218-10 was for point-to-point applications where the edge equipment received SST, packetized the PCM as TMoIP, and then converted back to SST. As telemetry equipment has evolved, especially with the influx of Spectrum Reallocation Funds (SRF) investments, the transport model has evolved as well. While the requirements of the traditional model still exist, other models have emerged. One such case is the output of TMoIP directly from receivers, bit syncs, and best source encapsulators. Another model exists where an endpoint ingests SST, produces TMoIP, and delivers it to a software decom. Lastly, the case can be made for a model that eliminates SST completely, where a receiver produces TMoIP and delivers it to a software decom. Based on the varied requirements of flight testing that involve encryption and best source selection, the equipment in each of the models can vary, resulting in the greater importance of interoperability and reliance on metadata.

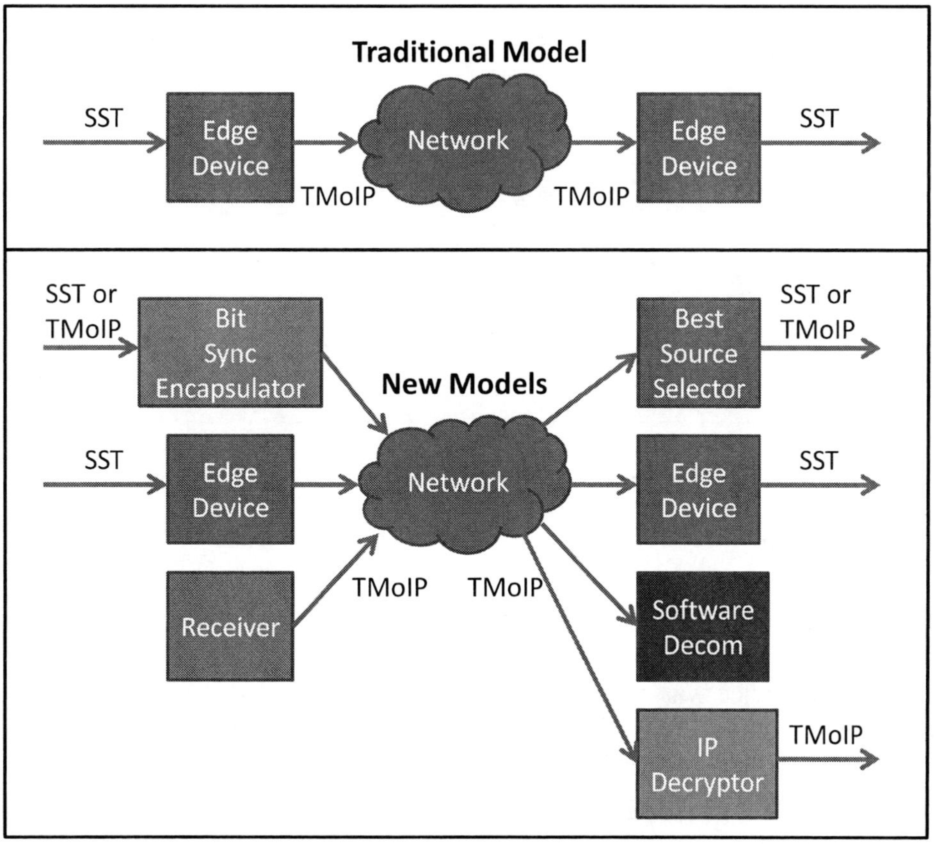

The primary deficiency identified was the lack of a timestamp. Proposals for resolving this requirement were all unique. There was some interest in using the RTP protocol mentioned in 218-10 for future exploration, but those solicited for input overwhelmingly preferred adding the timestamp as an expansion of the TMoIP control word rather than as an accompanying RTP packet. An RTP packet would carry its own unique overhead and would require correlation to the 218 base packet. Thus, the TTG has decided to add the timestamp to the control word of the 218 packet. The question then emerged of how to define the timestamp. Traditional telemetry uses the timestamp identified in IRIG 106 Ch 4.

4.7 Time Words

The following paragraphs describe the formatting of time words within a PCM stream. A 16-bit standardized time word format and a method to insert time words into PCM word sizes other than 16-bits are described.

In 16-bit standardized time word format, there shall be three words dedicated to providing timing information. These words are designated high order time, low order time, and microsecond time. High and low order time words shall be binary or binary coded decimal (BCD) weighted, and microsecond words shall be binary weighted. Time word construction examples are shown in Figure 4-3 and Figure 4-4.

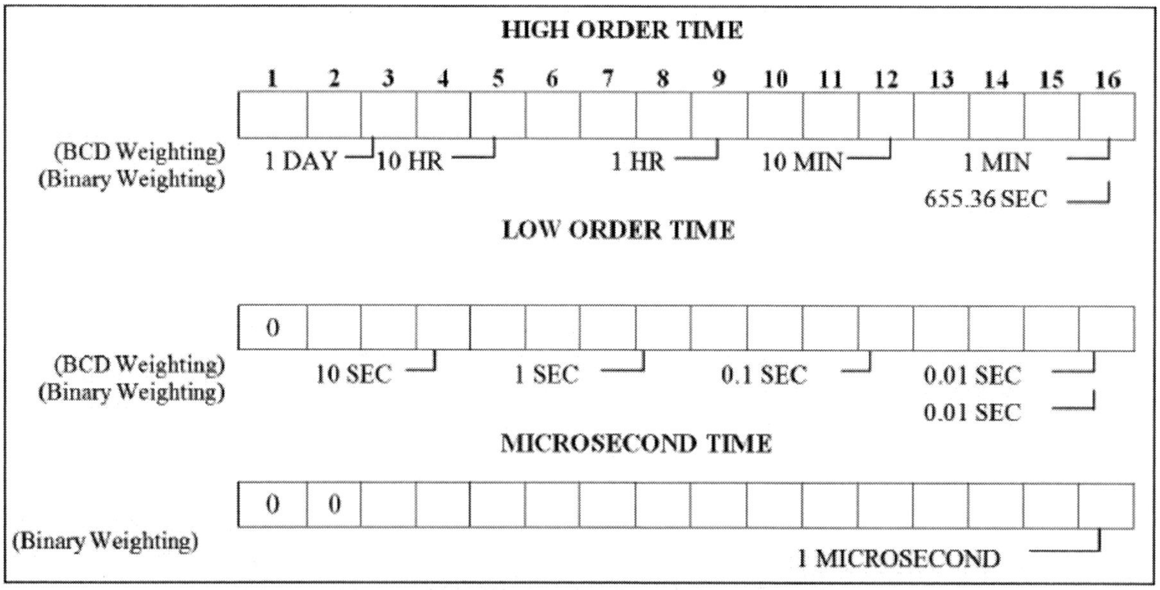

Figure 4-3. 16-Bit Standardized Time Word Format

One of the Range requirements was to provide a timestamp as close to the test article as possible. In this case, the test article lacked embedded time information in the PCM stream, the PCM is unencrypted, and the end device was a software decom. While the Ch 4 time format would be sufficient for this case, the millisecond resolution is insufficient for some of the features manufacturers hope to provide on next generation hardware. These features include minimized channel-to-channel skew and faster acquisition following changes in bit rate. To achieve the nanosecond resolution identified in the requirements request, a 64-bit timestamp will be implemented. The next stage in defining the 64-bit timestamp was to identify the time requirements. Results to this question were also varied. Currently, this remains unresolved, but the TTG is leaning towards a absolute time solution that includes year. Another question was whether to use UTC as legacy protocols do or use TAI as in 1588 PTP. The TTG decided that since there is already a lot of 218 hardware deployed that would likely use IRIG or NTP as a time source, the easiest path for those systems to become 218-18 compliant would be to use UTC. Another consideration was that few Ranges have implemented 1588 at this time, so more equipment would have to convert from UTC to TAI rather than TAI to UTC. New systems that choose to use 1588 as a timing source should be able to convert to UTC relatively easily.

Interoperability issues stem from optional criteria in the 218-10 standard that were implemented by manufacturers as proprietary 218 format variants. The TMoIP Industry Day that will be conducted at ITC hopes to identify the variations that lead to incompatibility. One of the major concerns is how vendors relay bit rate information. The Ranges have a desire for vendors to be able to communicate with one another with minimal circuit reconfiguration. For example, equipment should be able to calculate bit rate changes from mission to mission without the need for reconfiguration of the circuit. Payload sizes of the packets should be able to autoconfigure based on performance criteria (e.g. minimal latency) while still be recognizable by another vendor's product. As with the other requirements, the TTG's goal is to implement a format that allows for the continual operation of 218-10 equipment and minimal impact to 218-10 deployed systems that wish to implement 218-18, while respecting the desires of next generation products that will lead to next generation Range functionality and O&M cost savings.

Some of this next generation functionality involves bit sync and best source selection metadata. Traditionally, 218 is payload agnostic where it fills the payload of the packet, based on user preferences and network engineering, as bits are ingested from the SST. The newer models of telemetry delivery would benefit from the ability to shape the payload to either a PCM frame (see IRIG 106 Ch4) or a DQE (see IRIG 106 Ch 2 Appendix G) frame. The PCM frame would additionally benefit from carrying the bit sync status bits.

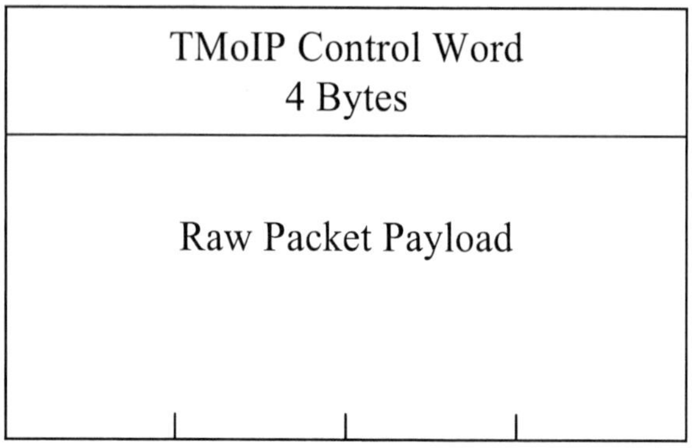

TABLE 3-3. TMoIP CONTROL WORD		
Field	**Bits**	**Description**
VER	4	Version identifier "0000" indicates 218-10 "0001" is reserved for proprietary modified 218 formats "0010" indicates 218-18 "0011" through "1111" reserved for future versions
VDB	12	Version Defined Bits
SEQ NUMBER	16	Sequence Number

Notes

218-10 remains unchanged.

218-P version was added to maintain functionality of several proprietary modified 218-10 variants in use.

218-18 version removes the unused alarm bits, adds the ability to designate that a payload is sized to a PCM or DQE frame, adds bit sync status for PCM frame aligned payloads, and adds a 64-bit timestamp.

TABLE 3-3.1. TMoIP 218-10 CONTROL WORD		
Field	**Bits**	**Description**
VER	4	Version identifier "0000" indicates 218-10
L	1	Local Defect Alarm, indicates local circuit fault in the TM stream
R	1	Remote Defect Alarm, indicates remote circuit fault in the TM stream
M	2	Local Defect Alarm Modifier
RES	2	Reserved
LEN	6	If non-zero, LEN indicates TMoIP Payload Length, defined as the TMoIP Control Word + Raw Packet Payload If zero, LEN indicates TMoIP Payload Length greater than 63 bytes. In this case the TMoIP payload length is determined via length fields in lower protocol layers.
SEQ NUMBER	16	Sequence Number

Notes

Req The TMoIP raw packet size shall be user configurable.
Opt The TMoIP raw payload size may be auto-configurable, based on user priorities (e.g. stream/delay characteristics).
Req The minimum TMoIP raw packet size = 1 byte.

Notes:
a. To limit the effects of Ethernet fragmentation, the final Layer 2/3/4/6 packet size should be less than the Ethernet Maximum Transmission Unit (MTU).
b. Padding may be required to meet the minimum Ethernet MTU size.

TABLE 3-3.2. TMoIP 218-P CONTROL WORD		
Field	**Bits**	**Description**
VER	4	Version identifier "0001" indicates Proprietary variants of 218-10
PDB	12	Proprietary Defined Bits
SEQ NUMBER	16	Sequence Number

Notes

This version allows vendors who created modified versions of 218-10 to maintain functionality. Due to variances in implementation, not recommended in mixed vendor environments.

TABLE 3-3.3. TMoIP 218-18 CONTROL WORD

Field	Bits	Description
VER	4	Version identifier "0010" indicates 218-18
PLD	2	Payload type "00" indicates no frame alignment "01" indicates PCM frame aligned "10" indicates DQE frame aligned "11" Reserved
FSS	2	Frame Sync Status (used for PLD = "01" only) "00" indicates Search "01" indicates Check "10" indicates Lock "11" indicates Flywheel
RES	2	Reserved
LEN	6	If non-zero, LEN indicates TMoIP Payload Length, defined as the TMoIP Control Word + Raw Packet Payload If zero, LEN indicates TMoIP Payload Length greater than 63 bytes. In this case the TMoIP payload length is determined via length fields in lower protocol layers.
SEQ NUMBER	16	Sequence Number
TIMESTAMP	64	Currently undefined 64-bit UTC-based timestamp

Notes

Req The TMoIP raw packet size shall be user configurable.

Req The TMoIP raw payload size shall be auto-configurable, based on user priorities (e.g. stream/delay characteristics).

Req The minimum TMoIP raw packet size = 1 byte.

Notes:
a. To limit the effects of Ethernet fragmentation, the final Layer 2/3/4/6 packet size should be less than the Ethernet Maximum Transmission Unit (MTU).
b. Padding may be required to meet the minimum Ethernet MTU size.

CONCLUSION

The TTG is working on updating the 218-10 standard to a format that satisfies identified requirements, minimizes impact to currently deployed TMoIP infrastructure, minimizes cost to equipment manufacturers, and produces features that will help Ranges satisfy next generation testing requirements at a lower cost. The plan forward for the TTG is to reassess the 218 standard on a two year cycle similar to the IRIG 106 process and implement future versions as needed. All are encouraged to contact the TTG directly or participate in the TSCC listserv to propose future requirements for consideration.

INSTRUMENTING AN ETHERNET AVIONICS BUS FOR APACHE TESTING

Vannesa Cuevas, Michael Diehl, and Tab Wilcox
Air Combat Systems Directorate, U.S. Army Yuma Proving Ground, Yuma AZ 85367

ABSTRACT

Last year, the U.S. Army Yuma Proving Ground (YPG) faced a short suspense requirement to instrument an Apache's Ethernet bus. YPG was able to implement an effective method to satisfy this requirement; however, more capable and effective methodologies could not be developed and utilized due to time constraints. While continuing to support ongoing Apache testing, YPG is working to implement more efficient methods to capture and utilize the Ethernet data. This paper will discuss YPG's initial implementation and the follow-on efforts being pursued.

KEYWORDS

1. Ethernet bus
2. Chapter 7
3. Streaming User Datagram Protocol (UDP)

INTRODUCTION

YPG, a subordinate command of the U.S. Army Test and Evaluation Command, is one of the largest military installations in the world. Located in southwestern Arizona, it encompasses 1,308 square miles. YPG personnel conduct tests on nearly every weapon system or piece of military equipment in the ground combat arsenal. With a mission to provide premier test services to the U.S. Government and her allies, YPG conducts, reports, and supports developmental tests, experiments, production tests, integrated developmental/operational tests, as well as provides training support. In 2016, YPG fired over 212,000 rounds of artillery, mortar, missile, and small arms munitions, flew over 2,900 aircraft sorties, drove over 68,000 test miles, and conducted over 1,500 airdrops. These numbers more than double when events conducted in support of training are counted, and for the last 7 years, YPG has been the busiest Army proving ground.

The Aviation Systems and Electronic Test Division at YPG is responsible for testing aviation ballistic weapons and missiles, unmanned aerial systems, aircraft systems, precision guided and unguided air delivered systems, personnel parachutes, sensors and surveillance systems, and electronic warfare systems. As a developmental test activity, the goal is to help the developer field their systems and provide the Soldier with the safest, most lethal equipment. Instrumentation is critical to capturing the necessary data that is generated by the item under test or capturing the "true" performance of the system. It is also important to display real-time critical information from the system under test so the test team can operate efficiently and safely.

BACKGROUND

One the systems tested at YPG that benefitted tremendously from the real-time display of critical data is the AH-64 Apache Helicopter Fire Control Radar (FCR). The FCR allows the aircrew the ability to quickly survey the battlefield by detecting and displaying both stationary and moving wheeled, tracked, and airborne vehicles. The FCR is capable of detecting and prioritizing hundreds of vehicles, allowing the aircrew to efficiently process all the provided information.

The FCR performance is based on Probability of Detection (the ratio of vehicle detections versus detection opportunities) as well as the False Alarm Rate (the number of times a false detection occurs). When testing the FCR, it is common to utilize 20 or more vehicles of various types for detection trials. During a typical 2-hour test flight, the FCR will scan a vehicle array hundreds of times, resulting in thousands of vehicle detection opportunities. Given the large number of detection opportunities and resulting detections (both real and false), trying to determine how the FCR is performing during the test flight quickly becomes impossible for the test team if the FCR/aircraft and ground truth data is not effectively presented.

After supporting FCR developmental testing for over 25 years, YPG developed an effective way to collect, merge, and display the available data in a way that allowed the test team to conduct real-time analysis of the radar's performance as well as provide enhanced situational awareness to optimize data collection. Figure 1 shows an overview of data flow.

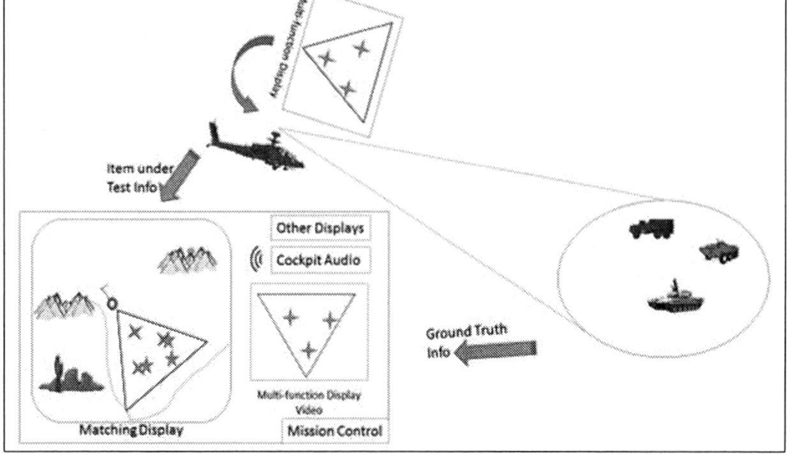

Figure 1. Data Flow Overview

To enable a real-time analysis capability for the test team, information from the Apache helicopter and ground truth information from the actual vehicles are sent to Mission Control to be displayed. The vehicle ground truth information was gathered using existing YPG infrastructure. Figure 2 shows the instrumentation scheme for the Apache.

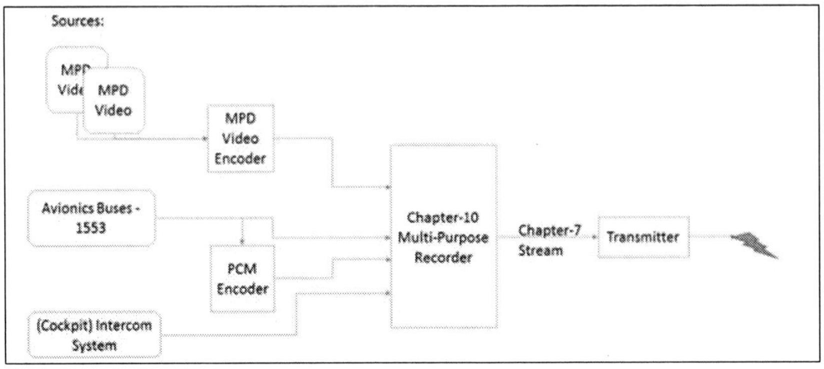

Figure 2. Apache Instrumentation Scheme

The Mission Control room was set up with large projector screens and workstations to display the key information provided to the test teams. In order to gain awareness of what the aircrew was seeing, the cockpit Multi-function display(s) (MFD) was shown with corresponding audio from the cockpit intercom system. Whenever possible, the various pages were displayed on the instrumented MFDs to allow the entire test team to witness the aircrew utilizing FCR to perform the planned test points. It also allowed the various SMEs in Mission Control access to any of the diagnostic or setup pages that could be brought up on the MFDs. This greatly increased the SMEs and aircrews ability to interactively perform different diagnostic and error isolation steps when anomalies were encountered during testing.

Although there were numerous benefits to displaying the cockpit MFDs, there were two main shortcomings for the test team with respect to analyzing FCR performance. The first shortcoming was that the MFDs were oriented with respect to aircraft heading, i.e. out the front of the aircraft. Multiple different headings were used during any particular flight test, which required the test team to reorient themselves when comparing the MFD video to other displays. The second shortcoming was the absence of any ground truth information on the MFDs as the MFDs only show the vehicles detected by the FCR. When FCR scans were performed, it was difficult to validate the detections shown on the MFD as true detections because the actual vehicle array was displayed elsewhere in a traditional map orientation, i.e. the top of the map was north. The aircraft's heading was rarely north so it was very difficult to match the FCR detections depicted on the MFD to the actual vehicle location displayed. The net result of these shortcoming was that the test team could not determine FCR performance from just the MFD video.

Based on the limitations of the displayed MFDs, an integrated display that showed both FCR detections and actual vehicle location was developed. This display became known as the "Matching Display" since both the FCR detections and actual vehicle location were superimposed on a common map. In order to validate an FCR detection, the test team only had to look at the display to see if the detection matched an actual vehicle location. The Matching Display also showed key geographical information such as terrain features and roads in a traditional map orientation (top of the display is north). All these feature combined on a single display allowed the test team to determine how well the FCR was performing. This display also helped the test team verify that the desired test point was actually executed by displaying information, such as radar field-of-view size, center of scan, terrain mode, etc.

Other displays utilized included a meteorological display that presented temperature, pressure, humidity, wind speed, and direction from various points on the test range, as well as a custom FCR Engineering Display. This Engineering used special instrumentation messages that were enabled on a MIL-STD-1553 bus to present important information to the FCR SMEs in real-time that was not usually shown on the MFDs during test.

INITIAL INSTRUMENTATION IMPLEMENTATION

The AH-64E model is the latest variant of the Apache. Currently, the U.S. Army is developing a Version 6 update to the AH-64E Apache that includes improvements to the FCR. With respect to instrumentation, many features were the same as previous models/versions; however, with the addition of a second Ethernet Routing Device (ERD) with Version 6 (ref 1), the FCR data was split with the FCR mode commands/responses being sent via a MIL-STD-1553 bus and all the vehicle detection information being sent via Ethernet bus, shown in Figure 3.

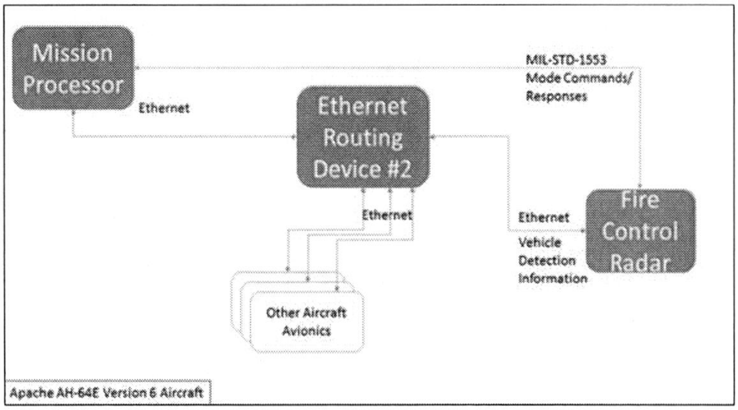

Figure 3. Apache Version 6 Overview

In the summer of 2016, YPG was requested to support a series of FCR developmental flights in support of the Version 6 improvements to the FCR. This was the first time YPG instrumented an Apache Version 6 aircraft, and due to the changes with Version 6, a significantly different instrumentation approach was required. Furthermore, the flight tests were schedule to begin in late summer 2016, which meant the new instrumentation suite would have to be designed, checked out, and installed on the aircraft in approximately 10 weeks.

Given the short time available, the initial efforts were focused on what modifications could be made to the existing instrumentation in order to capture the necessary Ethernet data. The data flow from the aircraft to mission control was broken down into six major areas and each area was examined for any necessary changes. A discussion for each of the six areas is provided below.

1. Aircraft Instrumentation: Perhaps the biggest impact to the existing instrumentation suite was to the aircraft instrumentation. Gaining access to the Apache's Ethernet bus was a primary concern. The Ethernet bus on the Apache is a hub and spoke design and the necessary data for the FCR test was passed on a single spoke to the Ethernet Routing Device (ERD) number 2 (Figure 3). Since it had not been previously instrumented, it was not clear if any new instrument components or specialized pieces such as connectors would be required. The short time necessitated that any new items be identified as soon as possible to ensure that they could be procured in the time available.

   Assuming that the necessary Ethernet access would be possible, the next question was how the Ethernet data would be sampled. Since a Chapter 10 (ref 2) multipurpose recorder was already being utilized and data was being transmitted utilizing Chapter 7 (ref 3) in the existing instrumentation suite, the Ethernet data could be sampled either on the aircraft or on the ground. Regardless of where the Ethernet data was sampled, an on-board recording—by virtue of its lack of transmission errors—was needed for any post-flight data reduction.

   Finally, the transition to Ethernet substantially increased the amount of data that could be sent at one time to or from the FCR. An entire block of FCR detections were now sent in a single Ethernet packet, where previously a block FCR detections required a series of multiple 1553 messages. Also, due to the greater throughput available with Ethernet, the amount sent with each block also increased. It was immediately apparent that these changes resulted in more data that would need to be telemetered in order to maintain the same real-time analysis capability that was provided in previous test events.

2. Downlink: It was understood that more data would be telemetered. At the time of this initial assessment, it was still unclear how much more data would need to be telemetered, but it was believed that it would be well within the capability of the transmitters currently in use. The transmitter utilized on the Apache instrumentation suite had already been used on another effort at a 20 Mbps, and the previous Apache data rate had been 5 Mbps. The only concern was that frequency diversity that was employed on Apache combat antenna masking (ref 4), and the increase in data rate would also result in a 2-fold increase of the spectrum requirements.

3. Receiving Sites: Similar to the transmitters, the impact to the tracking antennas and receivers was assessed as minimal. The components were capable of much higher data rates and had previous been utilized at a 20 Mbit per second rate in support of another test effort.

4. Ground Station: The ground station is a vital component when utilizing a Chapter 7 data stream as it breaks out the different signals that were combined when creating the Chapter 7 stream. The impact to the ground station was assessed as minimal, as the worst case anticipated data stream was well within the capabilities of the ground station, to include Ethernet data streams.

5. Decom: The impacts to the Decom were significant and on par with the impacts to aircraft instrumentation: the need to send more data to ground resulting in a larger data frame; the restructuring of the FCR data to include addition parameters; and, changes in data types meant that little of the existing set-up could be used. Furthermore, previously in support of the Matching Display additional processing had been done to FCR detection data to facilitate the display of data. Because of the significant changes to the FCR detection message, it was unclear what modifications to these processes were required but there would surely be many.

   Finally, as previous discussed, the sampling/parsing of the Ethernet data could potentially be done on the ground with the Decom rather than with the aircraft instrumentation. This brought to light an interesting concern, because although the ability existed to do the sampling/parsing, to do so would require the use of a new Decom. The new Decom had been available for a while; however, the test support workload, some integration issues, and other initiatives kept it from being incorporated into a mission support role.

6. Displays: With all the changes to the data sources, it was assumed that there would be changes needed. Both the Matching Display and the FCR Engineering Display had been developed over many years and had functioned well, so there was very little recent experience with the code behind the display. When trying to decide a path forward, a good deal of thought was put into the impact to the displays and trying to minimize the amount of changes required by doing additional changes upstream of the display. This decision was based on the fact that since other processes upstream would already have to be changed, additional changes would be less of an impact than trying to change the specific displays. Since the data for the FCR Engineering Display remained on the same MIL-STD-1553 bus for Version 6 aircraft as it had previously, efforts were made to preserve the input to the display to allow it to remain unchanged. Unfortunately with increase in data, changes in data types, and other data changes, it was likely that modifications to the matching display would be unavoidable.

IMPLEMENTATION

The resulting implementation is shown in Figure 4.

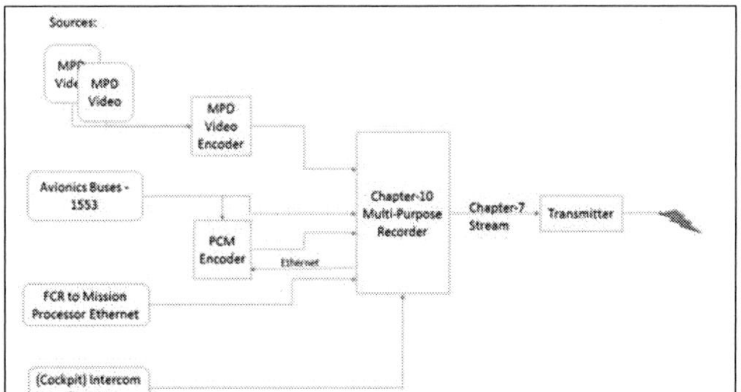

Figure 4. Resulting Aircraft Instrumentation Overview

The Apache Army Program Manager tasked Boeing Military Aircraft (BMA) to provide, among other things, instrumentation and maintenance support for the Version 6 test aircraft. After YPG completed its assessment, one of the first orders of business was to meet with BMA to discuss instrumenting the Ethernet bus. A meeting was held at BMA facility that was supported by key personnel who had extensive knowledge of instrumentation, the Version 6 aircraft, and FCR. This group decided that the best course of action to gain access to the Ethernet bus was to fabricate a custom harness to allow the installation of an Ethernet tap in the FCR to ERD-2 Ethernet path. Once installed, the Ethernet data could be sent to both ERD-2 and YPG's instrumentation with no impact to the aircraft's functionality. BMA agreed to take on the task of building the custom harness, installing the Ethernet tap, and routing the Ethernet signal to YPG's instrumentation pod.

An additional benefit from the meeting at BMA was that the FCR cognizant engineer provided YPG with some samples of the Ethernet traffic between the FCR and the Mission Processor. This combined with Interface Control Document (ICD) information was used to build up the process to sample the Ethernet data.

Armed with these data samples, some initial exploration of processing Ethernet data with the new DECOM was performed. In the end, it was decided to continue to use the legacy DECOM and sample the Ethernet data at the aircraft using the existing PCM Encoder. The primary reason for this decision was the lack of an available Ethernet parser on either DECOM. Although all the necessary tools and functions existed on the new DECOM, a custom parser would have to be written. This would put additional burden on the available resources and since the new DECOM had not yet been utilized, other integration issues would inevitably be encountered so it was decided not to use the new Decom for this effort.

As luck would have it, YPG had an Ethernet module for the PCM encoder that was already being used on Apache. This module was capable of sampling the Ethernet data down to the data word level. Although the native Ethernet structure would be lost, all of the selected Ethernet data would be in a PCM frame and manipulation of a PCM frame using the legacy DECOM was well understood. Even though YPG had the necessary Ethernet module, it had not been used for many years and only then in a much different application. The PCM encoder would now be used to capture all the 1553 and Ethernet parameters needed to drive the displays in mission control, so programing the PCM encoder was not a trivial task. YPG received outstanding customer support from the Encoder manufacture, which saved us precious time getting this critical component ready.

A strategic approach was taken when programing the encoder and building the corresponding PCM Frame. With the inclusion of the Ethernet data, the resulting frame needed to be much bigger. In the previous instrumentation scheme, the encoder output a 1-Mbps stream containing 1248 parameters, which provided a 50Hz sampling rate. With the inclusion of the Ethernet parameters, the encoder bit rate had to be increased to 2-mbits/sec stream containing 2496 parameters to maintain the 50Hz sample rate. A pragmatic approach was taken when creating the new PCM frame with a focus on the time available rather than the engineering aesthetics of design. The new PCM frame was constructed to preserve as many of the data word locations used in the previous frame as possible. Although this resulted in a slightly larger frame, it significantly minimized the number of changes the Decom or displays needed. This approach worked better than expected as the FCR Engineering display ended up needing virtually no changes.

The final consideration for the aircraft instrumentation was onboard recording of the Ethernet data, with the goal to record the Ethernet directly to the multipurpose Chapter-10 recorder being utilized. To accomplish this, the Ethernet data flow needed to take the following circuitous route. The Ethernet data was sent to YPG's instrumentation via the BMA Ethernet tap. Next, the Ethernet data was sent to the Chapter-10 recorder, which both recorded and passed the data to the PCM encoder. The PCM encoder ingested the Ethernet data, sampled the required parameters, and then sent the resulting PCM stream (with 1553 data) back to the recorder where the encoder's PCM stream was recorded. Finally, the Chapter-7 stream was created and sent to the transmitters.

Once the Ethernet data flow was finalized, it was realized that an additional multipurpose recorder module not previously identified was needed. This was a serious issue because by this time, there was not enough time to procure and receive the necessary module prior to the start of the test. YPG explained their situation to the recorder vendor and gratefully accepted an offer from the vendor to utilize one of their demo cards to support the upcoming test.

The resulting Chapter-7 stream was 7.5 Mbps and contained the PCM stream from the encoder, which now included both 1553 and Ethernet parameters, the 2 MFD encoded videos, and cockpit intercom. This was an increase from the previously implemented 5 Mbps Chapter-7 stream, but as anticipated no extraordinary efforts were needed to program the transmitters, receivers, and ground station for this higher bit rate. The spectrum requirement did increase, but fortunately any negative impacts were able to be mitigated by careful coordination of all the test programs on the range.

Even though all the necessary data was now accessible at mission control and efforts had been made to minimize the impact to the DECOM, there was more work to be done. When it came to programing the DECOM, the new PCM data frame was twice the size of the previous data frame and more than half of the parameters were completely new. Ultimately, all the data words, new and old, would have to be mapped from the PCM data frame to specific places in data blocks that would be sent to the displays. Also, any engineering unit conversions, discrete bits, multiple data word manipulations, and any additional data processing would have to be done. Once completed, the painstaking task of checkout began. The good news was that with respect to data words for the FCR Engineering Display and many of the FCR mode commands/responses, much of the previous work could be quickly modified to work in the new program since they remained on the MIL-STD-1553 bus and their data word locations from the previous data frame were preserved in the new frame; unfortunately, all the Ethernet data words were new and all had to be programed from scratch.

With respect to the displays, the careful plan to preserve has much of the previous structure paid dividends as the FCR Engineering required only minor changes to be fully functional. The Matching Display required a significant amount of reprogramming and check-out was difficult because of the lack of a contiguous data set that included both Ethernet and 1553 data. Included in the Matching Display development was the efforts for the programmers to re-familiarize themselves with the display since it had been more than 5 years since a significant update had been performed. This is not uncommon for specialized displays that are tied to an Army procurement cycle, i.e. feast or a long famine, but can be a challenge when working on a tight schedule. Standards and documentation help but even in the best cases, some re-familiarization will always be needed.

RESULTS

The preliminary build up was completed, checked out to the greatest extent possible, and 2 weeks prior to the start of flight testing the aircraft instrumentation was sent to the BMA for installation on the aircraft. Installation and a 3-day ground check-out on the aircraft was a joint effort between YPG and BMA personnel. The data flow on the aircraft was replicated fairly completely, to include the transmitting and receiving data from the aircraft. The data flow on the ground side was incomplete, because it was impractical to transport all necessary personnel and equipment to the BMA facility to simulate the ground side data flow.

The check-out included utilizing the FCR, and benefitted greatly from a built-in mode in the FCR that could simulate a complete radar scan, including target detections. A series for simulated scans was conducted and recorded so that a complete check-out of the displays could be done at YPG prior to aircraft arrival. Overall, the check-out went better than expected and all the modifications done by BMA and YPG seemed to be functioning as advertised. With the completion of a successful ground check, the YPG team packed up and headed home to continue the checkout process with the data captured.

The actual flight test went well, with only minor problems and corrections needed to provide the minimum support required by the test team. Although this seems a bit anticlimactic, given the large number of changes implemented in a short period of time, the flight test was a great success. Also, because this was a typical instrumentation effort, a series of anomalies with the existing components caused some added excitement to the test support. Similar to the Ethernet instrumentation, the effects of these anomalies were mitigated enough to allow the minimum requirements to be met. The problems encountered reminds us that no matter how many pre-flight checkouts are conducted, complex instrumentation efforts still need some flight time for check-out.

Future Goals: Based on the results of this initial implementation, several areas were identified for improvements either to provide more capability or eliminate older components. With respect to the aircraft instrumentation, the improvements focused on the multi-purpose recorder. The multi-purpose recorder can be configured to, in addition to recording, filter the incoming Ethernet and 1553 messages for transmission to the ground through the Chapter-7 data link; therefore, making the PCM encoder currently being used redundant, and removable. This PCM encoder is by the oldest component utilized on the aircraft and although very reliable it is long past its service life.

Removing the PCM encoder from the Apache would result in a lighter more compact instrumentation suite that would draw less power. It would also force us to handle the received data differently as the ground station would now output a Chapter 10 User Datagram Protocol (UDP) data stream containing the Ethernet and 1553 data of interest. The DECOM utilized for this test is not capable of receiving this UDP stream; however, the beauty of this UDP stream is the possibility to DECOM data with only software, i.e. no specialized hardware. In the near term the newer DECOM originally passed over for this test effort will be brought on line, eliminating the proven but aged DECOM.

The final improvement would be an effort to streamline connecting to the Apache's Ethernet buses. The current method of using a modified harness and an Ethernet tap works well, but is both intrusive and limited to a single Ethernet bus. A better approach would be to connect directly to the ERD-2 (see Figure 3) utilizing an available port. This would eliminate the need for any modification or custom harness. The challenge with this approach is that once connected to the ERD-2, the proper subscription/publish handshake would have to be established to gain access to the data. In this approach, the multipurpose recorder's Ethernet module would be modified to provide the appropriate handshake, and request information from any of the Ethernet buses. Initial investigation indicates the multipurpose recorder is capable of this, but much more coordination with the recorder manufacture, BMA, and the Apache PM will be needed to implement.

CONCLUSION

It would be easy, but not realistic to conclude that more time would have allowed YPG to implement a better approach than what was ultimately flown. The configuration flown met all the requirements, including schedule constraints, and improvements based on careful analysis of the results can and will be made for upcoming tests. Although the short suspense resulted in a hectic schedule and more than a little stress, it also resulted in increased instrumentation capability. Though requirements for the particular test were not completely anticipated, the capability to instrument Ethernet has been part of YPG instrumentation investment planning for years. That fact coupled with very good working relationships with several vendors plus a little luck resulted in a successful endeavor. Test support will always have the highest priority, but time for investment planning as well as interacting with industry is also needed for long term success.

ACKNOWLEDGMENTS

YPG would like to thank the technical representative from Curtiss-Wright Avionics & Electronics, Zodiac Data Systems, Boeing Military Aircraft, and Northrup-Grumman that provided valuable and much needed information in a very timely fashion.

REFERENCES

1. Haggar COL Jeffrey E. and Cofer, COL (Ret.) Robin D, "Apache Helicopter Project Office Update" ARMY AVIATION Magazine.com, 2 March, 2016, http://www.armyaviationmagazine.com/index.php/archive/not-so-current/974-apache-helicopter-project-office-update.

2. Telemetry Group, Telemetry Standards, IRIG Standard 106–13 (Part 1), Chapter 10, Secretariat, Range Commanders Council, WSMR, New Mexico, June 2013.

3. Telemetry Group, Telemetry Standards, IRIG Standard 106-15 (Part 1), Chapter 7, Secretariat, Range Commanders Council, WSMR, New Mexico, June 2013.

4. Diehl, Michael, Swain, Jason, and Wilcox, Tab, Rotary-Wing Flight Tests to Determine the Benefits of Frequency and Spatial Diversity at the Yuma Proving Ground, Air Combat Systems Directorate, U.S. Army Yuma Proving Ground, Yuma AZ, August 2016.

PUREmodules: An IoT Development Method to Simplify Prototype Hardware

Furqaan Afzal[1], Michael Rodriguez[1], Kyle Lee[1], Sashi Ono[1], Hua Lee[2], and Vincent R. Radzicki[2]

[1]Pure Engineering,
Santa Barbara, California
[2]Department of Electrical and Computer Engineering
University of California, Santa Barbara

ABSTRACT

Conventional hardware development of sensor systems can be complex and tedious due to the time-consuming procedures of prototyping the physical hardware and the lack of firmware examples for testing and evaluation. In this paper, we introduce the concept of a new development process in the format of one single core module with a standardized interface so that a large number of sensors may be connected easily in an ad-hoc manner. The sensors can then be programmed with standard development tools and processes. The concept of the design modules enables effective and rapid prototyping of hardware implementations without prior knowledge of the low-level details of the hardware components. With this capability, sensors can be connected to a core module for rapid development of low-cost high-performance devices.

INTRODUCTION

In recent years, semi-conductor fabrication technology has improved at a rapid pace making high-performance integrated circuit technology commercially available at extremely low cost and in very small packages. Microcontroller Integrated Circuits (ICs) and Micro-Electro-Mechanical Sensors (MEMS) specifically have greatly benefited from these improvements in process and fabrication engineering technology, yielding devices with extremely small form factors as well as increased power performance and functionality. Figure 1 displays the integrated circuit die of the PIC16C505 microcontroller and a MEMS device illustrating the extremely small-scale of this technology. The main costs of designs that utilize these components are now mainly driven by the hardware and software engineering development work required to produce functional sensor systems.

As a result of developments, the controls and sensor electronic components that are commercially sold today are often available with very low Size-Weight-and-Power (SWAP) specification. It is therefore desirable that fully-developed sensor systems are designed to minimize the overall SWAP of the system as well. There are many applications both military and commercial ones that would benefit greatly from reduced SWAP sensor systems, which is a main focus of industry work today. One of the largest market forces driving the commercial demand is the development of Internet-of-Things (IoT) products [2]. IoT products generally cover any sensor that can communicate either directly or indirectly with networked computers to transmit its sensor data for further centralized processing, such as automobile sensors that transfer vehicle performance data to the manufacturer for reliability tracking.

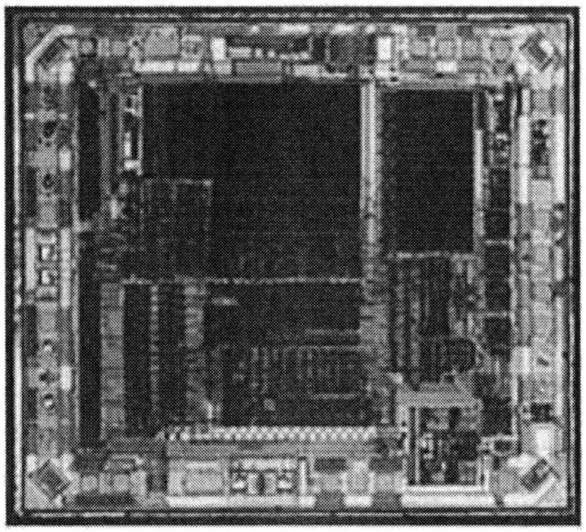

(a.)

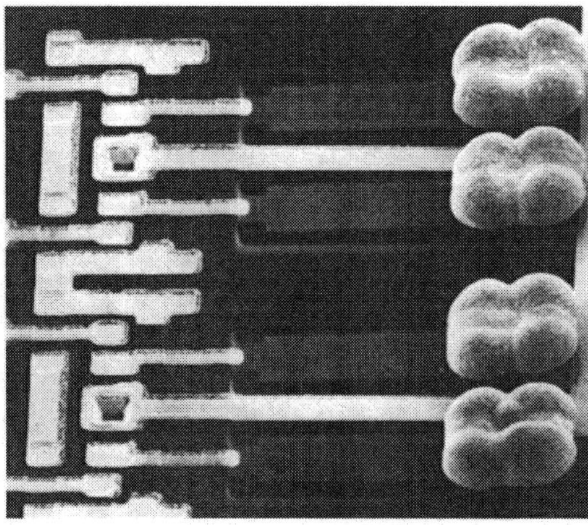

(b.)

Figure 1. PIC16C505 microcontroller die with 1200nm fabrication technology (a.) and silicon capacitive MEMS accelerometer (b.) [1]

Engineering and development of sensor systems in the past typically involved prototyping the system design with large rack mounted test and measurement equipment. In many cases the final product would be of similar size and shape of the large test-equipment although more tightly integrated. Designing small sensor systems for potential IoT applications, however, poses a new set of challenges to successful implementation of the system. Software complexity, electronic noise coupling, timing and synchronization problems, and electromagnetic interference (EMI) can all become serious issues when integrated multiple sensor packages and micro-processors into a small integrated embedded system. To save on development costs and minimize risk for miniaturized sensor systems, it would be desirable to start prototyping with the embedded sensor, communications, and microprocessor components immediately to identify the most problematic issues that could be encountered in the system design. Oftentimes, however, the initial software and hardware work required to construct a development environment to even begin testing the embedded sensors can be time-consuming and arduous. Once full-scale system testing is begun, it may be discovered that key sensor components are not adequate to meet the required system performance specifications. Suitable replacement parts may require completely different communication and software interfaces invalidating much of the design and development work.

A modular sensor architecture that can quickly begin acquiring data from a variety of commercially available sensor components with minimal development work would provide significant value to the engineering design process of embedded sensor systems. The motivation behind PUREmodules was to address this need by simplifying the development of prototype hardware mainly by building an adaptable interface for sensors around tested and robust hardware designs. The PUREmodule system is designed to be modular so that different sensors can be attached and detached from the system through simple mechanical interfaces. The benefit of this approach minimizes the need for constantly re-building custom printed circuit boards for similar sensors so that the risk and costs are drastically reduced. The overall goals of this project are to organize robustly tested hardware and software into modular components, provide a quick way to instantiate and experiment with new sensor designs, allow custom hardware to be easily added, and ensure that design resources are spent on the highest priority needs of the system.

SENSOR MODULE ARCHITECTURE

Careful consideration and design work was required to construct a modular hardware/software platform that is general enough to interface with a wide variety of sensors while also providing enough specialized functionality to be of utility to engineers. To meet these goals Pure Engineering has developed the PUREmodules architecture, which aims to develop an open and easy way to make plug-and-play sensors for developing hardware prototypes. By standardizing on a single common pinout and making it openly available, the modules can be made compatible with each other, and any interested parties can also add to the ecosystem of sensors by contributing their designs. A combination of the right set of boards can produce a custom hardware solution that can be assembled very rapidly to serve as the prototype sensor system. The key set of architecture design decisions consisted of determining the hardware connectivity and software protocols the sensor modules would utilize to communicate with one another.

The printed circuit board (PCB) layout and software work redundant to multiple sensor architectures was compiled into a single general purpose board that could emulate many embedded sensor systems. To allow for prototypes to be small and low cost a 62mil PCB edge connector was chosen as the hardware connection interface. Using a PCB edge connector keeps the sensor costs low and can be added to nearly any development board with no additional cost since no physical connector requires soldering. The 10 pin AVX Open Ended Card Edge (BTB) was chosen as the default interface to connect two PCBs together. The connection is very robust and allows for several insertions and can carry higher currents if needed, and provides polarity tabs to reduce the risk of accidental reversal. Lastly, the pin density is relatively high so that it does not require large amounts of PCB space.

The digital communications interface was the other key decision faced in the architecture design. For this task, the Inter-Integrated Circuit (I^2C) protocol was chosen to be the primary interface for communicating with new sensors added to the system. I^2C is a popular digital interface used in embedded systems today and was chosen mainly for the simplicity of its two wire interface and multi drop architecture. The protocol is typically used for low power hardware sensors, and for most common sensors no interconnect logic is required. Each sensor module has space for two BTB edge connectors, meaning one can be used as an input and the other as an output for the I^2C communications bus allowing for multiple sensor modules to be daisy chained together. The connection is very robust, and can carry high currents if needed. The other major digital interface used in embedded systems design is Serial Peripheral Interface (SPI) communications. For sensors that require an SPI connection, the General Purpose Input-Output (GPIO) can be used with the sensor to act as an SPI communications bus.

For sensors capable of low power sensing wherein the sensor is triggered before making a measurement, it useful to run an interrupt to the main microcontroller. An open pin is reserved for this function where the interrupt is modeled as an open collector, so that more than one sensor can trigger the microcontroller without resulting in contentions of the one interrupt line. If the sensor is in a push-pull configuration for its interrupt an Negative-Field Effect Transistor (N-FET) can be used to convert the configuration to open-collector input-output (IO).

The architecture proposed here is a balance between low-cost hardware manufacturing with flexible communications functionality so that the PUREmodules can achieve their stated goals from the previous section. This multi-layer interconnected system allows for daisy chaining multiple sensors while communicating over the main axis connectivity busses. With SPI and I^2C communications enabled, PUREmodules offer a low-cost prototyping platform that can communicate with the vast majority of embedded sensors commercially available today.

HARDWARE RESULTS

An early prototype of the PUREmodule has been designed and fabricated to determine its utility to the design process. A computer-aided design (CAD) model of the product can be seen in Figure 2. The small form factor and the general-purpose BTB edge connectors can be seen here

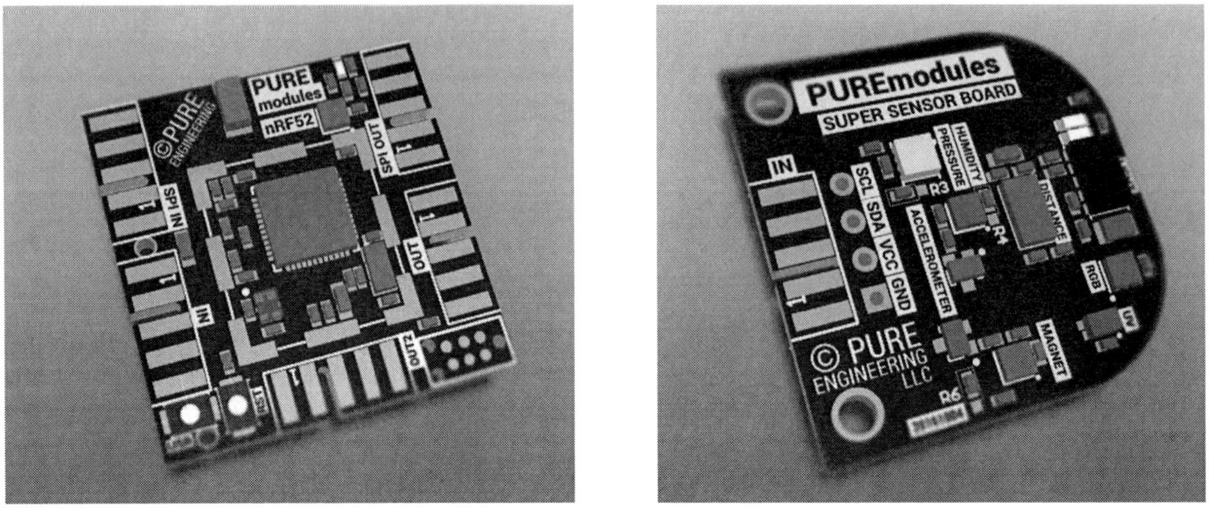

(a.) **(b.)**

Figure 2. PUREmodule core sensor module (a.) and super sensor board (b.)

The device in Fig. 2a is the CoreModule board which houses a Bluetooth capable Nrf52 microprocessor and an accelerometer. This CoreModule board is responsible for all the processing and computation required in any project. The sensor in this configuration in Fig. 2b is the PureModules SuperSensor which can fit multiple sensor components that can all communication with the CoreModule board directly. This system allows for rapid hardware prototyping to determine how multiple will sensors work together in a single integrated system.

Tables 1 and 2 demonstrate how to set the pin-out of the CoreModule boards to enable I^2C and SPI communications respectively. The I^2C connection simply requires setting pins 5 and 6 to the appropriate I^2C required data lines and preserving the remainder GPIO pins for other tasks. For SPI sensors, the GPIO pins 2-5 are reserved for SPI communication lines. To prevent accidental plugging into the I2C sensor port, the connector will be reversed. Additionally the recommended placement of the SPI port is to be adjacent to the standard I^2C sensor expansion port. This would allow both ports to run to a sensor board if needed.

The common pin-out is as follows

1	VCC system	2-3.3V range. Typically could be directly tied to a lithium coin cell or two AA's in series
2	RAW battery	3.3V-15V range. Cound be a li-ion battery, wall wart, solar cells, etc. this raw battery is typically regulated down to 3V-3.3V
3	GND	Common ground reference point
4	SDA	
5	SCK	
6	GPIO	(typically would be the uart TX from the micro-controller)
7	GPIO	(typically would be the uart RX to the micro-controller)
8	GPIO	
9	GPIO	
10	GPIO	(typically this connects to an interrupt open collector bus to the micro-controller

Table 1. I^2C Configured Pinout of the PUREmodule

1	VCC system	2-3.3V range. typically could be directly tied to a lithium coin cell or two AA's in series
2	SCLK	
3	GND	Common ground reference point
4	MISO	
5	MOSI	
6	GPIO	Chip Select pin
7	GPIO	Chip Select pin
8	GPIO	Chip Select pin
9	GPIO	Chip Select pin
10	GPIO	Chip Select pin

Table 2. SPI configured pinout of the PUREmodule

A hardware prototype of the design can be seen in Figure 3. Here the CoreModule is attached to the Super Sensor board through the BTB edge connector. The sensor board combines 8 different sensors into a very small form factor, and preserves communication fidelity and the measurement accuracy of each component.

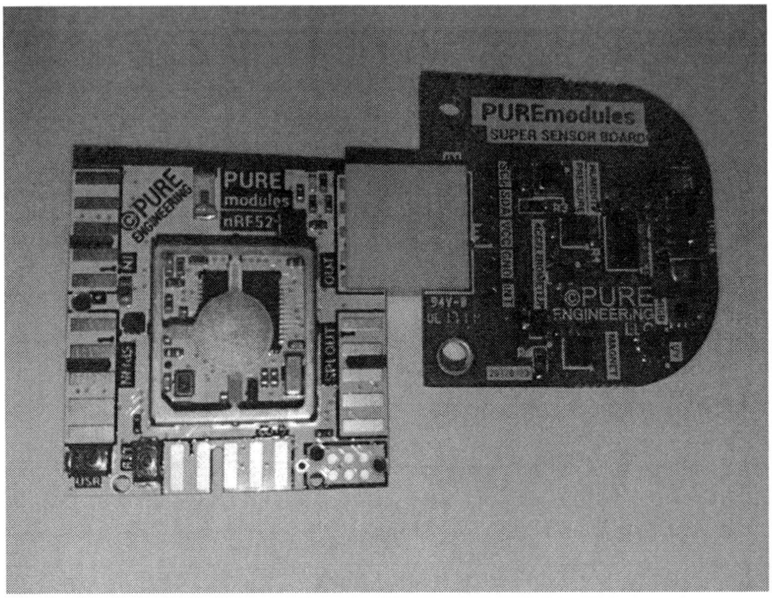

Figure 3. Hardware Prototype of PUREmodule Sensor System.

CONCLUSIONS

The goal of this project was to make a simple, low cost, low power sensor prototyping system, which we believe was achieved by PUREmodules. The system supports a variety of firmware platforms and offers modularity to keep prototyping cost down, and thus the overall development and implementation costs of sensor system products. It also introduces different ways to reduce the power of a sensor system. As the market for IoT products grow and price competition drives to keep the development costs low, the PUREmodules system offers an efficient and robust solution. One of the motivations of this paper is to encourage other parties interested in engineering and development of embedded sensor systems to contribute to the PUREModule system by contributing designs for various sensors of interest. It is the hope that a community of IoT product developers can grow from this effort and a wide variety of sensor will be available to quickly test and prototype with for any desired application. Future work is focused on further reducing the module shape and size for applications that require minimal sensor size.

REFERENCES

[1] Angell, James B., P. W. Barth, and Stephen C. Terry. "Silicon micromechanical devices." Scientific American 248 (1983): 44-55.

[2] I. Yaqoob et al., "Internet of Things Architecture: Recent Advances, Taxonomy, Requirements, and Open Challenges," in IEEE Wireless Communications, vol. 24, no. 3, pp. 10-16, 2017.

[3] PUREmodules… for dreamers, tinkerers, hackers and designers, Kickstarter "Why did we make PUREmodules " Nov. 2016.

[4] Simplifying development of low power wireless IoT hardware and software sensor prototypes with a modular sensor architecture, Hackaday, Sept. 2016

[5] https://www.linkedin.com/pulse/7-advantages-prototyping-modular-hardware-sashi-ono

[6] https://github.com/PureEngineering/PUREmodules

OPTIC FIBER SENSOR FOR STRAIN MEASUREMENTS IN HIGH TEMPERATURE SENSING APPLICATIONS

Julia White
Department of Electrical and Computer Engineering
Missouri University of Science and Technology
Rolla, MO, 65401
jwk8f@mst.edu

Faculty Advisors:
Jie Huang, Kurt Kosbar

ABSTRACT

Optic fiber sensors are employed in a variety of applications for the remote measurement of various parameters such as strain, pressure, or temperature. These sensors offer an array of benefits as well including light weight, compactness, and high resolution. In particular, Fabry-Perot interferometers (FPIs) maintain these benefits and can also be made to withstand extremely high temperatures. This advantage of the FPI allows it to be used in harsh environments where many other tools for parameter measurement could not survive. An FPI strain sensor is constructed and tested which has the capabilities to be used at high temperatures of over 1000°C for applications in gas turbine engine testing. This paper discusses the need for high temperature strain sensors in engine testing and this sensor's capabilities in this application.

INTRODUCTION

Strain measurements are required in various applications with harsh, high temperature, environments. While high temperature strain gauges exist on the market to meet these needs, most have limits ranging from 400 to 800 °C which can't withstand temperatures encountered in certain cases of engine testing. However, the Fabry-Perot interferometer (FPI) can be used as a strain sensor in these environments due to its ability to withstand high temperatures of over 1000 °C.

FPIs are well suited for applications in engine testing due to the benefits of optic fiber sensors (light weight, compactness, and high resolution [1]), and also for their high temperature thresholds [2]. The purpose of this paper is to show the capabilities of FPIs in terms of strain measurement for the high temperature conditions that occur in engine testing. This paper will demonstrate these capabilities through an explanation of the theory behind using FPIs for gas turbine engine testing and an explanation of the construction and testing of a FPI.

A SUMMARY OF THE FABRY-PEROT INTERFEROMETER

A FPI consists of two parallel reflective surfaces that can be used to measure various physical parameters such as strain, pressure, or temperature [3,4]. These measurements are based off of an interference pattern. The interference pattern is created when light waves reflected off of the two surfaces interfere with each other due to their different delays. This pattern can be analyzed to remotely determine the values of the aforementioned physical parameters by observing changes in the resulting waveform's phase, light intensity, or wavelength. In particular, changes in strain on an FPI result in a phase shift caused by distortion of the light being reflected off of the two parallel surfaces. Strain on the FPI physically lengthens the space between the reflective surfaces which causes this distortion. This is modeled by the formula for strain, ε, given in Equation (1).

$$\varepsilon = \Delta_L/L = \Delta_\lambda/\lambda \qquad (1)$$

Where l is distance, Δ_L is the change in distance, λ is wavelength at a certain point on the wave when there is no strain, and Δ_λ is the change in wavelength (phase shift) of this point when there is ε strain. Using (1), we can determine a value of strain with only the initial waveform where strain is zero and a secondary waveform where strain is ε. These waveforms are illustrated in Figure 1.

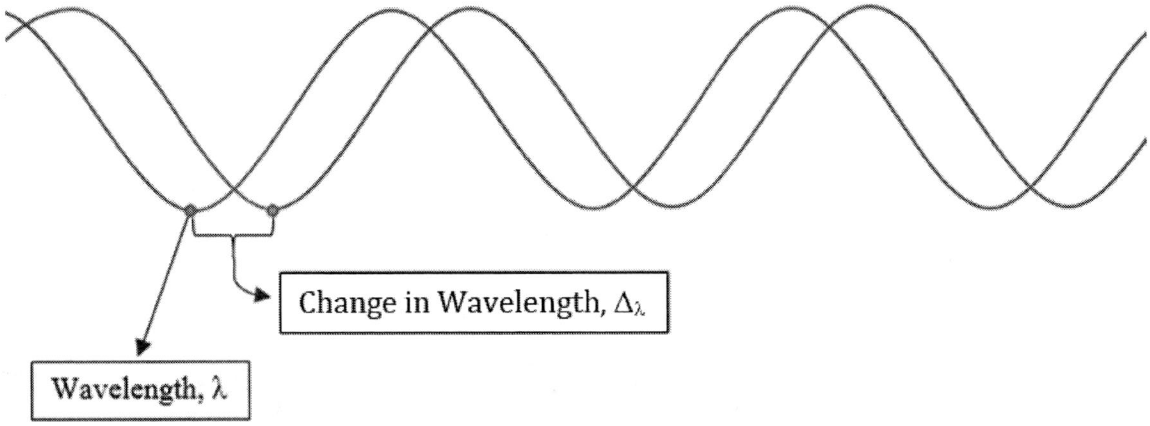

Figure 1 Waveforms Illustration

In addition to strain the FPI can also be used to measure other parameters such as temperature (which has interrelated effects on the interference pattern). Because of this, FPIs have been designed to measure and withstand extremely high temperatures of over 1000 °C using various methods of construction [2]. In the case of a FPI made with glass tubing, since glass fibers have extremely high melting points, a particular fiber could withstand up to 1200 °C before softening.

THEORY BEHIND IMPLEMENTATION IN GAS TURBINE ENGINE TESTING

Gas turbine engines reach temperatures of up to 2000 °C at their core due to the high temperatures produced by burning fuel. However, since the metals used in most jet engines begin to melt at 1300 °C, cooling mechanisms are implemented in order to keep temperatures below that threshold [5,6]. These temperatures are manageable for many variations of the FPI model.

FPIs can be constructed to withstand temperatures above 1000 °C, which makes them a good solution for gathering strain measurements in high temperature gas turbine engines where most strain gauges are not suitable. Temperature does have certain effects on the interference pattern of an FPI though. FPIs designed for measuring temperature typically have a cavity material with a temperature-dependent refractive index. This needs to be accounted for when taking strain measurements because temperature will alter the phase shift produced by a certain strain (i.e. the same strain value will show different wavelengths for a given peak at different temperatures) [7,8]. In order to accurately use the phase shift of the interference signal to measure strain, the FPI must be calibrated before measuring strain at different temperatures by using a temperature sensor in conjunction near the FPI. The measurements can then be used to account for the temperature strain cross-sensitivity of the FPI (which is a value in units of microstrain/°C and will vary for different constructions) and calculate strain.

FPI CONSTRUCTION

The FPI implemented in this report is a simple design which consists of single mode fiber and glass tubing. The FPI was prepared by first fusing a single mode fiber with a section of glass tubing. Then, the glass segment was cleaved under a microscope to the length of approximately 100 microns. After that, the cleaved end of the glass tubing was fused to another single mode fiber as shown in Figure 2.

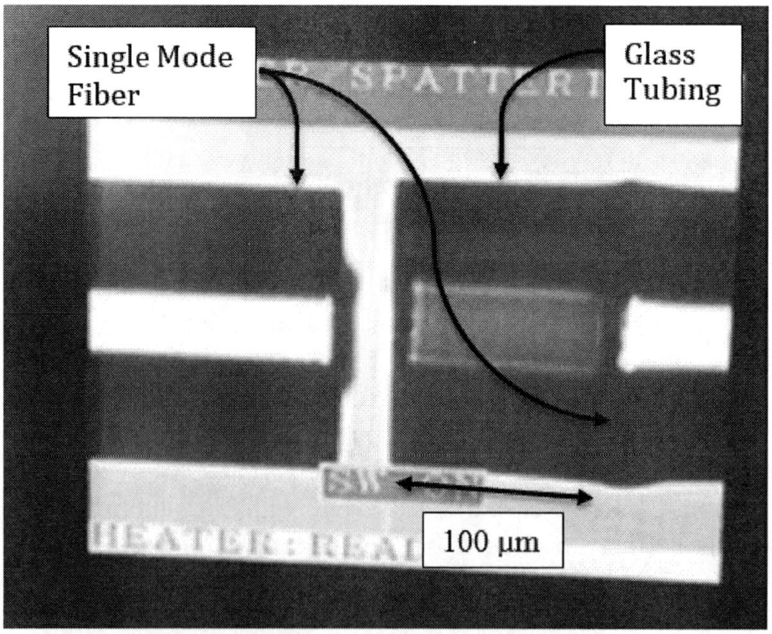

Figure 2 Fusing Single Mode Fiber with Glass Tubing

The surfaces where the fiber and glass tubing were fused together act as reflective surfaces. These surfaces reflect light with different delays which results in the interference pattern that is used to measure strain. Figure 3 depicts this construction.

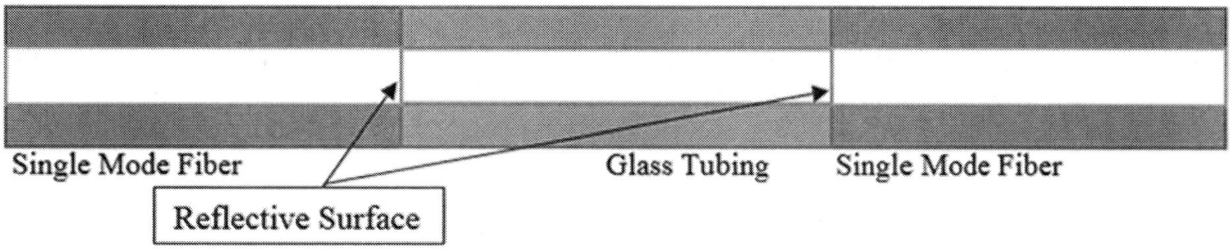

Figure 3 Fabry-Perot Interferometer

STRAIN TESTING

Strain testing of the FPI was performed with the use of an optical spectrum analyzer and a broadband ASE (Amplified Spontaneous Emission) light source. The single mode fiber on one end of the FPI was connected to the light source and the other end was connected with the spectrum analyzer. The resulting waveform displayed on the spectrum analyzer was irregular due to the varying amplitudes of the ASE light source. In order to normalize the waveform used for analysis, the FPI's output waveform was subtracted by the light source's waveform in the spectrum analyzer. This produced a stable sine wave which was recorded and used for calculations. The spectrum analyzer and light source connections are shown in Figure 4.

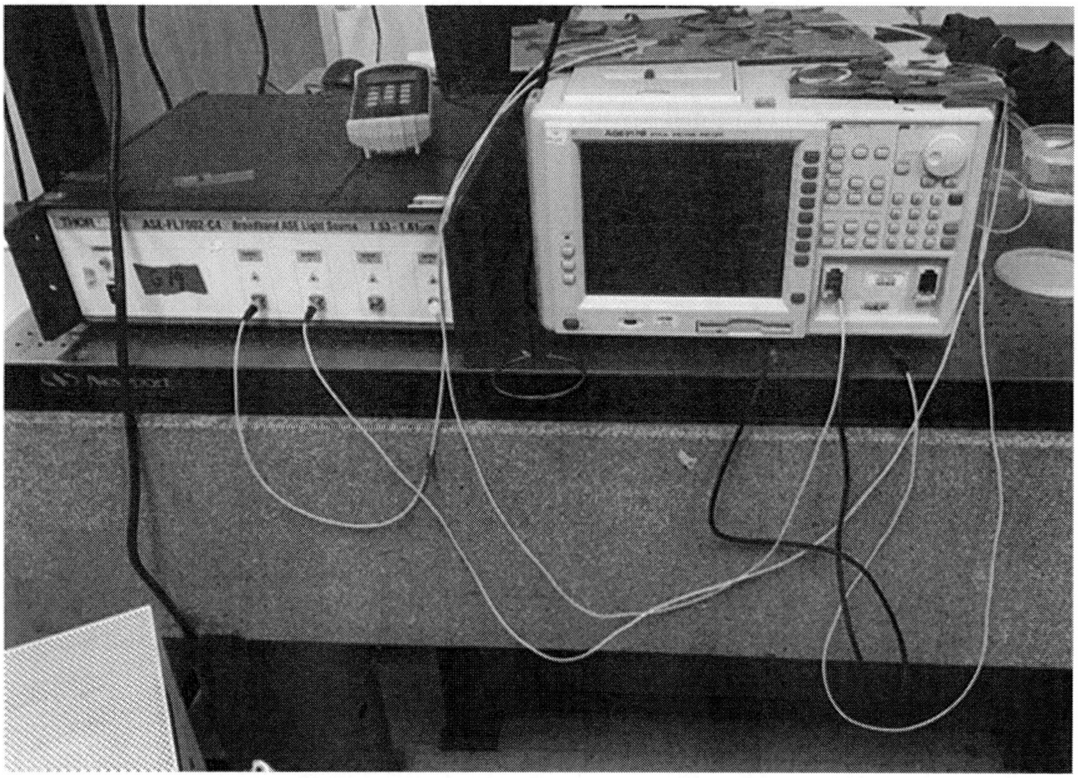

Figure 4 Spectrum Analyzer Connections

With the spectrum analyzer set up, the FPI was then glued with an epoxy resin onto two moveable plates as shown in Figure 5. After drying the epoxy, the waveforms were recorded as the plates were shifted apart in increments of approximately 0.025mm, slowly stretching the FPI. This stretching produced strain which could be measured by both the length of the FPI and wavelength measurements gathered on the spectrum analyzer.

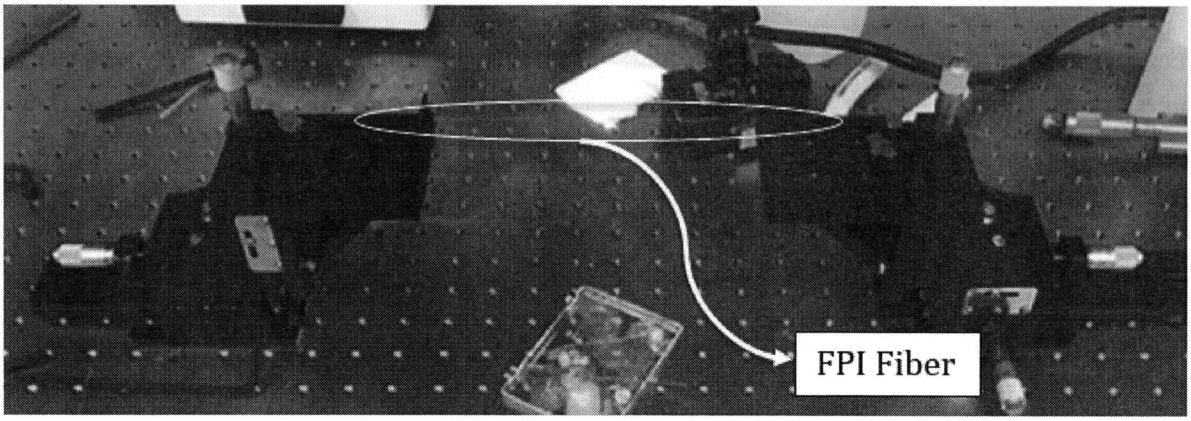

Figure 5 FPI Fixed to Moveable Plates

RESULTS AND ANALYSIS

The output waveforms gathered from the spectrum analyzer were recorded as the strain was increased. Four of the eleven waveforms gathered over eleven trials with varying strains are shown in Figure 6 to demonstrate the phase shift observed as strain increases. The trials depicted below (trials 0, 3, 6, and 10) are labeled with the FPI's change in length (Δ_L).

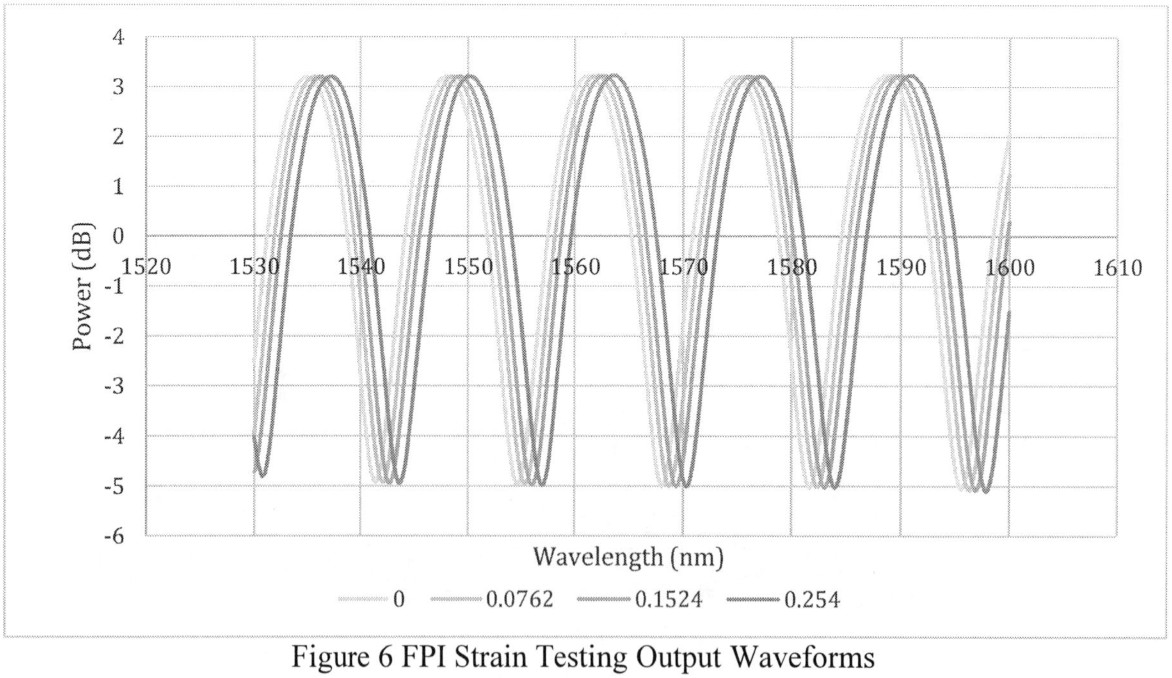

Figure 6 FPI Strain Testing Output Waveforms

A simple zero crossing technique was used to determine the phase shift of the waveform as strain increased. This method, rather than some more complicated signal processing techniques for calculating phase shift, was sufficient since the data presented in this report has a high signal to noise ratio. The third zero-crossing of the trial 0 (no strain) waveform was used as a reference point and its change in wavelength (Δ_λ) was tracked through the following ten trials.

Table 1 FPI Strain Testing Derived Values

Trial	Δ_L (mm)	Δ_L/L	Δ_λ (nm)	Δ_λ/λ	Percent Error (%)
0	0	0	0	0	0
1	0.0000254	0.000141	0.21	0.000136	3.623442
2	0.0000508	0.000282	0.455	0.000295	-4.40794
3	0.0000762	0.000423	0.665	0.000431	-1.73081
4	0.0001016	0.000564	0.91	0.000589	-4.40794
5	0.000127	0.000706	1.05	0.00068	3.623442
6	0.0001524	0.000847	1.295	0.000839	0.946316
7	0.0001778	0.000988	1.505	0.000975	1.328762
8	0.0002032	0.001129	1.75	0.001133	-0.39225
9	0.0002286	0.00127	1.995	0.001292	-1.73081
10	0.000254	0.001411	2.24	0.001451	-2.80166
					λ = 1544.14 nm
					L = 180 mm

Table 1 gives values for the change in wavelength of the third lower peak (Δ_λ) and length the FPI was stretched (Δ_L) in all 11 trials as well as the initial value for wavelength (λ) and length (L) before any stretching. Values for Δ_L/L and Δ_λ/λ are also given and, by Equation (1), these values both represent strain. These two columns (Δ_L/L and Δ_λ/λ) when compared give the error values listed in the sixth column. With Δ_L/L taken as the theoretical strain value and Δ_λ/λ taken as the experimental strain value, percent error was calculated by Equation (2).

$$\%_E = ((A - E)/A)*100\% \qquad (2)$$

Where $\%_E$ is percent error, A is the actual value for strain determined from the FPI length, and E is the experimental value for strain determined by the wavelength. The errors calculated (which range from -4.40794% to 3.623442%) are attributed to random error because the average value for percent error of all 11 trials is within (-)1% of zero at -0.54086% and therefore the error was not consistently in a single direction (which would indicate systematic error). The discrepancies between the two strain calculations are a product of precision limitations that can be attributed to imprecision of the measurements gathered by the spectrum analyzer. This could be minimized with better, more precise, equipment. However, even with random error the accuracy of the strain measurements given by $\Delta\lambda/\lambda$ are consistently within 5% of the actual strain values which shows that the phase shift of these waveforms could be used to relatively accurately (within 95% of the true value) predict strain at constant temperature.

CONCLUSIONS

The Fabry-Perot Interferometer is an alternative solution to taking high temperature strain measurements required in gas turbine engine testing without the use of a strain gage (most of which can't operate at high enough temperatures to accommodate the harsh environments encountered in certain cases of engine testing). The FPI's ability to withstand extreme temperatures and accurately, remotely, measure strain through changes in the interference pattern make it useful in these applications. This paper discussed the theory behind the implementation of FPIs in the high temperature environments which occur in gas turbine engine testing. The strain sensing capabilities of the FPI were also demonstrated as a FPI was constructed and tested for various values of strain. Future work could be performed to demonstrate potential variance in strain measurement limits at different temperatures.

ACKNOWLEDGEMENTS

The laboratory work discussed in this paper was performed with the assistance Yiyang Zhuang, graduate student at Missouri University of Science and Technology in the Electrical and Computer Engineering Department.

REFERENCES

[1] K. Grattan and T. Sun, "Fiber optic sensor technology: an overview," Sensors and Actuators A Physical, vol. 82, iss. 1-3, May, 2000, pp. 40–61.
[2] I. Rajibul, "Chronology of Fabry-Perot Interferometer Fiber-Optic Sensors and Their Applications: A Review," Multidisciplinary Digital Publishing Institute, 14, April, 2014, pp. 7451-7488.
[3] S. Tseng and C. Chen, "Optical fiber Fabry-Perot sensors," Applied Optics, vol. 27, iss. 3, Feb, 1988, pp. 547-551. 10.1364/AO.27.000547

[4] Q. Yu and X. Zhou, "Pressure sensor based on the fiber-optic extrinsic Fabry-Perot interferometer," Photonic Sensors, vol. 1, 2011, pp. 72-83.
[5] "The Jet Engine: A Historical Introduction," Stanford, 16, March, 2004.
[6] T. Jenkins, S. Allison, and J. Eldridge, "Measuring gas turbine engine component temperatures using thermographic phosphorus," SPIE, 20, March, 2013. 10.1117/2.1201303.004769
[7] G. Beheim, "Fiber-Optic Temperature Sensor Using a Thin-Film Fabry-Perot Interferometer," Case Western Reserve University, May, 1997.
[8] L. Yujie, H. Ming, and T. Jiajun, " Fiber-Optic Temperature Sensor Using a Fabry-Perot Cavity Filled With Gas of Variable Pressure," IEEE Photonics Technology Letters, vol. 26, iss. 8, 15, April, 2014, pp. 757-760.

TELEMETRY ENCODER MINIATURIZATION AND RUGGEDIZATION USING MULTI-CHIP MODULE AND SILICON DIE SCAVENGING TECHNOLOGY

Scott C. Wolfson
U.S. Army Redstone Test Center
Redstone Arsenal, AL 35898-8052
scott.c.wolfson.civ@mail.mil

ABSTRACT

The primary objective of this technical paper is to provide details pertaining to telemetry encoder size and weight reductions using Multi-Chip Module (MCM) design techniques and a silicon die scavenging process to decrease MCM development cost. Additional use cases relevant to the Department of Defense (DoD) community and recommendations for future efforts will also be presented.

INTRODUCTION

As military systems and subsystem technologies advance, the Test and Evaluation (T&E) community is constantly maturing and adapting methodologies and instrumentation strategies to collect the required information for evaluators and programmatic decision makers. Payload limited unmanned aerial systems and new missile technologies are just two examples that support an ever-expanding need to develop unobtrusive, miniaturized, and ruggedized on-system instrumentation. Meeting this need is further complicated by the evolving instrumentation requirements for larger data storage capacities, support for increased bandwidths, and improvements in power efficiency. Current methods for adding embedded instrumentation to these types of systems struggle to overcome the adverse effects to system performance caused by weight and center of gravity changes. With the current methods, the results include either: 1) accepting the system performance degradation or 2) creating a non-tactical configuration (i.e. tactical payload replacement telemetry kits). The following material documents a design technique that addresses this test technology gap for current and future integrated telemetry encoders, and includes some beneficial byproducts pertinent to interests within the instrumentation design community.

MULTI-CHIP MODULE DESIGN

Standard custom circuit design entails connecting plastic encapsulated or packaged Integrated Circuits (IC) via traces on a Printed Circuit Board (PCB) to perform a specific function. Each encapsulated or packaged IC typically contains a small silicon-based component, referred to as the silicon die, which provides all the functionality of the component, while the packaging material provides die protection, heat dissipation, and simplifies circuit board assembly [1]. The process of integrating multiple silicon die elements into a single package is referred to as an MCM [2].

As a point of reference, Figure 1 depicts the micro-focus X-ray image of a differential amplifier offered in an 8-pin Small Outline Integrated Circuit (SOIC) package and is intended to demonstrate the significant size difference between a plastic encapsulated integrated circuit and the silicon die. The size of the encapsulated area in Figure 1 is 5 x 4 mm, while the die is only 1.5 x 0.9 mm. This size difference is typical for ICs and can translate to a potential 50% to 75% reduction in PCB size when MCM techniques are implemented.

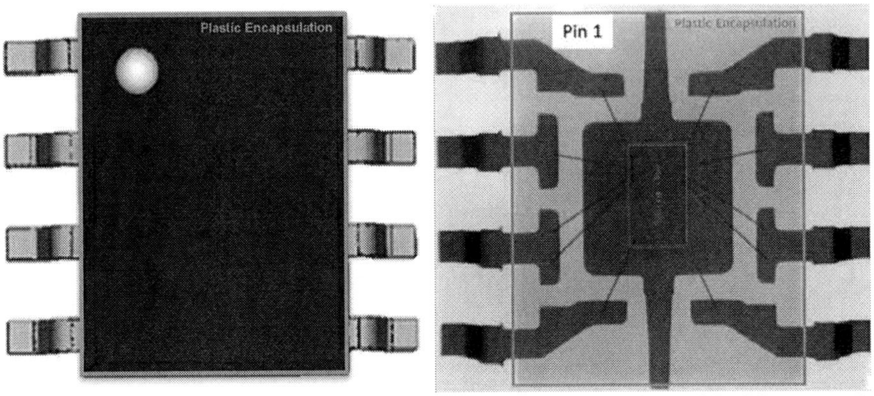

Figure 1 Component (left) and Micro-Focus X-ray (right) Images of an 8-Pin SOIC Integrated Circuit

For the miniaturized telemetry encoder, the foundational functional elements and components commonly used in previous designs were identified and evaluated to assist in locating viable candidates for MCM integration. These functional elements included a variety of serial interface types, analog signal conditioning, analog to digital conversion (ADC), memory (i.e. SRAM), programmable logic devices (i.e. FPGAs), and digital input/output (I/O) signals. The results of this system engineering effort revealed comparable components in die form that met or exceeded the performance of the components currently in use. The resulting 0.9 x 0.9 inch prototype Ball Grid Array (BGA) MCM printed circuit board and associated functional block diagram depicted in Figures 2 and 3 only requires ~30% of the board area used in previous designs.

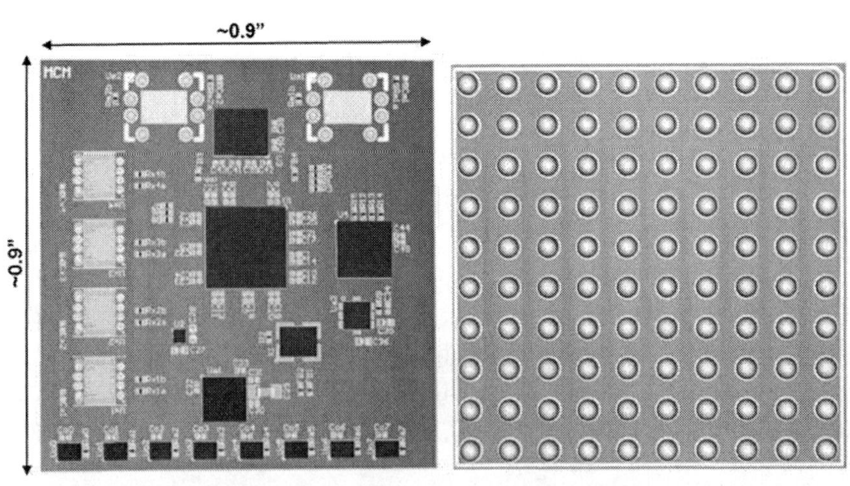

Figure 2 Top and Bottom View of Foundational Telemetry Encoder MCM

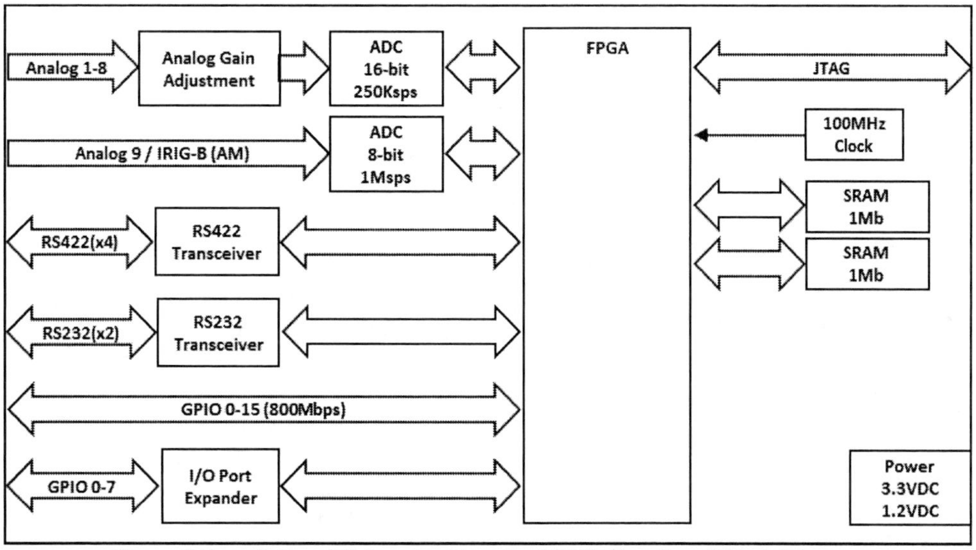

Figure 3 Foundational Telemetry Encoder MCM Functional Block Diagram

As with many modern digital circuit designs, incorporating a Field Programmable Gate Array (FPGA) has many advantages when compared to a microcontroller. These include process concurrency, flexible internal and I/O structures, and a large number of supported Intellectual Property (IP) cores [3]. When applied to this telemetry encoder, the end user can easily modify the frame format, redefine I/O logic levels (i.e. TTL, LVDS), and create/purchase IP cores (i.e. video compression, data encryption).

Functional verification testing and thermal characterization work is currently being performed on this design. This testing ranges from simple analog and digital stimuli to more complex interfaces such as inertial measurement units, magnetometers, accelerometers, secure digital cards, and IRIG-B (AM). Once complete, encapsulation and final acceptance tests will be performed.

SILICON DIE SCAVENGING PROCESS

A potentially complicated aspect of MCM design work is the difficulty and cost associated with procuring small quantities of desired dies for prototype and first article fabrication. This is due to the IC manufacturer's typical requirement of purchasing a full wafer (~500 to 10,000+ dies). Also, getting information such as die bond-pad to chip-pin assignments and die epoxy characteristics can be difficult. Because of this, joint efforts between the U.S. Army Redstone Test Center, academia, and private sector companies were initiated, and a process for scavenging dies from existing components was successfully demonstrated. The following list summarizes the scavenging process:

- Generate bond-pad to chip-pin map using micro-focus X-ray images
- Expose the encapsulated die using an acid based de-capsulation process
- Remove existing bond wires from die
- Remove die from remaining packaging material
- Re-bond a statistical sample into open-cavity packages

- Perform functional testing to determine scavenging yield

This scavenging process can reduce the cost of MCM prototyping by as much as 75% and was required for two of the components used in the miniaturized telemetry encoder design that were not available in die form. Figure 4 provides some images taken during a die scavenging and re-packaging process.

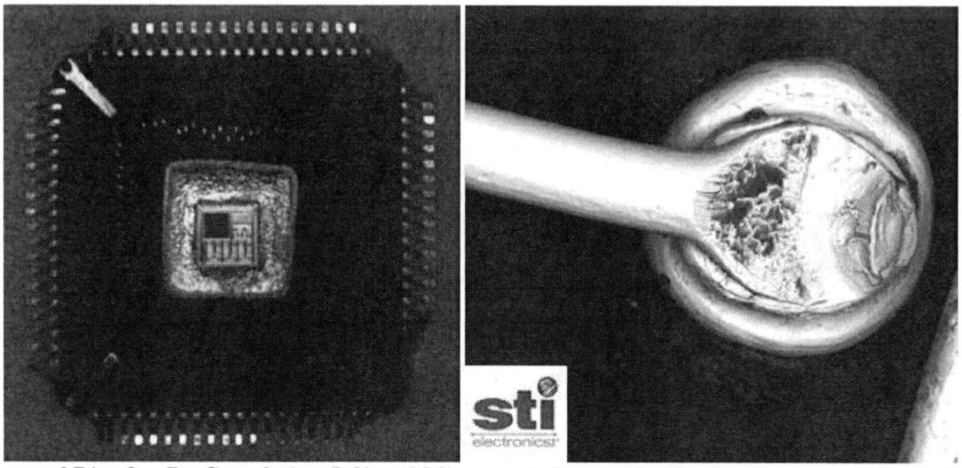

Figure 4 Exposed Die after De-Capsulation (left) and Microscopic Image after Re-Bonding to an Existing Bond (right)

ADDITIONAL USE CASES

As a result of this work, several potential secondary use cases exist that have relevance to current issues within the DoD and T&E communities:

- The ability to recover and repackage components provides an avenue for designers to improve the thermal and electromagnetic characteristics for ICs that meet functional requirements but are deficient in these areas. Thermal improvements are achieved by implementing a better heat dissipation strategy when designing the substrate. Similarly, the selection of metal type re-packaging material can aid in alleviating electromagnetic emission and susceptibility issues.
- Failure analysis is common throughout the T&E community, but isolating these failures to the IC level can be difficult without micro-focus X-ray and die exposing capabilities. Figure 5 illustrates two common component failure modes that can be revealed: 1) wire-bond separation resulting from a vibration or shock incident and 2) bond wire damage resulting from overcurrent conditions.

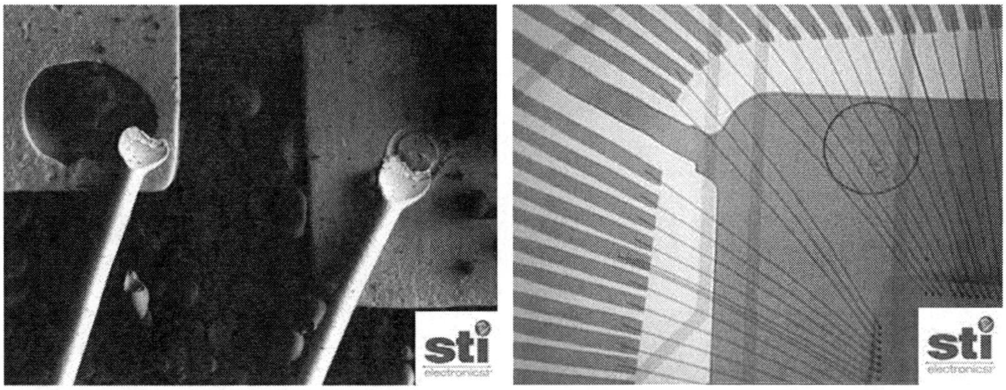

Figure 5 Separation of a Wedge Bond from Substrate Pad (left) and Bond Wire Damage (right)

- The process for harvesting dies from existing components can also be used to identify counterfeit components. The images in Figure 6 are an example where the silkscreen marking on the packaged component did not match the markings on the die after removing the encapsulation material. This is a common technique used by counterfeiters where the silkscreen of commercial grade components is replaced with the markings of the more expensive industrial or automotive grade versions.

Figure 6 Example of a Counterfeit IC. The Part Number Markings on the Component (left) Do Not Match the Markings on the Die it Contains (right)

CONCLUSIONS

In summary, any improvement to instrumentation size, weight, performance, and operational characteristics is highly desired in the Army/DoD test community. Implementing MCM and die scavenging techniques addresses these areas and advances the State of the Art in this specific area. Also, as this technique matures for T&E use cases, several opportunities exist for future refinement and improvement to the work presented in this paper. These include new methods to increase the yield of scavenged dies and the incorporation of vertical die stacking methods to further reduce the MCM size.

REFERENCES

[1] Jan M. Rabaey, Anantha Chandrakasan, and Borivoje Nikolic, "Digital Integrated Circuits (2nd Edition)," Pearson Education, Inc., 2003.
[2] R. Wayne Johnson, Mark Strickland, and David Gerke, "3-D Packaging: A Technology Review," NASA Technical Report, 2005.
[3] Joe Delaere, Stefano Zammattio, "Top 7 Reasons to Replace Your Microcontroller with a MAX 10 FPGA," Intel White Paper, 2017.

LOCALIZATION OF MEDICAL DEVICES BASED ON BLUETOOTH LOW ENERGY (BTLE)

Authors: Wondimu Zegeye, Lawrence Lee {wozeg1, lalee5}@morgan.edu

Advisors: Dr. Farzad Moazzami, Dr. Richard Dean

Morgan State University, Electrical and Computer Engineering Department

ABSTRACT

Wireless devices have invaded the medical industry with a wide range of capability as components of a wireless personal area network (WPAN) and Wireless Body Area Network (WBAN). The recent advances in Internet of Things (IoT) promises even larger contributions to the future of medical applications. This paper investigates the Bluetooth Low Energy (Bluetooth Smart- BTLE) for indoor localization of HealthCare devices used in medical telemetry applications and demonstrates the key role that localization plays in tracking of Bluetooth Low Energy enabled medical devices. Proper tracking of these devices provides better management which would directly reduce the transmission of infectious diseases which can result from improper sharing of these devices. This work will investigate the novel indoor localization technologies of BTLE devices with creative research strategies, and their applications as a basis for ultimately improving health standard using BTLE localization.

Key words: *Localization, BTLE, healthcare, Medical Devices*

1. INTRODUCTION

In Canada in 2007, there were 1077 reported cases of acute hepatitis B infection. In late 2006, five residents living in the same unit of a long-term care home (LTCH) in Toronto, Canada became infected with Hepatitis B Virus (HBV). This report describes the outbreak investigation and epidemiological analyses of this outbreak of acute HBV infection. All residents in the LTCH were serologically screened for hepatitis B markers. The infection control practices of the LTCH and of high-risk services provided to residents were reviewed. The risk factors for HBV transmission were investigated among residents and a case control analysis was conducted to identify associations with acute HBV infection. Results In total, five cases of acute HBV infection were identified in the same unit of the LTCH. The attack rate was 20.8% and the case-fatality rate was 60%. All five cases had diabetes mellitus, and HBV transmission was significantly associated with blood glucose monitoring. Conclusions results of this outbreak investigation and analysis demonstrated that hepatitis B transmission was associated with shared blood glucose monitoring equipment. To prevent hepatitis B transmission, it was recommended that a glucometer and finger-stick device be assigned to each diabetic resident requiring blood glucose monitoring in addition to following routine infection control practices [1]. Similar incidents have been investigated in [2].

Currently most of the medical devices such as a blood glucose monitoring devices are Bluetooth enabled. As a result these devices can be vulnerable to attacks related to this technology [3]. On the contrary this technology can be exploited to track these devices indoors in a medical facility. Proper tracking of the devices can enable to control or manage the sharing of these devices. Thus, we can reduce the transmission of infections associated with an improper sharing of these medical devices.

2. BACKGROUND

Bluetooth low energy is designed to enable new markets requiring low latency, low cost, low duty cycle and low power consumption data devices. These markets include healthcare, proximity, fitness, automotive, and smart grid applications. These may include device categories that are completely new to Bluetooth technology, or will use multiple features such as low energy and high speed Bluetooth technology within the same device.

Bluetooth Low Energy [4] devices operate in the 2.4 GHz license-free band, and so share the same indoor propagation characteristics as 2.4 GHz WiFi transceivers. The beaconing, or advertising mode, permitted in the BLE standard enables a very short, unsolicited message at very flexible update rates. These messages can be used to allow a device to detect close proximity to a specific location based on the Received Signal Strength (RSS). In this way, location specific triggers, adverts, vouchers and information can be provided to the user.

2.1 BLUETOOTH LE IN MEDICAL DEVICES

Embedding Bluetooth in medical devices has been transforming medical industry for the last decade. If we imagine the application of sensors in Healthcare which employ Bluetooth low energy, the possibilities and opportunities are diverse: continuous EEG monitoring with overlay of patient activities that would be triggers to epileptic activities (medical hat), 24 hour vital sign and ECG monitoring coupled with physical activity profile (medical underwear), simultaneous monitoring of airway resistance, environmental pollutants and actual medication delivery (medical inhaler), oximeter capable of adjusting the positive pressure to the changes in carbon dioxide concentration in the exhaled air of the user, ingestible cameras capable of real time transmission of the images to image processing module, and many others. These devices offer additional data that can be used to create more interactive, and health monitoring an activity of daily living.

2.2 BLUETOOTH LE BEACONS

BLE advertising beacons are particularly attractive to retailers because of the promise of long battery lives of many years, and so low maintenance requirements. Long battery lives are expected to require low radio power output and/or low beaconing rates. While this does not affect their use for proximity detection it does affect their usefulness for providing fingerprint-based positioning throughout an entire indoor environment.

Beacon device is a relatively new technology, first introduced by Apple in 2013. Since then, this technology has been put into tremendous use in the retail sales. Now Apple and Google have come up with their own beacon standards, "iBeacon" and "Eddystone". Beacons work on Bluetooth Low Energy – BLE (BTLE) which transmit a signal up to a certain distance, ranging from 15cm (~6 in) to 70m (~230 ft). Beacons provide a virtual region, when we scan the specific region of beacon then we can say that we are within that region. Beacon broadcasts Bluetooth signal which contains unique information of itself frequently, so the coverage area will be filled with beacon's signal; the area is called Beacon Region.

The beacons used in this work are Estimote beacons [5]. These beacons have broadcasting power (or transmitting power) which ranged between -40dBm to +4dBm, advertising interval of 100ms to 2000ms, RSSI.

3. BLUETOOTH LOW ENERGY LOCALIZATION FOR HEALTHCARE

Although there is no specific support for positioning service in Bluetooth technology yet the predominant technology used are signal strength measurement, link quality and bit error rate which rely on the services of the Host Controller Interface**.** Thus the Received Signal Strength Indicator (RSSI) value of the Bluetooth signal is used to get a correlation in the distance between sender and receiver in a network. RSSI value fingerprinting using Bluetooth Low Energy signals can be used for localization similar to the work in [6].

The work in this paper exploits BTLE signals detected from Estimote beacons to identify rooms in the LTCH scenario. The BTLE signals from medical devices (e.g. Glucometer) to identify the proper assignment of these devices to the correct room. The Glucometers used in this work are One Touch Verio Blood Glucose Monitoring Systems which are BTLE capable [7]**.**

The main reasons for using the Beacon technology are the following:

- Rely on BTLE which is low power consumption, can work 24 hours 7 days over a year with a button cell battery, very small and it can be installed very easily.
- High supported range.
- Cross-platform and support wide range of devices BLE is supported by different platforms including Android, IOS, and Windows… etc. In addition, according to Bluetooth SIG, they predict that "by 2018 more than 90 percent of Bluetooth enabled smartphones will support Bluetooth Smart" 2. Therefore, we can say BLE or Beacon is a trend in future.

3.1 ALGORITHM DESIGN

For the algorithm to work we need two data sets which store the offline data. The first data set is as specified in Table 1 (mapping of Room ID vs Medical Device) and the second one is as in Table 2 (mapping of Beacon Identifier vs Room ID).

To detect the identity of the rooms in the LTCH scenario, we apply the concept of proximity zones. These proximity zones are distance measurements correlations from the RSSI values of the BTLE signals of the beacons.

Proximity zones are defined as [5]:

- immediate (very close to the beacon)
- near (about 1-3 m from the beacon)
- far (further away or the signal is fluctuating too much to make a better estimate)
- unknown.

If the proximity to nearby beacon is immediate, we use the identifier of the beacon for mapping to the identity of the room.

Algorithm: Medical Device Localization

1. Scan Estimote Beacons
   Using proximity zones, Identify the Room Identity by mapping the immediate Estimote beacons identity to the room identity. This principle makes use of Table 2.
2. Scan BTLE devices and filter the closest devices based on range and map the values with the devices in Table 1 where we find the devices from the list of devices allocated in the health care.
3. By using the room identity from Step 1, we can use the Table 1 to identify the Glucometer assigned for the particular room.
4. Using the mapping from Step 2, we can detect the Glucometer detected in the room.
5. If the Glucometer detected at Step 4 matched with the Glucometer assigned in Step 2, the assignment will be labeled "correct" else "wrong".

Room ID	Glucometer (Mac Address)
141	F1:9F:7C:9F:0E:32
142	E6:1A:1A:6F:8C:43

Table 1: Room ID vs Medical Device

Estimote Beacon (Identifier)	Room ID
79a00688289bb9a713b897f8ef7e4d26	141
635ba5bb64ad0ea4e0db58d75f98162d	141
6abde74a4fae5a320ddb7d9978a45523	141
7ed22602334736ebdf9d7f49e7a39f2f	141
302b8565852e5f88014b4e34c123b71c	142
1c02b18fdaa395b11df1a5f2fdd7fe0a	142
32309c1d1eca0b12b118b1661056260b	142
f9ec6608dd55c381298b97629502c838	142

Table 2: Beacon Identifier vs Room ID

The closest BTLE devices are determined using attenuation model. The RSSI (in dBm) which is the ratio of the transmitter power and the received power decreases exponentially as the distance increases. This relation is given by equation (1)

$$RSSI_{dBm} = -10 * n \log_{10} d + A \quad (1)$$

The relationship between d and RSSI

$$d_m = 10^{\frac{RSSI-A}{-10*n}} \quad (2)$$

Where, n: signal propagation constant, also named propagation exponent (n=2 for free space) ; d: distance from the sender; A: received signal strength at a distance of one meter (in our case it is equivalent to Tx Power =-65dBm).

The distance (accuracy) property measures the correct RSSI and is calculated from the average RSSI over 9 samples.

3.2 EXPERIMENTAL SETUP: STANDALONE ARCHITECTURE

In case of the standalone architecture, the localization algorithm is implemented in a single Android App. In other words, one Android application is developed which makes use of the data sets (tables) to perform the BTLE medical devices localization. The tables of data sets are small in size. As a result, these values are coded as a hashmap. But future work will extend the implementation of these data sets as XML files which can be stored in the SD card of the device in order for the application to parse and use them accordingly.

The experiments conducted in this work are based on the standalone architecture. A single android application which can scan Estimote beacons and BTLE devices such as Glucometers. Based on the scan results the algorithm in section 3.1 is used to detect the correct assignment of a Glucometer to its respective room which it was assigned beforehand verifying the device is not misplaced.

3.3 EXPERIMENTAL SETUP: CLIENT-SERVER ARCHITECTURE

This section discusses the conceptual design to how the Cloud infrastructure of the Estimote can be used for BTLE localization. This design can be used equivalent to the design in section 3.1 and 3.2. This involves building our own App using the Estimote SDK.

With the Estimote SDK, we can easily implement Indoor Location in our own app. The Indoor Location SDK enables us to:

- Set up a location manually (draw a room with coordinates).
- Embed a built-in Location View - a map of location with the current position marked on it.
- The Location View is fully configurable – the appearance can be customized, show or hide labels or leave trace of the movement.

- Get raw positioning data – such as coordinates (x, y) and orientation.
- Save and load locations to and from Estimote Cloud.

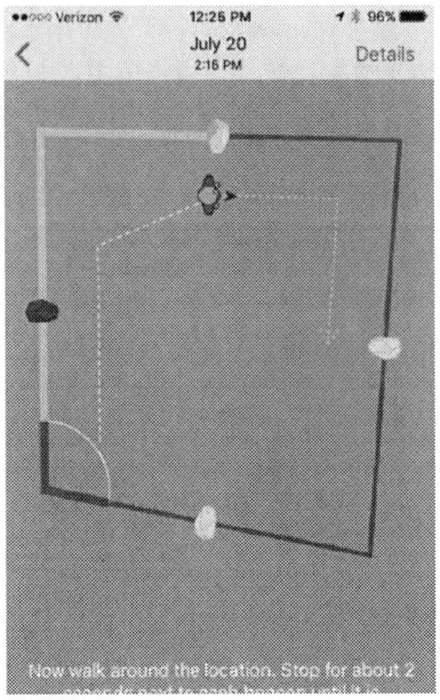

Figure 1: Estimote App for Beacons Placement

As can be seen from the Figure 1**,** we can scan walk around the room which we can to include in our localization and the app suggests the number and location of placement of the beacons properly. Once this phase is completed, we can use an application token to access our beacons for localization.

4. RESULTS

The experiment conducted in this work is based on the stand alone architecture described in section 3.2. The testing area used for the localization includes two adjacent rooms of the Schaffer engineering building of Morgan State University. These two rooms are as shown in Figure 2 and represent an LTCH scenario.

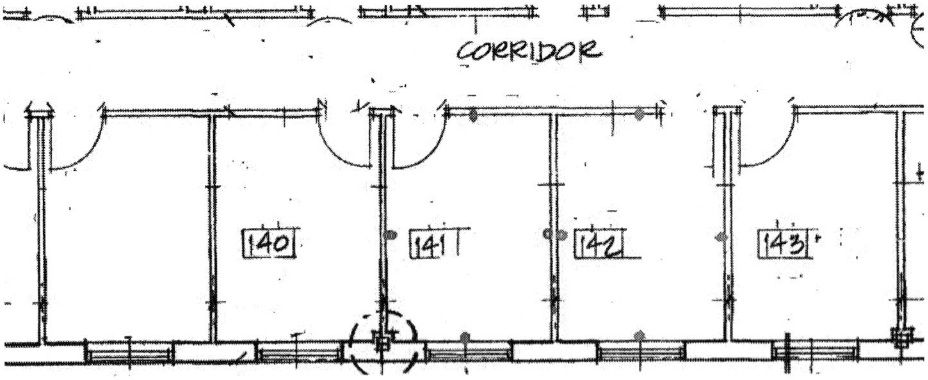

Figure 2: Layout of Testing Rooms

As can be seen from the Figure 2, 4 Estimote beacons are installed in each of the rooms (Room 141 and Room 142) for localization. As the Estimote beacons are powerful, it could be possible to identify a room identity using only a single beacon but we used four Estimote beacons for a

future extension of enabling the application navigation capability which can continuously display the user (device) location on the map of the building. The Estimotes are installed as shown with the red and green dots in each of the rooms.
The physical installation of the Estimote beacons is as shown below in Figure 3 which shows two Estimotes on their respective walls as an example.

Figure 3: Physical Location of Estimote Beacons

Using the Android App which runs the algorithm in Section 3.1 we can scan the Glucometer assigned room 141 as an example to show the working of this application. The result is as shown below in Figure 4.

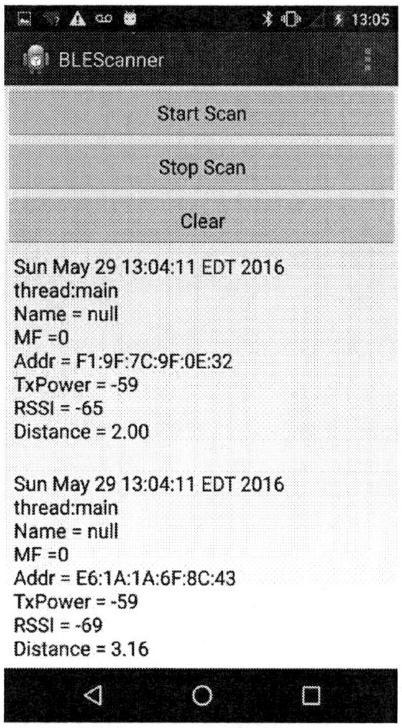

Figure 4: BTLE Device Scanner

The BTLE scanner App shows the scan result of reachable BTLE devices. For each of the detected devices the result includes the mac address, Tx Power, RSSI, and distance calculated.

From the scanned devices, candidates are filtered based on distance and compared to the device to room identity mapping table and the proper assignment of the device (Glucometer) is determined.

As can be seen from the first entry Figure 4, the first entry BTLE device has a mac address " F1:9F:7C:9F:0E:32" and it is located around 2m from the mobile device used for scanning. Using the Glucometer localization App in Figure 5, this BTLE device is detected when the "SHOW DEVICE" button is clicked. The detected device and the already assigned device in the table are matched. As a result the assignment status is set as "Correct". The also determines the identity of the room based on the proximity zone of the detected beacons.

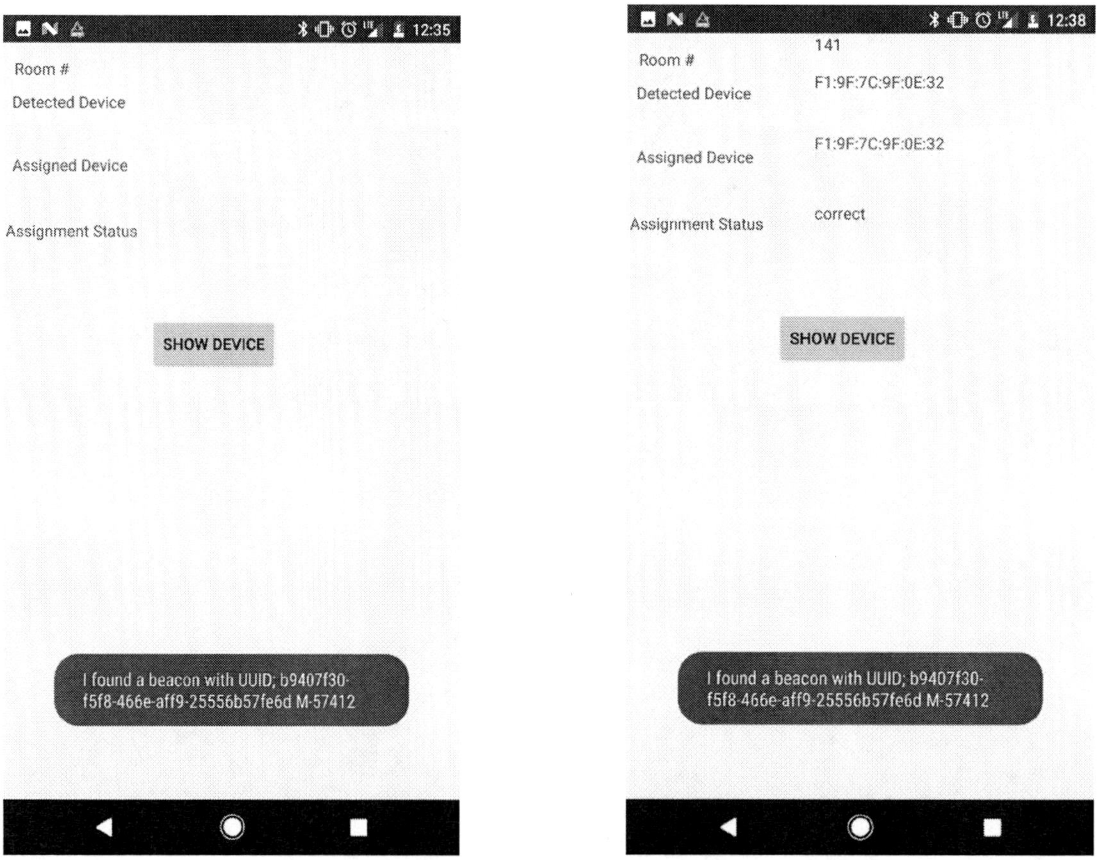

Figure 5: Glucometer Localization App

5. CONCLUSIONS AND FUTURE WORK

This work highlights the potential for use of commercial technology such as Bluetooth in telemetry applications in a long term health care. The use of wireless telemetry technologies such as Bluetooth Low Energy in medical environments brings major advantages to the existing

HealthCare medical services. The experimental work achieved shows how this Bluetooth Low energy technology which is currently available in many medical devices can be used to properly track these medical devices in a long term care home scenario. Using this localization technique can help to properly track these medical devices for better management which would directly reduce the transmission of infectious diseases which can result from improper sharing of these devices

Future work will investigate expansion of the indoor localization technique implemented here over a bigger testing area. In addition, it will integrate the navigation capability by loading the map of the testing area besides to locating the medical devices.

6. ACKNOWLEDGEMENT

The research reported in this publication was supported by the National Institute of General Medical Sciences of the National Institutes of Health under Award Number UL1GM118973. The content is solely the responsibility of the authors and does not necessarily represent the official views of the National Institutes of Health.

7. REFERENCES

[1] Consu Sachdeva, Herveen; Green, Corey I.; Arthur, Anne; Andonov, Anton P, "Risk factors associated with a nosocomial hepatitis B outbreak in a long-term care facility in Toronto", C*anadian Journal of Infection Control. Spring2012, Vol. 27 Issue 1, p9-14*

[2] Matthew E. Wise et.al., "Outbreak of acute hepatitis B virus infections associated with podiatric care at a psychiatric long-term care facility", *American Journal of Infection Control* Volume 40, Issue 1, February 2012, Pages 16-21

[3] W. K. Zegeye, F. Moazzami, R. Dean. Y. Astatke, "Exploiting Bluetooth Low Energy Pairing Vulnerability in Telemedicine," " *in proceedings of the 51st International Telemetry Conference (ITC)*, Oct. 2015, Las Vegas, NV.

[4] Bluetooth SIG. Bluetooth Specification Version 4.0, 06 2010.

[5] Estimote, retrieved from https://estimote.com/, August 6, 2017

[6] Wondimu K. Zegeye, Seifemichael Amsalu, Yacob Astatke, Farzad Moazzami, "WiFi RSS Fingerprinting Indoor Localization for Mobile Devices," The 7th IEEE Annual Ubiquitous Computing, Electronics & Mobile Communication Conference (IEEE UEMCON 2016), New York City, NY, USA 20 - 22 October 2016.

[7] OneTouch Verio® meter, retrieved from https://www.onetouch.com/products/glucose-meters/onetouch-verio, August 6, 2017

A SUMMARY OF DATA-AIDED EQUALIZER EXPERIMENTS AT EDWARDS AFB

Michael Rice, Chris Hogstrom, Chris Nash, Jeff Ravert
Brigham Young University, Provo, Utah, USA

Mohammad Saquib and Md. Shah Afran
University of Texas at Dallas, Richardson, Texas, USA

Arlene Cole-Rhodes and Farzad Moazzami
Morgan State University, Baltimore, Maryland, USA

Erik Perrins
University of Kansas, Lawrence, Kansas, USA

Kip Temple
412 Test Wing, Edwards AFB, California, USA

ABSTRACT

This paper summarizes the analysis of bit error rate data captured during flight tests designed to compare data-aided equalizers with SOQPSK-TG to unequalized and currently available blind, adaptive equalizers with SOQPSK-TG. The number of bit errors, on a second-by-second basis, are analyzed. The results are different for each test point. Given the uncertain behavior of the preamble detector for the data-aided equalizer and the differing channel conditions between the data-aided equalizer channel and the conventional serial streaming telemetry channel, we are unable to draw any firm comparative conclusions.

INTRODUCTION

Equalization is one of the methods employed to mitigate distortion due to multipath propagation. To date, commercially available equalization products for SOQPSK-TG are blind, adaptive filters usually based on the constant-modulus algorithm (CMA) [1, 2, 3, 4, 5]. A *data-aided* equalizer uses an estimate of the channel to compute equalizer filter coefficients that optimize some criterion. The open question is how well data-aided equalization might perform in aeronautical telemetry. To answer this question, flight experiments were conducted on 21 April 2017 at the Air Force Test Center, Edwards AFB, using SOQPSK-TG with periodically inserted pilot bits to enable channel estimation on the ground. This paper outlines the results of the flight test.

A number of signal processing tasks form the basis of data-aided equalization: the start of the received samples corresponding to the preamble must be identified. The samples corresponding to the preamble are used to estimate the frequency offset, the impulse response of the equivalent discrete-time channel, and noise variance. The frequency offset estimate is used to "de-rotate" the data. The channel and noise variance estimates are used to compute five different equalizers (for side-by-side comparisons). The equalized output is used by a modified SOQPSK-TG detector to create the bit decisions. Due to space constraints, detailed descriptions are not given here. The reader is referred to the following references: The sampling and downconversion are described in [6]; the preamble detector is described in [7, 8, 9, 10]; the frequency estimator is described in [10, 11]; the channel and noise variance estimator described in [10, 12] and analyzed in [10, 13]. The five equalizers considered in this project are the zero-forcing (ZF) equalizer described in [10, 14, 15], the minimum mean-squared error (MMSE) equalizer described in [10, 14, 15, 16], the MMSE-initialized constant modulus algorithm (CMA) equalizer described in [10, 14, 15, 17], and two frequency-domain equalizers described in [18]. The detection technique used here is described in [10, 19].

Some initial results from this project were described in [20]. Further analysis revealed a hardware problem that was corrected. A new round of flight tests were conducted in April 2017. This paper describes results from the April 2017 flight tests.

EXPERIMENTAL CONFIGURATION

The air-borne configuration used for the flight test is shown in Figure 1. The equipment was placed in a rack in a C-12 aircraft for the test. Two transmitters were used. The first transmitter transmitted a specially formatted data stream for the equalization tests: The iNET preamble (128 bits) and ASM (64 bits) were periodically inserted every 6144 bits as shown in the figure and used to modulate an SOQPSK-TG carrier at 1830.5 MHz with differential encoding disabled. This is called the "PAQ" link in the following. The iNET preamble and ASM bits formed a known bit sequence that was used by the estimators described in the references listed in the previous section. The data was a length-2047 PN sequence (called a PN11 sequence because $2^{11} = 2048$) at 10 Mbits/s. The periodic insertion of 192 bits every 6144 bits produced an over-the-air rate of 10.3125 Mbits/s.

The second transmitter transmitted the 10 Mbit/s PN11 sequence using differentially-encoded SOQPSK-TG without inserting the 192 preamble and ASM bits, thus mimicking the usual serial streaming telemetry (SST) link. The SST link was centered at 1800.5 MHz and served as a baseline for measuring any potential improvements realized by data-aided equalizers.

The signals were combined and transmitted from the lower aft antenna on the C-12 aircraft. Earlier experiments transmitted the same signal from both the upper and lower antennas (dual antenna transmission). Unfortunately, dual-antenna transmission caused problems with the preamble detector. So, dual-antenna transmission was abandoned for the flight test reported here.

The telemetry site, designated Building 4795 at the AFTC, was used as the ground station. A block diagram summarizing the ground station configuration is illustrated in Figure 2. Antenna 6 (a 10-foot parabolic reflector with conscan autotracking) was used as the receive antenna. The signal from the antenna feed was amplified and split between three receivers. The top receiver in the figure was tuned to the PAQ frequency and translated the RF signal to a 70 MHz IF signal that was sampled by the data acquisition card. The sampled IF signal was transferred to a host computer where IF-to-baseband downconversion, resampling, estimation, equalization, and detection were performed in three graphics processing units (GPUs) as shown. The ZF, MMSE, MMSE-initialized CMA, FDE1 and FDE2 data-aided equalizers were implemented in the GPUs. The five parallel bit streams corresponding to the five data-aided equalizers were output via an FPGA card and TTL driver circuit to the channels 1–5 of the bit error rate tester (BERT).

The middle receiver was tuned to the SST frequency and operated as a standard telemetry receiver (differentially-encoded SOQPSK-TG with no bit insertions). Its output bit stream was connected to channel 6 of the BERT. The lower receiver was tuned to the SST frequency and operated as a standard telemetry receiver with the equalizer described on pages 5–6 of the data sheet [21] enabled. The corresponding output bit stream was connected to channel 7 of the BERT. A spectrum analyzer was also attached for monitoring the two received RF signals.

Four test points are considered in this paper. All four test points are part of the larger AFTC complex at Edwards AFB. The test points are Cords Road (both east-to-west and west-to-east paths) and Black Mountain (both east-to-west and west-to-east).

RESULTS

The effectiveness of data-aided equalization is assessed by comparing its performance to that of the SST link without any equalization and to a proprietary adaptive equalizer described on pages 5–6 of the data sheet [21]. Because the latter two do not require periodic insertion of known bits, a separate link was used for these two signals (designated the SST link in Figures 1 and 2). The challenge with performing a true side-by-side comparison is the use of two separate carrier frequencies for the PAQ and SST signals. Because the two carriers were separated by 30 MHz, the two channels experienced simultaneous multipath only some of the time. Here, we restrict our attention to the points where both channels experienced simultaneous multipath propagation. The receiver AGC coupled with the bit error events were used to identify points along the test routes where multipath occurred.

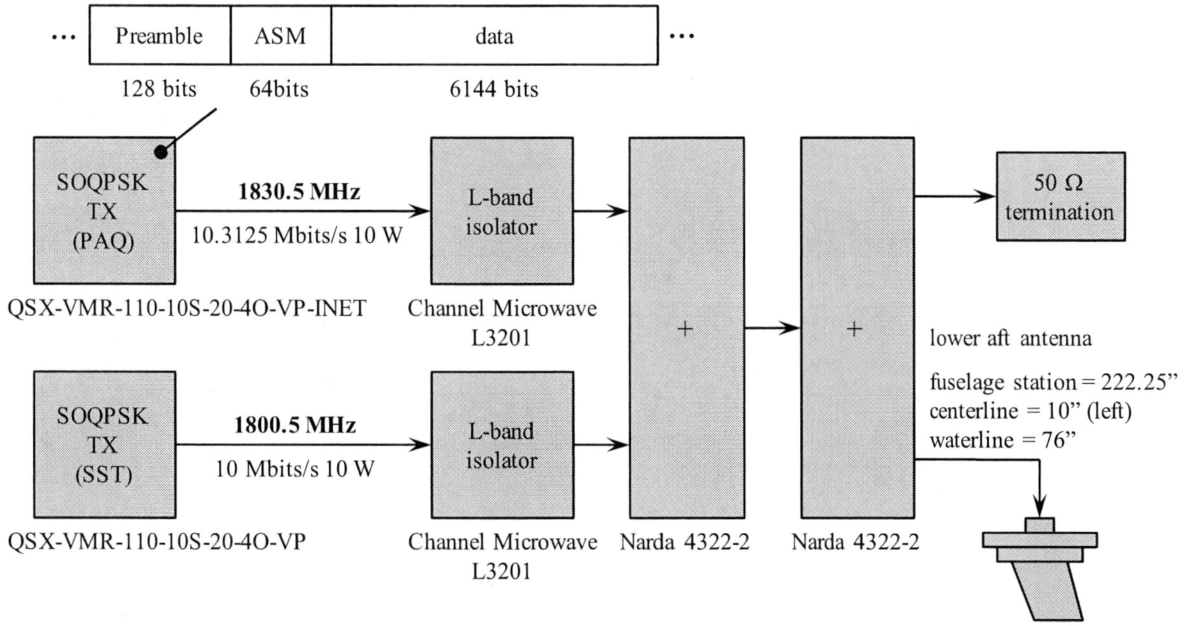

Figure 1: A block diagram of the airborne transmitter configuration.

A. CORDS ROAD EAST-TO-WEST

Plots of the AGC voltage vs. time and the measured bit error rate (BER) over one-second intervals are shown in Figures 3 for the Cords Road east-to-west test point. As is the usual case, multipath propagation (and hence, the occurrence of errors) occurs on the eastern end of Cords Road. There is a little multipath propagation on the western end of Cords Road as well. It is hard to tell from this plot how to order equalizer performance. Furthermore, for this test point, there were no intervals where both the PAQ and SST links experienced simultaneous multipath interference.

B. CORDS ROAD WEST-TO-EAST

Plots of the AGC voltage vs. time and the measured bit error rate (BER) over one-second intervals are shown in Figures 4 for the Cords Road west-to-east test point. Again note that most of the multipath propagation and accompanying errors occurred at the east end of Cords Road. A plot of the number of errors in each one-second interval for the intervals where simultaneous multipath occurred on the PAQ and SST links is shown in Figure 5. The total number of bit errors for each equalizer are listed in the plot. Sadly, only the ZF, FDE1, and FDE2 equalizers outperform the unequalized channel.

For the unequalized channel, there are a relatively large number of errored seconds with between 4 and 5×10^6 errors which indicates the case where the demodulator lost synchronization lock. The equalizers reduce the number of one-second intervals with sync loss, but there are still a relatively high number of such one-second intervals with the Quasonix equalizer. Relative to the unequalized

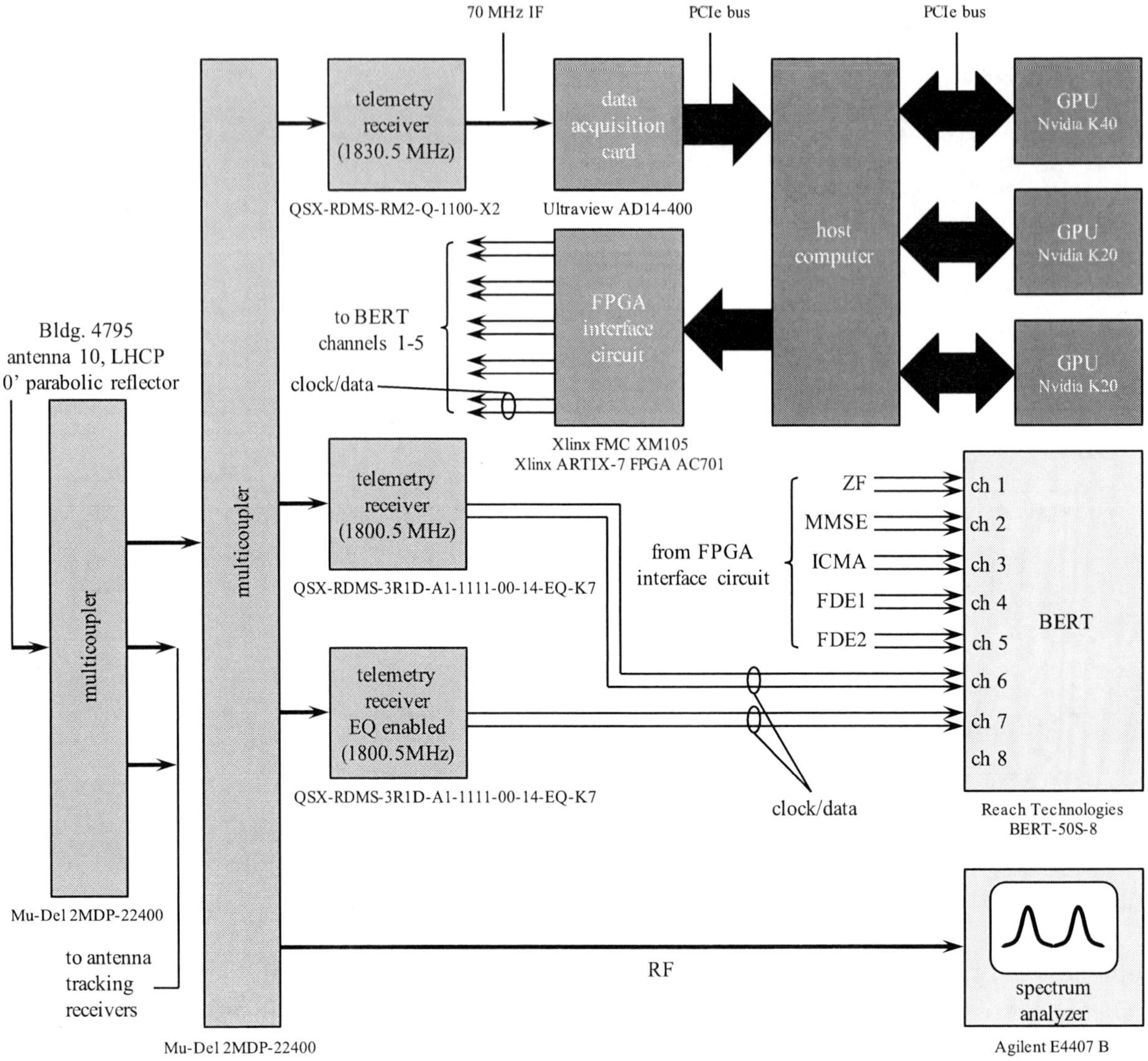

Figure 2: A block diagram of the ground-based receiving system.

channel, the equalizers also produce more one-second intervals with less than 10^6 errors.

C. BLACK MOUNTAIN EAST-TO-WEST

Plots of the AGC voltage vs. time and the measured bit error rate (BER) over one-second intervals are shown in Figure 6 for the Black Mountain east-to-west test point. As with Cords Road, note that most of the multipath propagation (and accompanying bit errors) occurs on the east end of the run. A plot of the number of errors in each one-second interval for the intervals where simultaneous multipath occurred on the PAQ and SST links is shown in Figure 7. The total number of bit errors for each equalizer are listed in the plot. Here *all* of the equalizers perform worse than the unequalized channel. Of the equalizers, the best performing is the Quasonix equalizer and the worst performing are the ZF, FDE1, and FDE2 equalizers — the exact opposite of the situation for the Cords Road west-to-east test point. Observe that during the last 100 seconds, the data-aided equalizers produce errors when the unequalized channel produces no errors. This was caused by the preamble detector "losing its way." When the preamble detector misidentifies the start of the preamble, all the estimates, including the channel estimate, are quite incorrect. And an equalizer based on an incorrect channel estimate produces a burst of errors for the corresponding packet.

D. BLACK MOUNTAIN WEST-TO-EAST

Plots of the AGC voltage vs. time and the measured bit error rate (BER) over one-second intervals are shown in Figure 8 for the Black Mountain west-to-east test point. As before note that most of the multipath propagation (and accompanying bit errors) occurs on the east end of the run. A plot of the number of errors in each one-second interval for the intervals where simultaneous multipath occurred on the PAQ and SST links is shown in Figure 9. The total number of bit errors for each equalizer are listed in the plot. Here, only the Quasonix equalizer outperforms the unequalized case (but just barely). But the time-distributions of the errors is the curious feature here. For the data-aided equalizers, the large error counts are confined to the 2-to-4 seconds interval near the beginning of the data set. The large error counts (relative to the unequalized case) are indicative of a malfunction of the preamble detector. A zoomed-in view of the plots in Figure 9 are shown in Figure 10. The zoomed-in view shows that, except for the preamble detector problem seconds, the pattern of errors for all equalizers is about the same. The number of bit errors for the data-aided equalizers is about half the number of errors for the unequalized and Quasonix-equalized channels.

CONCLUSION

The bit-error analysis of the four test runs displays some conflicting trends. The situation is confused by two factors:

1. When the preamble detector fails, the bit error rate for each packet with a misidentified preamble starting location is 0.5. Consequently, preamble detector failure produces a large number of errors that mask "what might have been" for the data-aided equalizers.

2. It is not clear if the multipath channel seen on the PAQ link is the same as that on the SST link. If channel equivalence could be guaranteed, more firm conclusions about the relative merits of blind equalization and data-aided equalization could be made.

Our observations during the flight test suggest that when the preamble detector is functioning properly, the data-aided equalizers seemed to perform better than blind, adaptive equalization. Unfortunately, it is difficult to verify this impression quantitatively.

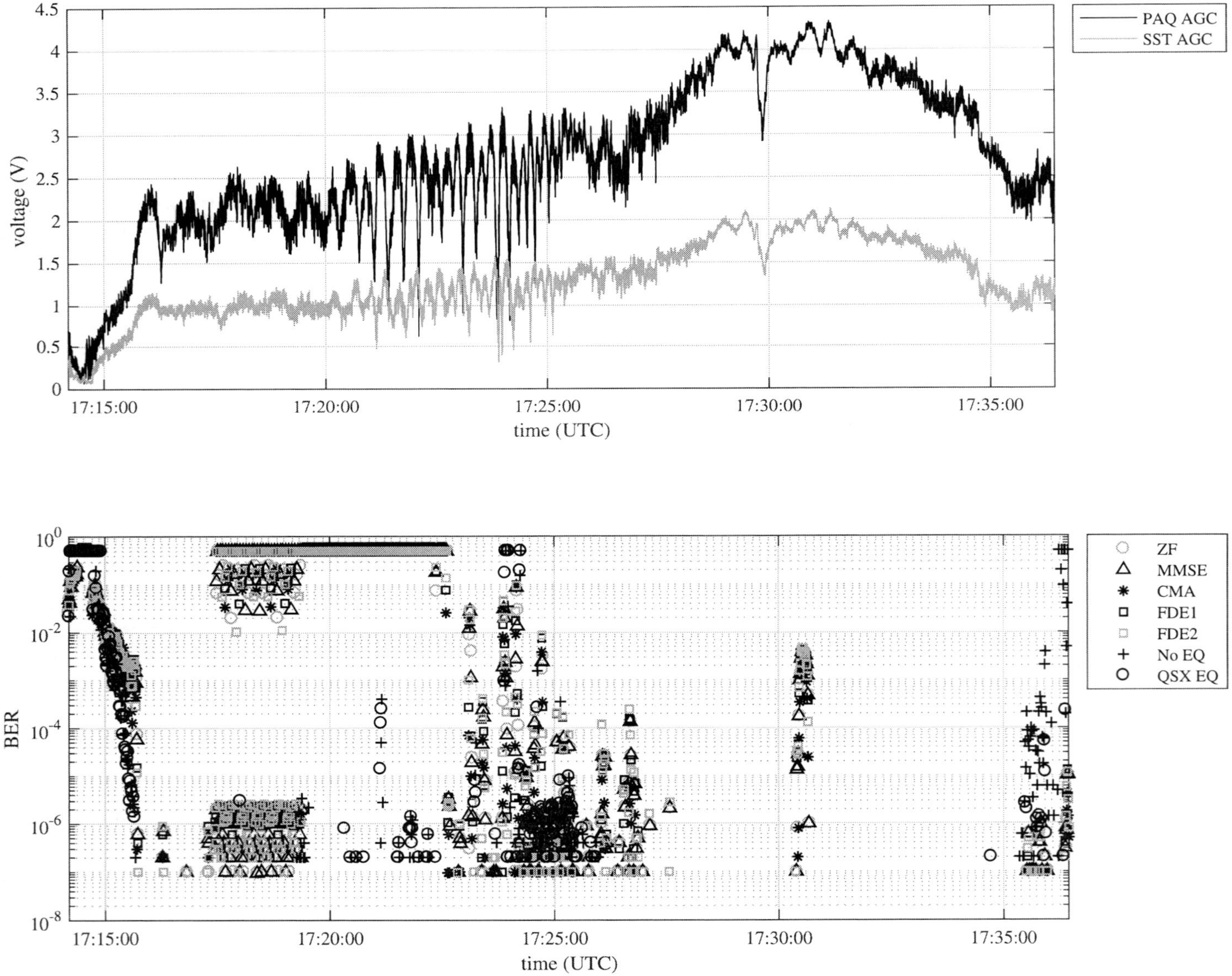

Figure 3: The AGC and BER for the Cords Road east-to-west test point.

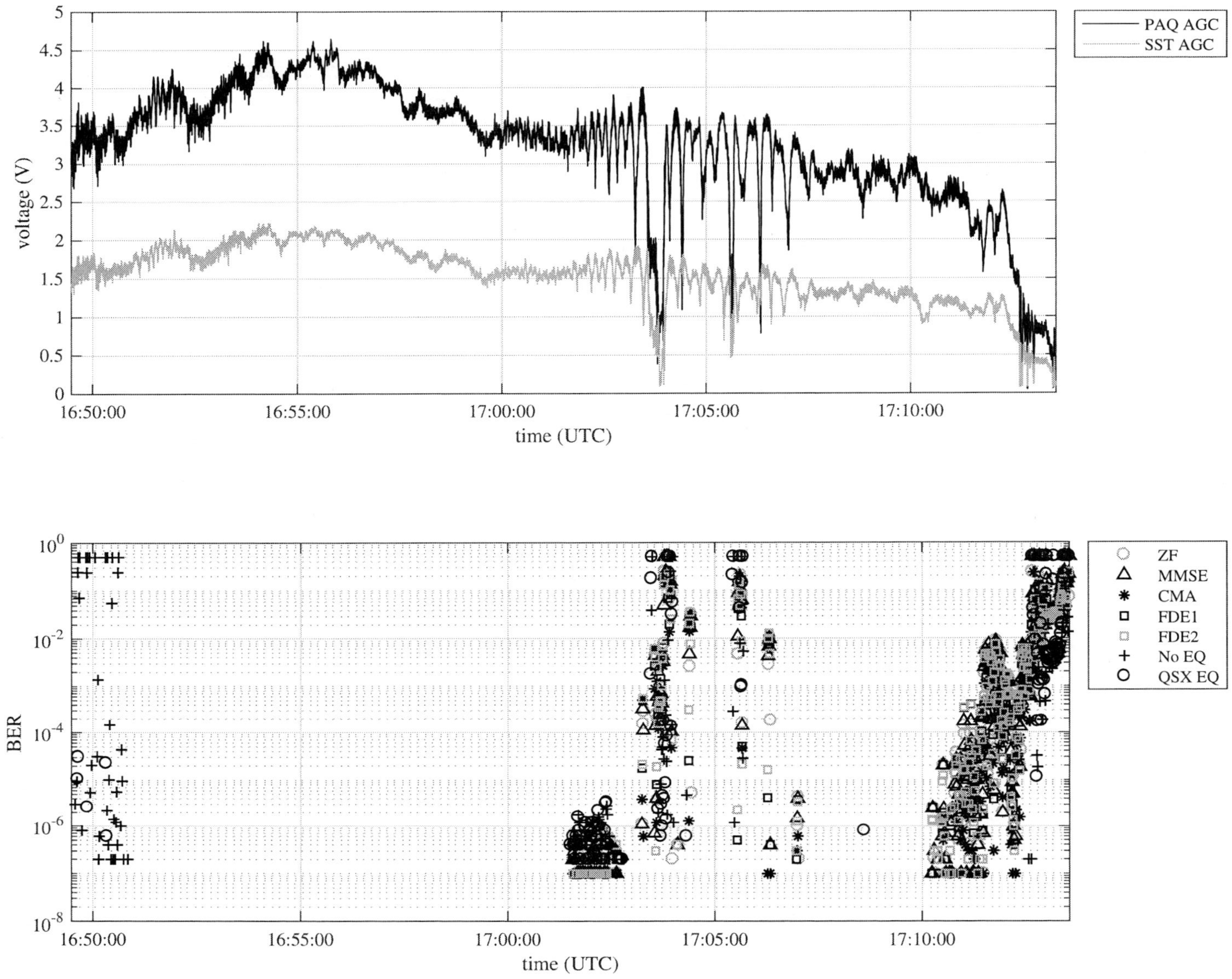

Figure 4: The AGC and BER for the Cords Road west-to-east test point.

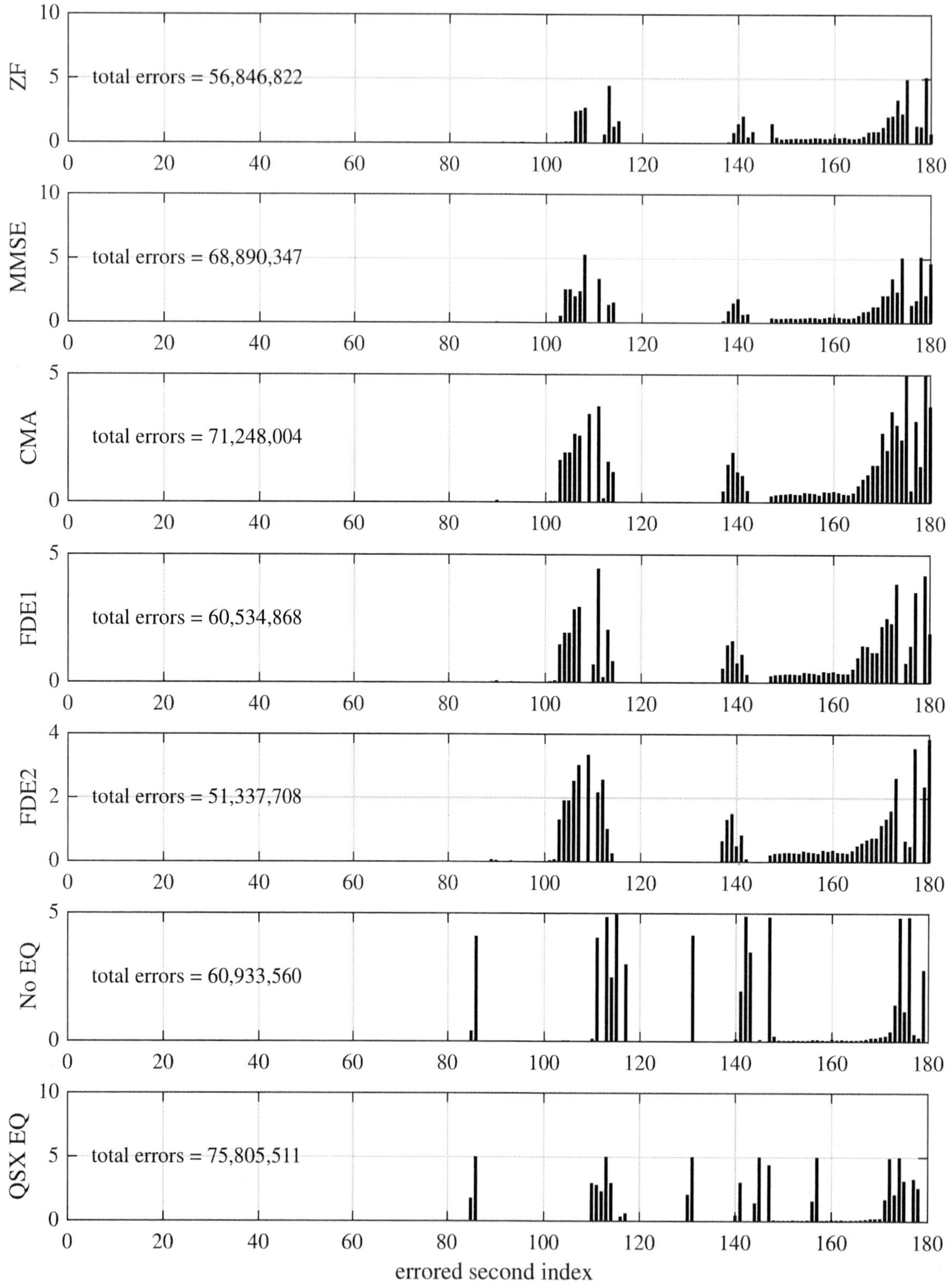

Figure 5: Bit errors ($\times 10^6$) vs. one-second intervals for the Cords Road west-to-east test point (cf. Figure 4).

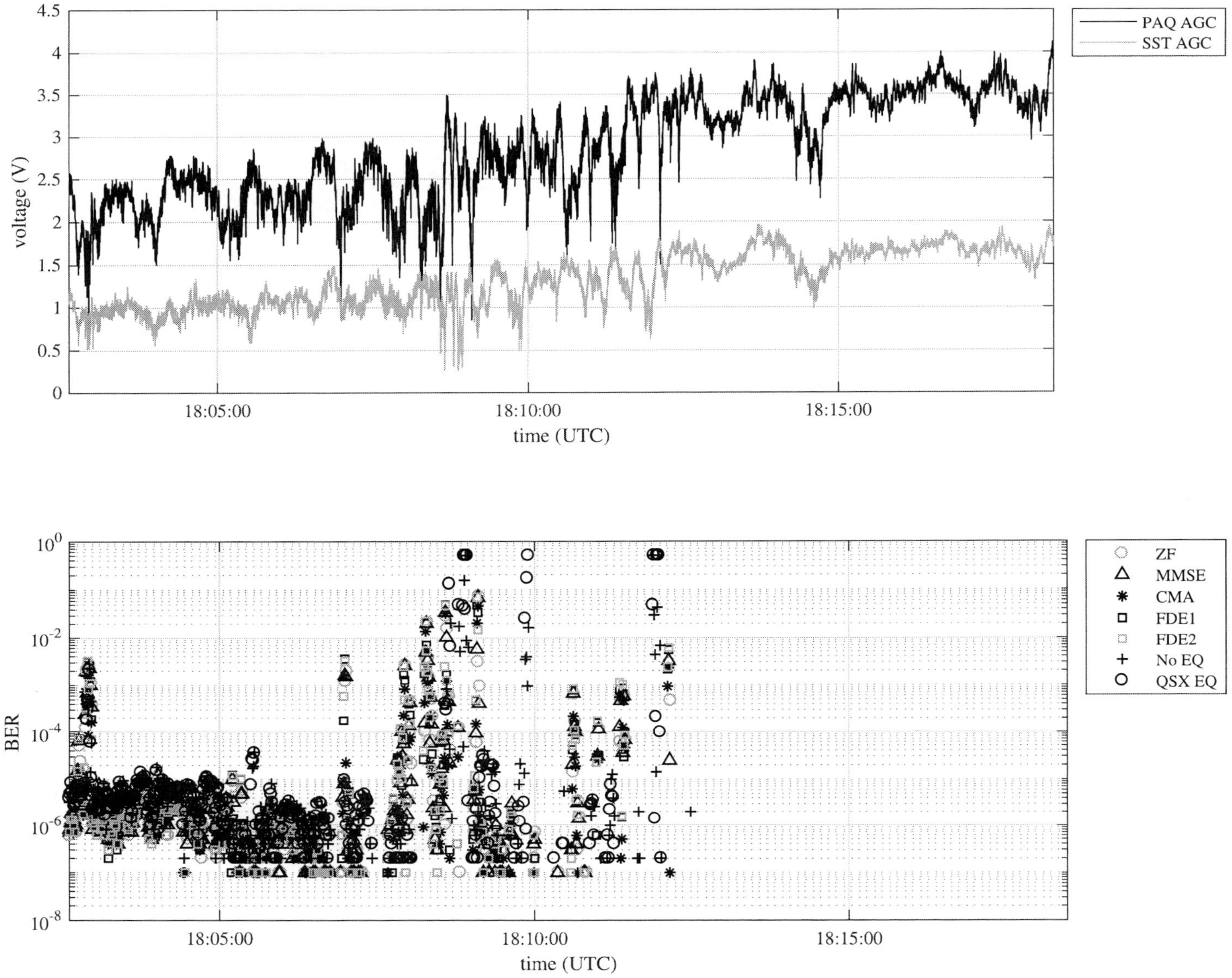

Figure 6: The AGC and BER for the Black Mountain east-to-west test point.

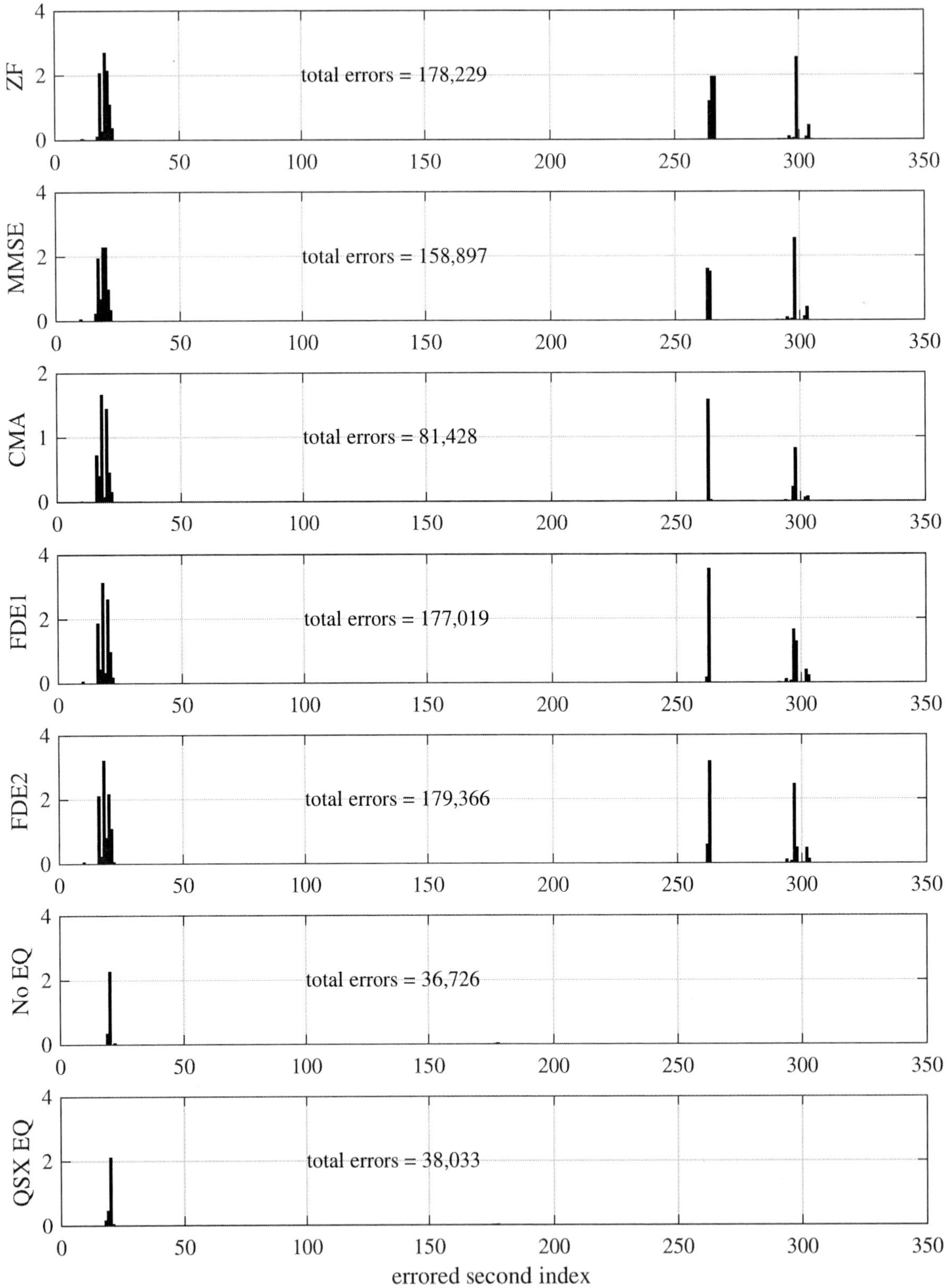

Figure 7: Bit errors ($\times 10^5$) vs. one-second intervals for the Black Mountain east-to-west test point (cf. Figure 6).

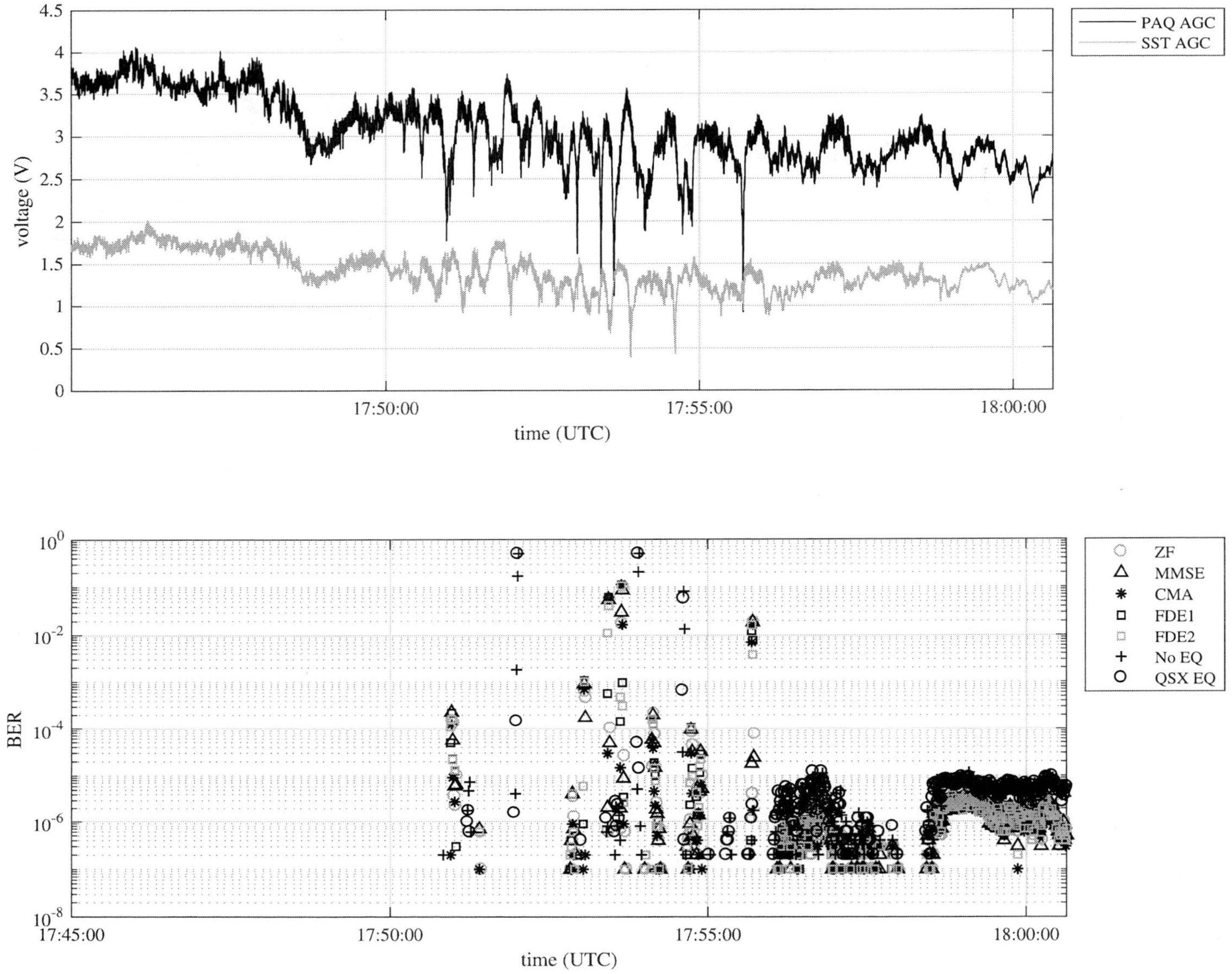

Figure 8: The AGC and BER for the Black Mountain west-to-east test point.

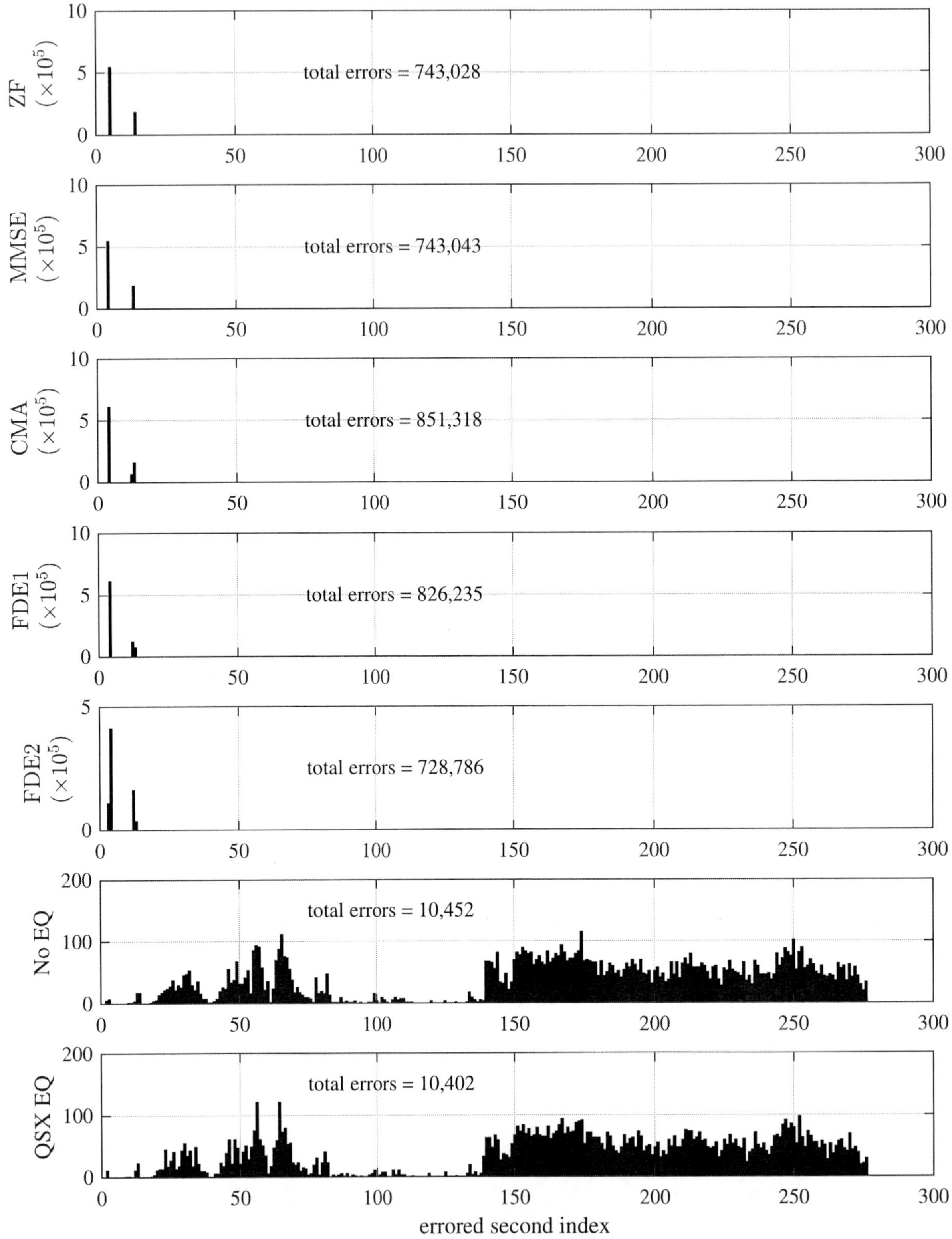

Figure 9: Bit errors vs. one-second intervals for the Black Mountain west-to-east test point (cf. Figure 8).

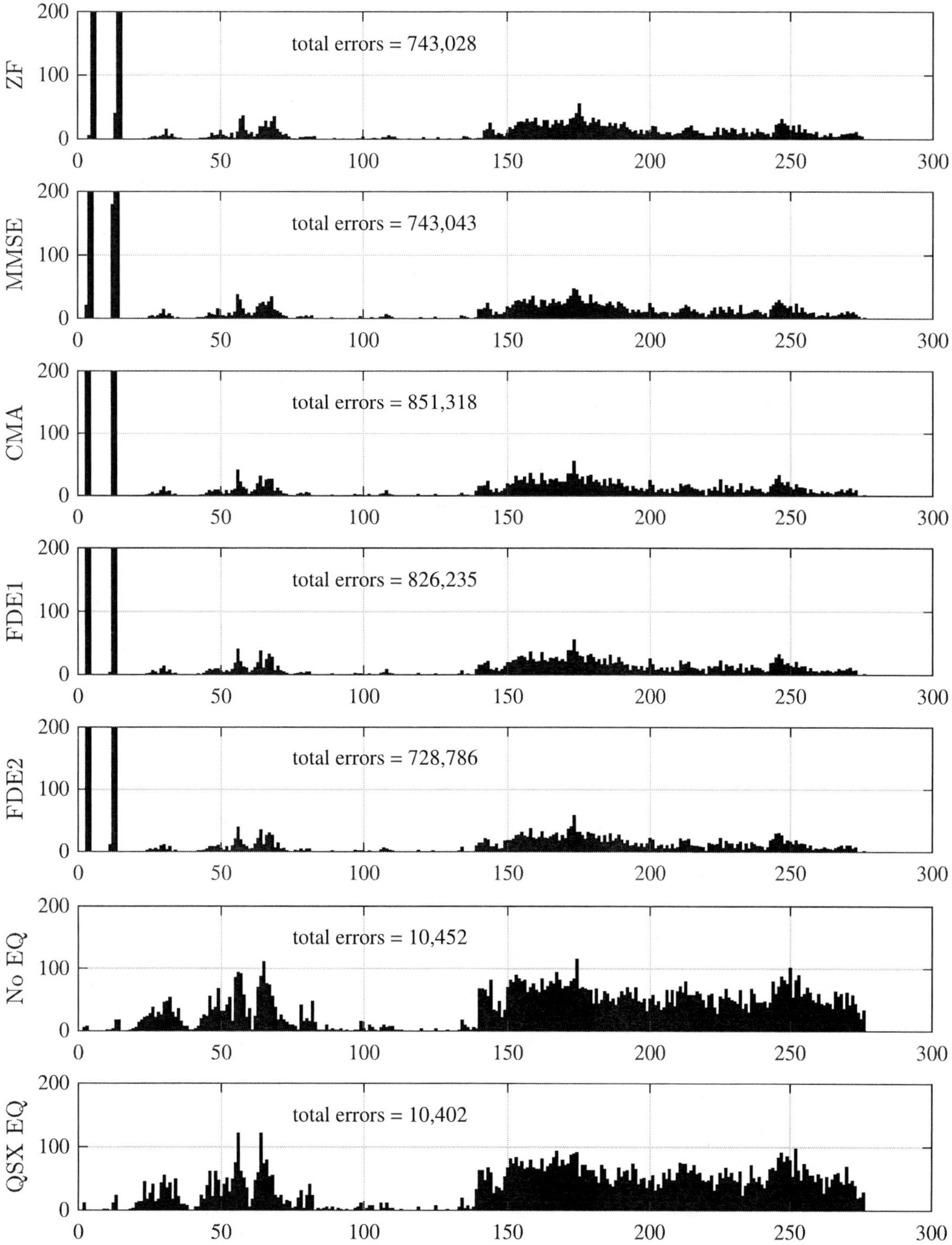

Figure 10: A close-up view of Figure 9: bit errors vs. one-second intervals for the Black Mountain west-to-east test point (cf. Figure 8).

ACKNOWLEDGEMENTS

This project is funded by the Test Resource Management Center (TRMC) Test and Evaluation, Science & Technology (T&E/S&T) Program through the U.S. Army Program Executive Office for Simulation, Training, and Instrumentation (PEO STRI) under Contract No. W900KK-13-C-0026 (PAQ). The Executing Agent and Program Manager work out of the AFTC.

REFERENCES

[1] T. Hill and M. Geoghegan, "A comparison of adaptively equalized PCM/FM, SOQPSK, and multi-h CPM in a multipath channel," in *Proceedings of the International Telemetering Conference*, (San Diego, CA), October 2002.

[2] M. Geoghegan, "Experimental results for PCM/FM, Tier I SOQPSK, and Tier II Multi-h CPM with CMA equalization," in *Proceedings of the International Telemetering Conference*, (Las Vegas, NV), October 2003.

[3] E. Law, "How well does a blind adaptive CMA equalizer work in a simulated telemetry multipath environment?," in *Proceedings of the International Telemetering Conference*, (San Diego, CA), October 2004.

[4] A. Guéguen and D. Auvray, "Multipath mitigation on an operational telemetry link," in *Proceedings of the International Telemetering Conference*, (Las Vegas, NV), October 2011.

[5] A. Skrzypczak, G. Blanc, and T. L. Bournault, "Combining a Reed-Solomon block code with a blind equalizer: Synchronization and bit error rate performance," in *Proceedings of the International Telemetering Conference*, (Las Vegas, NV), October 2015.

[6] M. Rice, C. Hogstrom, and C. Nash, "On IF-to-baseband translation and resampling in sampled-data receivers," in *Proceedings of the International Telemetering Conference*, (Las Vegas, NV), 23–26 October 2017.

[7] M. Rice, A. McMurdie, and E. Perrins, "A low-complexity preamble detector for iNET-formatted SOQPSK," in *Proceedings of the IEEE Military Communications Conference*, (Baltimore, MD), pp. 718–723, 6–8 October 2014.

[8] A. McMurdie, M. Rice, and E. Perrins, "iNET preamble detector performance in the presence of multipath interference," in *Proceedings of the International Telemetering Conference*, (Las Vegas, NV), 26–29 October 2015.

[9] M. Rice and A. McMurdie, "On frame synchronization in aeronautical telemetry," *IEEE Transactions on Aerospace and Electronic Systems*, vol. 52, pp. 2263–2280, October 2016.

[10] M. Rice, M. Saquib, A. Cole-Rhodes, F. Moazzami, and E. Perrins, "Preamble assisted equalization for aeronautical telemetry: Phase 1 final report," tech. rep., Brigham Young University, April 2014.

[11] M. Rice and E. Perrins, "On frequency offset estimation using the iNET preamble in frequency selective fading channels," in *Proceedings of the IEEE Military Communications Conference*, (Baltimore, MD), pp. 706–711, 6–8 October 2014.

[12] M. Rice, M. Saquib, and E. Perrins, "Estimators for iNET-formatted SOQPSK," in *Proceedings of the International Telemetering Conference*, (San Diego, CA), 20–23 October 2014.

[13] M. Rice, M. S. Afran, and M. Saquib, "Performance limits on joint estimation of frequency offset, channel, and noise variance in aeronautical telemetry," in *Proceedings of the International Telemetering Conference*, (Las Vegas, NV), 26–29 October 2015.

[14] M. Rice, M. Saquib, M. S. Afran, A. Cole-Rhodes, and F. Moazzami, "A comparison of three equalization techniques for iNET-formatted SOQPSK-TG," in *Proceedings of the International Telemetering Conference*, (San Diego, CA), 20–23 October 2014.

[15] M. Rice, M. Saquib, M. S. Afran, A. Cole-Rhodes, and F. Moazzami, "On the performance of equalization techniques for aeronautical telemetry," in *Proceedings of the IEEE Military Communications Conference*, (Baltimore, MD), pp. 456–461, 6–8 October 2014.

[16] M. Rice, M. Saquib, and M. S. Afran, "MMSE equalization for aeronautical telemetry channels," in *Proceedings of the International Telemetering Conference*, (San Diego, CA), 20–23 October 2014.

[17] A. Cole-Rhodes, H. Umuolo, F. Moazzami, and M. Rice, "Real-time CMA equalization for SOQPSK for aeronautical telemetry," in *Proceedings of the International Telemetering Conference*, (San Diego, CA), 20–23 October 2014.

[18] M. S. Afran, M. Saquib, and M. Rice, "Frequency domain equalizer for aeronautical telemetry," in *Proceedings of the International Telemetering Conference*, (Las Vegas, NV), 26–29 October 2015.

[19] J. Ravert and M. Rice, "On frequency offset compensation for equalized SOQPSK," in *Proceedings of the International Telemetering Conference*, (Glendale, AZ), 8–10 November 2016.

[20] M. Rice, C. Hogstrom, C. Nash, J. Ravert, A. Cole-Rhodes, F. Moazzami, M. Saquib, M. S. Afran, E. Perrins, and K. Temple, "Some initial results for data-aided equalizer experiments at Edwards AFB," in *Proceedings of the International Telemetering Conference*, (Glendale, AZ), 8–10 November 2016.

[21] Quasonix, "3rd Generation Rack-Mount RDMS^{TM} Telemetry Receiver." Data sheet. Available at http://www.quasonix.com/sites/default/files/qsx_rackmount_thirdgen_rdms_datasheet_0.pdf.

AN EXPANDED COMMUNICATIONS SYSTEM FOR OFF-GRID TELEMETRY

Kyle Boyer, Laura Brubaker, Kyle Everly, Richard Herriman, Paul Houston, Sean Ruckle, Rory Scobie, Ian Ulanday
Advisor: Dr. Michael Marcellin
The University of Arizona

ABSTRACT

The drivers of the University of Arizona Baja racing team must be intensely focused on tackling the jumps, boulders, mud bogs, and other challenges in the four-hour endurance race. These obstacles are just as demanding on the vehicle as the driver, so the pit crew needs effective ways of detecting problems within the vehicle before serious damage occurs. Our solution is a wide range, AES encrypted, Wi-Fi communications system that supports full IP protocol and live video feed, allowing our telemetry data to be accessed through a smartphone via a web interface. To provide a wide range of communication options, our system supports VOX and push-to-talk audio compatible with third-party radios, and has an on-board HDMI screen for text-based communications. Finally, our system is backwards compatible with prior generations to make repairs and replacements easy.

BACKGROUND

The University of Arizona's SAE Baja team designs, builds, and races a single seat off-road racecar at a yearly international collegiate competition. The race car's electrical system is subject to few regulations, leaving room to gain a competitive advantage through innovation and creativity.

DESIGN REQUIREMENTS

Voice

In the heat of a race it is crucial to be able to communicate clearly with the driver, whether to discuss strategy, report issues, or give encouragement and team updates. A clear communication line is important as it ensures messages are correctly understood and that excited drivers don't have to repeat themselves and risk distracted driving. It is also imperative to be able to communicate with other team members, as the competition venue is quite large and isolated, making cellphone communication unreliable at best. Communication with team members allows for physically distributed data sources and the ability to reach specific individuals in a timely manner.

VoIP

To get a higher fidelity for our voice communications, we chose to use a VoIP based system as our main communication system. A VoIP system has a higher bandwidth and with that comes better sound quality. A VoIP system requires a running server that clients can connect to and that facilitates the actual transfer of data between different devices. We chose to use Mumble because it is a free and open source VoIP system and was the easiest VoIP server for us to setup that allowed for a LAN based configuration as opposed to an internet based one. Mumble also has an app for both Android and iPhone which means that all team members can connect using just their smartphones, so there's no need for each person to have a dedicated radio to communicate (Mumble FAQ).

Radio

In the event of a catastrophic system failure, a simpler backup plan can be implemented. UHF radios are a reliable way to maintain communications if the Mumble app is unavailable or the IP based network fails. However, these radios come at the cost of lower fidelity. Radios are a last resort because a loud environment with multiple noise sources, such as engines, can seriously affect comprehension. Additionally, the radio band is very crowded because it is the preferred method of communication for other teams.

Access Controls

In order to keep the distractions for the driver to a minimum, we want to keep the number of people who are able to talk to the driver as low as possible. With Mumble, we are able to set up different channels and decide who has access to which channel, so we allow a certain few to talk to the driver and restrict everyone else from doing so. We are also able to set up separate channels for important competition information and casual chatting.

Privacy

Some teams inadvertently use the same channels for radio communication, which can lead to confusion. We have accidentally talked to members of other teams who were using the same radio channel as us, which created problems for both parties. Using VoIP ensures that our communications are exclusive to team members.

Video

One goal for our system is the ability to transmit live video from the car to the pit. As proof of concept, IP cameras were used to transmit video as a TCP stream (Figure 1). We tested this first by sending the stream from one 5.4 GHz Ubiquiti Bullet M5 to another, to a router, to a 2.4 GHz wireless access point, and finally to a computer (Ubiquiti Bullet M5 Datasheet). When this setup worked, we modified it to more closely match what would be used at competition, and sent the stream directly from the router to a Raspberry Pi. During competition, GoPros were used in place of IP cameras.

Eventually, the videos taken at competition will be used for route planning. The system will analyze the slightly different routes the driver took around the track and determine which one was the fastest, then suggest that route to the driver for future use. In addition to determining what radius of a turn is best, the system would be able to analyze which path is more efficient when the track diverges. These divergences are often created so that the driver has as difficult choice, whether to take a longer but smoother route, or a shorter but rockier one. Having a system to analyze times means that the driver can take each path once at the beginning of the race, and continue taking whichever is faster as the race progresses.

Figure 1 - Network Video Test Setup

Telemetry

A telemetry system can give a significant competitive advantage. By collecting and analyzing data from a suite of sensors, we can detect when something is going wrong on the car and even start to predict failures before they occur. In these cars, a small problem can commonly cause an even larger problem if left unattended. By predicting these failures, the small problems can be fixed quickly before they cause a larger one.

Acquisition

To acquire data, we use a distributed sensor network where each sensor has its own microcontroller which splits up the processing load and allows for every sensor to have interrupt enabled data acquisition. This means that we never miss a data point and have plenty of processing power to analyze that data.

Visualization

A web interface allows anyone connected to the network to view information about the car such as speed, position, and temperature readings. This lets team members easily monitor the status of the car at a glance from any location using their smartphones. The web interface is hosted by the control station in the pit.

The control station also has its own interface to visualize the data and interact with it. This station is able to map the car's position onto a satellite image, which allows those in the pit to see exactly where the car is. The interface also analyzes the data from the different sensors to alert the pit crew about possible problems.

A screen on the car lets the driver see important information about that status of the car, the main display being the current speed of the car. On top of that, the screen will show the driver when something is going wrong. For instance, if the brake pressure isn't what it's expected to be, that could mean that there is a leak in the break line and the screen will alert the driver of this potential problem. The screen also allows the pit to send text based communications in the event that both the VoIP and radio communications fail.

There are a few possible choices for the screen. We looked at LCD screens, which are fairly dim in direct sunlight but have a higher resolution, OLED which are brighter but have a lower resolution, and 7 segment displays which are the brightest option but also have the lowest resolution. Over several iterations, we have found that the added complexity and lack of visibility that comes with the LCD and OLED screens doesn't make up for the lack of resolution of the 7-segment display meaning that for our uses the 7-segment display is the best choice.

Other Design Considerations

Backwards compatibility

In an effort to maintain backwards compatibility and to allow for a backup plan, the driver's headset was chosen to be compatible with both standard 3.5 mm audio jacks and with the UHF radios. In addition to making the headset backwards compatible, all of the pins that were broken out on previous years boards were also broken out on this board. This meant that the new board is completely compatible with previous generations of hardware and software.

Range

5.4 GHz has less range than a single 900 MHz transmitter or UHF radios. It also has very little penetrating power, which ultimately renders it infeasible for use on a moving vehicle that regularly loses line-of-sight to the base station antenna. For this reason, we will be moving away from this frequency going forward, in favor of a lower frequency.

A 2.4 GHz access point in the pit enables people with cellphones to connect to our system and use the VoIP and Web Interface services over the entire pit area, and most of the track. As with 5.4 GHz, there is very little penetrating power, but this is easily resolved in the pit area by placing the AP on a flagpole.

Security

AES Encryption - the 2.4 GHz and 5.4 GHz links all use AES encryption

SSH - Software on the car can be remotely started, restarted and controlled via SSH. This includes remotely setting up Bluetooth devices while the car is in the pit, so the electronics can stay on the car. SSH provides protection against packet sniffing on the network, so we can have a guest network for demo purposes without worries.

HTTPS - self-signed certificate, since we are using a custom TLD and DNS hijacking to provide access to the visualization interface at a custom domain.

Reliability

The SAE Baja Competition lasts for four days. Such a short time frame can present unique challenges, as unforeseen environmental factors can play a large role in how well the system functions. Everything ran well during testing the first day, however, the competition organizers turned on their network the second day and our systems interfered with each other. To work around this issue, the plan is to go back to 900 MHz as was used at the 2016 competition. In order to keep the same data throughput, the system will have several transmitters running in unison. In the event our network entirely fails, the system stores telemetry data locally on the car to be offloaded and processed when possible

WHAT'S NEXT?

Looking at both what worked well this year and what could be improved, there are a few things we will be changing about our networking solution. The 2.4 GHz base station will be kept, to allow for individuals to connect to the network with mobile devices, view the telemetry data, and talk through Mumble. The 5.4 GHz connection that was used to connect from the pit to the car will be replaced with a 900 MHz system. This switch was brought about because the competition had its own network that interfered with our 5.4 GHz network while the 900 MHz band had no traffic. In order to maintain the data throughput that we need we will be using several 900 MHz connections in parallel.

CONCLUSION

This year we redesigned our entire communication system to increase the connectivity of our system. This was the most accessible system the UA Baja Racing Team has ever had, allowing anyone with a mobile device to view live data about the car. This system was a great learning experience for our team and will allow us to iterate upon it and make our system even better next year.

ACKNOWLEDGMENTS

Thank you to Dr. Michael Marcellin for making this project possible, our teammates on the Baja Racing Team for putting up with the billowing clouds of blue smoke and solder flux, and Josie Ruiz for always being there for us.

WORKS CITED

"Mumble FAQ" http://wiki.mumble.info/wiki/FAQ
"Ubiquiti Bullet M5 Datasheet" https://dl.ubnt.com/datasheets/bulletm/bm_ds_web.pdf

FEATURE SELECTION FOR CYCLOSTATIONARY-BASED SIGNAL CLASSIFICATION

Garrett Vanhoy, Noel Teku
Faculty Advisor: Dr. Tamal Bose
University of Arizona
Tucson, AZ, 85721
[gvanhoy, nteku1, tbose] @email.arizona.edu.

ABSTRACT

Cognitive radio (CR) is a concept that imagines a radio (wireless transceiver) that contains an embedded intelligent agent that can adapt to its spectral environment. Using a software defined radio (SDR), a radio can detect the presence of other users in the spectrum and adapt accordingly, but it is important in many applications to discern between individual transmitters and this can be done using signal classification. The use of cyclostationary features have been shown to be robust to many common channel conditions. One such cyclostationary feature, the spectral correlation density (SCD), has seen limited use in signal classification until now because it is a computationally intensive process. This work demonstrates how feature selection techniques can be used to enable real-time classification. The proposed technique is validated using 8 common modulation formats that are generated and collected over the air.

INTRODUCTION

Cognitive radio (CR) and software defined radio (SDR) are a pair of new technologies enabling rapid growth in the field of wireless communications. The concept of radio that can adapt to its environment was formalized by Joseph Mitola and was a revolutionary idea at a time where most radios could only perform in one way. This concept of Cognitive Radio has, at times, been simplified to applications in spectrum sharing where a radio can act as a secondary user to already licensed spectrum, but its applications go far beyond this. With the advent of SDR, which allows a radio to change fundamental transmission parameters such as center frequency and bandwidth, the concept of a fully Cognitive Radio has come closer to reality. SDR technology has developed over the last 20 years and has become a technology that is very necessary in many common devices such as cellular phones. The ultimate aim of CR is to be able to adapt a radio's operation to its environment, hence techniques that enable the understanding of its current operating environment must be developed.

Within an increasingly busy spectrum, a radio must be able to understand how the spectrum is currently being used. There are many existing techniques that allow a radio to determine whether a particular frequency is in use through measuring the amount of energy present at that frequency and comparing it to the amount of energy present when it is not in use. This technique enables spectrum sharing, but enabling more complex applications requires not only determining if a frequency is occupied, but *who* or *what* is occupying it. This can be accomplished through signal classification techniques. More specifically, the kind of signal classification that must be done here is both **blind** in the sense that no *a-priori* information is present, and **real-time**.

The task of determining the modulation of an incoming signal has been named modulation classification (MC), recognition, or identification [1]. Much of the existing work in this area has been on MC applications involving non-cooperative communications. Non-cooperative communications is when the signal of interest is coming from a transmitter that does not intend for its data to be interpreted by the observing radio. This is opposed to the normal wireless system that is designed to relay the modulation through the use of bits in the beginning of a packet that are of a known modulation. Non-cooperative communications is in general a more challenging scheme than cooperative communications as it is possible to exploit other known properties of the incoming signal.

A. *Likelihood-Based Modulation Classification*

Modulation classification methods can be largely separated into two types: likelihood-based and feature-based. Likelihood-based methods can be considered optimal in the sense that they achieve the least probability of misclassification [2]. However, this comes at the cost of a computational complexity that eludes real-time implementation in many cases [1]. Likelihood-based methods come in three categories including the Average Likelihood Ratio Test (ALRT), the Generalized Likelihood Ratio Test (GLRT), and the Hybrid Likelihood Ratio Test (HLRT).

The ALRT is named as such because it attempts to look at the averaged distribution of unknown parameters such as noise power and carrier phase offset with known probability density functions (PDF) conditioned on each modulation. This is the most accurate of the likelihood-based approaches and is often used as a theoretical upper bound for the probability of correct classification for other methods, but it is also the most computationally complex. Other types of likelihood-based approaches include the Generalized Likelihood-Ratio Test (GLRT) and the Hybrid Likelihood-Ratio Test (HLRT). The GLRT is less computationally complex than the ALRT, but is less accurate than the ALRT and suffers from being unable to differentiate nested constellations such as 16-QAM and 64-QAM entirely. These two algorithms can be combined to create the HLRT. The GLRT and HLRT computationally more tractable than the ALRT by design, but are still too computationally complex to use for many real-time applications.

B. *Feature-Based Modulation Classification*

A feature-based method extracts a set of descriptive values from the signal that differentiates each signal from each other. This approach is sub-optimal in terms of probability of correct classification, but substantially reduces the computational complexity. These features can include cumulants, time and frequency domain statistics, wavelet transform coefficients, or a combination of them [3]. Finding the best set of features to accurately identify modulation schemes from each other has been the subject of the bulk of research in the area of MC. The classification stage employs a variety of machine learning techniques to various degrees of success. The most prominent of these techniques include artificial neural networks, support vector machines, and decision trees.

C. Contributions

Likelihood-based MC establishes a framework capable of performing blind MC, but because of its computational complexity, it ultimately fails to lead to a real-time implementation. Feature-based MC significantly simplifies the signal classification process without sacrificing too much performance and thus enables real-time implementations. However, much of the work in feature-based MC makes prohibitive assumptions about the state of the received signal such as the need for system synchronization. This work proposes an MC design that does not require prohibitive assumptions and thus is both truly blind and enables real-time implementations.

In Section D., the spectral correlation density (SCD) is introduced as a feature by which several signal modulations can be classified is introduced. The method by which the computational complexity of classifying with SCD is significantly reduced is proposed is then proposed in F.. Classifiers that will be used in this work are detailed in Section G.. The results of using feature selection techniques using multiple classifiers as compared to not using feature selection techniques is presented in Sections H. and I..

BACKGROUND

D. Classification of Cyclostationary Signals

D..1 Signal Model

Signal models used in MC vary in their level of detail, but a model which incorporates all of the important parameters that are discussed in this work are presented here. The noiseless complex baseband signal is commonly represented by (1) from [1]

$$r(t) = \sum_{n=1}^{N} e^{j2\pi\Delta f_n t} e^{j\theta_n} \alpha_n s_k g(t-(k-1)T-\epsilon_n T) \qquad 0 \leq t \leq KT \tag{1}$$

with Δf_n, θ_n, and ϵ_n, and α being the carrier frequency offset, the time-invariant carrier phase, and timing offset of the nth signal with respect to the receiver's reference clock, and amplitude of the nth signal respectively. Each of the equi-probable complex data symbols s_k have symbol period T for the modulation of order P. The channel response to the transmitted pulse shape p_{TX} is the convolution between $h(t)$ and p_{TX} and is denoted by $g(t)$. The nature of channel impulse response $h(t)$ and the p_{TX} is assumed to be such that frequency-flat, slow Rayleigh fading takes place.

Attempting to classify the modulation of the nth signal can be broken down into three stages that each have its own challenges. First, a signal goes through a pre-processing stage in which common signal parameters such as the center frequency offset Δf_n or symbol rate can be estimated, channel effects such as multi-path fading can be compensated through equalization, and multiple signals are isolated using filtering techniques. This is done to maximize the effectiveness of the second stage, the feature extraction stage, in which a set of features are extracted from the signal using a variety of algorithms. Choosing a set of features that will most effectively describe the incoming signal has been the primary subject of research in MC. Lastly the classification stage implements a decision structure to best differentiate the incoming signals based on the extracted features. A few common decision structures include a decision tree, an artificial neural network, or a support vector machine. These three stages are not always treated as disjoint, but details about the pre-processing stage often omitted or lacking in detail in work in this area. For this reason, much of the work in MC can be summarized while focusing on the feature extraction and classification stages.

Many signals and systems have been modelled as wide-sense stationary stochastic processes where second-order statistics of the signal remain constant with time, but whose autocorrelation is independent of time. However, many man-made signals exhibit a periodic or an almost-periodic autorrelation function because they contain various periodic structures in time. These signals are called *cyclostationary*. Over the years, cyclostationarity has been studied rigorously in continuous and discrete, real and complex, and stochastic and non-stochastic contexts [4]. The cyclic spectrum at cycle frequency α of the process $x(t)$ can be written

$$S_x^{\alpha}(f) = \int_{-\infty}^{\infty} R_x^{\alpha}(\tau) e^{-j2\pi f \tau} d\tau, \tag{2}$$

which can interpreted as the time-averaged statistical correlation of two spectral components separated by cycle frequency α as the bandwidth of each spectral component approaches zero. For this reason, the cyclic spectrum can also be called the spectral correlation density (SCD). According to this definition $S^0(f)$ is actually the traditional power spectral density (PSD) of the process $x(t)$. The SCD has been used for analysis and classification of signals in many areas and this primarily stems from its ability to detect and characterize the presence of cyclic features such as cyclic prefix length, symbol period, or carrier frequency even in the presence of noise and other channel effects. Peaks in the SCD describe correlation in time between each pair of frequencies in the PSD. Naturally, analysis of the SCD permits the identification and characterization of underlying periodicities of the signal. The SCD is also granted a level of noise immunity because there are no inherent periodicities in noise. Thus, one might think of noise as spreading energy across all possible periodicities evenly.

E. Estimation of the SCD

The SCD is readily derivable in closed form for many continuous forms of communications signals and a substantial efforts has been made decades ago to estimate this quantity for a finite-duration digital signal. For a digital signal, estimating the SCD is commonly estimated using the FFT Accumulation Method (FAM) and is calculated as follows:

$$X_{N'}(n,k) = \sum_{r=-N'/2}^{r=N'/2} a[r]x[n-r]e^{-j2\pi k(n-r)T_s} \tag{3}$$

$$S_x^{\alpha}(n,k) = \frac{1}{N}\sum_{n=0}^{N-1} \frac{1}{N'} X_{N'}\left(n, k+\frac{\alpha}{2}\right) X_{N'}^{*}\left(n, k-\frac{\alpha}{2}\right) \tag{4}$$

where N' and N together determine a resolution in both the time and frequency domains and $a[n]$ is an arbitrary windowing function. Equation (3) is the sliding window discrete Fourier transform (DFT) with window $a[n]$. For this work $N' = 8$, $N = 512$, and $a[n]$ is a hamming window of length N'.

The SCD itself contains enough information to determine particular properties of each signal, however it contains far too many points to be used directly for classification and the number of points can be significantly reduced using feature selection techniques talked about in Section F.. A common way of reducing the SCD to a more reasonable number of points for classification is to take the maximum along the α axis to create an $\alpha-$profile and is thus used as a basis of comparison for some of the techniques presented in this work.

F. Feature Selection Algorithms

The objective of feature selection algorithms are to reduce a larger set of features to a smaller subset by removing features that are determined to be either redundant or irrelevant. There are three major types of feature selection algorithms: wrappers, embedded methods, and filters. Wrappers use the performance of classifiers to score the importance of features in a feature space. Embedded methods are similar to wrapper methods in that they both utilize the classifier when ranking features, but they "incorporate[s] variable selection as part of the training process" [5]. Filter methods function differently than embedded and wrapper methods by determining the importance of a feature using a classifier-independent scoring function [5]. For this project, filter and wrapper methods are used.

F..1 Score-based Feature Selection

Score-based feature selection techniques are an example of filter feature selection algorithms. These algorithms reduce an initial feature space of M features to a set of K features (where $K < M$) by passing all M features to an appropriate scoring function and selecting the K best-scoring features. This scoring function is independent of the classifier being used on the features which makes this technique a filtering algorithm. There are various scoring functions that can be used, but for this project, the analysis of variance (ANOVA) and mutual information feature selection techniques were implemented. The ANOVA scoring function determines the importance of each feature by comparing 1) the variance of a set of features $\boldsymbol{X}$ within samples of a common label (group) and 2) the variance of the set of features $\boldsymbol{Y}$ across all groups. The ANOVA test takes the ratio of these two values to produce a score known as the F-value as shown below [6]:

$$F_{rat} = \frac{MS_{bn}}{MS_{wn}} = \frac{\sum(\boldsymbol{X} - \bar{\boldsymbol{X}})^2}{\sum(\boldsymbol{Y} - \bar{\boldsymbol{Y}})^2} \tag{5}$$

If F_{rat} is closer to 1 this means that the variances are similar implying that the groups are similar looking only at one feature. The farther from 1, the more likely that this feature can be used to diff For a more detailed discussion on such experiments and one-way ANOVA, the reader is encouraged to refer to [6].

Another scoring function that can be used is based on mutual information. Mutual information is another statistical based on entropy that attempts to describe how much information is gained about one variable when another is known. The mutual information between several variables requires knowing joint probability density functions (PDFs) for each pair of variables and thus requires a substantial amount of samples. Instead, the joint PDF can be estimated using the provided samples with the k-nearest neighbors density estimation technique. An expression for this implementation of mutual information is provided in equation 6 [7]:

$$I(X, Y) \approx \phi(N) - \phi(m) + \log(V_{m;y}) - \phi(N_x) + \phi(k) - \log(V_{k;y|x}), \quad (6)$$

where X and Y are random variables, V is the volume, $\phi(n)$ is the digamma function, and k and m denote different neighbors.

F..2 Recursive Feature Elimination

Recursive feature elimination (RFE) is an example of a wrapper feature selection algorithm. RFE determines a feature's measure of importance by observing how the incorporation of certain features affects the classifier's performance. At each iteration of RFE, a certain number of features are removed. This new space is then used to train and test the classifier. Based on the classifier's performance on the testing data, a new set of features are removed from the feature space, and the procedure repeats until the desired number of features is obtained [8]. The advantage of this algorithm is that it weights the features based on performance rather than a statistical measure of redundancy. However, this algorithm has a large computational cost, as it requires a substantial amount of time to be completed for high-dimensional data sets.

G. *Classification Algorithms*

SVM: Support vector machines (SVM) are generally used to classify between two classes that are linearly separable. If classes are linearly separable, then a hyper-plane can be chosen such that points that have a negative distance to the hyper-plane are of one class and those that have a positive distance are the other [9]. Finding such an optimal hyper-plane involves solving an optimization problem using quadratic programming of which there are many tools available. Most data-sets are not linearly separable and often contain noise. So to counteract this, SVMs use a linear transformation with a chosen kernel to transform low-dimensional problem into a higher dimensional space where data is more likely separable.
Boosting: Boosting is an algorithm for training an ensemble of classifiers. Ensemble classifiers are used in widely in machine learning and is based on the principle that it is easier to generate weakly performing classifiers than creating one strong classifier. A boosting algorithm can be implemented in several ways, but in general, an ensemble is first trained on an initial data-set. Then, samples that are classified incorrectly are given heavier penalties to the training algorithm in hopes that the next time it is trained, these samples will be correctly classified. This algorithm is repeated iteratively until a chosen condition (perhaps performance related) is met. [10]
Bagging: Bagging stands for **b**ootstrap **agg**regat**ing**. Bagging is also an algorithm for training ensemble classifiers. It is known that an ensemble classifier performs well when the constituent classfiers are diverse in terms of their classifications. Bagging aims to create a diverse set of classifiers by training the constituent classifiers on different subsets of the original training set.[10]

Multi-layer Perception: The multi-layer perceptron (MLP) is essentially a feed-forward neural network. The MLP is a machine learning technique that attempts to model the human brain to make classification decisions. Its structure consists of connections between neurons, with each connection having an assigned weight. A standard neural network contains an input, output, and hidden layer. When training the MLP, the objective is to update the weights of each connection after calculating the error. A common method of training the MLP is called the back-propagation algorithm, which provides a method of updating the weights by estimating the local gradient of a neuron, which could be accomplished using stochastic gradient descent (SGD).

METHODOLOGY

The SCD calculation presented in section E. were implemented in Python and two sets of features were produced. The first set of features was the α-profile and the second was the 'flattened' featured. The 'flattened' features were produced by estimating the SCD and flattening the result (a matrix) into a one-dimensional vector. This could also be called the full SCD. Eight different modulations were generated on a Universal Software Radio Peripheral (USRP) N210 using GNU Radio, transmitted over the air and captured using GNU Radio on another USRP. The noise level in this process was kept as low as possible because the purpose was to emulate a signal being received by a non-cooperative receiver. Hence, the signal being received was not synchronized in frequency or time. The 8 modulations were BPSK, QPSK, 16-QAM, CPFSK, GMSK, OFDM, FM, and AM. Complex circularly symmetric additive white Gaussian noise was then added to each signal to simulate an AWGN channel with signal to noise ratios (SNRs) varying from -5 to 15 dB. This SNR was calculated by taking the ratio of the power in the signal to the power in the noise on a decibel scale. Note that this SNR is not equivalent to the SNR seen by a cooperative receiver because the non-cooperative receiver receives twice the necessary bandwidth to receive the signal. Hence, a cooperative receiver would see half as much noise as the non-cooperative receiver which is effectively 6 dB higher in SNR than the non-cooperative receiver. Receiving twice as much bandwidth as necessary from a classification standpoint will unlikely have an effect as the SCD measures the correlation of a pair of frequencies.

One-thousand independent samples were taken from each of the 8 modulations to be used for classification. Half of these samples are to be used for training classifiers and the other half are to be used for testing the classifiers. The α-profle and full SCD sets were separately reduced using feature selection algorithms discussed in F.. Implementations of these feature selection algorithms were obtained from the scikit-learn library, which hosts a wide array of machine learning algorithms programmed in python. The *SelectKBest* function was used as an implementation of the score-based feature selection. The ANOVA and mutual information scoring functions were implemented by calling *f_classif* and *mutual_info_classif* respectively, with each function finding their aforementioned score between features and labels. Each feature selection algorithm was used to reduce the set of features used for classification to 105 features based on the training set.

Once the training phase was complete, the classifiers were then given the testing data, and the percent errors were recorded for each configuration (i.e. feature selection algorithm and classifier). This procedure was repeated 10 times with the average error calculated over all 10 iterations. Scikit-learn's implementation of the four classifiers described in section G. were used. (**Note:** Scikit-learn's RFE implementation did not support Neural Networks and Bagging. Results obtained from RFE are only provided from using SVMs and Bagging) Their own specific set of parameters were used are as follows [8]:

SVM: The one vs one algorithm was selected for multi-class classification. A linear kernel was also chosen for all experiments.

Bagging: One hundred decision trees were chosen as the base classifiers. *max_features* was set to 0.5.

Boosting: One hundred decision trees were similarly chosen as the base classifiers. The learning rate was set to 0.5.

Multi-layer Perceptron: This consisted of 2 hidden layers (each with 100 neurons), using a rectified linear unit function as the activation function and a stochastic gradient-based optimizer to estimate the local gradient.

NUMERICAL RESULTS

H. Results by Feature Selection Algorithm

The results for percent correction classification for using an SVM while using different feature selection algorithms for both the flattened set of features and the α-profile are shown in 1. Each of the three algorithms shown in this Figure 1 reduced the feature set to 105 features. The 'baseline' method does not use any feature selection method and serves as a basis of comparison for the feature selection methods. Looking at both Figure 1b and Figure 1a, the 'baseline' algorithm generally performs the best as it uses all available features to classify each of the modulations. The 'RFE' method of feature selection has the best performance among the three feature selection methods and performs close to the baseline performance except for higher SNRs. At 15 dB SNR, the 'RFE' has a decrease of 7% percent correct classification. This is a meager reduction in terms of correct classification considering that less than 10% of the original features remain when 105 features are kept in the case of the α-profile and less than 3% of the features for the flattened features. This supports the idea that proposed approach to classification can be made real-time by reducing the feature set significantly.

I. Results by Classifier

The results for percent correction classification for using the mutual information based feature selection and different classifiers for both the flattened set of features and the α-profile are shown in 2. The number of features in Figure 2 are the same as the previous section 105 features. All classifiers appear to preform roughly the same using this feature selection algorithm except for Boosting. We suspect this is because Boosting used a decision tree classifier as its base classifier which is subject to over-fitting. What is interesting to note here is that the flattened features generally perform better than the α-profile for low SNRs At -5 dB SNR, the SVM classifier has an accuracy of 45% for the α-profile and 60% for the flattened features. This suggests that the flattened features are a better set of features to consider for future work than just the α-profile.

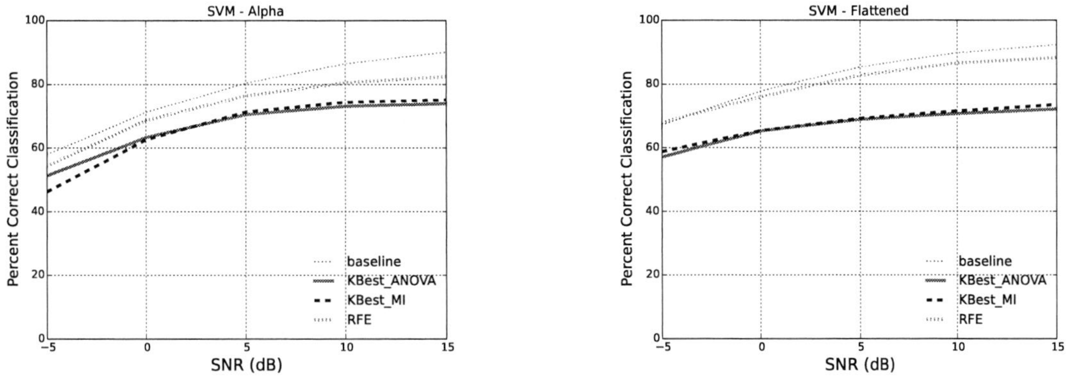

(a) (b)

Figure 1: Percent Correct Classification using SVM and α-profile **1a**) and flattened features **1b**)

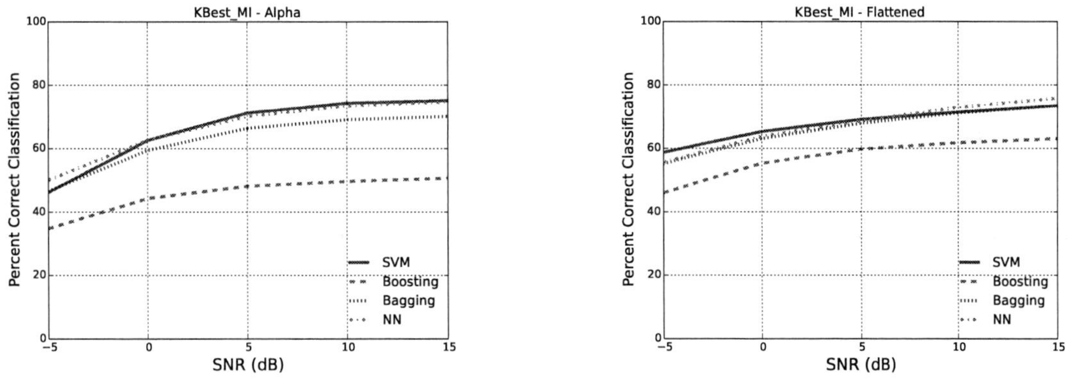

(a) (b)

Figure 2: Percent Correct Classification using mutual information based feature selection for α-profile **2a**) and flattened features **2b**)

CONCLUSIONS

In Cognitive Radio, signal classification is an important task in enabling higher levels of awareness. Modulation classification in non-cooperative scenarios requires that no *a-priori* information is used in the process of capturing a signal and classifying it from others. Until now, many modulation classification techniques have used features that require system synchronization to classify and hence assume some level of *a-priori* knowledge. The spectral correlation density (SCD) is a feature that can be used to classify signals without the need for system synchronization, but it generally too computationally complex for real-time use. Using feature selection techniques that have developed thoroughly in the machine learning community, only a small portion of the SCD needed to be calculated to properly classify 8 different modulations captured over the air in real time. Results show this technique is a promising baseline technique for classifying signals in non-cooperative scenarios with SNR ranges from -5 dB to 15 dB.

ACKNOWLEDGEMENTS

This project was partially supported by the Broadband Wireless Access and Applications Center (BWAC); NSF Award No. 1265960.

REFERENCES

[1] O. a. Dobre and R. Inkol, "Blind signal identification: Achievements, trends, and challenges," *2012 9th International Conference on Communications (COMM)*, pp. 349–352, jun 2012.

[2] F. Hameed, "On the Likelihood-Based Approach to Modulation Classification," *IEEE Transactions on Wireless Communications*, vol. 8, no. 12, pp. 5884–5892, 2009.

[3] A. Hazza and M. Shoaib, "An Overview of Feature-Based Methods for Digital Modulation Classification," *Conference on Communications, Signal Processing, and their Applications*, vol. 1, no. 08, 2013.

[4] W. A. Gardner, "Cyclostationarity: Half a century of research," *Signal Processing*, vol. 86, no. 4, pp. 639–697, 2006.

[5] I. Guyon and A. Elisseeff, "An Introduction to Variable and Feature Selection," *Journal of Machine Learning Research (JMLR)*, vol. 3, no. 3, pp. 1157–1182, 2003.

[6] G. Heiman, *Basic Statistics for the Behavioral Sciences*. Cengage Learning, 2010.

[7] B. C. Ross, "Mutual information between discrete and continuous data sets," *PLoS ONE*, vol. 9, no. 2, 2014.

[8] F. Pedregosa, G. Varoquaux, A. Gramfort, V. Michel, B. Thirion, O. Grisel, M. Blondel, P. Prettenhofer, R. Weiss, V. Dubourg, J. Vanderplas, A. Passos, D. Cournapeau, M. Brucher, M. Perrot, and E. Duchesnay, "Scikit-learn: Machine learning in Python," *Journal of Machine Learning Research*, vol. 12, pp. 2825–2830, 2011.

[9] E. Alpaydin, *Introduction to Machine Learning*. The MIT Press, 2nd ed., 2010.

[10] R. Polikar, "Ensemble based systems in decision making," *Circuits and Systems Magazine, IEEE*, vol. 6, no. 3, pp. 21–45, 2006.

LTE SYSTEM ARCHITECTURE FOR COVERAGE AND DOPPLER REDUCTION IN RANGE TELEMETRY

Achilles Kogiantis, Kiran Rege, Anthony A. Triolo
Vencore Labs, Red Bank, NJ 07701
Public Release No: 412TW-PA-17385

Abstract: A novel approach employing 4G LTE Cellular Technology for Test Range Telemetry is presented. Providing aeronautical mobile telemetry using commercial off the shelf (COTS) cellular equipment poses many challenges, including: Three-dimensional (3D) coverage, need for uninterrupted high data throughputs, and very high Doppler speeds of the Test Articles (TA). Each of these requirements is difficult to meet with a standard cellular approach. We present a novel architecture that provides 3D coverage over the span of a test range, allowing the TA to establish a radio link with base stations that have a manageable Doppler due to the reduced projected TA speed on the radio link line. Preliminary results illustrate that a variety of flight plans can be accommodated with commercial LTE technology by employing LTE's mobility mechanisms and adding centralized control. The resulting network architecture and Radio Access Network topology allow very high throughputs to be delivered throughout the test range with a judicious placement of base stations.

Keywords: Test, Range, Telemetry, Aeronautical, Mobile, Cellular, LTE, 4G, Handover, Doppler Shift, Frequency Offset, Compensation, X2, S1, 3D Coverage

I. Introduction

Currently, Aeronautical Mobile Telemetry (AMT) for test ranges is handled by point-to-point radio links with a wide beamwidth antenna on the Test Article (TA) paired with a high gain parabolic ground station antenna pointing to and tracking the TA. The spectrum used by the radio link is dedicated to that link for the duration of the test. Potential re-use of the same spectrum by another radio link at the same time anywhere in the test range requires coordination for maintaining orthogonality with the other radio link in the three dimensional (3D) space to control the potential interference. As a result, the spectrum efficiency of current telemetry practice is low and it requires intense human intervention for programming, scheduling and coordinating the tests. Moreover, test range telemetry is becoming increasingly complex with multiple radio links to serve each TA's telemetry needs, with test sorties of

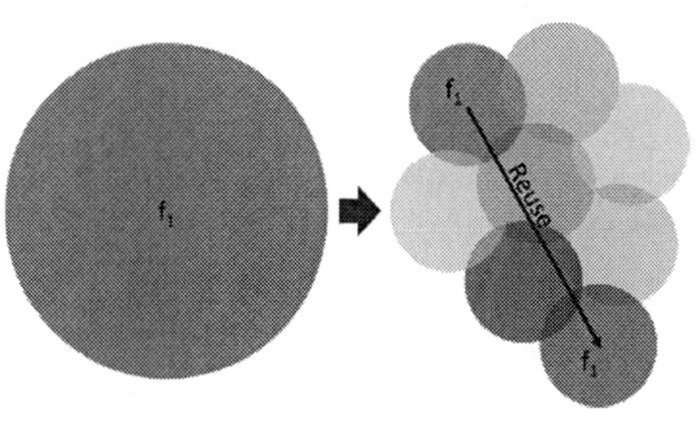

Figure 1 Test Range Telemetry transition to a cellular frequency reuse model

multiple coordinating TAs, and with multiple simultaneous tests occurring in the range. The overall complexity increase translates to difficulties in maintaining high spectrum efficiency, scheduling flexibility of the tests and logistical costs of the tests' execution.

A paradigm shift is attempted in this work to demonstrate that test range telemetry can utilize cellular communications concepts. The cellular approach is based on the localized use of the spectrum allowing reuse of the same block of spectrum in multiple and adjacent areas, as shown in Figure 1. Cellular providers have deployed extremely large numbers of fourth generation (4G) systems based on the 3GPP Long Term Evolution (LTE) standard. A cellular approach using LTE equipment for AMT applications allows leveraging of this low-cost, widely deployed COTS technology.

II. 4G LTE: A Candidate for Test Range Telemetry

The LTE standard was developed by the 3rd Generation Partnership Project (3GPP) a worldwide wireless mobile specifications body, with its first release (Rel-8) in 2008. 3GPP is the body that also developed the CDMA-based 3G UMTS and HSPA specifications, which were adopted widely by the regional Standards Development Organizations (SDOs). LTE has been built on the experience gained by the 3G on packet-switched optimized air-interface for multi-user environments. LTE enhanced 3G sophistication with support for numerous multiple antenna transmission modes, more optimized control channels, improved reverse link, support for diverse bearer QoS with a very wideband signal structure that is based on Orthogonal Frequency Division Multiplexing (OFDM). OFDM provides very flexible scalability in the frequency domain and enables a relatively simplified receiver structure that can handle channels with large amounts of multipath.

The LTE air-interface has been designed to accommodate up to 350 km/hr speeds, with some modes supporting up to 500 km/hr in anticipation of its use in high-speed train applications. LTE has been demonstrated to support air-to-ground links with aircraft speeds at 700 km/hr in field tests as well as in laboratory test environments, [1], [2], [3]. LTE's robustness is due to a narrow, 15 kHz wide, OFDM subcarrier structure populated densely with pilot subcarriers that make it robust to frequency selective, and very fast, fading.

LTE also expands on the already sophisticated mobility management of the 3G core network with a more streamlined network architecture and efficient signaling schemes. LTE supports mobility through two handover mechanisms, handover over the X2 interface, which relies on a direct exchange of messages between base stations (eNodeBs, or eNBs for short), or over the S1 interface, which relies on the base stations communicating through the Mobility Management Entity (MME). In both cases, the handovers are executed quickly, typically taking less than 200msec (100msec) if executed via the S1 (X2) interface. Any possible interruption of the air-interface link does not incur a loss of data as unacknowledged data are stored at the source eNB, and then forwarded to the target eNB for retransmission when the handover is complete.

LTE is currently the dominant global cellular technology marked by the widest and most rapid global adoption compared to any other cellular technology, e.g. WiMax, UMTS/HSPA, 1xEV-DO, or cdma2000. LTE is a living standard that is evolving through periodic releases (enhancements of the standard) by the 3GPP, with a continuous addition of functionality and features. The ecosystem of vendors is of a global scale, and is rapidly expanding. As a result, it offers economies of scale in cost, and wide availability of software, hardware and associated tools. Adopting LTE for range telemetry provides a transition path to constantly improving hardware and software releases and also an interoperable path into the 5G standards that are currently being defined.

III. Challenges of LTE for Range Telemetry

Although LTE is the best candidate for AMT amongst the available commercial cellular technologies, its application to test range telemetry is a challenging task that places LTE outside its normal envelope of operational environments. Specifically, two key challenges are present in AMT environments: Extreme Doppler shift and very high mobility.

Extreme Doppler Shift: LTE is designed to handle speeds up to 350 km/hr (189 Knots Ground Speed), or even up to 500 km/hr (270 KGS) under some channel conditions. The LTE speed limitation arises from the need to obtain accurate channel estimates for each symbol in a resource block (RB) in the time or frequency domain. In multipath environments with high user mobility, the RF channel frequency selectivity in combination with the rapid time variability between symbols makes channel estimation unreliable beyond some speeds. In terrestrial applications, the radio link experiences a rich multipath environment characterized by a large Doppler spread. The resulting smearing of the 15 kHz LTE subcarriers destroys the orthogonality between adjacent OFDM sub-carriers and introduces severe Inter-Symbol Interference (ISI). A metric of ISI is the ratio of max Doppler shift to the subcarrier spacing (f_d / Δ f) and is shown for the telemetry bands in Table 1.

Frequency Band	Max Doppler Shift f_d at 830 km/h (450 KGS)	f_d /Δf (%) Δf = 15Khz
L Band 1435 – 1525MHz	1.16 kHz	7.74%
S-Band 2200 - 2400MHz	1.84 kHz	12.5%
C-Band 4400 – 4940 MHz	3.8 kHz	25.31%
C-band 5090 – 5150 MHz	3.96 kHz	26.39%

Table 1 Doppler Shift relation to subcarrier spacing and carrier frequency

For the reasons stated above, if AMT radio channels were characterized by a rich multipath environment, it would be impossible for LTE to operate at the AMT speeds. However, in the AMT environment there is typically one dominant, line-of-sight, path between a TA and the ground station, and, possibly, much weaker multipaths typically caused by reflections from nearby mountain ranges. Thus the AMT channel experiences a significant Doppler Shift (DS) and possibly a weak Doppler spread. This is somewhat analogous to the case marked by a carrier frequency offset. Thus, even though the TAs are expected to operate at maximum speeds of thousand(s) of km/hr, this represents a relatively benign channel.

In LTE receiver implementations, the DS is compensated by first locking onto the Primary Synchronization Sequence (PSS) broadcasted by the eNB. PSS can be acquired before any Fast Fourier Transform (FFT) operation and at a very broad range of Doppler shifts. PSS lock provides the first, coarse, DS estimate that the receiver compensates for. Note that the receiver cannot differentiate between its own local oscillator (LO) ambiguity and the frequency offset associated with the received signal. Therefore, the receiver only locks on to what it estimates to be aggregate carrier frequency (inclusive of the DS) associated with the received signal. The receiver also performs additional, finer, frequency compensation offsets (FCO) to lock on to the aggregate carrier frequency. In an LOS channel such as most of those associated with the AMT environment, a receiver that can freely lock on to an arbitrary DS should not be impacted by the

TA's speed. In reality, however, commercial LTE handsets or eNB receivers have preprogrammed limits on the search window for possible DS estimation so they can optimize the fixed-point arithmetic of those operations. Determination of those limits on commercial products can be made through a receiver implementation design review or by testing the capability of devices to lock on to arbitrary DSs under lab conditions.

A second way to mitigate the extreme Doppler shifts in AMT is as follows: With cellular telemetry an opportunity exists to reduce the Doppler shift in the operational radio link by forcing the TA to connect only to that base station for which the projection of the velocity vector on the radio link line is small. This flexibility will be utilized in our solution.

Extreme Mobility: An AMT system that is based on a cellular architecture will have to make use of handovers to maintain seamless data connectivity with the ground as the TA traverses the test range. In LTE, connectivity during mobility is achieved by a standardized handover (HO) process, which is a break-before-make mechanism [4]: The UE reports to the connected (source) base station (eNB) the quality of potential radio links with the other candidate eNBs. The source eNB makes a decision as to which candidate eNB the TA should be handed over to. The TA disconnects from the source and attempts a random access procedure with the selected candidate eNB, now referred to as the target eNB. The number of signaling steps between the TA and the target and source eNBs are shown in Figure 2. In typical cellular applications with mobility speeds below 120 km/hr, a handover can be executed in a few hundreds of msec (200msec for S1 and less than 100msec for X2, typically), and in commercial systems it depends on the load of the nodes involved in the process. The MME or eNB nodes can be over-loaded as they have to maintain states or buffers and handle signaling and data traffic for a large number of users that are currently connected to or are trying to enter the network.

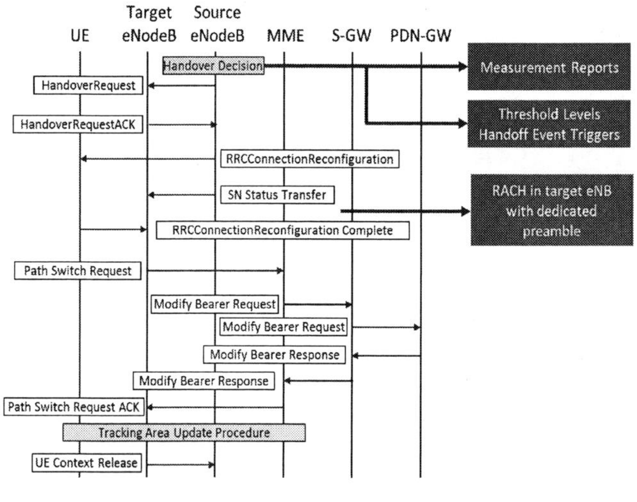

Figure 2 X2 Handover Call Flow for LTE per 3GPP standards specifications

In the AMT environment and at the high velocities to be encountered, the successful execution of all the handover steps will be more challenging when the velocities are high. Instructing the TA to connect with selected eNBs at lower apparent Doppler speeds will also aid the successful completion of handovers.

High Data Throughput Needs: The AMT data throughput needs are relatively high. A typical requirement may be 20Mbps in the TA to ground direction, and 2Mbps in the opposite, ground to TA, direction. The LTE standard allows in its basic formats 150Mbps peak data rate (with 64QAM or 16QAM and 2x2 MIMO) in the ground to TA direction (downlink in LTE terminology). It also allows 75Mbps (with 16QAM and single stream) in the TA to ground (uplink in LTE terminology) direction. Due to the benign nature of the AMT radio link characterized by almost free-space propagation, the link budget has sufficient room to yield high SNR operating points so that high-order modulation and coding schemes can be used. Therefore, the targeted data rates will be feasible with non-directional antennas (70-90 degree beamwidth in azimuth and elevation) in both directions.

IV. Proposed Cellular Architecture

In order to meet the two key requirements of extreme Doppler shifts and extreme mobility, we propose the following architecture with the name CeRTN (Cellular Range Telemetry Network) that is supported by a relatively dense deployment of base-stations so that most of the points on the likely flight paths of the TAs are within the coverage areas of multiple base-stations; moreover, the lines of sight to these base-stations are nearly orthogonal. The CeRTN system architecture contains an Integrated Cellular Network Controller (ICNC) to coordinate operation and handovers within the cellular network to minimize Doppler shift and optimize handoffs.

As shown in Figure 3 the eNBs in CeRTN are interconnected with X2 interfaces (when available). Moreover, each eNB is connected to the MME through an S1 interface on the control plane and to the Serving GateWay (SGW) with a S1-U interface on the user plane. The group of eNBs comprises the LTE Radio Access Network (RAN), while the rest of the functional elements other than the Network Orchestration and Management System (NOMS) and ICNC comprise the LTE Evolved Packet Core (EPC). The LTE RAN and the LTE EPC together are referred to as the LTE network.

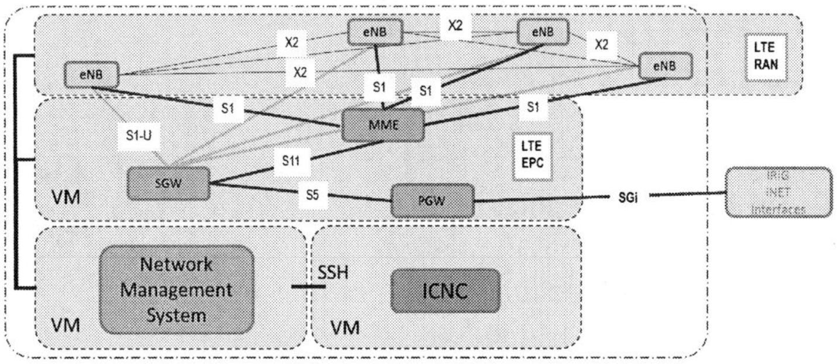

Figure 3 CeRTN Network Architecture showing EPC, NOMS, ICNC and interface to external networks

It should be highlighted that the interfaces are IP based and are logical. That is, no direct eNB-eNB physical connection is required. The control plane's physical connection links can be routed through the MME, if the MME is located at a central location. The X2 interface implies that packets originating in one eNB terminate at another eNB without the MME processing these packets. The Evolved Packet Core provides data to external elements, like the existing range telemetry system, through the SGi interface. The entire LTE network is managed through the NOMS. Each vendor's implementation of the NOMS is different, since it is not a standardized element. However, the functionality remains the same; that is, all NOMS provide facilities to manage parameters and settings on all of the LTE network elements, and to obtain key performance indicators (KPIs).

The key element of the proposed CeRTN system is the Integrated Cellular and Network Controller, (ICNC). The ICNC interacts with the elements of the LTE network to actively force handovers to desired eNBs, the mechanism to do so is through the NOMS. Parameters of the LTE network can be modified in real-time through the interface to the NOMS. Physically, the EPC, the NOMS, and the ICNC can be collocated on the same server, with each element existing within its own Virtual Machine.

Handover Control Scheme

The CeRTN system exercises handover control to enable the TA to connect to the base-station (eNB) that provides a communication link that not only has a high SINR, but also a manageable Doppler shift. Note that since the TAs making use of CeRTN are airborne, the LTE network must be designed to provide good coverage at all the altitudes at which the TAs are likely to fly within the test area. Thus, the coverage map of the test area will have to be 3-dimensional as shown in Figure 4 by way of an illustration. The message and control flow within the system as a TA takes off and follows its flight path is outlined as follows:

At some point during or after take-off, a TA establishes a connection with the LTE network using standard LTE procedures. For example, it establishes connection with the base-station with the strongest received signal strength (RSSI). The base-station connected to the TA is referred to as its serving base-station. The TA also receives a list of neighboring base-stations from the serving base-station. The TA periodically measures the RSSI associated with each of the base-stations in the neighbor list and reports them to the serving base-station. The reporting period may be set at a suitable level to accommodate the TA speed. The RSSI measurements are received by the serving base-station and forwarded to ICNC via the NOMS.

Location Estimation

ICNC uses the RSSI measurements reported by the TA to estimate its location. Additionally, ICNC may use (in addition to the RSSI) the timing advance reported by the TA to estimate its location. To that end, it uses a triangulation method in conjunction with a suitable path-loss model to estimate the TA's location. For most of the locations on a TA's flight-path that are at fairly high altitudes, a free-space path-loss model is accurate. Alternatively, other suitable models such as the Gierhart-Johnson propagation model may be used for these locations. Since a 3-D location estimate is sought, RSSI measurements associated with a minimum of four base-stations are required.

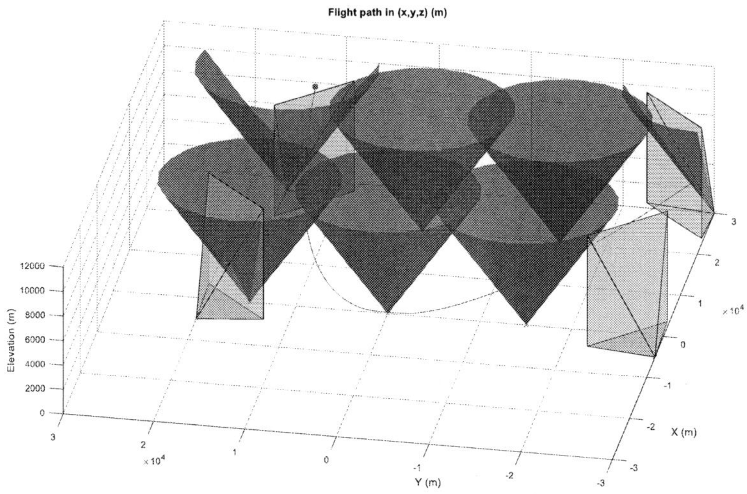

Figure 4 An Illustration of a 3-D Cellular Topology to provide coverage in the test range

Location Prediction

Once an estimate of the TA's current location is calculated, ICNC proceeds to predict the TA's location at a suitable point in time in the future using a suite of tools under the Flight Path Prediction Algorithm (FPPA) module, as shown in Figure 7. For example, if the current time is t seconds, ICNC predicts the TA's location at $t+\tau$ seconds, where τ is a suitable prediction horizon. Different methods can be used for location prediction. For example, one practical type of location prediction algorithm is based on "spline" functions. Due to kinematic constraints, the path between discrete samples of a flight-path is generally not a straight line. A common way to model the flight trajectory is a 3rd order spline function, which is a parameterized, piece-wise cubic function of time. Since ICNC does not have access to actual points on the flight-path, it uses the predicted TA locations in their place. Thus, given the most recent location estimates, say the current one as well as the past three estimates, ICNC can fit a 3rd order spline function to model the trajectory followed by the TA over the most recent segment of its flight-path, and assuming that its speed does not change significantly over the prediction horizon, it uses a continuation of the spline function to predict the locations the TA is likely to pass through in the immediate future.

While spline functions provide one way to model segments of the trajectory followed by a TA, they are, by no means, the only way to model the flight-paths. For instance, since the turning maneuvers of TAs usually follow a circular path, one can use a circular path prediction method to model segments of flight paths. In this method, ICNC selects the most recent location estimates for a TA and fits a circle to these points to model the path followed by the TA. In order to predict the TA's location at time $t+\tau$, it uses a continuation of this circle with the additional assumption that the speed of the TA does not change much from its current value.

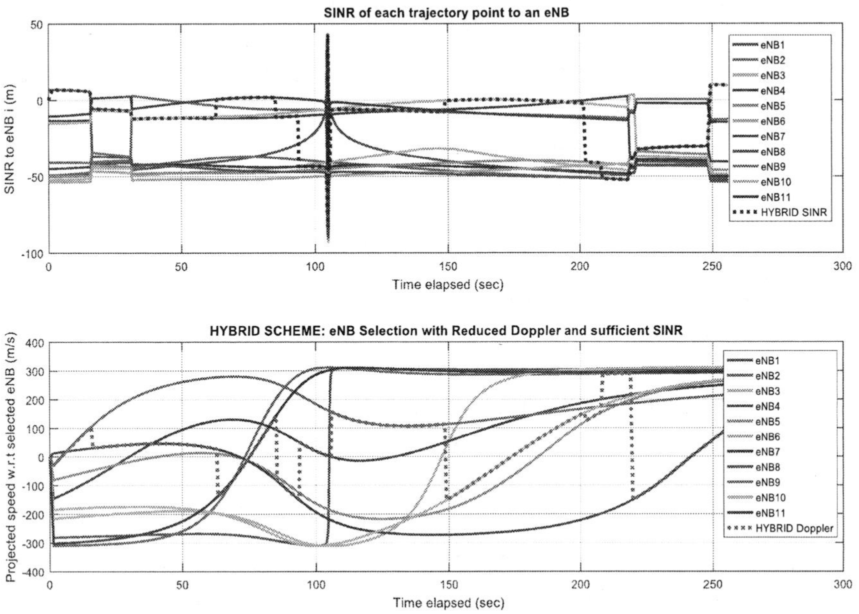

Figure 5 Illustration time series of Ground-to-TA SINR from each of 11 eNBs and the corresponding apparent doppler speed of each radio link. TA near parabolic trajectory is at 700mph constant speed starting and ending at 12Km elevation and performing a fly-by at 50m above control tower.

Determining the Best Base-Station to Handover to

After the previous step, ICNC has an estimate not only of the location of the TA at time $t+\tau$, but also its velocity vector. Using this velocity estimate, ICNC computes the Doppler shift that signals from the TA are likely to experience at each of the base-stations in the neighbor list. The Doppler shift for a base-station is computed by simply projecting the predicted velocity along the position vector between that base-station and the TA's predicted location. Also, given the TA's predicted location at time $t+\tau$ and the coverage maps of each of the base-stations on the neighbor list, ICNC can compute an estimate of the SINR associated with each of those base-stations.

Thus, at this stage, ICNC has, for each base-station on the neighbor list, an estimate of the SINR and the Doppler shift the TA is likely to experience if it is handed over to that base-station. Using the same method, ICNC can also compute estimates of these quantities if it remains connected to its current serving base-station. Among all these base-stations, ICNC selects the best base-station for the TA by considering the SINR associated with the radio link to each candidate base-station as well as the corresponding Doppler shift. The idea is to select, for handover, that base-station that not only has a decent SINR but also a manageable Doppler. Thus, a base-station with, say, a 10 dB SINR and a 50 Hz Doppler may be preferred over another base-station that has, say, a 25 dB SINR, but 2 kHz Doppler. If the selected base-station is the same as the TA's current serving base-station, no handover is needed. This means that ICNC does not have to do anything at this stage. On the other hand, if the selected base-station is different from the TA's serving base-station, a handover needs to be effected. Contrast the logic of this scheme with that used in typical commercial LTE networks where the choice of the base-station is based entirely on the SINR or received signal strength, irrespective of the Doppler shift.

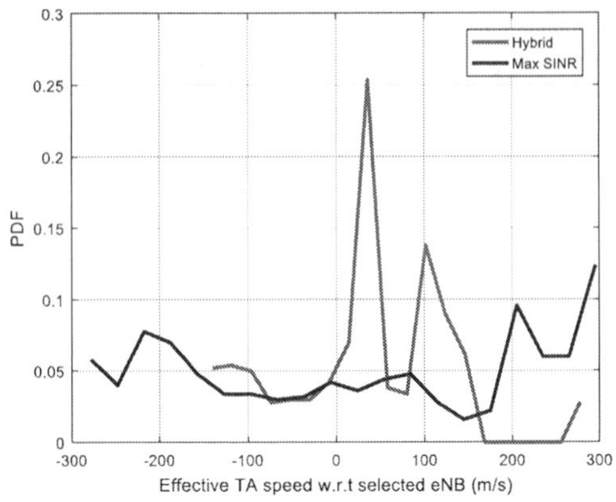

Figure 6 Histogram of reduced effective speeds of the hybrid scheme vs those incurred by maximum SINR eNB selection

The operation of this scheme is illustrated in Figure 5 using results from a simulation of a CeRTN network example shown in Figure 4. In this example, the TA follows a flight plan with a near parabolic trajectory and a constant speed of 608KGS. The TA initiates and terminates the flight segment at 12,000meters elevation, and during this segment it performs a fly-by at the origin at an elevation of 50meters. The radio link SINR and effective dopplers of the TA with 11 eNBs, each with wide beamwidth antennas pointing upwards or above the horizon, are recorded in a time-series as shown in Figure 5. A dashed line on each plot shows the resulting SINR and velocity (in m/sec) of the Hybrid scheme. Moreover, histograms of the Doppler shifts for the two handover schemes, Hybrid and Max-SINR, are compared in Figure 6, where it is evident that the hybrid scheme manages to reduce the Doppler to half of that in the Max-SINR scheme. It should be noted that in commercial LTE networks, HO mechanisms work similarly to the Max-SINR case.

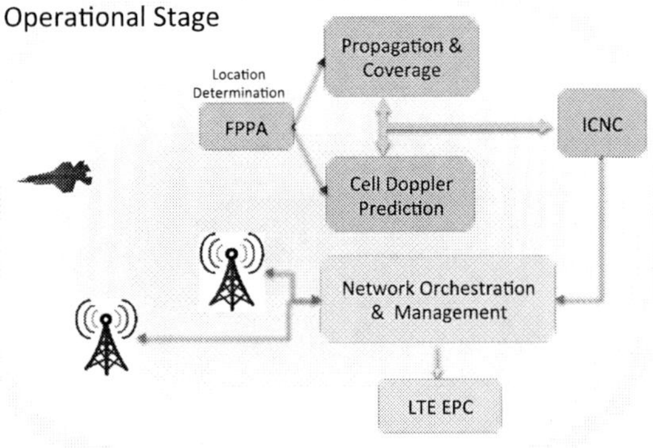

Figure 7 Operational view of the CERTN system. Information collected from the TAs triggers changes to the network through the NOMS, EPC and ICNC nodes.

Effecting Handovers

In those LTE hardware systems where blind handover commands (to handover a specific UE to a specific base-station independent of any measurement reports) can be sent to base-stations from the NOMS, ICNC can simply send a directive to the NOMS to issue the desired blind handover command. In those cases where the interface between the NOMS and the base-stations does not allow such commands, ICNC sends directives to the NOMS to change handover parameters such as thresholds, biases and hysteresis values. Specifically, ICNC changes these parameters in such a manner that the desired handover will happen naturally through the normal measurement reporting process. Conflicting handover requirements in the case of multiple TAs being on their respective flight-paths at the same time can be handled by assigning the TAs to different classes and crafting the handover parameters for each class to effect the desired handovers.

V. Conclusions

A novel approach for AMT employing cellular-based range telemetry has been introduced. It addresses coverage needs in the three dimensions, something that is typically far from what commercial cellular networks are designed for. The proposed scheme dynamically modifies network controls to tailor its behavior to the extreme Doppler and mobility needs of the AMT environments. The solution is based on near real-time information obtained from the network measurements and is immediately reflected in proactive actions involving TA handovers. Preliminary results indicate the scheme to be very effective in controlling Doppler shifts while maintaining reasonable radio link quality. A complete CeRTN system is currently being developed and evaluated in our lab and will be tested with outdoor airborne links over the course of this program.

Acknowledgments

Effort sponsored by the U.S. Government under Other Transaction number W15QKN-15-9-1004 between the NSC, and the Government. The US Government is authorized to reproduce and distribute reprints for Governmental purposes notwithstanding any copyright notation thereon.

The views and conclusions contained herein are those of the authors and should not be interpreted as necessarily representing the official policies or endorsements, either expressed or implied, of the U.S. Government.

References

[1] S. Caban, R. Nissel, M. Lerch, and M. Rupp. Controlled OFDM measurements at extreme velocities. In Proc. of 6th Extreme Conference on Communication and Computing (ExtremeCom), San Cristobal, Galapagos, Ecuador, August 2014.

[2] J. Rodríguez-Piñeiro, M. Lerch, J. A. García Naya, S. Caban, M. Rupp, L. Castedo. Emulating extreme velocities of mobile LTE receivers in the downlink. EURASIP Journal on Wireless Communications and Networking, Experimental Evaluation in Wireless Communications (2015), vol. 1; pp. 1 - 14.

[3] Ericsson Tests LTE in Extreme Conditions (2012) [ONLINE]. Available: https://www.ericsson.com/news/121101-ericsson-tests-lte-in-extreme-conditions_244159017_c

[4] 3GPP TS 23.401 v13.7.0 "General Packet Radio Service (GPRS) enhancements for Evolved Universal Terrestrial Radio Access Network (E-UTRAN) access," June 2016

ADVANCES IN A LOW-COST SOFTWARE-DEFINED TELEMETRY SYSTEM

Michael Don and Mark Ilg
U.S. Army Research Laboratory
Aberdeen Proving Grounds, MD, 21005
michael.l.don2.civ@mail.mil
mark.d.ilg.civ@mail.mil

ABSTRACT

The United States Army Research Laboratory (ARL) has developed a Pulse Code Modulated (PCM)/Binary Frequency Shift Key (BFSK) telemetry system using a low-cost commercial Software Defined Radio (SDR). Whereas traditional radio systems are implemented in hardware, much of the functionality of SDR is defined in software. This gives them the flexibility to accommodate military telemetry standards as well as other specialized functions. After a summary of previous development, new telemetry advances are described, focusing on telemetry transmitter, Advanced Encryption Standard (AES) encryption, and layered protocol design. Many of these functions were proven in a field experiment at Yuma Proving Ground, AZ in March of 2017 where data was successfully encrypted, transmitted, decoded, and processed in real-time.

INTRODUCTION

Traditional military radio systems implemented much of the processing in dedicated hardware equipment whereas much of the functionality of SDR is defined in software. This makes them a highly versatile platform that can support military telemetry standards as well as research into new telemetry schemes. As technology advances, SDR will become faster, smaller, more power efficient, and less expensive. Already, system-on-chip technology is incorporating multiple SDR components into a single Integrated Cirtuit (IC). This makes SDR not only a platform for receivers, but also a practical option for higher volume telemetry transmitters. Due to the versatile nature of SDR, telemetry research can be adapted to future munitions applications to include sensing [1], cognitive networks [2], and robust communications.

ARL has been utilizing telemetry links since the the mid 1990's to evaluate the aerodynamic flight behavior and diagnostic analysis of ballistic munition systems. Early methods used FPGAs to sample a small set of analog sensors, frame the data, and broadcast in the S-Band of the wireless spectrum using PCM over a BFSK channel [3]. A complex set of ground support equipment including antennas, receivers, bit-synchronizers, recorders, and data visualization systems provided the collection capability to receive the data from the systems under test. To aid in the development of smart guided weapons new telemetry systems have been developed to support interfaces including Global Position System (GPS) receivers and digital Micro Electro Mechanical Sensors (MEMSs) which have the ability to survive the harsh conditions of gun-launch. These systems relied on a Field Programmable Gate Array (FPGA) to frame the sensor data and comply with Inter-Range Instrumentation Group (IRIG) standards.

Using a FPGA for the framing and data processing were often time consuming and challenging due to the strict timing requirements and programming model. There are three possible solutions to this problem. The first is to implement a soft processor in the FPGA fabric which allows for higher level programming abstraction [4]. The disavantage of this method is that soft processors utilize valuable FPGA resources to process simple serial streams (ex. GPS receivers) or digital inertial sensors. Alternatively, tools such as Xilinx High-Level Synthesis tool can convert C/C++ to Register Transfer Level (RTL) [5], however FPGA high-level synthesis is not as straight forward as real microprocessor programming. The third method is for the FPGA to offload the protocol layer to a co-processor. In this paper, the later approach is taken, using a Digital Signal Processor (DSP) as the co-processor to ensure the most flexibility and longevity of the design.

In many munitions applications Size, Weight, and Power (SWAP) is an issue with any electronics system. Therefore, the problem is scoped into two classes, a SDR only solution and a hybrid approach of using SDR as a base-station and a custom transmitter solution embedded into the system under test. A full SDR solution would provide additional capability to include bidirectional radio links and implementation of commercial standards [6]. The hybrid solution allows for a more transitional adoption and integration with legacy proven and validated Radio Frequency (RF) front-ends. Focusing on a hybrid solution allows for laboratory evaluations of systems under test to transmit to a SDR base station which does not require costly equipment and allows for flexible data analysis tools. Whereas, during range testing, traditional telemetry equipment may be used for the RF reception and interface to the SDR back-end. For example, using range tracking antennas, RF receivers, and bit synchronization equipment and bypassing the radio portion of the SDR hardware.

In this paper we will describe the design of several components of a hybrid solution to transmit encrypted data from a guided munition. First, previous SDR telemetry development is summarized. Next, work on a SDR telemetry transmitter is presented, followed by an overview of an AES encryption scheme. Finally, the co-processor implementation of the protocol layers and processing pipeline are described.

PREVIOUS WORK

Previously, we presented initial research developing a SDR telemetry receiver suitable for our current Frequency Modulation (FM)/PCM Lower S-Band (2200-2290 MHz) [7] transmitters, which employ BFSK modulation [8]. Two SDR receiver implementations were presented. The first used a SDR to acquire Inphase/Quadrature (I/Q) data and a LabVIEW program on a host computer to implement the BFSK receiver. This design worked successfully, but was too slow for typical ARL applications. The second design moved the radio algorithms to the SDR's onboard FPGA.

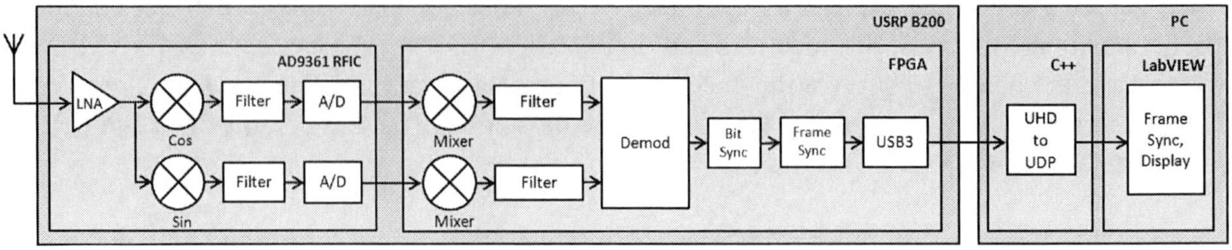

Figure 1: SDR telemetry receiver diagram.

This resulted in increased performance supporting data rates up to 10 Megabits per second (Mbs). Figure 1 shows a diagram of the SDR telemetry receiver. The Universal Software Radio Peripheral (USRP)'s Radio Frequency Integrated Circuit (RFIC) implements an I/Q demodulator, providing IF I/Q data to the FPGA. Ordinarily, the FPGA would downsample and convert the data to baseband before passing it to a Personal Computer (PC) for further analysis. In order to accelerate processing; demodulation, bit synchronization, and frame synchronization were added to the FPGA. Frame data was then passed to a PC where an intermediate C++ program received the data and sent it via User Datagram Protocol (UDP) to a LabVIEW program for display and archiving. This basic receiver architecture is ideally suited to form a basis for other custom applications.

SDR TELEMETRY TRANSMITTER

Several enhancements have been added to the SDR telemetry system including transmission capability, AES decryption, GPS time-stamping, spectrum monitoring, and an improved graphical user interface. An account of all of these improvements is out of the scope of this paper, but an overview of the AES decryption and SDR transmitter are presented here.

The USRP uses I/Q modulation for transmission given by

$$s(t) = I(t)\cos(2\pi f_c t) - Q(t)\sin(2\pi f_c t), \tag{1}$$

where $s(t)$ is the transmitted RF signal, $I(t)$ and $Q(t)$ is the I/Q data, and f_c is the carrier frequency. A BFSK transmitter modulates the frequency of carrier for symbols 1 and 0 by $+f_d$ and $-f_d$ respectively.

$$s_1(t) = \cos(2\pi t(f_c + f_d)) \tag{2}$$

$$s_0(t) = \cos(2\pi t(f_c - f_d)) \tag{3}$$

Using the angle sum identity and the fact that cosine and sine are even and odd functions respectively, $s_1(t)$ and $s_0(t)$ can be rewritten in the I/Q format from Eq. (1) with $I(t)$ and $Q(t)$ for the two symbols defined as

$$I_1(t) = \cos(-2\pi t f_d), \quad Q_1(t) = \sin(-2\pi t f_d) \tag{4}$$

$$I_0(t) = \cos(2\pi t f_d), \quad Q_0(t) = \sin(2\pi t f_d). \tag{5}$$

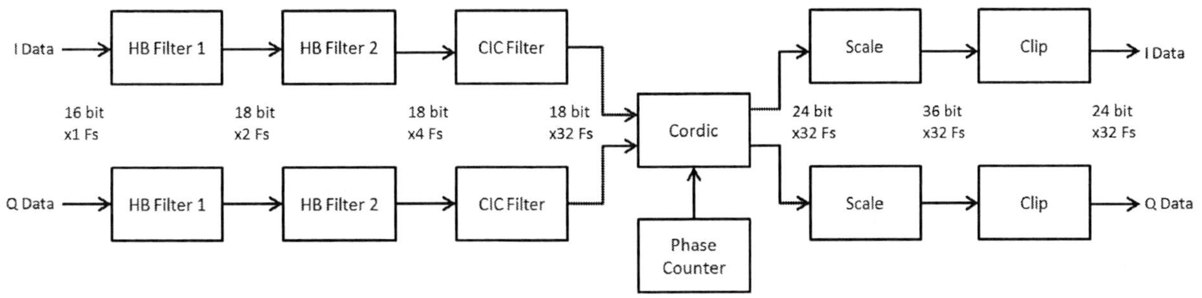

Figure 2: DUC diagram of USRP B200.

Typically, a SDR takes low-rate I/Q data generated on a PC, increases the sampling rate, and then shifts it to Intermediate Frequency (IF). Figure 2 shows a block diagram of the USRP's Digital Up Converter (DUC). Aside from the functional blocks, resolution and example sampling rate information is noted at each step for a case of upsampling by a factor of 32. The I/Q data travels through parallel paths, starting with two half-band filters that each increase the sampling rate by a factor of two. A Cascaded integrated-comb filter is used for the final upsampling. The I/Q data is then converted to IF through a Cordic block, and finally scaled and clipped before being sent to the RFIC for transmission.

For high data rate applications, however, it is more advantageous to generate the I/Q data onboard the SDR's FPGA. This has two benefits. First, it reduces requirements on the necessary transfer speed between the PC and SDR. Second, it offloads processing from the PC to the FPGA. The new transmitter diagram is shown in Figure 3, replacing the old DUC. Now 4 Mbs digital data is sent directly to the FPGA. The clock rate is set to the maximum value of 56 MHz in order to create the highest fidelity waveforms. The I/Q signals in Eqs. (4) and (5) are generated by Numerically Controlled Oscillators (NCO) implemented as Read-Only Memories (ROM). These contain 28 values representing one cosine period in order to support a frequency deviation of 2 MHz. Using the format of the original DUC at the Cordic input, these were structured as two's-complement Q2.15 values, where Qm.n is fixed point "Q" notation indicating m integer bits, n fractional bits, and one sign bit. The unsigned decimal values of the NCO were then

$$NCO = \begin{cases} [2^n x] + 2^{m+n+1}, & \text{if } x < 0 \\ [2^n x], & \text{otherwise,} \end{cases} \quad (6)$$

where x is the cosine sample value and $[x]$ denotes the nearest integer function. The I and Q phases are indexes to the ROMs, incrementing for a data bit of 0, and decrementing for a 1. Sine values for the Q data are simply obtained by shifting 90°, i.e. starting at index 7. FIR filters block the higher frequency components before using the rest of the original DUC blocks to complete the signal processing. In this manner, instead of generating and sending 56 Megasamples per second (Msps) I/Q data to the SDR, only 4 Mbs digital data is sent, the equivalent of 125 kilosamples per second (ksps) I/Q data. Figure 4 shows the spectrum of an example output of the BFSK transmitter.

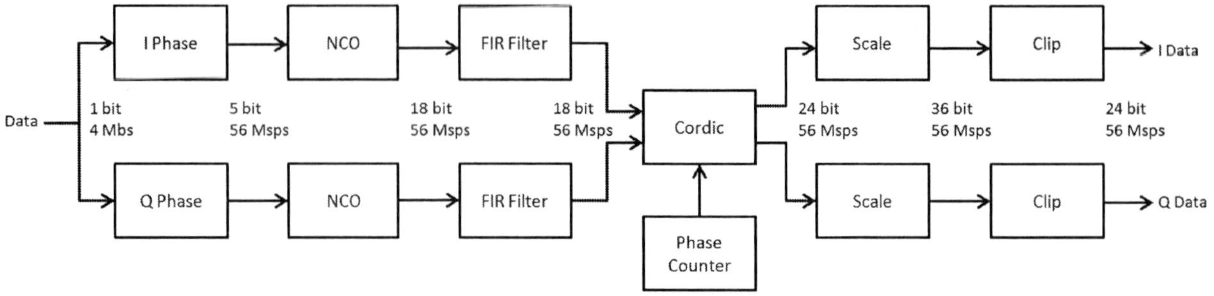

Figure 3: Modified FPGA transmitter block diagram.

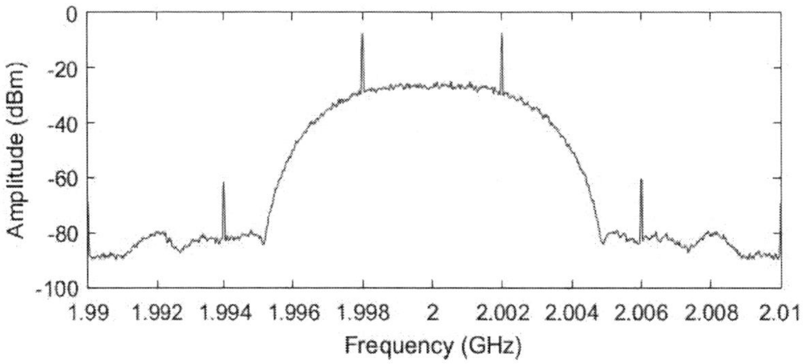

Figure 4: Spectrum measurement of transmit signal.

AES ENCRYPTION

Before presenting the encryption design, a short overview of AES is nessassary. The AES standard was established by the U.S. National Institute of Standards and Technology in 2001 [9] and approved for classified information by the NSA in 2003 [10]. It is a block cipher algorithm, i.e. encryption and decryption are applied to blocks of data as shown in Figure 5 on the left. AES uses 128 bit blocks, with 128, 196, or 256 bit keys. Blocks of data can be encrypted sequentially in order to encrypt an entire data stream. This simple application of a block cipher, called the electronic cookbook (ECB) mode, has inherent insecurities. Patterns can be detected in the data since identical data will always be encrypted the same way. This phenomenon is illustrated in Figure 6. The plaintext image on the left is encrypted using ECB mode to produce the ciphertext in the middle, where the ARL logo can still be detected. This problem can be circumvented by using the counter (CTR) mode shown in Figure 5 on the right [11]. The diagram on the lcft illustrates the encryption process. A counter is encrypted and xored with the plaintext to create the ciphertext. The same process is used to decrypt the chipertext, as shown on the right. The counter value increments by one for each block of data, with a random value chosen for the most significant bits called the nounce. Since it is the counter that is encrypted, and it is guaranteed to change with each block of data, there will not be any repetitive patterns in the ciphertext, as shown in Figure 6 on the right.

Using AES CTR mode with 256 bit keys, a simple encryption scheme was imposed on our typical frame format composed of 48, 16 bit words. The first two words comprise a 32 bit syncword, while the last three words are a 32 bit frame counter and a 16 bit Cyclic Redundancy Check (CRC). These

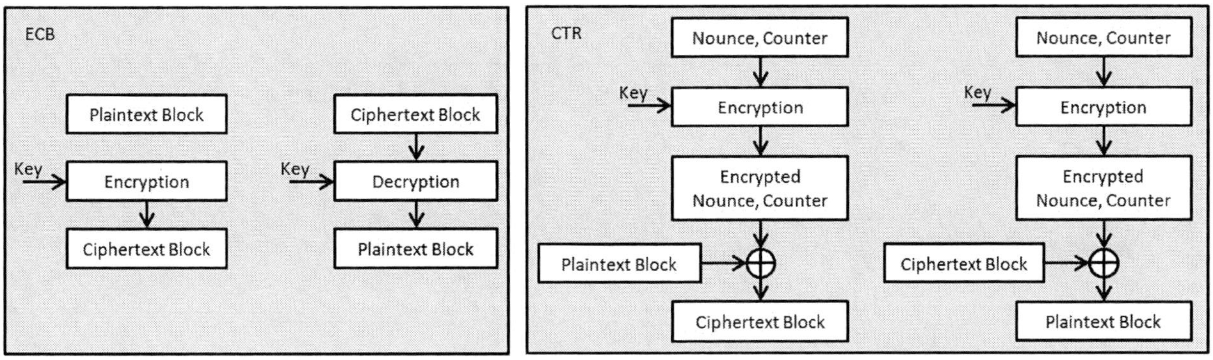

Figure 5: Block cipher encryption diagram using ECB and CTR modes.

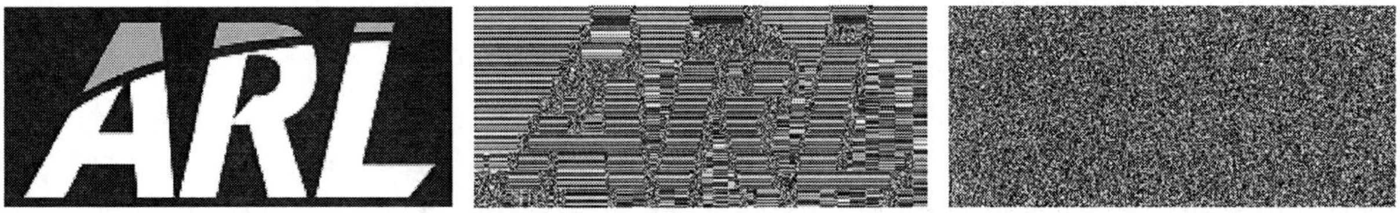

Figure 6: An image (left) encrypted using ECB mode (middle) and CTR mode (right).

words are left unencrypted, leaving 43 words of data for encryption. The first 40 data words are divided into 5, 128 bit cipher blocks. The last 3 data words are considered the most significant bits of a sixth cipher block, which can be encrypted normally, ignoring the least significant bits. The first block of 8 words is encrypted using a 96 bit nounce concatenated to the 32 bit frame counter from the previous frame. The counter value is incremented to encrypt each successive block, resulting in a final frame counter that increments by 6 for each frame.

Table 1 outlines an example frame encryption. The table starts with the last three words of the previous frame, showing the 32 bit frame count of x00000001. The first block of data in words 2 through 9 is encrypted using the previous frame count combined with a nounce. The counter continues to increment to encrypt each block of data, until the final block in words 42 though 44. The counter is incremented once more and inserted as the frame counter, which will be used to encrypt the first data block in the next frame.

This encryption scheme was implemented as a stand-alone board for encryption, and in the fabric of the SDR receiver's FPGA for decryption. Figure 7 shows a simplified block diagram of the encryption board. Unencrypted serial PCM data is input into the board, where frames are identified and words are extracted in the Frame Sync block. Detection of valid frames is indicated by asserting the Lock output. Words with their indexes are passed to into the frame encryption module, where a state machine in the Control block performs the encryption process. The state machine begins by waiting for a key load frame, which will load the AES 256 bit key and 96 bit nounce used for encryption. In order to simplify security, no data is output from the board before the key load frame is received. After it is received, all future frames are encrypted until a zeroize frame is received, or until the 32 bit frame counter reaches it's maximum value and the entire module is

Table 1: Frame encryption.

Word Index	Word Value	Encryption Counter
45,46	x00000001	
47	CRC	
0,1	SYNC	
2-9	Block1	x00000001
10-17	Block2	x00000002
...	...	...
42-44	Block6	x00000006
45,46	x00000007	
47	CRC	

reset. This allows for 38 hours of continuous transmission before reset using our typical 48 word frames and 4 Mbs data rate. A CRC module calculates a 16 bit CRC of the incoming frames, and is used to verify the key load and zeroize command frames. An Encrypting output provides feedback, indicating that the key load frame was successfully received and processed and the module is currently encrypting data. The Control block interfaces with the AES module[1] through a memory-mapped interface to encrypt the input data. Once a key load frame is received, the AES module is initialized with the key and encryption can begin. Input frames are encrypted as summarized in Table 1, with the unencrypted frame counter and CRC Out inserted at the end of the frames. Encrypted frames are serialized and passed through a 15-state RNRZ-L randomizer before being output for transmission. The final encryption board was a 1.5 inch diameter circular board using a Zilinx Artix-7 XC7A45T FPGA. Operation was verified up to 12.5 Mbs, but timing analysis indicates a maximum data rate of 152 Mbs.

For decryption, the FPGA of the SDR receiver in Figure 1 was modified to provide the additional capabilities shown in Figure 8. After the bitsync, a 15-state RNRZ-L derandomizer prepares the serial stream for frame synchronization. The CRC block then checks the frame CRC, replacing it with a 1 if correct or a 0 if incorrect. This is useful as a general telemetry system function that can be enabled independently of frame decryption. Key, nounce, and decryption enable settings are loaded into control registers from a PC through the interface program. As soon as enabled, decryption begins using the same procedure as encryption, except that instead of generating the frame count, the frame count is detected and used to decrypt the blocks of data in each frame.

PROTOCOL LAYER AND PROCESSING PIPELINE

A. TM STREAM GENERATION

The approach of using a DSP connected to a FPGA through a high speed synchronous serial interface provides a good balance between programmability and performance. This allowed the design

[1]The AES module was based on the avs_AES open-source code developed by Thomas Ruschival at www.opencores.org.

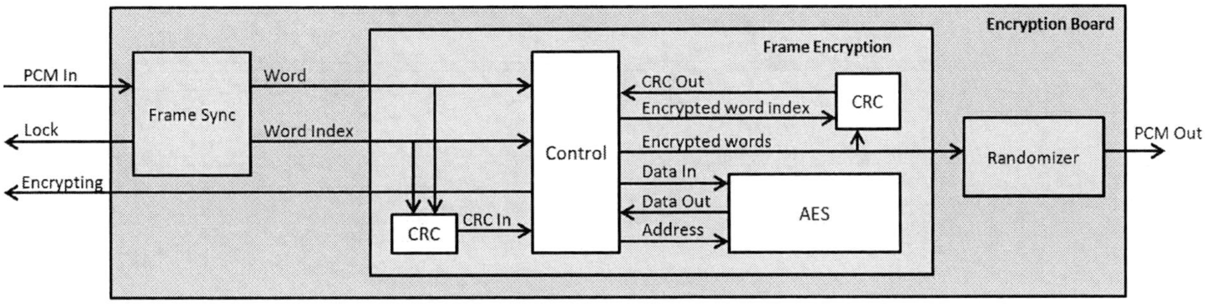

Figure 7: Encryption board block diagram.

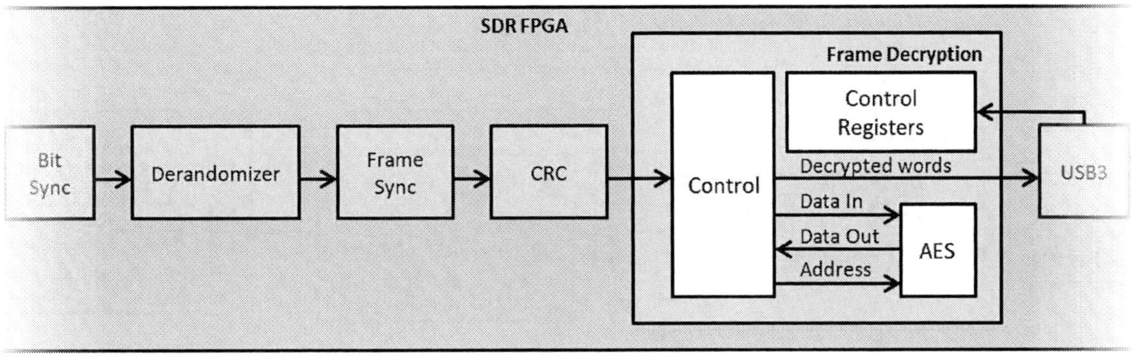

Figure 8: Decryption block diagram.

of all data handling and interfacing to external sensors through standard electrical interfaces. The DSP also allows for significant signal processing and estimation algorithms for real-time computations in the projectile for data compression and other data processing functions. It was determined the offloading of the AES encryption to the FPGA would allow for much higher throughputs of the telemetry data stream.

The DSP was then responsible for framing the data and processing incoming data payloads to generate the PCM stream. According to the IRIG standard [7], the PCM data is required to be placed into fixed length frames and embedded serial streams had to comply to a limiting set of requirements. Theses limitations are often burdensome and require significant frame and data planning to address the available channel bandwidth. Therefore, it is proposed to utilize a layered protocol where the top most protocol is the IRIG framing layer and subsequent protocols are embedded into this fixed length block. Figure 9 shows the structure of the telemetry stream. Each IRIG frame has a sync pattern as described in the previous sections with a Sub-Frame Identifyer (SFID) to identify the payload type, a frame counter, and checksum calculated using a CRC algorithm. This structure allows for creating unique SFIDs for specifically identifying fixed length frames which could comply with traditional standards. Utilizing a SFID also provides a universal framework for communicating with the SDR where standard fixed length frames may be added to the TM Stream. Each payload section is created using an asynchronous stream of messages.

The DSP creates the variable length message payloads through a First In First Out (FIFO) system

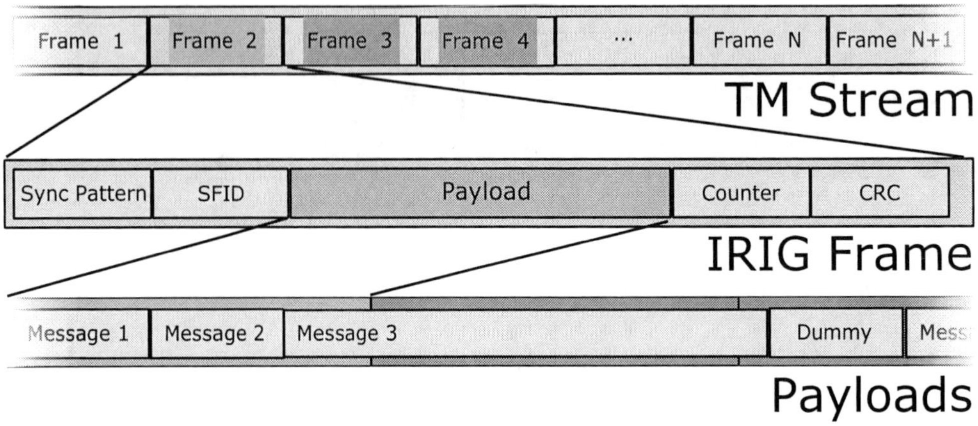

Figure 9: Protocol layer diagram.

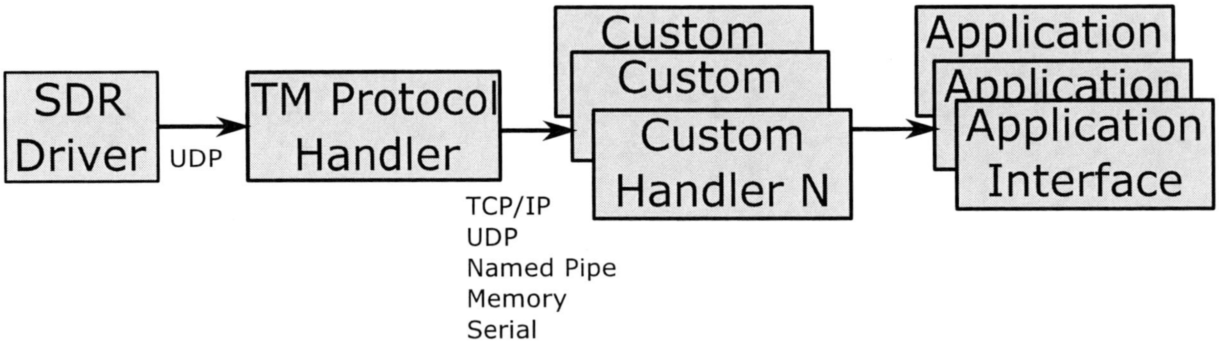

Figure 10: TM stream decoding.

and preserves the bitrate of the telemetry system through the injection of "Dummy" payloads when the FIFO is empty. Control of the payload is handled through a Direct Memory Access (DMA) channel to minimize the impact to the main CPU. Each message may have their own protocol and format. For example, a GPS message may have a uBlox protocol as described in the protocol specification [12]. Messages are thus only required to be defined as having a length and byte payload. For decoding the data, it is recommended to use higher level structuring of the data packet to ensure proper decoding.

B. TM STREAM DECODING

Figure 10 illustrates the Telemetry (TM) stream decoding process. The telemetry from the SDR first arrives through the SDR main driver to a UDP endpoint. This endpoint provides packets consisting of the fixed frames of data in Figure 9. The SDR ensures each frame has passed any decryption and CRC checks. A TM protocol handler then parses the standard IRIG frame and converts the remaining payload sections into a byte stream. The byte stream is then available on a variety of interfaces including Transmission Control Protocol/Internet Protocol (TCP/IP), UDP, Serial, Named Pipes, Memory, etc. These interfaces are then fed into a custom protocol handling software for decrypting to any application.

CONCLUSIONS

In this paper, we presented a SDR based telemetry for encrypted data transmission across a standard IRIG compliant S-Band channel. The telemetry link utilized multiple protocols for transmitting a variety of data sets using variable length messaging. The system was proven in a field experiment using range equipment at Yuma Proving Ground, AZ in March of 2017 where data was successfully encrypted, transmitted, decoded, and processed in real-time for data display.

-

REFERENCES

[1] H.-C. Chen, T.-H. Lin, H. Kung, C.-K. Lin, and Y. Gwon, "Determining rf angle of arrival using cots antenna arrays: a field evaluation," in *MILITARY COMMUNICATIONS CONFERENCE, 2012-MILCOM 2012*, pp. 1–6, IEEE, 2012.

[2] B. A. Fette, *Cognitive radio technology*. Academic Press, 2009.

[3] B. Davis, "Telemetry-based instrumentation for mortar systems," in *Proc. 2003 Mortars Conf*, pp. 4–8, 2003.

[4] M. L. Don, "Advances in telemetry capability as demonstrated on an affordable precision mortar," tech. rep., ARMY RESEARCH LAB ABERDEEN PROVING GROUND MD WEAPONS AND MATERIALS RESEARCH DIRECTORATE, 2012.

[5] Xilinx, "Vivado high-level synthesis," 2017.

[6] M. Iedema, *Getting Started with OpenBTS: Build Open Source Mobile Networks.* " O'Reilly Media, Inc.", 2014.

[7] Range Commanders Council Telemetry Group, Range Commanders Council, White Sands Missile Range, New Mexico, *IRIG Standard 106-04: Telemetry Standards*, 2004. (Available on-line at http://www.ntia.doc.gov/osmhome/106.pdf).

[8] M. L. Don, "A low-cost software-defined telemetry receiver," in *International Telemetering Conference Proceedings*, International Foundation for Telemetering, 2015.

[9] United States National Institute of Standards and Technology (NIST), "Announcing the ADVANCED ENCRYPTION STANDARD (AES)," *Federal Information Processing Standards Publication 197*, Nov. 2012.

[10] Committee on National Security Systems, "CNSS Policy No. 15, Fact Sheet No. 1: National policy on the use of the advanced encryption standard (AES) to protect national security systems and national security information," 2003.

[11] W. Diffie and M. E. Hellman, "Privacy and authentication: An introduction to cryptography," *Proceedings of the IEEE*, vol. 67, no. 3, pp. 397–427, 1979.

[12] *GPS.G6-SW-10018-F: μ-blox 6 Receiver Description Including Protocol Specification.*

Graph Theoretic Modeling and Energy Analysis of Wireless Telemetry Networks

Tristan A. Shatto, Kurt L. Kosbar, and Egemen K. Çetinkaya
Telemetry Learning Center
Department of Electrical and Computer Engineering
Missouri University of Science and Technology
Rolla, MO 65409, USA
{taszrc, kosbar, cetinkayae}@mst.edu

Faculty Advisors:
Egemen K. Çetinkaya and Kurt L. Kosbar

ABSTRACT

Network science provides essential tools to model and analyze topology and structure of dynamic wireless telemetry networks. In this paper, we model wireless telemetry networks using three well-known graph models: Gilbert random graph, Erdős-Rényi random graph, and random geometric graph models. Next, we analyze the connectivity of synthetically generated topologies using graph energy, which is the sum of absolute values of eigenvalues. Our results indicate second-order curves for adjacency and Laplacian energies as the connectivity of synthetically generated networks improve. The normalized Laplacian energy decreases, converging to the theoretical lower bound as the connectivity reaches to a maximum.

I. INTRODUCTION AND MOTIVATION

Network science is instrumental in analysis, design, and prediction of many military applications [1, 2]. Understanding the network behavior and improving the performance of networks is one of the essential areas of developing a strong network science capability for future military scenarios. Some applications areas include: the Darknet analysis of terror networks [3] and planning and developing future scenarios [4]. Accurate, realistic, and predictive modeling of a variety of military scenarios is still in the forefront of the network science [1, 2].

Temporal networks (also called dynamic or time-varying networks) are types of network whose structural properties vary over time [5]. Examples include social networks in which relations between entities change over time, biological networks in which protein interactions are dynamic, and wireless networks in which nodes might not have stable connectivity with their neighboring nodes as wired networks do. With respect to the time-varying wireless telemetry military application, a wireless airborne telemetry network scenario has been modeled and simulated using a predictive routing algorithm [6].

Capturing the topological and structural properties of networks have been studied widely [7]. Recently, we have been looking into the rigorous mathematical property of networks by means of eigenvalues, which are the roots of the characteristic polynomial. We studied eigenvalue multiplicities of the normalized Laplacian as the nodes and links are removed from backbone networks intelligently and found that eigenvalues approach to 0 when networks are under targeted attacks [8]. We also studied the graph energy, which is the sum of absolute values of eigenvalues, of backbone networks and found that as nodes are removed graph energy decreases with the exception of normalized Laplacian [9, 10].

In this paper, we simulate wireless telemetry networks using the well-known Gilbert random graph [11], Erdős-Rényi random graph [12], and random geometric graph models [13]. We then analyze the connectivity of these synthetically generated topologies using adjacency energy [14], Laplacian energy [15], and normalized Laplacian energy [16] measures. Our results indicate that adjacency and Laplacian energies reach a peak value before converging to the stable energy level for the full-mesh type of graph. The normalized Laplacian energy consistently shows a decreasing value before merging to the theoretical lower bound.

The rest of the paper is organized as follows: The background, including definitions and simple example about the energy of graphs is presented in Section II. The description of random graph models used in this paper is presented in Section III. In Section IV, we analyze three types of graph energies for the random graphs for varying node numbers. We conclude and showcase future work in Section V.

II. GRAPH ENERGY

Let $G = (V, E)$ be an unweighted, undirected graph with n vertices and l edges. The vertex set is denoted by $V = \{v_0, v_1, \ldots, v_{n-1}\}$ and the edge set is denoted by $E = \{e_0, e_1, \ldots, e_{l-1}\}$ for G. The graph connectivity can be represented by several structures such as an adjacency matrix, Laplacian matrix, and normalized Laplacian matrix. $A(G)$ is the symmetric adjacency matrix with no self-loops, where $a_{ii} = 0$, $a_{ij} = a_{ji} = 1$ if there is a link between $\{v_i,v_j\}$, and $a_{ij} = a_{ji} = 0$ if there is no link between $\{v_i,v_j\}$. The Laplacian matrix of G is: $L(G) = D(G) - A(G)$ where $D(G)$ is the diagonal matrix of node degrees, $d_{ii} = \deg(v_i)$. Given the degree of a node is $d_i = d(v_i)$, the normalized Laplacian matrix $\mathcal{L}(G)$ is denoted:

$$\mathcal{L}(G)(i,j) = \begin{cases} 1, & \text{if } i = j \text{ and } d_i \neq 0 \\ -\dfrac{1}{\sqrt{d_i d_j}}, & \text{if } v_i \text{ and } v_j \text{ are adjacent} \\ 0, & \text{otherwise} \end{cases}$$

Let M be a symmetric matrix of order n and I be the identity matrix of order n. Then, *eigenvalues* (λ) and the *eigenvector* ($\mathbf{x}$) of M satisfy $M\mathbf{x} = \lambda\mathbf{x}$ for $\mathbf{x} \neq 0$, viz., eigenvalues are the roots of the characteristic polynomial, $\det(M - \lambda I) = 0$. The set of eigenvalues $\{\lambda_1, \lambda_2, \ldots, \lambda_n\}$ together with their multiplicities (number of occurrences of an eigenvalue λ_i) define the *spectrum* of M.

A. *Graph Energy Definitions*

Energy of a graph, $\mathcal{E}$, is captured as the sum of absolute values of its eigenvalues [14, 17]. Given a graph, its adjacency energy, $\mathcal{E}_A$, Laplacian energy, $\mathcal{E}_L$, and normalized Laplacian energy, $\mathcal{E}_{\mathcal{L}}$, differ slightly for each type of graph representation's respective eigenvalues $(\lambda(A), \lambda(L), \lambda(\mathcal{L}))$. Given an adjacency matrix of a graph, $A(G)$, the graph energy is $\mathcal{E}_A(G)$ [14]:

$$\mathcal{E}_A(G) = \sum_{i=1}^{n} |\lambda_i(A)| \tag{1}$$

The graph energy of the Laplacian matrix, $L(G)$, is $\mathcal{E}_L(G)$ [15]:

$$\mathcal{E}_L(G) = \sum_{i=1}^{n} |\lambda_i(L) - 2l/n| \tag{2}$$

in which l is the number of links and n is the number of nodes. Lastly, given the normalized Laplacian graph, $\mathcal{L}(G)$, its energy $\mathcal{E}_{\mathcal{L}}(G)$ [16] is:

$$\mathcal{E}_{\mathcal{L}}(G) = \sum_{i=1}^{n} |\lambda_i(\mathcal{L}) - 1| \tag{3}$$

Our objective in this paper is to evaluate all three types of energies when forming random graphs representing wireless networks.

B. *Energy Analysis of Regular Networks*

In an effort to understand the energy characteristics, we study several regular graphs including star, full-mesh, bus, ring, and tree with node number 10 and 100. Among these regular network structures being analyzed, we only present (due to space) three cases – star, mesh, and ring – with 10 nodes and links removal scenarios as shown in Figure 1.

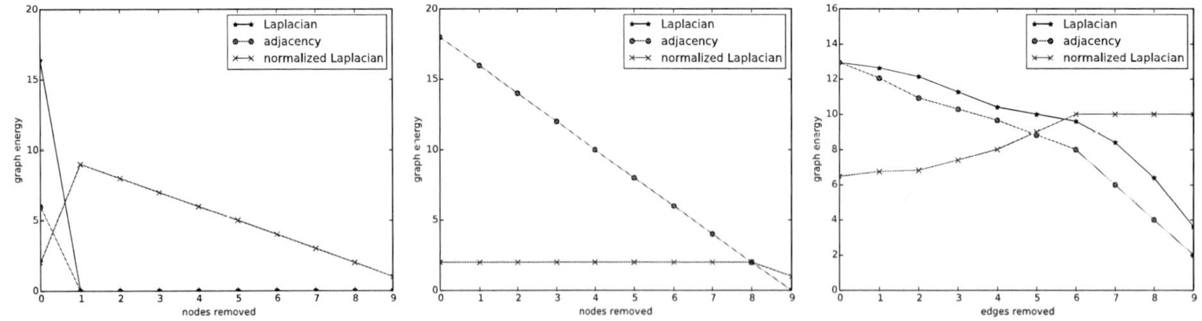

(a) Star with degree-based node removals (b) Mesh with betweenness-based node removals (c) Ring with betweenness-based edge removals

Figure 1: Energy characteristics of regular graphs (star, full-mesh, and ring) with 10 nodes

Before explaining the energies of regular graphs, we note that the eigenvalue multiplicities for adjacency, Laplacian, and normalized Laplacian matrices of star and full-mesh type graphs are shown in Table 1. The adjacency and Laplacian eigenvalue multiplicities [18], as well as the normalized Laplacian eigenvalue multiplicities [19] vary across different graph structures.

Table 1: Eigenvalue multiplicities of star and full-mesh graph types

Matrix types		**Multiplicities**
$A(G)$	star	$\pm\sqrt{n-1}, 0^{n-2}$
	mesh	$(-1)^{n-1}, (n-1)^1$
$L(G)$	star	$0^1, 1^{n-2}, n^1$
	mesh	$0^1, n^{n-1}$
$\mathcal{L}(G)$	star	$0^1, 1^{n-2}, 2^1$
	mesh	$0^1, (\frac{n}{(n-1)})^{n-1}$

Figure 1a shows the three types of energies for a 10-node star graph (1 root node and 9 leaf nodes) as the highest degree nodes are removed iteratively. From Table 1, a star graph has its eigenvalues at $\{-\sqrt{10-1}, 0, 0, 0, 0, 0, 0, 0, 0, \sqrt{10-1}\}$. When these eigenvalues are plugged in to Equation 1, we end up with an initial adjacency energy of 6. Once the highest degree node is removed, the empty graph with isolated edges have eigenvalues of 0, thus the adjacency energy becomes 0. While a similar decreasing pattern can be observed for the Laplacian energy, the normalized Laplacian energy starts from 2, peaks to 9, and then starts a decreasing energy pattern as shown in Figure 1a. The eigenvalues of a 10-node star graph are: $\{0, 1.11, 1.11, 1.11, 1.11, 1.11, 1.11, 1.11, 1.11, 1.11\}$, and when we plug these eigenvalues to Equation 3, we end up with a starting energy of 2. Once the highest degree node is removed, the eigenvalues become 0, but due to the -1 term in Equation 3, the empty graph's energy converges to $n-1$. Adjacency and Laplacian energies behave the same in a full-mesh graph and drop at an equal rate as nodes are removed (we note that $\mathcal{E}_A(G) = \mathcal{E}_L(G)$ for regular graphs [17]), whereas the normalized Laplacian energy remains the same at 2, as shown in Figure 1b. Finally, we look into the removal of links for a ring graph (intuitively, the connectivity of a ring lies between a star and a full-mesh) as shown in Figure 1c. In the case of ring network, the starting adjacency energy (i.e. 13 from Figure 1c) lies between a star (i.e. 6 from Figure 1a) and full-mesh (i.e. 18 from Figure 1b). The normalized Laplacian energy starts around 6 and ends with an energy of 10. As the links are removed, eventually the graph becomes an empty graph with isolated nodes, each node having an eigenvalue of 0. Due to the -1 term in Equation 3, the normalized Laplacian energy converges to n for an empty graph with all isolated nodes.

III. RANDOM GRAPH MODELS

We use the following well-known graph models to generate synthetic graphs: Gilbert graph model, Erdős-Rényi graph model, and random geometric graph model. Our assumption is that these random graphs capture the probabilistic nature of wireless telemetry network connectivity. We use the Python NetworkX library to generate these synthetic graphs [20].

A. *Gilbert Graph Model (GNP)*

In the Gilbert graph model, $G(n, p)$, given n nodes, these nodes are connected via a probability of p [11]. When the $p = 0$, then the graph generated will be an empty graph (i.e. graph with no edges). When the $p = 1$, then the graph generated will be a complete graph (i.e. full-mesh graph).

B. *Erdős-Rényi Graph Model (GNL)*

The Erdős-Rényi random graph, $G(n, l)$, takes number of nodes, n, and number of links, l as input parameters and outputs a graph with random l links between pair of nodes [12]. The maximum number of links in an undirected and bidirectional graph is $\frac{n(n-1)}{2}$. For example, in this model $G(5, 10)$ will output a full-mesh graph, whereas $G(5, 8)$ will result in a graph with 5 nodes and 8 links placed between pair of nodes randomly.

C. *Random Geometric Graph Model (GEO)*

In the random geometric graph model, $G(n, r)$, n nodes are placed on a unit surface at random and then two nodes are connected if the Euclidian distance between two nodes is less then a threshold value, r [13]. For example, $G(5, 0.1)$ will place 5 nodes uniformly at random on a unit square surface. If the distance between two nodes is less than 0.1, then these two nodes will be connected. Note that in a unit square surface, maximum possible distance between any pair of nodes is $\sqrt{2}$.

IV. GRAPH ENERGY ANALYSIS OF SYNTHETIC NETWORKS

To visualize how the connectivity of wireless telemetry networks can be improved by adding links between nodes, we examine the Laplacian, adjacency, and normalized Laplacian energy across synthetic networks of varying levels of connectivity. We generate synthetic networks with 10 nodes and 100 nodes using the random graph models. For the GNP model, we vary the p value in the [0, 1] range with increments of 0.1 for both $n = 10$ and $n = 100$ cases. For the GNL model, we increment the number of links by 1 for the $n = 10$ case and 100 for the $n = 100$ case. For the GEO model, we vary the r threshold value in the [0, 2] range with increments of 0.1 for both $n = 10$ and $n = 100$ cases. We average each point in the plots 100 times and plot with the 95% confidence intervals. By using three different random graph generators, and averaging the results of our energy calculations across numerous randomly generated networks, we can make observations into the connectivity tendencies of different sized telemetry networks. The 10 node graph energy analysis of random networks is shown in Figure 2 and the 100 node graph energy analysis of random networks is shown in Figure 3.

Using a GNP random graph generator (which places connections between n nodes based on a specified probability, p), we can notice that the Laplacian and adjacency energies peak near an edge placement probability of 0.6 for both 10 node networks and 100 node networks (seen in Figure 2a and Figure 3a, respectively). After peaking, the Laplacian and adjacency energies converge at a probability of 1, where each node is connected to every other node (i.e. a full mesh network). We may notice that due to the second-order curves of the Laplacian and adjacency plot lines, that the Laplacian and adjacency energies reach the same convergence value at multiple probability

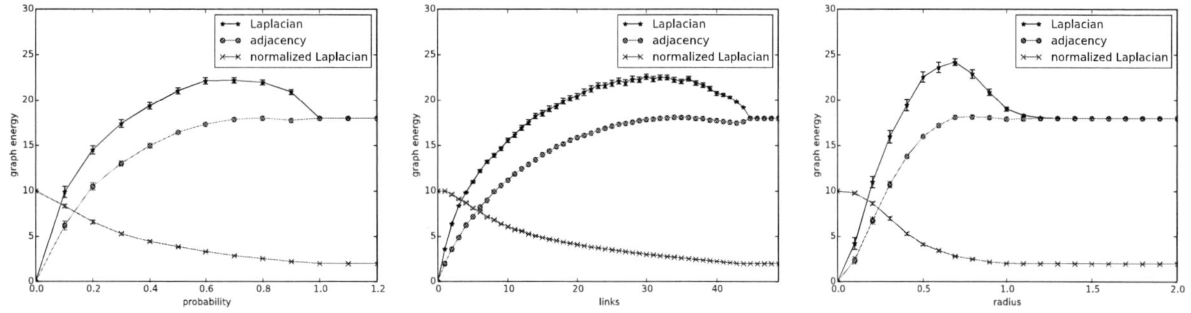

(a) 10 node GNP (b) 10 node GNL (c) 10 node GEO

Figure 2: Energy of random graphs with 10 nodes

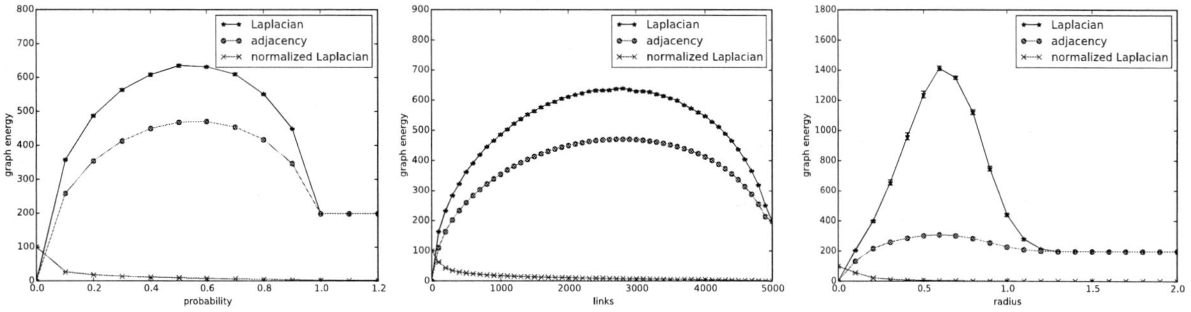

(a) 100 node GNP (b) 100 node GNL (c) 100 node GEO

Figure 3: Energy of random graphs with 100 nodes

values (e.g., 1.0 and 0.1 probability in Figure 3a). This is due to regular graphs having the same adjacency and Laplacian energy values, as explained in Section II.B. Seen most notably in the 100 node GNP plot (Figure 3a), while partially visible in the 10 node GNP plot (1.0 and 0.3 probability in Figure 2a), these matching points appear to signify varying levels of connectivity, with the lower probability point being the minimum probability for a connected graph and the larger probability (i.e., the probability when the Laplacian and adjacency energies converge) signifying full mesh connectivity. We can also observe that the normalized Laplacian energy exhibits a trend for both the 10 node networks and 100 node networks (Figure 2a and Figure 3a, respectively) where the energy gradually decreases from a peak value of n (for an n node network) at probability zero, before converging to a value at probability 1. Furthermore, we may note that the Laplacian and adjacency energies reach significantly higher peak values than the normalized Laplacian energy for both 10 node, and 100 node networks.

The GNL random graph generator produces plots similar to the GNP generator, with a notable difference being the sharpness of the Laplacian and adjacency energy peaks (seen in Figure 2b and Figure 3b); the peak value for the Laplacian and adjacency energies occurs near 30 links added for the 10 node networks, and near 3000 links added for the 100 node networks. Furthermore, we can see that the Laplacian and adjacency energies converge to a value near 18 after 45 links have been added to the 10 node networks, and converge to a value near 200 after 4950 links have been added to the 100 node networks, which are the respective points where the networks become fully connected as full mesh networks. The normalized Laplacian energy is also very similar to the GNL generated plots (shown in Figure 2b and Figure 3b), gradually decreasing from a value of n, for a n node network, and converging after reaching full mesh connection. Moreover, we can observe that the same trend of matching points seen on the Laplacian and adjacency curves in the GNP plots (i.e., two x-axis values where the Laplacian and adjacency energies are at their convergence value) are also present in the GNL plots: the lower point being the minimum number of added links necessary for a connected graph (near 10 links added and 200 links added for 10 nodes and 100 nodes, respectively), and the upper point being when the Laplacian and adjacency energies converge at full mesh connection (near 45 links added and 5000 links added for 10 nodes and 100 nodes, respectively).

The GEO graph generator notably results in a large peak of Laplacian energy near a radius of 0.6, for both 10 node and 100 node graphs (seen in Figure 2c and Figure 3c, respectively). The adjacency energy also peaks slightly near 0.6 for both 10 node and 100 node plots, but much less prominently than the Laplacian energy peak. Additionally, at a radius of approximately 1.4, we see that the Laplacian and adjacency energies appear to converge to the same value. Similar to the GNP and GNL plots, matching points can be observed where the Laplacian and adjacency energies reach their convergence value: the lower point (radius near 0.4 and 0.2 for 10 node and 100 node graphs, respectively) being the minimum radius for a connected graph, and the upper point (radius near 1.4 for both 10 node and 100 node graphs) being the radius at which the Laplacian and adjacency energies converge and reach full mesh connection. Furthermore, we can observe that the normalized Laplacian energy behaves similarly across all three graph generators, decreasing from n (for an n-node network) to its convergence value, which occurs near radius 1.4 for the GEO generator plots.

V. CONCLUSIONS

Network science tools enable analysis of the structure and predict the behavior of dynamic wireless networks. In this paper, we modeled the wireless telemetry network scenario using three random graph models: Gilbert, Erdős-Rényi, and random geometric graph models. We analyze the structure of these randomly generated graphs using the graph energy metric, which captures the sum of absolute values of eigenvalues. The synthetically generated Gilbert and Erdős-Rényi random graphs resulted in similar graph energy measurements. Among the graph energy measurements we conduct, adjacency and Laplacian energy measurements are second-order curves as the graphs converge to being complete graphs. The normalized Laplacian energy consistently decreases and approaching to the theoretical lower bound.

In our future work, we will investigate the reasoning of the second-order curves for adjacency and Laplacian energy relating to the connectivity of the graphs. We also plan to use realistic mobility models to model and study mobile wireless networks.

Acknowledgment

We would like to acknowledge members of the Complex Networks and Systems (CoNetS) group for discussions on this work. Tristan A. Shatto is in part supported by Missouri University of Science and Technology (Missouri S&T) Opportunities for Undergraduate Research Experiences (OURE) Program. This work was also funded in part by the International Foundation for Telemetering (IFT).

REFERENCES

[1] National Research Council, *Strategy for an Army Center for Network Science, Technology, and Experimentation.* Washington, DC: The National Academies Press, 2007.

[2] L. Gharai, K. Guan, and R. Ghanadan, "Military Scenarios and Solutions from a Network Science Perspective," in *Proceedings of the IEEE Military Communications Conference (MILCOM)*, pp. 1–6, October 2009.

[3] P. M. Dudas, "Cooperative, Dynamic Twitter Parsing and Visualization for Dark Network Analysis," in *Proceedings of the IEEE 2nd Network Science Workshop (NSW)*, (West Point, NY), pp. 172–176, April 2013.

[4] C. Arney, K. Coronges, H. Fletcher, J. Hagen, K. Hutchinson, A. Moss, and C. Thomas, "Using Rare Event Modeling & Networking to Build Scenarios and Forecast the Future," in *Proceedings of the IEEE 2nd Network Science Workshop (NSW)*, (West Point, NY), pp. 31–36, April 2013.

[5] P. Holme and J. Saramäki, "Temporal networks," *Physics Reports*, vol. 519, pp. 97–125, October 2012.

[6] J. P. Rohrer, A. Jabbar, E. K. Çetinkaya, and J. P. Sterbenz, "Airborne telemetry networks: Challenges and solutions in the ANTP suite," in *Proceedings of the IEEE Military Communications Conference (MILCOM)*, (San Jose, CA), pp. 74–79, November 2010.

[7] L. d. F. Costa, F. A. Rodrigues, G. Travieso, and P. R. Villas Boas, "Characterization of complex networks: A survey of measurements," *Advances in Physics*, vol. 56, no. 1, pp. 167–242, 2007.

[8] T. A. Shatto and E. K. Çetinkaya, "Spectral Analysis of Backbone Networks Against Targeted Attacks," in *Proceedings of the 13th IEEE/IFIP International Conference on the Design of Reliable Communication Networks (DRCN)*, (Munich), March 2017.

[9] T. A. Shatto and E. K. Çetinkaya, "Variations in Graph Energy: A Measure for Network Resilience," in *9th IEEE/IFIP International Workshop on Reliable Networks Design and Modeling (RNDM)*, (Sardinia), May 2017. Submitted.

[10] E. K. Çetinkaya and T. A. Shatto, "Eigenvalues for Resilience Analysis of Backbone Networks," in *Proceedings of the SIAM Workshop on Network Science (NS)*, (Pittsburgh, PA), July 2017.

[11] E. N. Gilbert, "Random Graphs," *The Annals of Mathematical Statistics*, vol. 30, no. 4, pp. 1141–1144, 1959.

[12] P. Erdős and A. Rényi, "On random graphs I.," *Publicationes Mathematicae Debrecen*, vol. 6, pp. 290–297, 1959.

[13] M. Penrose, *Random Geometric Graphs.* Oxford Studies in Probability 5, 2003.

[14] I. Gutman, "The Energy of a Graph: Old and New Results," in *Proceedings of the Euroconference Algebraic Combinatorics and Applications (ALCOMA)*, (Gößweinstein, Germany), pp. 196–211, September 1999.

[15] I. Gutman and B. Zhou, "Laplacian energy of a graph," *Linear Algebra and its Applications*, vol. 414, no. 1, pp. 29–37, 2006.

[16] M. Cavers, S. Fallat, and S. Kirkland, "On the normalized Laplacian energy and general Randić index R_{-1} of graphs," *Linear Algebra and its Applications*, vol. 433, no. 1, pp. 172–190, 2010.

[17] X. Li, Y. Shi, and I. Gutman, *Graph Energy.* Springer New York, 2012.

[18] A. E. Brouwer and W. H. Haemers, *Spectra of Graphs.* Springer New York, 2012.

[19] F. R. K. Chung, *Spectral Graph Theory.* American Mathematical Society, 1997.

[20] A. A. Hagberg, D. A. Schult, and P. J. Swart, "Exploring Network Structure, Dynamics, and Function using NetworkX," in *Proceedings of the 7th Python in Science Conference (SciPy)*, (Pasadena, CA), pp. 11–15, August 2008.

Virtual Reality and Motion Controlled Data Analysis

John Miscevich, Liam Burke and
Elmer Grubbs, Advisor

Abstract

In this technologically advanced society, statistics and data analytics are becoming increasingly important. As technology progresses, so should our methodologies of accessing and analyzing large data sets. By integrating the emerging field of virtual reality with data analysis, we can view and better understand data in a more engaging, hands on, and exciting way. Our project seeks to incorporate motion tracking technologies to truly immerse the user in the information that they are exploring. Using an Oculus Rift virtual reality headset, a Microsoft Kinect motion tracking sensor, and Unity Engine game development suite (using C#), we were able to make data representations that could be modified via physical motions. By giving users a firsthand view of their data in a virtual environment, and allowing them to manipulate graphical representations via intuitive physical gestures, they are more engaged and more willing and better able to understand the information that they are seeing. This new technique could be very valuable in allowing analysists to view telemetry data from test vehicles in a way to improve efficiency in finding the correct data files and then viewing the actual data.

Introduction

For years, the amount of telemetry data collected for test vehicles has increased almost exponentially. Even in the early years eight to ten signals displayed on a strip chart proved difficult to examine and then to be able to pull out the relevant information. As more signals have been added over the years, the huge amount of data collected has finally reached a point where it is extremely difficult to find the exact data that you would like to investigate. The era of "Big Data" is here and many different industries are now struggling to pull out the specific bits of data that they want to look at. The technique to be introduced here, using Virtual Reality, could be used for several different types of big data, but has been optimized for use where there is a large collection of data files, each containing a physical waveform. This easily lends itself to use in the field of telemetry. Virtual Reality (VR), using 3D glasses or a head mounted display to represent a three dimensional virtual world to the user, has been around for several decades now, but so far has not progressed much beyond its use for entertainment. It remains an untapped resource for scientific use other than perhaps in medicine, where it is used for training surgeons. Early versions of the glasses or head mounted displays were expensive and had very narrow fields of view. Fortunately, prices have lowered considerably over the last ten years and fields of view have increased.

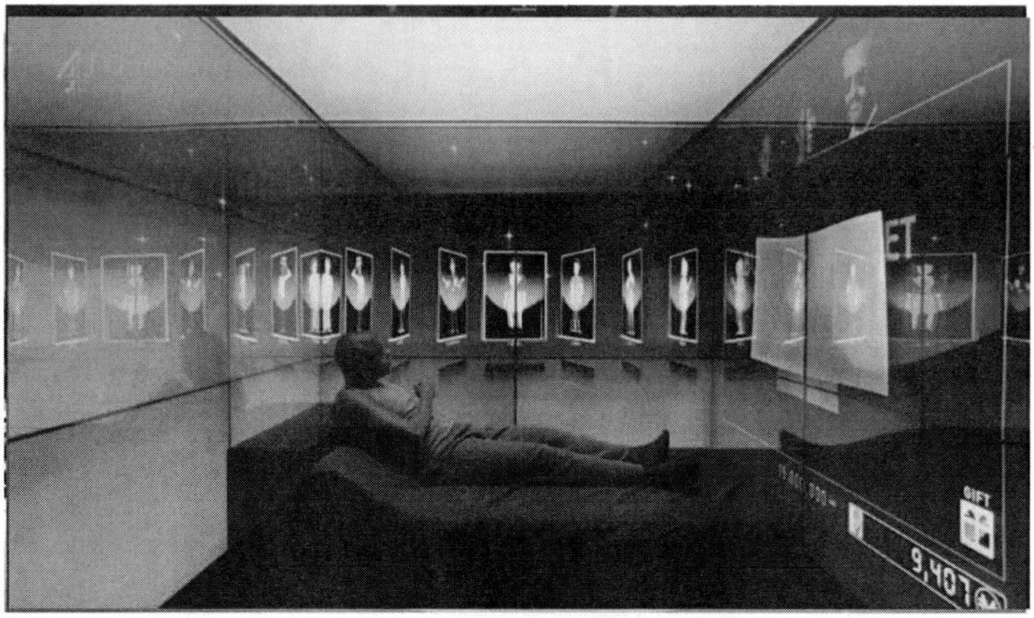

Figure 1

Figure 1 is an illustration of “The Cave”, a virtual reality experience that does not require a head mounted display or 3-D glasses.

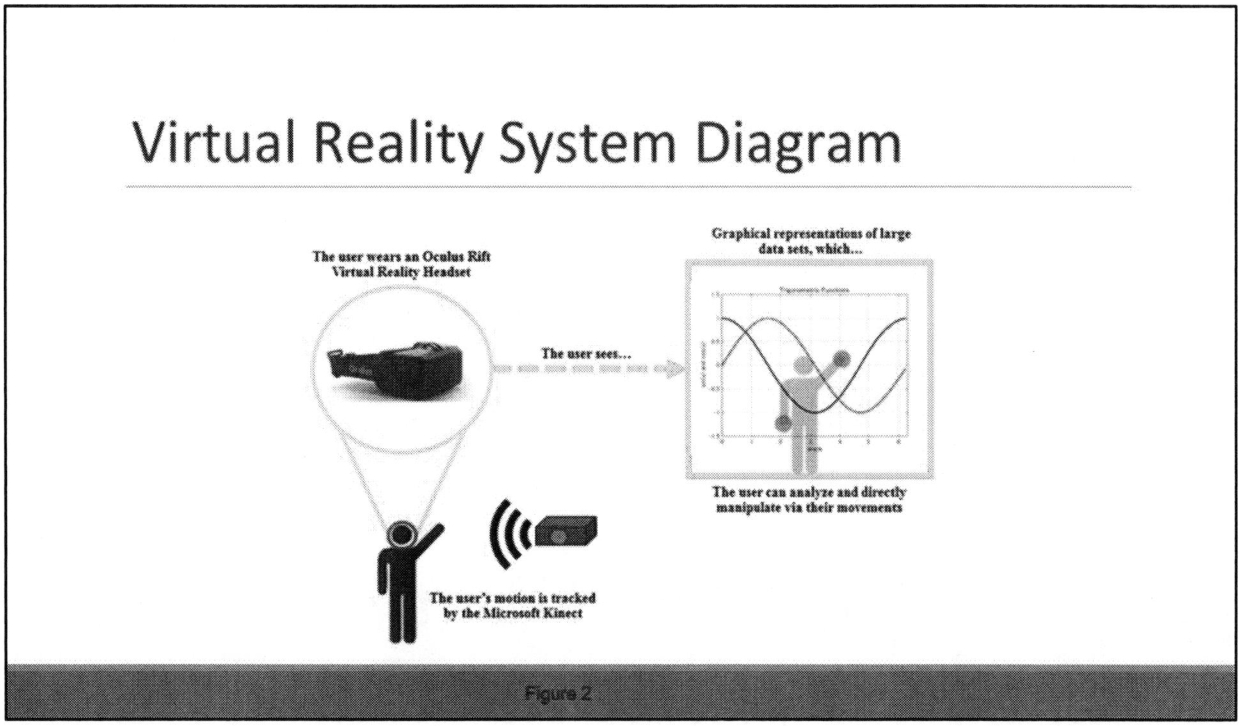

Figure 2

See Figure 2. These are the basic components of this Virtual Reality system, not including the computer system. It shows a user with a head mounted display and a Kinect motion tracker looking at and manipulating a graph similar to what one might see from a vehicle test signal.

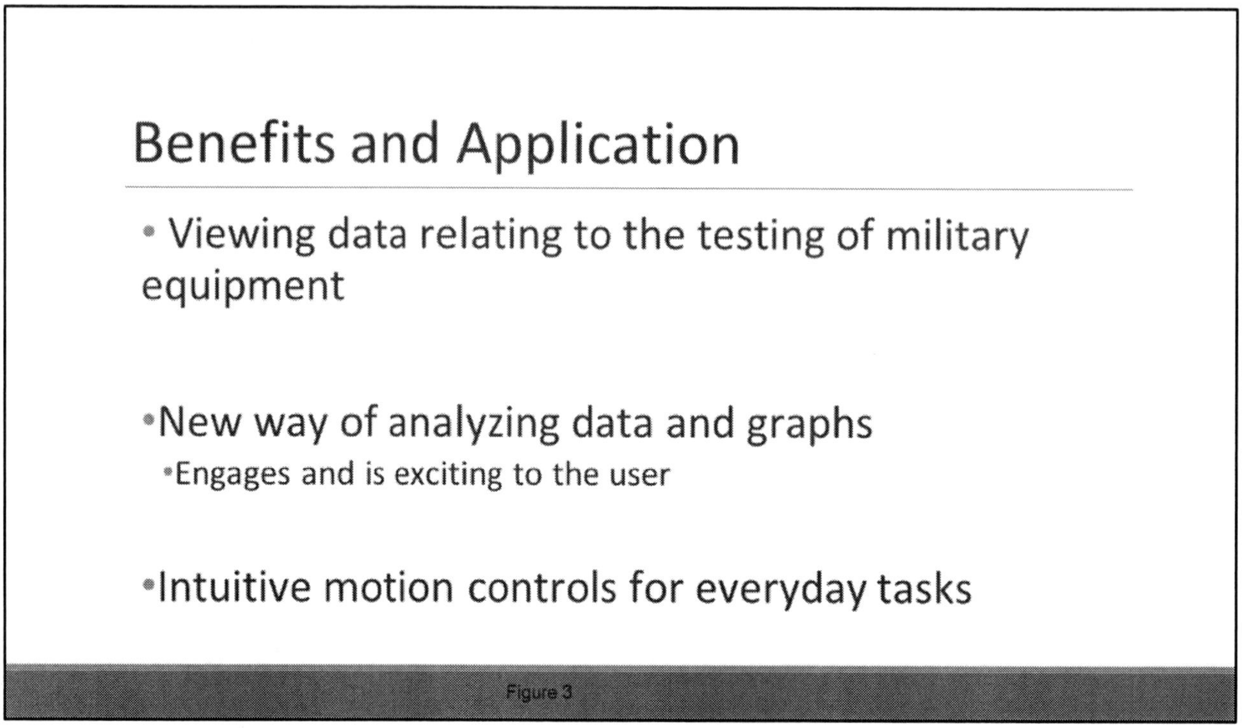

Figure 3

See Figure 3. This method of viewing test data includes adding innovative ways to the techniques currently being used, which should make it easier to analyze and view this data. Using gestures to view the data makes the process more similar to the techniques all of us use to process our data on cell phones and touch screens.

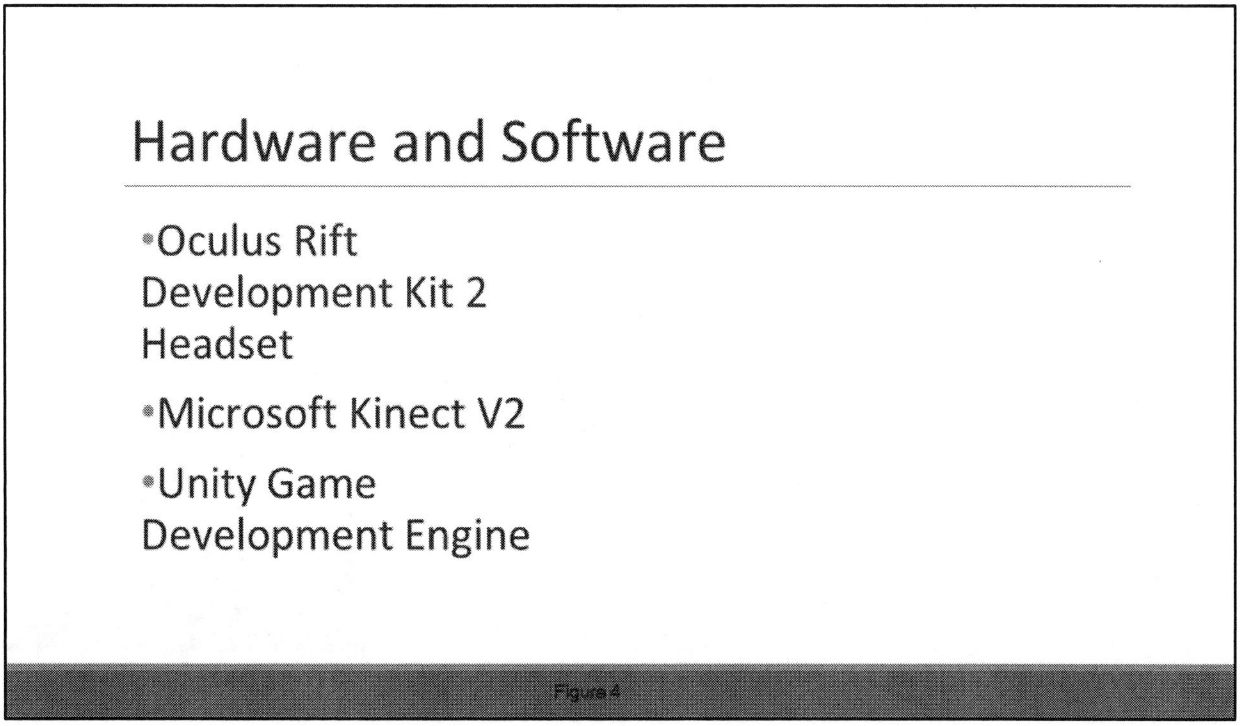

Figure 4

See Figure 4. In addition to the computer system itself, the system uses the Oculus Rift to process the location of the data and to see the data itself. It also uses the Microsoft Kinect to detect the user's gestures in order to control what is being seen. The software for the system is a game system called Unity, which produces the images seen by the user.

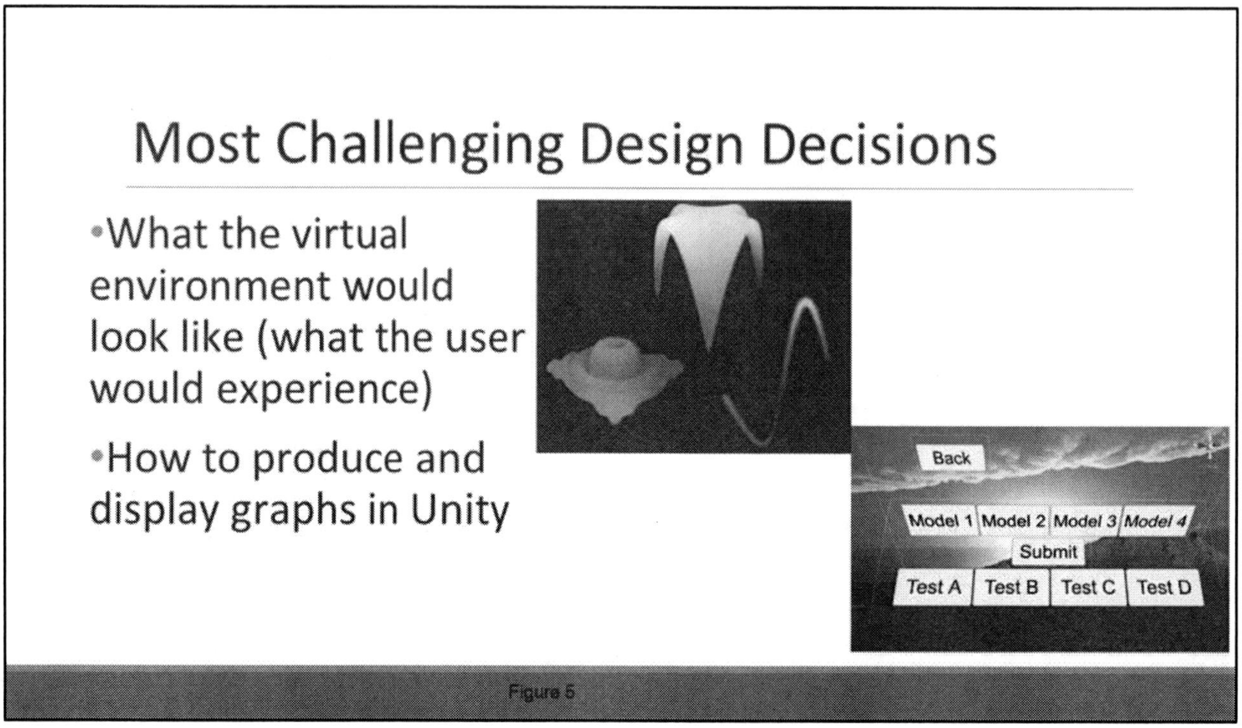

Figure 5

Figure 5 shows some of the design techniques that the authors used in building this system. Our design decisions involved mostly software issues. The image on the lower right shows how the user would choose a particular model of, for example a missile, and go to a particular type of test, such as flight test, hover test, etc.

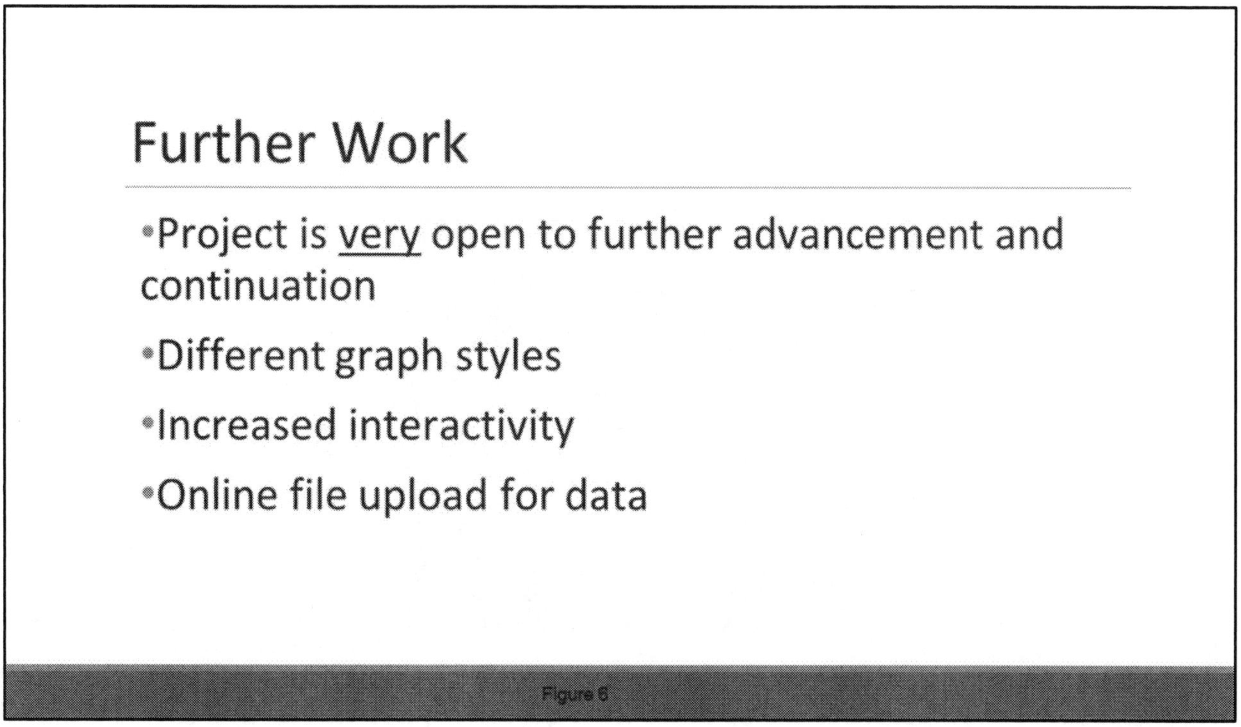

Figure 6

Figure 6 shows ideas for potential further work which could be done on the project. Of particular interest is number three, which allows data sets to be loaded onto a website for analysis using this technique. Foe test purposes, accelerometer data from moving an I-phone back and forth was loaded onto the website and is used to illustrate the graphing capabilities.

Graphing Software

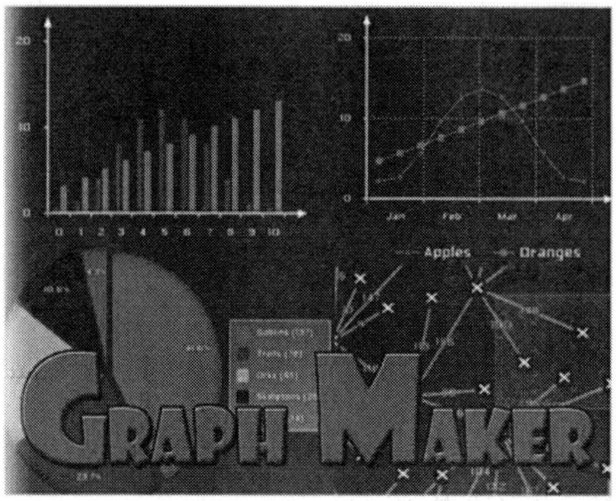

Figure 7

Figure 7 is an illustration of the software that was used to graph the data.

Figure 8

Figure 8 shows two important elements of the program's opening scene. First, there is a table with a tank, a plane and a missile. On the bottom right side is an image from the Microsoft Kinect, showing the user and his two arms. As he puts his right hand on the aircraft, a light illuminates it. By closing his hand as if to pick it up, the scenes switches to an image of several different types of aircraft and several different types of tests. Again, by grasping the type of aircraft and type of test, another image opens up with the dates of those tests. A scene which has not yet been implemented allows the user to choose a particular signal from the test to look at.

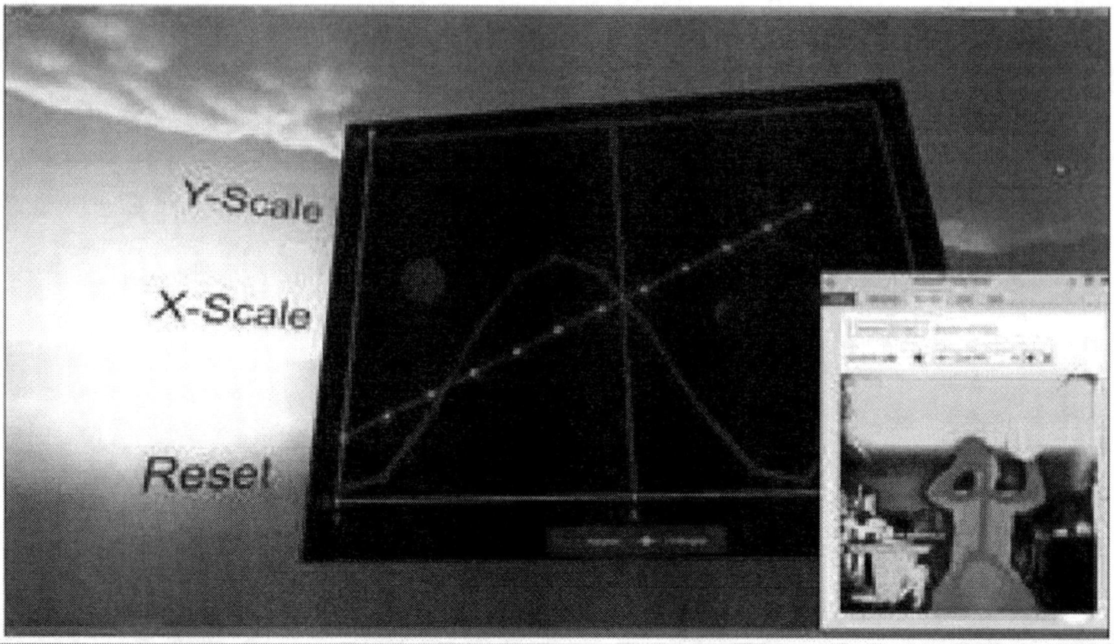

Figure 9

See Figure 9. An image then appears of the signal vs. time as shown here. As illustrated on the bottom right, the user can change the scale using gestures to adjust the X or Y axis values to investigate the particular waveform which is being viewed. Each axis can be expanded or contracted as desired by the user.

Conclusion

In conclusion, this is a new technique, using Virtual Reality, that allows the user to look at and manipulate data in a very user friendly way. Please visit http://virtualrealitynau.weebly.com/ for more detailed information, as well as more video demonstrations

TRACKING INFORMATION IN SAR IMAGE FORMATION AND CLASSIFICATION ALGORITHMS

Abhejit Rajagopal, Vincent Radzicki, Shivkumar Chandrasekaran, Hua Lee
Department of Electrical and Computer Engineering
University of California, Santa Barbra, USA 93106
{abhejit, vradzicki}@ece.ucsb.edu

ABSTRACT

Traditional target detection pipelines involve two sequential steps: the formation of a range-profile or likely-image, and the classification of likely targets within that image. Although it has been shown that target tracking in the RaDAR image-domain can be unnecessarily noisy, with more accurate and efficient implementations involving a direct analysis of the measured wavefield, image formation remains a desirable output in many applications due to its highly descriptive and interpretable nature. In this paper, we outline a mechanism for formalizing and accelerating this procedure in application-specific use cases. Enabled by recent advances in deep learning, we present a pipeline for automatically selecting an "optimal" filtered back-projection model, forming a likely-image, and performing target recognition and classification. The architecture allows practitioners to track and optimize the flow of information throughout the pipeline, enabling applications that utilize only intermediate outputs of the algorithm.

INTRODUCTION

The advent of radio detection and ranging (RaDAR) technology in the early 20th century was driven primarily by the need of allied strategic planners for an early aircraft warning system to provide them with important military intelligence [1]. The initial goal of early RaDAR systems was therefore detecting and tracking of both friendly and hostile aircraft. The resounding success that RaDAR systems found in this application generated widespread interest in the new technology and a search for other applications. Concurrently with these developments, research work on electromagnetic imaging techniques for scientific and medical instrumentation progressed rapidly [2]. The theoretical foundations between the two research areas were quickly recognized and work on developing radar imaging systems began soon afterwards [3, 4]. Radar imaging refers to estimating and forming spatial maps of the electromagnetic and other physical properties of objects and scenes probed by illuminating radar signals. The field of radar imaging covers a wide-variety of important engineering and scientific disciplines, and is utilized in a range of civil, medical, and military applications today.

As the capabilities of radar technology advanced from simple target detection and ranging to full-scale imaging, radar systems evolved to perform more complicated signal processing and analysis tasks. Tasks such as as automated pattern recognition and classification of radar data, for example, became highly desirable in a number of applications. In remote sensing airborne Synthetic Aperture Radar (SAR) systems, scene classification can be particularly useful for determining the environmental characteristics and topography of a surveyed area, where the scenes could be classified as forest, desert, mountainous, urban, or rural [5]. In radar array imaging, automated target recognition (ATR) is a useful tool to detect specific targets, such as hostile aircraft, and classify the target into a set of pre-defined categories (i.e. aircraft model, size, etc). In high-resolution radar tomography, where objects are imaged from multiple positions over a virtual or physical array, classifying the material composition of objects and propagation media is an important problem for many civil engineering applications.

The methodologies used to perform these recognition and classification tasks has been an interesting development as well. Radar imaging systems will typically make measurements in the time or frequency domains, and the raw data will undergo significant processing to be mapped to the spatial domain for visual image display. Early techniques focused on interpreting the raw data for patterns to help classify features, such as the harmonic resonances of structures at specific wavelengths of electromagnetic illumination, and frequency dependent scattering parameters. Other techniques explored time-dependent features such as the micro-doppler effects of rotating or moving structures fixed to moving targets [6, 7]. In general, however, these early techniques have been limited to scenarios where targets are known and their scattering profiles can be explicitly modeled in the linear, Fourier domain [8, 9, 10]. While this is a reasonable "first-order" approach, these algorithms have historically failed to transition well to real-world and real-time scenarios where the number and distribution of targets cannot be guaranteed [11, 12].

As such, rather than processing the raw wavefield pattern, recent approaches have shifted their focus instead to processing and characterizing materials in the reconstructed image or inverse domain [13, 14]. In these typical SAR imaging systems, back-projection or filtered back-projection is used to reconstruct a scene from the measured complex wavefield pattern [10, 15]. With machine learning and computer vision algorithms, including deep neural networks, having demonstrated remarkable performance in detection and classification tasks when applied to optical imagery, it was expected that the application of these same algorithms to radar images would be straightforward [16]. However, many of these algorithms tend to focus on morphological and spatially coherent properties of objects that are not present in radar images, thereby limiting their success. Furthermore, the properties of radar images, which are intimately dependent on the image reconstruction algorithm, often exhibit imaging artifacts that can lead to misinterpretation by humans and algorithms alike. Still, the application of machine learning and classification algorithms to SAR imagery shows much promise in uncovering hidden patterns in data and measurements, which could significantly improve both classification performance and our ability to understand the information content of radar images quantitatively. Therefore, to enable high accuracy classification of targets in SAR imagery, it is essential to track and understand the flow of information throughout the image formation and classification pipeline.

In this paper, we discuss popular image formation algorithms, subsequent processing using neural networks, and detail an architecture which incorporates the two into a comprehensive learning framework. The presented network architecture and parametrized objective function enables practitioners to weigh the benefits of forming human-interpretable images against high-accuracy classification. In short, this work suggest that radar images that are appear semantically "focussed" to humans may not be the optimal representation for signal discrimination by a machine. By removing this constraint on the image representation, we hope to achieve high accuracy ATR and target localization algorithms that can be tailored to user-specified applications and environments.

ADJOINT IMAGE FORMATION

In this section we briefly review some classical techniques for image formation using the adjoint operator. In particular, with appropriate assumptions, many electromagnetic imaging problems can be reduced to solving the follow linear system (Eq. 1):

$$d = Gm \tag{1}$$

where m captures the geometry and attenuation model of an environment, G represents the electromagnetic Green's function of the media, and d contains the known or measured data. In the case of active reflection-mode SAR, d contains both the known transmitted pulse or waveform and the received pulse (usually as a function of frequency), and m is the desired unknown that represents the reflection or absorption profile of the surveyed scene. From this linear-systems perspective, a reasonable estimate of m can be given by the least-squares solution (Eq. 2),

$$m = (G^H G)^{-1} G^H d \tag{2}$$

although in many cases $G^H G$ is invariably singular, ill-conditioned, and too large to easily invert. Classically, such conditions have encouraged practitioners to find suitable approximations of this equation. One common substitution is to eliminate the leading term by applying the approximation:

$$G^H G \approx I \tag{3}$$

which allows m to be computed easily as:

$$m \approx G^H d \tag{4}$$

where G^H is called the adjoint operator.

A. *Back-Projection*

Image formation traditionally involves using the measured data d to back-project the measured wavefield over either 1, 2, or 3 dimensions to construct a representative image of the scene m. The back-projection operator is typically selected as the inverse of the Green's function that defines forward-wave propagation in a homogeneous medium (i.e. backwards-wave propagation). For example, in Equation 4 the back-projection operator is chosen as the complex-conjugate of the Green's functions G and, in 3 spatial dimensions (x,y,z), takes the form:

$$G_i^H(r, \lambda) = \frac{1}{+j \cdot r \cdot \lambda} \cdot exp(+j \cdot 2\pi \cdot \frac{r}{\lambda}) \tag{5}$$

where r is the propagation distance of a coherent waveform, λ is the wavelength of light in an assumed medium, and j is the imaginary number $j^2 = -1$. This formulation permits a fast, simple, and vectorizable algorithm that can be tuned for real-time applications, as outlined below (Alg. 1).

It is worth noting that there are many ways to implement and speed up back-projection, with the presented pseudo-code being only one of many formulations that are optimized for parallelism and simplicity in understanding. In general, the complexity of back-projection stems from the fact that the Greens kernel changes for each "radial" point and operating frequency, with the pattern of how these functions are re-used being slightly more complicated. Typically, the explicit computation of these Green's functions is a bottleneck for the application of the adjoint operator, even if done in the frequency domain. As such, in practice the Greens kernel is often pre-computed with respect to a relevant antenna geometry, and re-used in all subsequent image formation tasks.

Algorithm 1 Image Formation using Back-Projection

```
procedure BP(grid, ant, d)            ▷ Given an evaluation grid, antenna array, and wavefield data
    img ← zeros(grid.shape)
    for x ∈ grid.X do
        for y ∈ grid.Y do
            for z ∈ grid.Z do
                pt ← (x, y, z)                                        ▷ Point in scene
                r ← ||ant.TX − pt||_2 + ||ant.RX − pt||_2             ▷ Path-length between TX/RX
                g ← 1/(jrλ̄) · exp(2πjrλ̄^(−1))                        ▷ Green's kernel for each λ̄_j
                img(x, y, z) ← sum(g(r, λ̄) · d)                       ▷ Apply adjoint operator
            end for
        end for
    end for
    return img
end procedure
```

B. Filtered Back-Projection

Although back-projection works well with limited modeling and computational effort, it is known that the resultant images are often unreliable for target identification. It is common, for example, for small scatters in particular geometries to be overly emphasized in the reconstructed image, particularly when imaging in the presence of multiple dielectric mediums. In particular, one of the issues with the back-projection model, besides its limitations in representing nonlinear systems, is that a single Greens kernel (i.e. derived for homogenous media) is applied throughout the reconstructed scene. For multi-layered heterogeneous media, this issue can greatly degrade the final image quality. One technique to overcome this is to modify the model to include environment-specific parameters, or spatial filters, that account for this inhomogeneity in a natural way. In this view, filtered back-projection can be viewed as a simple modification of the standard adjoint model to a system with the solution:

$$m \approx LG^H d \tag{6}$$

where the rows of L can be thought of as an application-specific environment or characterization model. Of course, the choice of L can be a difficult problem, but in general this allows for more degrees of freedom in the imaging model without sacrificing benefit of a fast algorithm. One such algorithm, which follows the structure of Algorithm 1 is presented on the next page in Algorithm 2.

Algorithm 2 Image Formation using Filtered Back-Projection

```
procedure FBP(grid, ant, d, L)   ▷ Given an evaluation grid, antenna array, wavefield data, and FBP model
    img ← zeros(grid.shape)
    for x ∈ grid.X do
        for y ∈ grid.Y do
            for z ∈ grid.Z do
                pt ← (x, y, z)                                  ▷ Point in scene
                r ← ||ant.TX − pt||_2 + ||ant.RX − pt||_2       ▷ Path-length between TX/RX
                g ← 1/(jrλ̄) · exp(2πjrλ̄^{-1})                   ▷ Green's kernel for each λ̄_j
                img(x, y, z) ← sum(L · g(r, λ̄) · d)              ▷ Apply modified adjoint operator
            end for
        end for
    end for
    return img
end procedure
```

C. *FFT Technique*

A main driver in the computational complexity of this back-projection algorithm is the complex operations involved in computing the Green's function. By discretizing the Green's kernel over propagation distance r and ignoring the amplitude attenuation component, a more efficient version of the algorithm can be computed. This possible modification involves altering the Green's kernel itself. In particular, practitioners have observed that the homogeneous Green's kernel (Eq. 5) closely resembles the commonly used Discrete Fourier Transform (DFT) basis functions:

$$f_i(\cdot) = exp(+j \cdot 2\pi \cdot k \cdot \frac{n}{N}) \tag{7}$$

where k represents the wave number, n is the sample index (e.g. in the time or spatial domain), and N is the total number of basis functions in the Fourier decomposition of the measured signal. Moreover, if N is restricted to be a power of 2, the fast Fourier Transform (FFT) can be utilized as a computationally efficient alternative formulation of back-projection, used to compute the reconstruction at each grid point. Furthermore, in this view the FFT formulation of SAR image reconstruction can be seen as an approximation of homogeneous Green's kernel by the Fourier basis, albeit one that enables a much faster computation, especially when the FFT is implmented directly in hardware.

PATTERN RECOGNITION

As discussed previously, it is both desirable and common for practitioners to localize, identify, and classify targets and materials using reconstructed images, rather than using the original wavefield data, even though mathematically these representations possess similar information. One reason for this might be that human operators can rely on geometric and spatial intuition to segment and identify objects when the information is represented spatially in a reconstructed image.

Specifically, the application of the previously described adjoint operator can be viewed as a low-pass filtering of the collected SAR wavefield data that spreads the measured information over the unambiguous range of a scene.

In this vein, various machine learning algorithms have been applied to pattern recognition tasks in SAR imagery, often incorporating intuition and priors from human operators that have been trained to interpret these images to produce application-specific analysis, such as military intelligence. While these tasks have historically utilized digital image-processing based denoising, segmentation, and classification algorithms [16], recently there has been significant interest in applying deep neural networks to these tasks [17]. Neural networks, which have emerged as a tractable approach to universal function approximation, with numerous applications in pattern recognition and image understanding, have been applied to a number of SAR datasets with varying levels of performance [18, 19]. One drawback, however, of current neural network architectures is that a majority of the existing machinery is built for real-valued inputs (e.g. intensities), rather than complex-valued inputs (e.g. wavefield data) [20, 21]. As such, a majority of neural network investigations utilize reconstructed magnitude images, which immediately reduces the amount of information that can be used to discriminate materials and targets in a scene.

Moreover, it should be noted that because the selected back-projection operator is not always invertible, the standard image-domain representation is not always optimal. As such, in addition to the various imaging artifacts that are introduced by simplifying assumptions in the system model, the reconstructed image itself is not always reliable for identification or characterization, perhaps due to the lack of information in this representation. That is, selected back-projection or filtered back-projection models that are semantically agreeable to human operators may not be the optimal intermediate representation of the wavefield data for characterization and classification tasks. By instead viewing the previously described image formation and pattern recognition algorithms (e.g. neural networks) as part of the same computation graph, it becomes clear that both the filter model and the neural network weights should be jointly optimized. An end-to-end classification and filtered-back projection algorithmic architecture is presented in Figure 1, on the following page.

In this architecture (Fig. 1), the image formation and pattern recognition task are optimized jointly using an objective function that values the interpretability of the image to both human operators and downstream classification networks. One such objective function could be parametrized as:

$$f_{image} = E(x, y) + \lambda V(y) \tag{8}$$

$$f_{detect} = ||z - z_{true}|| \tag{9}$$

$$f_{overall} = \alpha f_{image} + \beta f_{detect} \tag{10}$$

where x is either the output of standard back-projection or a known spatial map of the scene, y is the output of a learned filtered back-projection model, and z is the either the pixel-wise or collective classification result (e.g. the location and/or classification of a scatter in the scene).

The presented detection objective (Eq. 9) is fairly standard, and could be traded for a number of different objective (binary cross-entropy, mean-square error, log-liklihood, etc) depending on the application. Similarly, the image formation objective (Eq. 8) can match that of total-variation (TV) denoising, or other similar constraints commonly used in under-sampled image formation. For example, a popular choice is to pick (with an appropriate norm) a measure of anisotropic variance

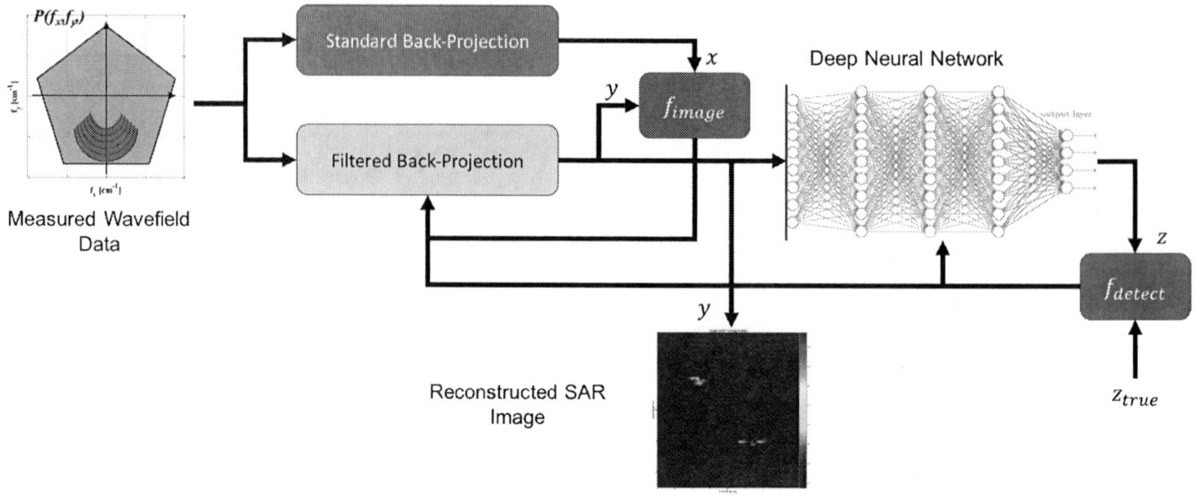

Figure 1: A deep neural network architecture that incorporates complex-valued image formation and complex-valued deep neural networks. The neural network at the end can be either a convolutional or fully-connected network that performs either pixel-wise or patch-wise classification and detections.

as:

$$E(x, y) = ||x - y|| \tag{11}$$

$$V(y) = \sum_{i,j} |y_{i+1,j} - y_{i,j}| + |y_{i,j+1} - y_{i,j}| \tag{12}$$

In this way, deep learning can be utilized to learn feature representations of the wavefield data that are semantically pleasing to human operators (e.g. via images), but still maintain a high level of discriminability to support their use in high-accuracy classification and target localization tasks. The degree to which this is achieved is then left to the user, who may select these parameters (α, β, λ) in real-time depending on their application requirements. For example, operators that require high-confidence classification may pick $\alpha = 0$ at the expense of losing visual interpretability in the image, while operators that require high-accuracy image formation may more carefully balance their selection of parameters.

CONCLUSIONS

We have reviewed the formulation of several popular image formation algorithms, and introduced an algorithmic architecture that suggests filter models for image formation should be tuned jointly with classification models. In particular, we present deep neural networks as an attractive candidate for the classification portion of the network, preceded by more-traditional filtered back-projection layers that can offer human-interpretable visual SAR imagery as a highly-descriptive intermediate output. Future work includes the incorporation of additional image formation and inverse-scattering models, the development of complex-valued neural network machinery, and the application of these to real-world terrain classification, material characterization, and target recognition tasks.

REFERENCES

[1] M. I. Skolnik, "Introduction to radar," *Radar Handbook*, vol. 2, 1962.

[2] S. Webb, *The physics of medical imaging*. CRC Press, 1988.

[3] A. Olte and S. Silver, "New results in backscattering from cones and spheroids," *IRE Transactions on Antennas and Propagation*, vol. 7, no. 5, pp. 61–67, 1959.

[4] D. L. Mensa, "High resolution radar cross-section imaging," *Boston, MA, Artech House, 1991*, vol. 1, 1991.

[5] L. Gomez-Chova, D. Tuia, G. Moser, and G. Camps-Valls, "Multimodal classification of remote sensing images: A review and future directions," *Proceedings of the IEEE*, vol. 103, no. 9, pp. 1560–1584, 2015.

[6] Y. Kim and H. Ling, "Human activity classification based on micro-doppler signatures using a support vector machine," *IEEE Transactions on Geoscience and Remote Sensing*, vol. 47, no. 5, pp. 1328–1337, 2009.

[7] V. R. Radzicki, "Detection and imaging of micro-periodic motion with fmcw sensing systems," in *International Telemetering Conference Proceedings*, International Foundation for Telemetering, 2015.

[8] N. E. Huang, "Survey of remote sensing techniques for wave measurement," in *Measuring Ocean Waves: Proceedings of a Symposium and Workshop on Wave Measurement Technology; April 22-24, 1981; Washington*, p. 38, National Academies, 1982.

[9] E. A. Kasischke, D. R. Lyzenga, R. Shuchman, Y. Tseng, and B. S. Termaat, "The use of synthetic aperture radar to detect and chart submerged navigation hazards.," tech. rep., DTIC Document, 1982.

[10] J. W. Goodman, *Introduction to Fourier optics*. Roberts and Company Publishers, 2005.

[11] E. R. Stofan, D. L. Evans, C. Schmullius, B. Holt, J. J. Plaut, J. van Zyl, S. D. Wall, and J. Way, "Overview of results of spaceborne imaging radar-c, x-band synthetic aperture radar (sir-c/x-sar)," *IEEE Transactions on Geoscience and Remote Sensing*, vol. 33, no. 4, pp. 817–828, 1995.

[12] D. Lu and Q. Weng, "A survey of image classification methods and techniques for improving classification performance," *International journal of Remote sensing*, vol. 28, no. 5, pp. 823–870, 2007.

[13] T. R. Reed and J. H. Dubuf, "A review of recent texture segmentation and feature extraction techniques," *CVGIP: Image understanding*, vol. 57, no. 3, pp. 359–372, 1993.

[14] V. Dey, Y. Zhang, and M. Zhong, *A review on image segmentation techniques with remote sensing perspective*. na, 2010.

[15] H. Bischof, W. Schneider, and A. J. Pinz, "Multispectral classification of landsat-images using neural networks," *IEEE transactions on Geoscience and Remote Sensing*, vol. 30, no. 3, pp. 482–490, 1992.

[16] L. Capineri, P. Grande, and J. Temple, "Advanced image-processing technique for real-time interpretation of ground-penetrating radar images," *International Journal of Imaging Systems and Technology*, vol. 9, no. 1, pp. 51–59, 1998.

[17] M. Wilmanski, C. Kreucher, and J. Lauer, "Modern approaches in deep learning for sar atr," in *SPIE Defense+ Security*, pp. 98430N–98430N, International Society for Optics and Photonics, 2016.

[18] K.-S. Chen, W. Huang, D. Tsay, and F. Amar, "Classification of multifrequency polarimetric sar imagery using a dynamic learning neural network," *IEEE Transactions on Geoscience and Remote Sensing*, vol. 34, no. 3, pp. 814–820, 1996.

[19] N. Kussul, M. Lavreniuk, A. Shelestov, and B. Yailymov, "Along the season crop classification in ukraine based on time series of optical and sar images using ensemble of neural network classifiers," in *Geoscience and Remote Sensing Symposium (IGARSS), 2016 IEEE International*, pp. 7145–7148, IEEE, 2016.

[20] D. A. Morgan, "Deep convolutional neural networks for atr from sar imagery," *Proceedings of the Algorithms for Synthetic Aperture Radar Imagery XXII, Baltimore, MD, USA*, vol. 23, p. 94750F, 2015.

[21] M. Wilmanski, C. Kreucher, and A. Hero, "Complex input convolutional neural networks for wide angle sar atr," in *Signal and Information Processing (GlobalSIP), 2016 IEEE Global Conference on*, pp. 1037–1041, IEEE, 2016.

EFFICIENT ARTIFACT-FREE VIDEO SYNOPSIS FOR TARGET DETECTION AND TRACKING

Jing-Ming Guo[1], Cheng-Hsin Chang[1], Hua Lee[2], and Vincent Radzicki[2]
[1]Department of Electrical Engineering,
National Taiwan University of Science and Technology
[2]Department of Electrical and Computer Engineering
University of California, Santa Barbara

ABSTRACT

Target detection and tracking is one of the most common applications of video systems. The requirement of processing time due to the computation complexity has been the most challenging element for these operations. To achieve computation efficiency, video synopsis is one of the most promising approaches. Video synopsis is a technique that focuses on the removal or reduction of spatial and temporal redundancies, while maintaining important activities in the original video. In this paper, a new video synopsis scheme is introduced. Major characteristics of this technique include efficient background subtraction, superior processing speed, and effective object comparison.

INTRODUCTION

The large volume of image data from the lengthy videos produced by surveillance systems introduces two basic challenges. One is the effective use of memory space and the other one is effective human interface for the utilization of the condensed video sequences, especially in target detection and tracking [1-15].

To address these technical elements, this new approach utilizes high efficient background subtraction algorithms with codebook models to extract objects from a video sequence precisely and efficiently [1-3]. Subsequently, the three-dimensional histogram descriptor is employed to describe the appearance of objects, and the artifact-free synopsis is applied to the tracking of the multiple targets.

This paper includes mathematical modeling and analysis, full-scale experiments, and organized comparison to conventional techniques. This technique can be modified for microwave, acoustic, and infrared sensing systems.

MULTI-LAYER CODEBOOK BACKGROUND MODEL

The main consideration in background subtraction for video synopsis is the effective detection of moving objects. This includes the possibility of dynamic background which is common in practical applications. Figure (1) illustrates the proposed flowchart. It can be partitioned into two phases based on the hierarchical strategy to build codebook models. The first phase is multi-layer background construction as the training period, and the second phase is the multi-layer background subtraction. Both of them employ the hierarchical strategy to build codebook models. The procedure can be summarized in four major steps. The first is the multi-layer codebook background model construction, which is followed by background subtraction. The third step is pixel classification, which is used to refine the block-based result to enhance precision. The pixel classification procedure is adopted to further classify the attribute of a pixel as real foreground, shadow, or highlight. To separate the shadow and highlight from the pixel-based background subtraction result, the Carmona algorithm is applied to improve the color model [16].

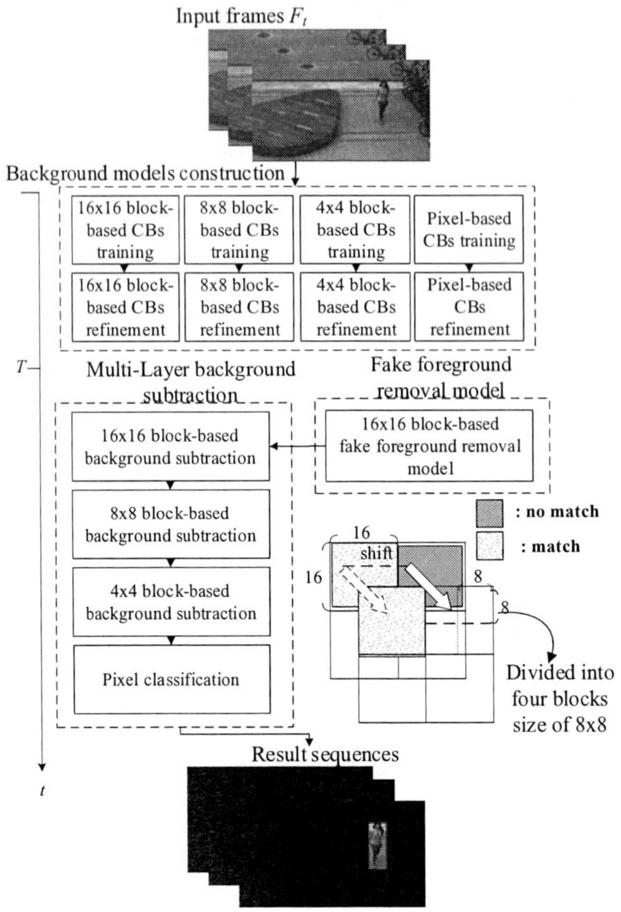

Figure (1): Multi-layer codebook model for background subtraction

The fourth step of fake foreground removal model (FFRM) is an important addition in practical applications, which is designed to merge the stationary foreground to the background model when the objects remain still for a long period of time. Figure (2) shows the flowchart of the FFRM which is utilized to record non-background information, and the construction procedure is identical to the background codebook model.

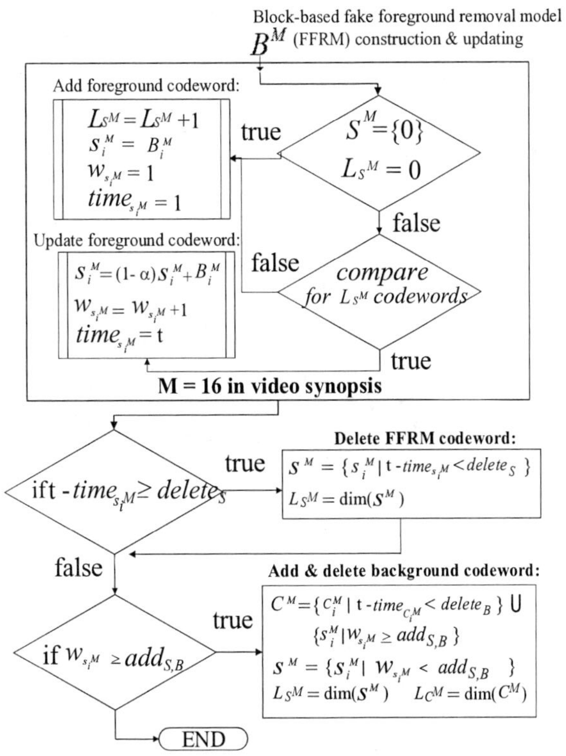

Figure (2): Fake foreground removal model

HIGH SPEED ARTIFACT-FREE VIDEO SYNOPSIS

Subsequent to background model construction and foreground detection, post-processing schemes such as two dimensional morphological operations can also be applied for the further refinement of the appearance of objects [17]. The next step in video synopsis is the formation of the object tubes, which involves the crucial factor of processing speed. Consequently, a color-histogram based object appearance descriptor is applied, which modifies the typical histogram descriptor. In addition to high computational efficiency, the advantage of using the color histogram as a descriptor is to avoid collisions, scaling, and rotation.

EXPERIMENTAL RESULTS

In this section, the performance of this technique is evaluated in terms of the FPS and condensation ratio. In this study, the testing platform was a Intel Core i7 3.40 GHz and 4GB of RAM. Figure (3) shows the testing surveillance videos captured from both indoor and outdoor environments.

(a) (b) (c) (d)

(e) (f) (g)

Figure (3, a-g): Sample surveillance videos in this study

Figures (4) and (5) show examples of the artifacts. Figure (6) shows the results processed by the new technique, which clearly illustrates the removal of artifacts. Table II documents the FPS and condensation ratio. The sixth column is the average FPS and the results, showing that the processing speed of this technique is 150 FPS in average, and the redundancy removal ratio reaches up to 87%, without artifacts.

CONCLUSIONS

Video synopsis is an algorithm to condense video by removing temporal and spatial redundancy. For this purpose, an efficient artifact-free video synopsis is introduced. To deliver correct information contents, redundancy in the temporal domain is removed and the final length of synopsis result ensuring that crucial contents can be preserved. This new technique removes the artifacts while providing a high processing efficiency for generating synopsis video. The temporal redundancy can be removed up to 90%, and the spatial position can be maintained for the further behavior analysis. Moreover, the relationship between object and synopsis background are fully considered during synopsis, and thus it significantly improves the accuracy of the information presented to the users.

Figure (4): Synopsis video artifacts

Figure (5): Synopsis video artifacts

TABLE I: EXPERIMENTAL RESULT AND TEST SEQUENCE INFORMATION.

Video Name	Resolution	Number of Object Tubes	Number of Frames in Original Video	Number of Frames in Synopsis Video	Average Processing Speed (FPS)	Redundancy Removal Ratio
Fig .11. (a)	320X240	10	6267	960	153.365	84.681%
Fig .11. (b)	320X240	12	6432	600	166.589	90.671%
Fig .11. (c)	320X240	22	5607	1560	137.974	72.17%
Fig .11. (d)	320X240	45	14787	1920	145.18	87.016%
Fig .11. (e)	320X240	19	9642	960	157.113	92.04%
Fig .11. (f)	320X240	27	6840	990	130.408	85.526%
Fig .11. (g)	320X240	6	36783	1800	167.618	95.104%

Figure (6): Synopsis results without artifacts

REFERENCES

[1] K. Kim, T.H. Chalidabhongse, D. Harwood, and L. Davis, "Real-time foreground-background segmentation using codebook model," Real-Time Imaging, vol. 11, no. 3, pp. 172-185, June. 2005.

[2] J.-M. Guo, Y.-F. Liu, C.-H. Hsia, M.-H. Shih, and C.-S. Hsu, "Hierarchical method for foreground detection using codebook model," IEEE Transaction on Circuits and Systems for Video Technology, vol. 21, no.6, pp. 804-815, June. 2011.

[3] J.-M. Guo, C.-H. Hsia, Y.-F. Liu, M.-H. Shih, C.H Chang, and J.-Y Wu, "Fast Background Subtraction Based on a Multilayer Codebook Model for Moving Object Detection", IEEE Transaction on Circuits and Systems for Video Technology, vol. 23, no.10, pp. 1809-1821, Oct. 2013.

[4] A. Rav-Acha, Y. Pritch, and S. Peleg, "Making a Long Video Short: Dynamic Video Synopsis", IEEE Computer Society Conference on Computer Vision and Pattern Recognition, vol. 1, pp.

435-441, June. 2006.

[5] Y. Pritch, A. Rav-Acha, A. Gutman, S. Peleg, "Webcam Synopsis: Peeking Around the World", IEEE 11th International Conference on Computer Vision, pp. 1-8, Oct. 2007.

[6] Y. Pritch, A. Rav-Acha, and S. Peleg, "Nonchronological Video Synopsis and Indexing," IEEE Trans. Pattern Analysis and Machine Intelligence, vol. 30, no. 11, pp. 1971-1987, Nov. 2008.

[7] Y. Pritch, S. Ratovitch, A. Hendel, and S. Peleg, "Clustered synopsis of surveillance video", Sixth IEEE International Conference on Advanced Video and Signal Based Surveillance, 2009. AVSS '09. pp. 195-200, Sept. 2009.

[8] Yongwei Nie, Chunxia Xiao, Hanqiu Sun, and Ping Li, "Compact Video Synopsis via Global Spatiotemporal Optimization", IEEE Transactions on Visualization and Computer Graphics, vol. 19, no. 10, pp. 1664-1676, Oct. 2013.

[9] Xiaobin Zhu, Jing Liu, Jinqiao Wang, and Hanqing Lu, "Key observation selection for effective video synopsis", 21st International Conference on Pattern Recognition (ICPR), pp. 2528-2531, Nov. 2012.

[10] C.-R Huang, H.-C Chen, and P,-C Chung, "Online Surveillance Video Synopsis", IEEE International Symposium on Circuits and Systems (ISCAS), pp. 1843-1846, May. 2012.

[11] Shikun Feng, Zhen Lei, Dong Yi, and S.-Z Li, "Online Content-aware Video Condensation", IEEE Conference on Computer Vision and Pattern Recognition (CVPR), pp. 2082-2087, June. 2012.

[12] Lei Sun, Junliang Xing, Haizhou Ai, and Shihong Lao, "A Tracking Based Fast Online Complete Video Synopsis Approach", 21st International Conference on Pattern Recognition (ICPR), pp. 1956-1959, Nov. 2012.

[13] Shizheng Wang, Jianwei Yang, Yanyun Zhao, Anni Cai, and Stan Z.Li, "A surveillance Video Analysis and Storage Scheme for Scalable Synopsis Browsing", IEEE International Conference on Computer Vision Workshops (ICCV Workshops), pp. 1947-1954, Nov. 2011.

[14] Shikun Feng, Shengcai Liao, Zhiyong Yuan and S.-Z Li, "Online Principal Background Selection for Video Synopsis", 20th International Conference on Pattern Recognition (ICPR), pp. 17-20, Aug. 2010.

[15] A. Yildiz, A. Ozgur, Y.S. Akgul, "Fast Non-Linear Video Synopsis", 23rd International Symposium on Computer and Information Sciences, ISCIS '08. pp. 1-6, Oct. 2008.

[16] E. J. Carmona, J. Martinez, and J. Mira, "A new video segmentation method of moving objects based on blob-level knowledge," *Pattern Recognition Letters*, vol. 29, no. 3, pp. 272–285, Feb. 2008.

[17] Satoshi Suzuki, Keiichi Abe, "Topological structural analysis of digitized binary images by border following", Computer Vision, Graphics, and Image Processing, vol. 30, no. 1, pp. 32-46, June. 1985.

NON-TRADITIONAL FLIGHT TEST SENSING SYSTEMS

Stephen A. Kilpatrick, Austin J. Whittington,
Myron L. Moodie, Ben A. Abbott
Southwest Research Institute®
San Antonio, Texas
stephen.kilpatrick@swri.org, austin.whittington@swri.org,
myron .moodie@swri.org, ben.abbott@swri.org

ABSTRACT

Traditional flight test sensing applications require installation of not only the sensor but also supporting cabling and interfacing infrastructure. The cost of this supporting infrastructure increases when it must cross pressure vessel boundaries, extend long distances, or interfere with operation of the aircraft. The continuing cost and schedule pressures on flight test programs demand approaches that minimize installation complexity and reduce the need to modify the aircraft under test. Some emerging approaches have leveraged wireless techniques for data transmission but this can only be used in certain circumstances and does not address the problem of power distribution. This paper describes ongoing research into alternative sensing approaches that utilize a mix of video processing, distributed processing, and power harvesting to provide additional solutions.

KEYWORDS

Flight Test Instrumentation, Sensors, Video Processing

INTRODUCTION

Full flight test programs require a significant investment in instrumentation hardware, made up of DAUs, sensors, and infrastructure. Current-generation DAUs are costly, interconnected pieces of hardware specialized for particular tasks, typically built on decades of legacy code. Often, they are unable to adapt to changing flight test needs without extensive rework and possibly the introduction of new, expensive hardware modules.

Smaller flight test programs may not have the benefit of having an aircraft complete with a full data acquisition and telemetry system. Instead, these programs rely on getting the data required for their tests by physically tapping into existing avionics buses on the aircraft. The work required to perform these activities is invasive, time-intensive, and costly.

Both scenarios leave open the opportunity for alternative sensor approaches, using hardware built from commodity parts that can bypass legacy acquisition methods and communication standards in favor of adaptability and agility, while maintaining the sensing accuracy and reliability requirements necessary to supply usable data. This paper explores methods that Southwest Research Institute (SwRI) is actively researching that can exploit the use of these commodity parts to develop a non-traditional DAU.

COMMODITY SENSOR DAU

The commodity parts we are evaluating have been in the hands of hobbyists pushing their boundaries since their initial release, especially the Raspberry Pi, originally released in February 2012. The Raspberry Pi is a low-cost option for managing or performing data acquisition. It is a full computer, the size of a credit card, and comes with both 802.11n capabilities and Bluetooth 4.1, giving multiple options for wireless communications [1]. It can run a standard Linux Operating System, including the ability to run cross-platform languages such as Python, which gives multiplatform options for the software framework.

These boards have been showing up in areas from weather sensors [2] to automobiles [3], which is very promising for our approach. Figure 1 shows a sighting of one board in a Ukrainian missile prototype at the Bezpeka Security Trade Show [4], which shows that even military applications are not out of the realm of possibility for this class of device.

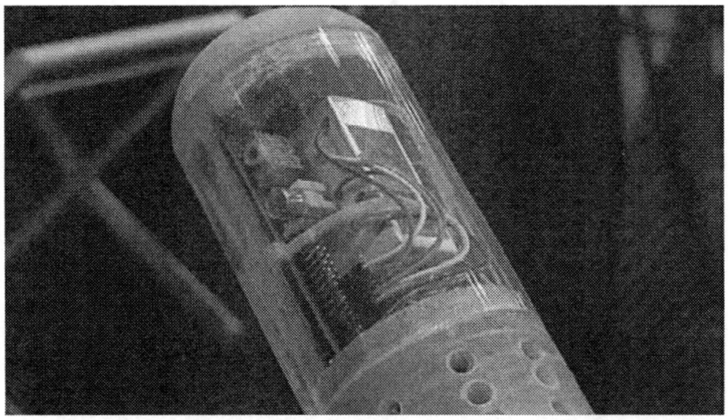

Figure 1 – Raspberry Pi in Ukrainian Missile Prototype

Sensors, DAUs, and associated wiring for carrying power and information are also a source of significant weight during flight test. Worse, when mounted in the wings or other non-easily accessible areas, they either require aircraft disassembly to remove, or are simply left as an expensive ballast when the test article is delivered to the customer. DAUs outside the pressure vessel, whether using legacy PCM or modern network-based data transport, must pierce the pressure vessel to carry data back to the central test infrastructure within the plane, requiring expensive modifications to aircraft structure.

There are several ways to mitigate these problems. First, the sensors can be made wireless, as has been shown in emerging products in the industry previously [5]. Curtiss-Wright has developed a protocol built on the IEEE 802.15.4 standard for Lossless Extended Range Synchronized wireless data [6]. By building on their techniques, we can create a working wireless implementation for the flight test environment. This removes the wiring necessary between the DAU and the sensor for carrying information, but does not alleviate the need for supplying power to the sensor.

To solve the power issue, we plan to explore energy harvesting techniques alongside our wireless solution. Use of a piezoelectric transducer can capture power on the order of 0.5 Watts of practical power per 1.5" x 2.5" by .0075" sheet [7], which is not enough to power a device by itself. However, it can augment power stored in supercapacitors or rechargeable batteries which are hardened for the target conditions. Induction charging can be used before the flight test to charge the devices without the necessity of physical access. The Qi Standard [8] for wireless power transfer provides for inductive charging across distances of up to 7 mm and resonant charging up to 45 mm, significantly more than the airframe skin on a Boeing 737 of .04 in (1.02 mm) [9], which is aluminum and thus not ferromagnetic, and so should not interfere with the charging process. By mounting devices (or only the induction coils) near the edges of the airframe skin, power reservoirs can be charged without any internal cabling needing to be laid.

The DAU must also be able to survive in harsh environmental conditions and temperatures. We plan to use a similar approach used in a previous research project titled "Next Generation Neutrally Buoyant Sensors" that developed prototype low-cost neutrally buoyant sensors [18]. Though the water sensor faced different challenges, specifically high pressure and water exposure, where the flight test environment presents an environment of high vibration, low pressure, and low temperature, the techniques used for hardening a commodity board will be similar. Commodity approaches to hardening these types of devices already exist (an example of which is shown in Figure 2). Also, with the advent of the availability of 3D printing to hobbyists, the community has been exploring commodity methods of hardening for environmental conditions, and we can leverage this previous work in the development of our approach.

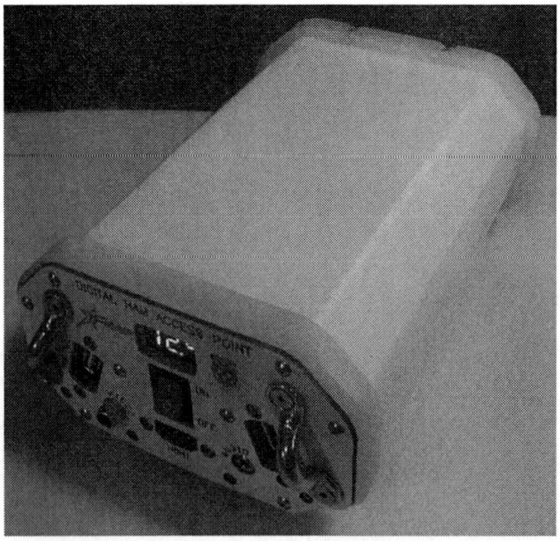

Figure 2 – D□HAP (D□Star Ham Access Point) enclosure/power supply for a Raspberry Pi

VIDEO PROCESSOR DAU

Some of the measurement data in which the test analysts are interested is available, but only to pilots on the cockpit displays. There is a wealth of information present on these displays in a wide variety of gauge shapes and formats, but extracting accurate measurement information from these displays using cameras is not a trivial task. A few existing patents describe methods of reading analog dial gauges in industrial plants [10] or cockpit displays [11, 12], but these approaches only address mechanical gauges and not the virtual displays of modern "glass" cockpits. None of these implementations use the data as input into a flight test DAU.

By using commodity cameras and processors (e.g. CPUs and GPUs), a flight test DAU that outputs cockpit display information can be created that exists alongside traditional DAUs. This method would provide a drop-in solution for companies with existing flight test data acquisition systems, where they could leverage all of their existing tools and processing systems without needing third-party software to interpret the cockpit display information.

While our end goal is to use cameras and computer vision techniques to create a "cockpit display" data acquisition unit, there are a number of technical challenges in proving the feasibility of this approach. The objective of our research is to prove that we can deal with the lighting, vibration, and image distortion problems that present challenges to any computer vision application. There is already some existing research in video processing for cockpit displays [13, 14, 15] that attempts to solve these problems, but each has its shortcomings.

For example, one solution [15] did not deal with the image distortion (skew) that occurs with cameras placed at angles relative to the gauge and required a gauge to be centered directly on the camera. In order to remove this gauge centering requirement, we plan to apply known geometric correction techniques [16] to the gauge images acquired at an angle to restore the view perspective as if it was observed with the camera centered and perpendicular to the imaged plane (see Figure 3 for an example of distorted and corrected images).

Figure 3 – Geometric Correction Techniques

Upon successful correction, the image can be assumed to be acquired with a perfectly centered camera and the proposed approach for angle detection of the needle can be performed. Other issues of vibration and minimal displacements can be targeted by applying image processing techniques that involve image stabilization [17]. Due to the nature of the problem, the rigid image registration should be sufficient to correct for the small displacements, which requires less computational power than the non-rigid one, thus, promising to be implementable for real-time image processing.

Although the experiments described in [15] include specular reflection effects, the claimed immunity to light reflection is questionable. None of the reported images show the specular reflections in the region of interest, which in this case, is the pivoting center of the needle. Therefore, the approach is expected to fail if the specular reflection happens to be within the region of interest.

EXPECTED RESULTS

The achievable capabilities of these non-traditional DAUs are described in Table 1. From our initial testing, these values are what we think is possible for these types of non-traditional DAUs.

Table 1 – Capabilities

Latency through data acquisition pipeline	50 ms
Time stamping accuracy	100 ns
Wireless fidelity	100% successfully delivered
Power Consumption	1 W
Operational temperature	-55 to +85 °C
Operational vibration	5g RMS
Installation time	75% of conventional DAU

The first capability is concerned with the amount of time a given sample needs, be that bus message or binary discrete value, until it is available for display to a user. This is not simply the processing time, as a 50 ms delay in processing a single bus message would mean that the device would be unable to keep up with the stream of data, but rather that a particular sample would enter the acquisition pipeline and be communicated out of the device and through the other FTI elements within the latency requirement.

The second capability covers the time stamping accuracy of the device, recording the exact time the signal was received. The third capability addresses the reliability of the wireless transmissions, versus those of a wired setup. This capability can be tested by storing the captured data on the device, and comparing the stored data to the data received over the wireless connection. The fourth capability involves using supercapacitors and induction charging to provide sufficient power to the small processor boards and sensors.

The fifth and sixth capabilities address the survivability of these types of DAUs. Environmental requirements for airborne electronics are controlled by the RTCA DO-160 specification. While this specification is targeted for production avionics that are intended to perform over the life of the airframe, some flight test organizations have adopted portions of this specification for requirements on their FTI. However, this has driven the cost and inflexibility of the current FTI offerings. We have set our capability below that required by DO-160, but high enough that we expect will be seen sufficient by flight test organizations.

The last capability looks at the usability of the device by measuring the installation time. There are many factors that determine the usability of a device, but installation time and cost are always issues in FTI scenarios. We expect that a measure of it is important to customer acceptance and a major factor driving the community's interest in wireless technologies.

CONCLUSIONS

This paper described ongoing SwRI research into alternative sensing approaches that utilize a mix of video processing, distributed processing, and power harvesting to provide additional solutions. The basic techniques behind this approach have been explored in previous SwRI projects, including the Remote Neutrally Buoyant Sensors effort which was the recipient of an R&D 100 Award [18], but have not yet been applied to the flight test instrumentation field. We believe that a commodity-based approach to flight test is not only feasible, but necessary as the continuing cost and schedule pressures on flight test programs demand approaches that minimize installation complexity and reduce the need to modify the aircraft under test.

REFERENCES

[1] https://www.raspberrypi.org/products/raspberry-pi-3-model-b/
[2] https://www.raspberrypi.org/education/weather-station/
[3] https://www.raspberrypi.org/blog/carputers-some-ideas-to-get-you-started/
[4] http://www.popularmechanics.com/military/weapons/a23774/raspberry-pi-ukrainian-weapon-system/
[5] http://www.lord.com/our-company/newsroom/curtiss-wright-collaborates-with-lord-corporation-to-bringwireless-networking-to-flight-test-instrumentation
[6]http://files.microstrain.com/Wireless-Sensor-Networks-LXRS-Data-Communication-Protocol.pdf
[7]http://www.piezo.com/tech3faq.html#app3
[8] https://www.wirelesspowerconsortium.com/about/benefits.html
[9] http://www.shotpeener.com/library/pdf/1998008.pdf
[10] M. K. Gellaboina, G. Swaminathan, V. G. Venkoparao, "Image based dial gauge reading." U.S. Patent 2013/0070099 A1, issued March 21, 2013.
[11] R. V. Weinmann, J. N. Gelinske, R. M. Allen, J. A. Wiig, "Optical image monitoring system and method for vehicles." U.S. Patent 2010/0214411 A1, issued August 26, 2010.
[12] J. Gur, I. Seror, "Method and system for the acquisition of data and for the display of data." U.S. Patent 2007/0236366 A1, issued October 11, 2007.
[13] D. Tappan, M. Hempleman, "Image Processing for Data Acquisition and Machine Learning of Helicopter Flight Dynamics," The 20th International Conference on Image Processing, Computer Vision, and Pattern Recognition, Las Vegas, NV, 2016.

[14] R. Hamza, W. Au, M. Johnson and T. Henderson, "Video to information (V2I) system enabling old commercial cockpit instruments to participate in the advanced information retrieval and safety services," Proceedings. The 21st Digital Avionics Systems Conference, 2002, pp. 12C4-1-12C4-10 vol.2.
[15] W. Khan, D. Ansell, K. Kuru and M. Amina, "Automated aircraft instrument reading using real time video analysis," 2016 IEEE 8th International Conference on Intelligent Systems (IS), Sofia, 2016, pp. 416-420.
[16] L. Lucchese, S. K. Mitra, "Correction of Geometric Lens Distortion Through Image Warping," Proceedings. 3rd International Symposium on Image and Signal Processing and Analysis, 2003, pp. 516-521 vol.1.
[17] S. Chakraborty, S. Ghosh, S. Chatterjee, S. Chowdhuri, R. Ray and N. Dey, "Rigid image registration using parallel processing," International Conference on Circuits, Communication, Control and Computing, Bangalore, 2014, pp. 237-241.
[18] http://www.swri.org/4org/d10/applied-sensing/special-sensing-tech.htm

HYBRID AUTOMATED UAV TARGET RECOGNITION SYSTEM

Undergraduate Students: David Hung, Kennon McKeever, Ricardo Ramirez
Advisor: Dr. Michael W. Marcellin
Department of Electrical and Computer Engineering
The University of Arizona, Tucson, AZ 85721

ABSTRACT

Accurate image classification is one of the core challenges in computer vision. At the annual AUVSI SUAS competition, this challenge is faced in the form of ground target classification from an unmanned aerial vehicle (UAV). Additionally, due to the constraints imposed by the UAV platform, the system design must consider factors such as size, weight, and power consumption. To meet performance requirements while respecting such limitations, the system was broken into two subsystems: an onboard subsystem and a ground based subsystem. This design allows the onboard subsystem, comprised of a DSLR camera and single-board computer, to capture ground target images and perform rudimentary target detection and localization. For further processing and to ultimately classify the targets in each image, data packets are sent to the ground-based subsystem via a 5 GHz wireless link. Convolutional networks are utilized on the ground to achieve state-of-the-art accuracy in classification.

INTRODUCTION

Background

The annual AUVSI Student Unmanned Aerial Systems (SUAS) competition is an event for which teams of students are required to design an unmanned aerial vehicle (UAV) capable of autonomous flight and navigation, remote sensing, and execution of specific tasks during a challenging flight mission. One of these tasks is vision based object detection, localization, and classification, and localization. In order to achieve this, the UAV must be capable of taking pictures of the search area, detecting objects of interest in images, determining the object's geographical position, classifying the object's characteristics, and uploading these results to an external server for scoring [1]. Examples of targets are shown in figure 1, below.

Figure 1 - Targets Displayed with various shapes, colors, and alphanumerics

In this paper, a vision system used for the 2017 AUVSI SUAS competition will be presented in detail. This vision system was designed as a senior capstone project in cooperation with the University of Arizona's Autonomous Vehicles Club. This paper will discuss the vision based mission tasks, the hardware and software designs produced to meet the mission requirements, and the results achieved by the system. It should be noted that the other aspects of the UAV are not covered in this paper, but are instead discussed in a separate paper by the Autonomous Vehicles Club.

Problem Statement

In the field of computer vision, object recognition is the ability to find and identify objects in an image or video sequence. Given competition targets that vary in viewpoint, size, scale, and rotation, developing a robust computer vision system is a multi-faceted challenge. The AUVSI SUAS competition poses this same challenge in the context of UAVs, which introduces an additional suite of complications. Current approaches to target identification for this competition are greatly limited, with many teams opting for manual (human) identification of targets. The few teams that attempt autonomous target recognition deploy dedicated Graphics Processing Unit (GPUs) at their remote system or stream full resolution photos down to their base station. These approaches have very high cost, as onboard GPUs are expensive and greatly increase power and heat management requirements, while high resolution image streaming demands large bandwidth and high powered transmitters.

Objectives

The vision system must be capable of high accuracy object recognition and characteristic extraction while respecting the constraints imposed by a limited size, weight, power, and thermal budget on-board a UAV. This requires some form of compromise between on-board system cost and overall system performance. This project seeks to find and implement an ideal balance between these two variables.

DESIGN SOLUTIONS

Hardware Design

In order to meet performance requirements while minimizing the system's impact on the UAV's flight characteristics, a 'hybrid' on-board/off-board computing architecture was used. The on-board subsystem interfaces with the UAV's gimbal stabilized DSLR camera via USB 2.0.

The On Board Computer (OBC), an Odroid XU4, is used to perform preliminary image processing on board before transmitting extracted regions of interest (ROIs) to the ground based subsystem. Images are securely transferred over a 5 GHz link using the Ubiquiti Bullet M5 wireless transceiver. Excluding the image capture subsystem, the onboard solution weighs less than 0.5 pounds and consumes less than 10 W of power. Figure 2 (below) shows the full onboard subsystem configuration used for this year's competition. The overall system weighed less than 5 pounds including camera, gimbal, and a 4S 5000 mAh LiFE battery.

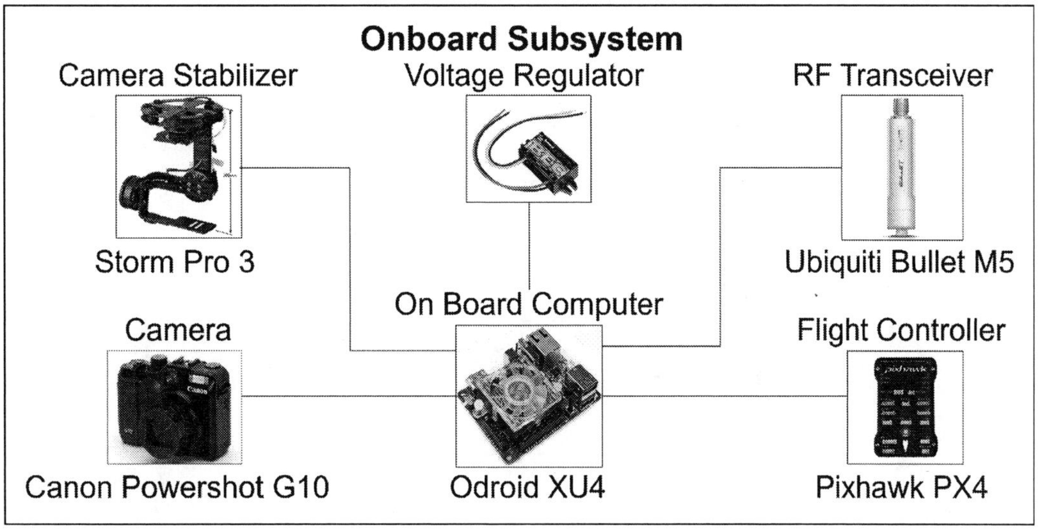

Figure 2 - On-board Subsystem Hardware Overview

The ground-based subsystem leveraged the much larger allowances on power, size, and weight consumption in order to shore up the computational weaknesses of the on-board subsystem. The core of the subsystem is a GPU accelerated computer, comprised of an AMD Ryzen 5 1600 CPU, Nvidia GTX 1060 GPU, and 8 GB of DDR4 RAM. The computer is further integrated with a network switch, monitor, and peripherals, and the entire package is housed inside a weatherproofed Pelican case for transport and deployment in the field. An Ubiquiti Nanostation Loco M5 wireless transceiver is mounted externally and maintains the data link with the on-board subsystem. Figure 3 (below) shows a block diagram of this hardware setup.

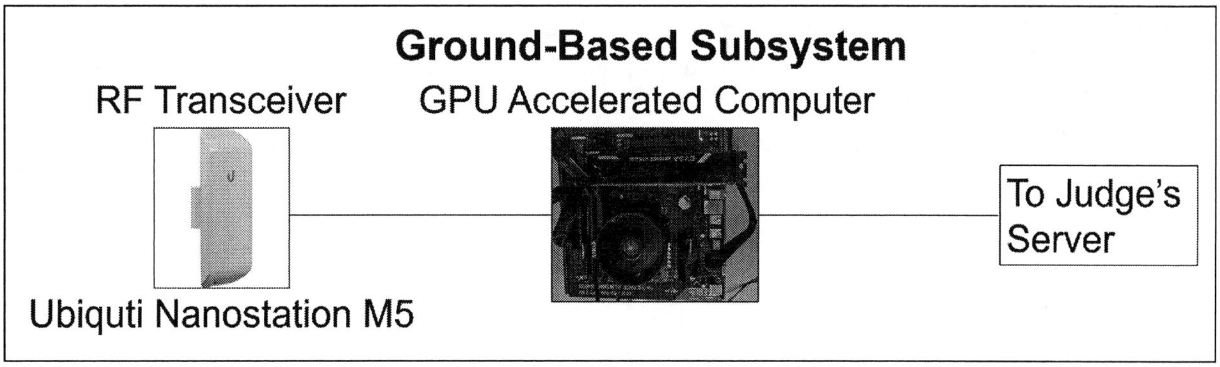

Figure 3 - Classification Ground Station Hardware Overview

Software Design

This on-board/off-board hardware architecture lends itself to a natural division in the image processing stack between image segmentation and feature extraction. In turn, this on-board image segmentation step dramatically reduces the bandwidth requirements of the system by transmitting only the detected ROIs and discarding the extraneous data. Figure 4 (below) presents an overview of the full software architecture, including the division between on-board and off-board processing.

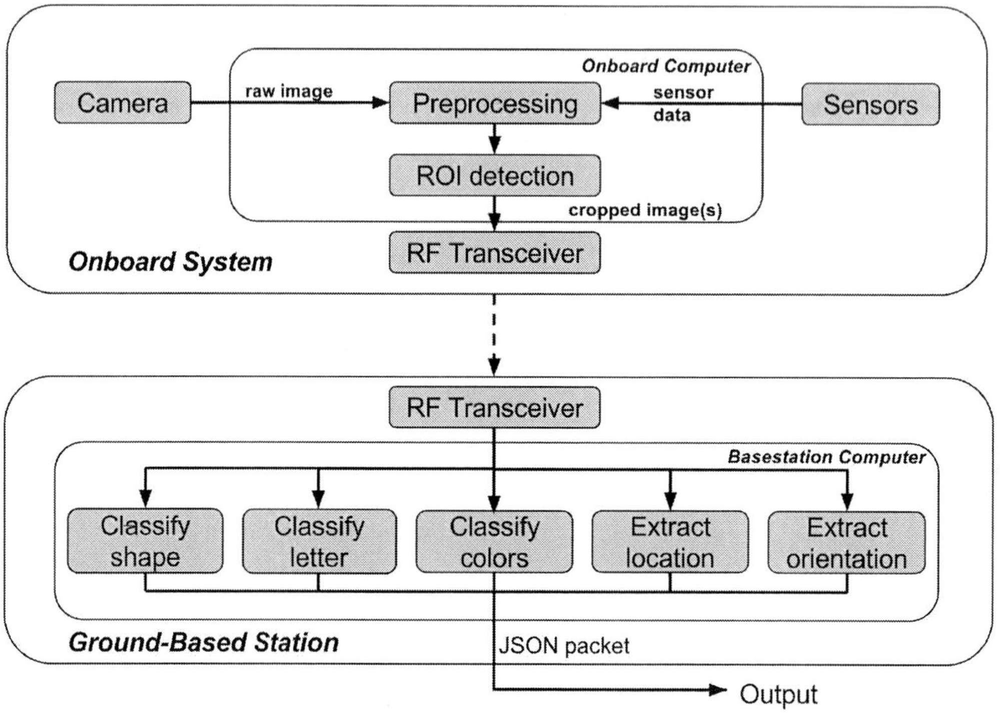

Figure 4 - Image processing software architecture overview

Images are programmatically captured by the OBC using the Canon Hack Development Kit (CHDK) Photo Transfer Protocol, an open source toolkit for programmatic control of Canon cameras. Captured images are dumped into a file directory in the OBC and are consumed by *Dispatchers*, which stamp the image with GPS location data and pushes the stamped image into the processing pipeline. Further down the pipeline are *Processor* nodes, which perform Canny edge detection [2] on the various color spaces of each image. The union of detections in each color space are fused into a single intermediate image, which is then smoothed and dilated in order to mitigate artifacts from noise. The resulting contours are finally passed to a bounding-box drawing function and filtered based on expected target size. Finally, the locations of these bounding boxes are used to crop potential regions of interest from the original image, and the result is sent to the ground-based subsystem for classification.

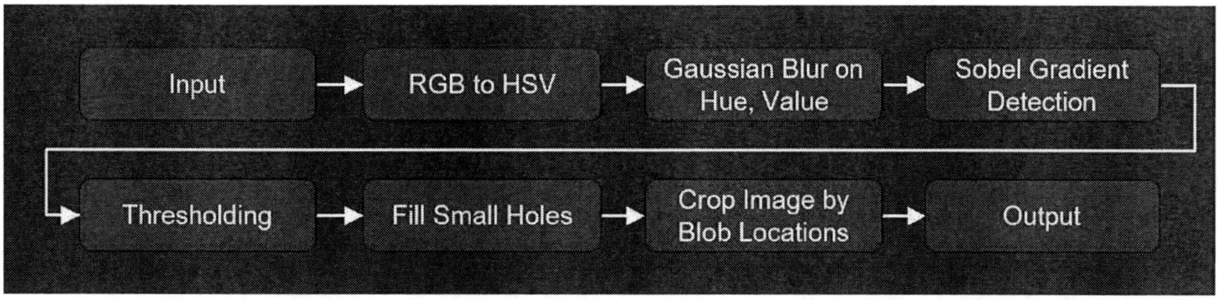

Figure 5 - ROI Detection Algorithm used in the Processor nodes

The off-board subsystem utilizes Convolutional Neural Networks (CNNs) in order to characterize target images. CNNs are a type of machine learning algorithms that were first pioneered in modern image classification by Krizhevsky, et al [3]. CNNs consist of core building blocks called "layers" which are used to construct a predictive model that can be trained on various datasets and deployed for state of the art classification accuracy. The main layers used by CNNs are as follows:

- Input Layer: Dataset of images. In the context of this paper, the input images are preprocessed aerial images captured during the mission.
- Batch Normalization Layer: This layer is used to normalize the inputs in order to achieve zero mean and unity variance within the neural network. These layers were found to significantly perform network performance as well as reduce the need for other techniques such as dropout and data preprocessing.
- Convolutional Layer: Convolutional layers slide a small window or filter across an input volume in a manner akin to convolution in order to detect various features. Filters employ a technique known as parameter sharing, which dramatically reduces the memory requirements for the network and also grants the network a degree of translational invariance. These are the core building block of CNNs.
- Activation Function: Nonlinear activation functions have been shown to be crucial in increasing the complexity of features recognized by neural networks.
- Pooling Layer: Pooling operations are generally used to reduce the dimensionality of layers in neural networks, which in turn can significantly reduce the memory footprint of the network.
- Fully Connected Layer: Unlike the convolutional layer which only perceives a small window of its inputs, fully connected layers are connected to every element in its input. Fully connected layers are nearly identical to traditional neural network layers and are generally used to compute the final output of the model.

For the system described in this paper, four separate classifiers were created and utilized. In order to classify target shape and alphanumeric, the Wide Residual Network (WRN) architecture [4], outlined in Figure 6 (below), was utilized. Target background color and alphanumeric color were much simpler features and thus complex models such as WRN were found to overfit, resulting in poor real world performance. Instead, a shallow, 3-layer multilayer perceptron was used. During mission runtime, the four models are evaluated for each received image and the classification results are exported for scoring.

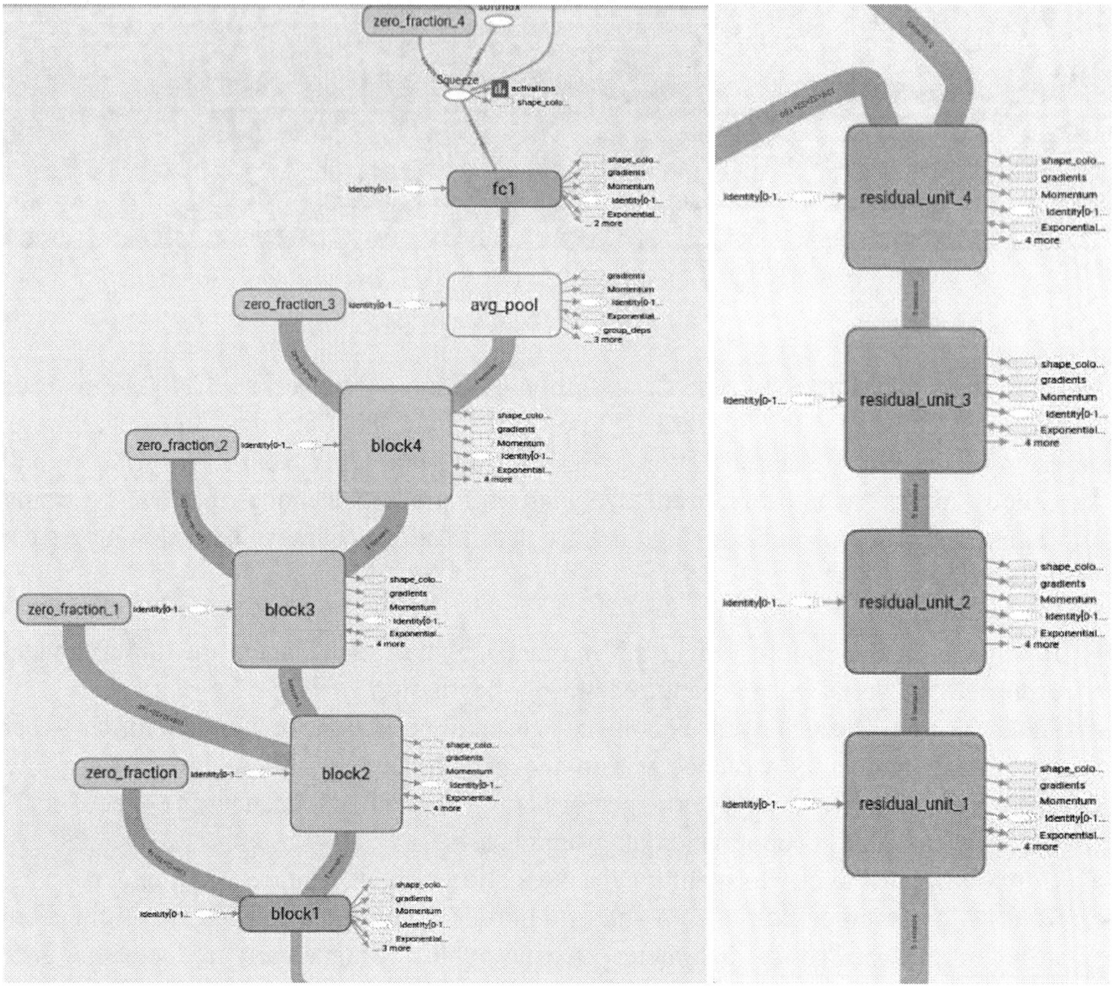

Figure 6a (left) - Overview of the convolutional neural network deployed on the ground station. Block1 consists of a single convolutional layer while blocks 2-4 contain stacked residual layers. Figure 6b (right) - an expanded view of the blocks shown in Figure 6a

CONCLUSION

The hybrid on-board/off-board processing architecture was an excellent compromise between on-board system impact and overall system capability. The region of interest extraction had a fairly high false positive rate of roughly 75%, but the rate still resulted in a major reduction in transmission bandwidth. Additionally, the image classifiers at the ground station are robust to background/non-target input images and appropriately identify them as such. Evaluation of the classifiers on synthetic data produced in classification accuracies ranging from a minimum of 90.1% for alphanumeric classification to 99.7% for shape classification. Unfortunately, evaluation of the classifiers on real world data is currently inhibited by resource limitations and is pending manual labeling of collected data with ground truth values. Overall, the system architecture is highly promising as a solution to target detection and classification from a UAV.

ACKNOWLEDGMENTS

This research was supported by our sponsor Professor Michael Marcellin. Dr. Marcellin has helped us open many doors of opportunity as we progressed throughout the school year, and for that, we are grateful. We would like to thank our colleagues from the University of Arizona who provided us with insight and expertise that greatly assisted us in our research. We would also like to show our gratitude to the University of Arizona for their support.

REFERENCES

1. "2017 Rules for the 15th Annual Student UAS (SUAS) Competition." *AUVSI SUAS Competition*. AUVSI Seafarer Chapter, 12 June 2017. Web. <http://www.auvsi-suas.org/static/competitions/2017/auvsi_suas-2017-rules.pdf>.
2. Canny, John, "A Computational Approach to Edge Detection," IEEE Transactions on Pattern Analysis and Machine Intelligence, vol. PAMI-8, Institute of Electrical and Electronics Engineers, Nov. 1986, pp. 679-698
3. Krizhevsky, Alex et al, "ImageNet Classification with Deep Convolutional Neural Networks," *Advances in Neural Information Processing Systems*, Lake Tahoe, Nevada, December 2012.
4. Zagoruyko, Sergey, Komodakis, Nikos, "Wide Residual Networks," *Proceedings of the British Machine Vision Conference,* York, United Kingdom, September 2016.

A DISTRIBUTED VIDEO ACQUISITION SYSTEM

Jian Song

Beijing Zoweetech Ltd.
Beijing, PR China
songj@zoweetech.cn

ABSTRACT

Camera has been used in telemetry system over decades now. Normally, the video signal outputted from cameras would be acquired by video module(s) located in general data acquisition equipment. In this paper, a distributed video acquisition system is presented. It is a synchronous acquisition system consisted of a record and control unit and several video acquisition units. The synchronous shutter control for cameras and data flow synchronous transportation inside the system is added as well. The accuracy of timing and synchronous acquisition of the system is less than 50ns.

KEY WORDS

Distributed Data Acquirement, Video Acquisition

INTRODUCTION

Cameras are used in variable ways in flight tests, such as eyes for ground or as actual measuring instruments, etc. The common practice for acquiring video and voice signals outputted from cameras is to mount video-acquiring modules in a data acquiring device. However, the video and voice outputted from cameras are usually analog signals which will be attenuated during transmitting from cameras to the data acquiring device. The longer the transmitting distance, the more the signal attenuation would be. Normally, it is difficult to arrange all the cameras near the data acquiring device in a flight test. Therefore, it is not quite satisfactory when the video from cameras are replayed later.

In this paper, a distributed video acquisition system is presented, which is composed by several video acquisition units along with a central control and record unit. In a flight test, each video acquisition unit will be mounted near a camera and connect to it by a cable as short as possible. Thus, the less attenuation of signal would result a better replay of the recorded video data.

Furthermore, a special synchronous mechanism [1] and IEEE 1588 PTP protocol [2] is applied in this system to improve the precision of shutter control for multiple cameras and to achieve a smooth playback when real-time monitor is needed.

SYSTEM OVERALL STRUCTURE

The distributed video acquisition system introduced in this paper is consisted of one central control and record unit and several video acquisition units. Sometimes a radio device might be added into the system to transport video stream to the ground for real-time monitoring. Each video acquisition unit could connect to one or two cameras and is responsible for acquiring video and voice signals. The central control and record unit is used for collect and record the video and voice data streams sent from all the video acquisition units. Figure 1 showed the structure and interconnections between devices inside the system.

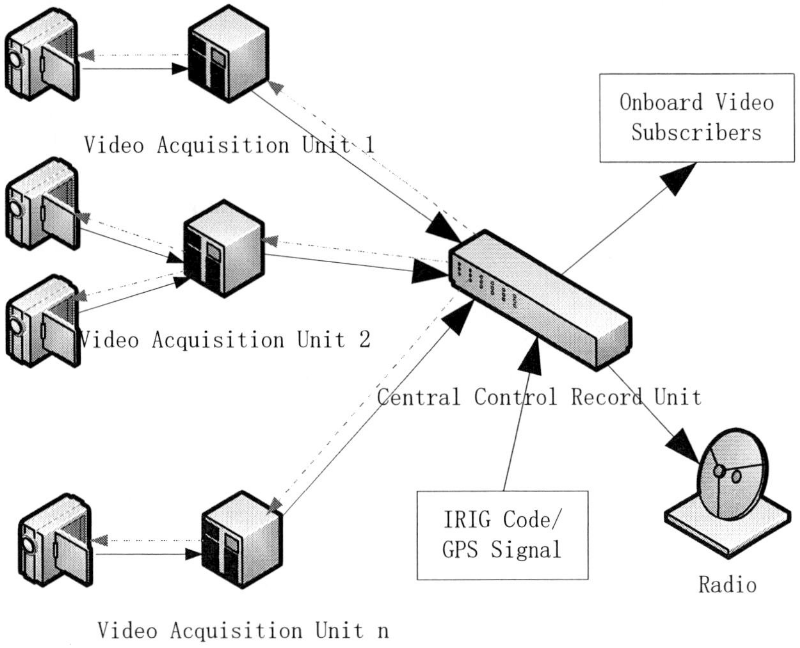

Figure 1 Structure of the Distributed Video Acquisition System

In this system, we have configured five kinds of components in the central control/record unit, including one main control module, several communication module, one record module (with one flash disk inside), one timing module, and one PCM output module. The communication modules will receive all the coded video and voice stream sent from the video acquisition module, and at same time, they will pass timing and controlling signals to the video acquisition unit connected. The record module will write the video data stream into the flash disk, and the GPS timing module will receive IRIG timing signals and resolve standard time from GPS signals, then manage the synchronous time for the whole system.

The video acquisition units in the system are responsible for capturing pictures from the cameras connected and coding the pictures into compressed data stream. At same time, the units are also responsible for passing the synchronous shutter control signal to the cameras as well.

Usually, Ethernet could be used for the interconnection between central control/record unit and each video acquisition unit. However, due to following reasons we decide to adopt direct Ethernet connection combining time division multiplexing technique [1] (DEC-TDM in short) plus improved IEEE 1588 PTP protocol instead.

a) Uncertainty of temporal latency in Ethernet transportation
The uncertainty of time delay in Ethernet transportation is caused by the CSMA/CD mechanics in Ethernet switch, which means that much time is needed for caching a full frame of video picture.

b) Precision of clock synchronization assured
IEEE 1588 PTP protocol is commonly used for clock synchronization for Ethernet. The latency of the timing signal transferred over Ethernet could be estimated by using this protocol. For a connection using DEC-TDM, this latency could be actually calculated precisely.

c) Precise control for shutters of multiple cameras
As a virtual channel for synchronous acquiring control signal is enabled in DEC-TDM, it easy for synchronous sampling control and synchronous shuttle control for multiple cameras. The precision of the control signals in the system is less than 50nm.

d) Transportation efficiency improved
Parallel process has been implemented in central control/record unit, so that the total data rate of video streams outputted from all the video acquisition units might reach nearly 10Gbps.

Finally, there are two output channels in the system, which are PCM output channel at the PCM output module and Ethernet output channel at the main control module. The PCM output module will code the video data and output to a radio device for ground real-time monitoring. Ethernet compatible output could be enabled by the configuration setup file stored at the flash disk for distributing videos to airborne subscribers.

VIDEO ACQUISITION UNIT

The video acquisition unit is made up of three modules, including one image acquisition and compression module, one main control module and one power supply module. Figure 2 shows the structure design of the video acquisition unit.

Among all three modules of the video acquisition unit, the power supply module is only a single function module, just in charge of supplying electricity for other modules. The acquisition/compression module plays the leading role, which will capture pictures from the output signals of the camera connected and then code the acquired data into compressed video data stream by using traditional H.264 coding algorithm, and then pass them to the main control module, which is actually a combine of communication, timing and management. A timestamp will be added in the user defined field in each video frame. The video data flow will then be sent to the central control and record unit.

The shutter control signal is also sent via the acquisition/compression module. System management information, including timing code and synchronous acquiring control signal, is received by the main control module. Following the instruction from the central/record unit, the synchronous acquiring control signal could be converted to shutter control signals and transmitted to the cameras through the acquisition/compression module.

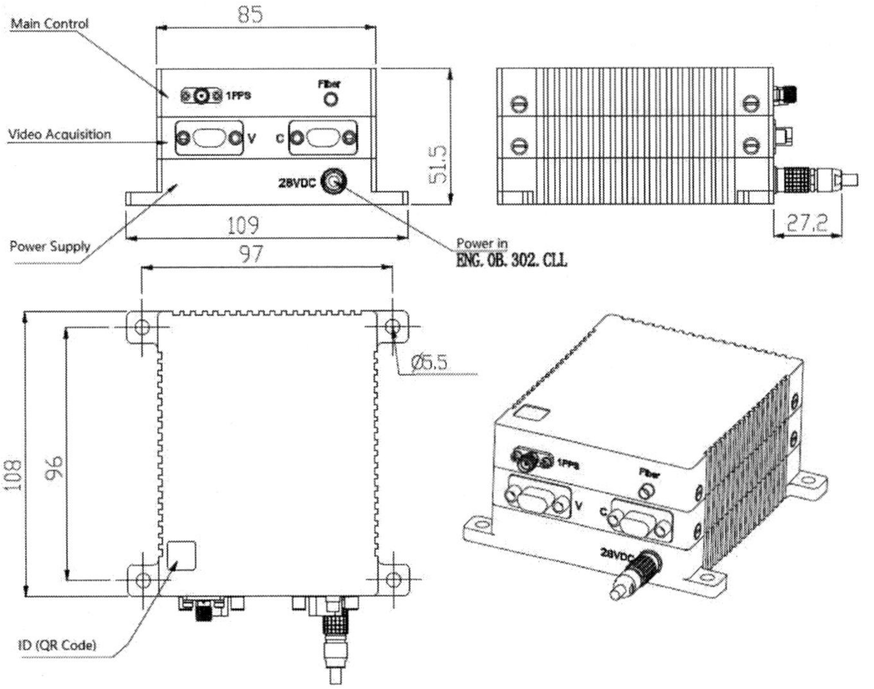

Figure 2. Video Acquisition Unit Structure Design

For the purpose of compatibility, in addition to ordinary video stream dispatch by Ethernet, there is an option that the outputted video data stream could comply with the iNET conventions.

If the original video source should be kept for later usage, the video acquisition unit should be capable of outputting the source to the central control/record unit. The GigE could be utilized for transporting the video source.

SYSTEM OPERATION MECHANISM

During an actual video acquisition task, one or two nearby camera(s) will be connected to one video acquisition unit, which will be connected to the central control/record unit with Ethernet cable. A radio device will be connected if real-time monitor needed on the ground.

A system clock is used for synchronization of video capture and control for multiple cameras, and this clock will be corrected periodically by IRIGB connection or standard time resolved from GPS signals. The timing module in the main control/record unit will use system clock to issue timing signals (codes) and synchronous acquisition control signals (codes) for synchronizing the local clocks in all video acquisition units and for controlling the shutters of the cameras connected. As direct Ethernet connection has been employed, the timing code and control code will not be delayed while these codes are transported from the central control/record unit to all the video acquisition units.

Cameras will start to take pictures as soon as the synchronous control signal was received, and then each camera will output video flow to connected video acquisition unit which will capture every frame of video and compress it with timestamp in each frame. All the fames compressed will be coded into a data flow which will be sent to the central control/record unit through the Ethernet connection interface.

The central control/record unit will receive all the data flow (video stream) from different video acquisition units and then record them into the flash disk. In the case some video sources are needed for measurement, the data rate for writing to the flash disk may exceed the writing speed of a single disk. So we had also tried a parallel recording structure in which multiple flash disk were used and the whole data flow needed to be recorded were divided into multiple data stream, and each disk could easily record one of the steams.

According to video acquisition task configuration, the central control/record unit will select one or more video stream to the onboard video subscribers through Ethernet interface located in its communication module, or send selected video stream by the PCM module to the ground via a radio set.

CONCLUSION

Distributed video acquisition system is better if multiple cameras had been used in a flight test, because the video acquisition unit could be mounted as close to the camera as possible and signal attenuation would be less correspondingly. In addition, for a video acquisition system, it is better to have synchronous control for all cameras connected to the system. It is very useful to get same moment pictures from different cameras if the pictures taken will be used as measuring resources.

As DEC-TDM has been implemented for transporting, both the timing and the synchronous acquisition control have been improved greatly. The laboratory experiments showed that the precision of timing synchronization and shutters synchronized control reached up to 50ns.

Parallel recording would be a solution for fast recording (data rate over 2Gb/s) when high-definition video cameras used in the system and the video source needed for further usage.

Furthermore, the video acquisition unit mentioned above could also be used as the front end for other iNet universal acquisition devices or recorders.

REFERENCES

[1]. Wang Zhongjie, Song Jian, etc. Synchronized data transportation for distributed acquirement system through direct Ethernet connection and TDM [C] // 51th Annual International Telemetry Conference. Las Vegas. USA：International Foundation for Telemetering. 2015.

[2]. IEEE. IEC 61588:2009, Precision Clock Synchronization Protocol for Networked Measurement and Control Systems [M]. Geneva: International Electrotechnical Commission, 2009, Edition 2.0.

DEFENDING T&E SPECTRUM THROUGH AUTOMATED FREQUENCY MANAGEMENT METRICS CALCULATION

Michael K. Painter, P.E.
Ronald Fernandes, PhD
Karthic Madanagopal
Knowledge Based Systems, Inc.
College Station, TX 77840
mpainter@kbsi.com

Charles H. Jones, PhD
C. H. Jones Consulting, LLC
Eugene, OR 97405
chjonesconsulting@gmail.com

ABSTRACT

Because of its economic value, there has been growing pressure to sell off spectrum currently allocated for defense purposes. These pressures come at a time when Department of Defense (DoD) spectrum needs are growing at an exponential pace, thus prompting heightened efforts to clearly demonstrate both the need and the responsible, efficient use of electromagnetic spectrum. In response, the DoD has developed a baseline set of standard metrics to measure spectrum utilization, demand, efficiency, and operational effectiveness. The focus has now shifted toward developing the automated tools to calculate, plot, and display these metrics. The purpose of this paper is to describe progress toward developing a Spectrum Management Metrics Toolkit (SMMT) to fill this need and its potential role in helping to analyze and defend Test and Evaluation (T&E) spectrum usage and needs.

INTRODUCTION

On June 7, 1995, a bill proposing the reallocation of 275 Megahertz (MHz) of government spectrum was brought to the floor of the U.S. Senate for a vote. The spectrum targeted for reallocation included the entire 225-400 MHz band, which the DoD had declared two years earlier to be the single most critical spectrum resource of our military tactical forces, both nationally and within the North Atlantic Treaty Organization (NATO). Focused primarily on the potential windfall that an auction for this spectrum would provide for the U.S. Treasury, the Senate passed this legislation by a voice vote. The DoD had been given no notice that the bill was even under consideration until the day of the vote. Had the bill not been altered at the last minute to exclude this band, the DoD would have had nine months to vacate the band [1].

Recurring efforts to reallocate DoD's electromagnetic spectrum for commercial applications pose an unwitting threat to our national security interests. This paper outlines the nature of the spectrum access challenge and the rising pressure for more widespread reallocation of this critical resource. It then describes the potential impact of spectrum loss in the Test and Evaluation (T&E) environment—as this application domain is a representative microcosm of the DoD spectrum management challenge. The paper concludes by describing recent efforts to develop metrics and tools designed to help the DoD establish its telemetry spectrum needs and demonstrate responsible, efficient use of this resource.

GROWING PRESS FOR SPECTRUM ACCESS

Effective spectrum and frequency management is essential to the success of United States (U.S.) military operations. Military operations increasingly rely on the ability to maintain full access and reliable control of the radio frequency (RF) spectrum for communications, radar, electronic warfare, remote fires, avionics, global positioning, logistics, medical support, and signals intelligence uses. Information dominance cannot be achieved without it. Consequently, effective spectrum and frequency management is central to present and future warfare plans, including Force 21, the Army After Next, the Expeditionary Aerospace Forces, and the Marine Expeditionary Forces.

The DoD's needs for electromagnetic spectrum are already significant and continue to grow at an exponential pace. The 2000 Defense Science Board report, Tactical Battlefield Communications, for example, estimates an anticipated growth rate in demand exceeding 18 percent per year [33333333]. Another report projected military spectrum demands to grow 70 percent in the decade leading up to 2010 [8].

Meanwhile, there is growing pressure to further reduce the DoD's allocation of spectrum. Much of this pressure is driven by a rapid growth in public demand for broadband wireless electronics, which is creating new private sector demands for spectrum access. Unquestionably, access to spectrum is central to a broad range of business and consumer communication, research and development (R&D) and information technology (IT) purposes, such as private and public telecommunication operations (e.g., mobile phone networks, wireless Internet communication, aviation, shipping, defense, public safety), broadcasting, radar, astronomy and various other applications including countless short-range, low-power wireless devices [11]. Additional pressure stems from perceptions that the DoD is hoarding spectrum, that it no longer needs as much spectrum, and that it is an inefficient user.

Outside the continental United States (U.S.), where U.S. forces require spectrum coordination with host and surrounding nations, spectrum-related issues are even more pronounced. Many nations view wireless communications services as the backbone for their future economic development, which means they are increasingly asserting their sovereign spectrum rights. Their spectrum allocations differ significantly from those used in the U.S., often making it difficult to conduct planned operations efficiently.

These and other factors have inevitably created recurring cycles of legislative activity to sell off portions of the DoD spectrum.

In 2006, for example, the FCC auctioned spectrum licenses in the 1710-1755 Megahertz (MHz) band that had previously been allocated for federal use. To meet the continued demand for commercial wireless services, NTIA assessed the viability of reallocating the 1755-1850 MHz band for private sector applications. This band was assigned to more than 20 federal users, including the DoD [12].

More recent legislative trends are equally troubling. Title VI of the Middle Class Tax Relief and Job Creation Act of 2012—also known as the Public Safety and Spectrum Act, or Spectrum Act, for short—seeks to expedite increased access to spectrum for commercial mobile broadband through spectrum reallocation, reassignment of spectrum rights, and changes in procedures to repurpose spectrum used by the federal government. Among other things, the act also includes provisions to apply spectrum license auction revenues toward deficit reduction. Further legislation of this kind clearly establishes a trend toward reducing federal government control of spectrum in favor of competing concerns. For example, the Efficient Use of Government Spectrum Act of 2013 (H.R. 2739, Representative Matsui) required the release of spectrum that was in use at the time by the DoD for shared use or auction. The Federal Spectrum Incentive Act of 2013 (H.R. 3674, Representative Guthrie) established that federal spectrum users are eligible to participate in incentive auctions like those being implemented for television broadcasters.

Further efforts to relinquish spectrum will almost certainly continue to rise with the emergence of new technologies like telemedicine, driverless vehicles, advanced robotics, cloud computing, and machine-to-machine communications (the Internet of Things).

Meanwhile, historical notions of spectrum allocation, assignment, and management are also changing dramatically. What was once a static, one-dimensional property is becoming a dynamic, multifaceted commodity. The original framework of spectrum allocation was based on the idea owning parts of the spectrum for a pre-defined purpose. The notion of spectrum ownership is shifting in response to recent developments in software-programmable radios and radars, spread-spectrum waveforms, digital signals, spectrum-sharing technology, and dynamically assigned frequency that allow users to share common spectrum, separated by frequency, location, time and waveform under the control of supervisory systems. The cost of upgrading, or even to move operations from the current band to some other frequency band, can be daunting. Twenty years ago, the military had already invested heavily in spectrum-using equipment including more than 800,000 radio frequency (RF) emitters worth over $100B [7].

These pressures prompt heightened concerns that rapidly growing DoD spectrum needs will soon reach a point where they can no longer be met—ultimately precipitating a corresponding erosion in our military's technological edge.

IMPACT ON THE T&E ENVIRONMENT

Today's T&E environment is a living microcosm of the larger spectrum management challenge facing DoD. The T&E environment requires war-like access to spectrum under the "train as you fight" and "test as you use" doctrines and involves the highly coordinated application of various resource types—including telemetry spectrum. This orchestration process and the numerous constraints involved present special problems in spectrum management. According to one source, the spectrum available for telemetry and instrumentation is inadequate to support the pace of activities in crowded areas like the R-2508 complex [3]. Lack of sufficient access to the needed spectrum comes at a cost. Historically speaking, flight testing accounts for as much as 15 percent of the cost of developing a new aircraft [4]. Test programs often involve hundreds of workers and can last months or even years. For large test programs, unplanned test delays can cost somewhere between $1 million to $3 million [5]. Taken together, an estimated $60 million a year is lost on test ranges today due to delays caused by telemetry spectrum shortages [6].

If the DoD hopes to maintain the spectrum it needs for T&E purposes, it will have to clearly demonstrate both the need for that spectrum and the responsible, efficient use of the spectrum. It must also dramatically improve its ability to quantitatively establish the technical, cost, schedule, and safety risks of reduced access and control.

Much of the difficulty in doing so stems from the lack of well-defined metrics and tools to:

- Accurately estimate current and future spectrum needs;
- Account for *actual* versus *scheduled utilization* of the spectrum that is allocated, so as to demonstrate responsible stewardship of the spectrum; and
- Quantify the cost and schedule implications of the loss of needed spectrum.

There has been precious little in the way of methods, metrics, and tools available to make these kinds of assessments. Consequently, those charged with securing and defending spectrum needed for T&E purposes have had little beyond anecdotal evidence to support their argument.

In response to these challenges, the Defense Science Board recommended that specific research and development (R&D) efforts be undertaken [3], including those aimed at developing improved operational and technical methods and advanced tools to promote more efficient spectrum use.

The Spectrum Efficiency Through Metrics (SETM) effort directly supports these goals by applying frequency metrics standards to monitor, assess, and improve the efficient use of spectrum while simultaneously working to give leaders the tools needed to defend access to T&E spectrum. The effort leverages many years' worth of research invested to define a baseline set of standard metrics for spectrum utilization, demand, efficiency, and operational effectiveness. These metrics are formalized in the RCC FMG Spectrum Management Metrics standard [2222]. The RCC FMG standard will provide the springboard for developing automated tools to calculate, plot, and display frequency management metrics.

SPECTRUM MANAGEMENT METRICS TOOLKIT OVERVIEW

Perhaps the most visible product of the SETM effort will be a set of tools designed to calculate, plot, and display frequency management metrics. Collectively, we refer to this set of tools as the Spectrum Management Metrics Toolkit (SMMT).

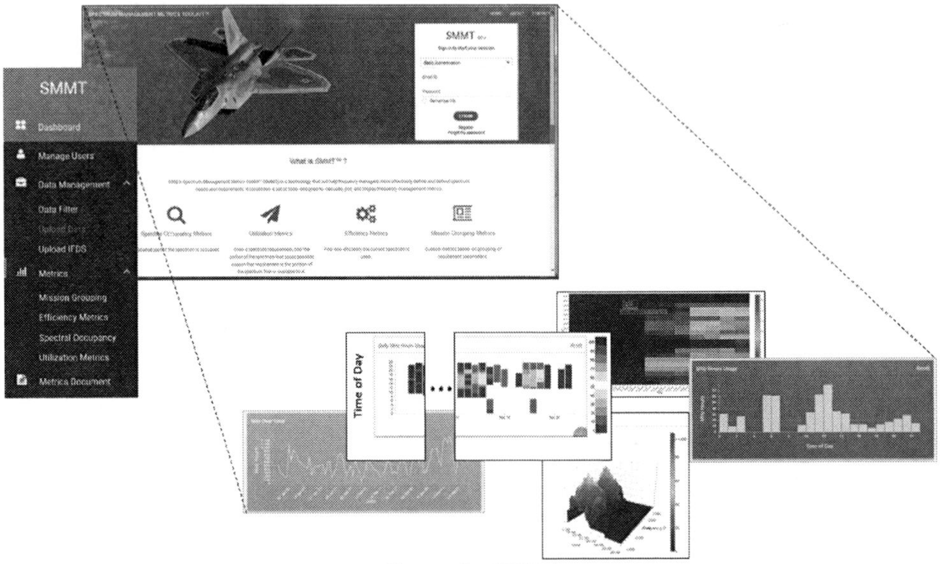

Figure 1. SMMT

The SMMT is architected as an enterprise level web application designed to ensure confidentiality and integrity of the data and the analysis results. As such, it provides a baseline set of enterprise system functions including SSO[1], RBAC, etc. As an interim measure to support end user testing, the SMMT also supports secure deployment using https:\\ with basic authentication. The toolset features an Angular 2.0 front end and a Node JS back end that uses a SQLite or MS-SQL database engine. The SMMT also provides a basic menu structure, a configurable dashboard interface, and an extensible library of metrics calculation and display "widgets" based on those found in the RCC FMG standard.

Some of the SMMT's key capabilities include the following:

- Provides a manual data import capability allowing users to load analysis data from various data sources ingested through a "drag-and-drop" interface.
- Allows users to define persistable filters to be applied to the data for analysis purposes.
- Provides a set of metrics calculation and display "widgets" using various chart types.
- Supports "cross-filtering" or "brushing and linking" across metrics widgets, meaning that a change made in the filtering criteria for one widget is automatically propagated to related metrics widgets.
- Supports the composition of personalized dashboards spanning metrics calculation types.

---

[1] The SMMT supports authorized access to the web application using either basic authentication (custom) or single-sign-on (Active Directory).

Based on the RCC standard, the SMMT is designed to support calculation and display of frequency management metrics in the following categories:

- Spectrum Occupancy
- Utilization
- Spectrum Reuse
- Efficiency
- Frequency Scheduling Operational Metrics
- Predictive and "What-if" Metrics
- Frequency Request Groupings

The paragraphs below provide a brief, representative overview of metrics in some of the more familiar metrics categories.

Spectral occupancy metrics are based on a time-frequency grid. These metrics in this category operate much like measuring occupancy in an apartment building. Consider an apartment building with four apartments that are up for lease this month. Percent Occupancy (PO) measures the extent to which the apartments available are leased out to residents. If there is one occupant in each apartment, PO is 100%. If only two of the four apartments are leased out to an occupant, PO is 50%. The PO metric disregards reuse, or concurrent occupancy. Thus, if each apartment had six occupants, PO would still be 100%. Percent Occupancy with Reuse (POWR) measures reuse of the spectrum. Reuse, or concurrent occupancy by more than one user of the spectrum, may be made possible through maintaining sufficient spatial separation to avoid interference, or by using multiple access methods. Returning to our apartment building example, if three of the apartments had one occupant while the fourth apartment had two occupants, POWR is defined to be 125%. Other metrics in this category measure how densely cells on a time-frequency grid are packed that are occupied (i.e., ignoring unused cells), the percentage of cells that are occupied with more than one occupant (regardless of how densely they are packed), etc.

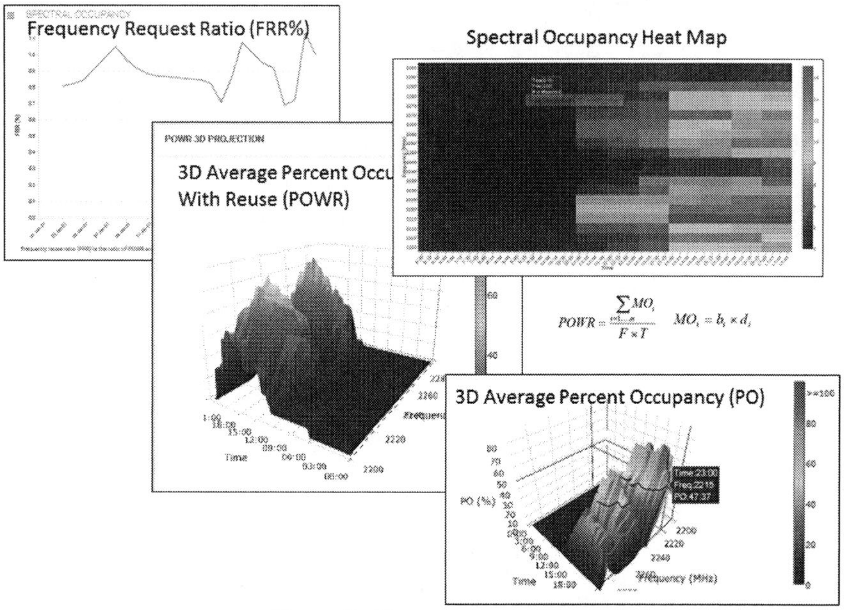

Figure 2. Occupancy Metrics

Efficiency metrics measure how efficiently the spectrum is being used and the extent to which technology and process change affect efficiency. Two forms of efficiency improvement are addressed: (1) advances in transmitter technology which allow more bits per second to be transmitted in the same bandwidth and (2) advances in scheduling technology (e.g., introduction of scheduling optimization algorithms). The metrics in this category attempt to answer questions like, "How much data are we transmitting?", "How much data is being transmitted per unit of spectrum per unit time (i.e., cost)?", and "Is the amount data being transmitted increasing as expected by having invested in technology and/or process change?" For example, the Average Mission Modulation Efficiency (AMME), which measures bitrate per unit of spectrum, would be expected to be vary based on the chosen modulation method (e.g., PCM/FM, SOQPSK or FOPSK). Each method has a theoretical performance—just like the EPA estimates a car's miles per gallon for city versus highway driving. But we know that the ideal performance may be affected by other factors, such as operating around mountains or imposing power restrictions. Charts that plot AMME as a function of modulation method help to surface what is actually possible.

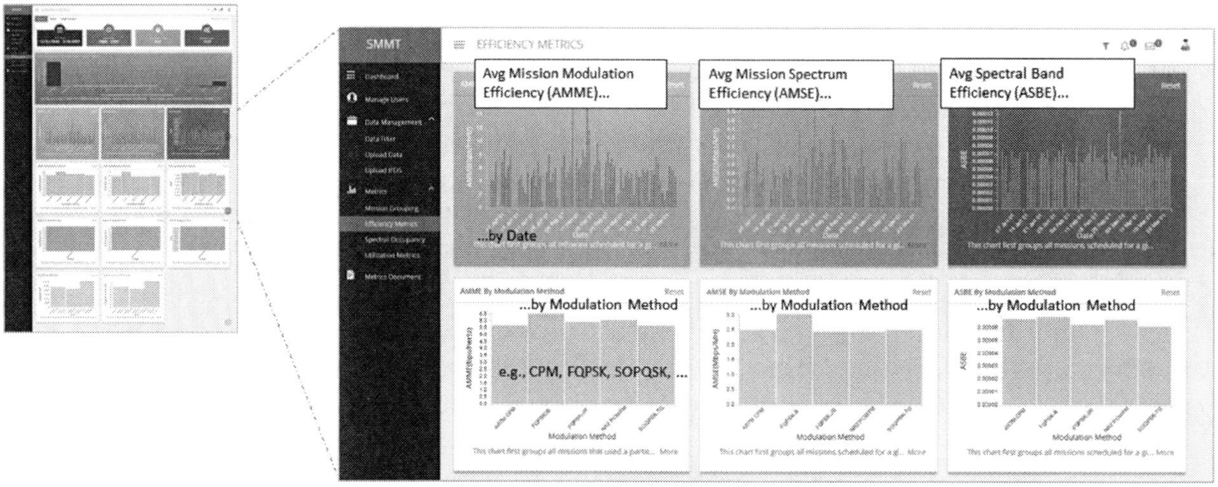

Figure 3. Efficiency Metrics

As viewed by the standard, Spectrum Utilization is defined in terms of spectrum availability. That is, in addition to actual occupancy in time, space, and frequency, access to spectrum may be denied to others for contingency purposes or as a consequence of the scheduled missions creating spectral fragmentation to a degree that unused spectrum cannot be scheduled. An example of the former case includes scheduling (but rarely actually using) a channel to execute in-flight termination of a missile in flight (i.e., reserving a frequency to initiate a self-destruct command, if required). Correspondingly, Spectrum Utilization metrics in the standard seek to answer questions like, "Can a given time-frequency request be accommodated given what has already been scheduled?", "What is the probability of being able to schedule a frequency request requiring a larger bandwidth?", "What does a typical frequency request look like?", "How has the likelihood of accommodating a typical frequency request changed with time?", etc. A heat map provided in the tool plots the number of frequency scheduling requests for a given bandwidth and duration in hours. In this example, we see that the most popular frequency request profile calls for 1.5 to 2.5 MHz of bandwidth for 1 to 2 hours. A companion heat map plots the likelihood that that a request calling for a certain bandwidth and duration can be scheduled. In this case, for the most popular request profile, the probability of it being scheduled is 70%. Of course, the more bandwidth required (moving vertically up the chart) and the longer the duration (moving horizontally to the right of the chart), the lower one's chances.

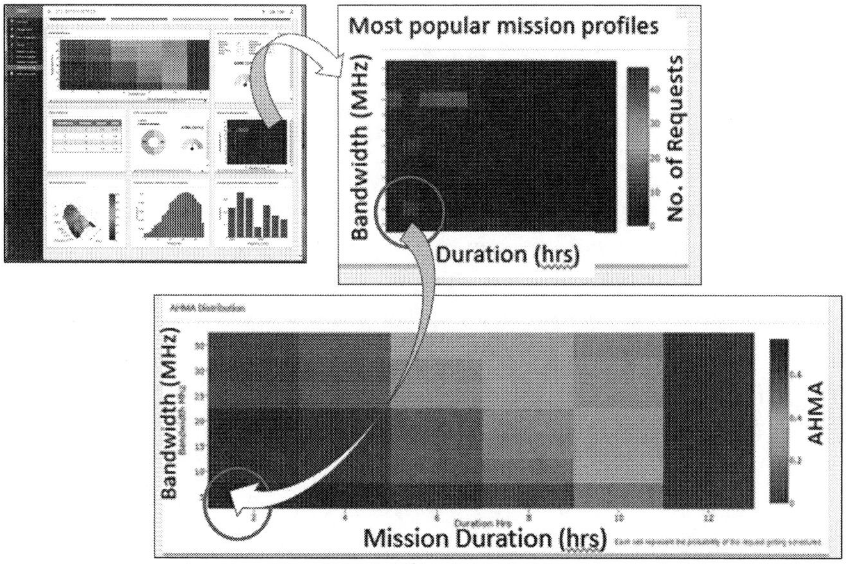

Figure 4. Utilization Metrics

CONCLUSIONS

Assured access to electronic spectrum is essential to the success of U.S. military operations both now and in the future. T&E spectrum needs are a vital component of that requirement. It is in that domain of spectrum application that we develop and maintain a decisive technological advantage. The two are inextricably linked. As more advanced technologies emerge, T&E spectrum needs will continue to grow. Many project that growth to be exponential. Meanwhile, the more crowded the spectrum becomes, the more it will become necessary to spread testing missions in time. Of course, doing so lengthens the development time between new generations of fielded weapons technology. Longer lead times, in turn, make it easier for enemy concerns to close the gap and achieve technological parity, or worse.

As evidenced by the past, economic pressures to increase the scope of commercial spectrum use will continue to drive further reallocations of DoD spectrum. If the DoD hopes to maintain the spectrum it needs for T&E purposes, it will have to clearly demonstrate both the need for that spectrum and the responsible, efficient use of the spectrum. It must also dramatically improve its ability to quantitatively establish the technical, cost, schedule, and safety implications of reduced access and control.

The emergence of the RCC's frequency management metrics standard coupled with tools like the one presented in this paper seek to provide the means to better manage and defend needed T&E spectrum.

REFERENCES

[1] Ryan, Mikel R., "Telemetry Spectrum Encroachment—Lessons Learned," European Telemetry Conference, May 2000, ADA377986.

[2] RCC 707-14 Spectrum Management Metrics Standards, Secretariat, Range Commanders Council, White Sands Missile Range, Mew Mexico. http://www.wsmr.army.mil/RCCsite/Documents/707-14_Spectrum_Management_Metrics_Standard/707-14_Spectrum_Management_Metrics_Standard.pdf

[3] Coping with Change: Managing RF Spectrum to Meet DoD Needs, Report of the Defense Science Board Task Force on DoD Frequency Spectrum Issues, Office of the Under Secretary of Defense for Acquisition and Technology, Washington, D.C., November 2000.

[4] Keane, Ken, "The Importance of WRC Agenda Item 1.5 to the Nations of the Americas," 2007.

[5] [Ernst 2007] Ernst, Darrell E., Carolyn A. Kahn, and David L. Portigal, "The Economic Importance of Adequate Aeronautical Telemetry Spectrum," The MITRE Corporation, February 2007.

[6] Kahn, C., "The Economic Importance of WRC Agenda Item 1.5 to Countries Worldwide," The MITRE Corporation, May 2007.

[7] Report of the Defense Science Board 1996 Summer Study Task Force on Tactics and Technology for 21st Century Military Superiority, Office of the Secretary of Defense, October 1996.

BIBLIOGRAPHY

[8] Federal Radar Spectrum Requirements, US Department of Commerce, NTIA Special Publication 00-40, May 2000.

[9] Chalfant, Timothy A. Telemetry Spectrum Encroachment: Taking Steps to Ensure the Future, in Proceedings of the Systems Engineering, Test & Evaluation Conference, October 2002.

[10] Ernst, D., Hoh, Y., Portigal, D., "The Growth of Data Rates for Aeronautical Telemetering and the Implications for the Radio Spectrum," MITRE Technical Report MTR 04W0000032, May 2004.

[11] Ben Freyens, "The Economics of Spectrum Management: A Review", School of Economics, Australian National University, 2007.

[12] "Spectrum Management: Federal Relocation Costs and Auction Revenues", Government Accountability Office (GAO) report to the Committee on Armed Services, U.S. Senate (GAO-13-472), May 2013.

Acknowledgment of Support: This project is funded by the Test Resource Management Center (TRMC) Test and Evaluation/Science & Technology (T&E/S&T) Program through the U.S. Army Program Executive Office for Simulation, Training, and Instrumentation (PEO STRI)

under Contract No. W900KK-16-C-0018. The Executing Agent and Program Manager work out of the AFTC.

Disclaimer: Any opinions, findings and conclusions or recommendations expressed in this material are those of the author(s) and do not necessarily reflect the views of the Test Resource Management Center (TRMC) and Evaluation/Science & Technology (T&E/S&T) Program and/or the U.S. Army Program Executive Office for Simulation, Training, & Instrumentation (PEO STRI).

Public Affairs Clearance: 412TW-PA-17391: Defending T&E Spectrum through Automated Frequency Management Metrics Calculation

DISTRIBUTION A. Approved for public release: distribution unlimited

COMPARING COMMERCIAL AND T&E SPECTRUM METRICS

Charles H. Jones, PhD
C. H. Jones Consulting, LLC
Eugene, OR, 97405
chjonesconsulting@gmail.com

Michael K. Painter, P.E.
Knowledge Based Systems, Inc.
College Station, TX 77840-2335
mpainter@kbsi.com

ABSTRACT

There is an ongoing need to understand how spectrum is used. In the context of defending T&E use against commercial encroachment, there is a recurring question of how commercial use compares to T&E use. This comparison is complicated by the lack of consistent definitions of the terms "use" and "efficiency" as well as different requirements of these distinctly different applications.

The International Telecommunications Union (ITU) recommendations lay out a foundation of spectrum metrics as applied to different applications. The RCC 707 Spectrum Management Metrics Standard provides a T&E perspective. Some insight about commercial metrics can be obtained through an internet search. This paper combines these resources to provide a coherent approach to understanding how to compare spectrum use within these different applications and how this affects the ability to defend against telemetry spectrum encroachment.

INTRODUCTION

There is an ongoing need to defend T&E spectrum use from commercial encroachment. Part of the approach to doing so is to better document T&E use and understand how this compares to commercial use (most notably cell phone use.) To this end, the T&E Science and Technology (S&T) Spectrum Efficient Technology (SET) focus area has funded the Spectrum Efficiency Through Metrics (SETM) project. As part of this project, a literature review of spectrum metrics was conducted on over 50 related documents.

A reasonable place to start this review is to look at existing standards for spectrum metrics. Within T&E this is mostly RCC 707 [1]. (See also [2].) This standard provides spectrum metrics for utilization, occupancy, reuse, and efficiency. In the broader telecommunications world, the

International Telecommunications Union (ITU) provides a series of recommendations some of which address spectrum metrics [3][4][5][6][7]. In particular, they define occupancy and spectrum utilization efficiency (SUE). No document explicitly defining spectrum metrics for the commercial world was found. However, there are some commercial spectrum metrics available on the internet. These are quite different from metrics in RCC 707 but give insight into commercial priorities. One of the values of the ITU recommendations is that they provide high level definitions that allow derivation of both T&E and commercial metrics. That is, they give us a foundation to analyze the relation between the two uses.

Simply from an initial look at the standards, we see that the concept of spectrum "use" is not well defined or, perhaps more accurately, it is multiply defined. This paper attempts to navigate these different definitions to provide insight into the differences in trying to measure spectrum use among the distinctly different applications of T&E and cell phones.

OCCUPANCY, USE, AND DENIAL TO OTHERS

Spectrum occupancy and use occur in a combination of frequency, space, and time (Hz x m^3 x s). As ITU-R SM.1046-2 [4] points out, occupancy is often thought of in terms of an active transmitter. However, a transmitter in no way prevents another transmitter from transmitting. Receivers, on the other hand, must have an interference free area of reception. So that receivers use the spectrum in the sense that transmitters that are not the target of reception can be denied the right to transmit. Thus, the ITU documents are explicit in defining spectrum use as including the denial of spectrum to others.

The RCC 707 spectrum metrics were developed by Dr. Jones primarily based on available data, which means data archived by the Integrated Frequency Deconfliction System (IFDS). This system records when frequency was *scheduled.* (Spectrum is scheduled in rectangles within a time-frequency grid based on min and max frequency and beginning and ending times.) Consistent with ITU definitions, RCC 707 defines "use" as "denial to others". One of the difficulties in presenting T&E spectrum usage has been the question of whether "scheduled" implies "used". *ITU-R SM.1046-2* [4] points out that "authorities deny licenses to transmitters in an attempt to guarantee interference-free reception." This is the sense in which "scheduled" means "used". T&E spectrum managers legally deny use to others by assigning frequencies to specific users for a scheduled time and place. This, of course, goes both ways. T&E cannot legally use commercially licensed spectrum (although there are exceptions both ways.)

There are currently efforts under way to be able to monitor spectrum use in the sense of transmitted RF power. These efforts will certainly provide useful data and the ITU provides several recommendations on how to monitor spectrum [5][7] including what data to collect [6]. But, a key point here is that "use" does not necessarily imply "occupied" in the sense of a transmission of RF power. On the other hand, the existence of RF power does not imply efficiency – transmitting static may occupy the spectrum, but it usually without value to the transmission. So, perhaps, a better way to look at spectrum use is in terms of efficiency rather than occupancy.

SPECTRUM EFFICIENCY

Efficiency comes in (at least) three forms: technical, economic, and functional [8][9]. In the current context, technical efficiency means minimizing the amount of spectrum used. Economic efficiency regards costs such as equipment or return on investment. Functional efficiency relates to the value of the service. For example, a 911 call is usually of more value than calling home to say "hi". It is perhaps useful to relate these different types of efficiency to the classic project management triangle of cost, schedule, and performance. The main difference being to replace schedule with spectrum. As with the project management triangle, it is only possible to optimize two, and not three, of the sides of the triangle.

The ITU defines Spectrum Utilization Efficiency (SUE) [4] which is generic enough to cover all three forms of efficiency.

$$SUE = M/U$$

where:

M: the useful effect obtained with the system in question; and
U: the spectrum utilization factor for that system.

Breaking down U we have

$$U = B \times S \times T$$

Where:

B: frequency bandwidth
S: geometric space, and
T: denied time

Giving units of (Hz $\times$ m^3 $\times$ sec). That is, the five dimensional segment of time, space and frequencies under consideration.

The definition of useful effect, *M*, is a bit vaguer. It can represent anything that the user considers useful. (In fact, this vagueness makes virtually every metric looked at to be derivable from this ITU definition – including those in RCC 707.) We'll consider economic and functional useful effects shortly, but let's start by looking at useful effect in terms of spectrum. At its simplest, we might define technical spectrum efficiency in terms of bits sent. (And this is one of the metrics in RCC 707. However, one of the weaknesses of RCC 707 is that it does not explicitly include geographic area in any of the metrics.) But, it isn't quite that simple when we look at different applications.

<u>SUE for T&E Point-to-Point Telemetry</u>

The *Spectrum Efficiency Metrics White Paper* [10] provides the following calculation for point-to-point telemetry**Error! Reference source not found.**. The key difference is the introduction of transmitted distance.

$$SUE_{p-p} = \frac{\text{info rate (bps/Hz) x Transmitted Distance}}{\text{Occupied Area}}$$

The unit of measure is thus: Bits/sec/Hz*km/km^2.

SUE For Cell Phones

This example is to provide a contrast to the point-to-point example. The example is from ITU-R SM.1046 [4] for a pico-cellular system covering a building. A primary difference is the use of number of channels. But it also emphasizes that "area" to cell phones can be very local, such as a building, vs. the large land areas flight test uses.

In order to calculate the total bandwidth required for the whole building, the vertical re-use distance in terms of the number of floors is required. This parameter is dependent on the floor losses and is different for different types of buildings. The total number of half duplex channels required for the building can then be calculated and is equal to:

$$2 \times \text{No. of channels per cell} \times \text{No. of cells per floor} \times \text{No. of floors of separation}$$

The factor 2 is needed here to reflect the number of channels needed for two-way communications. The spectrum efficiency, $SUE_{building}$, of the system providing coverage in the building can then be calculated:

$$\text{SUE}_{\text{building}} = \frac{\text{Total traffic carried in the entire building}}{\text{Total No. of channels} \times \text{channel bandwidth} \times \text{total floor area}}$$

After working through the units for each of these, the unit of measure is: Bits/sec/Hz/km^2.

COMMERCIAL METRICS

Formal definitions of metrics for the commercial communications industry don't seem to be discussed much outside of references to formal organizations such as the ITU, FCC, or IEEE. The market driven metrics they use are relatively intuitive and have a distinctly different emphasis then the ITU's approach. In particular, there was no data found on specific frequency use. Most likely, the industry has such data but companies probably keep them proprietary.

The industry standard for both identifying metrics and finding data appears to be the *Cisco Visual Networking Index: Global Mobile Data Traffic Forecast Update* [11]. Partly because it is a forecast, there is a lot of trend data and it quickly becomes clear that what they focus on is how much data is sent and how many users there are. This is broken down into types of data, devices used, number of devices, number of connections, and the like. The following (partial) list of charts in the forecast is provided to illustrate this difference in market perspective from the more technical emphasis of the ITU and RCC metrics. The scales of the axis are of some interest: billions of devices and exabytes per month. Several other references were found during the literature search that provide more data, but the data is similar in type to the Cisco forecast.

Example Chart Types for Cisco Forecast

Chart Title	Axis
Global Mobile Data Traffic	Exabytes per month x year
Global Mobile Devices and Connections Growth	Billions of devices by year and device
Global Mobile Traffic Growth by Device Type	Exabytes per month x year and device
Global Growth of Smart Mobile Devices and Connections	Billions of devices by year and device
Regional Share of Smart Devices and Connections	Region and Percent of the Regional Total
Effect of Smart Mobile Devices and Connections Growth on Traffic	Exabytes per month x year and smart vs. nonsmart device
Global IPv6-Capable Smartphones and Tablets	Number of devices by year
Global Mobile Devices and Connections by 2G, 3G, and 4G	Billions of devices or connections x year and connection type
Global Mobile Traffic by Connection Type	Exabytes per month x year and connection type
Global Machine-to-Machine Growth and Migration from 2G to 3G and 4G	Billions of M2M connections x year and connection type
Global Connected Wearable Devices	Millions of connected wearables x year
Mobile Voice Minutes of Use—VoWiFi, VoLTE, and VoIP	Minutes of use (billions) x year
Growth of types of data (file sharing, video ...)	Exabytes per month x year and type of data
Busy-Hour vs. Average-Hour	Tbps x year

SCHEDULED VS. ON-DEMAND VS. SHARING

Another way of comparing applications is via the underlying methods used by the different industries. One aspect of this is modulation method (e.g., PCM or CDMA) which can affect performance. But modulation techniques have changed for both T&E and cell phones. A more important aspect is how the service is provided. In particular, T&E spectrum is scheduled whereas cell phones are ad hoc and on-demand. This is, perhaps, the fundamental reason why the two forms of spectrum usage are difficult to compare. Scheduling is very specific about when spectrum is used and when it is not. In some sense, spectrum is always being used for cell phones. However, the fact that you can make a phone call at any moment means there was some portion of the spectrum not being used when you decided to make the call. Trying to analyze this form of non-usage is extremely difficult and can be dependent on the modulation method.

A current focus is the idea of sharing spectrum. Arguably the different methods of scheduling and on-demand usage are inherently incompatible from a sharing perspective. Fundamentally, when T&E schedules spectrum, any on-demand capability would have to be stopped for the duration. This would effectively force commercial companies to implement a form of scheduling.

So how can spectrum be shared? Potential solutions include networks, spread spectrum techniques, and cognitive radio techniques. Spread spectrum techniques (such as CDMA) require every user to coordinate keys. This is being phased out of use by cell phones because LTE is more efficient in terms of throughput. Cognitive radio (which monitors the spectrum and jumps in when a frequency is not being used) can use spectrum that might otherwise not be used but it does not guarantee a user gets as much spectrum as they need. So, networking it is.

Ethernet is a widely-used networking protocol which, effectively, allows on-demand usage. The iNET standards implement a form of Ethernet. This has the potential to move T&E from a scheduled approach to an on-demand approach. (This also potentially addresses some of the objections to metrics based on scheduling. In particular, it reduces spectrum fragmentation as described in RCC 707.) But this doesn't allow sharing with other users not part of the network. From the point of view of someone outside the network, this still looks like scheduled spectrum, albeit in larger chunks.

Cell phone methods (such as LTE and some earlier methods) are essentially complex network schemes. Every user gets (many small) dedicated slices of spectrum (in contrast to spread spectrum) but via complex monitoring and coordination protocols. There are currently some S&T funded efforts looking at using cell phone technologies for use in T&E. It is at least theoretically possible, but there are questions of cost and, if T&E and cell phones truly share the same infrastructure, security issues.

A final note is that a couple of documents found in the literature search address metrics for spectrum sharing. In particular, "Spectrum Sharing Metrics" [12] attempts to address this stating: "Useful metrics – e.g. true spectrum utilization efficiency – combine throughput and geographical area under the assumption that other systems are present in the area and use the same RF channel." The mathematics provided by this document are not simple.

COMPARISON

It is difficult to compare T&E and commercial spectrum usage. Indeed, in the case of SUE, it is logical impossible to do so because SUE derives different units for different applications as shown in the table. In particular, notice that (T&E) point-to-point and (cell phone) personal communications have different units. It is also noteworthy that SUE allows for introduction of users as part of the unit of measure.

SUE Units by Application

Service Classification	Unit
Satellite Broadcast System	Bits/sec/Hz
Satellite Point-to-Point	Bits/sec/hz*km^2
Terrestrial Broadcast	Bits/sec/Hz*users
Terrestrial Personal Communications	**Bits/sec/Hz/km^2**
Terrestrial Point-to-Point	**Bits/sec/Hz*km/km^2**
Terrestrial Hybrid	Bits/sec/Hz*km/km^2***users**

Besides SUE, there are some metrics that can be compared. However, these tend to show that T&E usage, from a technical efficiency point of view, are orders of magnitudes different from commercial use. For example, let's consider number of users and bits sent. For both metrics, let's calculate some estimated maximums for T&E.

Let's assume all military members and associated civilians are "users" of T&E spectrum. Using Wikipedia to provide approximate numbers, it lists about 1.5 million military members (all services) about 0.8 million reserves and about 0.7 million civilians. For completeness, throw in maybe 0.7 million contractors So, at most, we can talk about 3 to 4 million military related users. Compare this to the number of cell phone users: roughly half the world's population of 7 billion or ¾ of the 320 million people in the US.

On the other hand, because of its relation to national security, it is possible to argue that all U.S. citizens are users of T&E Spectrum. Or, even further, that all people in countries that are U.S. allies are users.

For bits sent let's calculate a rough upper bound based on maximum throughput in all frequency bands at every test and training range 24/7.

1. MHz available: 90+10+90+50+35+540+60+775 = 1650 MHz for 8 bands listed in IRIG 106. About 1.7 x 10^3
2. Transmission rate (upper end): 3/8 Bytes/Hz or 375,000 bytes per MHz. About 3.8 x 10^5
3. Seconds per month 60*60*24*30 = 2,592,000 secs. About 2.6 x 10^6
4. Say 100 test and training ranges (about 60 in IFDS). About 1 X 10^2

This calculates out to about 0.17 exabytes maximum transmission per month. (Realistically, T&E uses a tiny fraction of this.) Consider that some of the Cisco forecast charts have axes of exabytes per month. In other words, even based on outlandish assumptions, T&E maximum usage wouldn't be noticeable on Cisco's charts.

Let us return now to the three types of efficiency. SUE is, essentially, a measure of technical efficiency and, as has been shown, SUE does not allow for direct comparison. In terms of economic efficiency, it would be possible to calculate and compare total costs of spectrum equipment for the two industries and somehow normalize to users or bits sent. But this would take some effort to collect data and it's not clear it would provide a useful result. A simpler thought experiment is simply to note that the commercial world revolves around return on investment. In contrast, the T&E world spends tax money with no direct monetary return.

This leaves functional efficiency as the main useful point of comparison. So, we have national security vs. cell phone use for emergency response, business communications, and calling friends and family. Trying to compare functional value between these is extremely difficult and well beyond the scope of this paper.

Summary Comparison

Metric	T&E	Commercial
Spectrum Occupancy	**Scheduled Time Slots**	**On-Demand 24/7**
Spectrum Efficiency	**Not Comparable – Different Units**	
	Point to Point	**Personal Comm**
Economic Efficiency	**No monetary payback**	**Return on Investment**
Functional Efficiency	**National Security**	**Billions of Users**

SPECTRUM DEFENSE VS. SPECTRUM MANAGEMENT

Just as there are different applications for spectrum, there are different applications for spectrum metrics. And, different applications mean different approaches. Spectrum defense requires providing data to a different set of people than for spectrum management. When defending spectrum, the data is being presented to high level management (e.g., Congress) who are not always technically oriented. Data used for spectrum management is presented to people "in the trenches" and who, generally, have a technical understanding of spectrum issues.

Spectrum defense requires reduced data that might be thought of as a "one number" or a "one chart" approach to presentation. This is as much due to a lack of time of managers to research details as for any other reason. Spectrum managers need to look at details such as are provided by some of the 3-D charts produced by RCC 707 metrics. In fact, by RCC 707 emphasizing these 3-D charts, the data presented emphasizes how some portions of the spectrum are used more than others. This, in some sense, dilutes the usage message or, at the very least, distracts from the overall usage when presented to higher management. Part of the focus of the SETM project is to provide better presentation of data based on these two distinct uses.

CONCLUSIONS

It is very difficult to compare T&E and commercial spectrum usage. This is partly due to the fundamental difference of scheduled vs. on-demand services. But it is also due to significantly different applications.

Reporting spectrum occupancy, in the sense of transmitted RF energy, does not provide a complete picture of usage. And, in fact, understanding spectrum occupancy requires some technical background which makes it difficult to present to nontechnical people.

A better approach is to consider efficiency. But even then, as exemplified by SUE, efficiency is not necessarily comparable across applications and, again, requires some technical understanding. Economic efficiency is problematic in that T&E uses tax dollars and industry focuses on return on investment. Functional efficiency again raises difficulties. First, at its simplest level (number of users or bits sent) T&E is orders of magnitude less than cell phone use. This leaves national security as the primary argument for T&E spectrum use. But putting a value on this is extremely difficult.

REFERENCES

[1] RCC 707-14 Spectrum Management Metrics Standards, Secretariat, Range Commanders Council, White Sands Missile Range, Mew Mexico. http://www.wsmr.army.mil/RCCsite/Documents/707-14_Spectrum_Management_Metrics_Standard/707-14_Spectrum_Management_Metrics_Standard.pdf

[2] Charles H. Jones, "Average Typical Mission Availability: A Frequency Management Metric", Proc. International Telemetering Conf., Vol. XXXX, (2004) Paper 04-22-03

[3] Handbook on National Spectrum Management, 2005, International Telecommunications Union, https://www.itu.int/dms_pub/itu-r/opb/hdb/R-HDB-21-2005-R1-PDF-E.pdf

[4] Recommendation ITU-R SM.1046-2 Definition of Spectrum Use and efficiency of a radio system http://www.itu.int/dms_pubrec/itu-r/rec/sm/R-REC-SM.1046-2-200605-I!!PDF-E.pdf

[5] Recommendation ITU-R SM.1793 Measuring frequency channel occupancy using the technique used for frequency band measurement (2007) https://www.itu.int/dms_pubrec/itu-r/rec/sm/R-REC-SM.1793-0-200702-W!!PDF-E.pdf

[6] Recommendation ITU-R SM.1809 Standard data exchange format for frequency band registrations and measurements at monitoring stations https://www.itu.int/dms_pubrec/itu-r/rec/sm/R-REC-SM.1809-0-200704-I!!PDF-E.pdf

[7] Recommendation ITU-R SM.1880 (02/2011) Spectrum occupancy measurement https://www.itu.int/dms_pubrec/itu-r/rec/sm/R-REC-SM.1880-1-201508-I!!PDF-E.pdf

[8] Federal Communications Commission's Spectrum Policy Task Force. Report of the Spectrum Efficiency Working Group https://transition.fcc.gov/sptf/files/SEWGFinalReport_1.pdf

[9] J.W Burns, "Measuring Spectrum Efficiency – The Art of Spectrum Utilisation Metrics", Aegis Systems Ltd, http://www.aegis-systems.co.uk/download/article/ieespectrum1.pdf

[10] "Spectrum Efficiency Metrics", White Paper, Sharing Work Group, FCC Technological Advisory Council https://transition.fcc.gov/oet/tac/tacdocs/meeting92711/Spectrum_Efficiency_Metrics_White_Paper_by_TAC_Sharing_Working_Group_25Sep2011.doc

[11] Cisco Visual Networking Index: Global Mobile Data Traffic Forecast Update, 2015–2020 White Paper http://www.cisco.com/c/en/us/solutions/collateral/service-provider/visual-networking-index-vni/mobile-white-paper-c11-520862.html

[12] Spectrum Sharing Metrics (Aug 2014) http://www.spectrumconsult.net/wp-content/uploads/2014/07/Spectrum%20Sharing%20Metrics-postprint-jpkr.pdf

Acknowledgment of Support: This project is funded by the Test Resource Management Center (TRMC) Test and Evaluation/Science & Technology (T&E/S&T) Program through the U.S. Army Program Executive Office for Simulation, Training, and Instrumentation (PEO STRI) under Contract No. W900KK-16-C-0018. The Executing Agent and Program Manager work out of the AFTC.

Public Affairs Clearance: 412TW-PA-17362: Comparing Commercial and T&E Spectrum Metrics

EFFECTS OF AERONAUTICAL CHANNEL DYNAMICS ON THE PERFORMANCE OF LINK DEPENDENT ADAPTIVE RADIO (LDAR)

Author: Tasmeer Alam

Master of Science in Electrical Engineering, Department of Electrical and Computer Engineering, Morgan State University

Faculty Advisor: Dr. Farzad Moazzami, Dr. Richard Dean

ABSTRACT

This paper includes aeronautical channel dynamics in the flight simulations of the Link Dependent Adaptive Radio (LDAR) for realistic measurement of the throughput gain with the adaption of the modulation and coding parameters for aeronautical telemetry. Previously, the LDAR flight simulator used a fixed channel for different phases of the flight. This paper shows the performance of LDAR using a dynamic channel created by the customized two ray channel model considering delay spread, Doppler and other channel dynamics. The merger of such dynamic channel simulator with the LDAR flight simulator could help the telemetering community by providing reliable simulation results before test flights.

INTRODUCTION

Wireless technology is developing continuously for better communication. Improved quality of communication has always been a priority in wireless telemetry. Generally, wireless telemetry uses radio channels for transmitting test article data. But radio channels are unpredictable and dynamic and poor channel conditions affect performance in several ways such as noise, multipath, and Doppler effects. Channel dynamics, especially frequency selective fading of the channel becomes a bigger issue in highly dynamic environments such as aeronautical communication where the speed of the aircraft exceed mach1 [1-3].

Dedicated telemetry links between the ground station and test article has proven to be an inefficient bandwidth utilization method. The integrated network enhanced telemetry (iNET) project – supported by Test Resource Management Centre (TRMC) – has aimed at abandoning the point to point link dedication and moving toward networked telemetry. Georgia Tech Research Institute (GTRI) and Morgan State University have undertaken a project within iNET aimed at a bandwidth efficient and adaptive system for aeronautical communication. Link Dependent Adaptive Radio (LDAR) is an effort to maximize the throughput of telemetry links while ensuring an acceptable level of data quality and reliability [4-6].

LDAR

The next frontier of telemetry systems is creating a self-modulating coding rate based radio which adapts to the channel condition "on the fly". And Link Dependent Adaptive Radio (LDAR) does exactly that by choosing the best modulation scheme while analyzing the current channel condition based on the telemetry environment. This produces maximum throughput with mixed levels of link quality. Depending on the SNR or delay spread, if the communication channel quality is high between the Ground Station and Test Article then LDAR adapts to a higher data rate [4]. Conversely, having a low-quality channel (or when the error level crosses a predefined performance threshold) the adaptive system provides a feedback system which lowers the data rate during the successive communication phases. This adaptation is achieved by altering the modulation scheme or truncating the coding rate of the transmission [3-6].

The basic Link Dependent Adaptive Radio (LDAR) system model consists of a transmitter, a receiver, and a telemetry channel model. The transmitter has an encoder and a modulation which consists both Orthogonal Frequency Division Multiplexing (OFDM) and Shaped-Offset Quadrature Phase Shift Keying (SOQPSK). The receiver consists a demodulator and decoder [3-4]. The adaptation of modulation scheme and code rate takes place after the Signal to Noise Ratio (SNR) or delay spread is computed. Based on the predetermined table, LDAR decides on the next set of parameters for transmission.

To make optimum decision for parameter selection or for mode adaption depending on channel quality, different code rate and modulation has been tested previously. Figure 1 shows a comprehensive set of results with two different coding schemes for various Quadrature Amplitude Modulation (QAM) sizes with Additive White Gaussian Noise (AWGN) channel.

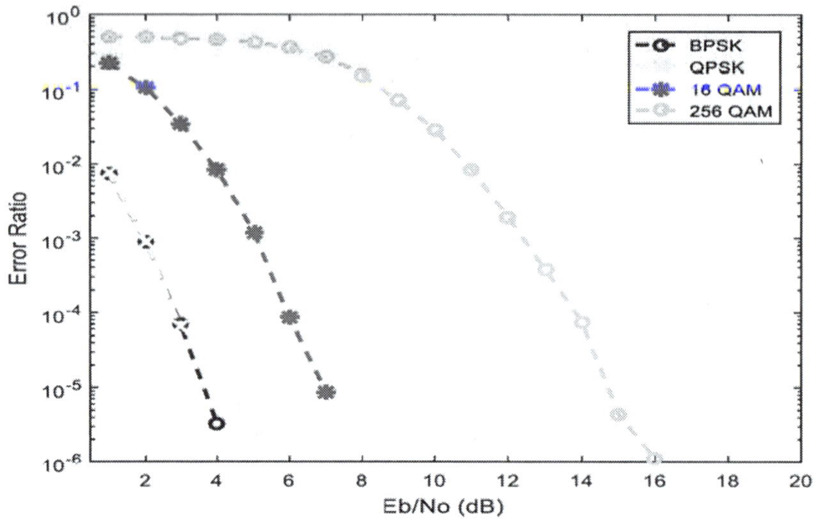

Figure 1: Error performance of variable QAM with half code rate

These curves result in the creation of LDAR tables for noisy channel. After the Error threshold is satisfied, we chose the highest possible throughput above the threshold line. A theoretical approach for Mode adaption and Transmission Table has been developed from this error performance graph of different modulation scheme.

FLIGHT SIMULATION OF LDAR

In our previous work, a flight path simulation has been developed where flight scenarios have been demonstrated to simulate the flight path for the Test Article (TA) and selects modulation schemes based on channel condition. The flight path demo had three different phases: Taxi, take off and Cruise as shown in Figure 2 below. The flight path simulator used a constant channel for different phases of the flight.

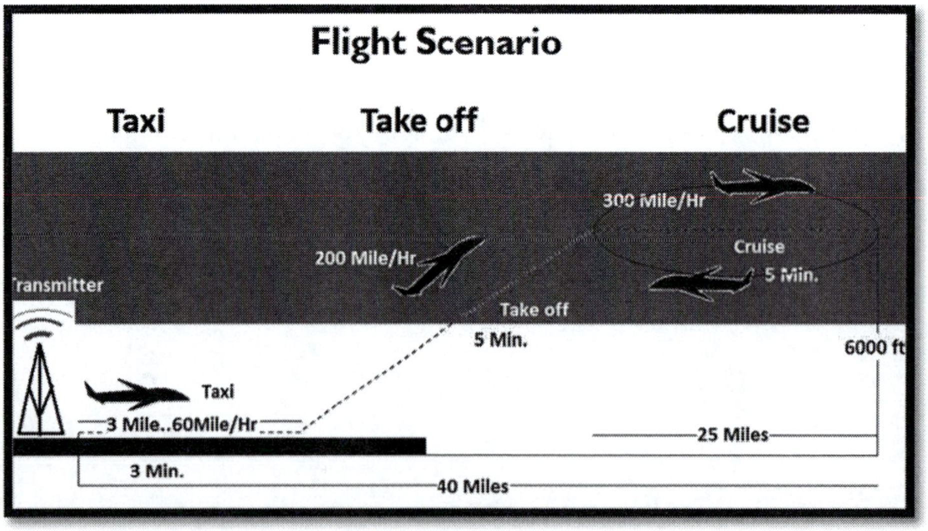

Figure 2: Flight path scenario of LDAR

The static channel impulse response and frequency response during Taxi, Take-off and Cruise has shown in figure below:

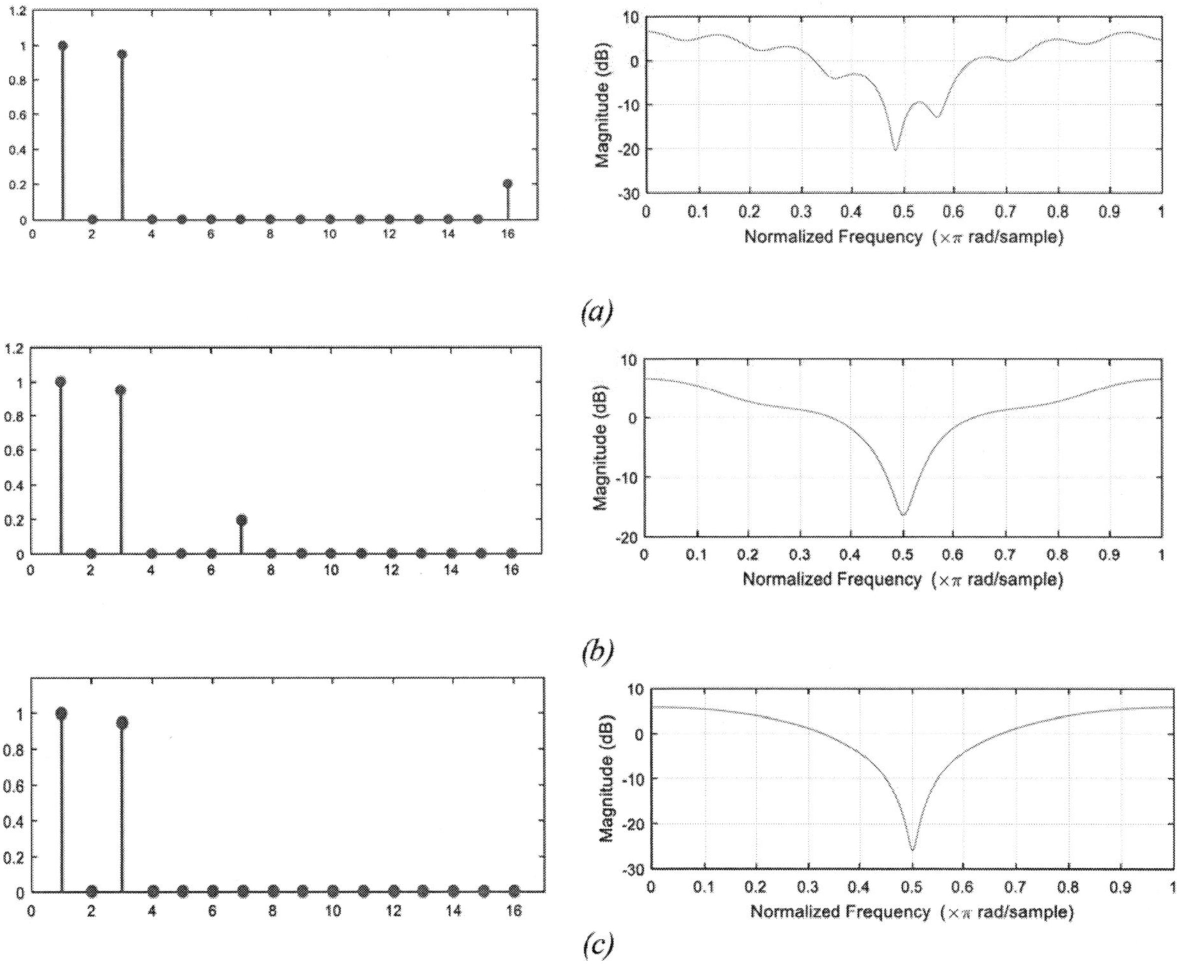

Figure 3: Impulse Response and Frequency Response of (a) Taxi Channel, (b) Take off Channel and (c) Cruise Channel

In taxi channel, there are three significant impulses where the first and second impulses denote the direct path and ground bounce respectively. One at the sixteenth tap represents the clutter perhaps from a nearby building in the test range. In the take-off channel, the first two impulses shown in the figure-3(a) are direct line of sight and reflected path respectively along with another reflecting object. And in the cruise channel, only line of sight and ground bounce is observed.

Using the Log distance path loss model, the Signal to Noise Ratio values for the entire flight path is computed. The duration of this flight simulation is 1560 seconds and sampled for each second. SNR values have been generated and stored prior to the test. Figure 4 below shows the SNR profile of the flight path for all phases including outbound and inbound.

Taxi phase lasts for 180 seconds and is 3 miles long. The path loss exponent used for taxi phase is n=4.

The exponent is assumed 4 because test article is close to the ground station and usually ground station is surrounded by other buildings and other test articles. First 180 seconds on figure 4 shows SNR values for taxi phase. The Taxi phase has a high SNR at the beginning and as the test article reaches to the end of the runway, SNR drops sharply. Takeoff phase lasts for 300 seconds and spans 12 miles. Note the rapid increase in SNR as the path loss drops with elevation. The path loss exponent considered for this phase is 2.5. This results in a smoother SNR compared to the taxi phase.

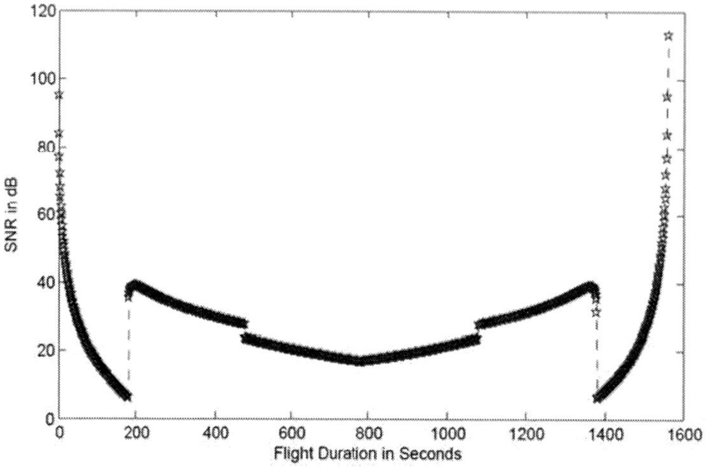

Figure 4: SNR profile of the channel for the flight path

During takeoff, the distance is increasing and SNR drops again. In the cruise phase n=2 results in smooth drop and at around 800 second, the test article starts to return to the base via the same flight path. So, the SNR profile is symmetric to the outbound path in reverse [6].

The flight is simulated with several modulation schemes such as BPSK, QPSK, 16 QAM, and 256 QAM. The Symbol Error Rate (SER) for all the modulation sizes in the flight path are shown in figure 5 below.

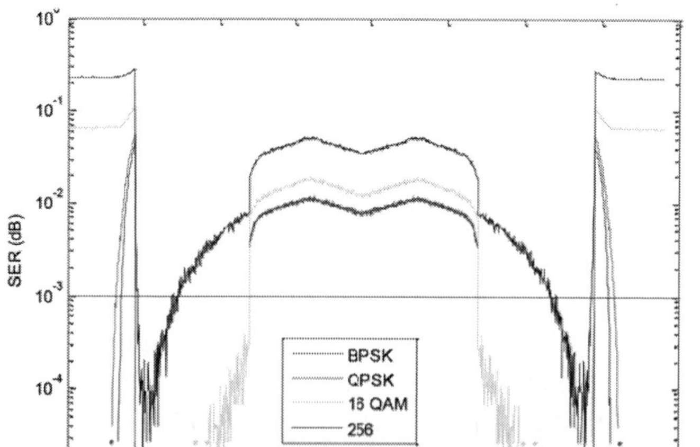

Figure 5: SER for the entire flight path with different modulation sizes

The modulation is dynamically selected by the LDAR controller given the tolerable error threshold and the previously ranked table for modes of operation. The final product is a live simulation of decision making in LDAR flight simulator. Figure below shows several snapshots of the simulator.

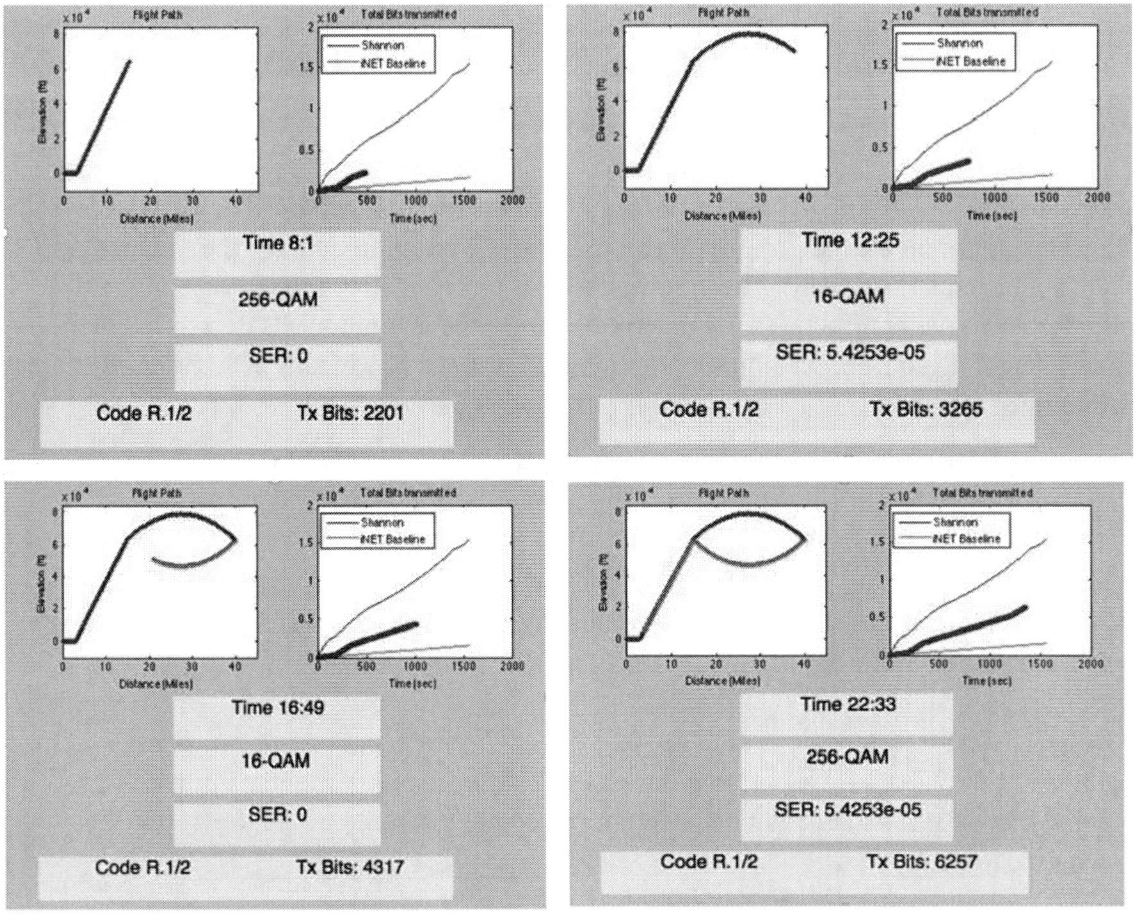

Figure 6: Snapshots of flight path simulator

The Total transmitted bits are increased by a factor of 5.5 over the iNET baseline. This is a promising result, however the channels within this simulator are static and averaged over the entire path which excludes the Doppler spread and delay spread changes during the flight [6].

The channels are selected from the results of Dr. Michael Rice's channel sounding test flights at EAFB. However, in the simulations only a snapshot of the channel is used for taxi, take off, and cruise phases without considering the delay spread and other dynamics of the aeronautical channel. We find this to be a common practice in flight simulations which could potentially result in considerable error levels in computer simulation of the test flights compared to the real flights.

The final goal of this prototype system (Link Dependent Adaptive Radio) is to ensure maximum throughput in a telemetry communication by adapting modulation scheme and error correction depending on the channel condition in real time. In a real time scenario channel is not static. Real time channel is dynamic. To increase the accuracy of this simulator, channel dynamics must be considered in this simulation. Using the two-ray ground reflection model, Doppler spread, and delay spread characteristics of such flight scenario, dynamic channels could be generated.

DYNAMIC CHANNEL SIMULATION

Dynamic channel simulation has been developed to generate dynamic channels and then to incorporate this in our LDAR flight simulation to ensure accuracy. The channel simulation process starts by identifying the number of taps from the impulse response of Dr. Rice's channel [7, 8]. Then, approximate the position of the taps and their amplitude from the impulse response graph (Tap_tau and Tap_mean). The amplitudes then normalized to the tap with highest amplitude. The other amplitudes were determined from the amount of dB decrease which normalized subtracted from the highest amplitude. Assume a bandwidth of 15 MHz thus the sampling rate is,

$$SR = \frac{1}{2 \times 15 * 10^6} = 33\ nsec$$

The next step is then to divide tap position values by sampling rate.

Effects added to the channel are: 1. The Doppler and frequency offset and 2. Delta radian per sample. Both has been described below.

The Doppler and frequency offset (Tap_fd): Given that each path or taps possesses its specific Doppler, each Doppler were added to the frequency offset between the transmitter and the receiver, which was calculated with an oscillation recovery located between ±15000 Hz, following the logic:

$$Frequency\ Offset = \cos(2\pi f_0 t)$$

$$Frequency\ Offset\ (FO) = 15000 \times 2 \times random\ input$$

$$FO\&Doppler = FO + doppler$$

The Doppler values varied between the following values: 1000, 100, 2.

Delta radian per sample (Tap_dTH): The calculation of the angle increment per sample was performed through the following formula:

$$Tap_{dTH} = \frac{2\pi\ (Tap\_fd)}{SR}$$

After setting the channels, a convolution of the input per sample was performed using a random created input of 100000 samples.

For example, a two-ray airborne channel from Dr. Rice's experiment. Snapshots of the frequency response of Dr. Rice's Channel are shown below [9]. The simulated dynamic channel responses for the same channel are shown below.

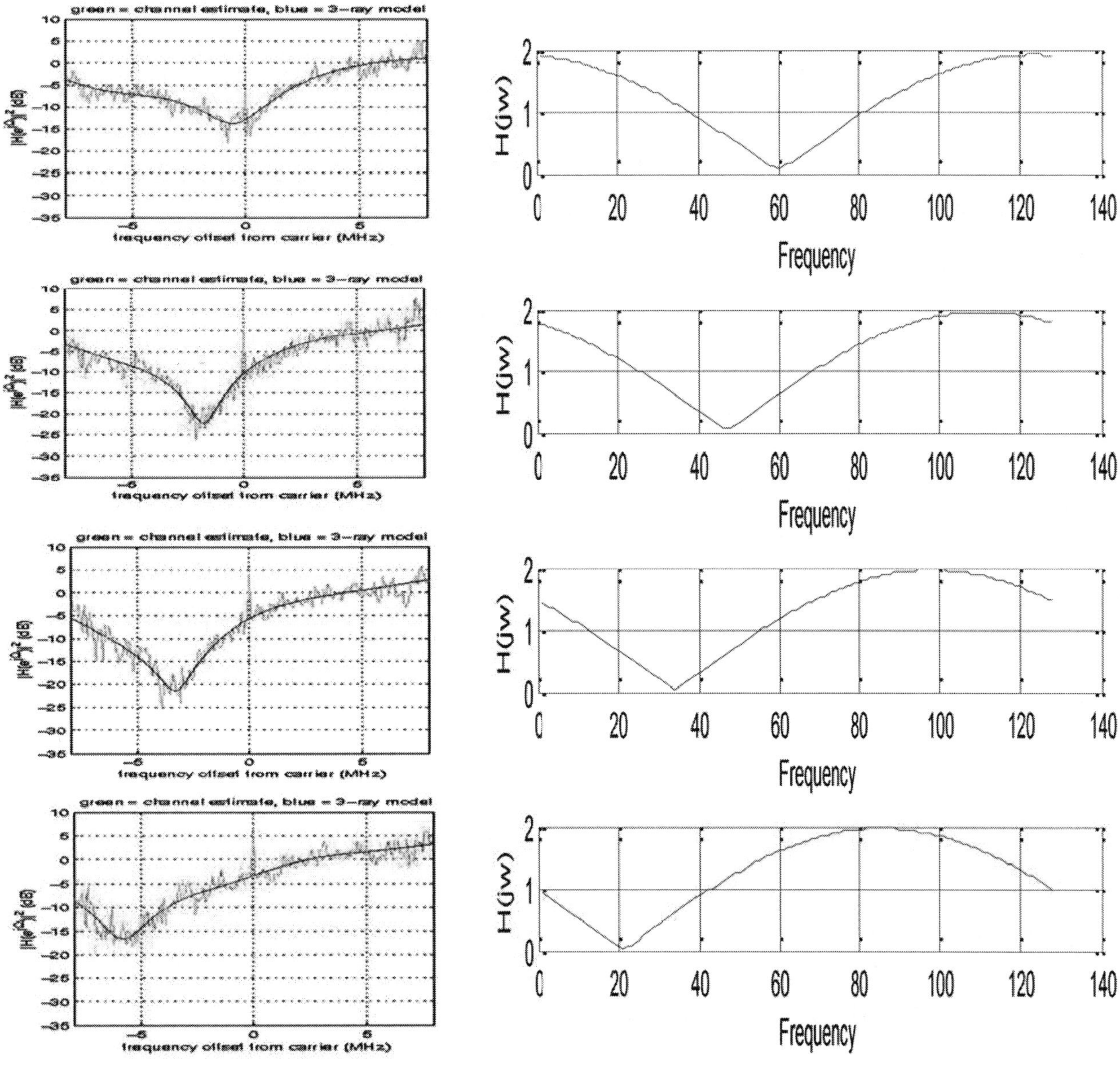

(a) *(b)*

Figure 7: (a) Snapshots of the frequency response of Dr. Rice's experiment in 2 ray airborne case; (b) Frequency response of dynamic channel simulator in 2 ray airborne case

The moving null in the frequency response of the channels are observed in figure7. The similarity of the actual and simulated dynamic channel responses are result of the inclusion on the simulated channel dynamics at the presense of two strong rayleigh paths. In addition, The four graphs illustrate that the doppler effect over time through the manifestation of the Nulls which are happening at different frequencies as the Dr. Rice's channel would behave.

NEXT STEP

Next step in this work is to incorporate the dynamic channel simulator into LDAR flight simulation and compare the throughput gain that is computed with the static and dynamic channels in the simulator. Now that the behavior of the channel with flight dynamics are simulated, the inclusion of this channel simulator in the LDAR simulator will provide more realistic set of results in a dynamic aeronautical environment. These dynamics could be critical in any modeling of aeronautical telemetry and provide support for testing new communications payloads in simulation prior to test flight. The impact of creating accurate simulation results with this dynaic channel simulator reaches beyond LDAR and will help the telemetry community to improve the accuracy of computer simulation in the design and pre-test stages.

ACKNOWLEDGEMENT

This work is funded by a grant from International Foundation for Telemetering (IFT). Author and other WiNetS laboratory members in the School of Engineering at Morgan State University thank IFT for their support of this research.

REFERENCES

[1] T. Rappaport, Wireless Communications, Principles and Practices, 2nd Edition, Prentice Hall PTR, Upper Saddle River, New Jersey 07458

[2] P. Jacob, Rajendra P. Sirigina, A.S. Madhukumar, Vinod A. A. Prasad, "Cognitive Radio for Aeronautical communications: A survey", IEEE 2016.

[3] Taser Alamo, Farzad Moazzami, Richard A. Dean, "Command Signaling for Link Dependent Adaptive Radio" , ASEE Mid-Atlantic Conference, Spring 2017, Baltimore, MD.

[4] Enkuang D. Wang, Brett T. Walkenhorst, Jieying Han, "A simulation Testbed for Adaptive Modulation and Coding in Airborne Telemetry", ITC 2014, San Diego, CA.

[5] Tara Pun, Deepak Giri, "Link Dependent Adaptive Radio Simulation", ITC 2014, San Diego, CA.

[6] Deepak Giri, "Flight Path Simulation in LabVIEW using LDAR", ITC 2015, Las Vegas, NV.

[7] Michael Rice and Adam Davis, "A Wideband Channel Model for Aeronautical Telemetry - Part 1: Geometric Considerations and Experimental Configuration," in Proceedings of the International Telemetering Conference, San Diego, CA, October 2002.

[8] Michael Rice and Adam Davis, "A Wideband Channel Model for Aeronautical Telemetry - Part 2: Modeling Results," in Proceedings of the International Telemetering Conference, San Diego, CA, October 2002.

[9] BYU Telemetry Laboratory: Wideband Channel Modelling, Animation: Flight 18, Time stamp: 15:18:18, captured on 07/20/2017.

WRITING BETWEEN THE LINES – THE STORY OF SPECTRUM AGGREGATION

Mr. Mark Wigent
Laulima Systems
Lihue, HI 96766
mark@laulimasystems.com

ABSTRACT

In order to meet the increasing demand for spectrum to support future DoD test and evaluation (T&E) requirements in an environment of decreasing spectrum availability, the DoD requires new spectrum aggregation and spectrum management technologies. Advanced radio technologies that sense and aggregate non-contiguous blocks of spectrum into a larger communication channel capable of supporting higher user data rates are needed. Moreover, the T&E community must more effectively manage and use the spectrum that is available to it, and utilize new paradigms in which spectrum is dynamically allocated on a non-interference basis to multiple, concurrent users and in a way that meets both planned and unplanned changes to mission requirements, and real time channel conditions, factoring in both Federal and non-Federal users of spectrum.

The objective of the Adaptive Spectrum Aggregation and Management (ASAM) project is to develop technologies that will sense available spectrum, aggregate white space within fragmented frequency bands to create communication channels, and dynamically allocate those channels across multiple range users using algorithms that increase spectrum utilization and efficiency while applying policy constraints. The ASAM project is sponsored by the Test Resource Management Center (TRMC) and was awarded in 2016 through the National Spectrum Consortium.

This paper will describe the ASAM project and technical challenges associated with its development.

INTRODUCTION

In 2017, it has become common knowledge among members of the test and evaluation (T&E) community that two trends are occurring. First, there is an increasing demand for spectrum both inside and outside the DoD. Second, usable electromagnetic spectrum is finite, and as a result these two distinct user groups are competing for the same limited resource. The voice of the telecommunications industry grows louder, and the needs of the T&E community, while growing in scope, are seemingly becoming less important. Spectrum relocation is a frequently discussed topic in our conferences, and the DoD is making significant investment in spectrum relocation as well as in developing new technologies which enable the DoD to operate using not only new frequency bands, but new paradigms, including those of greater levels of spectrum re-use and spectrum sharing. This paper introduces one of these new paradigms, one which is already ubiquitous in the commercial telecommunications sector and in our daily personal lives, whether we are aware of it or not – spectrum aggregation. The first part of this paper describes a new R&D effort focused on implementing spectrum aggregation technology as well as dynamic spectrum access in a way that supports telemetry. The second part of this paper presents an especially compelling use case for spectrum aggregation and illustrates how accessing white spaces - small unused portions – in spectrum gives us access to more of this precious resource and ultimately allows us to communicate more information. It is much like reading, and writing, between the lines on a page. This paper tells that story.

THE BODY

In order to meet the increasing demand for spectrum to support future DoD T&E requirements in an environment of decreasing spectrum availability, the DoD requires new spectrum aggregation and spectrum management technologies. Advanced radio technologies that sense and aggregate non-contiguous blocks of spectrum into a larger communication channel capable of supporting higher user data rates are needed. Moreover, the T&E community must more effectively manage and use the spectrum that is available to it, and utilize new paradigms in which spectrum is dynamically allocated on a non-interference basis to multiple, concurrent users and in a way that meets both planned and unplanned changes to mission requirements, and real time channel conditions, factoring in both Federal and non-Federal users of spectrum. The Test Resource Management Center (TRMC is sponsoring the Adaptive Spectrum Aggregation and Management (ASAM) project, which will combine carrier aggregation techniques, such as those applied in commercial cellular Long Term Evolution – A (LTE-A) implementations, with cognitive radio technologies for dynamic, policy-based spectrum allocation and management. Together these technologies will form the building blocks of a communications system which meets the T&E community's need to operate within new bands that have historically been too fragmented to support T&E requirements in a spectrum environment which is dynamic. The ASAM system is being designed to sense available spectrum, aggregate white space within fragmented frequency bands to create communication channels, and dynamically allocate those channels across multiple range users using algorithms that increase spectrum utilization and efficiency while applying policy constraints.

A key feature of this new technology is dynamic management of spectrum to support concurrent test activities, with an emphasis on sensing and aggregating non-contiguous bands in a spectral environment that may be fragmented by multiple users. Another key feature of the technology is that it continually updates spectrum allocation to increase overall system spectrum efficiency. The spectrum allocation algorithms presented in this proposal maximize total system utility based on the user defined policy constraints.

The ASAM system will perform the following functions:

- Sense unused portions of spectrum, or white space, within telemetry bands available on a DoD ranges. The band(s) of interest may be highly fragmented by different Federal and non-Federal users. Sensing may be performed either by base stations (BS) on the ground or by the test articles (TA) themselves.

- Identify sub-channels to be aggregated to create a single, logical channel to support test requirements out of unused portions, or white space, in spectrum. Selection of actual portions of telemetry spectrum to be aggregated will be driven by a number of factors, including not only availability of white space in spectrum but also estimated signal-to-noise (SNR) on subchannels, known interference locations, required communications rates for users, relative priority of users, user location, and other constraints such as sub-channel reservations or blackouts as established by pre-defined or ad hoc policy. The overall selection and distribution of sub-channels across multiple test articles will be optimized at the system level. Selection of sub-channels by the ASAM system will be dynamic, and the system will automatically adapt to time varying changes in spectrum availability.
- Aggregate and access available white space to create a full duplex communication link at the required data rate for a given test article. The ASAM system will employ orthogonal frequency division multiplexing, as this modulation scheme affords flexible access to unused portions of spectrum.

ASAM Concept of Operations

The following scenario highlights key features of the ASAM system, and illustrates how ASAM benefits test operations on a range with multiple-users and sources of interference. In this scenario, five test articles and two sources of interference on a test range illustrate the ASAM capability.

Test Article 1 (TA1) is a legacy system and is not equipped with the ASAM technology, so test engineers assign it a fixed, contiguous frequency range for its test.

Test Article 2 (TA2) is equipped with ASAM technology, but for historical programmatic reasons, test engineers also pre-assign it a contiguous frequency range for its test. The ASAM system is running in the test, and the system is configured to allow dynamic changes to the spectrum used by TA2 if a need arises.

Test Article 3 (TA3) and Test Article 4 (TA4) are also equipped with ASAM technology, and their onboard ASAM modules are set up to sense and aggregate optimal frequency bands for the test without an initial assignment of bands. The ASAM system uses a combination of factors to arrive at this solution, including user-defined policy, required data rates, and test priority. The test missions of TA3 and TA4 are of equal priority, but TA3 requires a higher data rate for the series of tests to be conducted. Also, TA3 will be operating on part of the range subject to significant degradation due to channel effects such as multi-path. TA4, on the other hand, will be operating on part of the range subject to interference from Interferer1, an operational system that does not coordinate spectrum allocation with the range.

At the beginning of the day, test engineers load the ASAM ground system with data that specifies which frequencies are available to ASAM-equipped test articles and which frequencies have been allocated to other non-ASAM equipped test articles or to other systems, such as training or operational assets, in the vicinity. This information is entered into ASAM through the system Policy Editor in the control room. Test Article (TA) 1 has been allocated a fixed range of frequencies that will remain unchanged during the test, as TA1 is not equipped with ASAM.

Though equipped with ASAM, TA2 has also been pre-assigned a frequency range for the test, but the ASAM configuration allows changes to the frequencies used during the test if TA2 experiences interference or if a frequency allocation change is required for overall system optimization. TA3 and TA4 are configured to fully use the ASAM capability, and frequency allocation is not pre-assigned for those test articles. The ASAM system will arrive at an optimal frequency assignment for those test articles, given the policy constraints of the total system, including the off-limit frequencies used by TA1, known interferers such as Interferer1, the initial assignment of frequency to TA2, test priorities for all systems, anticipated channel model parameters, and required data rates.

TA1 and TA2 begin their tests using pre-assigned frequency bands. TA3 and TA4 move onto different parts of the range to conduct their tests and rely on the ASAM system to sense and aggregate the optimal frequency usage for those systems given the underlying policy constraints entered into the system. The ASAM system modules on TA3 and TA4 communicate with the ASAM ground node and at the same time determine the amount of interference they experience on various channels in their respective locations on range, and the ASAM system determines an optimal frequency allocation. Communications between the ASAM modules on TA3 and TA4 is not required. TA3 is assigned a contiguous frequency range that provides an optimal solution given its high data rate requirements and location on the range that is prone to multi-path effects. The ASAM system module on TA4, on the other hand, ends up using non-contiguous set of frequencies due to the interferers operating in the vicinity in which it is operating. The ASAM system aggregates these non-contiguous white spaces into a single communications channel for TA4.

Interferer2 from a nearby training-activity begins to transmit in some of the non-contiguous blocks of spectrum used by TA4 as well as in some of the frequencies that were pre-assigned to TA2. Spectrum used by TA1 and TA3 are not impacted by Interferer2. Fortunately, TA2 and TA4 are equipped with the ASAM capability. The ASAM modules onboard TA2 and TA4 detect the additional noise on channels they are using, and the ASAM modules onboard each of those systems identifies new sub-channels to use. This determination is governed by overarching policy, such as the frequency allocation to TA1, and measured interference on various sub-channels. TA2 and TA4 adapt to the change in channel conditions caused by Interferer2, and their tests continue unimpeded.

Finally, TA5 enters the range for testing. It is a lower priority test, and there are no contiguous blocks of spectrum available to meet its data throughput requirements. However, because it is ASAM-equipped, it utilizes a number of non-contiguous channels that when combined meet its throughput requirements.

Upper C-band Use Case – Benefits of Spectrum Aggregation

A particularly compelling application which highlights the benefits of spectrum aggregation can be found in the Upper C-band from 5925-6700 MHz. Currently, the 5925-6700 MHz band is dedicated to non-Federal users, and includes a combination of fixed-satellite earth stations in the band 5925-6425 MHz and to a lesser extent in the 6525-6700 MHz band, which are used for fixed-satellite Earth-to-space communications; fixed service stations used for terrestrial communications; and in the 6425-6525 MHz band mobile broadcast service used for electronic news gathering and event video distribution.

Resolution 416 from the 2017 World Radiocommunication Conference (WRC-07), entitled *Use of bands 4400–4940 MHz and 5925-6700 MHz by an aeronautical mobile telemetry application in the mobile service*, considers and recognizes the growing needs for spectrum for Aeronautical Mobile Telemetry (AMT) flight test purposes as well as the existing allocation of 5925–6700 MHz to fixed service (FS), mobile service (MS), and fixed-satellite service (FSS) users, and resolves that administrations authorizing use of 5925–6700 MHz shall use criteria that prevent or reduce potential interference from AMT to incumbent users to acceptable levels.

Techniques for sharing the band among AMT airborne transmitters and the incumbent users are addressed in detail in the International Telecommunications Union's (ITU) Report ITU-R M.2119, *Sharing between aeronautical mobile telemetry systems for flight testing and other systems operating in 4400-4940 and 5975-6700 MHz bands*. The conclusions from that Report are that if a set of criteria are applied to AMT, sharing with incumbents in the 5925-6700 MHz band can occur without further coordination and without risk of interference to incumbents.

One of the technical measures identified in the report is establishment of a 450 km x 24 km coordination zone, defined by the location and orientation of each fixed or mobile service receiver antenna, outside of which AMT can be conducted with no further coordination with the incumbent user. Within the described 450 km x 12 km zone, AMT may still be possible but requires further bilateral coordination.

By applying the 450 km x 24 km coordination zone criterion to all fixed service and fixed-satellite service receivers within 600 km of four test ranges, Edwards AFB, NAS Pax River, Eglin AFB, and Wichita Test Center, and also by considering the frequency assignment of those incumbent receivers, one can reach several key conclusions. First, there is a significant amount of unexploited spectrum available for AMT within the 5925-6700 MHz band. Second, this spectrum is highly fragmented, much of it in smaller than 5 MHz-wide blocks. In order make the most efficient use of this fragmented spectrum is to employ a communications technology which can aggregate non-contiguous white spaces to form a single communications channel. To harvest the most amount of spectrum, the AMT system would be able to access distinct 1-2 MHz portions of the band. Third, due to the physical locations of the incumbent receivers, each point in the test area may have a different set of 1 MHz blocks available for aggregation. As one would expect, some overlap in the set of white spaces available between adjacent physical locations on the range exists, but the AMT system must be capable of accessing different blocks of data dynamically, as a function of its location on the range.

A detailed analysis of Edwards AFB, following the methodology described above (i.e. applying the Resolution 416 coordination zone criteria to each point on the range, and excluding all 1 MHz sub-bands used by FS stations within the coordination zone of that point on the range), shows that 100% of the Edwards AFB test area has at least 30 MHz of spectrum available through aggregation within the 5925-6425 and 6525-6700 MHz sub-bands. The same analysis shows that 93% of the test area has at least 50 MHz of available spectrum; and 40% of the test area has at least 100 MHz of spectrum available through aggregation.

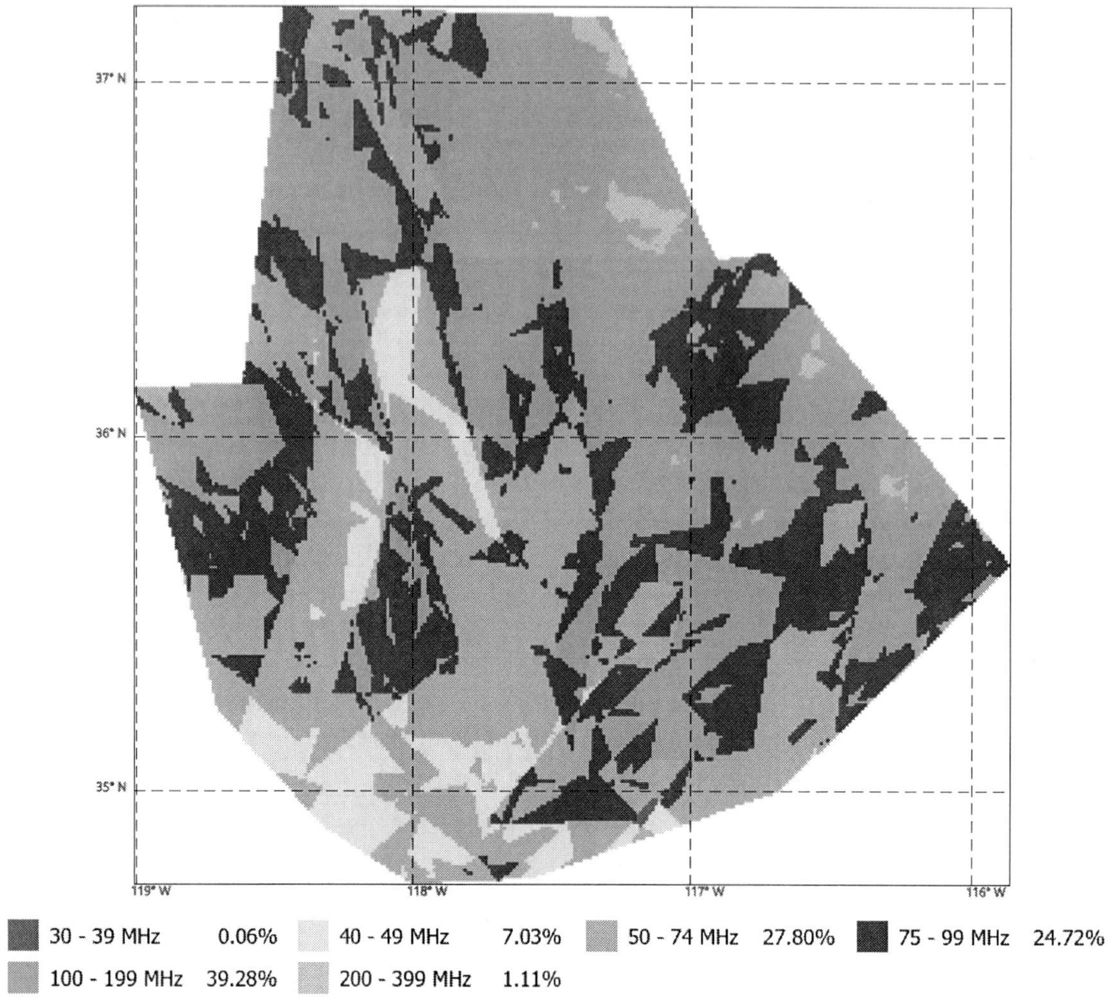

Figure **1** below illustrates the amount of spectrum in megahertz, that become available to support AMT at Edwards AFB through the use of spectrum aggregation techniques.

The same analysis conducted for the test areas at Eglin AFB, NAS Pax River, and Wichita Test Center yields similar results. At NAS Pax River, 100% of the test area has at least 25 MHz of aggregated spectrum available by applying the 450 km x 12 km criterion; 98% of the test area has at least 40 MHz of spectrum available; and 59% of the test area has at least 50 MHz of spectrum. At Eglin AFB, 100% of the test area has at least 40 MHz of available spectrum; and 90% of the test area has at least 100 MHz of available spectrum. At Wichita Test Center, 100% of the test area has at least 50 MHz of available spectrum; and 71% of the test area has at least 100 MHz of unused, aggregated spectrum available. Figure 2, Figure 3, and Figure 4 show total aggregated spectrum available for AMT at NAS Pax River, Eglin AFB, and Wichita Test Center, respectively.

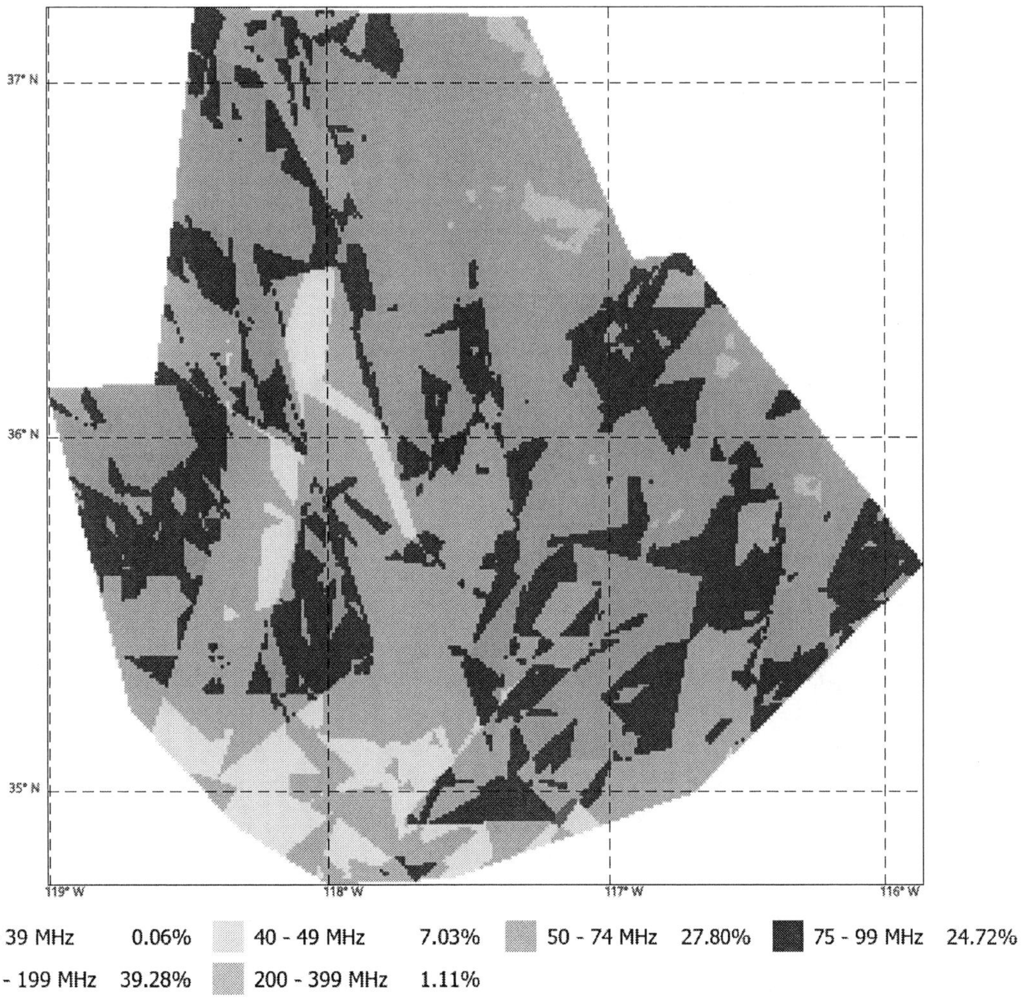

Figure 1. Aggregated Spectrum Available for AMT at Edwards AFB

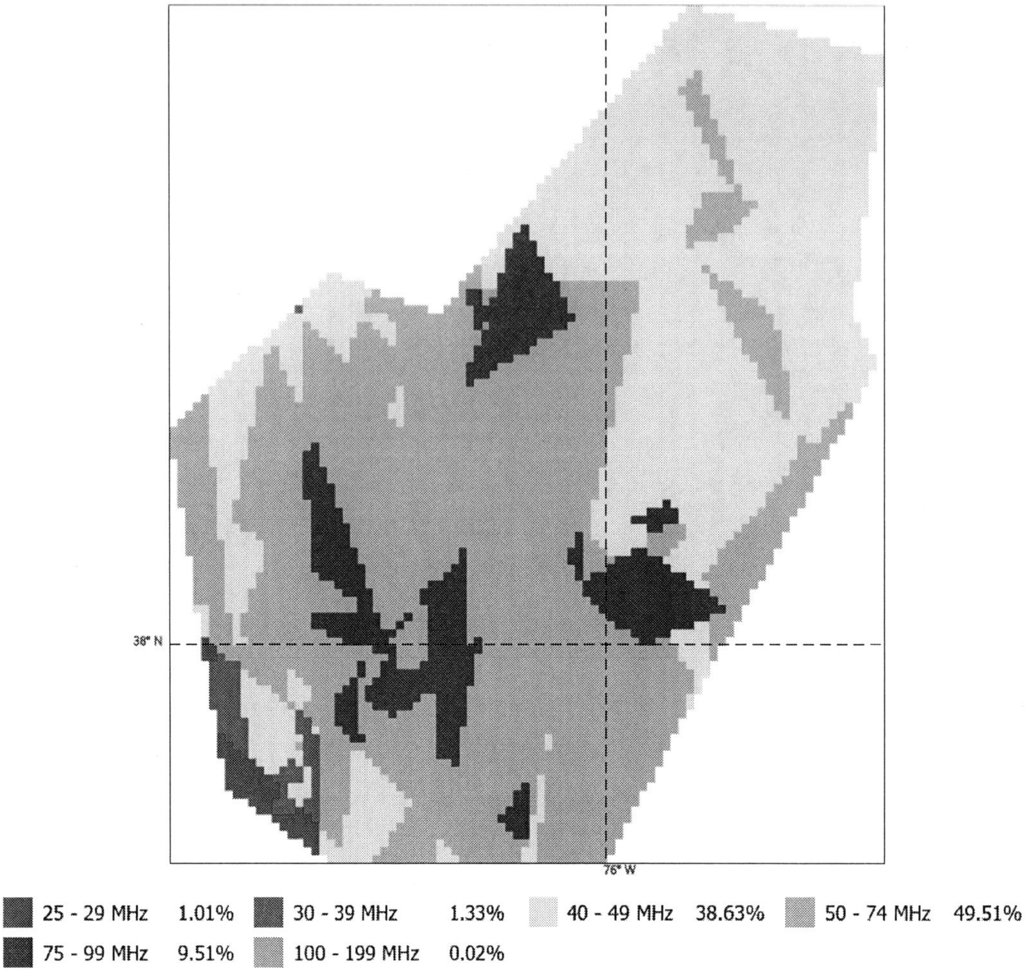

Figure 2. Aggregated Spectrum Available for AMT at NAS Pax River

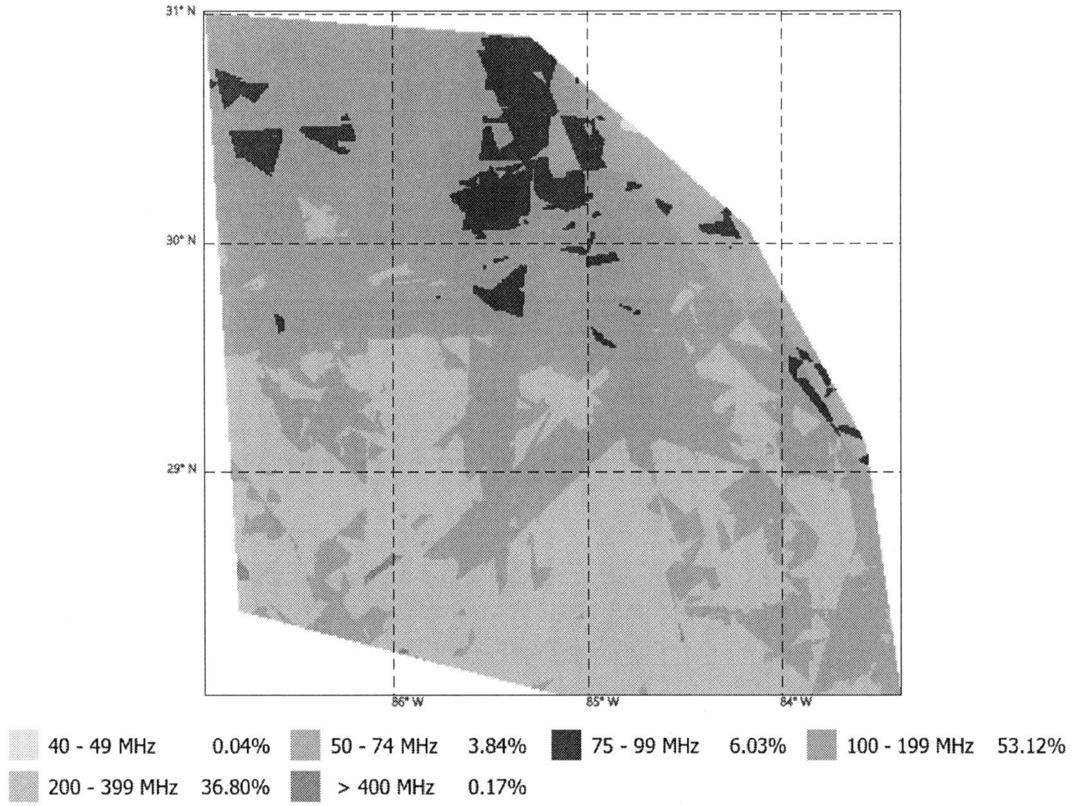

Figure 3. Aggregated Spectrum Available for AMT at Eglin AFB

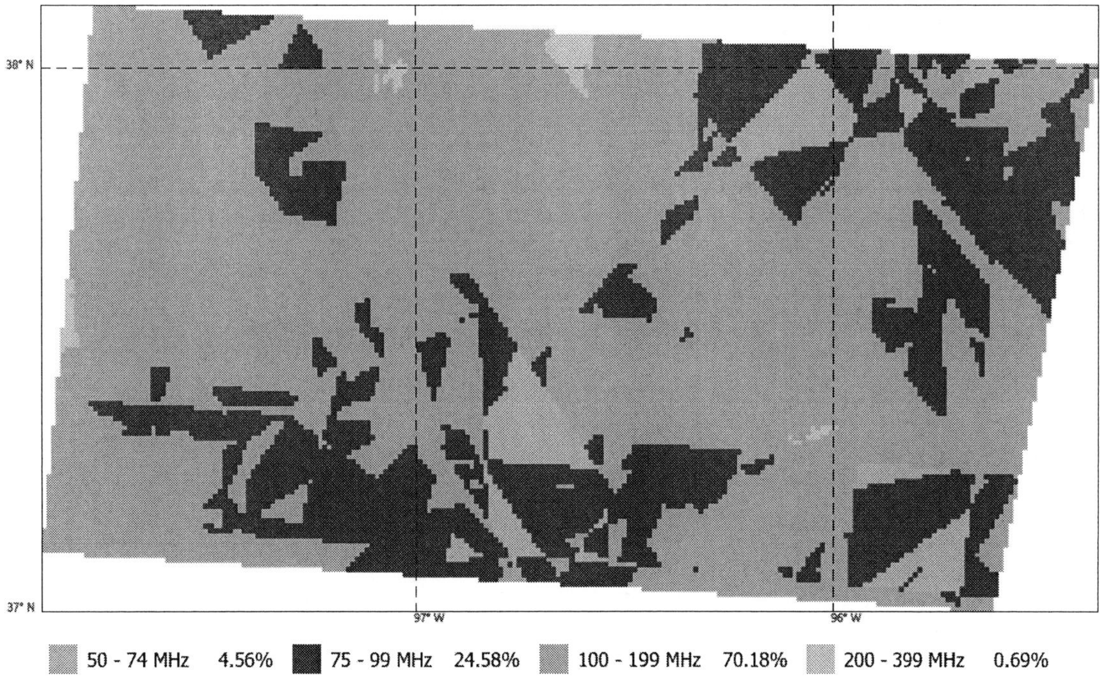

Figure 4. Aggregated Spectrum Available for AMT at Wichita Test Center

The keys to exploiting the whitespace segments lies in use of spectrum aggregation and dynamic spectrum access technologies. Because each point on the range has its own specific set of unused 1 MHz blocks of spectrum, an airborne AMT transmitter must have the capability to adjust the spectrum it accesses, as a function of its physical location on the range.

CONCLUSIONS

The benefits of spectrum aggregation are clear, particularly as we continue to move forward into a new era in which demands on spectrum – and its importance in our lives - will continue to grow. Spectrum aggregation, particularly when combined with technologies that effect dynamic access to spectrum, will enable more efficient use of existing telemetry bands by the T&E community and at the same time enable access to new bands, and to sharing among different sets of users, including those within and outside the DoD. This paper has presented a compelling example of the benefits of these technologies in the upper C-band, which is considered by its incumbent users to be too congested to support any additional use by the DoD. Certainly, with the advent of spectrum aggregating technologies and communications systems that have the ability to dynamically adjust the spectrum they access, our understanding and assumptions about how efficiently we can truly use this precious and finite resource we call spectrum can, and will, change.

ACKNOWLEDGMENT

This effort has been sponsored by the U.S. Government under Other Transaction number W15QKN-15-9-1004 between the NSC, and the Government. The US Government is authorized to reproduce and distribute reprints for Governmental purposes notwithstanding any copyright notation thereon.

The views and conclusions contained herein are those of the authors and should not be interpreted as necessarily representing the official policies or endorsements, either expressed or implied, of the U.S. Government

HIGH-G SURVIVABILITY OF AN UNPOTTED ON-BOARD RECORDER

Richard F. Granitzki, Douglas Weinhold
US Army Armament Research Development and Engineering Center, Telemetry Branch
Picatinny Arsenal, NJ, 07806

ABSTRACT

On-Board Recorders (OBRs) provide developers with data from sensors and computers by storing the data to memory devices contained within it's on board electronics system. At the expense of recovery, reusable OBRs provide projects with cost savings in terms of upfront non-recurring engineering, unit costs savings, and reduced field support setups. In this paper, the ARRT-158 OBRs used within artillery munitions systems to capture interior and exterior ballistics sensor and mission computer data will be discussed.

INTRODUCTION

On-Board Recorders (OBRs) provide developers with data from sensors and computers which are sampled and stored in on-board memory devices. At the expense of recovery, reusable OBRs provide projects with cost savings in terms of upfront non-recurring engineering, unit costs savings, and reduced field support requirements. In order to achieve a reusable, survivable onboard recorder, potting materials are often utilized to provide structural support to electronics. This paper discusses prior work done on potted OBRs as well as the experimentation of a pottingless OBR used within artillery munitions systems to capture interior and exterior ballistics sensor and mission computer data.

BACKGROUND

The ARRT-158 is a two rigid printed circuit board (PCB) data acquisition electronics design, interconnected by wires and secured to a 7075-T6 aluminum housing using mechanical standoffs, as shown in Figure 1. The entire assembly, within the housing is then potted using a reworkable potting compound to provide structural support, outside the load path of the acceleration forces, to the electronics board stack for high acceleration (high-g) environment survivability.

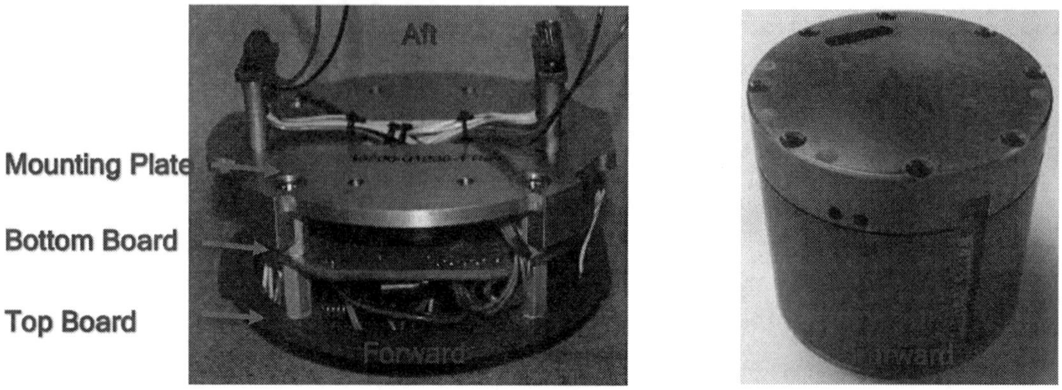

Figure 1 ARRT-158 Electronics Stack (Left); Fully Integrated ARRT-158 OBR (Right)

The ARRT-158's mission is to record interior and exterior ballistics acceleration environments for use in component or subassembly qualification. The operational environment for the ARRT-158 OBR is Picatinny Arsenal's 155mm Soft CATch (SCAT) Gun where test articles are subjected to accelerations up to 17,000G's during setback and balloting accelerations at muzzle exit and transition tube entry. The SCAT Gun is approaching 1,000 testing events with nearly each projectile tested an accompanying OBR is onboard recording the environment. Reusability of OBRs are vital to keeping testing costs low and ensuring operational readiness testing.

To achieve a level of reliability in such an extreme environment, potting materials are frequently used to fully support electronics to achieve survivability. However, potting materials themselves can lead to reliability problems resulting from improper cure and melting time and temperature which can lead to potting fracturing during testing, coefficient of thermal expansion and modulus mismatch during curing and testing, component stressing during potting removal, and dynamic component stresses caused by potting mass movement and fracturing during acceleration. It is important that fully coupled modeling and simulation environments be developed to evaluate the effects of potting materials. [1] Two of examples of reliability problems are shown in Figure 2.

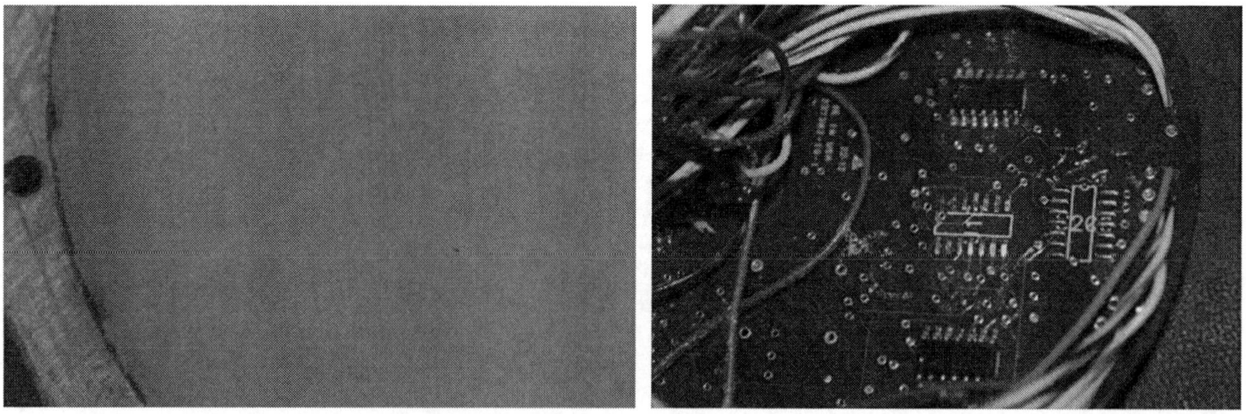

Figure 2 Cracked Potting (Left); Missing SOIC Package After Potting Removal (Right)

Extensive modeling and simulation of the electronics has been conducted in order to determine optimal electronics housing design as well as potting procedures and materials for survivability in high acceleration environments. A static deflection test was also conducted on the ARRT-158 digital board during a failure analysis effort.

Successful potting processes are important to reliability. If the process for the potting material is not strictly followed a repeatable process will be hard to achieve. This process variation can be detrimental to the cured material properties and lead to failures. Although modeling and simulation pointed to an acceptable level of board deflection within the potting structure, an effort was undertaken to eliminate potting materials so that the tensile and compression forces enacting on electronics components during potting preheating and curing processes were also eliminated. A pottingless solution is a valuable mechanical support alternative for any electronics assembly to achieve greater process repeatability and reliability as well as being able to be reworked.

PROOF OF CONCEPT

Before any changes were made to the ARRT-158 OBR to accommodate a pottingless design, two candidate rigid flex PCB designs were evaluated both of which had two 3.5" diameter 0.093" thick rigid PCBs with 22 independent traces run from one rigid board to the other through a flexible circuit, and is shown Figure 3.

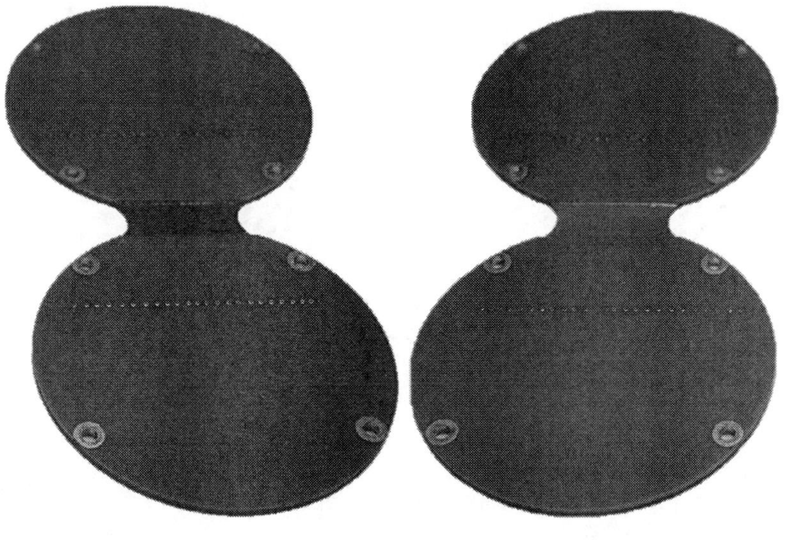

Figure 3 Rigid Flex Test PCBs: Flex Stack Up #2 (Left), Flex Stack Up #1 (Right)

The difference between the two rigid flex PCBs was the thickness of the flexible circuit dielectrics. An additional Dupont Pyralux dielectric was added to both sides of the flexible circuit making a second candidate deign. The flexible circuit material stack is shown in Figure 4 and the final integrated test article, without a lid, is shown in Figure 5.

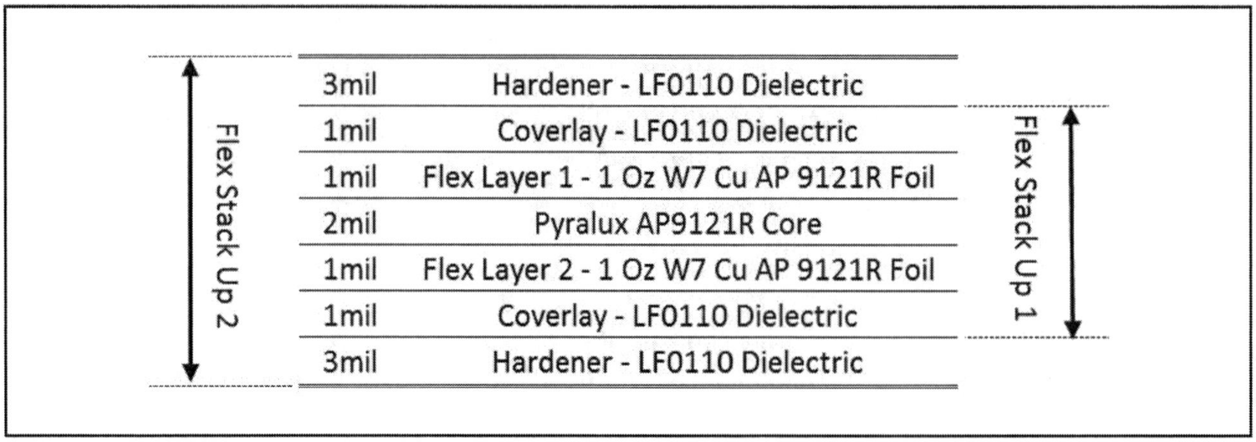

Figure 4 Proof of concept flex cross sections

Structural analysis predicted a maximum board deflection of 0.00535" [2] using a similar 155mm SCAT Gun MACS Zone 5 firing setback acceleration load curve as the analysis performed on the ARRT-158 OBR. With a reduction in PCB deflection of 0.0217" the team was confident in proceeding with testing the design.

Figure 5 Integrated Proof of Concept Test Article

The PCBs were inspected visually and via X-Ray before and after they were tested to determine if there was any damage to the flexible material or conductive traces when subjected to a high acceleration environment. The boards were tested at Picatinny Arsenals Soft CATch 155mm gun at MACS Zone 5 on March 13, 2014. No damage was observed to either test articles. Consequently the decision was made to choose the stiffer flex material design for added strength and protection.

PROOF OF DEMONSTRATION

With a successful Proof of Concept, the ARRT-158 OBR was redesigned to be a two board rigid flex PCB design taking on the new part number ARRT-167. During design of the new rigid flex OBR design, structural modeling and simulation was concurrently taking place and provided keep out areas on the PCB where structural spacers would reside for electronics survivability without potting. These cross hair style keep out areas are shown in electronics design artwork in Figure 6.

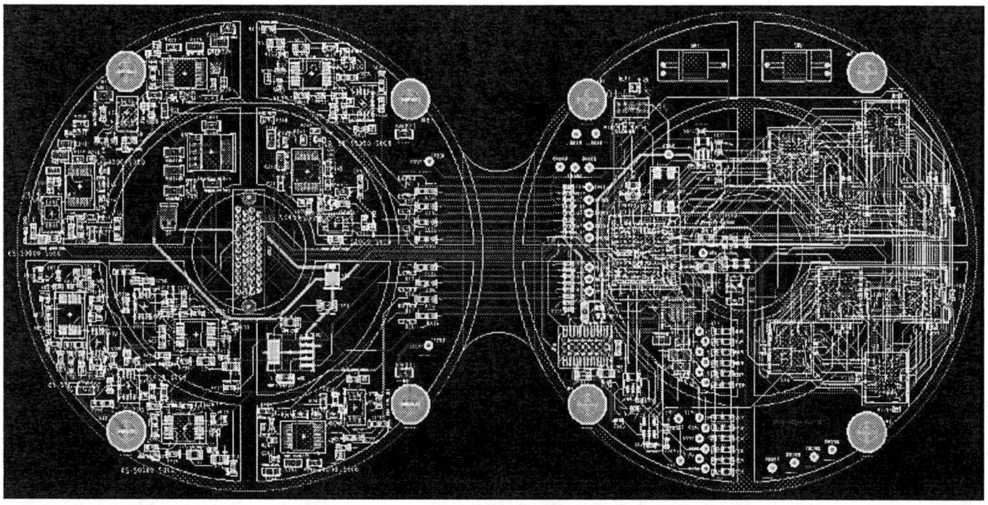

Figure 6 ARRT-167 PCB Layout with Spacer Keep Out Areas

The result was a 0.093" thick rigid PCB with the flexible section having two layers of DuPont Pyralux LF0110 composite coverlay for added flexible circuit support. The fabricated electronics board stack and structural spacers are shown in
Figure 7 and Figure 8.

Figure 7 ARRT-167 Rigid Flex PCB

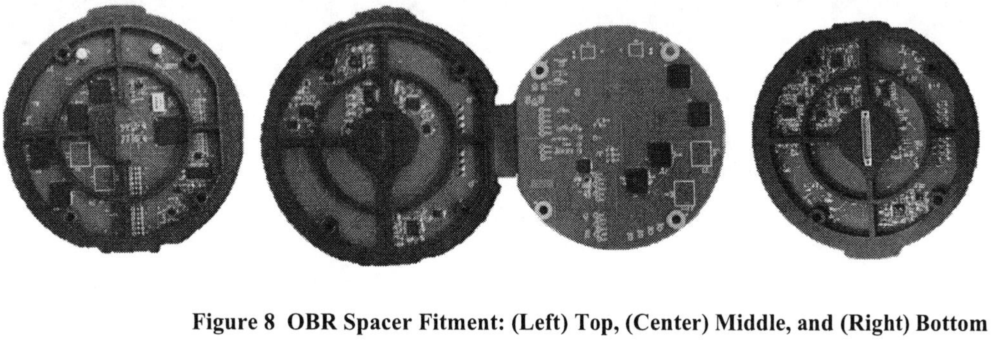

Figure 8 OBR Spacer Fitment: (Left) Top, (Center) Middle, and (Right) Bottom

The flexible material eliminated the need for wired connections between the PCBs unlike what is shown in Figure 1. Unsupported wires during high-g events, under their own mass, can impart enough stress on the solder joints to create intermittent or loss of electrical connection. Even in potted structures, wires that happen to be in a potting void can also impart stress on itself and solder joints. In this design, the rigid boards fit within a recess of the spacer such that the flexible section between the two circuit boards would be supported by a rounded groove in the spacer design as shown in Figure 9.

Figure 9 Flexible Circuit Interconnect Spacer Support

Component underfill material was applied to all PCB components to increase component adhesion to the PCB thereby reducing solder joint strain associated with component movement during accelerations. The purpose of this test was to evaluate the ability of the electronics to function during a high acceleration event without potting material, and while solder joints can structurally hold up to such accelerations this variable was removed to demonstrate the spacer performance in minimizing board deflections. A urethane conformal coat was also applied to seal the boards for added protection. Shown in Figure 10 are the complete electronics board stacks with and without spacers outside of its mechanical housing.

Figure 10 OBR in (Left) Board Form and (Right) with Spacers

While the schematic and the firmware were identical between the ARRT-158 and ARRT-167 OBRs the differences included: different PCB layout, flexible interconnect over thru hole wired connections, and 20,000 g range accelerometer (50 kHz frequency response) that were reused from previous systems over 60,000 g range accelerometer (100 kHz frequency response) that are replaced at the first sign of measurement anomalies.

TEST EVENT #1

ARRT-167 Serial Number 1 was tested alongside an ARRT-158 OBR during SCAT Gun Test #825. SCAT Gun Test #825 was fired at MACS Zone 5, had a projectile weight of 103.4 lbs, and an estimated muzzle velocity of 792.27 m/s. The data of both the Spacer OBR and Potted OBR are found in Figure 11.

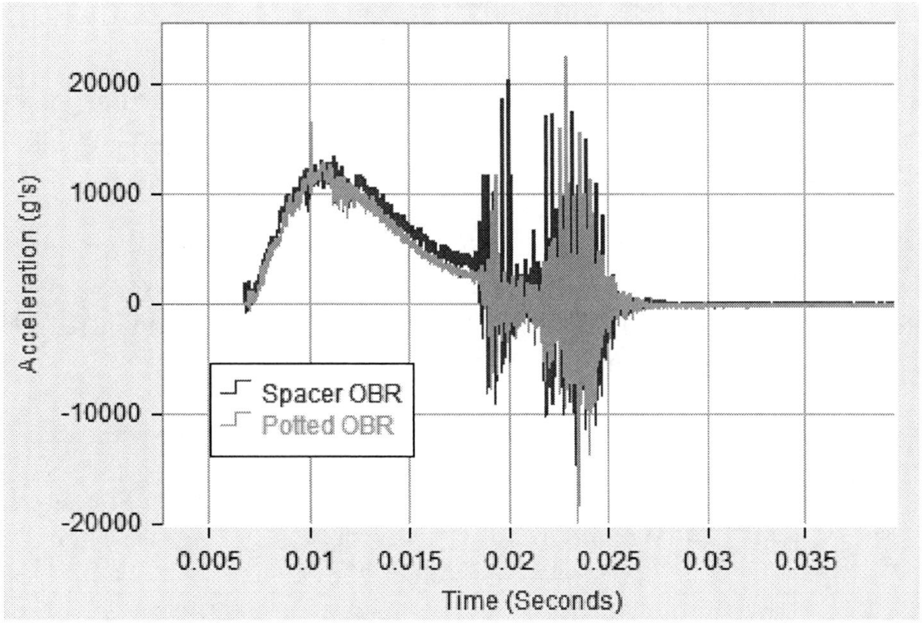

Figure 11 SCAT Gun Test #825, Spacer & Potted OBR Axial Acceleration Data

The data recorded on the Spacer OBR correlated to that of the Potted OBR significantly. Spacer damage was noted in two places. The first location is marked by a broken inner ring on the top spacer, and the second location is noted where one of the innermost supports around the connector separated at its thinnest section (shown in Figure 12). This spacer supported the bottom board to the mechanical housing. The damage was most likely attributed to the set-forward event and insufficient electronics stack preloading within the aluminum housing.

Figure 12 OBR Spacer Damage

The cracks developed in an area with a sharp edge, as highlighted in the images above. The spacers were redesigned (shown in Figure 8) to incorporate fillets and subsequent testing with these fixes have shown increased survivability.

TEST EVENT #2

The updated spacer design was tested in ARRT-167 Serial Number 2, in SCAT Gun #911, #912, and #914. Parameters for these test are shown in Table 1.

Table 1 SCAT Gun Test #'s 912 and 914 Firing Parameters

Shot Number	Charge (MACS)	Total Weight (lbs)	Estimated Muzzle Velocity (m/s)	Breech Pressure & Rise Time	Chamber Pressure
911	4	104.08	683	32.24ksi @ 4.88 ms	31.7ksi
912	4	96.84	701.55	30.4ksi @ 4.85 ms	29.9 ksi
914	4	97.18	698.57	29.99ksi @ 4.88 ms	29.69 ksi

The first shot in this event, SCAT Gun Test #911, involved shooting both the spacer OBR and a potted OBR for continued comparison. The electronics within both OBRs survived SCat Test #911, and captured identical data as shown in Figure 13. The post acceleration event bias shift is a result of common accelerometer die shifts that occur independent of the instrumentation electronics.

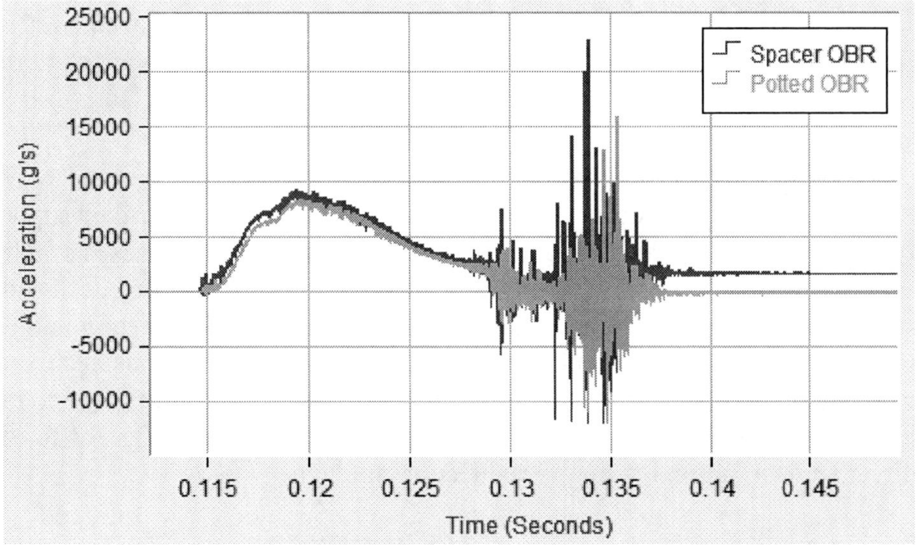

Figure 13 SCAT Gun Test #911, Spacer and Potted OBR Axial Acceleration Data

SCAT Gun Test #912 and #914's accelerations curves were over plotted due to their very similar firing parameters, and are shown in Figure 14.

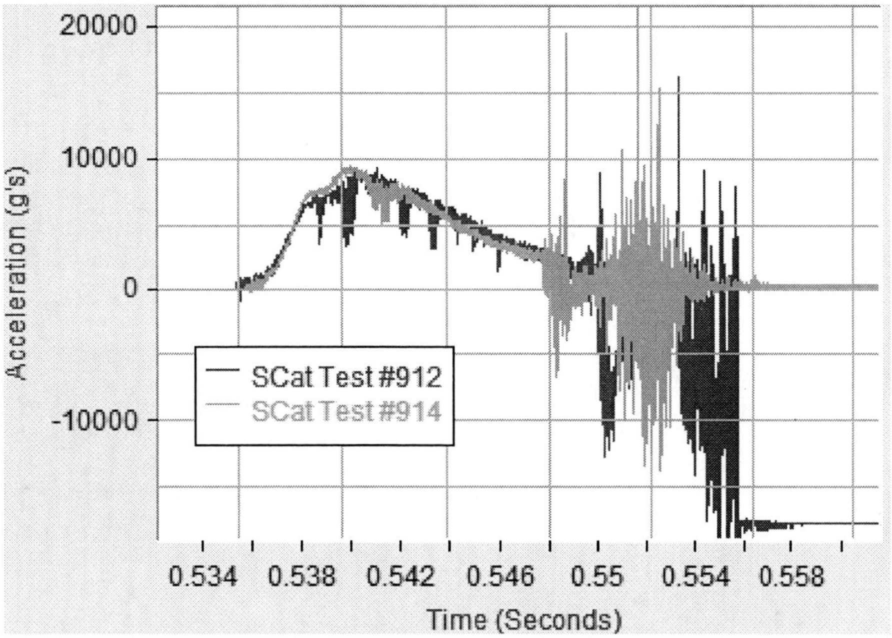

Figure 14 Spacer OBR SCat Test #'s 912 & 914 Axial Acceleration Data

From the data shown in Figure 14, the axial acceleration recorded from Shot #912 was very noisy and had a bias shift occurring after the set forward event. The accelerometer was not replaced before the next testing event and while it measured both set back and set forward acceleration phases in Shot #914, the data captured is questionable due to the fatigue of the gage incurred during Shot #912.

DISCUSSION

The ARRT-167 OBR's successfully demonstrated that the structural spacers supported the electronics board stack without potting materials during setback and set forward phases of a live fire 155mm artillery testing event. Additionally, with a significant reduction in board deflection, both the spacers and the electronics were able to be reused for continued survivability testing and reduced instrumentation cost. It is recommended that any electrical modifications necessary to update the design based on any revisions to the ARRT-158's electronics boards be made to move test articles away from the potted ARRT-158 OBR to the unpotted ARRT-167 design for further reliability and survivability characterization leading to reduced maintenance costs.

REFERENCES

[1] N.-H. Chao, J. Cordes, D. Carlucci, M. E. DeAngelis, S. Marhevka, J. Lee, L. Reinhardt and M. Tesla, "The Use of Potting Materials for Electronic-Packaging Survivability in Smart Muntions," in *Proceedings of the 2010 International Mechanical Engineering Conference and Exposition; IMECE2010-37433*, 2010.

[2] S. Manole, C. Stout, N. Baldwin, R. Granitzki, J. Caplinger, D. Weinhold and A. Rotundo, "Method for Designing Electronic Assemblies Without Potting for Gun Launched Applications Through the Use of Additive Manufacturing," *Technical Report - 16024,* US Army Armament Research, Development and Engineering Center, Picatinny Arsenal, NJ, In Publication.

USING MLC FLASH TO REDUCE SYSTEM COST IN INDUSTRIAL APPLICATIONS

Chris Budd
SMART High Reliability Solutions
Gilbert, AZ 85233
Chris.Budd@smartH.com

ABSTRACT

Storage devices based on Multi-Level Cell (MLC) NAND flash can be found in almost all computer systems except rugged, industrial systems; even though MLC is less expensive and more dense than devices based on standard Single-Level Cell (SLC) NAND flash, MLC's lower write endurance and lower retention has led system designers to avoid using it. This avoidance is unnecessary in many applications which will never come close to the endurance limits. Furthermore, new processes are leading to storage devices with higher write endurance. System designers should review the specific use-model for their systems and can select MLC-based storage devices when warranted. The result is lower system costs without worry of data loss due to write endurance.

KEYWORDS

SSD, SLC, MLC, endurance, retention, temperature, file system, ruggedized

INTRODUCTION

System designers make choices with various system components and usually the decision comes down to cost or quality: does the system absolutely need the quality afforded by the higher priced component? Since system designers are not necessarily experts with the details of each component, they can repeat the success of previous systems by choosing the same or similar components. This process works while subsequent generations of the components are relatively similar. However, when a technological shift occurs, system designers need to revisit the available products that satisfy a system requirement. One such requirement is storage. Almost all systems need storage, and designers of rugged, industrial systems generally choose solid state storage devices such as Solid-State Drives (SSDs) with NAND flash. Previously, only Single-Level Cell (SLC) NAND flash was compatible with the requirements of rugged systems. Now recent improvements in the NAND flash technology and the NAND flash controller have allowed Multi-Level Cell (MLC) NAND flash to work reliably throughout the industrial temperature range (typically -40 °C to +85 °C). While MLC has a lower cost than SLC, and MLC has a higher density than SLC, system designers are rightfully concerned about the reduced write endurance and data retention of MLC versus SLC. System designers may understand retention, but they may not fully understand the endurance requirements of their system. This paper will review the differences between MLC and SLC, the recent improvements with MLC, and provide guidance to better understand endurance requirements.

1. DIFFERENCE BETWEEN MLC AND SLC

Before making rash decisions, a system designer needs to understand the fundamental differences between the choices along with their tradeoffs. In the case of rugged SSDs, the two main choices are SSDs built with SLC or SSDs built with MLC. The traditional preference, if not requirement, has always been SLC since MLC was not considered appropriate for rugged systems.

1.1 DIFFERENCES IN VOLTAGE LEVELS

While the cell structure of MLC and SLC is fundamentally the same, the main difference is in the interpretation of the voltage level of the cell. Single-Level Cell NAND flash consists of a single threshold level; if the voltage in the cell is below the threshold level, logic in the NAND flash interprets the cell to contain the bit value "1"; if the voltage is above the threshold, then the cell contains a "0" bit. In Figure 1, the single threshold level in the SLC NAND flash cell divides the voltage along horizontal axis into two distinct ranges each with an associated bit value, 1 or 0. Multi-Level Cell NAND flash consists of multiple threshold levels; MLC is generally identified with three threshold levels producing four possible ranges for the voltage of the cell; each range has a pair of bits associated with it as also shown in Figure 1. In one cell, MLC stores two bits while SLC stores only one bit; therefore, MLC has twice the density of SLC. While this increased storage density is desirable, it comes with the drawbacks described in the following sections.

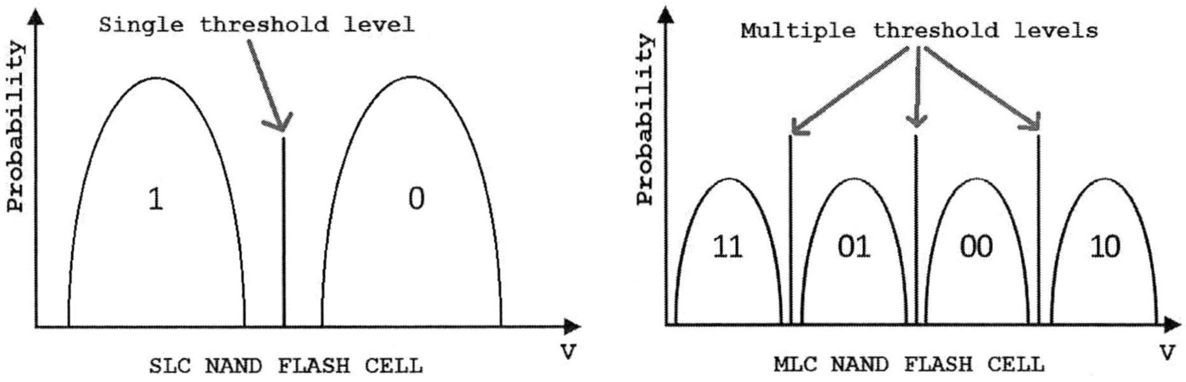

Figure 1: Threshold Levels in SLC and MLC

1.2 DIFFERENCES IN WRITE ENDURANCE AND RETENTION

The main drawback with MLC is its lower write endurance and reduced retention. The endurance of an SSD is generally defined to be the number of terabytes written (TBW) for a given capacity over the lifetime of the drive; retention is defined as the length of time the SSD can be powered off and still return the data without losing the data. A typical 240GB MLC drive has endurance of 100-400 TBW while a typical 240GB SLC has 20 times the endurance. Retention, on the other hand, varies depending upon drive age and temperature, but at 85 °C, an SLC drive can have well over 20 times the retention near the end of the MLC drive's endurance (73 days versus 2 days) [1]. Two difficulties with MLC cause the lower endurance and reduced retention.

As shown in Figure 1, there is essentially a 50% probability that an SLC NAND flash cell will have the voltage in the narrow range shown on horizontal axis below a particular curve. In an MLC NAND flash cell, the voltage range below each curve is still narrow, but the horizontal axis is more crowded with twice as many of these voltage ranges in the same cell structure. For any cells that end up near the thresholds, this crowding implies any minor variation of voltage in the cell could cause it to be interpreted as a different pair of bit values than was originally written. Since many of the modern NAND flash cells contain only a "handful of electrons" [2], it is easy to see how small disturbances could cause a small voltage variation and change the bits stored in the cell. In addition, after many program cycles, the curves tend to flatten out; with MLC, the curves cross not only the threshold levels, but the neighboring curves as well; therefore, as MLC wears out, it will have more bit errors without any external disturbance.

A second difficulty is that in order to write two bits to MLC, the flash controller must perform two program operations: once to set the least-significant bit, and the second to set the most-significant bit. This double program will greatly reduce the endurance of MLC compared to the single program required on SLC.

1.3 DIFFERENCES IN SENSITIVITY TO TEMPERATURE

All flash is subject to changes over temperature. For example, at hot temperatures, the endurance is greater since higher temperatures leads to fewer electrons getting trapped in the NAND flash cell, but consequently the retention is lower; the opposite is true for cold temperatures: reduced endurance, but longer retention [3]. Similarly, writing data to an SSD at one temperature extreme, and reading it back at a different temperature extreme causes problems for the SSD. Even though temperature affects all flash, it affects MLC more than SLC based on the reasons described previously.

2. RECENT IMPROVEMENTS

In spite of the difficulties with NAND flash, SSD users need not worry about them as the SSD and its NAND flash controller is designed to handle these difficulties and others. It is because of improvements to these NAND flash controllers, in addition to improvements to NAND flash technology itself, MLC drives exist at the industrial temperature range.

2.1 NAND FLASH TECHNOLOGY IMPROVEMENTS

When MLC was first introduced nearly 20 years ago, industrial temperature rated parts were not available. The goal of MLC was to increase the density of the NAND flash over SLC, and let SLC satisfy the other market segments such as those that need industrial temperatures. While NAND technology nodes get smaller and smaller, the densities of both MLC and SLC increase. New technologies like TLC and 3D NAND attract the attention and fabrication resources of the NAND flash vendors causing the vendors to be less interested in SLC and produce less SLC than they did previously [4]. Therefore, NAND flash vendors are now offering industrial temperature rated MLC parts.

2.2 NAND FLASH CONTROLLER IMPROVEMENTS

The early NAND flash controllers were not nearly as sophisticated as they are today. Improvements in the speed and density of these controllers have enabled them to accomplish more tasks in the same amount of time or shorter. Specifically, the NAND flash controllers have improved error correction algorithms and have added additional error correction bits, and the NAND flash has added additional space in which to store these extra bits; improved error correction helps to mitigate some of the problems discussed in previous sections including endurance and retention problems across the industrial temperature range. Some controllers include extraordinary error recovery techniques when the normal error correction algorithms fail. This additional and extraordinary error correction allows a controller to extend the endurance of the NAND flash. The controllers have also improved their wear-leveling algorithms so that they attempt to use all of the NAND flash blocks equally preventing a scenario where some blocks reach their endurance limit much sooner than other blocks. To help with retention, some controllers employ a read-scrubbing algorithm that periodically reads all the blocks and rewrites any block that is nearing the limit of the controller's error correction ability.

3. WRITE ENDURANCE REQUIREMENTS OF AN EMBEDDED APPLICATION

The improvements with MLC flash and NAND flash controllers are interesting, but a system designer still asks if an MLC SSD will have enough endurance for their application. The better question is, "How much endurance is needed?"

3.1 IMPACT OF A FILE SYSTEM

One important aspect of the endurance question is how the data is written. Will the system application write data through a file system, such as ext2 under a Linux operating system, or will it simply write data to the raw disk? A file system will increase the number of write operations to the SSD since the file system must manage its metadata like filenames and timestamps along with pointers to the files themselves. Some file systems are journaling, like ext3 and NTFS, meaning the file system writes a "journal" of its actions in addition to actually performing the action, which increases the number of write operations to the SSD even more. An alternative is to simply write to the raw disk without a file system; for example, if a system wants to record a stream of telemetry data, it can start at address 0 on the SSD and write to sequential addresses until the end of the steam or the end of the SSD. Not only does a raw write reduce the number of write operations to the SSD, it also helps the SSD manage its internal data mapping compared to the random nature of any file system; when using a file system, even if the application writes only one file sequentially, the file system must update its metadata and pointers causing the SSD controller to perform many additional internal read-modify-write operations to the NAND flash, and to perform frequent updates to its internal mapping tables.

3.2 TBW IN A REAL APPLICATION

The final answer to the endurance question depends on the system application. Below are three examples comparing the amount of data written against the endurance (TBW) of a typical MLC SSD.

- Map display: If the mapping information occupies most of a 240GB drive, the user could write a new map every day for 3 years before reaching 263 TBW, which would be near the endurance limit, 100-400 TBW, of a typical 240GB MLC SSD.

- Surveillance plane: If a surveillance plane records multiple streams of information to the SSD, the data access is more like random access than sequential access; because the SSD almost always has a slower random access data rate than its sequential access data rate, the maximum data rate of the SSD would be closer to 50 MB per second rather than 500 MB per second. During one 8-hour mission, where the application constantly writes 50 MB per second, the application would record close to 1.5 Terabytes (TB). Obviously, this application would require an SSD with a larger capacity, such as 1920GB, and one can expect the endurance to increase proportionally with the capacity producing a typical endurance of 800-3200 TBW. Therefore, one SSD could survive between 1.5 years to 5 years with one 8-hour mission per day.

- Ground-based transactional recorder such as in the base station of a cell phone tower: Although similar to the surveillance plane in recording small pieces of information randomly on the SSD, this recorder must be operational nonstop, 24 hours each day, 7 days each week; consequently, it will record 3 times more data per day than the plane did during its 8-hour mission. If this recorder has some mechanism to upload its data to an external server occasionally throughout the day, the drive could be written over and over at 50 MB per second – up to 4.5 TB per day. Using a large MLC drive, 1920GB, with an endurance rating of 3200 TBW, the replacement schedule for the drive should be just under 2 years.

Note that these examples assumed a worst-case amount of data written. A system designer would be wise to watch attributes from the drive, such as the attributes conveyed through Self-Monitoring, Analysis, and Reporting Technology (S.M.A.RT.), to determine a more accurate amount of data the application writes; many times the amount of data written is much less than the specified endurance of the SSD [5].

4. ADDITIONAL SELECTION CRITERIA FOR AN INDUSTRIAL APPLICATION

Endurance and retention are not the only requirements for an MLC SSD in a rugged environment such as those found with telemetry applications. These environments usually demand special features not found in commercial, consumer-class SSDs. Industrial temperature is an obvious difference, but other features could include a rugged enclosure capable of meeting MIL-STD-810G, special military erase sequences, write-protection, case isolation, interface connectors with 30 microinches of gold, etc.

4.1 DIFFERENT GRADES OF MLC

Another selection criterion for an industrial application could be the difference between MLC and enterprise-grade MLC (eMLC). The technology is the same between MLC and eMLC except that the eMLC flash is specially tuned to increase endurance. Even though greater endurance is welcome, it comes at the expense of retention. Typically in enterprise environments with uninterruptible power supplies, the application performs many small write operations leading to more emphasis on endurance than the length of time an SSD can retain data while powered off. Since most industrial applications cannot guarantee power, they require a reasonable balance between endurance and retention, and cannot use eMLC.

CONCLUSION

Historically, rugged, industrial systems required solid-state storage based on SLC. Even though SLC is more expensive than MLC, the reasons for this SLC exclusivity included temperature rating along with endurance and retention. However, recent improvements in NAND flash technology and NAND flash controllers allow MLC to exist in environments spanning the industrial temperature range. While retention is determined by the system designer, the endurance of MLC was the last obstacle to selecting MLC for these systems. Newer, larger capacity models allow for much more data to be written to the SSD than in years past, and most likely exceed the endurance requirements of many applications. This technological shift should encourage systems designers to review their specific application. They can not only compute the worst-case amount of data written, but also estimate the actual amount of data written using attributes from the drive. They may find the endurance they require is well below the endurance provided by SLC drives, and even below the endurance provided by MLC drives. Then the lower cost of MLC versus SLC allows the designer to save money on the overall system.

REFERENCES

[1] S. Phillips, "The Effects of Industrial Temperature on Embedded Storage," *RTC,* March 2015.

[2] P. Brace, "High-Performance and Large-Capacity Storage: A Winning Combination for Future Data Centers," in *Flash Memory Summit*, Santa Clara, CA, 2015.

[3] E. Tiomkin, "Industrial Temperature and NAND Flash in SSD Products," 24 July 2012. [Online]. Available: https://www.eeweb.com/blog/eli_tiomkin/industrial-temperature-and-nand-flash-in-ssd-products. [Accessed 25 May 2017].

[4] A. Lovinus, "SSD Storage: Is SLC NAND Extinct?," 28 September 2015. [Online]. Available: https://blog.neweggbusiness.com/components/ssd-storage-is-slc-nand-extinct/. [Accessed 26 May 2017].

[5] J. Handy, "Comparing DWPD to TBW," 1 June 2015. [Online]. Available: http://thessdguy.com/comparing-dwpd-to-tbw/. [Accessed 25 May 2017].

NOMENCLATURE

ATA or Advanced Technology Attachment: An interface standard from T13 (http://www.t13.org/) for storage devices typically using a parallel bus (PATA), or serial bus (SATA).

Flash: A non-volatile memory device using an array of transistors each with a floating gate to store a charge.

GB or Gigabyte: 10^9 bytes.

HDD or Hard-disk drive: Traditional mass storage device using a rotating, magnetic platter.

MB or Megabyte: 10^6 bytes.

MLC or Multi-Level Cell: A flash technology which stores more than one bit per transistor cell by using more than two voltage ranges on the floating gate; most commonly referred to two bits per cell using four distinct voltage ranges.

NAND: A high-density flash device usually with defect blocks marked by the factory; read and write operations must be done at a page level (several kilobytes), and erases must be done at an erase block level consisting of several hundred pages.

SATA or Serial ATA: A storage bus interface where the data is transferred serially rather than through parallel data wires as in previous ATA devices.

SED or Self-Encrypting Drive: A storage device that automatically encrypts all data written to the storage medium and decrypts all data read from the storage medium. No user or host intervention or support is necessary to manage the encryption key or perform the encryption.

SLC or Single-Level Cell: A flash technology which stores one bit per transistor cell by using two distinct voltage ranges on the floating gate.

S.M.A.R.T. or Self-Monitoring, Analysis, and Reporting Technology: A standard protocol for a host system to retrieve information about the drive status and health.

SSD or Solid-State Drive: A mass storage device typically using the same form factors as traditional hard-disk drives, but without the moving parts. An SSD typically stores data in SLC or MLC NAND flash.

TB or Terabyte: 10^{12} bytes

TBW or Terabytes Written: A measure of endurance of an SSD for a given capacity.

BOEING NW TELEMETRY SYSTEM

William C. Gregoire
Boeing Test & Evaluation
Seattle, WA, 98108
william.c.gregoire@boeing.com

INTRODUCTION

Most of flight/ground testing of Boeing Commercial airframes, and their military derivatives, are conducted with a full crew of analysts required to support the test scenario. This provides a degree of freedom to conduct multiple simultaneous flight tests anywhere in the world.

However some tests are high risk that require a minimum flight crew. During these types of tests, flight data is telemetered to ground station analysts who monitor test conditions and ensure safety of flight parameters are met. The Boeing NW Telemetry system is different from other experimental aircraft test programs as data is telemetered in packets rather than IRIG 106 Chapter 4 PCM format.

This paper will attempt to explain the methods and capabilities used in the NW test ranges. It will also explain the problems and limitations faced when flying near a major metropolitan area between two mountain ranges.

Why the Boeing NW Telemetry system differs from other experimental aircraft test programs

The data is telemetered in packets rather than IRIG 106 Chapter 4 PCM format

Using a conventional system based on IRIG-106 Chapter 4 PCM, would be giving up many of the advantages gained from formatting data within Ethernet packets used on the DAR (Data Acquisition and Recording) system that Boeing NW uses. An IEEE Std 1588 Precision Time Protocol (PTP) grand master time code generator is used by to provide a common time-of-day to all Data Acquisition Units (DAUs). The DAUs all share this time-of-day to within a microsecond. The Ethernet packet retains the timing information when telemetering avionics bus data. This is not practical with IRIG 106 compliant PCM frames which are intended to telemeter the synchronous analog data.

All of the NW Data Acquisition units output data over Ethernet. Ethernet based data is acquired with time stamping at the source of the data and is transmitted over the network in UDP (Universal Datagram Protocol) format. To be efficient and obtain maximum bandwidth, the Ethernet packet needs to be as large as possible. Each incoming message is placed in a structure which is referred to as a Data Segment. One or more data segments make up an Ethernet packet. DAUs are programmed to output Ethernet packets if they reach a predetermined size or exceed a maximum time latency threshold.

Like IRIG-106 Chapter 4 PCM, the telemetry system requires a continuous bit stream. However, unlike Chapter 4 PCM, the data is not in a fixed pattern of repeating frames. The continuous bit stream allows word and message synchronization to be obtained on the received data using conventional techniques. The onboard data system can have a total data network bandwidth of over 200 Mbps while TM is limited to 15 Mbps. To prevent the system from completely filling up all available TM bandwidth and to get as much data to the ground as possible, the format described in Figure 1 was developed.

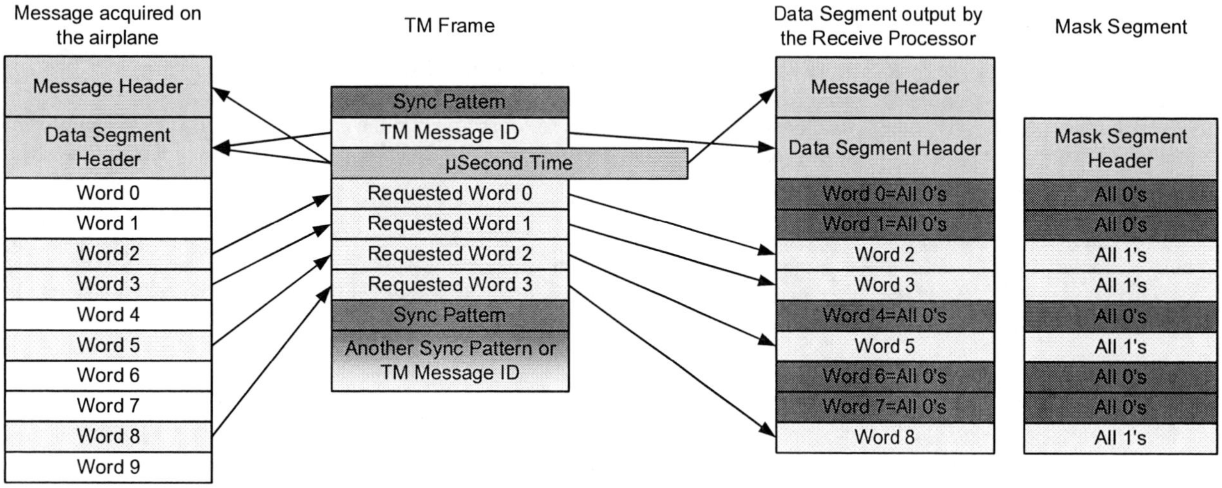

Figure 1: Telemetry data Transmission Format and Data Message

Telemetry Transmit Processor checks TM database for Message ID, Source, Label and Words and sends the data
Only the requested data is transmitted, instead of the whole message. Users identify measurements from the test setup database to be transmitted. The Telemetry Transmit Processor (TTP), reads the incoming messages to see if they have data that have been requested by the users to be telemetered. Both TTP and Telemetry Receive Processor (TRP) use the same test setup database to locate TM Message ID, message source, message label and the requested words to be telemetered. If the system cannot locate data to be transmitted, it fills in with sync patterns to maintain a continuous bit stream. Once per second, the TTP transmits airplane data system time for use by the TRP.

Receive Processor, synchronizes to incoming data stream, examines each incoming message, looks up its setup information and reconstructs the original message.
Before the TRP can output data on the ground network, it must first synchronize to the incoming TM data. The synchronization process is accomplished in two steps: 1) Data Sync; 2) Time Sync. When three valid messages in a row are received the TRP is in Data Sync. A single bad message causes loss of Data Sync. Once a valid time message is received, the TRP is in Time Sync and will begin to output data on the ground network. Time Sync is lost if more than two sequential time messages are missing. A loss of Data Sync forces a loss of Time Sync. The TRP will not output data on the ground network unless both are in sync.

Using the same database as is being used on the test airplane the TM room looks up the data source, message, the number of words and which words we requested then compares what is being transmitted to the database. Reconstruction of the message can then take place.

Since a minimum amount of data is telemetered, the TRP uses the TM Message ID to reference the test setup database in order to reconstruct the data of interest in the original message. When words are not requested for transmission, zeroes are placed where the data would have been. Words that are requested retain their position in the message. The TRP appends a mask segment to the output Data segment which delineates bits that have been telemetered from those that have not been.

Types of telemetry testing are supported in the NW

Flight Line Gauntlet testing – With a flight emulation test system connected to the aircraft on the ground mimic flight conditions.
High Speed Taxi – Check systems with aircraft on its own power for the first time.
First Flight – First flight test conditions can vary dependent on whether this airplane is a new major model or a derivative.
Initial Airworthiness – Aerodynamic Stability – Clearing a flight envelope of all required speeds and altitudes

Flutter – Structural Stability – Clearing a flight envelope of all required speeds and altitudes.

Problems and limitations faced when flying near a major metropolitan area between two mountain ranges (Seattle area)

Test airplane direct to Seattle Telemetry Analysis Station (TAS) room

A Line of Site (LOS) is required to provide all data and communications from the airplane RF transmitter to the roof antenna of the building collecting and viewing the data (Figure 2). Radio signals can bounce off ground support vehicles, buildings and terrain creating multi-path interference problems.

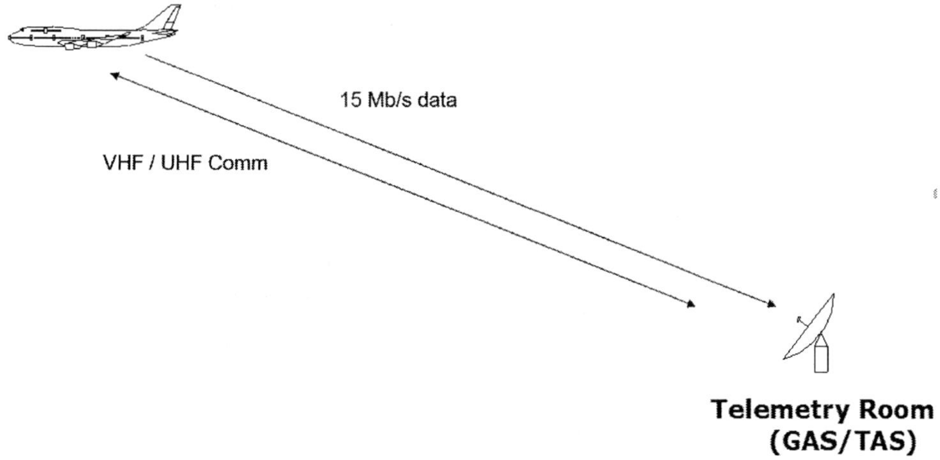

Figure 2: Direct Airplane line-of-sight to Telemetry room

Range/altitude limited

Conducting TM tests originating out of an airport located near a major metropolitan area between two mountain ranges, somewhat hampers the otherwise unobstructed signal range that could have been obtained without them. This restriction limits the "Telemetry Coverage" area to mostly over the Puget Sound waterways when testing at the lower altitudes.

Seattle Area No Flutter Test zone(s)
In order to protect high density populated areas from potential falling debris, a "No Flutter zone" was established that reduces the risk of occurrence. IAW (Initial Airworthiness), Flutter, and VMCA (Minimum Control Speed in the Air) are examples of testing that are not allowed in these zones. This restricted area consists of highly populated areas of western Washington/Oregon.

Telemetry interference issues with Wireless Communications Service (WCS)
In addition to the normal issues of Multi-Path Interference, WCS can also cause problems. Cell phone service has been deployed around the US for 4G service after approval in 1997. The proportion of the wireless service in the 2345-2360 MHz band is causing interference with the S Band telemetry system when pointing the telemetry receive antenna toward those sites. If the TM receive antenna is pointed away from a WCS antenna there is no interference with the TM signal. If the TM receive antenna is pointed at a WCS service antenna there is an overlap causing interference. The telemetry antennas used in the NW have wide bandwidth Low Noise Amplifiers (LNA's) that pull in the WCS signal and oversaturate the amplifier. WCS are operating legally and within FCC rules however the NW TM antenna designs do not block the WCS signal. Most of TM signal range locally north-west of the Telemetry room site are totally blocked by WCS. Similar issues future 5G wireless services could occur when moving to C-band frequencies.

Telemetry interference issues with AeroMACS at larger airports
AeroMACS (Aeronautical Mobile Airport Communications System) is a wireless system that supports the transmission of safety & regulatory flight data at airport surface facilities. The system is using the (5091-5150 MHz) frequency spectrum that Boeing NW plans to use for the C-band telemetry when operating at larger airports. The local large commercial airport (SeaTac) will have these systems operational. No problem should occur when using TM away from the larger commercial airports. Inference could occur during telemetry testing in takeoff or landing phases of flight.

Factors when operating TM System(s) remotely

Test airplane direct to Remote Test Station (RTS) Remote Test Station fixed or mobile ground station
Since direct LOS to Seattle is limited, testing at a remote location is required for most TM testing. Similar to operating locally, data must first be sent from the airplane to a ground station LOS, and therefore all other issues previously mentioned are still applicable.

Flutter tracks are utilized to support test airplanes during flutter. In an agreement with the FAA, areas of low population and adequate distance (150-200 straight miles) are utilized where flight maneuvers can be conducted without affecting other airline traffic. Several Flutter Tracks are located in Eastern Washington which require a RTS. Operating distances are difficult to specify as they depend on height of the test aircraft above the ground and how flatness of the terrain. The maximum operating distance from the RTS is approximately 150 Miles LOS at 15,000 feet and 200 Miles at 30,000 feet altitude.

From the RTS, data and communications are sent one of two ways back to the Seattle ground station monitoring team.
The first method is an uplink of the data and communications through a satellite in geosynchronous orbit. NW TM is currently licensed to operate anywhere within the Continental United States. Rental of satellite time is expensive and scheduling has to be coordinated a minimum of 1 week in advance and in blocks of 12 hours. TM over Satellite adds latency to the data that could be minimized through other routes.

The second method is to send data over the internet through a "Telemetry over Internet Protocol", a process called TMoIP. (Figure 3)

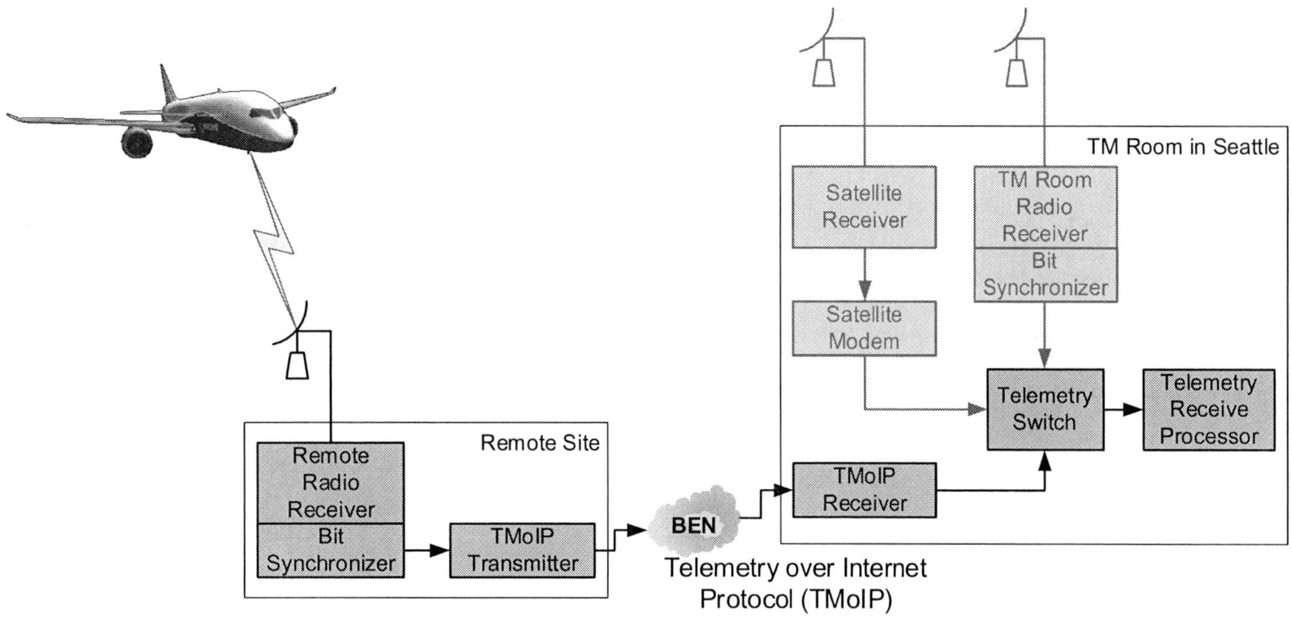

Figure 3: Telemetry over the Internet

The TMoIP transmitter takes the incoming serial bit stream and breaks it up into pieces. Each piece is then placed into the payload of a message that will be transmitted over the internet. The TMoIP receiver reassembles those messages to recreate the serial bit stream that was transmitted from the airplane. The output of this unit goes to the TRP.

CONCLUSIONS

These are a few of the reasons that make up why the Boeing NW Telemetry system is different from other experimental aircraft test programs.

1. Data is telemetered in packets
2. Only the requested data is transmitted, instead of the whole message.
3. Most other flight test programs telemeter their data all of the time.

Listing more reasons for differences would be time prohibitive. Some of which could be the amount rain fall in Seattle or the massive quantities of coffees that are consumed but you can form your own opinions on those areas.

REFERENCES

[1] "Feasibility of sharing the frequency band 5091 – 5150 MHz between Aeronautical Mobile Telemetry (AMT) and AeroMACS systems deployed at or near airports" (Dan Jablonski), July 10, 2015.

[2] "Telemetry Interference Issues with WCS (Wireless Communications Service) and AeroMACS" (Thomas F. Fisher)," March 14, 2017.

[3] Telemetry Standards, IRIG Standard 106-15 (Part 1)", July 2015. http://www.wsmr.army.mil/RCCsite/Documents/106-15_Telemetry_Standards/chapter4.pdf

[4] "Network-Based Data Acquisition System for Flight Test" (Lee H. Eccles, Larry Malchodi, Kenneth A. Wilhelm)," April 24, 2013.

[5] Telemetry Room Training, Course Number TR014503 (John Corrigan)," July 25, 2013.

[6] "BCA Telemetry System" (Lee H. Eccles), April 27, 2012.

[7] "WiMAX Forum –AeroMACS" -Copyright © 2001 – 2017 http://wimaxforum.org/Page/AeroMACS

THE TENA AND JMETC SOLUTION FOR TELEMETRY IN DISTRIBUTED RANGES AND FACILITIES

Gene Hudgins and Juana Secondine
TENA Software Development Activity (SDA)
Niceville, FL, 32578
gene.hudgins@tena-sda.org and juana.secondine@tena-sda.org

ABSTRACT

TENA and JMETC enable interoperability among Telemetry Instrumentaion (TM) ranges, facilities, and simulations in a timely and cost-efficient manner. TENA provides for real-time system interoperability, as well as interfacing existing range assets, C4ISR systems, and simulations; fostering reuse of range assets and future software systems. JMETC is a distributed, LVC capability which uses a hybrid network architecture; the JSN is used for secret testing and the JMN is the network solution for all classifications and cyber testing. JMETC provides readily available connectivity to the Services' distributed test and training capabilities and simulations, as well as industry resources. TENA and JMETC allow for the most efficient use of current and future TM range resources via range resource integration and remote control. This integration fosters interoperability and reuse within the test and training communities, critical to validate system performance in a highly cost-effective manner.

INTRODUCTION

Historically, the DoD's test and training ranges were developed in isolation, focusing on specific requirements, and constrained by aging techniques and technologies. As the U.S. military evolved from standalone Service branches to the concept of a "Joint" military, warfighters and operational systems now fight Jointly; meaning they must therefore be tested and trained in a tightly-coupled Joint environment as well. The existing DoD test and training range infrastructure was not designed to be interoperable, and rapidly became inadequate in this new era of Joint warfare. An urgent need was identified for the integration of thc complex and disparate test and training ranges.

The DoD's response to this crippling issue was to develop the Test and Training Enabling Architecture (TENA), an implemented architecture comprised of middleware, object models, tools / utilities, and collaboration capabilities. TENA provides real-time software system interoperability and the capability to interface existing range assets, systems, and simulations at distributed facilities. Government-owned and free for anyone to use, TENA allows the most efficient use of current and future range resources via range resource integration. This integration invariably fosters interoperability and reuse within the test and training communities – critical in validating system performance in a highly cost-effective manner.

TENA provides a middleware software component and can be used on any internet protocol (IP)-based range or distributed network, such as the Joint Mission Environment Test Capability (JMETC) networks and the Joint Staff (JS) J7 Deputy Director Joint Training (DDJT) Joint Training Enterprise Network (JTEN).

Upgrading an existing range system to TENA can be achieved in a drastically shorter time frame than traditional software integration efforts. Additional benefits include the cost-effective replacement of unique range protocols, enhanced exchange of mission data, and organic TENA-compliant capabilities at sites which can be leveraged for future events, enhancing both reuse and interoperability.

TENA AND THE TELEMETRY COMMUNITY

As in the past, present telemetry (TM) support requires operators to be on location with the TM receive system or at a remote TM console (with a remote TM antenna control unit). This often results in temporary duty (TDY) for operators and potentially an insufficient number of operators to support all scheduled operations. The capability to remotely operate the telemetry system (i.e., perform status monitoring, data distribution, and / or command and control from a centrally-located, manned site) greatly reduces operational costs of TDY to remote TM sites. A remote control capability could altogether eliminate the existing requirement for personnel at both the local TM system antenna site as well as the TM control facility, alleviating previous manpower issues.

Naval Air Station (NAS) Patuxent River, MD (Pax River) Atlantic Test Range (ATR) is an excellent example of how beneficial TENA can be for TM control. Before work began to develop and field an enterprise approach to remotely monitor and operate all components of remote ATR ground telemetry systems, Pax River was faced with four major, and incredibly common, TM range issues: operator proximity, lights-out operations, a generalized interface, and Information Assurance (IA) requirements.

The existing approach at Pax required TM operators to be on location with the TM Antenna Control Unit (ACU) during missions. Any near-term remote operations concepts required a one-to-one correlation between the remote ACU and remote TM Antenna, and no sub systems were supported. They also had no ability to fully power-on, configure, operate, or obtain the status of their remote Auto-Tracking Telemetry System (ATAS) and Mobile Telemetry Acquisition System (MTAS) systems, therefore requiring personnel on-site to perform power-on and to configure all systems with no distributed status available from TM system components.

Vendor-specific interfaces and data models were used, which meant operators had to gain proficiency on each system component. This generalized interface prohibited uniform operator consoles, and limited the ability to easily access and share relevant metrics and engineering data. Furthering the problem was that Pax River had a limited ability to meet evolving IA requirements and Security Technical Implementation Guides (STIG) on system components.

Working alongside members of the TENA Software Development Activity (SDA), NAS Pax River developed an enterprise approach to remotely manage and operate all components of remote ground telemetry systems. This method provides a common architecture (TENA) which interfaces system components, regardless of system manufacturer. Upon completion, this effort now provides for single operator control of several remote TM systems, therefore reducing travel and manning requirements at remote sites. It also allows TM status information, setup, and control to be distributed to appropriate destinations for system verification and operations.

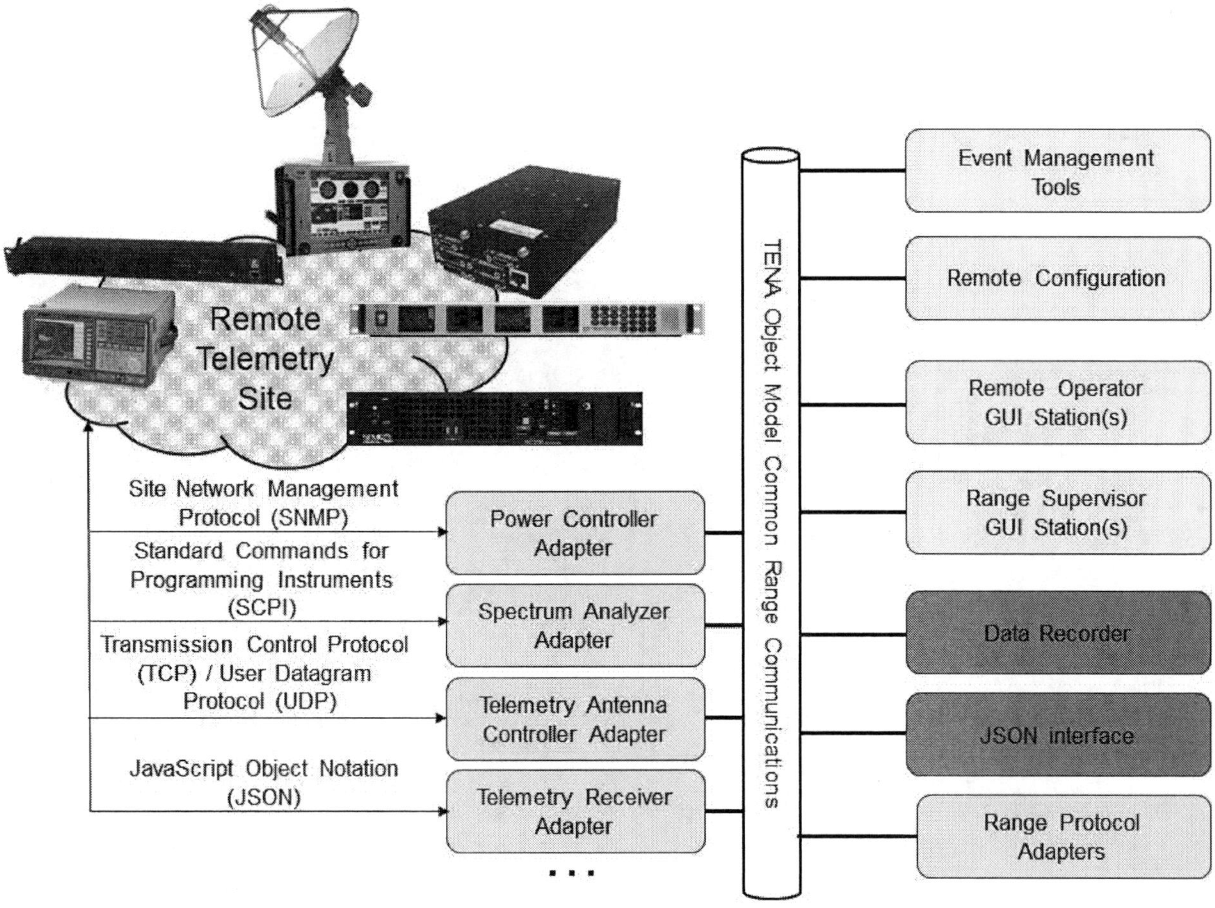

Figure 1 Architecture to a Remote TM Site.

Other recent implementations of TENA were at Yuma Proving Ground (YPG) in Yuma, Arizona and Redstone Test Center (RTC), in Huntsville, Alabama. YPG chose to add a TENA interface to their TCS Antenna Control Unit (ACU) model M1 used on their TM pedestals. Operational testing is currently underway on the remote monitoring and control capabilities of the telemetry antenna system using TENA.

RTC decided on the use of TENA to pull real-time Time, Space, Position Information (TSPI) data from the Smartronix Omega NeXT decommutation software. Engineers at Redstone developed a "Data Adapter Tool" which fuses other real-time TSPI sources such as inertial measurement unit (IMU) and differential Global Positioning System (GPS). The Data Adapter

Tool allows operators at RTC to now transport data via TENA using the Standard platform object model.

Other TM applications of TENA are ongoing at White Sands Missile Range (WSMR) in New Mexico, Vandenberg Air Force Base (AFB) in California, and Eglin AFB in Florida. WSMR reached out to the TENA SDA seeking a TENA-capable range interface unit (RIU) for existing radars; a TENA-capable Telemetry Tracker pointing data interface (as a modification to the existing RIU); and a persistent, distributed TENA capability through WSMR's Inter Range Control Center (IRCC). TENA is currently being used to connect FPS-16 radars, telemetry systems, and optics systems. Future plans at WSMR include the use of TENA for Real-Time Data Processing (RTDPS).

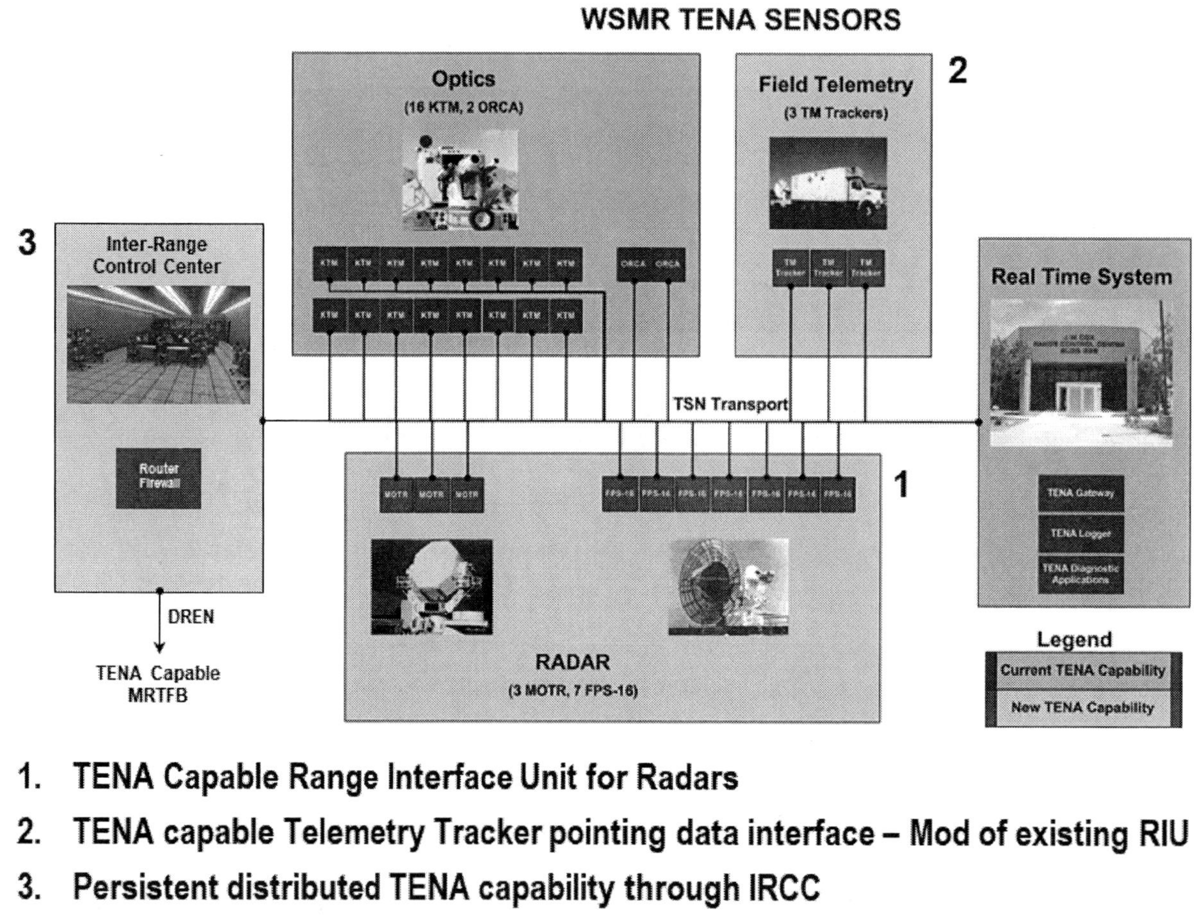

Figure 2 Use of TENA at WSMR.

PROVIDING INTEROPERABILITY AND RESOURCE REUSE

Interoperability is the characteristic of an independently-developed element to use commonality to work together with other elements toward a shared goal. Reuse is recognized as the ability to use an element in a context for which it was not originally designed; in essence focusing on the multiple uses of a single element and often requiring well-documented interfaces. In order to achieve interoperability and communicate meaningfully, a common architecture (including a common language and a common communication mechanism) and a common context (including environment and time) must be present. TENA was developed to bring the efficiency and economic advantages of interoperability and reuse to DoD test and training ranges. The initial interoperability and reuse efforts were completed in early Fiscal Year 2005, and the continuing interoperability and reuse refinement of TENA is managed by the TENA Software Development Activity (SDA).

Being successful in the development of any Joint testing capability requires a supporting and guiding activity, and in December 2005, the JMETC program element was formed. JMETC is a distributed live, virtual, and constructive (LVC) testing capability which supports the acquisition community throughout program development, developmental testing (DT), operational testing (OT), interoperability certification, and includes demonstration of net-ready key performance parameters (KPP) requirements in a customer-specific Joint mission environment (JME). JMETC provides readily available connectivity to the Services' distributed test capabilities and simulations, as well as that of Industry test resources. Initially developed as a testing capability, JMETC is aligned with and complemented by the JTEN, and together offer integration solutions fostering test, training, and experimental collaboration.

JMETC uses a hybrid network architecture. The JMETC Secret Network (JSN), which leverages the Secret Defense Research and Engineering Network (SDREN) for connectivity, is the T&E enterprise network solution for secret testing. The JMETC Multiple Independent Levels of Security (MILS) Network (JMN) supports higher classifications, as well as cyber testing.

The JMETC program uses TENA to improve its testing support infrastructure. With TENA, a live range instrumentation architecture and field-proven in exercises and numerous distributed test events since 2002, it provides JMETC a technology already deployed in the DoD. Where TENA serves as the middleware and software component, the persistent JMETC networks provide the connectivity for distributed testing.

UNDERSTANDING TENA

Understanding composability as the ability to rapidly assemble, initialize, test, and execute a system from members of a pool of reusable, interoperable elements, the TENA architecture is a technical blueprint for achieving an interoperable, composable set of geographically distributed range resources (both live and simulated) that can be rapidly combined to meet new testing and training missions in a realistic manner. TENA is made up of several components, including a domain-specific object model that supports information transfer throughout the event lifecycle, common real-time and non-real-time software infrastructures for manipulating objects, as well as standards, protocols, rules, supporting software, and other key components.

The TENA Middleware (currently at Release 6.0.5 and available for free download at the TRMC web site: https://www.trmc.osd.mil) combines distributed anonymous publish-subscribe, and model-driven, distributed, and object-oriented programming paradigms into a single distributed middleware system. This unique combination of high-level programming abstractions yields a powerful middleware system that enables TENA middleware users to rapidly develop complex yet reliable distributed applications.

The TENA object model consists of those object / data definitions derived from range instrumentation or other sources, which are used in a given execution to meet the immediate needs and requirements of a specific user for a specific range event. The object model is shared by all TENA resource applications in an execution. It may contain elements of the standard TENA object model although it is not required to do so. Each execution is semantically bound together by its object model.

Therefore, defining an object model for a particular execution is the most important task to be performed to integrate the separate range resource applications into a single event. In order to support the formal definition of TENA object models, a standard metamodel has been developed to specify the modeling constructs that are supported by TENA. This model is formally specified by the Extensible Markup Language (XML) Metadata Interchange standard and can be represented by Universal Markup Language (UML). Standards for representing metamodels are being developed under the Object Management Group Model Driven Architecture activities. The TENA Object Model Compiler is based on the formal representation of this metamodel, and TENA user-submitted object models are verified against the metamodel. However, it is important to recognize the difference between the TENA metamodel and a particular TENA object model. The object captures the formal definition of the particular object / data elements that are shared between TENA applications participating in a particular execution, while the object model is constrained by the features supported by the metamodel.

Another significant benefit for TENA users is auto-code generation. The TENA Middleware is designed to enable the rapid development of distributed applications which exchange data using the publish-subscribe paradigm. While many publish-subscribe systems exist, few possess the high-level programming abstractions presented by the TENA Middleware. The TENA Middleware provides these high-level abstractions by using auto-code generation to create complex applications, and these higher-level programming abstractions (combined with a framework designed to reduce programming errors) enable users to quickly and correctly express the concepts of their applications. Re-usable standardized object interfaces and implementations further simplify the application development process.

Through the use of auto code generation, other utilities, and a growing number of common tools, TENA provides an enhanced capability to accomplish the routine tasks performed on the test and training ranges in support of exercises. The steps in many of the tasks are automated, and thanks to the enhanced software interoperability provided by TENA, the information flow is streamlined between tools and the common infrastructure components.

TENA utilities facilitate the creation of TENA-compliant software and the installing, integrating, and testing of the software at each designated range. This complex task falls to the Logical Range Developer who, in this phase, performs the detailed activities described in the requirement definitions and event planning, as well as the event construction, setup, and rehearsal activities of the Logical Range Concept of Operations. While some manual exercise and event setup is required at ranges, TENA tools, as they are developed and become accepted across the range community, make exercise pre-event management easier.

SUPPORT FOR TENA USERS

The TENA SDA has developed a website that provides a wide range of support for the TENA user, including an easy process to download the Middleware, free of charge. The website also offers a helpdesk and user forums that will address any problems with the Middleware download and implementation. The TENA SDA is very aware of the need to inform range managers and train TENA users, and the TENA SDA presents regular training classes that are designed to meet attendees' needs; from an overview or technical introduction of TENA, all the way to a hands-on, computer lab class on the TENA Middleware.

TENA's continuing evolution in its support of the test and training range community is managed by an organization of users and developers. This collection of TENA stakeholders, called the Architecture Management Team (AMT), meets several times a year to be updated on TENA usage, problems, and advancements. Although the agenda involves briefings, it is open to wide-ranging discussions. This ensures the users' concerns and inputs are understood, recorded, and action items are made if necessary. Of equal importance, TENA developers and management have had a long and mutually beneficial relationship with the Range Commanders Council.

CONCLUSION

Although it was a technological and software evolution that was the impetus for TENA's growth in its enabling of range interoperability and resource reuse, the Middleware found its needed validation on the DoD test and training ranges. On those ranges, the U.S. Military evaluates the warfighting equipment, personnel, and concepts that are deployed in support of ongoing missions around the globe. Unfortunately, test and training events only provide the opportunity for evaluation. It is the data collection and analysis that determines the war-worthiness of the equipment or concept; this data can quickly and definitively illuminate any necessary improvements needed to ensure effective and safe weapon system operation and training. TENA and JMETC are time-tested, proven, integral parts of that equation.

JMETC reduces the time and cost to plan and prepare for distributed events by providing a persistent, readily-available network, and the TENA common integration software is easily-integrated into telemetry environments and applications. Even the remote control capability alone alleviates previous manpower issues and greatly reduces operating costs for the telemetry community. Using TENA and JMETC, Test Directors can put their focus back where it needs to be – on the warfighter and the task at hand.

For more information about TENA and JMETC, contact Ryan Norman, TENA SDA Program Manager (PM), AJ Pathmanathan, JMETC PM, or George Rumford, Test Resource Management Center (TRMC) Deputy Director – Major Initiatives Technical Analysis (MITA), E-mail: feedback@tena-sda.org or feedback@jmetc.org or go to the TRMC Web Site: https://www.trmc.osd.mil.

LINK DEPENDENT ADAPTIVE RADIO – DESIGN FOR iNET

Author: Tasmeer Alam

Master of Science in Electrical Engineering, Department of Electrical and Computer Engineering, Morgan State University

Faculty Advisors: Dr. Farzad Moazzami, Dr. Richard Dean

ABSTRACT

This paper focuses on the design of a simple Time Division Multiple Access (TDMA) signaling structure for the Link Dependent Adaptive Radio (LDAR) prototype wireless radio communication system to meet the timing requirements of the iNET standard. Built for aeronautical telemetry, LDAR adapts its modulation and coding scheme based on the channel condition in real time. In this paper, a simple protocol for transmission of Command Message and Data Message between Ground Station and Test Article is discussed. This protocol includes all analysis for the continuous adaptation of modulation scheme and coding rate to maximize throughput while ensuring a minimum level of link quality. This project was a collaborative effort between Morgan State University and Georgia Tech Research Institute and is a continuation of our previously published work on LDAR.

Keywords: *Link Dependent Adaptive Radio (LDAR), TDMA, iNET, OFDM, SOQPSK*

INTRODUCTION

Spectrum scarcity has become one of the major issues in wireless communication with the dramatic increase of data demand. With the growing demand, the aeronautical spectrum is expected to be congested [1,2,4]. Also wireless telemetry uses unpredictable and dynamic radio channels for transmitting information which affect performance in several ways including Doppler effect, noise, multipath etc. Channel dynamics, specially fading of channel becomes a bigger issue in highly dynamic environments such as aeronautical communication where the speed of aircraft exceeds mach1. Higher performance in data transformation requires improved channel [4]. Dedicated links between the ground station and test article of a telemetry communication has proven to be an inefficient bandwidth utilization method. The integrated network enhanced telemetry (iNET) project – supported by Test Resource Management Centre (TRMC) – has aimed at abandoning the point to point link dedication and moving toward networked telemetry. Georgia Tech Research Institute (GTRI) and Morgan State University have undertaken a project within iNET aimed at a bandwidth efficient and adaptive system for aeronautical communication. Link Dependent Adaptive Radio (LDAR) is an effort to maximize the throughput for telemetry links while ensuring an acceptable level of data quality and reliability [3-7].

LDAR

The main concept of Link Dependent Adaptive Radio (LDAR) is to develop a wireless radio communication system that adapts its modulation system and coding rate based on the channel condition in real time in a telemetry environment. Analyzing the current channel condition in real time, LDAR selects a modulation scheme (that suits best current channel condition) which maximizes throughput while ensuring a minimum level of link quality. If the quality of channel (SNR or delay spread) is improved, a control signal informs both Ground Station and Test Article of a telemetry communication to switch to a higher data rate [4]. But if the channel quality decreases or the performance error exceed the predefined acceptable threshold, the same adaptive mechanism ensures that next transmission happens in lower data rate, either by changing the modulation scheme or by reducing the coding rate [4,5,6].

SYSTEM MODEL

A system model for this prototype system is shown in figure 1 below.

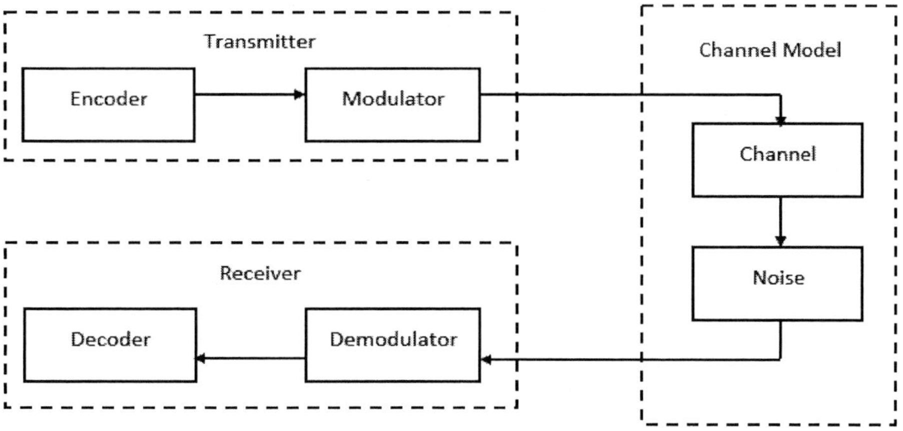

Figure 1: System Model

The system consists of a transmitter, a receiver, and a telemetry channel model. The transmitter has an encoder and a modulator which consists of Orthogonal Frequency Division Multiplexing (OFDM) and Shaped-Offset Quadrature Phase Shift Keying (SOQPSK) modulation. The receiver consists of a demodulator and decoder. The adaptation of modulation scheme and code rate takes place after the Signal to Noise Ratio (SNR) or delay spread is computed. Based on the predetermined table, LDAR decides on the next set of parameters for transmission.

CHANNEL PARAMETERS ADAPTATION

To make optimum decision for parameter selection or for mode adaption depending on channel quality, different code rate and modulation has been tested previously. Figure 2 shows a comprehensive set of results with two different coding schemes and all the Quadrature Amplitude Modulation (QAM) sizes under study of Additive White Gaussian Noise (AWGN) channel [6].

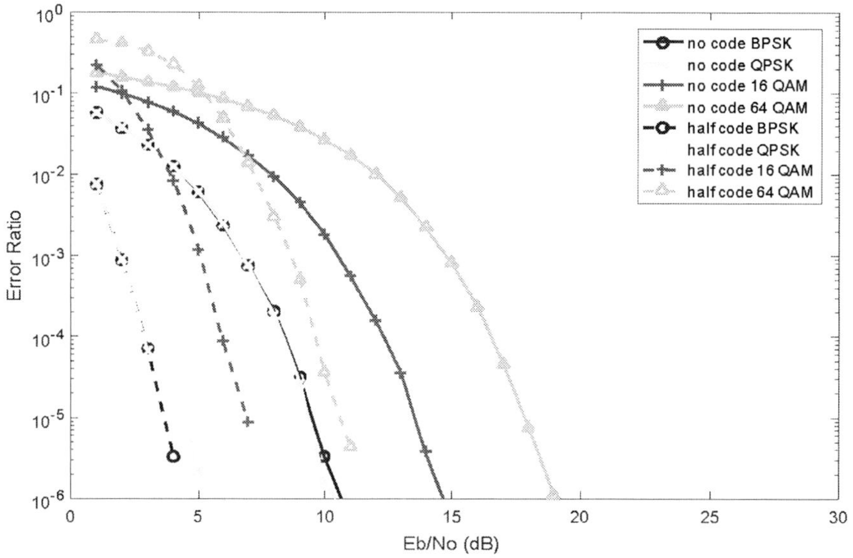

Figure 2: Error performance of AWGN channel with variable QAM and code rate

These curves result in creation of LDAR tables for noisy channel. After the Error threshold is satisfied, we chose the highest possible throughput above the threshold line. In our previous work, only OFDM-QAM modulation with different code rate were considered for LDAR and the above graph reflects those results. However, in iNET standard, two types of modulation are considered, SOQPSK and OFDM-QAM. Later, Both OFDM and SOQPSK modulation were incorporated in this project. A theoretical approach for Mode adaption and Transmission Table has been developed to be followed throughout the project. OFDM modulation incorporates QPSK, 16-QAM and 64-QAM modulation schemes with different Forward Error Correction (FEC) coding rates. Figure 3 shows the order of modulation schemes for LDAR with respect to the accepted delay spread and SNR of the channel. Table 1 shows different Modulation schemes with different coding rates which results in various values for throughput.

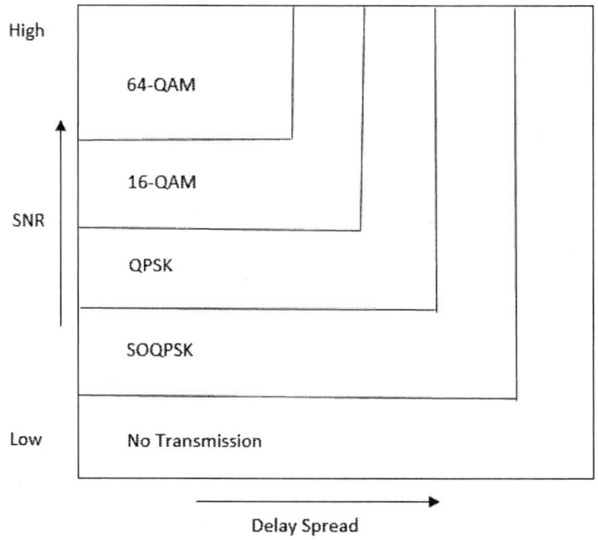

Figure 3: Link Dependent Adaption

Mode	Modulation	Code Rate, R	Data/symbol
1	SOQPSK	0.33	0.333
2	SOQPSK	0.50	0.500
3	SOQPSK	0.88	0.875
4	SOQPSK	1.00	1.000
5	QPSK	0.33	0.667
6	QPSK	0.50	1.000
7	QPSK	0.88	1.750
8	QPSK	1.00	2.000
9	16QAM	0.33	1.333
10	16QAM	0.50	2.000
11	16QAM	0.88	3.500
12	16QAM	1.00	4.000
13	64QAM	0.33	2.000
14	64QAM	0.50	3.000
15	64QAM	0.88	5.250
16	64QAM	1.00	6.000

Table 1: Different Transmission choices

In the next section, the two major Modulation schemes in iNET standard are introduced.

MODULATION SCHEMES

OFDM: Orthogonal Frequency Division Multiplexing (OFDM) is a powerful modulation technique that increases bandwidth efficiency and helps to mitigate inter-symbol-interference (ISI) and inter-channel interference (ICI). OFDM is a technique to combat frequency selective fading and has been shown to prove a good foundation for adaptable spectrally efficient modulation. OFDM divides spectrum into several sub-carriers. By dividing the bandwidth into smaller channel, narrowband channels are created from a wide-band environment. These narrowband subcarriers experience flat fading, which is an important characteristic for the aeronautical multipath problem. Each narrow band subcarrier supports an M-ary Quadrature Amplitude Modulation Scheme with M=4, 16, 24 or 64 [8].

In i^{th} block of OFDM, a set of N QAM symbols are transmitted over N sub-carriers and these frequency components are converted into time samples by performing FFT. After parallel to serial conversion, a cyclic redundancy of length v is added as a prefix and the signal is then serially transmitted over a noisy multipath channel with channel response h and noise vector w.

The output of the channel can be written as,

$$y^i = h_{circ}x^i + w$$

Where, h_{circ} is an $N \times N$ circulant matrix.

SOQPSK: The Shaped Offset Quadrature Phase Shift Keying (SOQPSK) modulation scheme is defined as a continuous phase modulation (CPM) with baseband modulated signal [3]. Different versions of SOQPSK differ by their respective frequency pulse. In this project, we use SOQPSK-TG (Telemetry Group SOQPSK).

The output of the channel can be written as,

$$y_n = h * x_n + w_n = \sum_{l=0}^{L-1} x_{n-l} h_l + w_n$$

Where, x_n is a discrete time SOQPSK waveform, h_l is the multipath channel impulse response, L is the channel length and w_n is an additive white Gaussian noise [9].

In the next section, LDAR modes of operation is discussed.

LDAR MODES OF OPERATION

This section includes an update to our prior publication on the signaling of LDAR [4]. The transmission of Control Messages over Link Dependent Adaptive Radio (LDAR) depends on four different modes between Ground Station and Test Article: i) Ground Station Transmitter ii) Test Article Receiver iii) Test Article Transmitter iv) Ground Station Receiver

The diagram of LDAR Modes of Operation has shown below in figure 4.

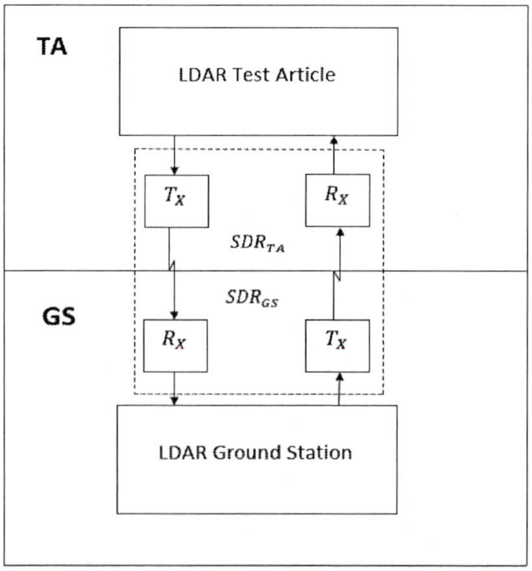

Figure 4: LDAR Modes of Operation

The process starts by sending a command message (CM) by ground station transmitter using a test sequence script through an Uplink radio channel. The receiver of the test article accepts this message to direct or command the Test Article (TA) to transmit the Data Message (DM). The initial command message includes information about the modulation scheme and coding rate on which the Data Message can be assembled. After receiving Command Message, test article receiver then decodes CM and sends the mode selection and coding rate to test article transmitter. The transmitter then processes data message according to the specification of mode and code rate. The TA transmitter conveys DM to ground station receiver through downlink radio channel.

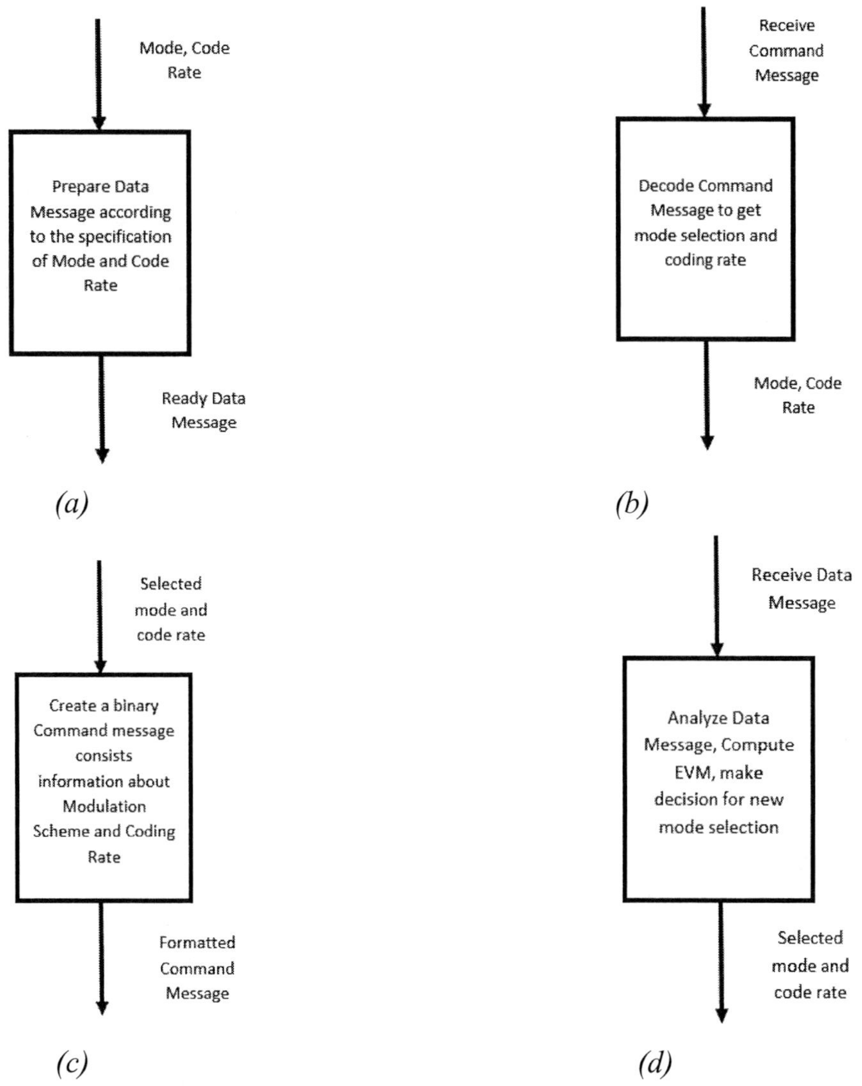

Figure 5: Modes of operation (a) Ground Station Transmitter (b) Test Article Receiver (c) Test Article Transmitter (d) Ground Station Receiver

COMMAND MESSAGE

The Command Message is the coded version of Mode Rate. This is a 32-bit message with 5 repetitions. It comprises of Mode Rate code as 16 bits and 16 bits of Frame Structure. 16-bit Mode Rate contains 8-bit data and 8-bit Cyclic Redundancy Check (CRC). 16-bit Frame Structure is also created with 8-bit data and 8-bit CRC. The 8-bit data for Mode rate includes the binary information of Modulation and Code Rate selection. And 8-bit data for Frame structure contains information for different frame structure for data packet to be sent.

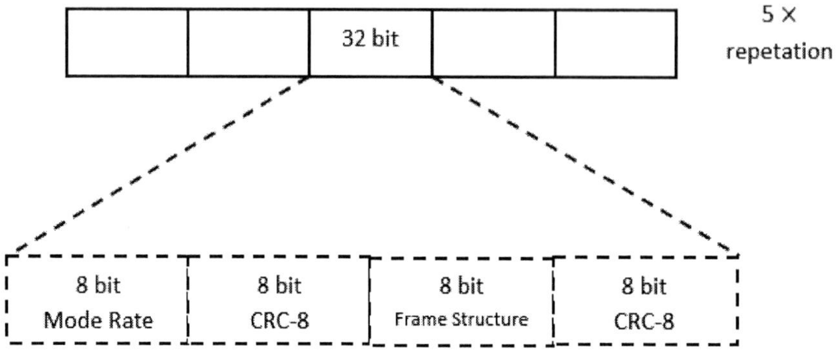

Figure 6: Command Message Structure

DATA MESSAGE

Based on the Mode Rate and variable frame structure, data message of our simulation is the 2047 PN sequence initialized as 00000000001 from polynomial $z^{11} + z^2 + z$ with a '1' appended to last bit. This Sequence is repeated multiple times to represent a complete frame. (sub-frame × framemultiplier) and truncated for the last packet as needed.

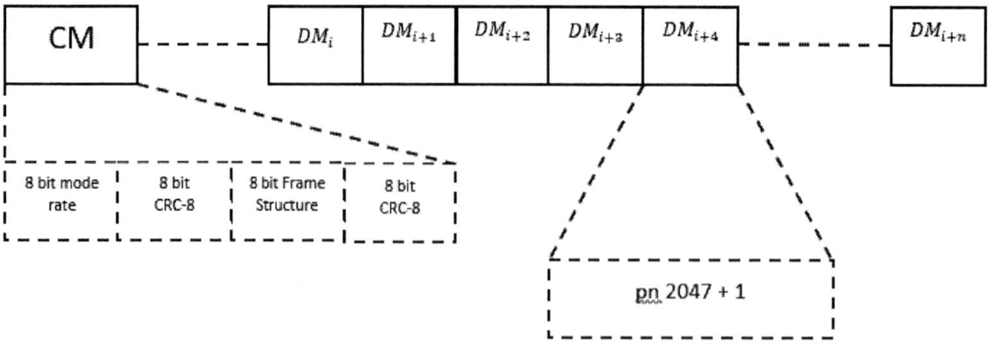

Figure 7: Data Message Structure

Table 2 and 3 show the specifications of Packet Structure and Frame Structure for command and data messages.

Mode	Modulation	bits/symbol, K	Code Rate, R	Data/symbol	Sub-Frame	Data/Sub-Frame
1	SOQPSK	1	0.33	0.333	1.4msec	2731
2	SOQPSK	1	0.50	0.500	1.4msec	4096
3	SOQPSK	1	0.88	0.875	1.4msec	7168
4	SOQPSK	1	1.00	1.000	1.4msec	8192
5	QPSK	2	0.33	0.667	1.4msec	8192
6	QPSK	2	0.50	1.000	1.4msec	12288
7	QPSK	2	0.88	1.750	1.4msec	21504
8	QPSK	2	1.00	2.000	1.4msec	24576
9	16QAM	4	0.33	1.333	1.4msec	16384
10	16QAM	4	0.50	2.000	1.4msec	24576
11	16QAM	4	0.88	3.500	1.4msec	43008
12	16QAM	4	1.00	4.000	1.4msec	49152
13	64QAM	6	0.33	2.000	1.4msec	24576
14	64QAM	6	0.50	3.000	1.4msec	36864
15	64QAM	6	0.88	5.250	1.4msec	64512
16	64QAM	6	1.00	6.000	1.4msec	73728

Table 2: Data/Packet Structure

Frame Multiple	Frame (ms)	Guardband (ms)	Frame Period (ms)
10	1.4	0.14	15.4
30	1.4	0.14	46.2
100	1.4	0.14	154
300	1.4	0.14	462
1000	1.4	0.14	1540
10	1.4	0.28	16.8
30	1.4	0.28	50.4
100	1.4	0.28	168
300	1.4	0.28	504
1000	1.4	0.28	1680
1000	1.4	0	1400

Table3: Frame Structure

TDMA STRUCTURE

The Uplink and Downlink channels have a simple Time Division Multiple Access (TDMA) structure in order to meet iNET specification [3].

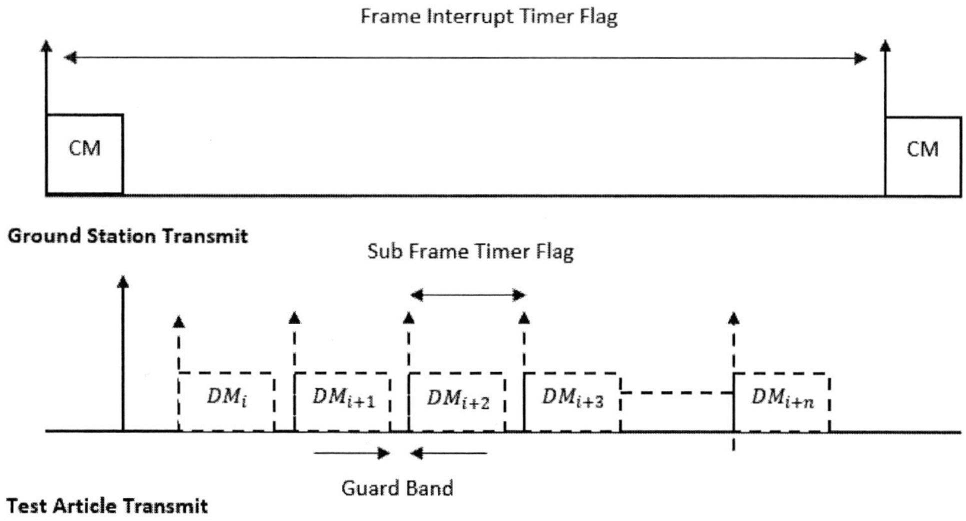

Figure 8: TDMA structure for uplink and downlink channel

DETERMINATION OF CHANNEL QUALITY

For LDAR to maximize the throughput of the telemetry link while ensuring the minimum level of reliability, the channel condition must be estimated to determine what mode and code rate needs to be selected. Georgia Tech Research Institute (GTRI) has developed a new approach of an adaptive link metrics that measures the degradation due to thermal noise and channel effects. They present the Error Vector Magnitude (EVM) for OFDM and second order Godard Dispersion ($D^{(2)}$) for SOQPSK and derived a mathematical relationship between these two metrics [9].

EVM measures the difference between the received symbols and their expected positions in the I/Q plane. EVM has no meaning for SOQPSK since there is no soft decision in the demodulation of the signal. Second Order Godard dispersion, defined as $D^{(2)} = E\,[(|y(n)|)^2 - R_2)^2]$ where $y(n)$ is the received signal and R_2 is the amplitude of the constant modulus signal, to measure the quality of a received SOQPSK signal. The mathematical relationship between EVM and Godard dispersion derived by GTRI enables to utilize a set of empirically-derived adaptation rules that incorporate both modulation schemes [9].

Using the mathematical equation provided in the mentioned paper [9], we can determine the channel condition by comparing the incoming Data message through the channel.

CONCLUSION

This paper shows the design of LDAR with a simplified iNET Time Division Multiple Access (TDMA) structure of Uplink and Downlink channel. This work maps the LDAR design into timing structure of iNET by creating command messages that governs the mode of operation with a simple TDMA structure. Its adaptation of Modulation depending on the channel condition ensures error performance is controlled by a specific command and control messaging that we showed in this paper.

REFERENCES

[1] T. Rappaport, Wireless Communications, Principles and Practices, 2nd Edition, Prentice Hall PTR, Upper Saddle River, New Jersey 07458

[2] P. Jacob, Rajendra P. Sirigina, A.S. Madhukumar, Vinod A. A. Prasad, "Cognitive Radio for Aeronautical communications: A survey", IEEE 2016.

[3] iNET, Communication Link Standard, version 0.7, February 2010.

[4] Tasmeer Alam, Farzad Moazzami, Dr. Richard A. Dean, "Command Signaling for Link Dependent Adaptive Radio", ASEE Mid-Atlantic Conference, Spring 2017, Baltimore, MD.

[5] Enkuang D. Wang, Brett T. Walkenhorst, Jieying Han, "A simulation Testbed for Adaptive Modulation and Coding in Airborne Telemetry", ITC 2014, San Diego, CA.

[6] Tara Pun, Deepak Giri, "Link Dependent Adaptive Radio Simulation", ITC 2014, San Diego, CA.

[7] Deepak Giri, "Flight Path Simulation in LabVIEW using LDAR", ITC 2015, Las Vegas, NV.

[8] M. Elrais, B. Mengiste, B. Gautam and E. Dambia, "Variable OFDM Performance on Aeronautical channel", ITC 2013, Las Vegas, NV.

[9] Jieying Han, Brett T. Walkenhorst, Enkuang D. Wang, "Adaptive Modulation Schemes for OFMD and SOQPSK using EVM and Godard Dispersion", ITC 2014, San Diego, CA.

INEXPENSIVE UHF TRANSCEIVER LEVERAGING COTS COMPONENTS

Owen Chiaventone, Kyle Avola, Stetson Tuschhoff
Electrical and Computer Engineering Department
Missouri University of Science and Technology
Rolla, MO 65401
Faculty Advisor:
Dr. Kurt Kosbar

ABSTRACT

This paper describes the design of an inexpensive UHF transceiver which leverages some of the recently developed commercial off-the-shelf (COTS) components. The initial goal is to implement digital voice transmit and receive function, although the design can accommodate a wide range of digital communication and telemetry applications. The handheld transceiver transmits 5 watts of power in the 430-435 MHz UHF band. A 1.2 kHz wide GFSK modulation format is used, generated by a Silicon Labs radio chip. The recently released Raspberry Pi Zero processor implements a low bit rate audio coding which conforms to the Codec2 standard. The transceiver fits in a 3 cm x 8 cm x 14 cm volume. It is powered by two 18650 lithium ion cells, and draws approximately 1 watt of power during receive, and 6 watts during transmission.

Key Words

UHF GFSK Codec2 DSP

INTRODUCTION

This paper describes a cost effective digital handheld two-way radio suitable for a broad range of applications. FM and AM two-way handheld radios have been around for a very long time, have batteries capable of handling a day of continuous use, and a real world range exceeding 30km on 5 watts of power. A digital mode UHF/VHF radio would have several advantages over an analog radio, but its primary benefit is signal quality. With a quality codec and implementation, the digital radio has a greater range (A real world example of greater than 45 km for the same power) for the power input. Instead of a fading out effect as the user goes beyond range, the digital radios exhibit consistent voice quality until the connection is dropped, giving the user time to finish their statements as they go out of range. This radio aims to provide a ready-to-use radio or a software-modifiable platform for modifying the unit however the end user requires.

To make this a viable project, the radio must include a microprocessor with sufficient processing power to decode the voice data stream and a radio of sufficient sensitivity to decode very weak signals, such as those -100dBm and below. These two requirements must then be weighed against the ever present restriction of the battery life able to be given to a hand held project, the ability to fit easily in one hand. The radio must have a volume no more than 3x8x14cm, and should not weigh much more than a typical cellular or current hand held device, also imposing restrictions on the battery. The radio must be able to decode weak signals as interruption in decoding capability will cause the cessation of communication, and should be sufficiently powerful to provide communications over the specified range. The microprocessor should be able to handle the continual conversion of the microphone data stream using the DSP-intensive codec.

Our design does not rely on a proprietary codec or encoding chip, focusing on using off-the-shelf technology and an open hardware and software design. Our radio focuses on an open source codec which is processed on the radio's processor, saving the cost and extra power draw of an encoding chip. Looking to other examples of radios built using this codec, this radio will use prepackaged ICs whenever possible, saving the time, development cost, and board area common with a discrete approach.[4]. In contrast, our team wished to diminish development time and get a mature product to the market faster at the cost of some additional features that were likely to be unused outside of the experimental community. Finally, for the experimental community (or for those who require more control over their radio), our radio is still easy to work with in terms of source availability and hardware compatibility.

Our approach centers around a ready built radio chip with a built in microcontroller being controlled by a low power (<500mW) ARM processor capable of providing the digital signal processing. An off the shelf approach allows us to keep our time investment very low. Instead of using discrete components, we were able to use a single, low cost chip with a small amount of supporting circuitry to conserve board space and component cost, versus a digital synthesizer and oscillator chip approach. With a ready made radio front end, our team was able to focus on programming the radio and ensuring the signal chain was correct from the microprocessor to the radio. Because of the decision to incorporate the codec into a standard, well documented ARM chip, our team could simplify the board layout, as well as conserving the battery by using one less chip. In addition, this provided one more opportunity for meeting the cost savings goal, as a free codec with sufficient ability greatly reduces the cost of the end unit, with the codec being a major expense in the radio hardware. Focusing on off-the-shelf and open source technology allowed the team to greatly compact and dramatically lower the cost of the final product.

SPECIFICATIONS AND REQUIREMENTS

In order to be a practical, viable radio in the market, it was agreed upon that the unit must be able to conform to the following criterion: Power, Sensitivity, Runtime, Clarity, Cost, and Dimensions. First and foremost, in order to compete, the radio had to approach or exceed 5W of power on the UHF 430 MHz ISM band. While an advanced codec is able to decode at a lower signal receive power and a higher signal to noise ratio[5], 5W was set at the target output to maximize the range while maintaining runtime. The radio also had to be capable of receiving signals fainter than existing products. The Si4464 chips used in the second and third prototypes are capable of decoding to -126dBm[3]. Runtime was a must have, and our target was for 120 minutes of continuous use (Assuming a 20% transmit usage cycle). In order to meet this goal, a pair of Lithium 18650 cells were used. With the power and range handled, the codec would be at its optimum performance, and we wished for clarity of speech to remain excellent to the very edge of its range. The open source Codec2 was selected for its excellent properties, including weak signal performance and low bandwidth constraints. [2] Our original target market, the local amateur radio community, was polled for the cost constraint. It was found that a sub-75 USD radio would be a very popular item, and every effort to remain below the 75 USD parts per unit price was made. Finally, the dimensions (both volume and weight) were in our minds as we made our prototypes, however the final prototype made to date does not yet follow these dimensions.

TECHNICAL APPROACH

SOFTWARE

Rowetel's Codec2 is at the heart of the radio design. The codec provides communications quality digital transmission at bitrates between 700 and 3200 bits per second. While not cellphone quality audio, a radio powered by Codec2 would allow for legible communications in many usage scenarios (such as construction, operations, or navigation) where users could reasonably expect noisy backgrounds and external interference to make perfect quality audio unnecessary. As communications requirements grow and frequencies become congested, efficient use of bandwidth is a requirement. The team selected a 1.2kHz channel with GFSK, providing more than the 700 bits per second requirement with room to fit forward error correction. A simulated channel was created with the ability to vary the noise floor, providing an opportunity to tune the codec before finishing the hardware (as seen in Figure 1).

PROCESSOR AND RADIO

The selection of the hardware was based around the codec and modulation scheme chosen to make sure no lock-in products were selected, leaving the decision as platform-agnostic as possible. The processor would handle the combination of the various user input sources (keypad, microphone, battery life remaining, USB charge/data) and output (Speakers, LCD Display, USB Data), as well as the radio I/O buffer. The processor would also encode and decode the microphone input stream, and the radio buffer input stream. The radio could then be any data chip with a byte I/O buffer.

USER I/O

To simplify the operation of the radio, a simple user interface was selected, composed of a frequency change mechanism, an on/off/volume potentiometer, and a USB charging system. The frequency change mechanism consists of a series of pushbuttons, a pair for up and down frequency select, and a pair for up and down band select. An LCD screen displays the users current frequency. The on/off rotary switch disconnects the battery from the regulation circuitry to save power, while

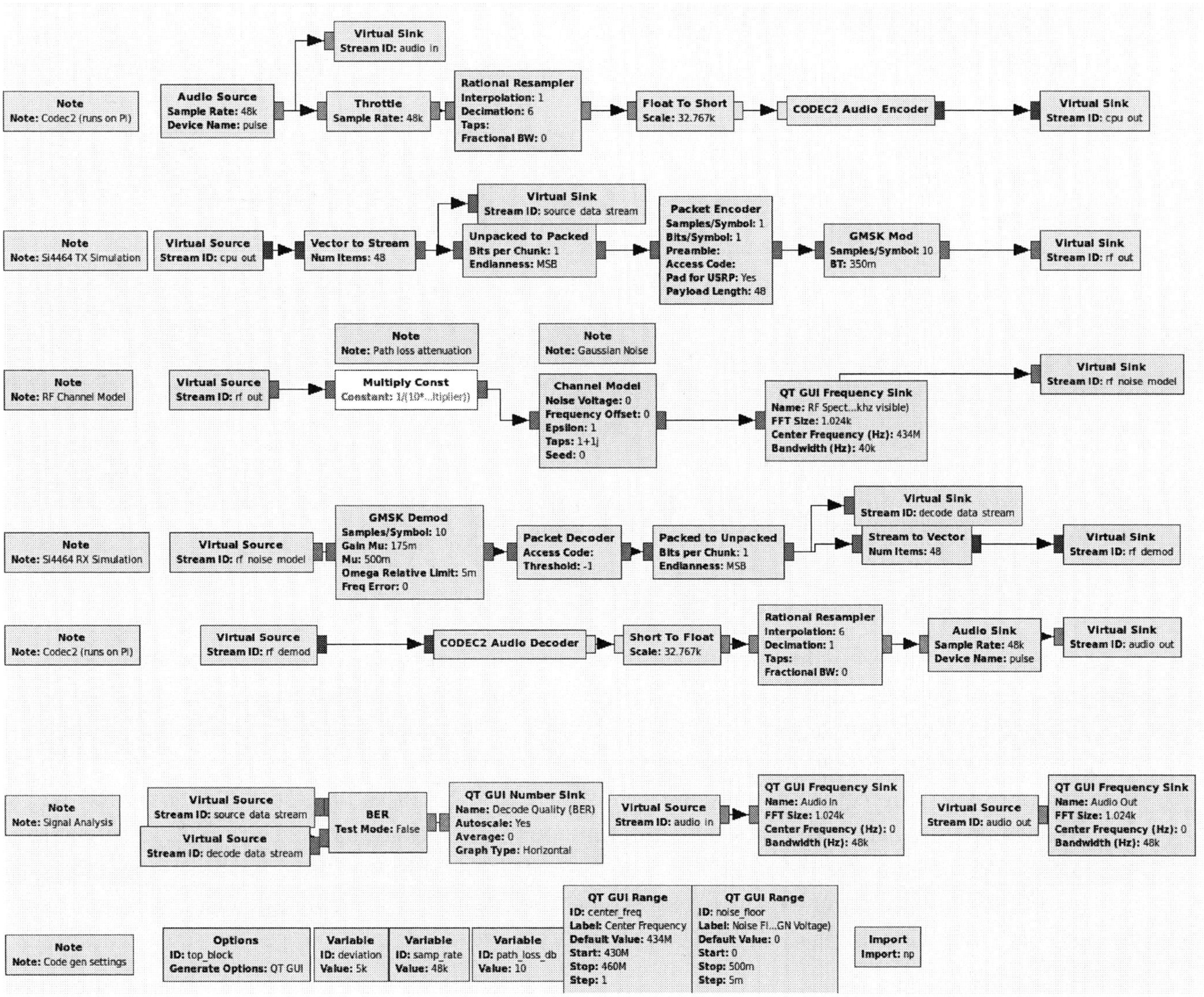

Figure 1: GNU Radio Flow Graph

the integrated volume potentiometer is connected to the processor via a 10 bit ADC. The USB port is capable of serial communication and fast charge for programming applications.

SUPPORTING CIRCUITRY

The battery charge and regulation system was selected to provide a constant 5V/3.3V source through a series of buck and boost converters. The USB port provides power to the charging buck converter to charge the battery, which is boosted from 3.2-4.2V to the running voltage of 5V, with a seperate 3.3V buck path.

FIRST REVISION AND REDESIGN

PROCESSOR, RADIO CHOICE

The initial task was the selection of the processor and radio. In order to get the signal to noise ratio required for long range communications, our choice fell to two modulation schemes: GFSK and QPSK. QPSK-16 was chosen due to its excellent performance in terms of power per bit ($\frac{E_b}{N_0}$), which left us with only one radio front end chip available at the price point and external interface complexity we were willing to expend. The On AX5043 offered excellent frequency range, very good sensitivity at a high data rate, and QPSK-16 support. An STM F4 series chip was original selected as our processor, providing low power draw and high performance. A development board was made for the radio, and a discover board was bought for the processor.

RESULTS

The results from this combination of hardware were not greatly encouraging in the first round. Two major issues prevented us from going further with this development track, time and complexity. Approximately 5 months was available to complete the project, leave a great short of time available for working through complex issues. The STM chip, while powerful, was a complex chip to work with, requiring either in-depth knowledge of the API and feature set exposed, or the time to develop low-level drivers for the chip, time our team did not have. The AX5043 shipped with a driver and configuration for the 8051, however, it had a difficult to work with feature set as well, and little if any community support for which to draw help. Eventually, the team was able to get a carrier from the combination, but the chip was seen as a lost cause to get the setup working within the time frame.

PITFALLS

We learned several lessons from this loss, but the most important being what to look for in selecting components. Instead of looking solely for a chip based on its technical capabilities, it was important to also look at support available, tools available, and overall time to complete a task for the given platform. If no tools were available, or were difficult to use, more time would be spent working with the toolset than developing for our application. If no support was available, the team would quickly find itself burning valuable time attempting to understand low level hardware than the issues which it set out to solve. If either of these grew to great, we would not be able to

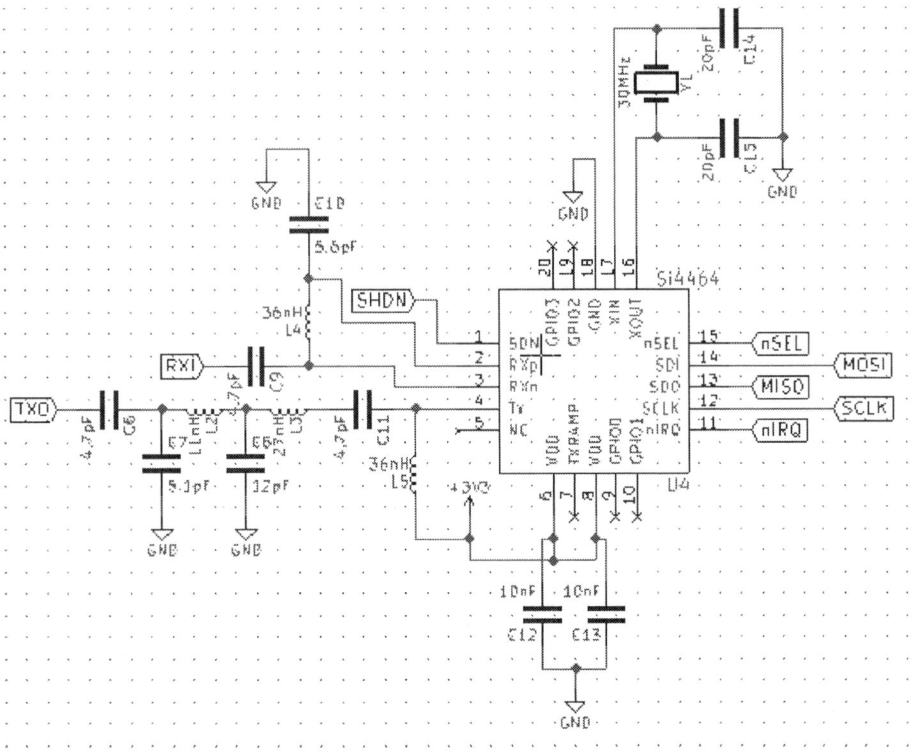

Figure 2: Radio and Matching Network

complete the project.

SECOND & THIRD REVISIONS

PROCESSOR, RADIO CHOICE

With the previous failure to make the radio and the processor communicate effectively, it was decided to simplify as much as possible to produce a working product. A Raspberry Pi was selected for its ease of GPIO control and heavy processing capability, which we hoped would allow for quick and easy turnaround on our software development. Our standby choice, an Si4464 radio, was selected to take over the radio front end. The Si4464 had excellent community support, prebuilt drivers for the raspberry Pi, and would be easy to build a radio platform for. Selecting the Pi, however, forced us to use a processor that was not power efficient for the task, and the Si4464 was not able to support QPSK-16. It was decided that these were worthwhile trade-offs to get a running prototype. In order to "breadboard" the radio in a way suitable for use with the Raspberry Pi, a development board was built, which included the radio, supporting circuitry and matching networks, and the volume control knob (Figure 2). In revision three, a preamplifier, final amplifier, and RX/TX path select switch was added to the matching network front end (Figure 3).

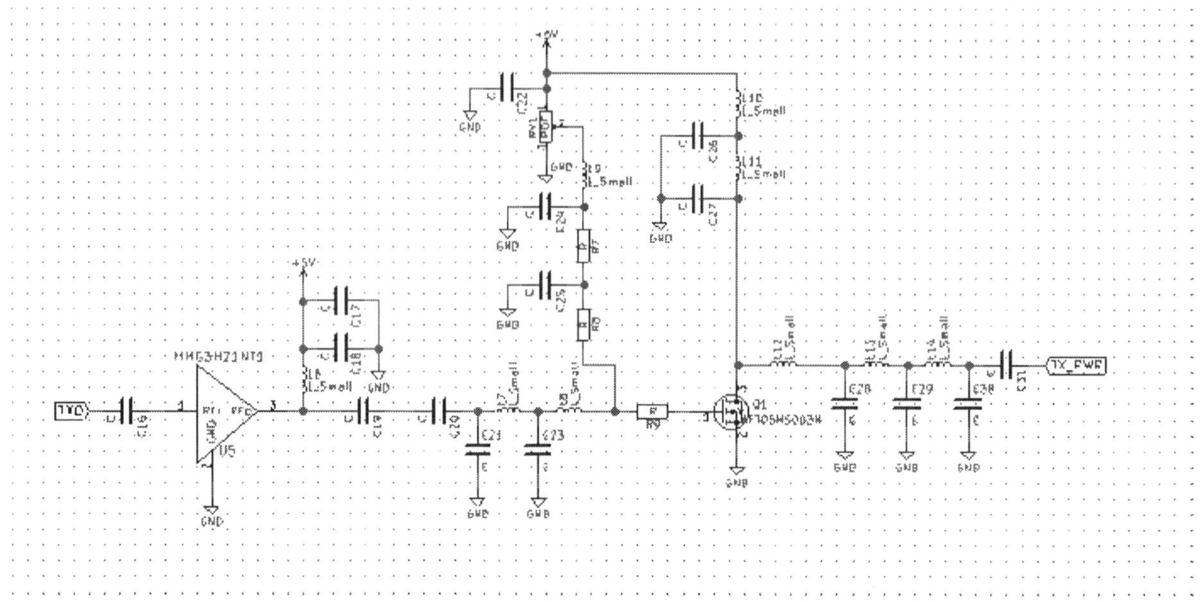

Figure 3: Amplifier Chain

BATTERY MANAGEMENT

The decision was made early on that the radio would be charged through a standard 5V USB micro B port. Such a charging method would enable quick, convenient charging with a charger that almost everyone would already have. To extend the capacity of the 18650 cells, the two 3.7V cells were placed in parallel. This allowed a standard buck converter to reduce the 5V input to a 4.2V charging voltage for the batteries (Figure 4). On the output side of the converter, a boost converter was used to supply 5V power to the rest of the system (An NPN MOSFET allowed for boost cutout when the USB charger was plugged in and the unit was operating at low load), followed by a buck converter to provide 3.3V to the rest of the system.

USER INTERFACE

The User Interface was comprised of a volume control knob, a series of pushbuttons for frequency selection, a Push-To-Talk switch and a LCD Screen. It was decided to keep the user interface simplistic to ensure the radio would not become too difficult to learn, a common problem with modern digital radios which include a great deal of features, filling never ending menus. Instead, volume was easily controlled, a frequency was able to be selected (and later, power output), and then one simply had to transmit.

RESULTS

The second and third prototype boards were greatly encouraging compared to the original choices. The user interface was complete, and a battery management system was demonstrated and capable of providing charging and output power to the system. While drivers were built for the Pi, a welcome improvement, they were for a previous version and had to be modified for the Pi

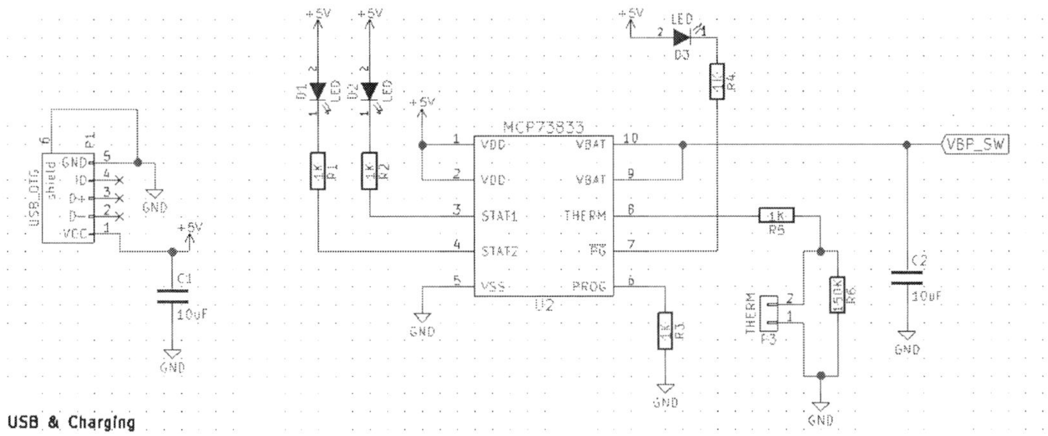

Figure 4: Battery Charging from USB

v3. Unfortunately, the Si4464 used an asynchronous serial select line, which is not handled by the Linux kernel. We were unable to get the Pi and the radio to talk with one another, requiring instead the use of an Arduino for testing purposes. As the Arduino was unable to handle codec processing of the microphone input, a demonstration was able to be made of a computer handling encoding of the data, passing the data to the Arduino, and then sending the data (received by an SDR).

FUTURE WORK

TRANSMISSION AND RECEPTION

As the team begins to wrap up its work, the next milestone is the successful transmission and reception of voice data packets. For this milestone to be met, a slight addition will need to be made to the radio to overcome the asynchronous serial issue met during testing of the third revision. Depending on microprocessor choice, the raspberry Pi processor will require an external processor which is capable of acting as a buffer and I/O controller for the Pi (Assuming we wish to continue with the Linux kernel), or a switch back to a dedicated low-power DSP capable ARM chip will have to be made. The benefit of the raspberry Pi processor is the vast array of drivers already included with the kernel, and the ease of accessing input and output for prototype use and changing. However, we eventually will have to migrate to a different microprocessor, and until then, the Linux kernel will act as an intermediary, slowing communications and requiring an IO processor to handle the radio. If the switch to a more complex chip were made at this time, effort would be spent early on to port drivers and handle more of the underlying hardware instead of providing a quick prototyping environment, but we would be more knowledgeable of the end product when the time comes to switch to the new chip.

FULL INTEGRATION

Following a successful transmission and reception of voice data, the radio must be integrated to form a seamless experience. At this time, the prototype is bulky and over size requirements. To finalize the product, the various subunits of the product must be designed to work with one another and the unit must be compacted. For the subunits to be designed together, the power management must be added to the processing board, which should be able to snap on in a low-profile manner to the radio board (which has been moved off the processing chip for concerns of signal and noise issues). Following the resolution of those items, the unit will be a full-featured radio.

CONCLUSIONS

While it is clear that there is a great deal of work necessary to get this radio in a state suitable for use in emergency and commercial services, our team has built a solid base on which to work on this project in the future. With a method of simulating encoding and decoding, adding noise and processing times, the future of the project will not rely as greatly on any one bottleneck issue. With the user interface in place and integrated with the radio drivers and processing front end, future work will be easier and quicker to implement, allowing the team to spend more time on adding features and cleaning up issues as they arise. Future work will also involve a better management of resources, including money, time, and personnel. A digital voice mode radio using the Codec2 codec is not far away.

REFERENCES

[1] "DVSI home page," in DVSI, 2016. [Online]. Available: http://www.dvsinc.com/index.htm. Accessed: Dec. 13, 2016.

[2] D. Rowe, "Codec 2," in Rowetel, 2016. [Online]. Available: http://www.rowetel.com/?page_id=452. Accessed: Dec. 13, 2016.

[3] "SI4464/63/61/60 Datasheet," in Silicon Labs, 2016. [Online]. Available: https://www.silabs.com/Support%20Documents/TechnicalDocs/Si4464-63-61-60.pdf. Accessed: Dec. 13, 2016.

[4] david rowe, "SM2000 Part 1 – Open VHF digital voice radio," in Rowetel, 2015. [Online]. Available: http://www.rowetel.com/?p=4694. Accessed: Dec. 13, 2016.

[5] Stallings, Data and Computer Communications, 10th edition, pgs 98-103

[6] RadioHead: RadioHead Packet Radio library for embedded microprocessors, RadioHead: RadioHead Packet Radio library for embedded microprocessors. [Online]. Available: http://www.airspayce.com/mikem/arduino/RadioHead/. [Accessed: 02-Mar-2017].

DESIGN OF A LONG RANGE COGNITIVE HF RADIO WITH A TUNED COMPACT ANTENNA

Noel Teku*, Gitansh Gulati*, Hamed Asadi*, Garrett Vanhoy*,
Ahmed H. Abdelrahman**, Kevin Morris*
Faculty Advisors: Dr. Tamal Bose* and Dr. Hao Xin*
*University of Arizona, Electrical and Computer Engineering Dept.
Tucson, AZ, 85721
**University of Colorado Boulder, Electrical, Computer and Energy Engineering Dept.
Boulder, CO, 80309
[nteku1,gitanshg,hasadi,gvanhoy,abdelrahman,kevinmorris,tbose,hxin]@email.arizona.edu
[ahmed.abdelrahman]@colorado.edu

ABSTRACT

High frequency (HF) communications, ranging from 3 to 30 MHz, are utilized by many radio enthusiasts to conduct transmissions with users across the globe. These communications depend on successfully reflecting signals off of the ionosphere. However, numerous factors (i.e. power level, coding, modulation, etc.), combined with the instability of the ionosphere, can make transmissions over this frequency band unreliable. Thus, an HF communication system design is proposed to offer more robust long range HF communications. The system has a cognitive engine that can determine transmission parameters (i.e. coding, modulation, etc.) capable of providing a high throughput and low bit error rate in various environments. The system also has a low-profile helical antenna that, combined with a matching circuit, is capable of receiving signals over different subsets of the HF band. These two components constitute a system capable of effectively transmitting and receiving signals over the HF band.

INTRODUCTION

High Frequency (HF) communications, occurring over the frequency range of 3-30 MHz, has become a popular medium for communications over the past few decades. They enable radio enthusiasts to conduct long-range communications across the globe without the use of a satellite. These communications are conducted using ionospheric reflections to relay the signals across large distances. These reflections can be made with low power and do not require the use of any external equipment (i.e. satellites, beacons, etc.) to transmit signals. Because of this lack of dependence on major communication architectures, in the event of major systematic failure, HF communications would be more reliable than most mainstream communication systems [1]. However, the conditions of the ionosphere are constantly fluctuating and dependent on numerous parameters (i.e. time of day, location, solar flux, etc.), making it a somewhat unreliable medium. In addition, transmissions made over the HF band contain very low data rates. Thus, it is desired to develop a system that can overcome these issues and be capable of making transmissions over the HF band.

This paper proposes a design for such a system with an electrically small helical antenna, matching network, and an embedded intelligent agent known as the cognitive engine. The helical antenna, which is 2 meters in height, is demonstrated employing passive electronically switched matching circuit at discrete frequencies for amateur, maritime, and broadcast applications. In addition, an active broadband matching circuit has been developed to achieve wideband improvement of received signal strength. Furthermore, a wideband unbalanced-to-unbalanced transformer has also been designed, tested, and measured for qualitative analysis in terms of received signal strength, signal-to-noise ratio (SNR), and signal intelligibility. A cognitive engine (CE) is defined as "an intelligent agent which observes the radio environment and chooses the best communication settings that best meet the application's goal" [2]. In other words, the cognitive engine enables the radio to determine the most optimal method for transmitting in a specific environment. As [3] explains, a radio should be capable of using "a large number of the plethora of current communication methods,...[and] be able to select the method that best meets its objective under the current operating environment." The CE is implemented in software and allows the radio to strategically explore different communication methods and utilize the methods deemed most effective by the algorithms. When a CE algorithm is applied to a radio, the end result is referred to as a cognitive radio (CR). The CR subsequently increases "radio performance and maximizes utilization of the available resources such as power and radio frequency (RF) spectrum" [2].

The following outlines the structure of the paper. First, the background section provides details on the objectives of the helical antenna and matching circuits as well as the CE algorithms that constitute the digital component. The development section provides an explanation regarding the implementation of the CE algorithms and how the current specifications of the helical antenna and matching circuits were obtained. The experimental results section shows verification of the performance of the helical antenna and matching networks in relaying signals from long ranges and the effectiveness of the CE algorithms in determining the best methods for transmitting when trying to maximize a certain parameter (i.e. throughput, power consumption, etc.). The conclusion section provides a summary of the work completed in this project and next steps.

BACKGROUND

A. Cognitive Engine

As stated in the introduction, the objective of the CE is to determine the best configuration for a radio to transmit in any environment. The main complication is that there are numerous parameters to examine; such as, modulation, coding schemes and transmitting power. It is often impractical for the radio to spend large amounts of time determining which combination of these parameters are the most effective for transmitting, especially in situations that require dire urgency [4]. Likewise, it is often not practical to select a single method that may not be as effective as other possible options. Thus, a balance has to be satisfied between experimenting with different parameters for the radio, and utilizing the parameters that have already been determined to be the most optimal to enable the radio to transmit effectively. This scenario is referred to as "exploration vs. exploitation". Exploration is the experimentation of various options without prior knowledge of each outcome, and exploitation is the usage of the most effective option out of the options already explored [4]. To find this balance, it is best to look at this problem through the perspective of reinforcement learning, algorithms that train a model by rewarding specific actions the model takes. Rewards are defined as "a measure of success" [5], and, in the context of reinforcement learning, are determined by the objective that is desired to be maximized or minimized based on the nature of the objective's relationship to the communication system (i.e. spectral efficiency, power consumption, etc.). Thus, the objectives of the CE algorithms are to utilize reinforcement learning to find the options that maximize a certain objective, and subsequently providing large rewards, by utilizing different "exploration vs. exploitation" techniques. To set up the general problem solved by the CE algorithms, the following assumptions were made [2]:

1. There are K possible communication methods and each has a potential reward R_k.
2. Each method k has a potential reward R_k.
3. The mean μ_k and standard deviation σ_k are determined from n samples of the reward distribution.
4. Each method has a belief state $\pi_k(n)$ that represents what is currently known about the underlying reward distribution at a time step n.

Three CE algorithms were implemented in this project. The first is known as the ϵ-Greedy Algorithm, where ϵ represents the probability that different options are explored. Likewise, $1 - \epsilon$ represents the probability that different options are exploited. The main complication with this algorithm is that ϵ is fixed, meaning that it will continue randomly exploring/exploiting options at the same rate, which can lead to the algorithm only using sub-optimal configurations or never converging on an appropriate configuration. This observation spurred the development of an annealing version of the ϵ-greedy algorithm, which primarily explores options at the beginning because it is initially unknown if one option is more effective than the others. As knowledge about the performance of different options is acquired over time, ϵ decreases - reducing the amount of exploration and increasing the amount of exploitation. The update equation for ϵ_d can be found in [2].

The second CE algorithm is known as the Softmax Strategy. Essentially, it incorporates information about the rewards of the present options in its exploration process [5]. This is done by controlling two parameters. The first is μ_k, the mean of the reward distribution, to influence the probability associated with selecting a particular method k [2]. In this way, options with a higher mean (i.e. higher reward) will be more likely to be chosen and vice versa. The second parameter is the temperature (T). Its usage is similar to how atoms behave in a material exposed to different temperatures. As the temperature increases, atoms engage in random motion, corresponding to random exploration of different options. Conversely, as temperature decreases, atoms tend to move in a slower and more orderly fashion, corresponding to a more structured approach of exploration of different options [5]. In the context of reinforcement learning, high temperatures encourage choosing any option while low temperatures encourage choosing options that give higher rewards. The Softmax expression and update equation for T can be found in [2].

The third algorithm is known as the Gittins Index Strategy. As [2] summarizes, it reduces the K-dimensional problem established at the beginning of this section to "K one-dimensional problems" by using a specific indexing technique. The distribution of R_k that is assumed significantly affects how the Gittins indices are determined. For this application, the Gittins index for the Normal Reward Process (NRP) is utilized, implying that in the event of a successfully received packet, the return will be equal to the rate of the current method chosen. If the packet is not successfully received there is no reward [2]. Because calculating these indices can be a complicated process, a look-up approach is utilized with the indices reported in [6]. To learn more about the Gittins Index Strategy, the reader is encouraged to refer to [3], [4], and [6].

B. Helical Antenna and Matching Circuits

To establish an effective communication link at HF-band wavelengths (i.e. 10-100 meters), large antennas are required; however, this significantly reduces portability. Thus, research has been conducted in developing electrically small antennas that can operate in the HF band. Albeit efforts have be made towards designing small antennas, there are fundamental limitations such as low radiation efficiency, reactive impedance matching, and low bandwidth. Several design approaches such as an electronically switchable broadband loaded antenna for 10.5-30 MHz band [7], a broadband bow-tie shaped HF antenna covering 6-30 MHz band [8], and naval structural antenna systems [9] have been implemented, but the height of the antennas in each application are not less than 15 meters.

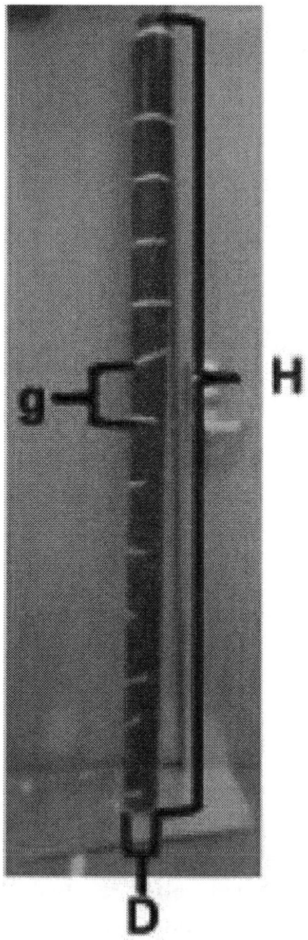

Figure 1: HF helical antenna

DEVELOPMENT

C. Cognitive Engine

For this project, the three CE algorithms discussed in the background section were developed in Python using GNU Radio and LiquidDSP. These two software frameworks are open-source, actively being developed, and contain multiple functions required to perform digital signal processing (DSP) techniques for various software-defined radio (SDR) applications. LiquidDSP enables the implementation of physical layers of various wireless standards (GSM, Wi-Fi, LTE, etc.) and allows for essential physical layer parameters (i.e. preamble length, header, contents, etc.) to be easily set [10]. GNU Radio's DSP functionality is implemented in various C/C++ blocks, while Python is used to connect these blocks in a very user-friendly way [11].

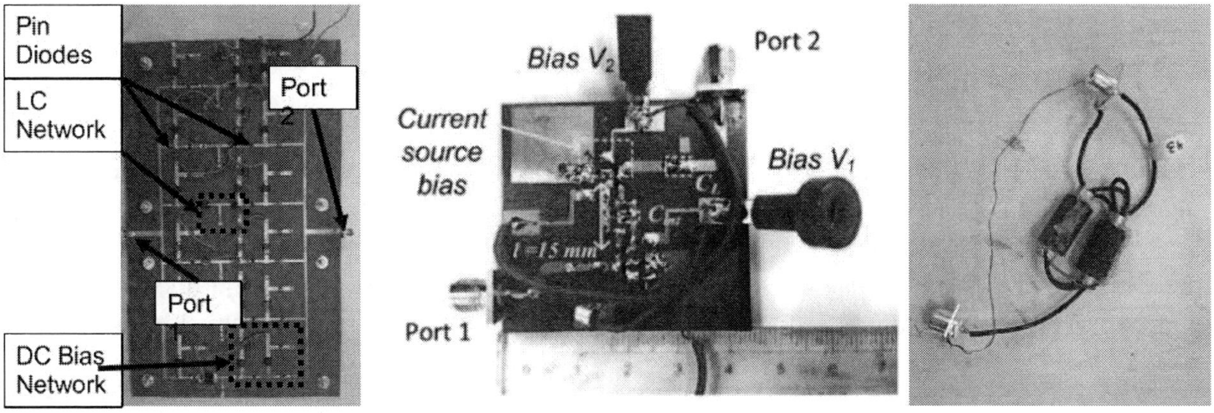

Figure 2: **Left)** Narrowband Switchable Matching Circuit **Middle)** Active Broadband Non-Foster Matching Circuit **Right)** Passive Broadband 2.5:1 UNUN Transformer

D. Helical Antenna and Matching Circuits

The vertically-polarized normal-mode helical antenna resonating at 24.5 MHz is designed, fabricated and tested as shown in Figure 1. The proposed antenna is 2 meters long ($H = 2$ m), such that it is $\frac{\lambda}{50}$ meters at the lowest frequency. The parameters for the antenna (i.e. helix diameter ($D = 10$ cm), number of turns and spacing between the turns ($g = 17$ cm) have been optimized to achieve an antenna with low-self resonance along with moderate radiation resistance within the HF band using a commercial full-wave simulation software called ANSYS HFSS.

A passive reconfigurable matching circuit, based on a T-matching network consisting of two inductors and one capacitor, has been proposed for discrete bands. To facilitate instantaneous switching between distinct bands, pin diodes have been used along with independent DC-bias networks as shown in Figure 2. In addition, to overcome the challenges associated with proposed reconfigurable passive matching network, an active broadband non-Foster matching circuit, employing a stable negative capacitor ($\approx$-40 pF) based on Linvills negative impedance convertor (NIC) is also designed, fabricated and tested as shown in Figure 2. Stability is an important concern for active non-Foster circuit. A normalized determinant function (NDF) analysis is a theoretical method used to figure out the stability of complex circuits, including multiple-feedback circuits. Hence, an NDF analysis has been carried out, using a circuit simulation software called SPICE, to ensure the stability of the designed matching circuit [12].

The non-Foster matching circuit uses transistors, forcing the design process to require a relatively low noise figure so that the system retains a low SNR. Therefore, transformer matching has also been implemented as it possesses an instantaneous broad bandwidth along with a improved SNR. A broadband 2.5:1 UNUN transformer was designed based on the complete equivalent circuit model of the ferrite transformer, taking into account parasitic reactances, suitable ferrite material, minimum core loss, high power sustainability, galvanic isolation and high load reactance cancellation. A schematic of the fabricated transformer is shown in Figure 2.

EXPERIMENTAL RESULTS

E. Cognitive Engine

The three CE algorithms were simulated using an AWGN channel with multipath, to effectively model an HF environment. The multipath effects were simulated by implementing a two-tap FIR filter in Python. The channel noise was varied to simulate different SNR levels. The objective of these experiments were to maximize throughput through the channel. Table 1 outlines the possible configuration options that could be chosen - consisting of different modulations, inner, and outer codes.

Modulation	Inner Code	Outer Code
BPSK	None	None
QPSK	Conv-V27 (1/2 rate)	Golay (24,12)
8-PSK	Conv-V27 (2/3 rate)	ReedSolomon-M8
16-PSK	Conv-V27 (4/5 rate)	Hamming (7,4)
DBPSK	Conv-V27 (5/6 rate)	Hamming (12,8)
DQPSK	Conv-V27 (6/7 rate)	SECDED (22,16)
8-DPSK	Conv-V27 (7/8 rate)	SECDED (39,32)
4-ASK		SECDED (72,64)
16-QAM		
32-QAM		
64-QAM		

Table 1: Possible Communication Methods

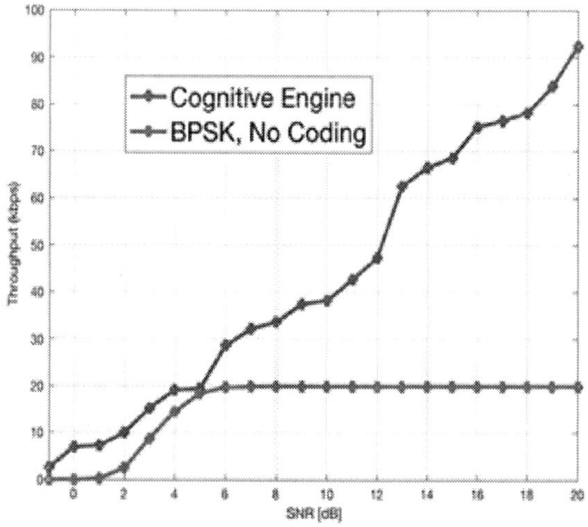

Figure 3: CE throughput (kbps) vs. SNR over AWGN

In the beginning, the CE algorithms have no prior knowledge regarding how effective each of the possible combinations shown in table 1 are. However, as the CE explores these options, it will begin to converge towards the combination of parameters that provide the highest reward with respect to the current objective, which in this experiment was to maximize throughput. For each trial, a packet with a payload of 64 bytes was sent until the 1000th packet was received. Then, the throughput of each packet was determined as shown in equation 1:

$$C = F * log(M) * B \tag{1}$$

where F is the combined coding rate of the inner and outer code, M is the number of symbols in the modulation, and B is the signal bandwidth which is assumed to be 20 kHz. The results of the experiment are shown in figure 3, with the simulated SNRs ranging from -1 to 20 dB. The throughput of the BPSK modulation option without coding is also included in figure 3 for comparison. Figure 3 indicates that throughput increases linearly with respect to SNR. This trend shows that the CE chooses the best modulation and coding for this HF channel. The configuration that

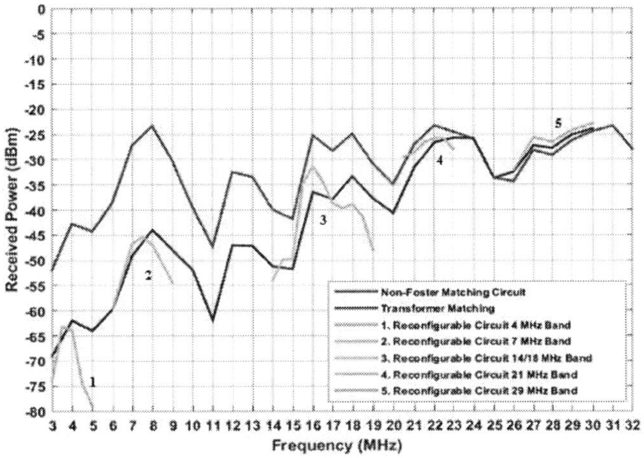

Figure 4: Received Signal Strength Comparison of Proposed Matching Circuits

provided the slowest throughput was achieved using BPSK with a $\frac{1}{2}$ rate code. The configuration that provided the highest throughput was achieved using 32-QAM with a $\frac{1}{2}$ rate code. However, it is expected that this linear trend would not exceed a certain SNR because the channel's capacity would eventually be reached.

F. Helical Antenna and Matching Circuits

Outdoor near-field test measurements were conducted for a qualitative analysis in terms of received signal power strength. A transmitter source transmitting CW signal power of 16 dBm, situated at a distance of 50 m from the receiving helical antenna with the non-Foster matching circuit, achieved significant improvement in received signal strength. This is in comparison with the case of the passive electronically switchable matching circuit and broadband transformer matching network, as shown in Figure 4.

To evaluate the approximate SNR as a result of using the helical antenna with the three matching circuits, an experiment involving voice transmission over multiple licensed bands was also performed. Single side band (SSB) modulated voice signals were transmitted, with a power of 50W, over a 650 meter line-of-sight distance using a FLEX-6500 software-defined radio (SDR) as a transmitter and the receiver setup shown in figure 5. The receiver setup is largely based on the setup proposed by Ettus Research [13]. It consists of the helical antenna, matching circuit,

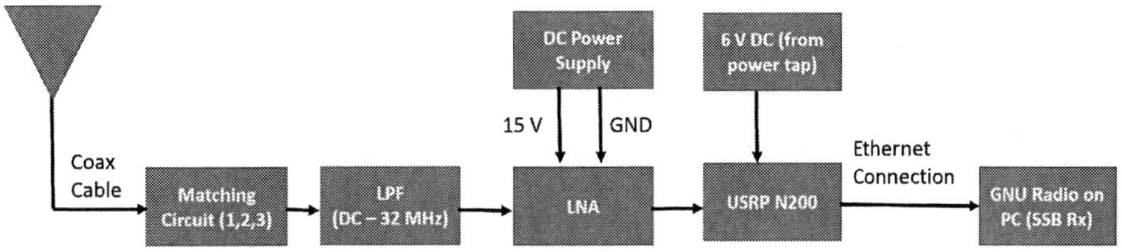

Figure 5: Antenna and Matching Circuit Testing Setup

Table 2: Receiver: GNU software-defined radio (SNR dB)

Frequency (MHz)	Passive Electronically Switched Matching Circuit	Active Non-Foster Matching Cricuit	Passive Transformer Matching
4.035	15	13	22.5
7.021	32	25	33
13.983	67	50	62
17.615	65	60	65
29.935	57	58	64

low pass filter, and low noise amplifier, all connected via coax cable. The low noise amplifier is then connected to a USRP N200 with a "BasicRX" daughterboard, which digitizes the signal and transmits it to a PC hosting GNU Radio. A single side band (SSB) flow graph for GNU Radio, also from the Ettus application note [13], is used to process the incoming signal. The flow graph can be adjusted to operate as either an upper or lower side band (i.e. USB or LSB) receiver. It also creates a GUI that allows a user to observe and tune to different frequencies on the spectrum. It was observed that despite the higher received signal strength of the active non-Foster circuit, its SNR performance was not superior to other passive matching circuit prototypes, as shown in Table 2. This is due to the overall increase in the system noise level. It was concluded that passive transformer matching can provide an adequate system performance with minimal transformer loss. Apart from competitive SNR, the ferrite transformer is also advantageous in terms of power handing capacity.

An experiment was also performed to test the receiver setup's ability to relay long-range communications. Figure 6 shows the approximate locations that could be observed with the setup, where each path is indicative of a different matching circuit used in the receiver. These locations were noted by either hearing the user state it during the transmission, or recording their call sign and reporting the location shown for that callsign on the FCC database. It is worth noting that the figures are not an accurate comparison of the ranges being received by the different matching circuits. The data for this figure was collected at different instances, making it unable to authoritatively assert whether one matching circuit outperformed the others. Such a comparison would also be dependent on the users' activity (or lack of) at each instance as well. Thus, the figure is only indicative of the general performance of the electrically small helical antenna and the three matching circuits in receiving long-range communications. The blue paths represent the usage of the Passive Broadband 2.5:1 UNUN Transformer. The black paths represent the usage of the active matching circuit. The red paths represent the usage of the reconfigurable matching circuit.

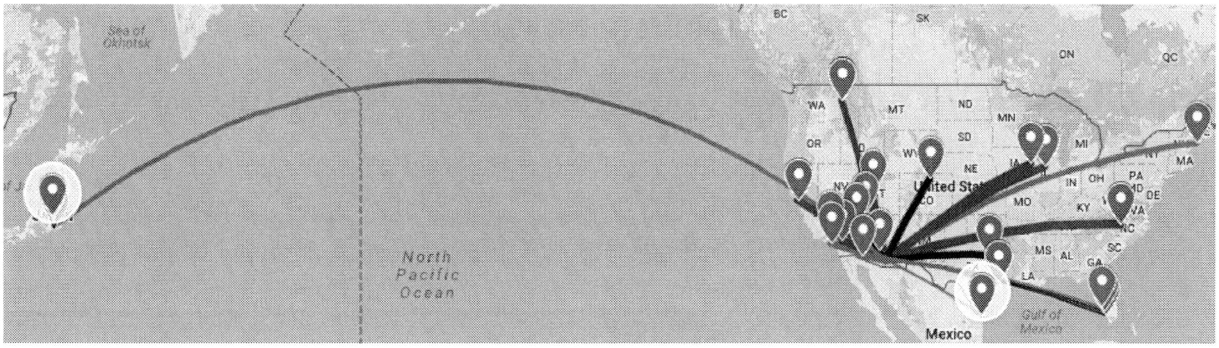

Figure 6: Pictorial Depiction of locations observed with Helical Antenna and Matching circuits. Two circled points are Japan and San Juan, Texas

CONCLUSIONS

The objective of this project was to implement a design capable of conducting HF communications. The proposed system consisted of an electrically small helical antenna, matching networks, and a digital component (i.e. CE). The CE was implemented using liquidDSP and GNU Radio, which are commonly used for SDR applications. The design of the helical antenna, resonating at 24.5 MHz, was simulated using ANSYS HFSS. This paper has shown that the components of the long-range HF communications system have been developed effectively. Figure 3 shows that as SNR increases, assuming an AWGN channel, the CE's performance in choosing optimal communication parameters that maximize throughput also increases (almost linearly). Figure 6 shows that the receiver setup shown in figure 5 is capable of relaying long-range communications over the HF band. One next step that will be pursued is reducing the height of the helical antenna to make it more optimal to carry and set up for a potential user. In addition, more experiments will be planned to verify the performance of the CE over-the-air for long-range HF channels. These experiments will provide further information on the CE's performance because the effectiveness of the configurations it selects will be tested as the signals experience actual ionospheric reflections. In addition, more advanced CE algorithms will be incorporated to form what is known as a meta-CE [14]. Once this is completed, the CE will be combined to the helical antenna to verify that the whole system can transmit and receive in various HF channels.

ACKNOWLEDGEMENTS

This project was partially supported by the Broadband Wireless Access and Applications Center (BWAC); NSF Award No. 1265960.

REFERENCES

[1] A. L. Saverino, A. Capria, and F. Berizzi, "Realization of a Flexible Technological Demonstrator for HF Sky-wave Data Links," pp. 2305–2309.

[2] H. Asadi, H. Volos, M. M. Marefat, and T. Bose, "On Quantifying the Experience Level of a Cognitive Engine," *SDR-WInnComm*, pp. 9–15, 2015.

[3] H. I. Volos and R. M. Buehrer, "Cognitive engine design for link adaptation: An application to multi-antenna systems," *IEEE Transactions on Wireless Communications*, vol. 9, no. 9, pp. 2902–2913, 2010.

[4] H. I. Volos and R. M. Buehrer, "On balancing exploration vs. exploitation in a cognitive engine for multi-antenna systems," *GLOBECOM - IEEE Global Telecommunications Conference*, pp. 0–5, 2009.

[5] J. White, *Bandit Algorithms for Website Optimization: Developing, Deploying, and Debugging*. O'Reilly Media, 2012.

[6] J. Gittins, *Multi-armed bandit allocation indices*. Wiley-Interscience series in systems and optimization, Wiley, 1989.

[7] K. Yegin and A. Q. Martin, "On the design of broad-band loaded wire antennas using the simplified real frequency technique and a genetic algorithm," *IEEE Transactions on Antennas and Propagation*, vol. 51, pp. 220–228, Feb 2003.

[8] W. Lin and Z. Shen, "Broadband horizontally polarized hf antenna with extremely low profile above conducting ground," in *2013 IEEE Antennas and Propagation Society International Symposium (APSURSI)*, pp. 688–689, July 2013.

[9] G. Marrocco and L. Mattioni, "Naval structural antenna systems for broadband hf communications," *IEEE Transactions on Antennas and Propagation*, vol. 54, pp. 1065–1073, April 2006.

[10] J. D. Gaeddert, *liquidsdr*. Available at http://liquidsdr.org/.

[11] *GNU Radio*. Available at http://gnuradio.org/.

[12] Q. Tang, *"Active metamaterial: gain and stability, and microfluidic chip for terahertz cell spectroscopy"*, 2016. Ph.D Dissertation, Dept. Elec and Comp. Eng., University of Arizona, Tucson, 2016.

[13] E. Research, *"Receiving HF Signals with a USRP Device"*.

[14] H. Asadi, H. Volos, M. M. Marefat, and T. Bose, "Metacognition and the next generation of cognitive radio engines," *IEEE Communications Magazine*, vol. 54, pp. 76–82, January 2016.

ELASTIC NET FOR CHANNEL ESTIMATION IN MASSIVE MIMO

Ture Peken, Ravi Tandon, and Tamal Bose
University of Arizona
Tucson, AZ, 85719
[turepeken,tandonr,tbose]@email.arizona.edu

ABSTRACT

Next generation wireless systems will support higher data rates, improved spectral efficiency, and less latency. Massive multiple-input multiple-output (MIMO) is proposed to satisfy these demands. In massive MIMO, many benefits come from employing hundreds of antennas at the base station (BS) and serving dozens of user terminals (UTs) per cell. As the number of antennas increases at the BS, the channel becomes sparse. By exploiting sparse channel in massive MIMO, compressive sensing (CS) methods can be implemented to estimate the channel. In CS methods, the length of pilot sequences can be shortened compared to pilot-based methods. In this paper, a novel channel estimation algorithm based on a CS method called elastic net is proposed. Channel estimation accuracy of pilot-based, lasso, and elastic-net based methods in massive MIMO are compared. It is shown that the elastic-net based method gives the best performance in terms of error for the less pilot symbols and SNR values.

Keywords - Elastic net, channel estimation, massive MIMO, compressive sensing

INTRODUCTION

Current wireless communication systems like IEEE 802.16M and 3GPP LTE/LTE-Advanced widely use MIMO technology [1]. MIMO technology in which multiple antennas serve more than one user terminal (UT) is called multi-user MIMO (MU-MIMO). MU-MIMO brings significant benefits such as higher data rate, improved reliability, and energy efficiency to the wireless systems of the future [2]. It is anticipated that the fifth generation (5G) wireless networks will be deployed by 2020. These future wireless networks are expected to support even higher spectral and energy efficiency, higher data rates, lower latency, and improved reliability. Massive MIMO is considered to be one of the fundamental technologies of 5G due to its potential for spectrum and energy efficiency, secure, and robust networks [2]. In massive MIMO, the BS employs a few hundreds of antennas to serve dozens of UTs in the same radio channel. Massive MIMO improves spectral efficiency and throughput by focusing the energy into much smaller regions with hundreds of antennas [2]. Energy efficiency can be achieved by adding the signals transmitted from the antennas of the BS constructively at the intended UTs and destructively at the other UTs. The spectral efficiency in massive MIMO is better than in conventional MIMO since dozens of UTs can be served simultaneously in the same frequency-time resource [2]. Consequently, massive MIMO is a very promising technology to build the wireless systems of the future.

In order to recover received signals accurately, channel state information (CSI) which is represented with a complex matrix needs to be estimated with minimum error. Channel estimation methods can be classified as pilot-based, blind, or semi-blind methods. In pilot-based methods, the complex channel matrix is evaluated by sending a set of training signal sequences, called pilot sequences, from each UT to the BS. Pilot sequences need to be orthogonal to prevent a specific type of inter-cellular interference called pilot contamination. The channel coherence time imposes an upper-bound on the number of symbols of each orthogonal pilot sequence [2]. Therefore, UTs in different cells reuse the same pilot sequences in a cellular network. Pilot contamination is caused by the interference between a desired UT and another UT which share common pilot sequences [2].

Blind channel estimation methods estimate the channel by using statistics of the received symbols after having been transformed by the channel without using any pilot sequences. Pilot contamination can be mitigated by using blind channel estimation methods, but they have higher computational complexity than pilot-based methods since they require advanced signal processing techniques. To compromise between the computational complexities of pilot-based and blind methods, semi-blind channel estimation methods can be used. Semi-blind methods use shorter pilot sequences and additional signal processing techniques to estimate the channel.

Pilot contamination is one of the biggest challenges in massive MIMO since it limits the gain that can be achieved. Using shorter pilot signal sequences to estimate the CSI gives promising solutions to the pilot contamination problem in massive MIMO. The channel response between the BS and the UT becomes sparse as the number of antennas of the BS increases [3]. In [4], they exploited the sparse nature of signals to reduce the pilot overhead for massive MIMO OFDM systems by using a CS-based technique. Different researchers have proposed compressive channel sensing (CCS) methods to estimate sparse channels. CS methods estimate the strong channel responses which correspond to desired channels while they treat the interfering channels as noise. Therefore, they provide very promising solutions to the pilot contamination problem. Channel estimation by using compressive sensing methods have been studied in the context of massive MIMO [5]-[6]. These compressive methods are based on different regularization and variable selection methods. One of the most popular compressive methods is least absolute shrinkage and selection operator (lasso) which is a L_1-penalized least squares (LS) method [7]. Since it shrinks some coefficients and sets others to 0, it provides a good estimation for the sparse channels. However, the lasso method does not perform well when the pairwise correlations among the group of variables are high and there are high correlations between predictors. Most of the CS based techniques for sparse channel estimation are based on L_1 penalization. Adaptive-lasso has been used to estimate the channel in [3]. Recently, the CS based techniques, which use Bayesian framework, have been studied. In [8], a Bayesian based CS method is proposed for pilot reduction in massive

MIMO systems. In [9], a new regularization technique is proposed that generally gives better performance than lasso called elastic net. With the additional regularizer (L2), the elastic net does automatic variable selection and continuous shrinkage at the same time. Therefore, it can select groups of correlated variables which improve its prediction accuracy compared to lasso. Since the correlations among the predictors are high in massive MIMO, we propose to use elastic net based channel estimation. To the best of our knowledge, elastic net method for channel estimation in massive MIMO has not been studied in the literature.

In this paper, we implement a massive MIMO system in MATLAB. Then, we compare the performance of the LS pilot-based channel estimation method, lasso channel estimation method and elastic net based channel estimation method in this massive MIMO system. We analyze the relative accuracy and computational complexity between these methods, with and without pilot contamination. We compare their accuracy for different number of pilots and values of SNR. Computational complexity is observed when the number of pilots changes. We show that elastic net outperforms LS and lasso with and without pilot contamination when the number of pilots and SNR decrease. Lasso has the worst performance among these methods. Moreover, we show that elastic net continues to estimate the channel successfully with less pilots. Therefore, we show that elastic net based channel estimation gives a promising solution to the pilot contamination problem. However, computational complexity of elastic net is higher than LS and lasso according to the simulation results.

SYSTEM MODEL

A massive MIMO uplink system in which the BS has M antennas and each UT has one antenna is considered in this paper. There are N UTs where $M \gg N$ in this system. We consider P number of pilot symbols per frame. The L received symbols at the M antennas of the BS can be represented as:

$$Y = SH + W, \tag{1}$$

where $Y \in \mathbb{C}^{L\times M}$ is a matrix of the L received symbols at the M receive antennas of the BS. $S \in \mathbb{C}^{L\times N}$ is a matrix of the transmitted L symbols. $H \in \mathbb{C}^{N\times M}$ is the CSI between N UTs and M antennas of the BS, and $W \in \mathbb{C}^{L\times M}$ is the additive white Gaussian noise (AWGN) matrix.

In massive MIMO, the radio propagation channel is composed of a superposition of different multipath components (MPCs). MPCs are caused by the interaction between the radio waves and the objects (or scatterers) in the environment. We can reduce the number of channel modeling parameters significantly by grouping MPCs with the same delay and directions into clusters. We extend the cluster-based COST 2100 channel model [10], which captures important massive MIMO channel characteristics.

PILOT BASED CHANNEL ESTIMATION

In pilot-based channel estimation methods, the BS estimates the complex channel matrix by using known pilot sequences transmitted from the UTs. They are preferable due to their low computational complexity. However, bandwidth is not used efficiently in pilot-based channel estimation methods since a predetermined section of the bandwidth is used for transmitting pilot sequences. Pilot contamination is another issue which occurs when pilot sequences are reused among different cells. One of the most common pilot-based channel estimation methods is the LS method. Let's assume the same pilot sequences are reused among K cells, i.e., $S_{p_1} = ... = S_{p_K}$. Here S_{p_i} stands for the pilot sequence of the ith cell. The LS estimate of the complex channel matrix at the BS of the cell i is defined as:

$$\hat{H}_i = (S_{p_i}^H S_{p_i})^{-1} S_{p_i} Y = H_i + \sum_{k \neq i, k=1}^{K} H_k + (S_{p_i}^H S_{p_i})^{-1} S_{p_i} W, \tag{2}$$

In (2), the interfering channels H_k, $k = 1, ..., K$, $k \neq i$ will contaminate the desired channel estimate H_i, which causes pilot contamination [11].

CHANNEL ESTIMATION VIA LASSO

Channel estimation based on lasso applies a L_1-penalized least squares (LS) method on each column vector of the complex channel matrix. By using lasso algorithm, the jth column vector of H, which is the complex channel matrix between N UTs and jth antenna of the BS, is given as:

$$\hat{h}_j = \arg\min_{h_j} ||y_j - S_p h_j||_2^2 + \lambda ||h_j||_1, \tag{3}$$

where S_p are the pilot sequences such that $S_p = (s_1|s_2|...|s_N)$ and $s_i = (s_{i_1}, s_{i_2}, ..., s_{iP})$, $i = 1, ..., N$ are the predictors. y_j is the vector of received symbols from N users at the jth antenna of the BS. λ is a fixed non-negative tuning parameter.

CHANNEL ESTIMATION VIA ELASTIC NET

A novel method for channel estimation based on elastic net is proposed in this paper. The jth column vector of H which corresponds the complex channel coefficients between N UTs and jth antenna of the BS is estimated by using elastic net:

$$\hat{h}_j = \left(1 + \frac{\lambda_2}{n}\right) \arg\min_{h_j} ||y_j - S_p h_j||_2^2 + \lambda_2 ||h_j||_2^2 + \lambda_1 ||h_j||_1, \tag{4}$$

λ_1 and λ_2 are any fixed non-negative tuning parameters.

An algorithm called LARS-EN is proposed to solve the entire elastic net path in [9]. This algorithm is based on the LARS algorithm which is proposed for solving the lasso problem in [12]. Given a fixed λ_2, the elastic net problem is equivalent to the lasso problem on the augmented data set. In this algorithm, all coefficients are set to zero initially. Then, the predictor which is most correlated with response y_j is found. Let's assume this predictor is s_{j1}. The largest possible step is taken in the direction of s_{j1} until another predictor s_{j2} has the same correlation with the current residual. The algorithm steps in the direction equiangular between s_{j1} and s_{j2} until another predictor s_{j3} has the most correlation with the current residual. The algorithm continues along the least angle direction between s_{j1}, s_{j2}, and s_{j3} until another predictor earns its way to the active set which is the most correlated set. It continues so on.

The predictors are simply the P pilot symbols transmitted from N UTs. Pilot symbols are known by the receiver. By applying an elastic net estimator to the P received pilot symbols at the jth antenna, the complex channel coefficients between N UTs and jth antenna can be estimated. Elastic net estimator is applied to each antenna of the BS separately.

SIMULATION RESULTS

A model of a multi-cellular massive MIMO system is constructed in MATLAB to test the relative performance of LS channel estimation and the proposed channel estimation via elastic net method. In the simulation, there are nine cells and each cell has a 10,000 square meter area. Each BS is equipped with 100 ideal dipole-type antennas and serves different number of UTs. There are 90 UTs in total. Each UT is equipped with one antenna and has a velocity less than 1 m/s. The channel is modeled as COST 2100 with Rayleigh fading and non-line of sight (NLOS) situation. The corresponding pseudo-color plot of the strength of complex channel coefficient matrix $H \in \mathbb{C}^{N \times M}$ is depicted in Figure 1. As expected, the channel matrix is approximately sparse. The cells are assumed to be synchronized during the time pilot symbols are transmitted. This is the worst-case scenario for pilot contamination. A carrier frequency of 2.4 GHz is assumed for this massive MIMO system.

First, RMSE between real and estimated complex channel coefficients matrices of LS, lasso, and elastic net methods for different number of pilots are compared when there is no pilot contamination. Figure 2 shows RMSE between real and estimated complex channel coefficients matrices of LS, lasso, and elastic net methods when the number of pilots increases from 18 to 30. The number of data symbols are decreased from 142 to 130 simultaneously. In this scenario, the SNR is 10 dB. Figure 2 shows that same RMSE is obtained with decreasing number of pilots for elastic net while performance of LS significantly decreases. It is seen that lasso gives the worst performance in terms of RMSE.

Then, RMSE between real and estimated complex channel coefficients matrices of LS, lasso, and elastic net methods for different number of pilots are compared when there is pilot contamination. Figure 3 shows that elastic net has nearly the same RMSE with LS for higher number of pilots because the performance of LS method gets worse with pilot contamination.

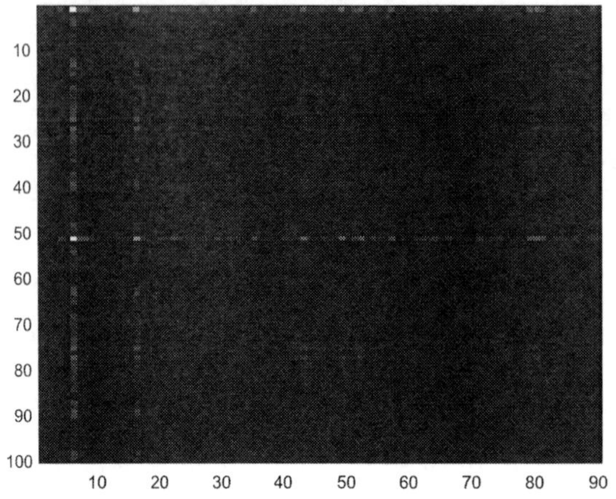

Figure 1: A pseudo-color strength of complex channel coefficient matrix

However, lasso performs worse than both LS and elastic net method.

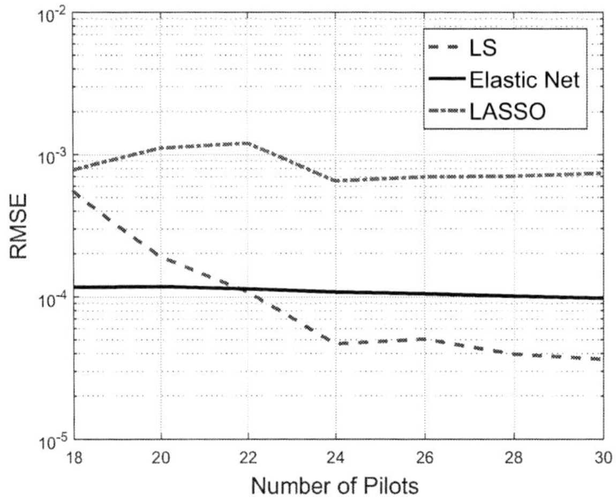

Figure 2: RMSE between real and estimated complex channel matrices of LS, lasso, and elastic net methods without pilot contamination for different number of pilots and data symbols

Figure 4 shows RMSE between real and estimated complex channel coefficients matrices of LS, lasso, and elastic net methods when SNR changes from 0 dB to 20 dB. The results in Figure 4 demonstrate that elastic net outperforms LS and lasso for decreasing values of SNR. Lasso has the worst performance for the all SNR values. Then, the pilot contamination effect on LS, lasso, and elastic net based channel estimation methods are compared in terms of RMSE when the SNR changes from 0 to 20 dB. It is shown in Figure 5 that elastic net gives less RMSE than LS. Lasso again has a greater RMSE than LS and elastic net for all the SNR values. It is shown in [13] that signals become less sparse while per element signal strength increases. Elastic net is affected adversely by the reduction in sparsity of signals.

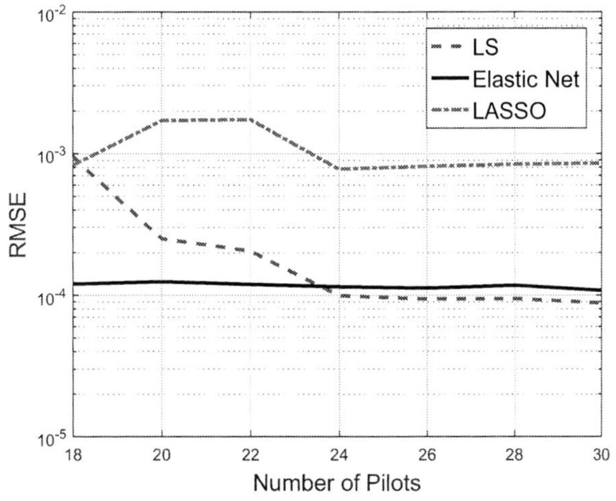

Figure 3: RMSE between real and estimated complex channel matrices of LS, lasso, and elastic net methods with pilot contamination for different number of pilots and data symbols

Even though the signal quality gets better with increasing values of SNR, less sparsity of signals causes a bottleneck. Therefore, elastic net does not perform better with increasing values of SNR.

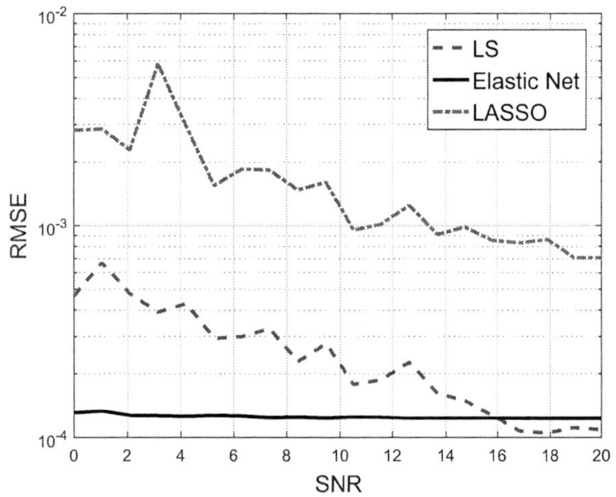

Figure 4: RMSE between real and estimated complex channel matrices of LS and elastic net methods without pilot contamination for different SNR values in dB

Then, we investigate the minimum number of pilots that is required to estimate the channel with lasso and elastic net methods when there is pilot contamination. Figure 6 shows RMSE between real and estimated complex channel coefficients matrices of lasso and elastic net methods when the number of pilots increases from 2 to 30. The number of data symbols are decreased from 158 to 130 in the meantime and the SNR is selected as 10 dB. It is seen that lasso and elastic net start to fail at estimating the channel when the number of pilots

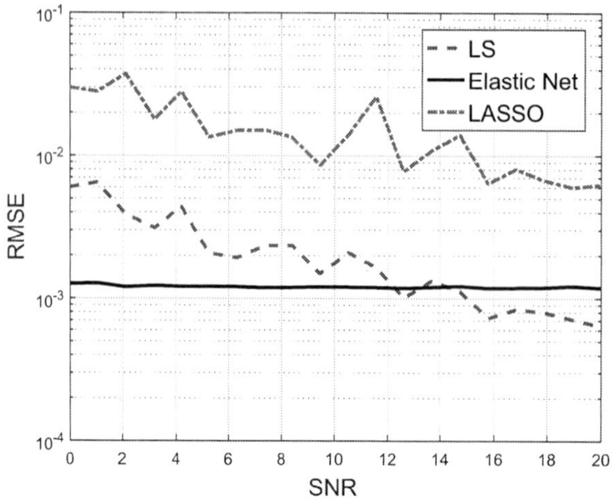

Figure 5: RMSE between real and estimated complex channel matrices of LS and elastic net methods with pilot contamination for different SNR values in dB

is below 10 and 8, respectively. According to these results, better performance than LS can be achieved by using elastic net even with a significantly less number of pilots compared to data symbols.

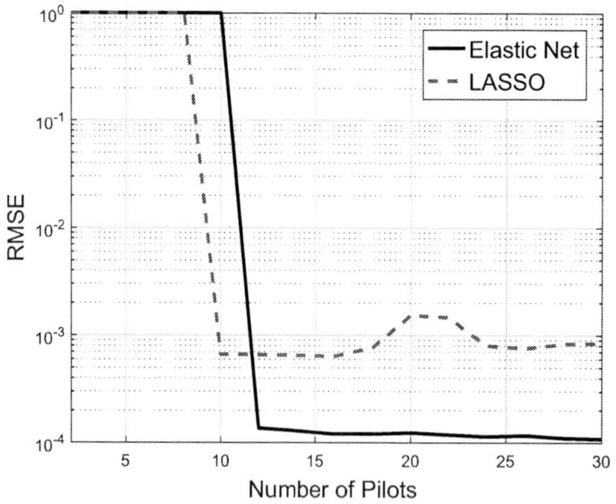

Figure 6: RMSE between real and estimated complex channel matrices of lasso and elastic net methods with pilot contamination for different number of pilots

Finally, computational complexity of LS, lasso, and elastic net are compared. We show the average run time of one frame for LS, lasso, and elastic net based channel estimation methods in Table 1. According to the results, lasso method has a higher runtime than LS. Elastic net has the highest runtime among these methods.

Table 1: Average run time of LS, lasso, and elastic net methods

Channel Estimation	Run Time
LS	0.0004 s
Lasso	0.16 s
Elastic Net	0.87 s

CONCLUSIONS

Massive MIMO is expected to be one of the key technologies in 5G wireless systems. Massive MIMO, in which significantly more BS antennas are used, has been proposed to improve spectral efficiency and data rates. Effective channel estimation for massive MIMO is one of the most important challenges in addressing pilot contamination. By exploiting sparse channel in massive MIMO, CS methods can be used to estimate the channel. Communication resources such as latency, bandwidth and energy can be used more efficiently by using CS methods. Moreover, pilot contamination can be overcome by reducing pilot signal sequences. In this paper, we propose a novel channel estimation method based on elastic net regularization and variable selection method. The pilot contamination effects on LS, lasso, and elastic net based channel estimation methods are compared by implementing a MATLAB simulation of massive MIMO. Both of the scenarios with and without pilot contamination are compared. Simulation results show that elastic net gives better performance than LS and lasso in terms of error rate for low values of SNR and less number of pilots. Therefore, we can reduce pilot contamination significantly with elastic net method. Moreover, communication resources can be used more efficiently in the elastic net method while achieving a better performance than the LS method with a decent number of pilot signal sequences. Computational complexity of these channel estimation methods are also compared by using MATLAB simulation. It was observed that LS has the least computational complexity. In the future, we plan to improve the elastic net based channel estimation algorithm in order to achieve less computational complexity.

ACKNOWLEDGEMENTS

This project was partially supported by the Broadband Wireless Access and Applications Center (BWAC); NSF Award No. 1265960.

REFERENCES

[1] Q. Li, G. Li, W. Lee, M. il Lee, D. Mazzarese, B. Clerckx, and Z. Li, "MIMO Techniques in WiMAX and LTE: A Feature Overview," *Communications Magazine, IEEE*, vol. 48, pp. 86–92, May 2010.

[2] E. Larsson, O. Edfors, F. Tufvesson, and T. Marzetta, "Massive MIMO for Next Generation Wireless Systems," *Communications Magazine, IEEE*, vol. 52, pp. 186–195, February 2014.

[3] G. Destino, M. Juntti, and S. Nagaraj, "Leveraging sparsity into massive MIMO channel estimation with the adaptive-LASSO," in *2015 IEEE Global Conference on Signal and Information Processing (GlobalSIP)*, pp. 166–170, Dec 2015.

[4] J. W. Choi, B. Shim, and S. H. Chang, "Downlink pilot reduction for massive mimo systems via compressed sensing," *IEEE Communications Letters*, vol. 19, pp. 1889–1892, Nov 2015.

[5] M. Masood, L. H. Afify, and T. Y. Al-Naffouri, "Efficient coordinated recovery of sparse channels in massive mimo," *IEEE Transactions on Signal Processing*, vol. 63, pp. 104–118, Jan 2015.

[6] W. U. Bajwa, J. Haupt, A. M. Sayeed, and R. Nowak, "Compressed channel sensing: A new approach to estimating sparse multipath channels," *Proceedings of the IEEE*, vol. 98, pp. 1058–1076, June 2010.

[7] R. Tibshirani, "Regression shrinkage and selection via the lasso," *Journal of the Royal Statistical Society, Series B*, vol. 58, pp. 267–288, 1994.

[8] J. W. Choi and B. Shim, "Compressive sensing based pilot reduction technique for massive mimo systems," in *2015 Information Theory and Applications Workshop (ITA)*, pp. 115–118, Feb 2015.

[9] H. Zou and T. Hastie, "Regularization and variable selection via the elastic net," *Journal of the Royal Statistical Society: Series B (Statistical Methodology)*, vol. 67, no. 2, pp. 301–320, 2005.

[10] L. Liu, C. Oestges, J. Poutanen, K. Haneda, P. Vainikainen, F. Quitin, F. Tufvesson, and P. Doncker, "The COST 2100 MIMO Channel Model," *Wireless Communications, IEEE*, vol. 19, pp. 92–99, December 2012.

[11] T. Marzetta, "Noncooperative Cellular Wireless with Unlimited Numbers of Base Station Antennas," *Wireless Communications, IEEE Transactions on*, vol. 9, pp. 3590–3600, November 2010.

[12] B. Efron, T. Hastie, I. Johnstone, and R. Tibshirani, "Least angle regression," *Ann. Statist.*, vol. 32, pp. 407–499, 04 2004.

[13] S. Reid, R. Tibshirani, and J. Friedman, "A Study of Error Variance Estimation in Lasso Regression," *ArXiv e-prints*, Nov. 2013.

PERFORMANCE EVALUATION FOR DECISION-FEEDBACK EQUALIZER WITH PARAMETER SELECTION ON UNDERWATER ACOUSTIC COMMUNICATION

Husam Nassr, Kurt Kosbar
Dept. of Electrical & Computer Engineering,
Missouri University of Science & Technology
Rolla, MO 65409.
hhn4k4@mst.edu,
kosbar@mst.edu

ABSTRACT

This paper investigates the effect of parameter selection for the decision feedback equalization (DFE) on communication performance through a dispersive underwater acoustic wireless channel (UAWC). A DFE based on minimum mean-square error (MMSE-DFE) criterion has been employed in the implementation for evaluation purposes. The output from the MMSE-DFE is input to the decoder to estimate the transmitted bit sequence. The main goal of this experimental simulation is to determine the best selection, such that the reduction in the computational overload is achieved without altering the performance of the system, where the computational complexity can be reduced by selecting an equalizer with a proper length. The system performance is tested for BPSK, QPSK, 8PSK and 16QAM modulation and a simulation for the system is carried out for Proakis channel A and real underwater wireless acoustic channel estimated during SPACE08 measurements to verify the selection.

KEY WORDS

Decision feedback equalizer(DFE), minimum mean-square error (MMSE).

INTRODUCTION

In the last few decades underwater acoustic communications has become important in many application that utilize remotely controlled equipment and vehicles for sensing and exchanging information between devices located under the surface in the oceans such as military and commercial fishing and oil exploration [1]. Acoustic waves are used in underwater wireless communication instead of electromagnetic waves employed in radio frequency (RF) communication due to the rapid attenuation of electro-magnetic signals through salt water. Transmission over a frequency selective channel which is characterized by its long spread time is a challenging, especially when the channel suffers, in addition to multipath degradation, from Doppler shift which cause a rapid change in the channel impulse response (CIR). Due to the multipath, the receiver receives multiple copies of the same signal which causes the inter-symbol interference (ISI) distortion. To deal with the Doppler effect, a channel estimation within the shorter block of symbols is necessary and on the other hand, coding and equalization can help with the symbol detection and to mitigate the ISI distortion caused by the channel. Minimum mean square error decision feedback equalizer (MMSE-DFE) is a very common equalization algorithm used to alleviate the problem of the ISI. An important step in the implementation is how to select the parameters of the equalizer based on the characteristics of the channel. The channel taps are classified into postcursor (causal) and the precursor (anti-causal) parts are important characteristics needed to be estimated during design of the equalizer. Usually the equalization process implies designing an equalizer with large number of coefficients such that the amount of distortion is minimized. This introduces a higher computational complexity to the design since it involves manipulating with large matrices.

To design a finite length MMSE-DFE, the main parameters that need to be determined are the coefficients of the feedforward filter (FFF), the feedback filter (FBF) and the decision delay. The length of the feedforward filters (K = K_1+K_2+1) where K_1 and K_2 represent the precursor and the post-cursor length respectively, and the length of the feedback filter (K_3) needs to be carefully selected in addition to the decision delay Δ. Some of the previous design methods are based on ad-hoc selection while some others use exhaustive search among these three parameters. Paul *et al.* [2] investigated the effect of the decision delay on the performance of the finite length DFE. An optimum solution for the FFF and FBF and the decision delay has been proposed for RF channels, but for Underwater wireless acoustic channel this solution might be too complicated to be implemented due to the large matrix manipulation, and the result could be suboptimal. Naofal *et al.*[3] proposed a simple order recursive algorithm that reduces the exhaustive search to one dimensional, instead of three dimensional, to select the parameters for the MMSE-DFE. Although the result is promising, it is not clear how to relate this parameter selection to the channel characteristics. William J *et al.* [4] mentioned that the length of the FBF (K_3) needs to be on the order of the postcursor part of the channel and length of the FFF should be greater than the channel length and the decision delay should be longer than the channel span, but there is still an issue of how to select these parameters. Yu Gong *et al.* [5] proposed an adaptive method that optimizes the structure based on the characteristics of the channel. It shows that the decision delay should be in the range between 0 and $K_1 - 1$.

The performance will not necessarily improve by selecting a large number of taps for the feedforward filter and feedback filter while keeping the decision delay constant. This could lead to increased noise, in addition to increasing the complexity. The system performance can be improved further by incorporating channel coding where the equalized hard/soft output sequence from the HD-DFE/SD-DFE is fed to the decoder to estimate the most likely sequence. This paper seeks to study the relationship between the channel length L and the length of the equalizer need to be used. A Proakis Channel A and an estimated SPACE08 channel have been used in the evaluation to discover the best design for a single input single output (SISO) equalizer in term of removing the distortion with minimum number of computations. On average, the estimated SPACE08 channels,

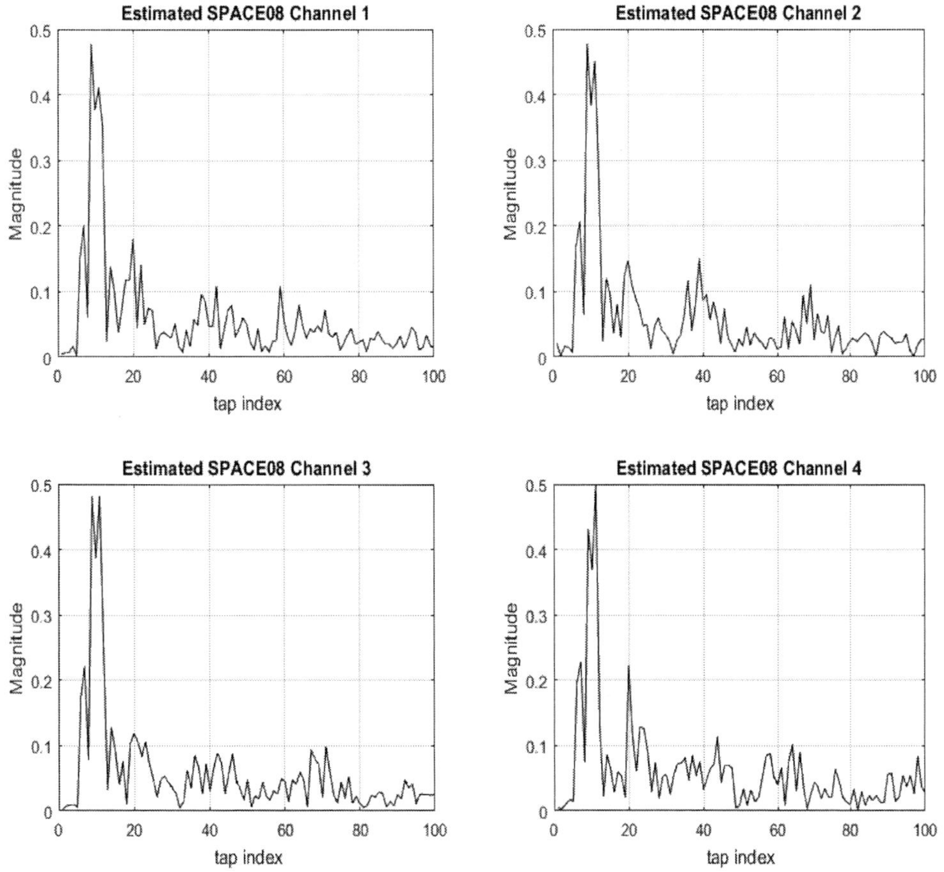

Figure 1: Tap Structure of the Magnitude of the Estimated SPACE08 Channel.

as shown in Figure 1, have around a 100 taps and none of them come from direct path (line of sight path). The length of the causal part for these channels L_2 is usually longer than the anti-causal part L_1 and in general these underwater acoustic channels are considered longer than the conventional RF channels. For these kinds of channels, minimizing the length of the equalizer is very important, especially when the channel is changing very fast, since channel estimation and equalization need to be repeated many times for each short block of the received data.

SYSTEM MODEL AND DATA FORMAT

Consider the single-input single-output (SISO) communication system in Figure 2, where the information bit stream is encoded and mapped such that the output of the encoder is a sequence of encoded bits stream $\mathbf{c} = [\mathbf{c}_1\mathbf{c}_2 \cdots \mathbf{c}_\mathrm{N}]$ where $\mathbf{c}_\mathrm{k}$ represents $[c_{k,1}c_{k,2} \cdots c_{k,q}]$ with bits $c_{k,j} \in \{0,1\}$, and N is the number of transmitted codewords and q is the number of bits per symbol. The mapper uses a Gray code to map each bit vector $\mathbf{c}_\mathrm{k}$ to a symbol x_k from the 2^q array constellation set $S = \{\alpha_1\alpha_2 \cdots \alpha_{2^q}\}$, where α_i corresponds to the deterministic bit pattern $\mathbf{s}_{i=} [s_{k,1}s_{k,2} \cdots s_{k,q}]$ with

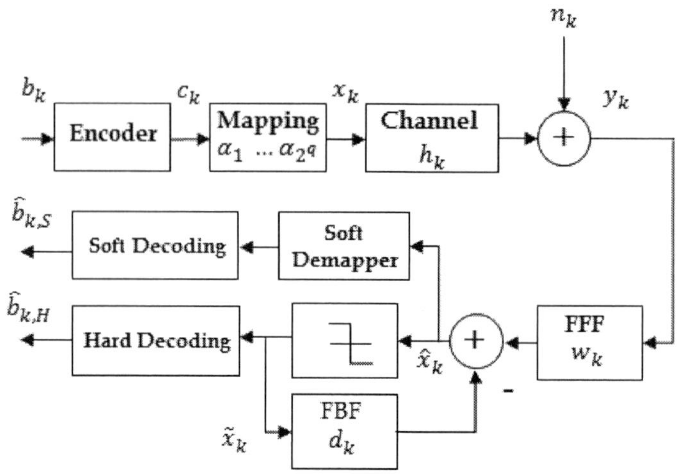

Figure 2: Simplified communication system.

$s_{i,j} \in \{0, 1\}$ which specifies the mapping between the encoded bits and the elements of the constellation as shown in Table I in [6]. Next, the generated symbols are modulated with a carrier and transmitted through the channel h_k. At the receiver, the received baseband signal can be written as

$$y_k = \sum_{l=-L_2}^{l=L_1} h_l x_{k-l} + n_k \tag{1}$$

here h_l is the lth tap of the channel and L_1 and L_2 are the lengths of the causal and anti-causal part of the channel respectively as shown in Figure 3, and L= L_1+L_2+1 is the total length of the channel. In addition, n_k represents an additive Gaussian white noise with zero mean and variance σ_n^2. Arranging equation (1) into a matrix form we have

$$\mathbf{y}_k = \mathbf{H}\mathbf{x}_k + \mathbf{n}_k \tag{2}$$
$$\mathbf{y}_k = \left[y_{k-K_1}\; y_{k-K_1+1} \cdots y_{k+K_2}\right]^T$$
$$\mathbf{x}_k = \left[x_{k-K_1-L_1}\; x_{k-K_1-L_1+1} \cdots x_{k+K_2+L2}\right]^T$$
$$\mathbf{n}_k = \left[n_{k-K_1}\; n_{k-K_1+1} \cdots n_{k+K_2}\right]^T$$
$$\mathbf{H} = \begin{bmatrix} h_{L_1} & \cdots & h_{-L_2} & \cdots & 0 \\ \vdots & \ddots & \ddots & \ddots & \vdots \\ 0 & \cdots & h_{L_1} & \cdots & h_{-L_2} \end{bmatrix}.$$

The bold letters represent a vectors and bold capital letters for a matrix. The variable k is used for the time index and $(\cdot)^*$, $(\cdot)^{-1}$, $(\cdot)^T$ and $(\cdot)^H$ are the conjugate, inverse, transpose and hermitian operators respectively. The signal $\mathbf{y}_k$ is input to the feedforward filter and the input of the feedback filter comes from the previously detected symbols. The output of the equalizer can be represented with the following equation.

$$\hat{x}_k = \sum_{j=-K_2}^{K_1} w_j y_{k-j} + \sum_{j=1}^{K_3} d_j \tilde{x}_{k-j}, \tag{3}$$

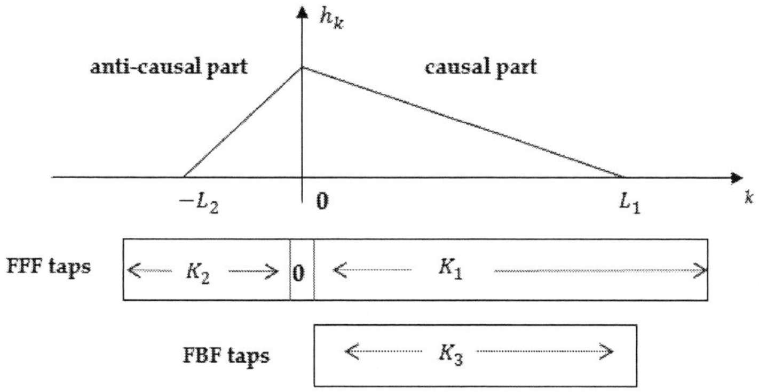

Figure 3: Parameter in relation to the channel.

where $\hat{x}_k$ and $\tilde{x}_{k-1}$ represent the estimated symbol and the previously detected symbol at time instance k respectively. $\{w_j\}$ and $\{d_j\}$ represents the the tap coefficients of the feedforward and feedback filter respectively. Assuming the detection of the previous symbol $\tilde{x}_{k-1}$was correct and equals to the previously transmitted symbol x_{k-1}, equation (3) can be rewritten at time instant k into a matrix form as:

$$\begin{aligned} \hat{x}_k &= \mathbf{w}\mathbf{y}_k + \mathbf{d}\mathbf{x}_k^d \\ \mathbf{x}_k^d &= [x_{k-K_3}\ x_{k-K_3+1} \cdots x_{k-1}]^T \\ \mathbf{w} &= [w_{K_2}\ w_{K_2-1} \cdots w_{-K_1}] \\ \mathbf{d} &= [d_{K_3}\ d_{K_3-1} \cdots d_1] \end{aligned} \tag{4}$$

where $\mathbf{x}_k^d$ is a vector contains the past K_3 decided symbols and $\mathbf{w}$ and $\mathbf{d}$ are feedforward and feedback coefficients at the time instance k respectively. Using the em orthogonality principle the values of $\mathbf{w}$ and $\mathbf{d}$ can be calculated as

$$\mathbf{w} = \mathbf{m}\mathbf{H}^h[\mathbf{H}\mathbf{H}^h + \sigma_n^2\mathbf{I}_{K\times K} - \mathbf{H}\mathbf{M}\mathbf{M}^h\mathbf{H}^h]^{-1} \tag{5}$$

$$\mathbf{d} = -\mathbf{w}\mathbf{H}\mathbf{M} \tag{6}$$

where $\mathbf{I}_{K\times K}$ represents the identity matrix of dimension K $\times$ K, and $\mathbf{m}$ and $\mathbf{M}$ are defined as

$$\mathbf{m} = [\mathbf{0}_{1\times(K_2+L-1)}\ \ 1\ \ \mathbf{0}_{1\times K_1}]$$

$$\mathbf{M} = \begin{bmatrix} \mathbf{0}_{(K_2+L-K3-1)\times K_3} \\ \mathbf{I}_{K_3\times K_3} \\ \mathbf{0}_{(K_1+1)\times K_3} \end{bmatrix}.$$

The equalized symbols are now ready for decoding. For hard decision decoding (HDD), a decision based on the euclidean distance from the received symbol and the all points in the signal constellation needs to be made. The coded bits can then be determined using the regular mapping. In contrast, for soft decision decoding (SDD) soft bits are needed, and that is the job of the soft

Decoder. For BPSK, it is easy to extract the soft information, but for higher modulation level a specific technique is needed. Jianing Su *et al.*[7] provided an efficient way to extract the soft information from the equalized symbols for QPSK and 8PSK which can be generalized for 16QAM or any other constellation.

SIMULATION AND EXPERIMENTAL RESULTS

The simulation uses Proakis channel A with channel taps $h = [\ 0.04, -0.05, 0.07, -0.21, -0.5, 0.72, 0.36, 0, 0.21, 0.03, 0.07]$ and underwater acoustic channel estimated from the SPACE08 experiment. In the simulation, the binary bits are encoded by 1/2 coding rate with generator polynomial G=[7,5]. The encoded bits then were mapped into BPSK, QPSK, 8PSK, and 16QAM and sent through the channel. As can be seen from Figure 1, the channel impulse response for the SPACE08 channel is very long and has common characteristics such as the length of the causal and anti-causal parts. During the equalization process, and for each received block of data, the lengths of equalizer parameters K_1, K_2 and K_3 have been selected within the range proportional to L_1 and L_2 of the estimated channel. Figure 4 and Figure 5 show the effect of K_1 on system performance for Proakis channel A while fixing the other parameters. As shown in Figure 4, the

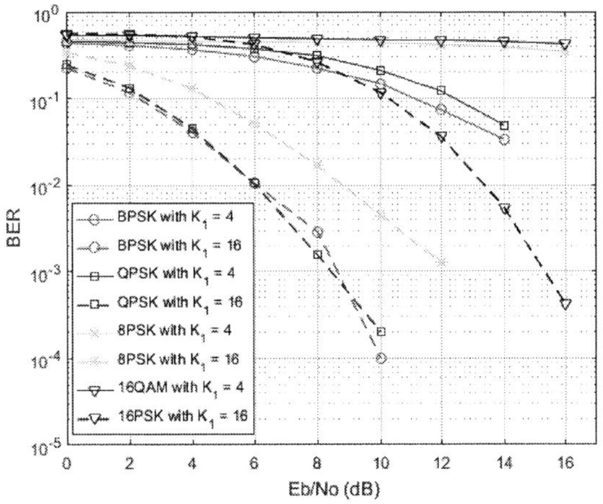

Figure 4: Impact of K_1 on system Performance for Proakis Channel A with K_2=4 and K_3=3.

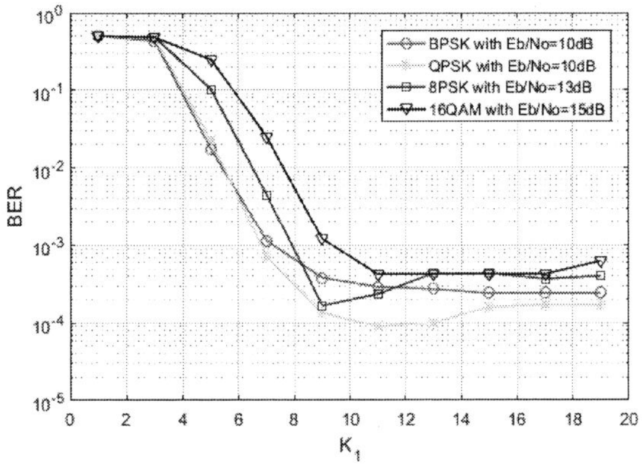

Figure 5: Changes of BER versus K_1 with K_2=4, K_3=3 for Proakis Channel A when Eb/No is fixed.

system performance can be substantially improved by proper selection of K_1. For example, at Eb/No of 16 dB the bit error rate (BER) can be reduced from around 0.4 to less than 10^{-3} by increasing K_1 from 4 to 16. On the other hand, and at 10 dB Eb/No, Figure 5 shows that increasing K_1 will not improve the system performance if it is increased beyond 10, which is about the length of the channel, where BER becomes almost constant. In such situations, increasing K_1 will add complexity to the implementation without significantly improving system performance. Moreover, it is clear form Figure 4 the amount of improvement in system performance depends on the Eb/No, the modulation order, and the values of the other two parameters K_2 and K_3 as well.

For the SPACE08 channel, and as shown in Figure 6 and Figure 7, the system performance at certain Eb/No increased by increasing K_1, but by an amount much smaller than what we had in the case of Proakis channel A at the same Eb/No. This is because of the length of the SPACE08 channel compare to the Proakis channel A. Also, the improvement of the system performance is proportional to the value of K_1 and when K_1 approaches length of the channel the performance becomes almost fixed. This is similar to Proakis channel A, where increasing K_1 will increase complexity to the implementation without any signficant improvement in performance. Figure 8

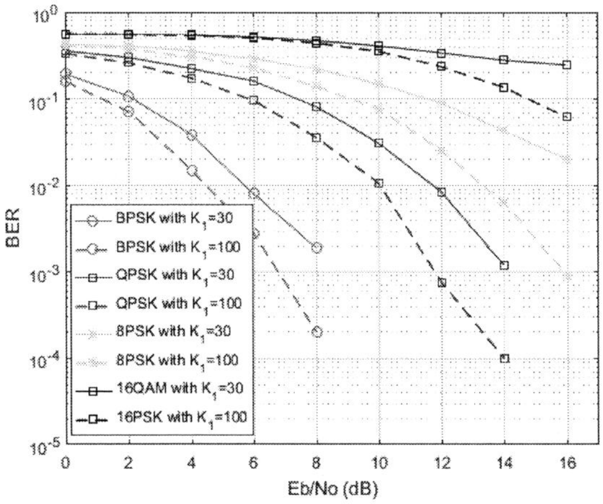

Figure 6: Impact of K_1 on system Performance for SPACE08 Channel with K_2=40 and K_3=40.

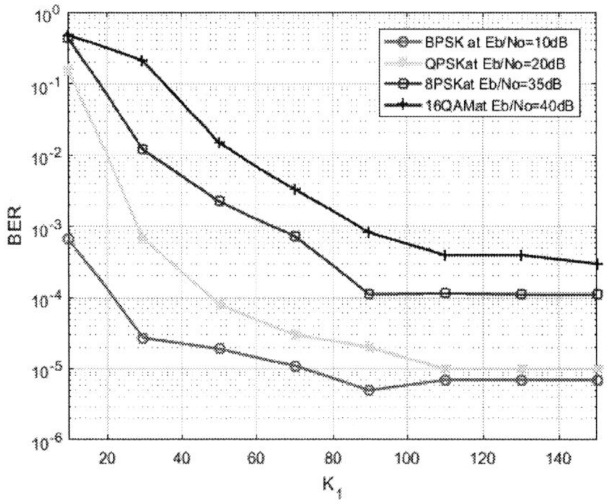

Figure 7: Changes of BER versus K_1 with K_2=30, K_3=30 for SPACE08 Channel when Eb/No is fixed.

and Figure 9 show the effect of K_2 while other parameters are fixed. As shown in Figure 8, at Eb/No less than 16 dB, changing k_2 will not provide improvement in performance for this short channel, because the job at this level of Eb/No can be done by K_1 and K_3 if they have been selected properly and this is clear also from Figure 9. For the SPACE08 channel and as shown in Figure 10

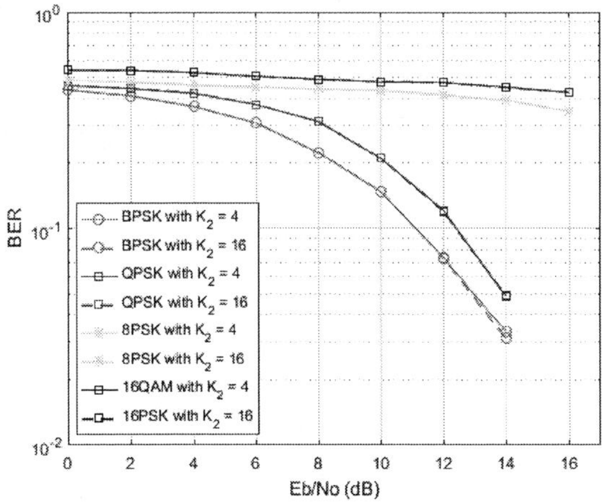

Figure 8: Impact of K_2 on system Performance for Proakis Channel A with K_1=4 and K_3=3.

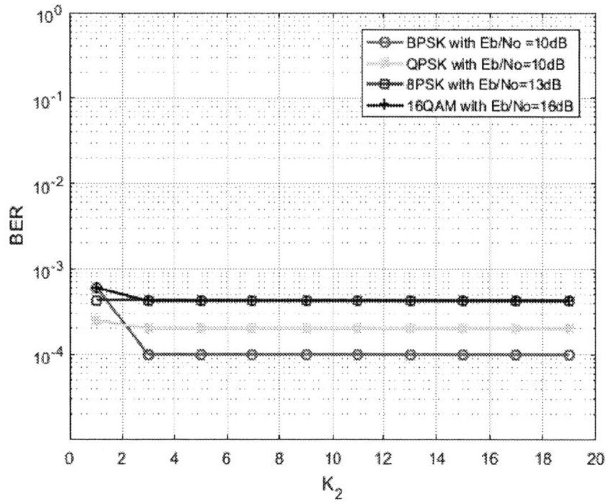

Figure 9: Changes of BER versus K_2 with K_1=4, K_3=3 for Proakis Channel A when Eb/No is fixed.

and Figure 11, K_2 still has an important effect on system performance and that is clear form Figure

10 where, for BPSK as an example, around 2 dB can be gained by increasing K_2 from 30 to 160. Such an improvement needs to be carefully evaluated because 160 in length of K_2 may add huge computational complexity. This result indicates that the selection of this parameter depends also on the length of the channel, and for such a long channel K_2 is important parameter and it could help improve the performance. The effect of the feedback section of the DFE is shown in Figure 12

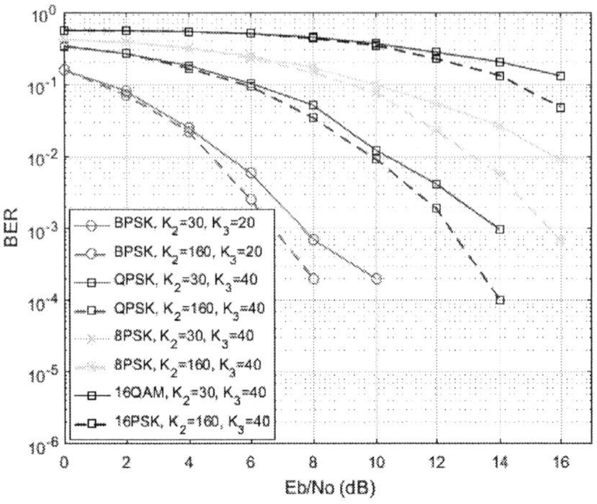

Figure 10: Impact of K_2 on system Performance for SPACE08 Channel with K_1=100.

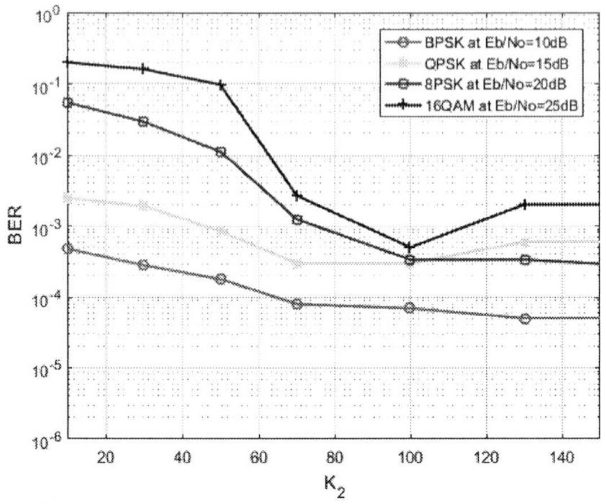

Figure 11: Changes of BER versus K_2 with K_1=50, K_3=30 for SPACE08 Channel when Eb/No is fixed.

through Figure 15 for both Proakis channel A and SPACE08 channel, where K_3 has been changed while the other parameters are held fixed. Similar to the previous results, this effect depends on the Eb/No and the values of other parameters and on the modulation order. For Proakis channel A and when K_1 is kept fixed, increasing K_3 will provide a large improvement in performance. This improvement will stop at a certain value of K_3 which is approximately the length of the causal part of the channel. In addition, the impact of changing K_3 on system performance is much larger in case of SPACE08 channel compare to Proakis channel A. This could be an another indication that for long channels, all parameters contribute to the improvement of the system when they selected properly. It is important to not increase the length of the parameter beyond the effective range, because at that moment we are adding only complexity to the system as in Figure 13 and Figure 15.

CONCLUSION

During this experimental simulation, the communication through Proakis channel A and SPSCE08 channel has been evaluated in relation to the equalizer parameter. For both channels, the equalizer parameters K_1 and K_2 are varied over a wide range around the length of channels L. The length of the feedback filter K_3 has been selected within a range around L_1 of each channel. As shown in the results, the parameter K_1 has a large impact on the system performance and it should be selected in the range of L_1 of the channel. For both channels, there will be minimal improvement in the BER when K_1 is chosen to be larger than L_1. Regarding the parameter K_2, the results showed that this parameter has a minor effect on the system performance, especially in Proakis channel A, while

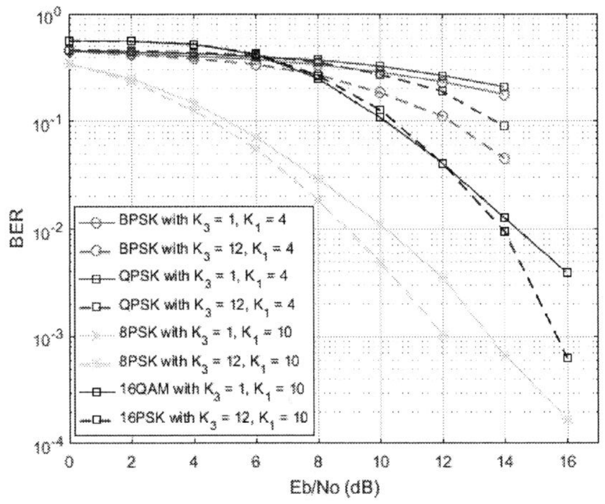

Figure 12: Impact of K_3 on system Performance for Proakis Channel A with K_2=4.

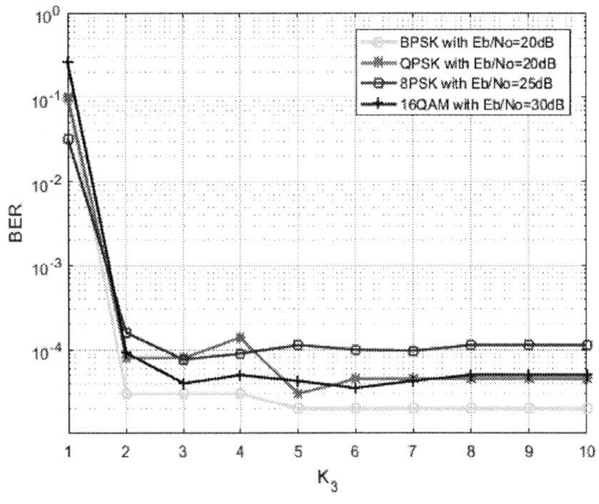

Figure 13: Changes of BER versus K_3 with K_1=4, K_2=4 for Proakis Channel A when Eb/No is fixed.

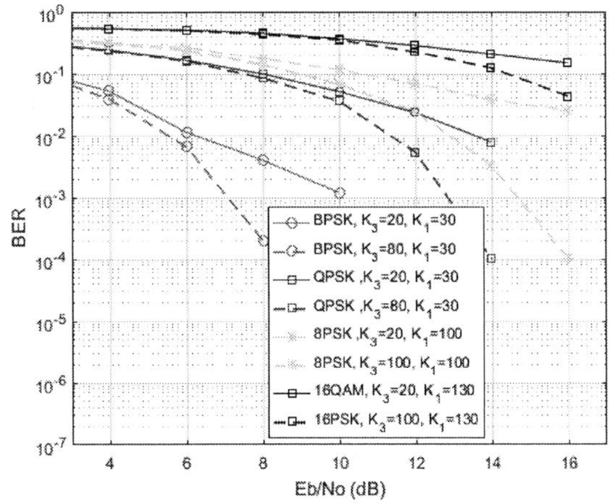

Figure 14: Impact of K_3 on system Performance for SPACE08 Channel with K_2=50.

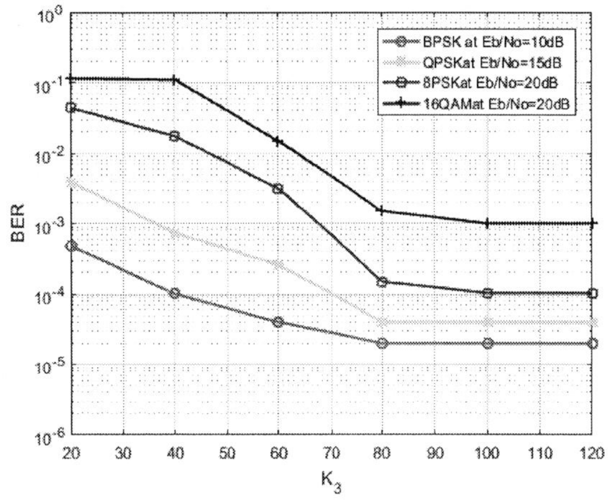

Figure 15: Changes of BER versus K_3 with K_1=40, K_3=40 for SPACE08 Channel when Eb/No is fixed.

it has limited impact in the case of the SPACE08 channel. On the other hand, K_3 is an important parameter and this makes DFE is preferable over Linear Equalizer.

REFERENCES

[1] A. C. Singer, J. K. Nelson, and S. S. Kozat, "Signal processing for underwater acoustic communications," *IEEE Communications Magazine*, vol. 47, pp. 90–96, January 2009.

[2] P. A. Voois, I. Lee, and J. M. Cioffi, "The effect of decision delay in finite-length decision feedback equalization," *IEEE Transactions on Information Theory*, vol. 42, pp. 618–621, Mar 1996.

[3] N. Al-Dhahir and C. Fragouli, “”How to choose the number of taps in a DFE,”,” in *Proc. 36th Annu. Conf. Information Science Systems*, (Princeton, NJ), Mar 2002.

[4] W. J. Hillery, M. D. Zoltowski, and M. Fimoff, “Decision feedback equalizer design for insensitivity to decision delay,” in *Acoustics, Speech, and Signal Processing, 2003. Proceedings. (ICASSP ’03). 2003 IEEE International Conference on*, vol. 4, pp. IV–505–8 vol.4, April 2003.

[5] Y. Gong and C. F. N. Cowan, “A self-structured adaptive decision feedback equalizer,” *IEEE Signal Processing Letters*, vol. 13, pp. 169–172, March 2006.

[6] A. Rafati, H. Lou, and C. Xiao, “Low-complexity soft-decision feedback turbo equalization for mimo systems with multilevel modulations,” *IEEE Transactions on Vehicular Technology*, vol. 60, pp. 3218–3227, Sept 2011.

[7] J. Su, Z. Lu, X. Yu, and C. Hu, “A novel low complexity soft-decision demapper for qpsk 8psk demodulation of dvb-s2 systems,” in *2011 IEEE International Conference of Electron Devices and Solid-State Circuits*, pp. 1–2, Nov 2011.

PERFORMANCE OF IEEE 802.11S FOR WIRELESS MESH TELEMETRY NETWORKS

Nathan D. Price, Arul Mathi Maran Chandran
Faculty Advisors:
Dr. Kurt Kosbar, Dr. Maciej Zawodniok
Missouri University of Science and Technology, Rolla, MO, USA

ABSTRACT

Wide area sensor networks have numerous agricultural, industrial, scientific, medical, and military applications. For many years now, such networks have been implemented through wireless mesh networks. As wireless mesh networks have come to maturity, several network standards have become readily available. The WiFi mesh extension 802.11s was finalized in 2008 and officially adopted into the 802.11 standard in 2012, is extremely interesting thanks to the wide spread adoption of WiFi and low cost. In this paper, we evaluate the performance of a small 802.11s network implemented on low-cost, off-the-shelf single board computers for the purpose of building a telemetry sensor network.

INTRODUCTION

A number of developments have lead to the realization of cost effective wireless mesh networks. A tremendous amount of wireless mesh research [1] over the last 20 years has culminated in a number of standardized wireless mesh protocols. The cost of computing has declined to the pint that requirements required for this project can be had for as little as $5. The widespread adoption of wireless computing such as WiFi has allowed for commodity pricing of the necessary wireless hardware. Additionally, the open source community generously provided an implementation built into the Linux operating system.

Despite those advancements, the adoption of mesh networking is slow. The main challenge is the understanding of limitations and practical constraints that a multi-hop, mesh network faces in realistic deployment. Some researchers [2–6] have evaluated WiFi based mesh test beds and identified some weaknesses including effects of TCP algorithms, network topology, and payload length. However, many of these findings are limited in that the performance data is only applicable to that test environment. For example, a loss of 1 Mbps in performance is significant if the maximum throughput of the channel is 11 Mbps but is insignificant if the maximum throughput is 300 Mbps. In contrast, this paper presents analysis of the performance from an experimental setup and compares them to benchmarks made in the same environment.
Wireless mesh networks (WMNs) have several advantages over traditional infrastructure based

wireless networks. The most notable advantage that a WMN has over a traditional wireless network like a cellular network or traditional infrastructure-mode WiFi is that it requires no central infrastructure to manage and relay communications. Each wireless device communicates directly in an ad-hoc or peer-to-peer fashion and can forward traffic on behalf of another device when no direct route is present. This makes WMNs cheap to operate as they don't require extra infrastructure beyond the client devices themselves. Should an individual route or neighbor device fail WMNs generally possess the ability to re-route traffic. This makes WMNs extremely robust as they generally don't have a single point of failure.

Background

One of the distinguishing features of WMNs is the ability to relay and route traffic. WMN routing schemes fall somewhere between two paradigms of routing: proactive routing and reactive routing [1]. In proactive routing each STA proactively creates and maintains routes. The benefit of proactive routing is that there is no time wasted setting up a route prior to its use, however there is the constant overhead of maintaining routes that scales with the size of the network. On the other hand, reactive (on-demand) routing creates routes as required. By contrast, reactive routing has the benefit of low overhead, but incurs a setup time penalty before a route is used.

IEEE 802.11S

802.11s adds new type of network, a Mesh Basic Service Set (MBSS), to the WiFi standard. Both the MBSS and Basic Service Set (BSS) commonly referred to as an ad-hoc mode WiFi network posses no central access point unlike traditional infrastructure-BSS networks. STAs in a BSS can communicate only with other STAs in range of direct communication. An MBSS network distinguishes itself from a BSS network in that its STAs posses the ability to route and forward traffic in the data link layer (i.e. layer 2 of the OSI model) [7,8]. In the past, multi-hop routing schemes such as AODV, OLSR, DSDV, DSR, B.A.T.M.A.N., etc [9–12]. have been designed and implemented for BSS networks though only through the use of network layer (i.e. layer 3 of the OSI model) routing and by violating the principal of strict layer separation [13].

MBSS networks contain three different types of STAs: mesh stations (mesh STAs), mesh access points (MAPs), and mesh gates. A mesh STA differs from a normal STA in that it has the ability to forward traffic to and from other STAs and perform path discovery. A MAP is a mesh STA and acts as a WiFi access point that can bridge standard STAs into the mesh network. A mesh gate is an STA that provides access to an external network (i.e. Internet access). None of the three new types of STAs are mutually exclusive meaning a mesh STA can serve multiple roles simultaneously.

A. *MAC LAYER*

Like traditional WiFi, 802.11s uses the distributed coordination function (DCF) mechanism for channel contention but adds a its own unique extension called mesh coordination function (MCF). Recall from 802.11 that DCF is a carrier sense multiple access with collision avoidance (CSMA/CA) based MAC layer. The MCF extension to DCF can operate operate in both a contention mode using the enhanced distributed channel access (EDCA) and a contention free mode called MCF controlled channel access (MCCA). In this contention free mode, mesh STAs use EDCA to bid for time slots. Then the mesh STAs can operate in a contention free mode similar to an infrastructure BSS.

B. *HYBRID WIRELESS MESH PROTOCOL*

For routing, 802.11s uses a routing scheme that is a combination of proactive and reactive routing techniques aptly named the hybrid wireless mesh protocol (HWMP). HWMP is most similar to the reactive routing scheme called the ad-hoc distance vector (AODV) protocol [7, 9], however it also incorporates proactive techniques.

The process of path discovery in AODV takes place via path requests (PREQ) and path replies (PREP) messages. To establish a route from a source STA to a destination STA, a source begins by broadcasting a PREQ. This PREQ contains the address of the original source and destination as well as a hop counter. When a PREQ is received by an STA that a is not the destination the hop count is incremented and the PREQ is rebroadcasted. Each forwarding STA records a reverse route (destination to original source) via the previous, PREQ-forwarding STA. In this way, a reverse route is created in steps from the destination to the original source as the PREQ is relayed forward.

Once the PREQ is received by the destination a reply process begins. The destination first responds with a path reply (PREP) message. The PREP gets relayed along the reverse route established by the PREQ. As the PREP propagates back toward the original source, a forward route (original source to destination) is established via the previous, PREP-forwarding STA. In an identical fashion to the PREQ, each STA increments a hop counter contained in the PREP message until it reaches the original source. Since a PREQ is broadcasted, it is likely that a destination STA will receive multiple copies of the PREQ each having traveled a different route to the destination. The destination STA replies to the PREQ with the lowest hop count.

Each STA tracks its routes via a timer to detect breaks. When a route is used the timer is reset. Should a route timer expire (a route break), a path error (PERR) message is sent backward to all STAs on the route before break. Path discovery must take place again before further communication with the destination STA of a broken route.

HWMP differs from AODV in the use of the Airtime Link Metric (ALM). The ALM takes the place of the hop counter in PREQ and PREP messages. The ALM is an estimation of amount of time required to send one frame between two STAs. Before forwarding a PREQ or a PREP, an STA will calculate the ALM between itself and the previous STA and add this number to accumulated sum in PREQ or PREP. The destination STA is likely to receive several copies of the same PREQ hence it will reply only to PREQs with the lowest accumulated ALM.

HWMP introduces three proactive routing modes and three types of proactive routing messages respectively. The new types of routing messages are the root announcement (RANN) message, the gateway announcement message (GANN), and proactive PREQ. STAs serving as a gateway can be configured to send a GANN on a periodic basis. The GANN is processed like a PREQ to create a route from a STAs to the mesh gate, but GANNs are not replied to. This allows STAs in the mesh to maintain an active route to external networks via gateway STAs and keep ALMs up to date. Alternatively, STAs (including gateways) can be configured as a root. Root STAs periodically broadcast RANNs to all STAs to maintain an active routes between themselves and all other STAs in the mesh. If an STA receives a RANN and the ALM contained in a received RANN is better than its stored metric, the STA will initiate path discovery with RANN-sending, root STA. Any STA, for example an STA with an important network resource, can be made a root STA. Finally, root nodes can be configured to send proactive PREQs periodically. Proactive PREQs are the same as PREQs except the destination address is set to all ones. STAs may optionally reply with a PREP, or they may be compelled to reply depending on configurable flags in the proactive PREQ.

C. AIRTIME LINK METRIC

$$\frac{1}{1-e_f} = \frac{1}{1-q} = \frac{1}{p} \tag{1}$$

The ALM uses frame error rate e_f to weight the estimated time needed to transmit each frame. If the success or failure of each transmission is thought of as a sequence of Bernoulli trials equation (1) resembles the expected value for a geometric distribution.

$$c_a = \left[O + \frac{B_t}{r}\right]\frac{1}{1-e_f} \tag{2}$$

Thus the ALM seen in equation (2) is simply the expected number of transmissions times the time needed for each transmission. The value O is the overhead time generated from frame headers, error checking, etc. relative the current configuration of the physical layer (e.g. bits per symbol, coding rate, etc.). The second term $\frac{B_t}{r}$ is simply the maximum frame size B_t (i.e. 8192 bits) divided by the current data rate r.

$$MIN \sum_{i\in I, j\in J} c_a(i,j)x_{ij} \tag{3}$$

Equation (3) represents the total ALM seen by an STA for a particular route. The x_{ij} is a boolean value indicating if the route made a single hop from from STA i to STA j thus the final ALM is the of the summation of ALMs across all single hops taken. HWMP keeps the route with lowest ALM for any particular destination. Since the ALM is calculated in units of time and since HWMP uses the route with the total ALM, ALM is therefore an attempt at a latency optimized routing.

TESTING METHODOLOGY AND EXPERIMENTAL SETUP

Our network consisted of inexpensive single board computers (SBC) and inexpensive WiFi adapters. Our single board computers consisted of 4 Raspberry Pi 2s and 2 Raspberry Pi 1s running the Raspbian Linux distribution (kernel version 4.4). Our WiFi adapters were USB connected and based on the Atheros AR9271 chipset. Choice of WiFi adapter was made carefully as not all adapters currently support 802.11s mode. The adapters were configured for 20 MHz channels in high throughput mode allowing for up to 75 Mbps of throughput theoretically. A summary can be found in Table 1

Table 1: Hardware and Software configurations

Hardware	Setup
SBC	Raspberry Pi 1 & 2
WiFi Adapter	Atheros AR9271 chipset
Channel	Ch1–2.412 GHz
Bandwidth	20 MHz
Mode	Hight Throughput (802.11n)
MIMO Configuration	1x1
Distribution	Raspbian Jessie
Kernel Version	v4.4
IPerf3	v3.0.7

The presented work focuses on experimental performance evaluation of WiFi-based mesh networking. A set of end-to-end metrics were gathered on our demo network for TCP and UDP performance using the open-source tool IPerf3. For TCP, throughput and retry count were recorded, and for UDP, throughput, jitter, and packet loss were recorded.

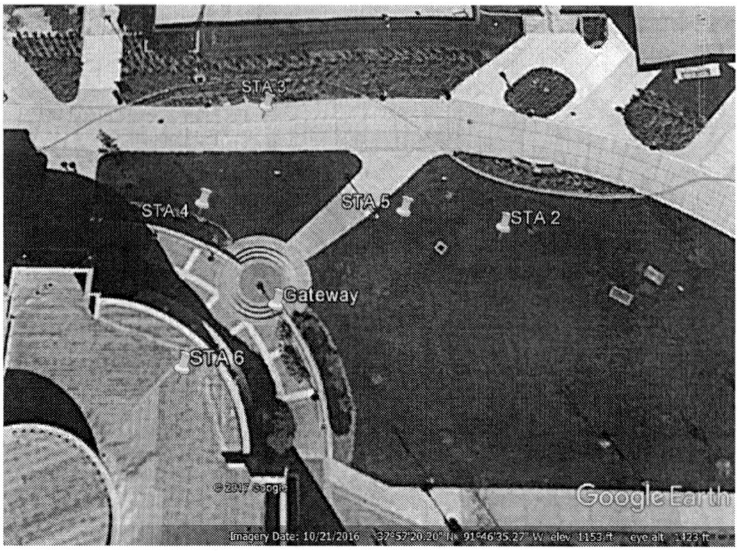

Figure 1: Mesh network in the Campus.

A random cluster topology was used for the primary experiment. The cluster features 5 data generating STAs running the IPerf3 client that were placed placed at various angles and distances relative to a gateway STA running the IPerf3 server. Depending on the application each STA potentially could contain a sensor platform gathering multiple channels of data. In this scenario, multiple flows of data (i.e. one per STA) were relayed to the gateway STA that could represent a local server, a connection to a cloud-server, etc. This experiment was conducted on our campus green with STAs placed directly on the ground.

A secondary, linear topology was used to gather performance figures that could be used as a benchmark. STAs were arranged in a line with equal spacings. Routing was set statically so that each node could communicate only with its one-hop neighbor(s). A single data flow was generated and relayed across varying numbers of hops. Testing was repeated with 10 m, 15 m, 20 m spacing. Results from this test can be used a relative ceiling for performance. This experiment was run once on the same campus green and once in a nearby park both with STAs placed directly on the ground.

Figure 2: Placement of nodes in the park to test individual hop metrics

RESULTS

This section presents the results of the experiments namely the cluster topography and the static-routing, multi-hop benchmark at both locations. Table 2 and Figures 3 and 4 show results from the mentioned experiments.

The performance observed in the cluster topography are reasonable despite low numbers. Figure 3a shows the maximum single-hop single-flow performance to be about 4.1 Mbps. After review routing tables on each STA, we observed that all STAs were using a single hop route to connect to the gateway. With that in mind the best expected performance for each STA is roughly $1/5^{th}$ of the maximum single-hop, single-flow speed (i.e. 0.82 Mbps). The throughput values in Table 2 are near 0.82 Mbps.

Table 2: Iperf3 TCP Results

STA	Distance from the Gateway (m)	Throughput (bits/s)	Retries
2	27.5	1.10 M	202
3	25	966 K	159
4	15	902 K	141
5	19	756 K	142
6	12	748 K	179

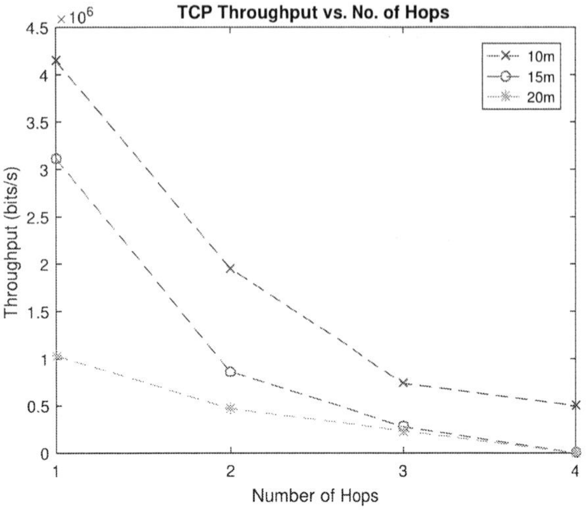

(a) Throughput vs Number of hops

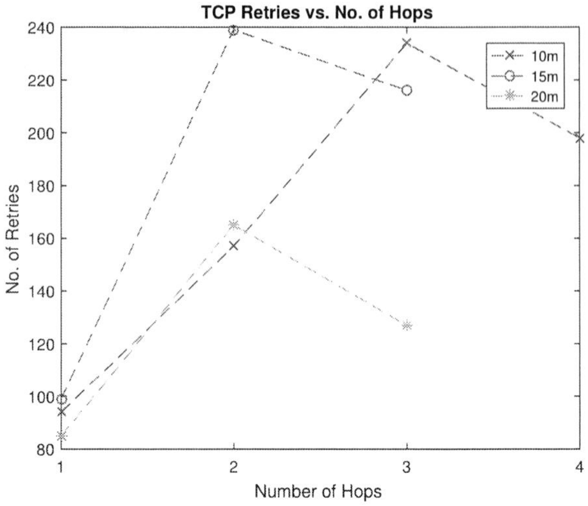

(b) Number of retries vs Number of hops

Figure 3: TCP results when operated the network with interference from other network

Plots from Figures 3 and 4 show similar trends across multiple hops. Maximum throughput was achieved in the park scenario in contrast to the campus scenario. The primary difference between these two scenarios was that there was no WiFi activity on channel 1 in the park whereas there was a campus network active on channel 1. The campus network likely introduced additional channel contention that reduced performance of the mesh. The number of retires in both scenarios show a peak at two hops. There are multiple factors that may have contributed to this trend including but no limited to the distance covered, the hop number, the topography, and/or WiFi's physical layer configuration.

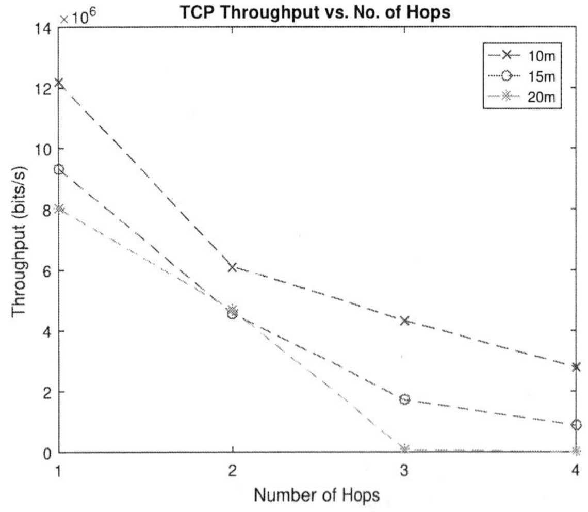

(a) Throughput vs Number of hops

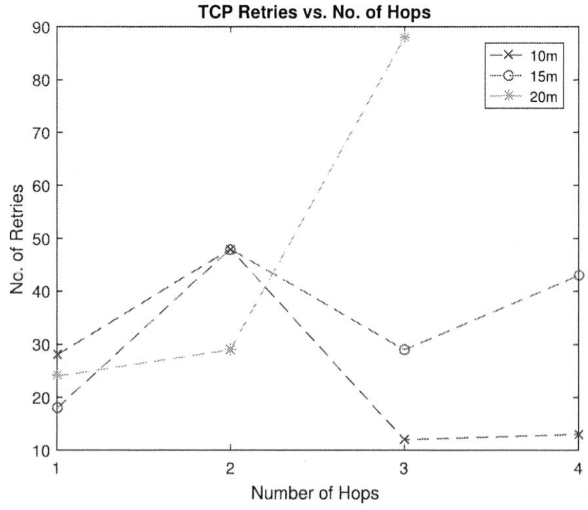

(b) Number of retries vs Number of hops

Figure 4: TCP results when operated the network with no interference from other network

During our testing, we found IPerf3 to be an unreliable tool for measuring UDP performance in a wireless mesh network. IPerf3 does not throttle UDP data generation automatically like TCP. This would require the user to manually identify the throughput maximum throughput by trial-and-error. An overestimation resulted in IPerf3 crashing whereas a underestimated value did not result in the maximum throughput. The trial-and-error result was not reliable across multiple trials. Additionally, IPerf3 failed to establish a link with the server when throughputs dipped below approximately 200 Kbps which is why there are data points missing in Figure 3b.

CONCLUSIONS

In this paper we evaluated the performance of a small demo mesh network built using 802.11s. Our goal for this paper was to provide context for such performance figures. The maximum throughput in our multi-flow scenario comprised of five flows of data was 1.10 Mbps and the minimum was 745 Kbps. The maximum throughput achieved in a single-flow, single-hop, activity-free, WiFi channel was 12.2 Mbps whereas 4.2 Mbps was achieved in an active WiFi channel. When compared against our single flow metric we found the multi-flow results to be reasonable. We believe 802.11s a network has place in telemetry systems.

In our paper we found IPerf3 to be an inadequate tool for measuring UDP performance in a mesh network. In future work, we hope to incorporate better tools. Due to the small size of our network, we did not observe any multi-hop routing. We would also like to increase the scale of the network so that we may make observations related to routing and routing techniques.

REFERENCES

[1] J. Sarangapani, *Wireless Ad Hoc and Sensor Networks: Protocols, Performance, and Control*, ser. Control Engineering Series. Boca Raton, FL: CRC Press Taylor & Francis Group, LLC, 2007.

[2] M. Rethfeldt, P. Danielis, B. Beichler, B. Konieczek, F. Uster, and D. Timmermann, "Evaluating Cross-Layer Cooperation of Congestion and Flow Control in IEEE 802.11s Networks," in *2016 IEEE 30th International Conference on Advanced Information Networking and Applications (AINA)*, Mar. 2016, pp. 181–188.

[3] J. C. P. Wang, B. Hagelstein, and M. Abolhasan, "Experimental evaluation of IEEE 802.11s path selection protocols in a mesh testbed," in *2010 4th International Conference on Signal Processing and Communication Systems*, Dec. 2010, pp. 1–3.

[4] M. N. Alam, R. Jntti, J. Kneckt, and J. Nieminen, "Performance Study of IEEE 802.11s PSM in FTP-TCP," in *2012 IEEE Vehicular Technology Conference (VTC Fall)*, Sep. 2012, pp. 1–5.

[5] E. Elahi, A. Qayyum, and S. Naz, "The effect of payload length on QoS in IEEE 802.11s Wireless Mesh Networks," in *2012 18th IEEE International Conference on Networks (ICON)*, Dec. 2012, pp. 70–73.

[6] T. Imboden, K. Akkaya, and Z. Moore, "Performance evaluation of wireless mesh networks using IEEE 802.11s and IEEE 802.11n," in *2012 IEEE International Conference on Communications (ICC)*, Jun. 2012, pp. 5675–5679.

[7] "IEEE Standard for Information TechnologyTelecommunications and information exchange between systemsLocal and metropolitan area networksSpecific requirements Part 11: Wireless LAN Medium Access Control (MAC) and Physical Layer (PHY) specifications Amendment 10: Mesh Networking," *IEEE Std 802.11s-2011 (Amendment to IEEE Std 802.11-2007 as amended by IEEE 802.11k-2008, IEEE 802.11r-2008, IEEE 802.11y-2008, IEEE 802.11w-2009, IEEE 802.11n-2009, IEEE 802.11p-2010, IEEE 802.11z-2010, IEEE 802.11v-2011, and IEEE 802.11u-2011)*, pp. 1–372, Sep. 2011.

[8] R. C. Carrano, L. C. S. Magalhaes, D. C. M. Saade, and C. V. N. Albuquerque, "IEEE 802.11s Multihop MAC: A Tutorial," *IEEE Communications Surveys Tutorials*, vol. 13, no. 1, pp. 52–67, 2011.

[9] "RFC 3561 - Ad hoc On-Demand Distance Vector (AODV) Routing." [Online]. Available: https://datatracker.ietf.org/doc/rfc3561/

[10] "RFC 4728 - The Dynamic Source Routing Protocol (DSR) for Mobile Ad Hoc Networks for IPv4." [Online]. Available: https://datatracker.ietf.org/doc/rfc4728/

[11] "RFC 3626 - Optimized Link State Routing Protocol (OLSR)." [Online]. Available: https://datatracker.ietf.org/doc/rfc3626/

[12] "Better approach to mobile ad-hoc networking (B.A.T.M.A.N.)." [Online]. Available: https://www.open-mesh.org/projects/open-mesh/wiki

[13] V. Kawadia, Y. Zhang, and B. Gupta, "System services for implementing ad-hoc routing protocols," in *Proceedings. International Conference on Parallel Processing Workshop*, 2002, pp. 135–142.

AN INTERNET-BASED REMOTE COMMAND AND TELEMETRY SYSTEM FOR A MICROWAVE PROPAGATION STUDY

Mario Colapelle, Brian Zamore, Brian Kopp
Electrical Engineering Department
University of North Florida
Jacksonville, Florida, 32224
Brian.kopp@unf.edu

Randy Pierce
Intelligent Transportation Systems Office
Florida Department of Transportation
Tallahassee, Florida, 32301
Randy.Pierce@dot.state.fl.us

ABSTRACT

A research project investigating microwave radio frequency propagation in a 500 mile link across the Gulf of Mexico requires a remote-control process to command microcontroller-based devices including power control modules and antenna feedhorn positioners, and to telemeter system parameters back to the operators. The solution that was developed is a simple, webserver-based user-interface that can be accessed both locally and remotely via the internet. To interface the webservers with the microcontroller-based devices, a polling protocol, based on MODBUS, was developed that provides an efficient command and telemetry link over a serial RS-485 interface.

KEY WORDS

Command, telemetry, MODBUS, webserver, internet.

INTRODUCTION

This paper discusses the remote command and telemetry process under development at the University of North Florida (UNF), in support of a joint-research project with the Intelligent Transportation Systems (ITS) office of the Florida Department of Transportation (FDOT). The background section discusses the purpose and design of the research project and the need for remote command and telemetry. The design of the test stands is presented next and then the Telemetry and Command (T&C) requirements are discussed. There are multiple devices that can be commanded and have their status telemetered. These are reviewed in the T&C section of the paper. A brief discussion on off-the-shelf protocols that were reviewed and considered for use in the project is presented next. This is followed by a discussion of the serial protocol that was ultimately developed by UNF and used in the project. The paper concludes with a discussion on the status of the project.

BACKGROUND

The University of North Florida and the ITS office of the FDOT are conducting a study to investigate the feasibility of using the atmospheric evaporative duct at sea-level to create a communications link across the Gulf of Mexico between Key West and Destin, Florida, a distance of 500 miles. It is hoped a reliable enough link can be established to offer some data network redundancy to the single points of failure that the FDOT's statewide ITS network has along the Florida Keys and Florida panhandle. The first phase of this project is underway and involves a propagation study at 5 and 10 Giga-Hertz (GHz) to assess the suitability of the evaporative duct over a long period of time.

To implement the propagation study, a transmitter is being constructed for deployment at Naval Air Station (NAS), Key West and a receiver is being constructed for deployment at Henderson Beach State Park in Destin Florida. The transmitter will be capable of transmitting Continuous Wave (CW) signals and modulated signals for the study. Software Defined Radios (SDR) will be used to transmit and receive the test signals. Microwave component up-conversion and down-conversion will permit propagation studies at 5 and 10 GHz. Large prime focus 2.4 meter dish antennas mounted at a height of approximately 30 feet will launch and capture the microwave signals into and out of the evaporative duct, at an elevation angle near zero degrees and an azimuth that matches Key West with Destin.

The propagation tests will be conducted over a long duration period of at least 6 months to evaluate seasonal and environmental effects on propagation. Remote operation of the transmitter and receiver will be necessary to support the testing in a cost effective manner. Students in the UNF Advanced Telecommunications Research Program (ATRP) will conduct the propagation tests from the ATRP lab at UNF in Jacksonville, Florida. Internet connectivity at the ATRP lab will be used to remotely operate the transmitter and receiver test stands. The test stands are being constructed on communications trailers provided by the FDOT and will use cellular modems for internet connectivity. Both test stands must be similar in design to reduce design risk and cost. Mains power is anticipated to be available at the NAS Key West transmitter site but will not be available at the Destin receiver site, at Henderson Beach State Park in Destin. Due to the power limitations at Destin, and the need to use similar designs, the overall design used by both test stands will therefore be power efficient.

TEST STAND DESIGN

In combination, the two test stands form a complete communications link. The transmitter test stand uses an SDR radio that is controlled by a mini Personal Computer (PC). The mini PC is remotely controlled through the internet using a commercial remote, computer-control, software package. A cellular modem/router provides the internet connection. The mini PC has a dual-port Network Interface Card (NIC) to interface directly with the SDR and also connect to the internet via Ethernet. Mathworks Simulink simulates the transmitter source and receiver sink on the mini PCs and controls the interfaces with the SDRs.

The SDRs use a first Intermediate Frequency (IF) of 270 Mega-Hertz (MHz). A second IF of 1020 MHz is used in between the first IF and the two selectable RF frequencies of either 5840 MHz or 10250 MHz. The SDRs and the IF and RF oscillators are all disciplined with a 10 MHz reference provided by a Global Positioning Satellite (GPS) receiver on each test stand. The SDR and mini PC equipment are housed in an insulated enclosure on the FDOT provided trailers. The second IF and RF microwave equipment are housed in a tower mounted enclosure behind the antennas to keep the RF cable runs short. The large antennas are mounted on combination azimuth and elevation rotors, controlled via an Ethernet connected controller housed with the mini PC and SDR at the bottom of the towers. Figure 1 contains a high level schematic of the communications architecture on the transmitter test stand. The receiver test stand communications architecture is nearly identical to the transmitter test stand, with the exception that there is no transmit power meter in the up-tower components.

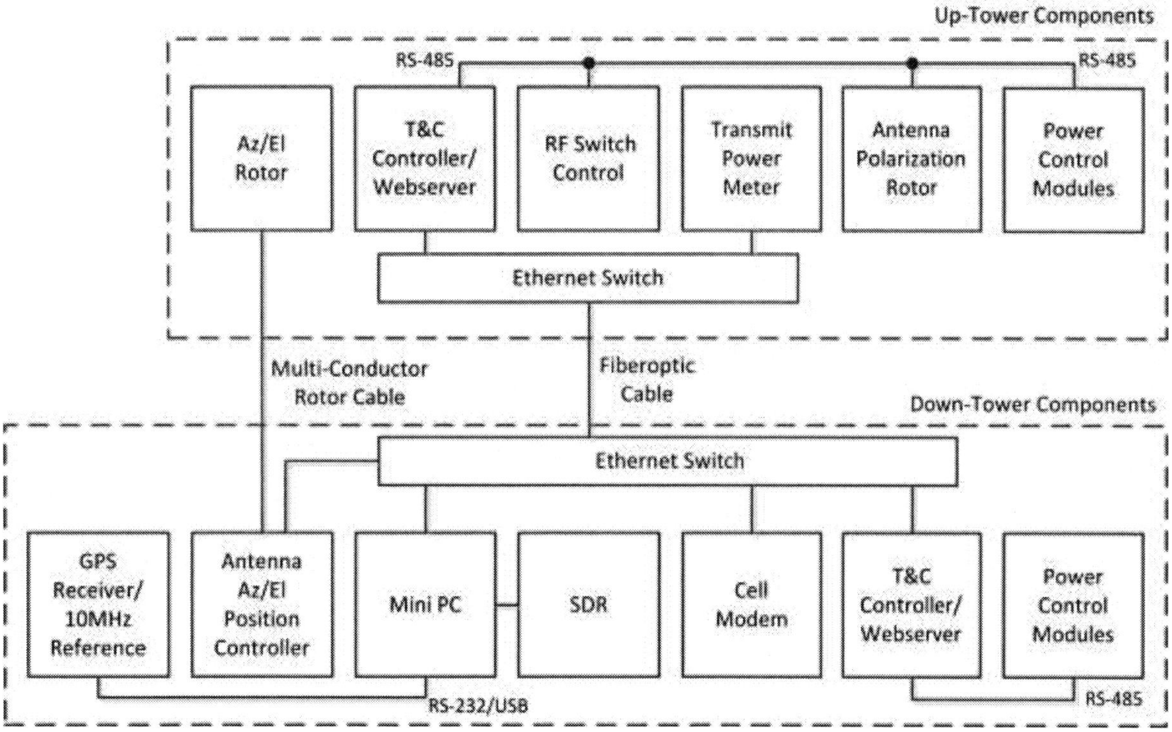

Figure 1. Transmitter test stand communications architecture.

The feedhorns on the prime focus antennas have dual linearly polarized feeds, one matched to 5840 MHz and the other to 10250 MHz. Polarization rotation by the evaporative duct is one parameter that will be studied in this project. Rotating the linear transmit and receive feedhorns will facilitate this investigation. The feeds will be mounted on a UNF designed polarization rotor assembly that uses a stepper motor. A motor control interface is being designed to connect the T&C controller to the polarization rotor assembly.

TELEMETRY AND COMMAND REQUIREMENTS

The use of the internet and the Ethernet Local Area Network (LAN) on the test stands will provide remote access from the ATRP lab at UNF to the mini PCs on the test stands. Most of the remaining devices on the test stands will be controlled via the LAN from the mini PC during tests. The mini PC will use vendor software to interface with the antenna azimuth and elevation position controller and the Transmit Power Meter via the LAN. To monitor and command many of the other devices on the test stand that do not have built in Ethernet capability a T&C controller is being developed.

The T&C controller serves two purposes. It interfaces with the user via Ethernet and it interfaces with the devices. To connect the T&C controllers to the users via Ethernet, webservers were chosen. This choice permits the user to interface with the T&C controllers from the mini PC (via remote PC access) or through the internet directly from the ATRP lab. A maintainer can also attach a laptop to the test stand LAN and have access to the T&C controllers. On the device side, the T&C controllers have several duties to perform. They must control power management by commanding power supplies to turn on and turn off and they must also telemeter bus voltages. The T&C controllers must also be able to command the RF switches on the test stands that route IF and RF signals between various microwave components. The most challenging task for the T&C controllers is the control of the polarization rotor by commanding it to move to a specific position. When queried, the current position of the rotor must be reported as well.

Each test stand has several dozen devices requiring T&C control. All of the devices are physically close to one another with only relatively short cable runs between them. No wireless connectivity is necessary. The data rate requirements of the T&C control operations are low because none of the devices have high speed telemetry or command needs. In addition, erroneous commands cannot cause damage or create an unrecoverable condition. The environment does contain RF energy of multiple watts so there is concern for Radio Frequency Interference (RFI) on the test stands. There is no perceived threat that dictates a strong need for communications security on the test stands.

OFF-THE-SHELF PROTOCOLS

The T&C controller requirements suggest a simple wired protocol will suffice for interfacing with the various devices on the test stand that do not already have an Ethernet interface. Several industrial and military protocols were initially reviewed. Ultimately, a custom protocol that closely models one of the reviewed protocols was developed.

From the industrial market two protocols that are typically used for process control and monitoring were reviewed: MODBUS RTU [1] and CANbus [2]. Other industrial protocols including Foundation Fieldbus H1 [3], Profibus DP [4], and Distributed Network Protocol 3

(DNP3) [5], are known to the authors but were judged to not offer anything significant to the relatively low protocol demands of this project beyond what MODBUS or CANbus offer.

MODBUS, is a query and response (master and slave) protocol that can be implemented in hardware using Ethernet, a high speed tokens scheme, or asynchronous serial communications. This protocol is mature and is very popular for simple network connections between host computers and Remote Terminal Units (RTUs) or industrial process devices. Its asynchronous serial communication version, known as MODBUS RTU, makes it a good candidate for low-speed, low-overhead, byte-oriented, communication schemes. The MODBUS protocol has been in use since the late 1970's.

MODBUS, like many protocols, uses addresses to identify which device is to respond to a query. However, the industrial CANbus protocol does not use any addresses. This pier to pier protocol uses message identifiers to indicate what type of device sent the message, not which specific device the message came from or is going to. A simple but very effective bus priority-arbitration scheme ensures that data collisions are resolved. This permits CANbus devices to communicate whenever they have something to say. CANbus is mature and used extensively in the automotive industry, where many modern cars have multiple CANbuses onboard. In the past 30 years CANbus has also spawned related protocols including Allen Bradley's DeviceNet, in the process control industry, and the National Marine Electronic's Association (NMEA) 2000 protocol in the marine industry.

One military protocol, the Military Standard 1553 protocol was considered. Like MODBUS it is a query and response protocol where a bus controller communicates with multiple remote terminals. The protocol operates at 1 Mega-bit-per-second (Mbps) and uses a dual-redundant bus architecture. Signaling of data uses Manchester coding which assists in synchronization. The Military Standard 1553 protocol is popular in military avionics because of its reliability. It is a mature protocol and has been used on many military avionics systems as well as the space shuttle.

UNF T&C CONTROL PROTOCOL

After reviewing the available off-the-shelf protocols it was decided to develop a custom protocol for this project that is very similar to the asynchronous MODBUS RTU protocol. The CANbus protocol could have been adapted for use but since the nature of the T&C controller function in this project is that of a query and response transaction from specific addressed devices, it was decided to use a more simple conforming protocol. The chosen physical layer is RS-485, which is supported by MODBUS RTU. The twisted-pair cabling and differential signaling of RS-485 will help mitigate any RFI and also any motor noise from the antenna azimuth, elevation, and feedhorn polarity rotors. The data link layer will consist of asynchronous byte oriented data transfers, which also conforms with MODBUS RTU. This will facilitate the use of simple hardware interfaces available in microcontroller peripherals. In this particular case, the mode chosen for asynchronous byte transmissions is to send 1 start bit, 8 data bits, an odd parity bit and one stop bit. The data frame formats for MODBUS RTU and for the UNF T&C control protocol are shown in Figure 2.

The UNF T&C control protocol introduces several minor changes to the data frame structure. The start and stop flags consist of 4 full bit times of silence, instead of a minimum of 3.5 bit times. This will simplify the timer computation requirements slightly in a typical microcontroller program. In addition, the Cyclic Redundancy Check (CRC) error detection field is optional in the UNF T&C control protocol. The reason for this requires a discussion on the general topic of T&C control error checking.

As mentioned above, the UNF T&C control protocol has been developed for a specific project that will control simple devices that are not operating on a time critical scale. An error in transmission of a command cannot cause an unrecoverable condition. If an erroneous command is executed it can be easily detected by direct means via the telemetry data in a status check command response, or indirectly by monitoring the other systems on the test stand, such as the RF power meter, mini PC, or SDR. Similarly, a suspected erroneous response can be easily checked with a repeated command query. Never the less, there are mandatory error checking processes built into the protocol. The first one has already been mentioned. Odd parity is required on all byte transmissions in the UNF T&C protocol. The use of parity is well known, easy to implement in hardware, and can detect any single bit errors in a byte (as well as any 3, 5, and 7 bit error patterns in the byte). A higher layer error checking process that is also mandatory involves the handling of the data frame. Delays between bytes, or reception of an incorrect number of data bytes for a particular function code, will trigger an error in the software reception process and this will be reported to the user as a failed (unexecuted) transaction. If the error occurs in the reception process of a field device, the response to the query will include an error code that indicates a communications failure.

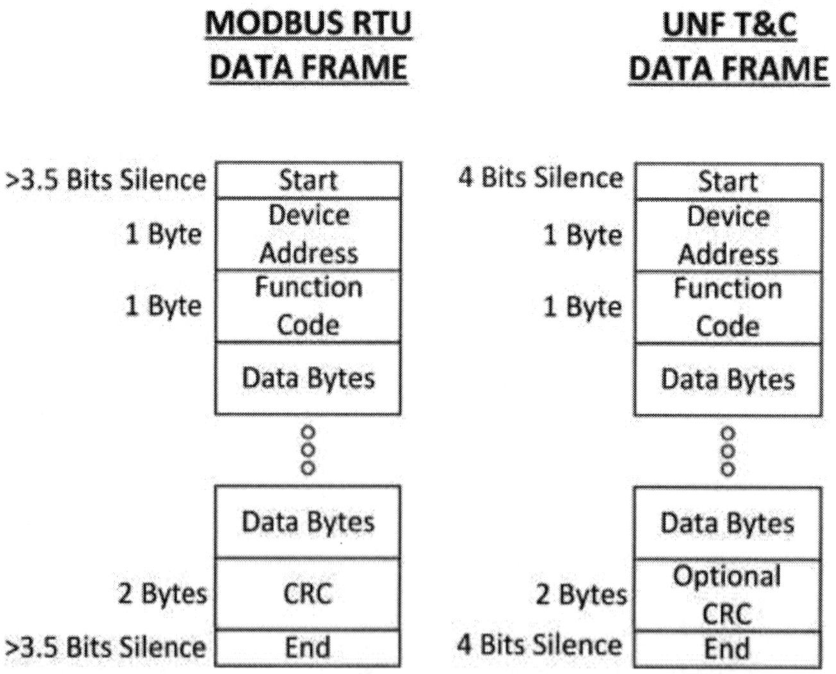

Figure 2. Comparison of MODBUS RTU data frame and UNF T&C data frame.

Given the use of the mandatory error checking processes built into the UNF T&C control protocol, coupled with the less demanding needs of the project in general, it was decided to make the use of the CRC error checking optional. The option to use the CRC is invoked through the function code. The MODBUS RTU function code is an 8-bit byte but there are only 128 recognized function codes. The Most Significant Bit (MSB) of the MODBUS RTU function code is used by field units to indicate an error either during command reception or command execution. However, it is unused on the outbound command query function code. In the UNF T&C data frame we set this bit in the command data frame to indicate that a transaction (query and response) does NOT use the optional CRC error checking. This effectively creates a second set of 128 possible identical function codes that are executed without a CRC check.

Through this point the UNF T&C control protocol has maintained compatibility with the MODBUS RTU protocol. Nothing in the UNF T&C control protocol would prevent a MODBUS RTU compatible field device from being added to a UNF project using the UNF T&C control protocol (with CRC enabled). The function code assignments and error handling processes in the MODBUS RTU protocol include public codes used universally, as well as vendor specific codes. Most of the public function codes in MODBUS are very flexible to facilitate use by various types of field devices. For instance, the MODBUS public function code to enable multiple field device discrete outputs is function code 15, known by the arcane name: "Force Multiple Coils" [1, pg. 29]. The data field in the command query of this function code contains as few as 6 bytes and as many as 251 bytes, permitting control of as few as 1 discrete output and as many as 1968 discrete outputs. Rather than adapt these public function codes for use on this relatively simple project, UNF chose to take advantage of the user defined function code assignments available in the MODBUS protocol. Figure 3 shows the MODBUS function code assignment space. Note that there are 17 addresses available for user defined function codes.

By employing these user defined function codes the UNF T&C protocol can maintain compatibility with MODBUS RTU for some future projects that are as yet undefined. Several user defined function codes have already been developed for the field units being used in the current project. They are shown in Table 1. More function codes will be developed as needed for the project.

PROJECT STATUS

The field device that will be used in this project to control either power modules or RF switches, has been developed. The combined device is referred to as a Power Control Module (PCM). Two PCMs are shown below in Figure 4. They use a PIC18F2580 microcontroller and support five discrete outputs. Three discrete outputs control the delivery of DC power to three power outputs via independent high-side load switches. The two remaining discrete outputs can be used to command two non-latching relays or a single latching relay, using the Pulse Output function code. The DC power bus used to supply the power outputs is monitored by the microcontroller and reported when queried with the Read ADC Value function code. The T&C controller and webserver is being developed on an Arduino platform with an Ethernet shield and an RS-485 shield to support the webserver and the T&C control interface, respectively. The T&C controller and webserver hardware is also shown in Figure 4. The polar rotor controller is still under development. The test stands and the entire command and telemetry system are expected to be operational by the fall of 2017.

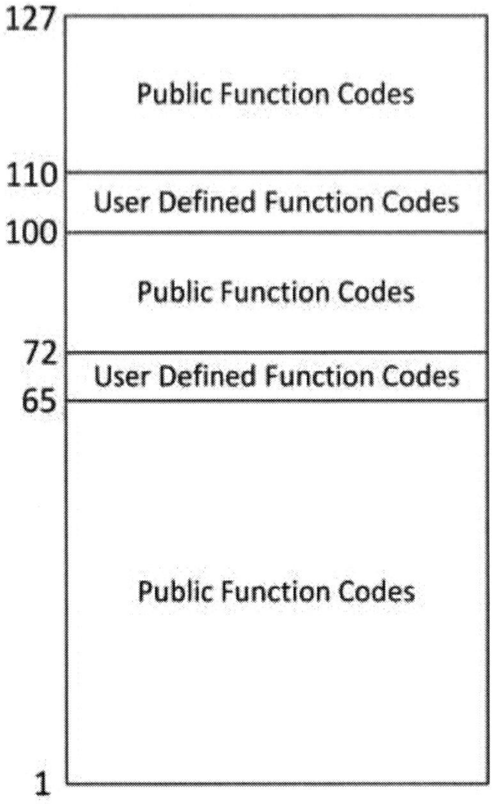

Figure 3. MODBUS Function code assignment space [1, pg. 10].

Table 1. UNF T&C control protocol user defined function codes.

Function Code	Function Code Without CRC	Name	Description
65	193	Do Outputs	Set and clear 1 – 16 discrete outputs
66	194	Get Outputs	Query status of 1 – 16 discrete outputs
67	195	Pulse Output	Apply pulse output to 1 of 16 discrete outputs (Used for latching relays. Pulse duration set by device)
68	196	Read ADC Value	Query for report of monitored bus voltage
100	228	Broadcast output clear	Clear all discrete outputs on all output devices
105	233	Rotor Position Set	Set the rotor position to a specific angle (0-360 deg.)
106	234	Read Rotor Position	Read the current rotor position

CONCLUSIONS

The development of a simple T&C control protocol has leveraged an available off-the-shelf industrial protocol to help develop a command and telemetry system for a joint UNF and FDOT research project in a cost effective and timely manner. By also using low-cost and easy-to-learn solutions for the T&C controller and webserver, students at the UNF ATRP lab have been able to participate in this research project as well. After this project is completed it is hoped other UNF research projects will be able to take advantage of the lessons learned and deploy the new UNF T&C control protocol.

REFERENCES

[1] The MODBUS Organization, "MODBUS Protocol Specification", V1.1b3, Available www.modbus.org/specs.php

[2] Robert Bosch GmbH , "CAN Specification", V 2.0, 1991, Stuggart, Available http://www. bosch-semiconductors.de/media/ubk_semiconductors/pdf_1/canliteratur/can2spec.pdf

[3] M. C. Carolan, "Foundation Fieldbus: Fieldbus Basics", Brisbane, 2011, Available www.fieldbus.org/images/stories/international/asiapacific/singapore/presentations/2011-04%20Vietnam/1_ff_basics_with_additional_points.pdf

[4] Profibus and Profinet International, "Profibus System Description", 2016, Available www.profibus.com/nc/download/downloads/ profibus-technology-and-application-system-description/download/21420/

[5] DNP User Group, "DNP3 Primer", Revision A, 2005, www.dnp.org/AboutUs/DNP3% 20Primer %20Rev%20A.pdf

Figure 4. UNF T&C control hardware for the propagation project.

The opinions, findings and conclusions expressed in this publication are those of the author(s) and not necessarily those of the Florida Department of Transportation or the U.S. Department of Transportation.

OVERVIEW OF SECURE COGNITIVE RADIO MAC PROTOCOL IN THE PROPOSED 3.5 GHZ BAND

Author: David Oyediran

Doctor of Engineering Candidate, Department of Electrical and Computer Engineering
Morgan State University

Faculty Advisors: Dr. Richard Dean, Dr. Farzad Moazzami

ABSTRACT

Spectrum sharing between federal and commercial users is proposed by the FCC and NTIA to open up the 3.5 GHz band for wireless broadband use. The proposed technology requires the detection and subsequent allocation of available licensed spectrum for temporary use by other users without compromising the privacy of the licensed user. DoD has a documented requirement of 865 MHz by 2025 to support telemetry but only 445 MHz is presently available. Research is presently on-going at DoD to realize, test and evaluate spectrum efficient technology with the aim to develop, demonstrate, and evaluate technology components required to enable flight and ground test telemetry operations. The use of cognitive radio (CR) in spectrum sharing has gained much popularity in that CR senses the unused spectrum at a specific time and location and dynamically allocates to users as required. This paper will provide an overview of a secured CR Media Access Protocols using the IEEE 802.22. Carrier Sense Multiple Access Collision Avoidance (CSMA/CA) will be utilized for protocol transmission in order to properly identify user's location and identity thereby providing a secured network against false alarm from external attack.

INTRODUCTION

The demand for wireless spectrum, particularly wireless and mobile broadband services is growing every day, because of increasing mobile connectivity, see figure 1. This growth represents a significant economic opportunity but also presents a challenge, as increased usage strains the capacity of the available. Unfortunately, this growing demand is not limited to commercial users. Government departments have increased test system complexity and resultant high data rates to support new and modern weapon technology. This requires about 865 MHz in the near future to support telemetry with only 445 MHz presently available [1].

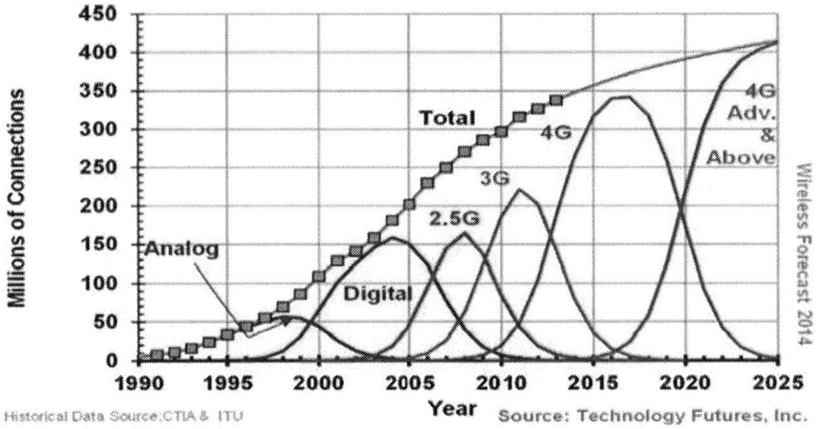

Figure 1: Technology Future Inc. Industry Wireless Generations Forecast
(https://tfi-ctfg.com/wireless-forecast/)

Spectrum by itself is non-depletable resources but limited in some constraint i.e. the same frequency, same technique cannot be used at the same time and same area. Securing additional spectrum has been difficult, partly because spectrum usable by today's mobile broadband technologies is currently designated for a variety of other uses [1]. Since spectrum is a finite resource, allocating more spectrum for some services inevitably requires consideration of re-allocation of spectrum from other uses, which can cause disruption and cost as spectrum is re-configured. As a result of this, the (Federal Communication Commission) FCC and the (National Telecommunications and Information Administration) NTIA are placing much greater emphasis on the efficiency with which spectrum is used. Presently, consideration is being given to far more innovative ways to use radio spectrum that has traditionally been considered – making available un-used gaps in spectrum [1].

To alleviate these challenges, efforts have turned to include spectrum sharing with incumbent users. One user may have priority over the other, but generally all users assigned to those frequencies can use them. Spectrum sharing between Federal and commercial users is a technique proposed by the FCC and NTIA to open up the 3.5 GHz band for wireless broadband use. The 3500-3650 MHz band is one of the candidate bands identified and recommended by the NTIA for reallocating 100 MHz of its 3550 – 3650 MHz for wireless broadband use within the shortest possible time [3, 4]. The band was selected by NTIA because WiMAX equipment has already been developed. NTIA understood this band to be used primarily for high power shipborne radars designed to operate in the 3500-3650 MHz band due to specific propagation and atmospheric conditions unique to this frequency range [5]. It is also used for communications with missile systems for data updates to the missile while in flight to its target. The radars in this band represent significant investment on the part of DoD and many are incorporated into ship and aircraft design. Redesigning for other frequency ranges to make this spectrum available for wireless broadband may require new technology, and significant redesign of their associated platforms.

However, the main challenge in spectrum sharing would be designing a system or network that protects the security of the licensed spectrum and users. The PU's must be protected from unacceptable interference from unlicensed users this could be achieved through the use of a proactive spectrum selection and interference avoidance processes.

To allow unlicensed users to access a radio spectrum when not in use by licensed users, there is need to develop the wireless technology. Cognitive Radio (CR) wireless network allows the unlicensed user to use only unused spectrum opportunistically at a particular time to prevent interference with the licensed user [2]. CR therefore is a wireless radio that uses machine learning techniques for the allocation of spectrum dynamically over a given period. Figure 2 shows CR spectrum sharing network.

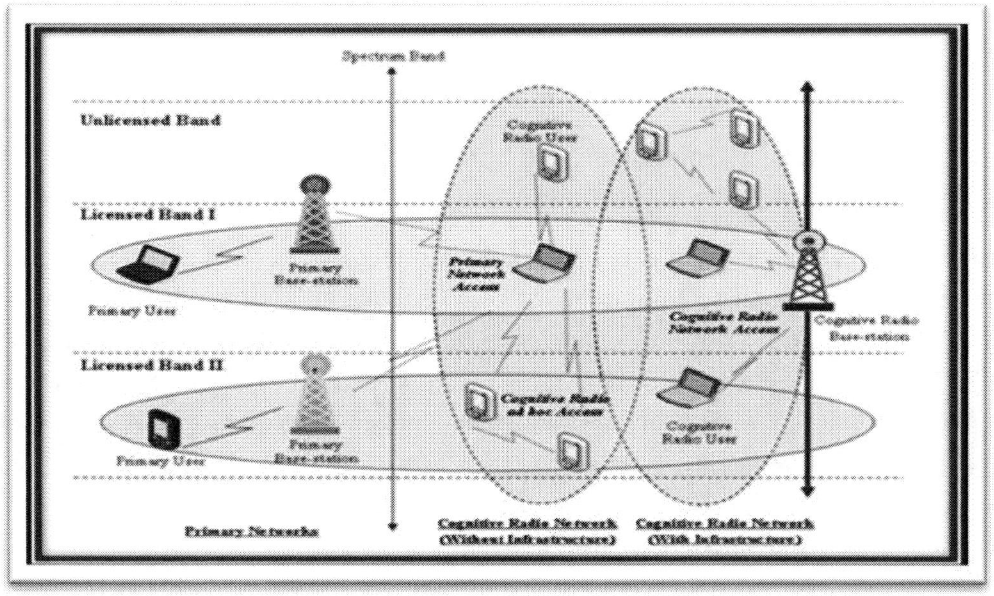

Figure 2: Cognitive Radio Network
(Image from http://bwn.ece.gatech.edu/CR/projectdescription.html)

To ensure proper coordination of information, spectrum sensing and interference free allocation between the licensed users also known as the primary users (PUs) and the unlicensed user, the secondary users (SUs), there is need to have a well-designed medium access control (MAC) protocol. The design must address how the SUs determine when and which channels they should use to transmit or receive SU packets without affecting communication among the primary users (PUs) [3].

For spectrum access, CR utilizes CSMA/CA protocol. In this protocol, after sensing, users back-off by a random time before transmission. If there is a collision of transmitted packets by any two users, the users double the back-off window and retransmit. This protocol entails appropriate scheduling algorithm to determine the right time to transmit over shared channel. The process of sharing could be briefly described as shown in figure 3.

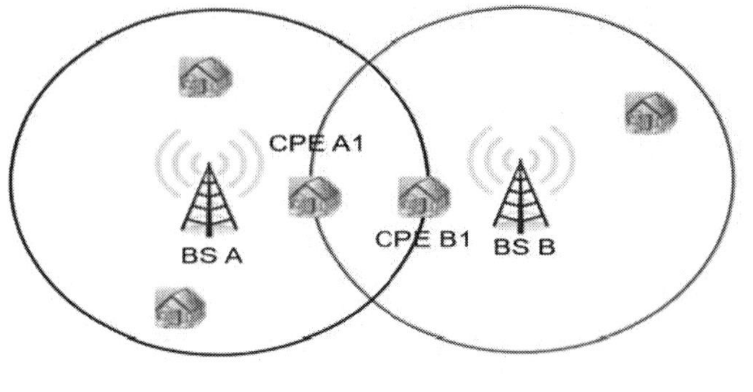

Figure 3: Cognitive Radio Spectrum Sharing Mechanism
(Diagram from https://jwcn-eurasipjournals.springeropen.com/articles/10.1186/1687-1499-2012-28)

In spectrum sharing, the spectrum is divided into a number of channels and suppose base station A (BS A) and customer premise equipment (CPE A) operate on a given channel N, when a new BS B and CPE B start operation, the channels are scanned to detect the available channels and CPE B1 eventually detects BS A's Super-frame Control Header (SCH) packet already transmitted by the BS A. Here a first coexistence mechanism is executed by BS B in order to avoid interference. The reallocated 100 MHz of the 3500-3650 MHz band could be divided into different channels for sharing purposes between the SUs and the PU.
The remaining part of the paper is organized as follows. Section 2 introduces the IEEE 802.22 standard, CR network infrastructure and the PHY/MAC layer. Section 3 provides the application of CSMA/CA protocol to CR and MATLAB implementation. In Section 4, the simulation results were analyzed while section 5 concludes the paper.

IEEE 802.22 STANDARD

The IEEE 802.22 working group provided the cognitive radio-based international standard for wireless regional area network (WRAN) with the purpose of providing broadband access in remote and rural areas by opportunistically exploiting the unused TV band. This is the first worldwide standard for Wireless Regional Area Network (WRAN) based on CR by exploiting the unused TV band. The WRAN systems can operate on vacant TV channels in VHF/UHF band ranging from 54 MHz to 862 MHz frequency and the average coverage area of WRAN cell is 33 km^2 with possibility for extension up to 100 km^2. This coverage area is larger than IEEE 802.11 based Wi-Fi and IEEE 802.16 based WiMAX. IEEE 802.22 incorporates advanced cognitive radio capabilities including dynamic spectrum access, incumbent database access, accurate geolocation techniques, spectrum sensing, regulatory domain dependent policies, spectrum etiquette, and coexistence for optimal use of the available spectrum. Figure 4a and 4b below shows the characteristics and network applications IEEE 802.22 standard relative to other IEEE 802 wireless data transmission standards.

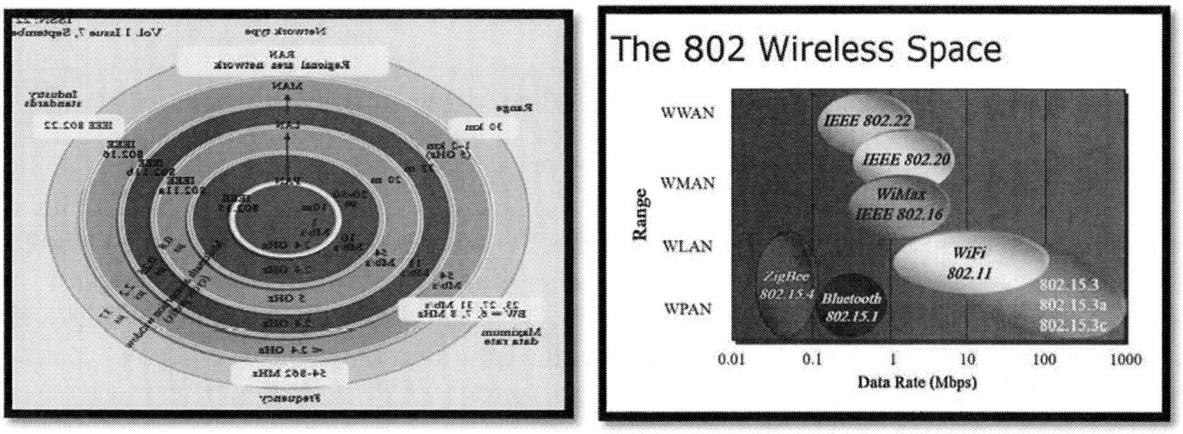

Figure 4: IEEE 802.22 standard relative to other IEEE 802 wireless data transmission standards
(Diagrams from http://amanuel-wran.blogspot.com/2011_06_01_archive.html and http://slideplayer.hu/slide/2129723/)

COGNITIVE RADIO NETWORK INFRASTRUCTURE

As shown in figure 5, the IEEE 802.22 network is based on master/slave architecture with single wireless regional area network base station (WRAN BS). This WRAN BS manages all the CPE through MAC. The BS controls all the communication in the cell, i.e. there is no peer-to-peer communication directly between the CPEs. Figures 5a and 5b below shows the basic infrastructural configurations of IEEE 802.22.

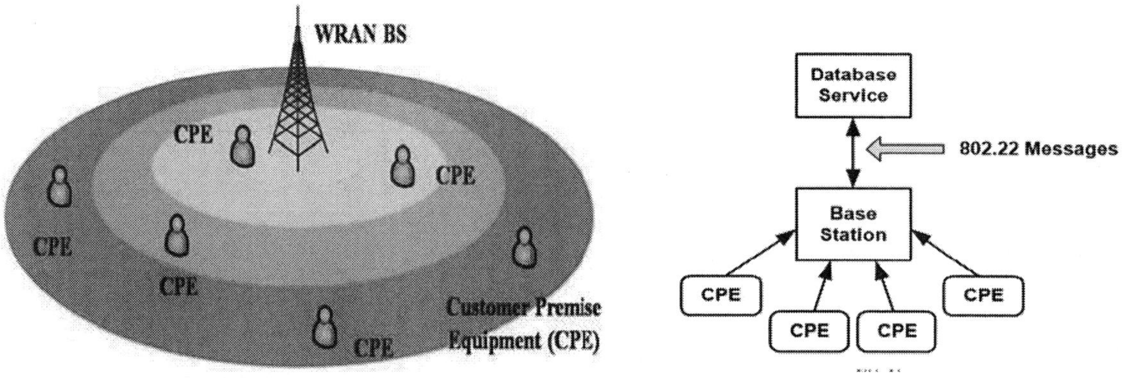

Figure 5: Cognitive Radio Infrastructure and 802.22 database interface model
(Diagrams from http://ecee.colorado.edu/~ecen4242/802_22/general_info. htmlhttp://slideplayer.com/slide/9906767/)

In CR network, no CPE is allowed to transmit without receiving any proper authorization from WRAN BS. The downstream transmission, where WRAN BS transmits and CPE receives, is based on time division multiplex (TDM) and the upstream transmissions are shared by CPEs on demand basis according to orthogonal frequency division multiplex (OFDMA) scheme [4].

PHY/MAC LAYER

The MAC and PHY specifications is a standard developed by the IEEE for designing how the SUs determine when and which channels they should use to transmit or receive SU packets without affecting communication among the primary users (PUs) [3] [4]. The IEEE 802.22 MAC provides mechanisms for flexible and efficient data transmission, and supports cognitive capabilities for both reliable protection of incumbent services in the TV band and self-coexistence among 802.22 systems. The MAC is a connection-oriented mechanism supporting unicast, multicast and broadcast services. Connection is defined by 12 bits; 9 bits are reserved for station ID (SID) and 3 bits are for flow ID (FID). The station being under the control of BS (CPE) is uniquely identified by SID. The FID represents a specific traffic flow assigned to a CPE. The combination of SID and FID defines a connection identifier (CID) for the particular CPE. The MAC facilitates up to 512 stations, each with a maximum of 8 service flows and each service flow requires a separate connection in a WRAN cell.

IEEE 802.22 PROTOCOL REFERENCE MODEL

The IEEE 802.22 protocol reference model defines the system architecture as shown in figure 6. The system is partitioned into data plane, management plane and cognitive plane. The data plane consists of the Physical Layer (PHY), the medium access control (MAC) layer and the convergence sublayer (CS). Service access points (SAPs) are added in between these layers to allow different components to communicate with each other. Security sublayers are appended between SAPs to provide data security functionalities and applications. The cognitive plane is composed of Spectrum Manager, Geolocation, Spectrum Sensing Automaton and Spectrum Sensing Function [5].

APPLICATION OF CSMA/CA PROTOCOL TO COGNITIVE RADIO NETWORK

To ensure interference free and to enhance QoS in spectrum sharing, the Carrier Sense Multiple Access with Collision Avoidance (CSMA/CA) protocol can be utilized by cognitive radios for spectrum access. In this protocol, after sensing, users back-off by a random time before transmission. If there is a collision of transmitted packets by any two users, the users double the back-off window and retransmit. The capacity of the WRAN cell as defined in IEEE 802.22 standard ranges from 1 to 512 CPEs. When CPE powers on or loses its association with the WRAN BS, it then attempts to register with the WRAN BS [5].

CSMA/CA PROTOCOL IMPLEMENTATION FOR SPECTRUM SHARING

From figure 4, like the 802.11 CSMA/CA MAC layer, we propose the 802.22 CSMA/CA MAC layer using the CSMA/CA with request-to-send (RTS) and clear-to-send (CTS) exchange technique. This strategy is important since it has a more efficient performance in term of average throughput [6] and solves the hidden node/terminal problem [7] in the network.

Assuming we have a network with many terminals, (set of node pairs, consisting of a sender and receiver), one access point (AP) and N number of channels through which the CPEs send their registration requests to the WRAN BS. We assume that each CPE sends its registration request on separate channel. If two or more CPEs send their requests on the same channel at the same time, then collision occurs. However, with the CSMA/CA with RTS/CTS technique, if the channel is busy for the CPEs, each CPEs chooses randomly a backoff time in the interval [0, CW-1], where CW is a contention window. The wait counter freezes when the channel is busy, and resumes when the channel is available again. The backoff counter is decremented by one each time the channel is detected to be available for a Distributed Inter-Frame Space (DIFS) duration. When the backoff counter attempts zero, the source CPE randomly chooses one band over the N available bands to send a RTS to the destination node. It waits for receiving a CTS from the AP before transmitting data. The AP listens simultaneously to all RTS bands and if one or more RTS is detected, the AP broadcasts CTS over all the bands indicating the authorized station to communicate.

In this technique, the PU is given priority over all the SUs by making the PU to have smaller interframe space (IFS), IFS is the period of time a node has to wait before sending a packet to the AP when a channel is found idle. We want high priority wait for shorter time where wait time is calculated by specific range for each priority. For demonstration purposes, we simulated a CSMA and the CSMA/CA and showed how much performance increase is found using CSMA/CA. Although, this simulation was done in the MATLAB environment, this is not the best simulator for CR network simulation it does not have cognitive capability applications. Network Simulator v9 and above (NetSim) would be the applicable software because the library contains the network/protocols compactible with the simulation of Wireless Regional Area Network (WRAN). NetSim functionalities for CR (as per IEEE 802.22 MAC and PHY) includes but not limited to detection and channel switching and the cognitive plane [8]. It contains spectrum manager, spectrum sensing function, spectrum sensing automaton and channel model. We don't have access to NetSim so we are using MATLAB for a modified simulation.

CSMA/CA MATLAB SIMULATION

The MATLAB program generates uniformly distributed mobile in 2 D space and computes the path loss between nodes. To ensure connectivity of all nodes, the program varies the power/range until full connections are established. After this, traffic is generated for this fully connected network and collisions associated with traffic are measured. We then increase the traffic and measures traffic throughput. The results are shown in the figures below. The simulation was done using an ideal physical layer (no path loss, no fading, no shadowing) is considered.

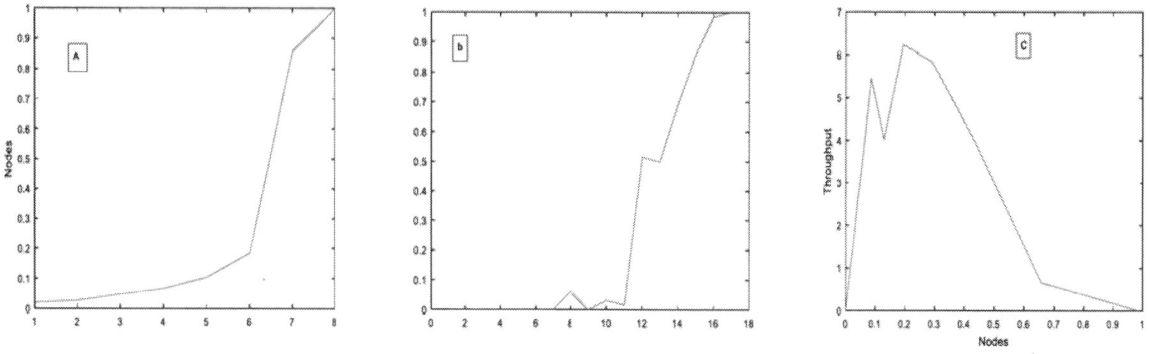

(a) (b) (c)

Figure 7: (a) and (b) Nodes connection and probability of collision with traffic without delay (c) throughput with all the nodes connected in the network with traffic

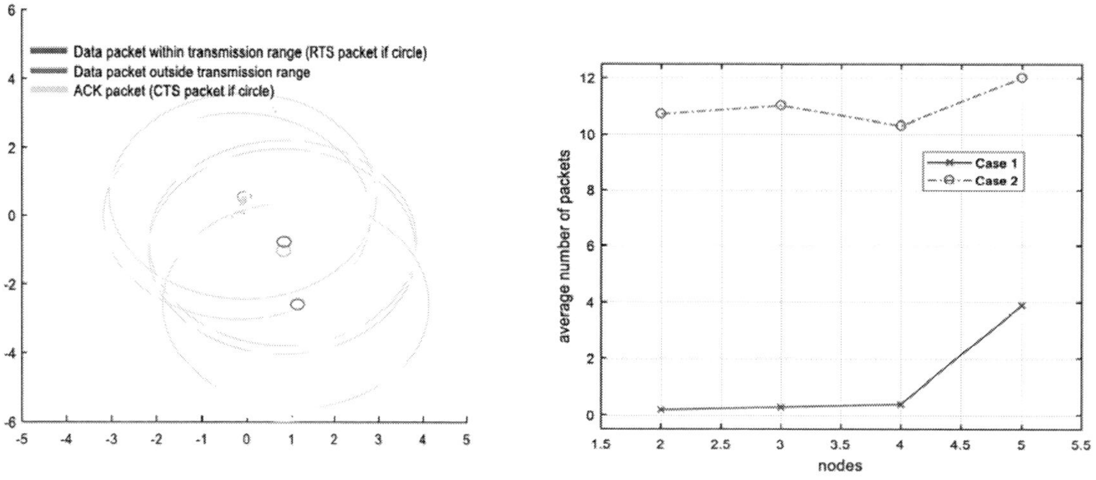

Figure 8: concentric circles showing RTS/CTS exchange and CSMA with without collision avoidance.

For example, assuming a node, A wants to transmit a data packet to another node B on the network, node A firstly broadcasts RTS packet then B broadcast CTS packet. Node A then send the data packet to node B and node B send an acknowledgement packet to node A upon successful completion of the data exchange. The probability of success, P_s = Data x PR_{nc} x PC_{nc} x PD_{nc}, where P_s is the probability of success, PR_{nc} is the probability of sending RTS without collision, PC_{nc} is the probability of sending CTS without collision and PD_{nc} is the probability of sending data without collision. This could be translated as $p(1-p)^{N-1}$ where N is the number of nodes in the network.

Assuming N nodes are trying to send data in the network, the probability that a node is successful (i.e. no collisions) is $Np\ (1-p)^{N-1}$. We considered five nodes in this simulation and our focus is to determine the successful packet transmission rate of the network.

ANALYSIS

Figure 7a is a plot showing the nodes connections when power is sufficient for connectivity and figure 7b demonstrates that the collision probability increases with the number of nodes using CSMA protocol without collision avoidance. More nodes lead to deadlocks in the network. Figure 7c shows that the throughput decreases with increased network connectivity.

In figure 8a, the RTS/CTS exchange is shown using concentric circles where the red concentric circles denotes an RTS packet and green circles representing a CTS packet. The graphics were set-up using pointers to graphical objects where the objects were initially circles. For example, W (i) = plot (exp (j*ph)*(1+r), 'EraseMode', 'xor'); The MATLAB file "pilot1.m" changes the properties of these objects to show whether a packet is in range. The circle eventually get changed to lines to show packet transmission. At the initial exchange of RTS/CTS packets between the sender node and the recipient node, the sender transmits a DATA packet, which is represented by a red line. The receiving node sends an ACK packet back to the sender once the packet arrives. A green line represents an ACK packet.

The case 1 in figure 8b represents collision CSMA while case 2 is successful transmission CSMA/CA. The impact of RTS/CTS protocol on the system performance shows in figure 8 that there is improvement in packet transmission with reduced collision rate. We found out that the number of successful transmissions increased and the number of collisions (both DATA and ACK packet collisions) has decreased using CSMA/CA. The simulation was done using an ideal physical layer (no path loss, no fading, no shadowing) is considered.

CONCLUSION

In this paper, we have provided an overview of a secured CR MAC protocol using the IEEE 802.22 standard and suggested the application of CSMA/CA to cognitive network in order to reduce interference between the Primary and Secondary Users to increase throughput. Although, MATLAB does not support the cognitive functionality and IEEE 802.22 standards, we implemented the 802.11 MAC protocol and suggested the application of the strategy to IEEE 802.22 since both standards supports applications in the wireless networks. CSMA/CA was utilized for protocol transmission in order to properly identify user's location and identity. We have found that the inclusion of RTS/CTS handshaking greatly improves the performance of our simulated wireless network.

ACKNOWLEDGEMENT

The authors express their gratitude to members of **WiNetS** (Wireless Networking and Security) laboratory of the Electrical and Computer Engineering Department at Morgan State University for their constant support.

REFERENCES

[1] D. Oyediran and F. Moazzami, "Spectrum Sharing: Overview and Challenges of Small Cells Inovation in the Proposed 3.5 GHz Band," in *International Telemetry Conference*, Las Vagas, 2015.

[2] I. Akyildiz, W. Lee and M. V. S. Mohanty, "NeXt Generation/Dynamic Spectrum Access/ Cognitive Radio Wireless Networks: A Survey," *Computer Networks Journal (Elsevier),* vol. 50, pp. 2127-2159, 2006.

[3] M. A. Raza, S. Park and a. H.-N. Lee, "Evolutionary Channel Sharing Algorithm for Heterogeneous Unlicensed Networks," *TRANSACTIONS ON WIRELESS COMMUNICATIONS,* vol. 16, 2017.

[4] C. Cordeiro, K. Challapali, D. Birru and S. S. N, "IEEE 802.22: An Introduction to the First Wireless Standard based on Cognitive Radios," *Journal of Communications,* vol. 1, pp. 38-47, 2006.

[5] IEEE, "IEEE 802.22 Working Group on Wireless Regional Area Networks," 2012. [Online]. Available: http://www.ieee802.org/22/.

[6] K.-L. Hung and B. Bensaou, "Throughput Analysis and Rate Control for IEEE 802.11 Wireless LAN with Hidden Terminals," ACM, New York, 2008.

[7] IEEE, "Wireless LAN Medium Access Control (MAC) and Physical Layer (PHY) specifications: Higher Speed Physical Layer (PHY) Extension in the 2.4 GHz band," IEEE Xplore, June 1999. [Online]. Available: http://ieeexplore.ieee.org/stamp/stamp.jsp?arnumber=1210624.

MULTIHOP ROUTING OF TELEMETRY DATA IN DRONE SWARMS

Alexis Pospischil (ampospis@nps.edu)
Department of Computer Science, Naval Postgraduate School, Monterey, CA, 93943
Faculty Advisor: Dr. Justin P. Rohrer (jprohrer@nps.edu)

ABSTRACT

In 2015, a group of Naval Postgraduate School (NPS) professors and students set the record for largest fixed-wing unmanned aerial vehicle (UAV) swarm flown at one time. The swarm had 50 vehicles flying simultaneously and successfully demonstrated distributed decision-making with all processing occurring on swarm vehicles rather than a centralized control station. Much of the decision-making is based on telemetry data that is continuously streamed from all the nodes. At that time all telemetry data was broadcast in a single-hop radio environment using 802.11 in AdHoc mode. In the future, drone swarm distribution and mobility patterns will necessitate multi-hop communications for this telemetry data. This paper models the network currently used by the NPS drone swarm as well as potential future topologies and evaluates candidate multihop routing protocols for this application.

INTRODUCTION

In the natural world around us, swarms have demonstrated a high degree of success. Bees, ants, termites, and naked mole rats maintain large groups that distribute tasks among individuals in order to achieve great things for the success of the colony. Fish, birds, and bats move and swirl in great numbers without colliding. Man has sought to reproduce these working models.

One key aspect that the natural models share is a form of reliable communication between individuals. Information at the small scale builds to achieve goals at the large scale. For a swarm of aerial vehicles, that communication is likely over some form of 802.11 network. This work models the network currently used by a swarm of fixed-wing aerial vehicles developed at the Naval Postgraduate School (NPS) and compares its performance to Delay (or Disruption) Tolerant Networks (DTNs) in order to observe tradeoffs between potential routing protocols for the current swarm configuration as well as future larger scale swarms.

In 2015, a group of NPS professors and students set the record for largest fixed-wing unmanned aerial vehicle (UAV) swarm flown at one time [1]. The swarm had 50 vehicles flying simultaneously and successfully demonstrated distributed decision-making with all processing occurring on swarm vehicles rather than a centralized control station. This swarm uses broadcast messaging, and no multi-hop routing is implemented.

This makes for a low latency networking scheme when all entities are within broadcast range. Where the swarm runs into problems is scaling and future sub-swarm missions. Fifty vehicles is the current highest concentration of vehicle achieved using broadcast over UDP with application layer messaging. It is not currently known how many vehicles will bring the network to saturation nor how far the vehicles may travel before losing the ability to communicate with the ground station. Vehicles have been flown as far as 2 kilometers (km) away and 850 m of elevation away from the ground station (at the limits of the testing grounds) without losing connectivity. As mission behaviors become more advanced, it may be necessary to test these limits. At this point, a routing protocol that does not overly stress the bandwidth but allows intermediary vehicles to relay messages through the swarm would prove beneficial.

This research constructs a model of the ad hoc communication network used by a large number of fixed-wing UAVs flown by NPS faculty and students in order to determine the best parameters by which to measure successful packet delivery, and compare its performance to other network protocols that might provide better performance.

BACKGROUND AND RELATED WORK

This section introduces the necessary concepts of UAV swarm research; how it has progressed, common issues, and research towards finding solutions to these issues. Very often, dynamic networks have properties unique to their environment and/or purpose. Solutions for one group may not work for another without changes to the routing protocol. As a result, there are many variations to the basic founding ideas to consider.

A. Swarm Networking Issues

Aerial swarms have the capacity to move quickly, not only in relation to the ground and base station, but also in relation to other vehicles. The swarm is composed of the NPS-designed and built Zephyr II UAV, which is a 2.5 kilogram (kg) fixed-wing vehicle based on a commercially available platform from Ritewing. It has a wingspan of 1.45 meters (m) and a cruising speed of 18 meters per second (m/s) [1]. The NPS swarm uses no routing protocol. Rather, it broadcasts all messages over the 802.11n wireless network and relies on the vehicles to determine from the application-level message header whether to drop a message not intended for that vehicle, or accept the message and respond accordingly.

R. Stefano *et al.* [2] implemented a fixed wing swarm with a cruising speed of about 16 m/s. This high mobility creates a highly dynamic topology wherein links are frequently formed and broken. Stefano *et al.* addressed this issue with a routing protocol that uses knowledge of Global Positioning System (GPS) data to predict the best routing path over the 802.11n wireless network.

Another issue that arises for mobile aerial vehicles is the ability for them to travel away from the base station and/or other vehicles. This allows for a larger operating area as vehicles disperse throughout an unbounded three-dimensional space, which may benefit the mission. The result is highly heterogeneous node dispersion and clustering. Power must be split between communications, flight, and sensors. In any case, until aerial swarms are able to scale well, they will have lower node density than sensor networks to which they are sometimes compared [3].

At the transport layer, Rohrer *et al.* [4] examine the shortfalls of using a TCP/UDP/IP in an aeronautical environment. The TCP routing protocol was designed for long-lasting connections along an established path. Attributes of TCP, like the three-way handshake, slow start algorithm, and congestion control algorithm, do not allow this routing protocol the flexibility to deliver packets quickly in an aeronautical node environment. UDP is not encumbered by the same restraints as TCP, but offers no acknowledgment that packets were delivered to the destination [5].

Bandwidth is yet another constraint in aerial vehicle swarms. When all the nodes are trying to send messages, and the number of nodes is significant, bandwidth can be overwhelmed. Message conveyance is not the only stress on bandwidth, though. Overhead from routing protocols can be the most significant taxation on bandwidth. Section C. provides more detail about the kinds of overhead produced by multi-hop routing protocols.

B. Current Swarm Networking Tactics and Models

The NPS swarm uses an in-house application for vehicle-to-vehicle and vehicle-to-ground messaging. User Datagram Protocol (UDP) broadcast is used to transmit these messages over an 802.11n ad-hoc wireless network. The application-layer is responsible for accepting messages and sending replies when appropriate. Several distinct message types are utilized, a number of which are transmitted at regular intervals. For safety reasons, a heartbeat message is sent from the ground station to the vehicles at an interval of 1 hertz (Hz). If a vehicle does not receive that message for a period of 30 seconds, it aborts its current mission and loiters at a pre-designated point to attempt to re-establish its link with the ground station. If the vehicle does not receive an update within two minutes, it executes its autonomous landing procedure [1].

Individual vehicles send flight status messages to the ground station at a rate of 2 Hz. Vehicles also send pose messages to update the ground station and other swarm vehicles as to their current state at a rate of 10 Hz [1]. The last synchronous message is the red pose message, transmitted by a special purpose ground station, the arbiter, as a means of providing 'virtual sensor' information during competitive multi-swarm events [6].

In addition to these synchronous messages, a number of asynchronous messages are provided. These messages address a number of special purpose requirements such as assignment of vehicles to sub-swarms for tasking, initiation and termination of swarm behaviors, direction or parametrization of individual vehicle actions, and the exchange of behavior-specific information between vehicles [1]. Asynchronous messages can be broadcast to the entire swarm or directed to a specific vehicle or ground station (although still transmitted as a UDP broadcast message).

Real-world vehicle-to-vehicle packet delivery rates for the 50-UAV swarm event were described in [1]. Given their frequency and contents, synchronous messages comprise the bulk of the NPS swarm communications requirement. This work will therefore focus on analysis of synchronous message traffic within the NPS swarm.

Examples of other swarms and their techniques include:

- 2004: Three quadcopter UAVs using GPS navigation and a Bluetooth connection with a ground station that is able to adjust positioning data, and send navigation commands as waypoints. Commands are given to each vehicle through an established end-to-end Bluetooth connection. Vehicles are not able to exchange information [7].

- 2013: Three fixed-wing UAVs in a leader-slave-slave relationship. Each vehicle is given updated information about the flight path throughout the flight from a ground station via a modem using a point-to-multipoint setting. Vehicles are not able to exchange information. Safety personnel on the ground stand by to take control with RC controllers (one person per vehicle) [8].

- 2014: Ten quadcopter UAVs in a decentralized swarm. Each vehicle is responsible for collision avoidance calculated with GPS data shared between vehicles via Xbee wireless radios.

These position reports are sent in a broadcast mode between relative neighbors without establishing an end-to-end connection. Flight status information is sent to the ground station for monitoring and record keeping. The vehicles receive no information from the ground station pertaining to flight or organization. One individual can safely run the swarm due to the high level of autonomy [9].

- 2016: Intel demonstrates a 500+ quadcopter UAV lightshow swarm. Little information is available on the communication methods used, other than that they are controlled from a single computer and the choreography is preprogramed. This allows the vehicles to execute their flight plan without communicating with other vehicles [10].
- 2017: Ehang demonstrates a 1000+ quadcopter UAV lightshow swarm. Little information is available on the communication methods used.
- Ongoing Research: The swarming micro air vehicle network (SMAVNET) uses up to ten small fixed-wing UAVs (420g, 80cm wingspan) to conduct swarming behavior and communication experiments. Collision avoidance is accomplished via vehicle-to-vehicle communications while in flight. Neither operator-provided information nor on-board sensors are used to detect the in-air location of other vehicles. The SMAVNET team uses commercially available wifi hardware on their vehicles. Swarming behaviors are either reverse-engineered controllers or modeled after biological examples such as ants leaving pheromone trails [11].

One potential solution to many issues described is the use of a Mobile ad hoc network (MANET). MANETs are groups of mobile nodes with the capability to receive, route, and transmit network traffic. These nodes are free to move which creates a dynamic topology. Mobility comes at the cost of being dependent upon a limited energy source as well as relying on wireless connections resulting in bandwidth constraints. These routing protocols fall in three different categories; proactive (table based), reactive, and hybrid schemes [12]. Originally designed to bridge the vast distances in space, DTN routing protocols use a store-and-forward tactic to deliver bundles of data when an end-to-end connection between source and destination is not possible [13]. There are many kinds of DTNs. Their design and performance are dependent upon their environment. As the NPS swarm grows in scale, multi-hop routing protocol will likely be necessary.

The NPS swarm's various messages have different reliability and latency requirements. Some, like the heartbeat message, may not require every message to reach a vehicle, as long as one does within a defined window. It is not vital that missed heartbeats be delivered once the next has been sent by the ground station. Others, like asynchronous messages carrying behavior commands, must reach their destination or be able to inform the ground station that it was unable to reach the destination or vehicles will not join the correct behavior. More than one protocol may be chosen to work within the swarm's communication network, each chosen for a message that uses that protocols strengths to the swarms advantage. Section C. reviews protocols that represent various tactics used in network routing. Some are well established, like the four applied in this study (DSDV, OLSR, AODV, and Epidemic). Others are relatively new as more variation is needed for unique wireless networks such as flying ad hoc networks (FANETs), or disruption-tolerant networks (DTNs) [14].

METHODOLOGY

This section provides a review of the simulation built to model the NPS swarm in ns-3. Ns-3 is a discrete-event network simulator designed by the members of the NS-3 Consortium [15]. It contains many helpful tools within its library that replicate the behavior of real-world networks.

Table 1: Message Parameters

Message	Data Rate	Packet Size	Frequency
Heartbeat	160 bps	20 Bytes	1 Hz
Flight Status	768 bps	48 Bytes	2 Hz
Pose	4480 bps	56 Bytes	10 Hz
Red Pose	3200 bps	40 Bytes	10 Hz
Asynchronous	as required	variable	as required

Individual NPS swarm messages each have unique characteristics affecting behavior at the link layer that must be captured in the model. Synchronous messages described in Section B. are broadcast at message-specific frequencies and have standard sizes. Within the simulation, each message type is also assigned a unique port over which to be received. All messages, regardless of type, include a standard 16 Byte header added at the application layer. Applications were started at staggered times in order to ensure that collisions were not generated by artificial synchronization.

Table 2: Swarm Configurations

Mobility Model	Ground Station	Sub-swarms	Sub-swarm nodes	Simulation nodes
6 v 6	1	2	6	12
10 v 10	1	2	10	20
15 v 15	1	2	15	30
25 v 25	1	2	25	50

Instead of using one of the synthetic mobility models in ns-3, the swarm nodes follow traces from real swarm flights. The NPS group can fly the swarm in a red versus blue scenario in which two competing groups attempt to target and engage all vehicles in the opposing swarm [6]. The vehicles are assigned a swarm ID, which places them in either sub-swarm one (red) or sub-swarm two (blue). After launch, each vehicle flies to its assigned altitude and begins circling a designated area. Once all vehicles are launched, there will be two columns of circling UAVs standing by for their next order, which will arrive via an asynchronous message. The two sub-swarms will come together and simulate an air-to-air engagement, then separate back into their designated columns. After all behavior scenarios are complete, individual UAVs will be told to land until all vehicles are safely on the ground. Table 2 illustrates the composition of the four swarm events used for the mobility model in this work.

C. Routing Protocols

Once the simulator reproduced message traffic in a broadcast configuration, code was added for additional routing protocols that are representative of the major categories traditionally used in ad hoc networking. The ns-3 library provides a number of useful ad hoc routing protocols:

- OLSR: a link-state proactive protocol is able to maintain tables for immediate addressing of the shortest path [16].

- DSDV: a distance-vector proactive protocol [17].
- AODV: a distance-vector reactive protocol [18].
- Epidemic: a simple flooding-based DTN routing protocol [19].

Though these protocols are not optimized for use in FANETs, they can be treated a representatives of fundamental routing tactics and are therefore useful in demonstrating strengths and weakness inherent to all protocols that fall within these respective categories. We vary the simulated radio antenna gains to emulate varying node densities and radio coverage ratios. Transmit and receive gains were always set to matching levels. Simulation experiments were conducted swarm trace data for twelve, twenty, thirty, and fifty vehicle events.

The following metrics are utilized for quantitative comparison of protocol performance:

- Packet Delivery Ratio (PDR): calculated by dividing the number of messages received by the number sent.
- Overhead: message copies, retransmissions, and routing control messages.

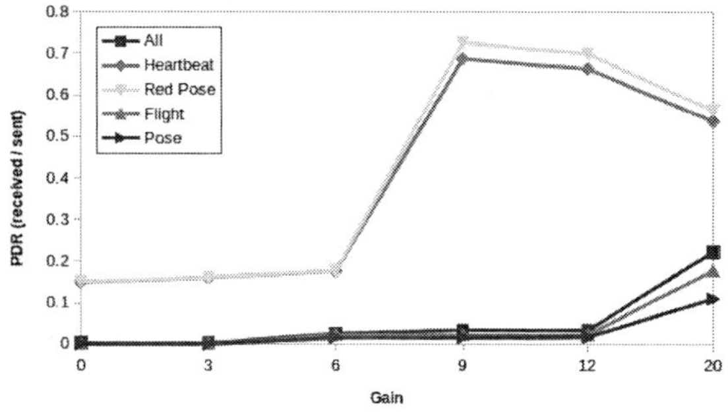

(a) Broadcast 6 v 6 Swarm

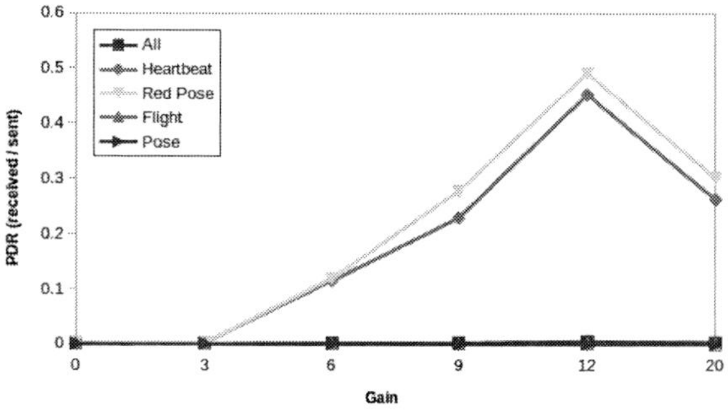

(b) Broadcast 10 v 10 Swarm

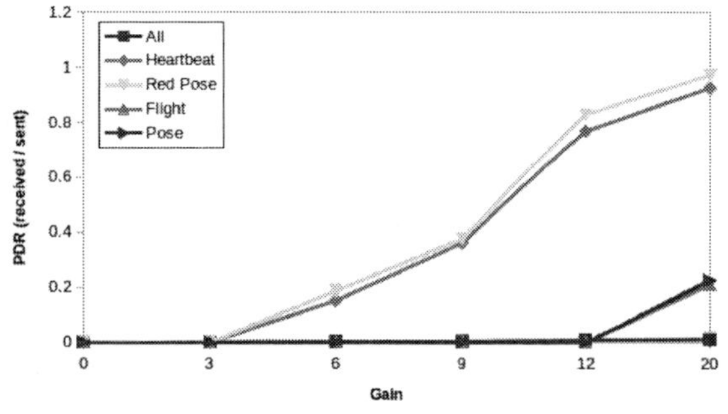

(c) Broadcast 15 v 15 Swarm

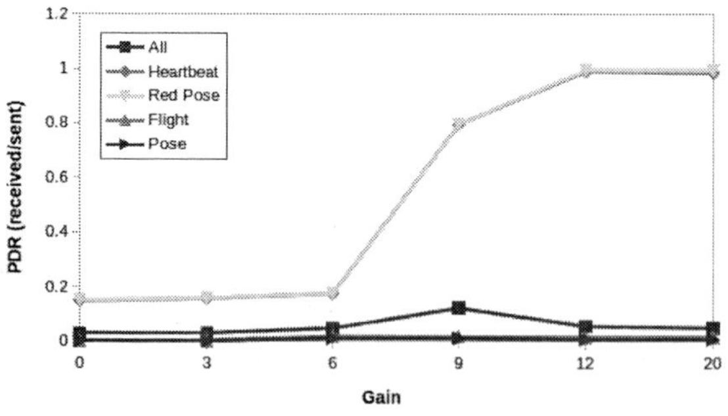

(d) Epidemic 6 v 6 Swarm

Figure 1: Packet Delivery Ratio

RESULTS AND ANALYSIS

This section reviews the results from the 120 simulations and highlight a number of illustrated trends and relationships.

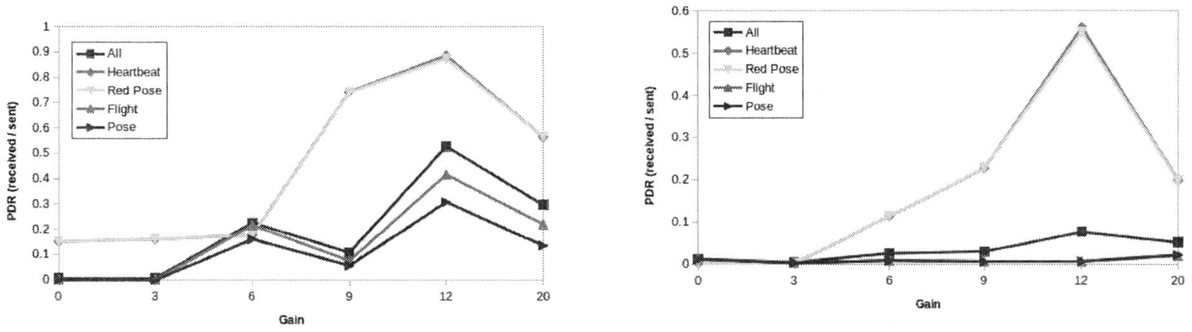

(a) 6 v 6 Swarm

(b) 10 v 10 Swarm

Figure 2: Packet Delivery Ratio Using DSDV

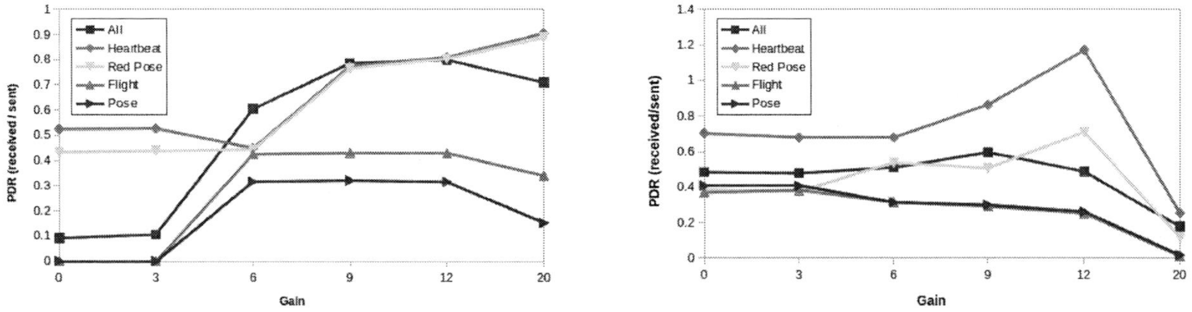

(a) 6 v 6 Swarm

(b) 25 v 25 Swarm

Figure 3: Packet Delivery Ratio Using AODV

Packet delivery rations (PDRs) for flight and pose messages are calculated by dividing total messages received by total messages sent.

In general, the broadcast method improves as the gain in the environment increases, as shown in Figures 1a, 1b, and 1c. Packets traveling from the vehicles to the ground achieved a higher packet delivery ratio than those traveling from the ground station to the vehicles. Heartbeat and red pose messages destined for the vehicles are close in values and patterns. Likewise, flight status and pose messages also behave in a close relationship.

DSDV's best performance was in the gain of 12 environment across all swarm sizes (2a and 2b).

AODV's PDR performance (Figures 3a – 3b) is best in the smaller swarm of 6 v 6 after reaching the gain of 6 environment. Delivery rates are low in the 10 v 10 and 15 v 15 swarms. The 25 v 25 swarm maintains around 50% of packets delivered, peaks in the gain of 12 environment and then crashes in the most inclusive gain environment. This crash is in contrast to the behavior in the 6 v 6 swarm, which performs closer to expectations.

OLSR's packet delivery rates (Figures 4a – 4d) are quite poor. Rarely are more than half of sent packets received. The vehicle-destined heartbeat and red pose messages, which have thus far outperformed all other messages using the other routing protocols, lay at the bottom of these graphs. It observed in the trace files that OLSR has a difficult time creating the routing tables for vehicle destinations in such a dynamic topology. Looking at the raw trace files generated by the simulator, one can see numerous routing table messages. They far outnumber actual message traffic. Their frequency creates many collisions at the data link layer.

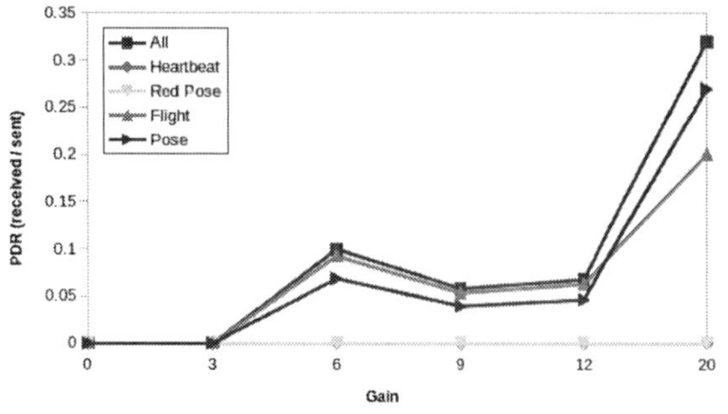

(a) 6 v 6 Swarm

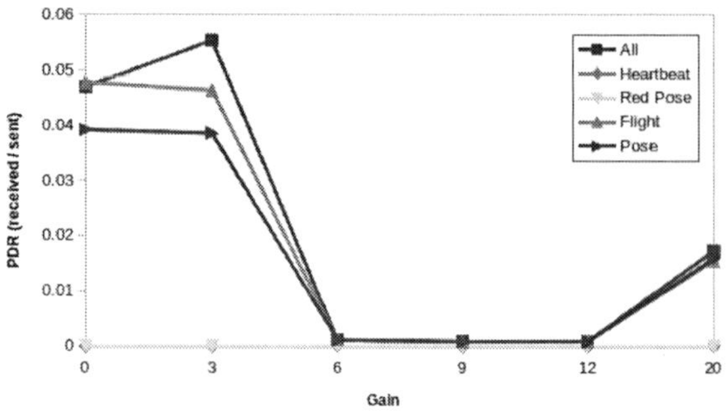

(b) 10 v 10 Swarm

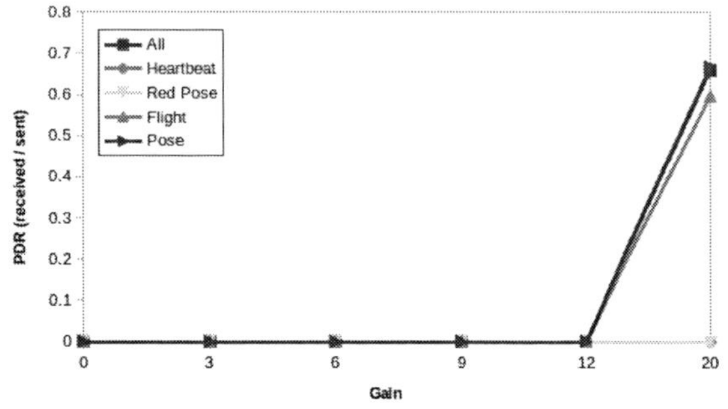

(c) 15 v 15 Swarm

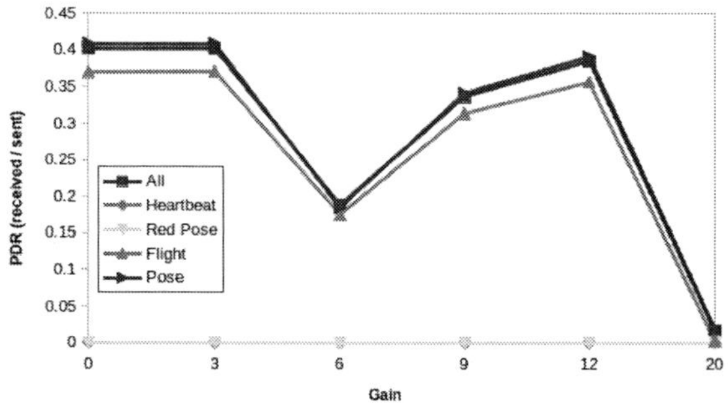

(d) 25 v 25 Swarm

Figure 4: Packet Delivery Ratio Using OLSR

Epidemic's PDR performance in the 6 v 6 swarm shows that it can deliver messages to the nodes as the environment becomes more inclusive (more nodes able to be touched by the flooding) but the nodes are never really able to deliver messages to the ground (Figure 1d).

Figures 5a and 5b give a final perspective of PDR by placing the routing protocols' all messages side by side per swarm size. AODV PDR remains above them all in every swarm. The recovery of stats in the 25 v 25 swarm is even more evident.

CONCLUSIONS AND FUTURE WORK

Some of the results were unexpected. No protocol consistently performed close to expected values. Those protocols that generated significant overhead created collisions or failed converge the network Not only did overhead create conditions for numerous retransmissions, but they had low message delivery metrics. The protocols with less overhead fared better.

AODV, as a reactive protocol, was able to discover multi-hop routes in restricted connectivity environments but was able to back off when nodes operated in a highly connective environment. It was able to perform across more environments than any other protocol both when delivering messages to the ground station and to nodes. Even in the 15 v 15 Swarm, where many protocols failed to deliver messages to the ground, it was able to deliver some messages.

DSDV had similar performance numbers to broadcast which implies it did not harm message delivery while updating its neighborhood routing tables but it also did not improve delivery in environments when single-hop routing was not available. OLSR, the other proactive routing protocol, was unable to deliver messages to the vehicles in all swarm sizes. The highly dynamic node topology caused OLSR to flood the network with routing table update messages.

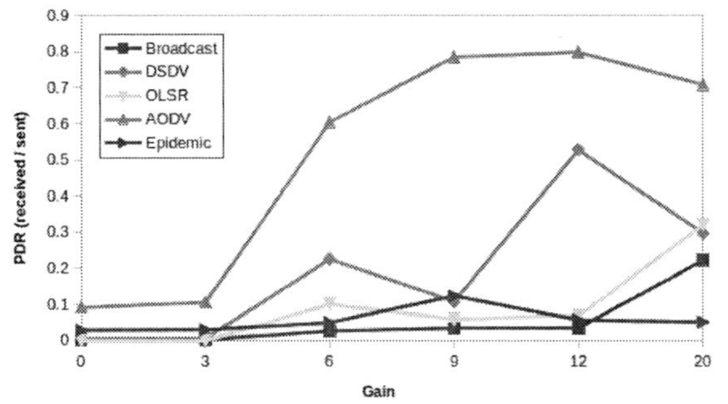

(a) 6 v 6 Swarm

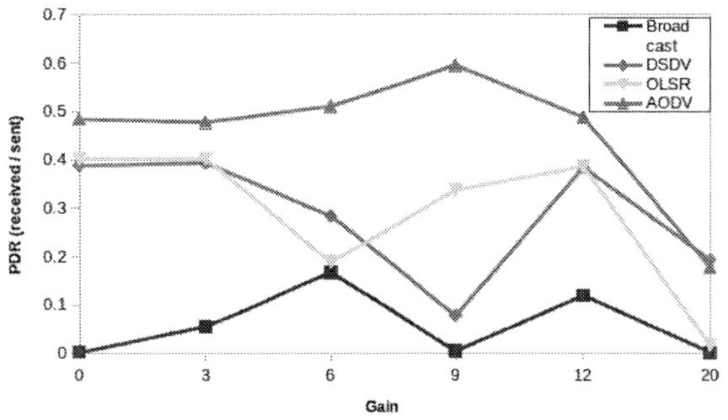

(b) 25 v 25 Swarm

Figure 5: Packet Delivery Ratio for all protocols

At this time, the 1-hop communications used by the NPS swarm is working consistently with 50 airborne nodes. As NPS swarm experimentation scales up, we recommend choosing a routing protocol that is either reactive in nature or is focused primarily on reducing overhead to avoid exhausting the limited bandwidth available. Future routing protocols may need different limits, rules, or settings, based on whether the message is traveling ground to node or node to ground.

In future it would be useful to perform the same simulation with additional mobility traces of the same numbers of nodes, in order to understand the generality of the results. Averaging PDRs at the same size but with different flight patterns could help factor out extreme cases based purely on lucky or unfortunate flight paths of certain vehicles. The data sets might also be more clear if take-off and landing flight data were excluded from the mobility models.

At this time, the ns-3 libraries are limited to traditional routing protocols for MANETs. Networking researchers have had more opportunity and motivation to work with them than to network UAVs. It would be extremely beneficial to develop and add to the libraries various protocols that are more feasible or specifically designed for FANETs and DTNs.

REFERENCES

[1] T. H. Chung, M. R. Clement, M. A. Day, K. D. Jones, D. Davis, and M. Jones, "Live-fly, large-scale field experimentation for large numbers of fixed-wing UAVs," in *IEEE International Conference on Robotics and Automation (ICRA)*, pp. 1255–1262, 2016.

[2] S. Rosati, K. Kruzelecki, L. Traynard, and B. Rimoldi, "Speed-aware routing for UAV ad-hoc networks," in *2013 IEEE Globecom Workshops (GC Wkshps)*, pp. 1367–1373, 2013.

[3] I. Bekmezci, O. K. Sahingoz, and Ş. Temel, "Flying ad-hoc networks (FANETs): A survey," *Ad Hoc Networks*, vol. 11, no. 3, pp. 1254–1270, 2013.

[4] J. P. Rohrer, A. Jabbar, E. Perrins, and J. P. Sterbenz, "Cross-layer architectural framework for highly-mobile multihop airborne telemetry networks," in *MILCOM 2008 - 2008 IEEE Military Communications Conference*, pp. 1–9, 2008.

[5] K. S. Pathapati, T. A. N. Nguyen, and J. P. Rohrer, "Performance analysis of the AeroTP transport protocol for highly-dynamic airborne telemetry networks," in *International Telemetering Conference Proceedings*, 2011.

[6] D. T. Davis, T. H. Chung, M. R. Clement, and M. A. Day, "Multi-swarm infrastructure for swarm versus swarm experimentation," in *MILCOM 2008 - 2008 IEEE Military Communications Conference*, pp. 1–9, 2016.

[7] G. Hoffmann, D. G. Rajnarayan, S. L. Waslander, D. Dostal, J. S. Jang, and C. J. Tomlin, "The stanford testbed of autonomous rotorcraft for multi agent control (starmac)," in *The 23rd Digital Avionics Systems Conference*, vol. 2, pp. 12–E, IEEE, 2004.

[8] J. Han, Y. Xu, L. Di, and Y. Chen, "Low-cost multi-UAV technologies for contour mapping of nuclear radiation field," *Journal of Intelligent & Robotic Systems*, vol. 70, pp. 401–410, April 2013.

[9] G. Vásárhelyi, C. Virágh, G. Somorjai, N. Tarcai, T. Szörényi, T. Nepusz, and T. Vicsek, "Outdoor flocking and formation flight with autonomous aerial robots," in *2014 IEEE/RSJ International Conference on Intelligent Robots and Systems (IROS 2014)*, pp. 3866–3873, IEEE, 2014.

[10] K. Kaplan, "Disney makes history with first U.S. light show powered by 300 intel drones," 2016. Accessed February 10, 2017.

[11] J. Z. S. Hauert, S. Leven and D. Floreano, "The swarming micro air vehicle network (SMAVNET) project." Accessed March 27, 2017.

[12] J. Macker, "Mobile ad hoc networking (MANET): Routing protocol performance issues and evaluation considerations," RFC 2501, January 1999.

[13] K. Fall, K. L. Scott, S. C. Burleigh, L. Torgerson, A. J. Hooke, H. S. Weiss, R. C. Durst, and V. Cerf, "Delay-tolerant networking architecture," RFC 4838, April 2007.

[14] J. P. Rohrer and K. M. Killeen, "Geolocation assisted routing protocols for vehicular networks," in *2016 International Conference on Connected Vehicles and Expo (ICCVE)*, pp. 1–6, IEEE, 2016.

[15] "The ns-3 simulator." Accessed March 21, 2017.

[16] T. Clausen and P. Jacquet, "Optimized link state routing protocol (olsr)," RFC 3626, October 2003.

[17] C. E. Perkins and P. Bhagwat, "Highly dynamic destination-sequenced distance-vector routing (DSDV) for mobile computers," in *ACM SIGCOMM computer communication review*, vol. 24, pp. 234–244, ACM, 1994.

[18] C. Perkins, E. Belding-Royer, and S. Das, "Ad hoc on-demand distance vector (AODV) routing," RFC 3561, July 2003.

[19] M. J. Alenazi, Y. Cheng, D. Zhang, and J. P. Sterbenz, "Epidemic routing protocol implementation in ns-3," in *Proceedings of the 2015 Workshop on ns-3*, pp. 83–90, ACM, 2015.

A TRADE STUDY TO DETERMINE THE BEST LOCATION FOR TM RECEIVERS

Michael Diehl, Ryan Fraser, Jonathan Green, and Jason Swain
Air Combat Systems Directorate, U.S. Army Yuma Proving Ground, Yuma AZ 85367

ABSTRACT

As part of a larger effort to improve telemetry link availability, the U.S. Army Yuma Proving Ground (YPG) is currently in the process of upgrading its telemetry receivers. As YPG begins integrating new receivers into existing range infrastructure, the question of where to place these receivers to provide maximum benefit must be considered. Should the receivers be placed at each of the remote antennas or should they be centrally located at the primary telemetry site? Although many Telemetry-over-Internet Protocol (TMoIP) and radio frequency (RF) over fiber solutions exist to transport these data, there are numerous concerns including network and Cybersecurity limitations to consider when implementing either of these approaches. This paper will document the trade study conducted at YPG to explore the benefits of each approach.

KEYWORDS

1. Telemetry over Internet Protocol
2. Radio Frequency over fiber
3. Radio Frequency over Internet Protocol (RFoIP)

INTRODUCTION

YPG is a subordinate command of the Army Test and Evaluation Command (ATEC) and is one of the largest military installations in the world. Located in southwestern Arizona, it encompasses 1,308 square miles. YPG personnel conduct tests on nearly every weapon system or piece of military equipment in the ground and air combat arsenal. With a mission to provide premier test services to the U.S. Government and her allies, YPG conducts, reports, and supports developmental tests, experiments, production tests, and integrated developmental/operational tests, as well as provides training support. In 2016, YPG fired over 212,000 rounds of artillery, mortar, missile, and small arms munitions, flew over 2,900 aircraft sorties, drove over 68,000 test miles, and conducted over 1,500 airdrops. These numbers more than double when events conducted in support of training are counted, and YPG was the busiest Army proving ground for the last 7 years.

The Aviation Systems and Electronic Test Division at YPG is responsible for testing of aviation ballistic weapons and missiles, unmanned aerial systems, aircraft systems, precision guided and unguided air delivered systems, personnel parachutes, sensors and surveillance systems, and electronic warfare systems. As a developmental test activity, the goal is to help the developer field their systems and provide the Soldier with the safest, most lethal equipment. Instrumentation is critical in order to capture the necessary data reported from the item under test or capture the "true" performance of the system. It is also important to display critical information from the system under test at remote locations in real-time so the test team can operate efficiently and safely.

In order to support Telemetry missions, data needs to be moved around the range efficiently and reliably. At YPG, that means moving data from three remotely controlled fixed antennas back to the main Telemetry Building (TM Bldg) and between the TM Bldg and the mission control center utilizing YPG's extensive fiber network. The locations of these assets are shown in Figure 1. On some occasions, data is also required to be moved from two mobile Telemetry vans to the TM Bldg, and data is also sent to other test ranges; however, while this is a rare requirement, it is one that is increasing in frequency.

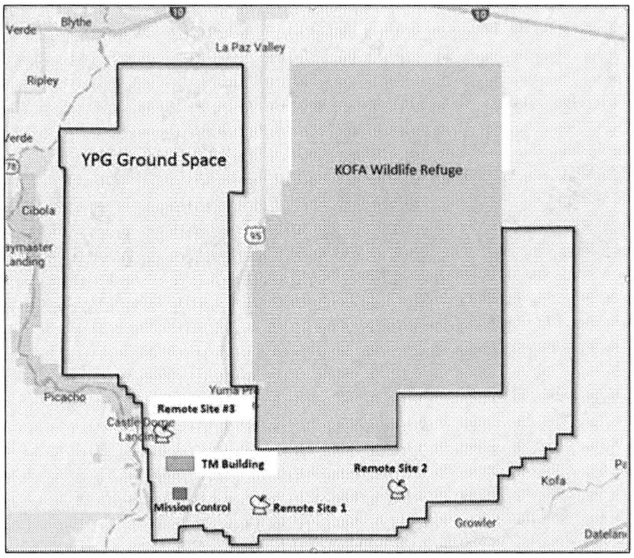

Figure 1. Map of YPG with Telemetry Assets

With respect to telemetry antennas and receivers, YPG's current setup is a typical one where telemetry receivers are located at each antenna location. This reduces RF loss and allows for receiver's digital Pulse Code Modulation (PCM) output to be used for data transfer, which is less prone to interference and noise. However, this approach requires that 2 to 4 receivers be located at each antenna site, which is both expensive and inefficient as each receiver costs between $65,000 and $85,000 and are idle when the antenna is not in use.

Alternatively, if all the Telemetry receivers could be kept at the TM Bldg, a less expensive, more efficient scheme may result. This scheme would likely require fewer total receivers that could be easily assigned to the various antennas as required, resulting in less idle time; however, this scheme is only possible if a low-loss RF path could be made between remote fixed antennas and the TM Bldg.

Currently, YPG is in the procurement phase of an investment cycle to replace its legacy receivers. These new receivers have advanced capabilities that will greatly increase link availability while also having the ability to support C-Band frequencies. As is always the case, the goal of this investment cycle is to maximize the Army's investment and provide YPG the greatest flexibility to support its diverse test mission. The purpose of this paper is to evaluate options for moving data throughout YPG as well as off YPG. Due to various mission requirements, multiple options may be required for different applications.

TECHNOLOGY OPTIONS

There are three technologies evaluated in this paper that can be used to transport telemetered data around YPG: TMoIP, RF over fiber, and RFoIP. All three would use YPG's extensive fiber network of buried fiber going to all improved locations or to lay above ground fiber cables to unimproved locations, as needed. Although Internet Protocol (IP) technologies do not require fiber, due to the availability of fiber at YPG, it will be used as transport medium for most IP traffic.

1. TMoIP: TMoIP takes the receiver's PCM digital data and packetizes it in an IP stream. An interface box is required on both ends, but the data can then travel on any network-capable medium, from a fiber to an Ethernet cable. When IP is sent via fiber, it shares the medium with other IP traffic, so it does not require a dedicated fiber path. There is minimal concern over packet delay when there is heavy IP traffic; however, IP does have a packet priority capability to help mitigate packet delay, which TMoIP hardware is capable of utilizing. YPG has already invested in TMoIP hardware and uses it regularly. For example, the TM Bldg and YPG's mission control center are approximately four miles apart, and TMoIP is currently used to transfer the received telemetered data over that gap.

   a. Pros:

   (1) Currently in use at YPG.
   (2) TMoIP is also the most convenient option for transferring data between different test ranges since it is IP-based. Within the last year, YPG has collected mission data on the range and sent it via TMoIP to Redstone Test Center (RTC) in Alabama and Edwards Air Force Base (AFB) in California with little effort and great success.
   (3) TMoIP is very mature and readily available from many vendors. When YPG first investigated TMoIP options in 2015, five vendors were evaluated on-site and all performed exceptionally well.
   (4) The Range Commander's Council (RCC) has begun discussion to start standardizing TMoIP.
   (5) Easily integrated into YPG's existing range network infrastructure. TMoIP allows shared use of fibers, does not significantly increase network traffic, and IP is preferred by YPG's Range Communications group.

b. Cons:

(1) Packet delay due to heavy IP traffic. This has not been an issue at YPG thus far, but as more and more instrumentation systems are starting to use YPG's range network, it could be in the future.
(2) TMoIP has not been standardized; consequently, cross compatibility between vendors equipment is not guaranteed. Although the standardization process has begun, it will take several years to complete, and matched sets of hardware will be required to ensure compatibility until it is completed.
(3) Receivers still need to be collocated at the remote sites so no reduction of required receiver assets is possible.
(4) TMoIP equipment will have to conform to Cybersecurity requirements.

c. Market Availability and Cost: Because all solutions have different functional capabilities, there is a wide cost delta of $7-54K for 4 to 16 channels. Some vendors in this marketspace include Apogee Labs, GDP Space Systems, IPtec, NetAcquire, and RT Logic.

2. RF Over Fiber: This solution also has an interface box on each end and takes the RF from the antenna and converts it to light. The RF is then reconstituted on the receiving end, ready to input into a telemetry receiver. Depending on the capability of the interface box, RF levels can be attenuated or gain added to mimic the original RF level. YPG has not previously utilized this technology for mission support but it is worth noting that RF over fiber does not seem to be as common as TMoIP at other test ranges; however, it is in use reliably at some, notably Point Mugu.

a. Pros:

(1) RF over fiber, unlike TMoIP, provides the capability to locate Telemetry receivers at a central location. This means that if all receivers were located at the TM Bldg, the total quantity required could be reduced to approximately 20% without loss of test support capability.
(2) RF over fiber technology is also very mature and readily available commercially from several vendors.
(3) There are no cybersecurity requirements for RF over fiber technology because the hardware never touches the network. Some units can be configured via a computer, but it would not be a daily mission requirement.

b. Cons:

(1) One of the drawbacks for this technology is it requires a dedicated fiber path, meaning no other instrumentation or signals can share the fiber line. At YPG, this is of some concern. Each fiber cable at YPG consists of 12 individual fibers, and the TM Bldg, being a major range communications hub, has multiple cables available. However, when fiber lines go bad, they are not dug up and repaired—technicians switch connection to one of the good lines. This means in some places, the number of fibers

is limited. For these reasons, YPG's Range Communications group, who are responsible for both fiber and networking, strongly encourage shared use of fibers, and particularly use of IP, by all users.

(2) Multiple fibers and interface boxes may be required to send right hand circular polarization (RHCP) and left hand circular polarization (LHCP) RF from the antenna to the receiver if the RF over fiber interface boxes were only single channel.

(3) RF over Fiber is susceptible to noise and dB loss. From a technical aspect, one needs to be careful when using RF over fiber because it can affect the RF dynamic power range. Any noise on the input from cables, multi-couplers, etc. increases the noise figure and reduces the RF link budget. An overall loss of 0.25-0.5 dB in gain to noise temperature (G/T) is possible when using RF over fiber equipment. Understanding the entire RF link path and careful planning can minimize the loss.

(4) Not all vendors advertise the ability to support C-Band.

c. Market Availability and Cost: Readily available commercially from several vendors, including some vendors not traditionally in the Telemetry industry. Vendors that produce RF over fiber hardware include but are not limited to Finisar (Optium), Glenair, Microwave Photonic Systems, Emcore (Ortel), Miteq (L3), and Universal Switching Corporation. Cost for this technology is $10-20K per link.

3. RF Over IP: Like the name suggests, it converts the RF signal into IP packets for transmission across any network. Like TMoIP and RF over Fiber, it also requires an interface box on each end. This technology takes the benefits of both TMoIP (uses IP, non-dedicated fiber) and RF over fiber (transfers RF). Currently, only RT Logic markets this capability. RT Logic's SpectralNet box utilizes IP standard for intermediate frequency (IF) ANSI/VITA 49.0 to transfer the data. It can only handle one channel up to 54 MHz spectrum; if two transmitted frequencies are further apart than 54 MHz, two boxes will be required. The SpectralNet allows for the user to configure the box to send only the spectrum required. For instance, one could configure the unit to send only 3 MHz spectrum centered on a 1 Mbps signal. YPG has not utilized this solution.

a. Pros:

(1) RFoIP, unlike TMoIP, provides the capability to locate Telemetry receivers at a central location. This means that if all receivers were located at the TM Bldg, the total quantity required could be reduced by approximately 20% without loss of test support capability.

(2) Easily integrated into YPG's existing range network infrastructure. RFoIP allows shared use of fibers and IP is preferred by YPG's Range Communications group.

b. Cons:

(1) The market for RFoIP is small, as only one vendor markets the capability, RT Logic.

(2) It also is only capable of supporting Telemetry's L and S bands, not C Band. RT Logic is developing a product that will handle 500 MHz and C Band in the future, but the timeline for completion is currently unknown.

(3) RFoIP can use large amounts of IP bandwidth in order to transfer the RF spectrum. As a mitigating option, the SpectralNet allows for the user to configure the box to send only the spectrum required. For instance, one could configure the unit to send only 3 MHz spectrum centered on a 1 Mbps signal.
(4) RFoIP equipment will have to conform to Cybersecurity requirements

c. Market Availability and Cost: Only one vendor, RT Logic, markets this capability. If market demand increases, then other vendors may develop products as well. Cost of a SpectralNet box is ~$15K.

Table 1 summarizes the pros and cons of the technologies listed in this paper.

Table 1. Pros and Cons of Technologies

Pro		Con
Currently own Hardware	Telemetry over IP	No reduction in receivers possible
Shares Fiber		Receivers remain at remotes sites
Easy to transmit off range		Matched sets of hardware required
Easy to Transport around range		Required to conform to cybersecurity rules
Mature technology		Future possibility of packet delay
IP preferred by YPG's Range Communications group		–
Receivers can be at central location	RF over Fiber	Dedicated Fiber is required
Fewer receivers required for operations		Possible requirement for multiple interface boxes and fibers at each remote site
Mature Technology		Susceptible to noise and dB loss
No cybersecurity requirements		Not all vendors advertise C-band capability
Receivers can be at central location	RF over IP	Only one viable supplier identified
Fewer receivers required for operations		Uses large amount of IP Bandwidth
IP preferred by YPG's Range Communications group		Required to conform to cybersecurity rules
Easy to Transport around range		Does not support C Band

CONCLUSION

The pros and cons of the different technologies lend themselves to different applications. At YPG, there are five different applications to consider: Remote Antenna Site(s) to the TM Bldg, the TM Bldg to Mission Control, YPG to another range, TM Van to/from the TM Bldg, and Airfield to the TM Bldg.

1. Remote Antenna Site(s): The most obvious application where transferring RF data could benefit YPG would be between remote antenna sites and the TM Bldg (see Figure 1). YPG

currently has three remotes site and future plans include the possible addition of more sites. The remote sites are already remotely controlled from the TM Bldg and have receivers collocated with the antennas. All data captured by the antenna and processed by the receivers is sent to the TM Bldg. When YPG looked at options in 2015 for transferring data from these remote antenna sites, YPG's Range Communications group was a strong proponent of an IP solution. The basis for this advocacy for IP solutions was the availability of fiber lines on the test range, which was limited and would only decrease with time as the fiber lines degraded. The addition of new fiber lines is not planned in the near future.

The RF over Fiber technology offers the potential to reduce the number of receivers needed by 20% and had no cybersecurity compliance requirement. Although these were significant strengths, the lack of available fiber lines ultimately ruled out RF over fiber for this application. RF over fiber would require too many dedicated fiber lines, especially if two fibers were required to send RHCP and LHCP. This concern is amplified when fiber availability runs out, because the cost and complexity of adding new fibers (dig permits, cost, environmental assessments, etc.) far outweighs any cost savings from using fewer receivers. This left RFoIP and TMoIP as options.

For YPG, the technical limitations of RFoIP do not make it a viable solution at this time. This becomes apparent when evaluating a typical Telemetry operation because tracking antennas have RHCP and LHCP feeds available; therefore two SpectralNet boxes would be required to send both polarizations. In order to support two frequencies greater than 54 MHz apart, and both RHCP and LHCP, then eight total SpectralNet boxes would be required (four on each end). If a third frequency is added, more than 54 MHz from the first two, then four more boxes would be required. The added complexity and hardware required makes this option cost-prohibitive for a typical operation. This leaves TMoIP as the best option for remote sites at YPG.

2. TM Bldg to Mission Control: For data transmission between the TM Bldg and Mission Control, the data has already been converted from RF to PCM by the telemetry receivers, so it makes sense to use TMoIP. This has been the standard at YPG since 2015 and has proven to be very capable data link.

3. YPG to Another Range: Similar to data transmission between the TM Bldg and Mission Control, data being transmitted between YPG and another Test Range will be converted to PCM at YPG prior to transmission. Digital data is better suited for long distance transmissions. There are many straight forward networking options available to establish an IP link between YPG and another Test Range, making TMoIP the obvious choice.

4. TM Van to/from TM Bldg: TM Vans and the TM Bldg are configured to be complete stand-alone ground stations as many customers want to see the received data in real-time. In order to be able to receive, process, and display data, telemetry receivers are installed at the TM Bldg and in all TM Vans. Since both the TM Bldg and the TM Vans have receivers, it makes sense to use TMoIP to transfer data between them because the data is already in PCM format.

5. Airfield to TM Bldg: The last application is a checkout antenna at the airfield used to provide a data link to the TM Bldg. The TM Bldg does not have good line-of-sight to the airfield: furthermore, buildings, terrain, and other aircraft cause multipath and masking issues. An antenna located at the airfield addresses all these issues and provides much better RF coverage. For this application, RFoIP and RF over Fiber are viable since the costs of either is less than the cost of a receiver. Only one dedicated fiber would be required for the RF over fiber solution. Although no dedicated fiber would be required for RFoIP solution, the market space is small, creating sustainability and obsolescence concerns for the user. Since this application is the least utilized and not directly tied to test support, it could serve as a test bed to gain more experience with RFoIP. At this point, it is a coin toss between the two, and a decision that YPG will have to discuss in a broader group. Results from this paper should help guide that discussion. Table 2 shows a summary of the applications.

Table 2. Summary of Applications

Application	Best Technology Solution
Remote Site to TM Bldg	TMoIP
TM Bldg to Mission Control	TMoIP
YPG to Another Range	TMoIP
TM Van to/from TM Bldg	TMoIP
Airfield to TM Bldg	RF over Fiber or RFoIP

LEGEND:
RF – Radio Frequency
TM – Telemetry
TMoIP – Telemetry-over-Internet Protocol
YPG – Yuma Proving Ground

ACKNOWLEDGEMENTS

YPG would like to thank individuals and vendors that were technical resources for this study: Kevin Bossoletti, PT MUGU NAWC; Josh Korson & Richard Stewart, Microwave Photonic Systems; Bill Asiano & David Soderholm, RT Logic; Jesse Woolridge, Universal Switching Corporation; and, Joe Milan, Apogee Labs.

REFERENCES

1. ITEA 2015 Paper, Considerations for Adapting RF/IF over IP Within Range Applications, ITEA, Las Vegas, Nevada, May 2015

2. Telemetry Group, Telemetry Standards, IRIG Standard 106–13 (Part 1), Chapter 10, Secretariat, Range Commanders Council, WSMR, New Mexico, June 2013.

AUTONOMOUS GROUND RECONNAISSANCE DRONE USING ROBOT OPERATING SYSTEM (ROS)

Undergraduate Students: David Hung, Cinthya Tang, Coby Allred, Kennon McKeever, James Murphy, Ricky Herriman
Faculty Advisors: Dr. Michael W. Marcellin
Department of Electrical and Computer Engineering
The University of Arizona, Tucson, AZ 85721

ABSTRACT

The Arizona Autonomous Club is a student organization at the University of Arizona which designs, builds, and competes with Unmanned Air Systems (UAS). This year, a 25% scale Xtreme Decathlon model aircraft was selected and successfully converted into a fully autonomous UAS for the AUVSI Student Unmanned Aerial Systems (SUAS) 2017 competition. The UAS utilizes a Pixhawk autopilot unit, which is an independent, open-hardware project aiming at providing high-end autopilot hardware at low costs and high availability. The Pixhawk runs an efficient real time operating system (RTOS) and includes sensors such as a GPS unit, IMUs, airspeed, etc. The UAS also includes an onboard imaging system, which is controlled by an onboard computer (OBC). The Pixhawk and OBC are interconnected with two ground control stations (GCS) using the Robot Operating System (ROS) framework, which is capable of extending overall system capabilities to include an expanded telemetry downlink, obstacle avoidance, and manual overrides.

INTRODUCTION

Background

The Association for Unmanned Vehicles Systems International (AUVSI) has hosted the Student Unmanned Aerial Systems (SUAS) Competition annually since 2002. The competition is centered around the design, fabrication, and deployment of unmanned aerial systems (UAS). The UAS are tested on various mission tasks such as autonomous flight, obstacle avoidance, ground target identification and classification, and payload delivery.

This year the University of Arizona Autonomous Vehicles Club was organized into three sub-teams: hardware, flight control software, and imaging. The hardware team was responsible for the end-to-end design and fabrication of the airframe and avionics. The flight control software team was responsible for all software aspects for flight, including autopilot setup, mission planning, and creating custom airframe control configurations. The imaging subteam was responsible for the entire imaging stack, including imaging hardware both onboard and offboard the system as well as all peripheral software such as image processing and wireless communications. The imaging subteam was also in charge of the networking interface with the AUVSI SUAS judges' server.

Problem Statement

Success at the AUVSI SUAS competition requires a robust platform capable of accomplishing a wide breadth of tasks. As first time competitors, Arizona Autonomous prioritized the production of a solid basic platform to act as a launchpad for future development. As a result, the primary development focus for the 2017 AUVSI SUAS competition was placed on deploying a minimally featured framework which would be added to in subsequent years.

Objectives

The competition poses several mission tasks to be completed including fully autonomous flight, target recognition and classification, and payload delivery. This year's platform was also designed to prove infrastructure for the first two tasks, while secondary tasks such as payload delivery were excluded. A complete list of mission tasks may be found in the official AUVSI SUAS competition rules [1].

DESIGN SOLUTIONS

Airframe

This year, a 25% scale Xtreme Decathlon RC Acrobatics aircraft was selected as the base for the airframe. It has a hobby plywood and balsa construction with Ultracote covering for aerodynamics. The airframe has an overall wingspan of 92.2 inches and a fuselage length of 66 inches. It is powered by a Desert Aircraft 50cc engine with a 23x8 inch two blade propeller and is controlled using flaps, ailerons, elevators, and a rudder. In order to fit within the custom interior, a custom servo mount was designed and 3D printed and installed onto the airframe.More information about the 25% scale Xtreme Decathlon RC Acrobatics aircraft can be found at the vendor's website [2].

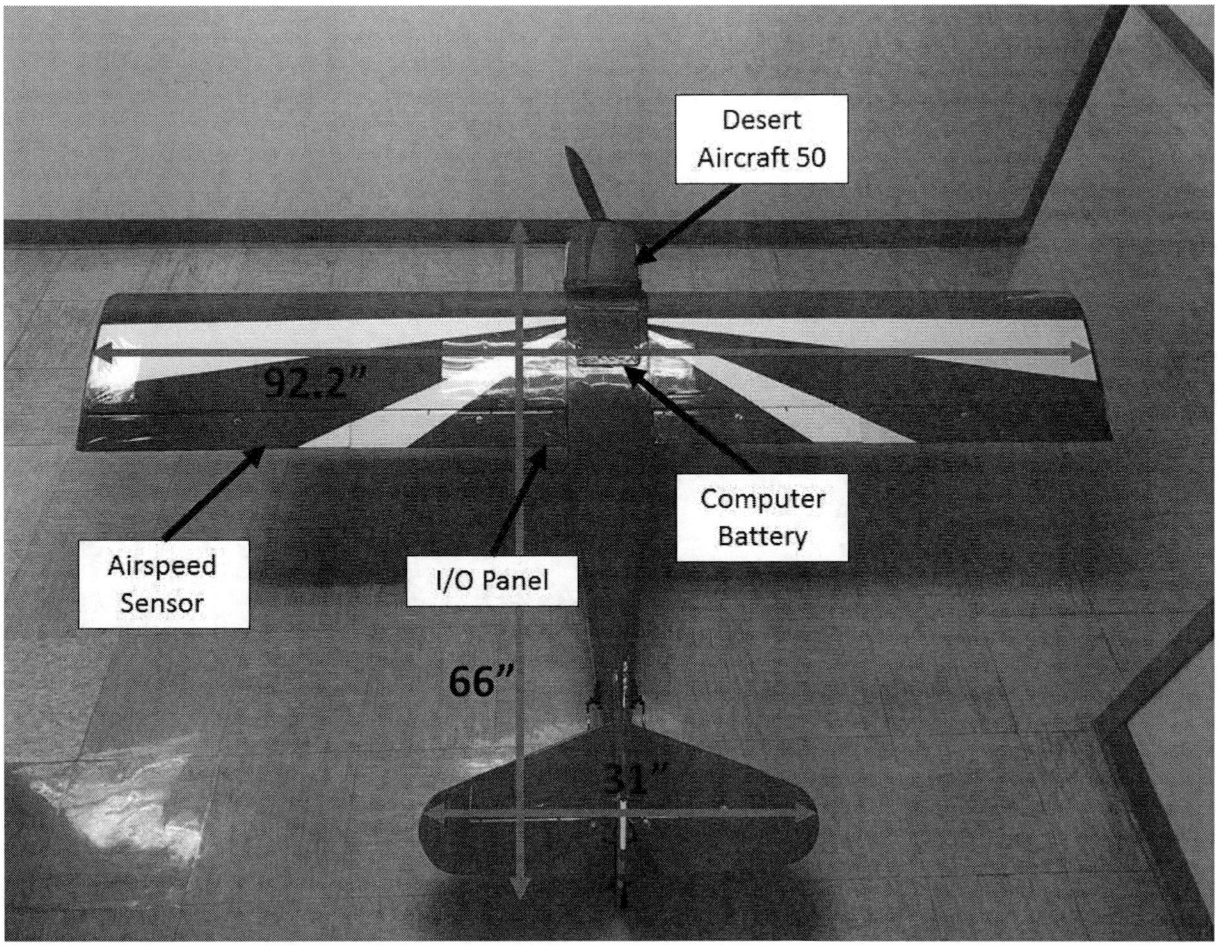

Figure 1. Top-down image of the UAS with measurements (in inches) and labels.

The airframe was modified with custom mounting solutions in order to support the various avionics and imaging equipment necessary to complete the mission tasks. Externally, a portion of the aircraft's belly was cut out and replaced with a flat piece of hobby plywood in order to act as a mounting location for the gimbal assembly. On the wings, a mounting position and routing path were created for an airspeed sensor with accompanying pitot tubes and electrical wiring. Internally, plywood plates were added to create a three layered interior structure. This structure was leveraged to mount all avionic and computational equipment, such as the Pixhawk autopilot, Odroid on-board computer, and three independent, battery power circuits. Finally, a custom I/O panel was added to the airframe and was mounted on the side of the aircraft to provide an easily accessible interface for regular maintenance and software updates. It houses three switches to power for each of the three independent power systems, three charging ports, five indicator LED's to indicate system power states and status, and various data ports. The addition of the I/O panel greatly decreased development overhead and streamlined the preflight setup process.

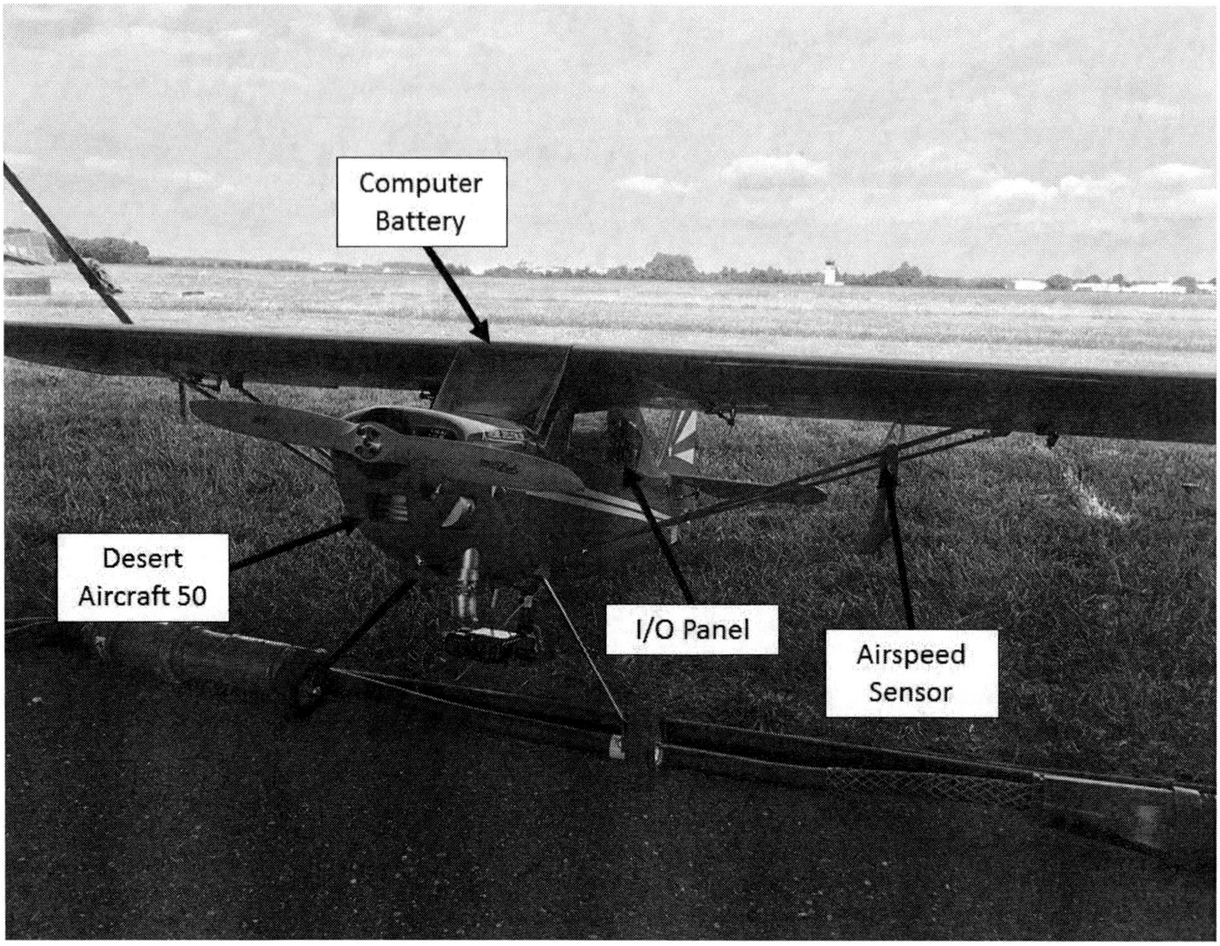

Figure 2. Front view image of the UAS with labels.

Flight Control Software

To control the UAS, the affordable, open-source Pixhawk autopilot platform was installed on the airframe with the PX4 flight stack. It utilizes a 32-bit ARM Cortex M4 core with 256KB of RAM, 2MB of flash memory, and a secondary 32-bit failsafe co-processor and is capable of automated, real-time control of all of the control surfaces on the airframe through the use of PWM servo outputs. The Pixhawk includes a variety of sensors including a 16 bit gyroscope, a 14-bit accelerometer and magnetometer, a 3-axis accelerometer and gyroscope, and a barometer. The Pixhawk platform also supports a variety of peripherals through the use of auxiliary servo outputs to control the mounted camera gimbal and other equipment. A custom airframe configuration was created to allow for the higher number of servos onboard the UAS than the included airframe options. The Pixhawk also supports offboard control which is accomplished using an ODROID XU4 running Ubuntu and ROS. When offboard control is desired, a MAVLink message is sent from the XU4 to the Pixhawk which gives control of the UAS to ROS from the Pixhawk [4].

The primary software used to control the Pixhawk during flight is the ground control software QGroundControl, an application specifically targeted at the Pixhawk and other similar platforms. This software is used for the initial configuration of the Pixhawk and calibration of its onboard sensors along with support for further parameter and control loop tuning. Telemetry data that is sent from the Pixhawk to QGroundControl is displayed in a GUI which shows airframe position, airspeed, mission waypoints, and other telemetry data from the UAS. QGroundControl also supports mission planning through the use of an interactive GUI that allows for waypoint navigation, takeoff and landing, and special mission triggers.

Figure 3. QGroundControl mission planning of 2017 AUVSI SUAS competition.

In addition to the core autopilot software, an additional utility named Path-Generator was developed. Path-Generator is used to parse input coordinates into waypoints, boundary vertices, and search area vertices in order to produce a sequence of waypoint coordinates usable by the Pixhawk autopilot. While input waypoint coordinates are trivial to translate into autopilot coordinates, search area vertices do not directly map to waypoints. In order to properly handle this, Path-Generator leverages QGroundControl's existing Survey command. By storing each of the search area vertices within a particular format QGroundControl is commanded to draw the proper ladder path within the mission's search area. Path-Generator also reads in and stores standard mission waypoints into a single list of waypoints which can be read by QGroundControl. The program will then prompt the user for a decision on whether the search area or the mission waypoints will be executed first. This function was desired in order to achieve the optimal mission time and minimum fuel usage after deciding which runway would be used for takeoff after evaluating wind conditions on the flight line. These two lists are then pushed to Path-Generator's formatting function in the appropriate order. Finally, the program will publish the final path to a .mission file in a JSON format which the flight software,

QGroundControl, is able to read. After successfully writing the mission, the mission is then pushed to the UAS where the Pixhawk will begin its processing of the flight.

Telemetry

To communicate with the UAS, three separate links were established. The primary communication method for the UAS employed the 915 MHz band which utilizes frequencies from 902-928 MHz and was responsible for relaying telemetry data between the GCS and the UAS. This data includes all of the telemetry collected from the Pixhawk's built-in sensors and other sensors onboard such as the equipment responsible for airspeed readings. This data is packaged as MAVLink messages, which is a lightweight communication method that converts data from the XML format to language specific source code. The 915 MHz band is also utilized as the primary communication method from the GCS to the UAS regarding mission objectives in the form of MAVLink waypoint and event messages created by QGroundControl. The second link established with the UAS is over the 2.4 GHz band which allows for manual takeover of the UAS by a safety pilot using the Futaba T10J in case of data link loss or autopilot failure. Also controlled by the 2.4 GHz link and the Futaba T10J is a fibre-optic killswitch for the UAS's engine in case of complete system failure.

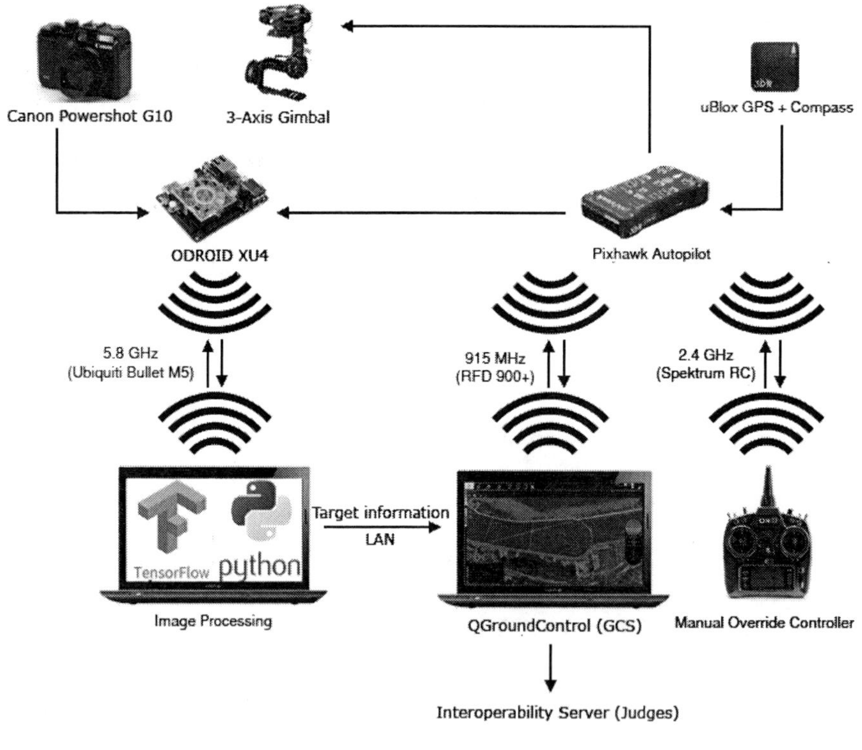

Figure 3. Communication system overview

A third communication link was established between the UAS's on-board imaging computer and the classification ground station. The connection consisted of an Ubiquiti Bullet M5 with an omni-directional antenna on-board and an Ubiquiti Nanostation loco M5 on the ground. These devices utilized the 5 GHz band with channels operating in the frequency range of 5170 to 5825

MHz. These transceivers connect to their respective machines via ethernet, so the established link will make the devices appear to be on the same Local Area Network.

Robot Operating System

Networking for all communications throughout and between the UAS and ground station were managed by the Robot Operating System (ROS), a framework that simplifies communication between processes distributed across different machines. In addition to abstracting most of the networking details away from the software developers and significantly reducing the imaging system's development time, the ability to automatically monitor and restart processes in case of failure, and the ability to easily communicate with the Pixhawk autopilot, ROS also provided a few features which were invaluable to this particular application.

One useful feature of ROS was its ability to automatically route data. Aside from simplifying the communication between the on-board computer and the classification ground station, it helped increase the on-board image processing stack's throughput by spawning multiple instances of the image processing routine and intelligently balancing workload between nodes. Aside from a slightly longer delay between any given input and its corresponding output, the end result was similar to enabling the use of multiple threads in the sense that multiple images could be processed simultaneously, but the developers never had to face the design challenges associated with multithreading. Furthermore, this communication structure required no additional bandwidth from the 5 GHz wireless data link due to ROS' peer-to-peer model for establishing routes for data [4].

Another useful feature that ROS provided was protection against brief network interruptions. In the case of short network interruptions, ROS can automatically store any data that failed to transmit correctly in a queue. Once the network becomes healthy, ROS then transmits all of the data to any machines affected by the outage. Due to the high likelihood of network interruptions in the imaging system, this helped to ensure that all packets containing target data would be received by the classification ground station.

RESULTS AND CONCLUSION

Arizona Autonomous successfully competed at this year's AUVSI SUAS competition, receiving an overall placement of 37 out of more than 50 teams internationally. The team's performance was negatively impacted by several unexpected technical complications, such as inhibitively high GPS noise, mechanical failure in the propellor mounting bolts, and a software synchronization error. With the exception of these technical difficulties, Arizona Autonomous' participation in this year's SUAS competition demonstrated some of the features offered by the team's ROS based framework, and the issues experienced require only nominal fixes for correction. Overall, the 2017 AUVSI SUAS competition proved to be a highly educational experience and allowed the team to collect valuable data for future development efforts.

ACKNOWLEDGEMENTS

The Arizona Autonomous Vehicles Club would like to thank our faculty advisor **Professor Michael Marcellin** for his endless support, patience, and guidance. We would also like to thank him for being our safety pilot throughout the year and for performing outstanding emergency landings on separate occasions. Additionally, we would like to thank all of our team members for their hard work, dedication, and perseverance. We would also like to thank the members who attended the 2017 AUVSI SUAS Competition. Lastly, we would like to thank the University of Arizona for their support.

REFERENCES

1. "2017 Rules for the 15th Annual Student UAS (SUAS) Competition." *AUVSI SUAS Competition*. AUVSI Seafarer Chapter, 12 June 2017. Web. <http://www.auvsi-suas.org/static/competitions/2017/auvsi_suas-2017-rules.pdf>.

2. "The New Bill Hempel's "Super Decathlon"." Bill Hempel's Team Edge! - Super Decathlon. R&D VISIONS, n.d. Web. 20 Aug. 2016. <http://www.billhempel.com/decathlon/>.

3. "Home." Home - Pixhawk Flight Controller Hardware Project. N.p., n.d. Web. 26 Nov. 2016. <https://pixhawk.org/>.

4. "ROS/Introduction." ROS/Introduction - ROS Wiki. *Open Source Robotics Foundation.* n.d. Web. 22 May 2014. <http://wiki.ros.org/ROS/Introduction>

Mobile Diagnostics for Personal Electrics Transportation Devices

Zachary A. Gilchrist
Electrical Engineering Design Laboratory
University of Kansas
Lawrence, KS 66044
zacharygil@ku.edu

Faculty Advisor:
Prof. Glenn Prescott

ABSTRACT

The ultimate purpose of this project was to improve on a new electric form of transportation. These methods of transportation have been gaining popularity for those who have relatively short commutes, and this project poises that group as the target audience. As a result of these trends and the practicality of small personal transportation, the design team made the purpose to attain usable real-world practicality and walking replacement ability. Other types of devices have been created before, however they usually have a short ride time and do not go much faster than the average walking speed. That being said, this design also uses telemetry to serve as a proof of concept that internal data can be sent to a receiver from a distant location. The end result was expected to be an enjoyable experience that gave a mobile diagnostics system.

INTRODUCTION

This paper is to discuss the research and development of the Telemetry involved in a mobile diagnostics system for a single wheeled electric transportation device. The device was referred to as Riding One-wheeled balanced electronic range transportation (ROBERT).

THE BODY

The telemetry system is made up of a few major components that come together to form one complete system. The complete system's function is to take in data, in this case temperature, and transmit that data to a computer wirelessly. The individual component/components are an Arduino Uno, Adafruit Temperature sensor, and two XBee transceivers. The components come together as seen in Figure 1. The temperature readings are taken into the Arduino that uses the XBee to transmit that data to an XBee receiver connected to a computer through USB. Whenever the Arduino based XBee transmits data through the extended antenna it flashes a red LED beacon. The XBees communicate through a 2.5-GHz frequency and have individual PAN IDs to allow for secure and uninterrupted communication.

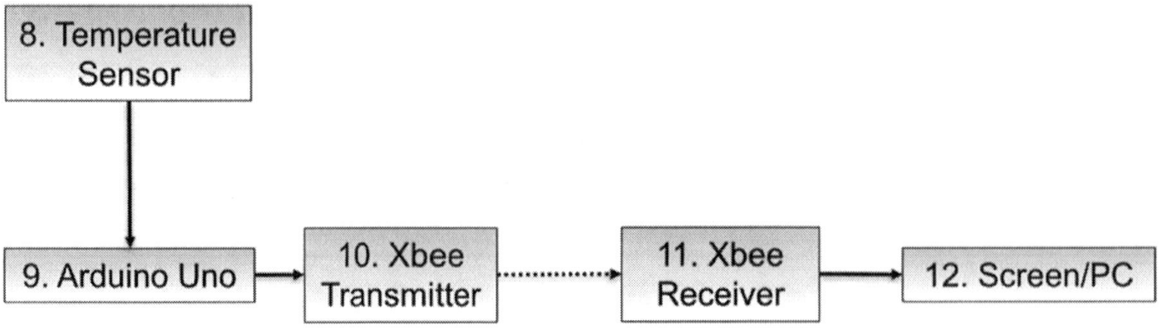

Figure 1 Telemetry Circuit functionality block diagram

Software aspects:

The software has two main areas, that are the Arduino IDE and then the XCTU program for the XBee configuration. First, the Arduino IDE is interesting in that it is written in Java and based on the Processing Project. Though the code used in programming the Arduino is a C and C++ mixture. Both languages will work, with the note that most of the standard libraries will NOT work. This is in part due to the fact that Arduinos have extremely low RAM compared to something like a PC. The code written for the Telemetry portion is C++ with imported hardware libraries for the sensor and XBee.

The Arduino code has a loop function that it continuously goes through. This is useful for microcontrollers because there isn't a required "wake-up" and "shut down" function, or a need to write code to create a live loop. Within this loop our code had two methods that were waiting on "YES" to come to their conditions. This can be seen in Figure 2. The first loop waited on data to be seen as being transmitted to it. More specifically, it waited for the XBee to say it had data being transmitted to it. In this code that data was irrelevant. For this code the data just showed that the user on the computer tapped some key and that data was sent to the Arduino based XBee. Once that condition was met, the data was read from the temperature sensor using methods within the downloaded library. That data was then algebraically converted to Fahrenheit and from hexadecimal form. Then transmitted back to the XBee that originally transmitted data to it. In essence, it showed that someone was asking for data, read the sensors, then sent data back to that someone.

The other of the two impeded loops was an LED switch. It waited for data to be transmitted. As soon as the data was transmitted from within the last loop, it set that digital PIN to high and turned on an LED. As soon as the data stopped transmitting, the LED is set to low. The LED is on for a very brief period of time relative to the data being transmitted. If the data was continuously transmitted (as tested) the LED simply stays on. This can be seen visually in Figure 4.

The other half of the software in the project is the program XCTU, which is a configuration and test utility software. In this case, we used a USB dongle to connect to the PC to be able write and read the XBee. Both XBees were first read and configured through the XCTU program, even though one was only used in the Arduino setup. XCTU allows for multiple parameters to be changed, but the most important and debatably easiest to change is the PAN ID. Ours is set to 1776 to separate it from any other XBees in the area. Then post calibration use, the XCTU program was used to monitor the XBee module connected to the Arduino. It sent a keystroke to the Arduino to signify that it wanted to receive data. That data was then received and displayed in the "Console Mode." It shows the data in two forms -- Hexadecimal and as received. As received in this case was the temperature in Celsius and Fahrenheit.

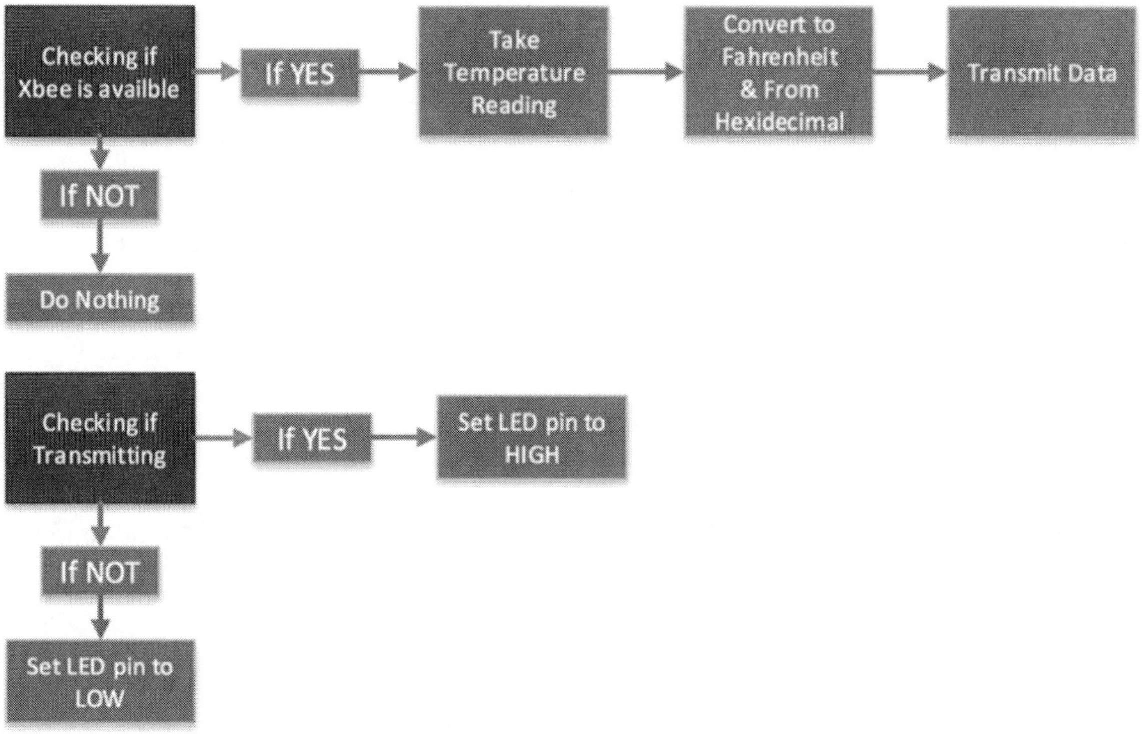

Figure 2: Arduino Software Flow Chart

How it was implemented:

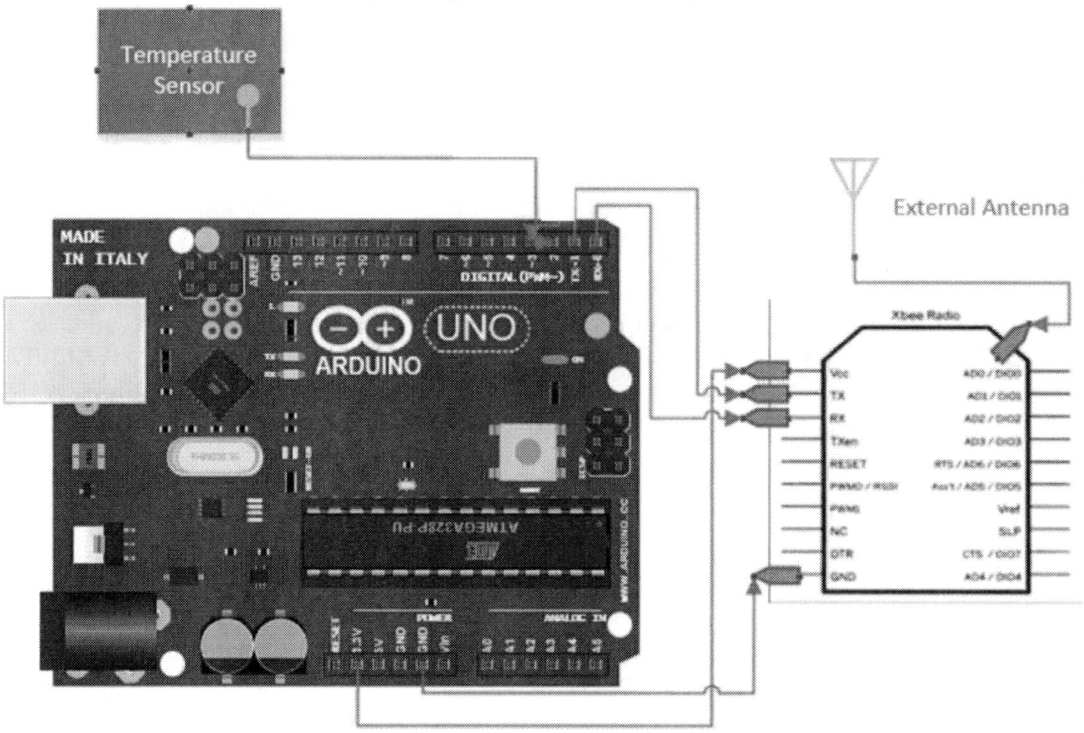

Figure 3: Component interconnections with Arduino for telemetry

The hardware connections are largely set up in a very direct and obvious way. The Xbee is connected to a custom shield for Arduino that lies directly on top of the Arduino Uno. The Xbee for the computer is connected directly to a USB dongle that was made for Xbees.

There are two Xbees; one with a SMA connector and one with a permanently attached wire antenna. While they are both interchangeable for placement with the USB dongle or Arduino Uno, the one with the SMA connector is to be used with the Arduino set up. This is because the 7.9-mm connector is to be used with an extended Wi-Fi antenna located outside of the board.

The temperature sensor is connected to two analog pins for data, and then 5-V power and ground. There are 4 unused pins on the actual temperature sensor for alerts and other unused options. The LED's anode is connected to a digital pin while the cathode is attached to a 220-W resistor that is attached to ground. This is all powered by a 9-V battery housed in a case. The case has a male 2.1-mm center positive barrel jack, that goes into the standard female 2.1- mm center positive barrel outlet of the Arduino Uno.

Lastly, all of these components are held together with industrial hook-and-loop fastener tape, Velcro, to insure a secure mount. It also allows for the Telemetry portion to be completely removed for sensor adjustment or added as an option.

Testing

The telemetry portion was evaluated through results and a distance test. The inflow of data can be seen in Figure 6. This was the data that was sent from the board to the XCTU program on the computer. The screen shot is of the console within the XCTU program after a "." keystroke was sent. It shows the output of the "." in blue, and the input of the temperature in red. Also, to be noted, the Arduino set the LED to be illuminated every time the data read that "." and transmitted back data. This can be seen in Figure 4. The returning data of the temperature shows not only that transmission was achieved, but also that the temperature readings were accurate. The temperature sensor was accurate when at room temperature within 1 degree. It was also able to rise accurately and quickly, but took over a minute to cool down if there was a quick influx of temperature. This is expected, as it is an *ambient* temperature sensor. We tested this aspect specifically with a heat gun.

After transmission was proved successful and the sensors were calibrated, the next aspect for evaluation was the maximum range of transmission. To do this, the telemetry Arduino module was attached to its 9-V power supply and taken away until the connection broke. Line of sight gave roughly 200-meters till no connection was available. The term roughly is used as the connection was in and out beginning at 178-meters, and completely out at 200-meters. Within a building, we were able to go from the fourth floor of Eaton Hall down to the second floor of Learned Hall's main hallway and still have connection for half of the length of the hallway. This direct line distance is 73.152-meters. It goes through 4 walls at minimum, and 2 stories.

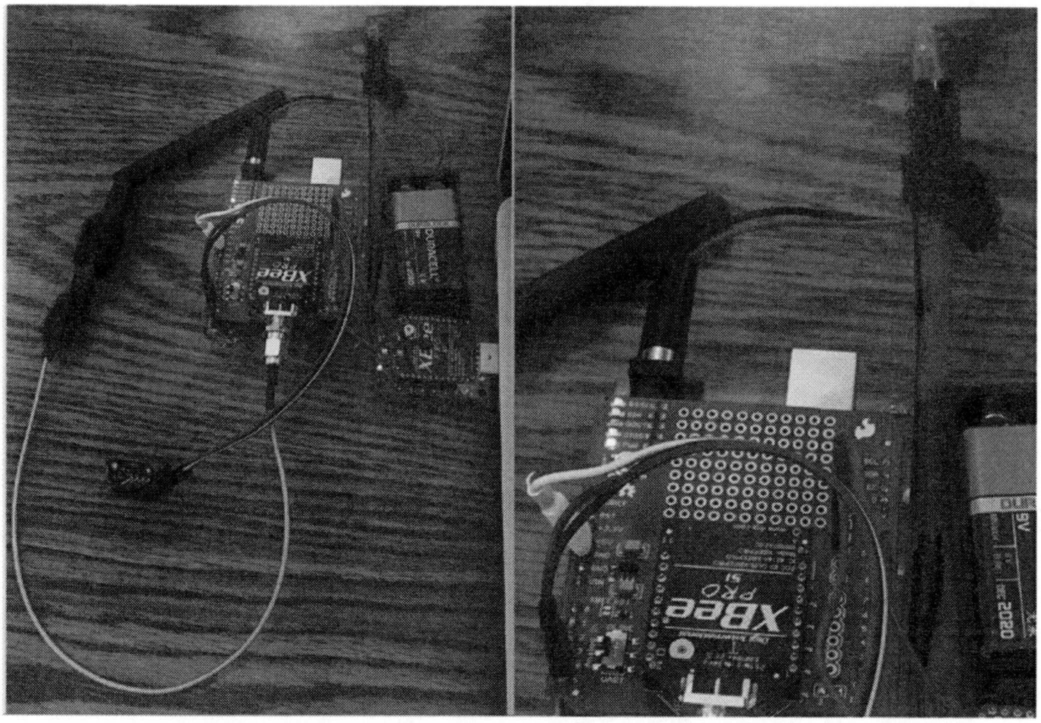

Figure 4: The complete circuit set up with illumination on right

Figure 5: The USB Dongle

```
. C: 22.12  F:71.71
. C: 22.06  F:71.82
. C: 22.12  F:71.71
. C: 22.06  F:71.71
. C: 22.06  F:71.71
. C: 22.06  F:71.71
. C: 22.06  F:74.97
. C: 23.87  F:79.59
. C: 26.44  F:81.61
. C: 27.56  F:83.07
. C: 28.37  F:83.97
. C: 28.87  F:84.65
. C: 29.25  F:85.21
. C: 29.56  F:85.66
. C: 29.81  F:86.00
. C: 30.00  F:86.34
. C: 30.19  F:86.68
. C: 30.37  F:86.34
. C: 30.19  F:85.44
. C: 29.69  F:85.10
. C: 29.50  F:84.87
. C: 29.37  F:84.65
. C: 29.25  F:84.43
. C: 29.12  F:84.20
. C: 29.00  F:84.09
. C: 28.94  F:83.75
. C: 28.75  F:83.52
```

Figure 6: The Transmission Data as seen from XCTU console

Evaluation of Test Data

The complete system worked as the data shows. An expected range for transmitting with proof of concept values was achieved. The temperature sensor was ambient so it rose relatively quickly, but was left to its silicon and metal makeup that restricted it from cooling down just as quickly. The Arduino was never pushed as far as capabilities go, so it performed as necessary. The XBees worked well for transmitting, and allowed for antenna extension. The antenna in our case, compared to other noise factors, added a dismissible gain advantage. Though it did provide an advantage of getting the antenna outside and away from the wheel and elements. The antenna could be shot or soaked, and it would not affect the rest of the transceiver components. More so, in a case where the housing interrupts a 2.4-2.5 GHz signal, the antenna proves mandatory. The cut off distances were expected and relatively close to the ones listed on the data sheet.

CONCLUSIONS

The complete system worked well and met all expectations. With 9 digital pins unused there is room for the addition of multiple other digital sensors. The extended antenna and portable battery allowed for safe and independent placement of the Arduino based XBee within the housing. If any modifications were to be made for the next model it would be within the capabilities of the microcontroller and XBee. A raspberry Pi matched with a XBee series 3 PRO would allow for much greater range and processing capabilities. Though the proof-of-concept mobile diagnostics system was a success.

REFERENCES

[1] Iswandi, H. T. Nastiti, I. E. Praditya and I. W. Mustika, "Evaluation of XBee-Pro transmission range for Wireless Sensor Network's node under forested environments based on Received Signal Strength Indicator (RSSI)," *2016 2nd International Conference on Science and Technology-Computer (ICST)*, Yogyakarta, 2016, pp. 56-60.
doi: 10.1109/ICSTC.2016.7877347
[2] H. Kumbhar, "Wireless sensor network using Xbee on Arduino Platform: An experimental study," *2016 International Conference on Computing Communication Control and automation (ICCUBEA)*, Pune, 2016, pp. 1-5.
doi: 10.1109/ICCUBEA.2016.7860081
[3] J. Á. Ariza, "A proposal for teaching programming languages through open hardware tools," *2016 IEEE 8th International Conference on Engineering Education (ICEED)*, Kuala Lumpur, 2016, pp. 202-207.
doi: 10.1109/ICEED.2016.7856072

A DISTRIBUTED SENSOR NETWORK FOR AN OFF-ROAD RACING VEHICLE

Kyle Boyer, Laura Brubaker, Kyle Everly, Richard Herriman, Paul Houston, Sean Ruckle, Rory Scobie, Ian Ulanday
Advisor: Dr. Michael Marcellin
The University of Arizona

ABSTRACT

The University of Arizona Baja Racing Team competes annually in an intense off-road racing competition. This year's car features a distributed sensor network capable of displaying useful data to the driver, the benefits and technical aspects of which are examined by this paper. Based on the ATmega2560 chip, the system is USB programmable, features hot-swappable batteries, and includes SMD components. Each sensor is custom designed, functions as an I2C slave, and contains its own ATtiny85 microcontroller allowing all the sensors to be addressable and enables them to be wired in parallel. The system also includes interrupts for almost every single sensor, which allows for more accurate data collection and guarantees that no important data will be missed. A custom-made board was created to connect these sensors and serve as a microcontroller data logger based on an Arduino reference design.

INTRODUCTION

In 2012, a few members had the idea that a speedometer would be a valuable addition to the vehicle. A speedometer would allow for constant monitoring of speed, which would enable the driver to make consistent turns and jumps. The following year, a new speedometer system was built, with the addition of a fuel gauge. The team also decided that receiving this data in real time would be extremely beneficial to the team. Therefore, the next system was designed to be capable of using radio communication to send speed and fuel data back to the pits, creating the team's first real time monitoring system. At the end of the year, it was decided that a separate subteam completely dedicated to the development of electronic and sensing equipment would be created. As the years have gone by, the Electrical Subteam has grown, gained experience, and increased its access to equipment. These resources have allowed for the system to develop into a completely custom sensor network capable of detecting and sensing dozens of parameters on the vehicle.

MOTIVATION

As a university club, we are first and foremost focused on education. As such, we approach each project with the goal of learning - and enabling others to learn - something new. At the same time, we strive with each project to improve our system in some way. As the complexity of our system increases, we are attempting to break up the project into a collection of smaller projects, each with reduced complexity, while also expanding the capabilities of the system.

DESIGN REQUIREMENTS

Durability

During any given race, the vehicle may encounter every type of terrain, from flat desert, to rock fields, to mud pits. To survive these conditions, all electronic device placed on the vehicle must be waterproof, dust proof, and strongly resistant to vibrations. To keep our devices waterproof and dustproof, we've implemented several different strategies. For anything we wanted to be accessible later, we designed housings that were either 3D printed or laser cut from polycarbonate, sealed with a silicone based gasket material. For devices that never need to be opened, permanent solutions are potting or conformal coating. In the interest of repairability, we preferentially use an ESD safe acrylic conformal coat over potting.

Connectors

In order to protect the internal electronics from dust and mud, all external connections in the system are made with IP68 rated (waterproof/dustproof) connectors. All wiring interconnects in the harness are achieved using either waterproof connectors or crimp connectors with integrated marine heat shrink tubing.

The kill switches on the vehicle work by shorting the spark plug contact to ground. When not engaged, unshielded kill-switch wire acts as an antenna. To avoid coupling between the kill switches and the data lines, we decided to use burial grade shielded CAT5e cable for all data transmissions on the vehicle.

While this does work, it turns out that ethernet insulation melts at a very low temperature. This is exacerbated by the copper data wires in the ethernet cable, whose high thermal conductivity cause the insulation to begin melting before the solder does. Because of this, we plan to restrict our use of ethernet cable to networking applications in the future, and use stranded wire for other data and power connections.

Modularity

We are constantly attempting to improve the system by adding additional features and removing bugs, so the ability to replace, add or remove entire subsystems without destabilizing the overall system is very useful. Further, by distributing the computational load across all devices in the system and using standard communication protocols wherever possible, we reduce the damage caused by a single component failing. This also reduces the additional development effort required to integrate additional sensors.

I2C

I2C allows all sensors on the vehicle (up to 112 - See Appendix 4) to operate in parallel on a single bus. This design decision was made after the electrical interface on the data logging and communications hub was finalized for this generation of the system. Using I2C reduces the number of pins required on the main board from 48 to 20, which will allow us to drastically reduce the footprint of the main board in future generations. I2C also opens up the possibility of integrating the microcontroller based data logger into a Raspberry Pi "Hat" as a means of further reducing the mechanical footprint of the board.

Unfortunately, the ATtiny85 does not natively support I2C (See Appendix 2). Using the ATtiny85 would require writing software to implement the protocol on GPIO pins. This functionality is not trivial to emulate in software, which is part of the reason we switched to the ATmega328 based Arduino Nano for the sensor implementations. We have since decided to switch from the ATmega328 chip to the ATmega32u4 for future implementations. This chip is used on the Arduino Leonardo, which is the same form factor as the Uno and has all of the same capabilities as the ATmega328. In addition to that, it has built in support for USB, meaning it does not require an external FTDI chip to be programed over USB.

Processing time

One of the major concerns with a growing number of sensors is the processing time required to poll them. While interrupts help by allowing some measurements to be made asynchronously, as the sensor load increases the interrupts begin to encroach on the functions that process and store data. Due to the small amount of memory available, data must be written to nonvolatile storage very regularly - roughly every 50 readings of the sensors. If this process is interrupted too often, it may not get a chance to complete and data samples may be lost. It also does not leave much room for calculations.

One solution is to use a microcontroller with more memory. Unfortunately, the ATmega2560 already the largest memory capacity of the Arduino supported 8-bit microcontrollers (See Appendix 1). Another solution - the one we decided to go with - is to use multiple microcontrollers to distribute the processing load and memory required to store multiple samples of data for averages and integrations. This data can then be polled at set intervals by a master device, which can then write the data to a microSD card or other nonvolatile storage.

Form Factor

The vehicle is a small scale single seat racecar, so form factor is a major concern. There is a limited amount of space available, especially with all of the standard racing equipment inside. There are only two small compartments that are available to place all of the electronics in. This forced us to minimize the PCB footprint.

SMD

In order to make the system more compact this year, almost all of the components on the circuit boards are surface mount, rather than through hole. Though SMD parts are more expensive and difficult to solder, we felt that it would be an excellent learning experience for our members. This did impose a learning curve, as no members of the team had experience in surface mount, but after a fair amount of research and practice, we were able to successfully mount all of the components. These included components that required reflowing or drag soldering, such as the main microcontroller which was a TQFP package ATmega2560 or the 28 pin SSOP package FTDI chip. Using these surface mount components made it possible to have more LED indicators and functionality, yet still maintain a relatively small form factor. We are in the process of sourcing inexpensive components and designing a board that is just to assist in training our members.

PCBA not shield

In previous years, an Arduino shield that was directly compatible with the Arduino Mega was designed and produced. As the system gained complexity, the shield's footprint exceeded the footprint of the Arduino Mega. Because Arduino is an open source hardware with open source schematics, it was possible to produce an Arduino compatible circuit board that includes the extra functionality that we previously added with a shield. This circuit board used the same microcontroller as an Arduino Mega, had all the circuitry necessary for the microcontroller to function, and all of the components that would have previously been added by a shield. One of the convenient features of developing a custom board, was eliminating the typical Arduino headers and instead being able to route all the outputs and inputs to a custom waterproof connector (Figure 1).

Figure 1 - Waterproof Connector Post-Race

Reliability

COTS batteries

This year the rules for the competition were updated to no longer exclude the use of lithium based battery chemistry. In previous years teams were restricted to sealed lead acid (SLA) batteries and this change allows for a much higher energy density. However, using lithium batteries comes with the added risks of overcharging or draining the batteries, which can result in the battery catching on fire or exploding. These risks are mitigated by using proper charge and discharge circuitry. In light of this, we chose was to use COTS lithium ion drill batteries as they are designed to be durable and have all the needed protection circuitry built in.

Power Supplies

Due to the high premium placed on weight in this competition, the batteries are necessarily quite small. To compensate, any unnecessary inefficiencies must go. For this reason, the linear regulators from the old system were replaced with higher efficiency switching regulators. To reduce weight, a single regulator for each voltage is mounted to the firewall and the output is routed to the rest of the vehicle through shielded wires.

Where switching regulators can achieve efficiencies above 90%, linear regulators can be approximated as voltage dividers for efficiency calculations, and frequently operate at less than 50% efficiency. The 12V to 5V converter on the board is intended solely as a backup in case the

5V power from the switching regulator fails. It operates at around 42% efficiency. The 3.3V regulator achieves 66% efficiency (See Appendix 3) by using a 5V input instead of a 12V input to reduce the voltage difference and therefore the wasted power.

Interrupts

One of the main reasons we created our own custom sensor modules was so that we could connect every sensor that was logically to an interrupt enabled pin. Using interrupts to collect the data means that no data points would be missed due to the system polling the sensors at the wrong time. When the sensor triggers the interrupt, the sensor module is able to store that data until the main board requests it. This means that the main board can poll the sensors without risk of losing any data points.

Availability

Battery swap

Last year, the telemetry system and brake lights both ran on a single SLA battery. Due to the small capacity of the battery, it had to be replaced every hour. Having the battery run down so quickly is a significant risk in this competition, as the car may be black-flagged (removed from the competition) if the brake light fails to function properly.

To help alleviate this concern, we have separated the electrical system into two distinct and isolated subsystems. One subsystem consists only of the brake lights and a small lithium ion drill battery. Based on the estimated power consumption of the lights, the battery should sustain several hours of continuous use. Since the brakes are only engaged intermittently, and the race is only 4 hours long, one battery is more than enough for the entire competition.

The second subsystem consumes far more power as it contains all of the telemetry and communications gear. For both ease of use and safety of the electrical system, we used 12V and 20V drill batteries for the brake system and telemetry system respectively. We also custom designed battery mounts so that the batteries could be removed and replaced quickly during pit stops.

Because the telemetry and communications system requires a non-trivial time (about two minutes to boot and reestablish connections) to come online after a power interruption, it is desirable to have a means of replacing the battery without a reboot. To that end, the system uses two batteries in parallel, with a diode in series with each to prevent them from back feeding into each other. Either battery may be removed without interrupting the flow of power to the load.

Compatibility

CAN

The CAN protocol is used in most commercial vehicles as well as a host of other products. In an effort to be compatible with as many devices as possible, the main board has a CAN controller and transceiver chip. The associated data pins are broken out to the board mount connector and can be used to connect any device that supports CAN.

USB Programmable

The ATmega2560 must be programmed before it is usable. Two options are available for this: ICSP and USB. Each has its own benefits.

Programming over ICSP is reliable and works with no modification to the system. The main drawback to this method is that it requires an external device to perform the programming operation. If the device must be programmed repeatedly, the ICSP header must be broken out to an accessible location.

Programming over USB requires burning a bootloader to the device, which reduces the system resources available for the program itself. The benefit of this method is that the microcontroller can be programmed over UART. Consequently, a host computer can both program the system and communicate with it easily after programming by using a USB to serial converter IC. A common USB to serial converter is the FT232RL by FTDI, which has freely available drivers.

The Arduino bootloader is the first thing that runs when the chip is powered and it listens for programming attempts on the UART receiver - if it detects one, it proceeds to program the microcontroller. If it doesn't find a device trying to program the microcontroller, the bootloader exits and the program that was previously loaded on the microcontroller runs. This means that the microcontroller can only be programmed immediately after it is powered.

There are a few ways to accomplish this. The method we used was to start uploading the program and hit the onboard reset button at the right time. This required exact timing, or the upload would fail. The other (more ideal) method allows the computer to reset the microcontroller. The programing sequence on the computer sends a command to the FTDI chip when the reset is needed and one of the FTDI's pins is pulled high, which causes the reset. On our first boards, the trace that connected that pin on the FTDI to the reset line was missing, which is why a manual reset was needed. We tested that adding that trace would fix the problem by soldering a jumper wire between FTDI pin and the reset line. With this modification, the upload process worked with no manual reset. We have added the trace on our new designs.

Microcontroller Choice

Arduino is one of the most popular and well-known platforms for rapid prototyping, and as such there are a wide variety of hardware and software components available for it. By maintaining electrical (but not mechanical) compatibility with the Arduino, the custom board can use libraries written for the Arduino with little or no modification.

ATmega2560

The ATmega2560 that we chose to use on the main data acquisition board is the same chip that is used on the Arduino Mega. We chose this microcontroller over others like the ATmega328 (used on the Arduino Uno and Nano) or the ATmega32u4 (used on Arduino Leonardo and Arduino Pro Micro) because it was the most versatile of the different chips we considered. With multiple UARTs, built-in support for I2C and SPI, and the largest memory capacity in its line, it provides a wide range of options for future expansion.

ATtiny85

This year the custom sensor boards used the ATtiny85 was the microcontroller. This is a small IC with only 8 pins in total. Its small size and commensurately low cost make it an attractive choice for this application. In addition to that, there are also existing open source designs that make use of it. The small package does come with some drawbacks, however. The ATtiny85 lacks a UART, reducing compatibility with devices. It also lacks native support for I2C, which meant to be used in our distributed sensor network we would have to emulate I2C in software, reducing the speed and reliability. I2C and SPI are actually both listed as supported on the product page, though third-party libraries are required to make use of either. We are looking at the ATmega32u4 as a more fully featured replacement with a similar form factor.

CONCLUSION

This year provided our team with a lot of learning opportunities. Designing our own microcontroller board was an entirely new challenge for our team, but each of our members came away from the experience with greater knowledge, skills, and appreciation for telemetry. This project has also opened doors for our team in terms of future advancement. Using 5.4 GHz was also a new advancement that taught many lessons, but also exposed the team to new ideas and methods.

ACKNOWLEDGEMENTS

Thank you to Dr. Michael Marcellin for making this project possible, our teammates on the Baja Racing Team for putting up with the billowing clouds of blue smoke and solder flux, and Josie Ruiz for always being there for us.

APPENDIX: DATASHEETS

1. Microchip Technology Atmel ATmega640/V - 1280/V - 1281/V - 2560/V - 2561/V
2. Microchip Technology Atmel ATtiny25/V - 45/V - 85/V
3. National Semiconductor LM1117
4. NXP Semiconductors UM10204 I2C Bus Specification and User Manual

Permanent Magnet Synchronous Motor Variable Frequency Drive System

Judah Schad, Cameron Nichols, Katelyn Brinker
Advised by: Dr. Kurt Kosbar
Missouri University of Science and Technology, Rolla MO, USA

ABSTRACT

This paper discusses a permanent magnet synchronous motor (PMSM) variable frequency drive (VFD) system developed for an all-terrain Wifi-HaLow connected (802.11ah, 900 MHz) modular electric vehicle that competed in the Mars University Rover Challenge (URC). The quadrature axis flux linkage for each motor was estimated using on-board voltage and current measurements. A synchronous control algorithm tracked the electromagnetic operating parameters, which are highly dependent on variations in motor construction and load conditions. A feed-forward model-driven observer solution calculated flux linkage angles by direct-quadrature-zero transformation of three-phase shunt currents using DSP processors.

INTRODUCTION

This paper is intended as a design-project/application-note for the VFD power inverter component of a modular IoT electric vehicle system, which was named Gryphon by the Mars Rover Design Team of Missouri S&T. Gryphon competed against other electric vehicles of similar nature in the URC, which tasks collegiate teams with developing astronaut assisting rovers. Rovers are required to retrieve and deliver objects in the field, deliver assistance to simulated astronauts, service complex equipment, engage in exobiological analysis, and traverse a wide variety of terrain, all under the assumption that the telerobots are being operated by astronauts, stationed at various base locations on Mars [1].

The six VFDs present in Gryphon were installed as one collective sub-system operated as one individual IoT endpoint with a "drive board," that was connected across a wide-area wireless backhaul access link. The system was used to monitor and control six permanent magnet synchronous motors, with the drive board acting as serial communications transmission master to aggregate, transmit, and receive remote vehicle telemetry and command target values for each motor over six independent two-wire asynchronous serial ports. Each of six VFDs independently receive their own serial command target values on separate duplex streams, and respond with both sensed and derived operating telemetry over each of the six serial interfaces to be aggregated by the Drive Board. The interaction between the user, drivetrain subsystem, and other vehicle subsystems was done over a WiFi HaLow (802.11ah) 6.5 watt radio, operating in the 33 centimeter band (900Mhz) [2].

Electromagnetic operating parameters must be independently monitored for each of the six PMSM's and are highly dependent upon variations in specific motor construction and load conditions []. In order to derive motor model transformations, each of the six individual VFDs utilized their own parallel-processing DSP core, to independently execute filter algorithms concurrently to each 32-bit floating-point real-time microcontroller co-processor. Voltage and current readings acquired by each DSP core corresponded to capture triggers of enhanced hardware peripheral sub-modules. Feed-forward model-driven observers then filtered real time data on each drive, and derived flux field angles by direct-quadrature-zero transformation of each motor's three-phase shunt currents.

VEHICLE SPECIFICATIONS

The MRDT purchased commercially available 24 slot stator, 8 pole rotor Neumotor 1907-480 PMSMs for Gryphon's drive motors. These motors maintain a 480Kv motor velocity constant (back EMF constant), and were originally marketed for large radio controlled (RC) hobby aircraft. When comparing PMSM for digital electric power systems, the Kv rating of a motor in RPM/Volt describes the "low end" response and intended operating power band. The brushless motor Kv is the ratio of the motor's unloaded RPM to the peak (not RMS) voltage on the wires connected to the coils (the back EMF). "Kv" is how many unloaded rpm the motor will run for each volt that is connected to it. High winding resistance, with two windings in series for each phase, provides high torque at low rpm, which is desirable for the terrain conditions faced by rovers at the URC.

The MRDT also fabricated a custom lithium manganese nickel cell 29.6V 25A nominal operation battery pack with a battery management and power distribution system that ran MDRT open source firmware, and connected to the the RoveComm network. The Drivetrain electronics were designed at 60V, 60A max ratings, in order to add electrical headroom, and create a reasonable margin of safety for competition operating conditions.

The drivetrain sub-system as well as other rover sub-systems are addressable over the 802.11ah Point-to-multipoint network. Wireless communication to the device is accomplished through the MRDT command/telemetry distributed communication software protocol, RoveComm, which acts as the IoT IP gateway for remotely publishing and subscribing telemetry and dispatching commands to individual subsystems over duplex UDP streams. RoveComm is an open source C++ library that is compatible with both the Arduino Wiring software framework and the Texas Instruments Launchpad Boosterpack hardware standard and is currently made available on the MRDT Github repository [2].

PMSM VFD CONSIDERATIONS

Electrical Machines convert electrical energy to mechanical energy using magnetic field interactions as the link between the two types of energy. Electric motors are unique when it comes to their power band, creating power across changes in this magnetic field linkage by different mechanisms with different transfer characteristics that vary greatly by motor type, power source, and available sensor feedback for various control schemes. PMSM's offer compelling improvements over other motor topologies, at the expense of added sensor feedback complexity and digital control requirements.

The advantages in PMSM's are well known when compared to other motor topologies, and include higher torque and speed bandwidths, lighter weight, smaller size, greater efficiency and increased dynamic response. The permanent magnets produce a high torque-to-inertia ratio, responding quickly to control-signal changes, making them useful in dynamic traction applications. Most motors produce their maximum torque at a particular speed range, called the torque peak. Above or below the torque peak, the motor produces less torque than that maximum value. PMSM's however, can operate at all speeds while maintaining the full torque output at the rated load. Assuming ideal flux alignment waveforms from the timing controller, system efficiency is high and any applied load within operating region will only change the phase angle of magnetic of coupling between rotor and stator to take the load, keeping the physical rotor speed constant. Any load attempting to slow the rotor will induce a slippage in the rotor field, and a corresponding current increase at the inverter, and as long as power output of the drive is not exceeded and the ideal flux alignment maintained, a compensating torque will be generated in order to maintain speed [3].

Disadvantages of the PMSM are primarily due to the choice of sensor used to fulfill data feedback requirement. When wired directly to any constant frequency power source, the PMSM will not spin. Sensor feedback and digital control are required to operate the motor. A PMSM is a synchronously commutated motor, meaning that its rotor spins at the same speed as the motor's internal rotating magnetic field. A timing controller must be employed to ensure the stator field and the rotor field maintain consistent linkage [4].

Three separate torus structures in the PMSM stator are wound in a symmetrical way, with three separate coil loops wound at an angle of 120 degrees from one loop to the next, wired together with respect to a shared system ground. The rotor is made of permanent magnets mounted around the rotor perimeter and aligned such that the field strength around the perimeter of the stationary rotor maintains a static magnetic field with a single true north polarity (negative) peak and a single true south polarity (positive) peak, at rest with respect to the rotor. For any position of the PMSM rotor, regardless of changes to load, speed, and supply frequency, there is a single direction of the stator field which maximizes torque, as well as a single direction which will produce no torque. When the rotor position is such that a pole of the rotor is near a pole of the stator, the poles repel each other and torque is produced. When the rotor position is such that a pole of the rotor is balanced between two poles of the stator, the poles balance each other, and no torque is produced [3]. In order to maintain flux linkage, the stator field must magnetically rotate in sync with the physical rotation of the rotor and rotor field. This dictates that the supply voltage requires knowledge of the rotor position, in order to determine which stator coils to energize, and leads to expense increases due to the sensor-connection cables and the electronic management needed to run synchronous commutation algorithms. A microcontroller is needed to energize the stator coils at the correct moment using input from sensors indicating the position and speed of the rotor [4]. Maintenance issues center on the lifetime and reliability of the sensors.

PMSM's with motor lead voltage ranges below 120 Volt AC are often also referred to as Brushless Direct Current Motors (BLDC) due to their intended applications, rather than their topology. In the lower voltage market, the PMSM is generally still referred to as a DC motor because it's coils are typically driven by a lower voltage DC power source and a crude trapezoidal waveform applied to the stator coils in a sequential pattern by some type of digital controller instead of the expected higher voltage continuous power available at the common three phase AC industrial installation.

Several techniques exist to operate the synchronous motor, all involving some form of digital control. The most popular technique commonly available in the low voltage market is to monitor the rotor position using Hall-effect sensors and apply a digital power commutation as a trapezoidal energizing sequence by switching the dc bus supply voltage in order to vary the current of the motor. By using Hall-effect sensors, motor flux linkage can be phase-found with zero initial movement. Bipolar Hall-effect sensors incorporate multiple magnets into a simple structure referred to as a ring magnet to measure voltage (back emf) on an internally applied current [].

In a typical low voltage permanent magnet motor, the three latching bipolar Hall sensors are mounted on the stator at 120 degree spacing with respect to the rotor axis of revolution. Latching Hall-effect sensors act as a sample and hold capture and require both positive and negative magnetic fields to operate. A magnet presenting a south polarity (positive) magnetic field of sufficient magnetic flux density will cause the device to latch to its on state, even after the magnetic field is removed. When a north polarity (negative) magnetic field of sufficient strength is presented, the device reverts again to an off state. Regardless of the number of stator coils and rotor poles, each particular possible position is only available once per rotation, describing one electrical cycle for each complete mechanical revolution. When taken together, the three binary outputs of the Hall-effect sensors form a three-bit word, resulting in 8 possible digitally coded position readings, with only 6 codes representing valid positions. Readings of "000" and "111" are typically considered illegal values, and may be ignored or flagged by the digital controller as a fault condition, indicative of a bad connector, a wire harness that has been cut, or the Hall-effect power supply has died. The motor's electronic controller may energize the three stator coils sequentially in response to the Hall readings, turning off on one phase and simultaneously switching on the other two phases at each state. This creates trapezoidal waveforms via a process known as commutation to generate a rotating electric field [4].

To produce torque on the shaft, the timing between control current flowing through the coil and the position of the shaft must be as accurate as possible. It can be easily seen that with a six-step sensor technique there is no possibility of keeping the angle between the rotor magnetic field and the stator magnetic field precise. The real angle varies from 60° to 120°, leading to inefficiencies and instabilities. Any error in the Hall-effect sensor switching point and any delay in the sensor's response to changes in the magnetic field leads to lower bandwidth and accuracy. Inaccuracy in the commutation in lower effective torque and introduces a torque ripple commonly referred to as cogging torque, that will impact stability, reliability, efficiency of the overall system performance [4].

Pulse Width Modulation, or PWM, generates a chopped output waveform from a steady voltage by periodically switching transistors on and off []. Each of the three motor terminals are switched, either at the voltage supply, or the system ground, for any given instant. By it's nature, a motor is effectively a passive integrator, or inductive low pass filter, and when the PWM is intrinsically filtered by the inductance in the motor, the power output waveform is the average of the applied voltage over time. By varying the duration of the periodic on state, a variable voltage may be applied, allowing voltage control of the motor's speed. Motor terminal PWM, however, is not generally viable on the three phase outputs of small, high speed synchronous motors, as they are typically designed with very low inductance and high resistance. If PWM is applied by the power transistors of the three-phase inverter to the motor terminals on such a motor, the current waveform will rapidly and frequently magnetize and demagnetize the metal, causing high thermal losses and magnetic hysteresis [3].

Electronic speed controllers intended for these motors, typically utilize a second PWM signal at a high frequency (above 10Khz) coupled to an inductor and the capacitor to set the DC supply voltage itself at the desired level, called variable DC link. The motor power stage still uses the same six power transistors to fully turn on and off the three-phase commutation, however the DC bus voltage level is controlled by the two additional transistors in the variable DC link. The variable DC link six-step inverter controls the base voltage on the motor by filtering high frequency PWM, while the discretely switched commutation is performed by the three-phase power phases of the inverter. The digital controller can buffer the captured the Hall-effect sensor readings and utilize them in the voltage scalar controller. Each new arriving edge of any Hall sensor signal calls the interrupt routine commutation algorithm. As there is a delay between the Hall sensor edge and the current commutation, the current is not symmetrical, and commutation advance is usually generated using a timer countdown in order to compensate. The timer countdown period is calculated using a lookup table of timing sequences for each commutation sector and the time between two particular commutation edges in the previous step. Two problems arise that usually necessitate additional sensors. First, as the Hall-effect sample rate is a function of the rotor rotation rate, and at low speeds, when the motor is going really slow, the signal gets contaminated with quantization noise, and the decrease in reaction time leads to complexity and instability. Second, in drive train applications, low end torque, which is proportional to current, not voltage, can be quite high, leading to large spikes in current draw. To address these issues, current sensors are usually added, in order to provide the required current mode [].

For motors with a high enough inductance and low enough resistance, motor terminal PWM may be employed. A PWM sinusoidal commutation sequence improves operation, but usually necessitates more sensor complexity. Sinusoidal commutation operates in much the same way as the traditional six-step trapezoidal commutation, but typically utilizes an higher resolution sensor or encoder to measure rotor position, and a lookup table of trigonometric values to determine the PWM commutation sequence [4].

One of the most economical techniques used to measure motor current is to use a precision resistor placed at a strategic location in the motor drive circuit, with one end of the shunt resistor connected to the ADC's ground reference, usually grounded to the negative rail of the drive's DC bus to eliminate any common mode signals. In a three phase motor, assuming no winding shorts, it is only necessary to read two of the motor currents, since the third current will be the negative sum of these two currents. Some drives use an optional third resistor shunt so they can detect winding shorts [].

Motor and Hall sensor geometry inaccuracies cause many unwanted conditions to occur. If high voltage builds up on the DC supply, the variable DC link inverter will switch to connect the bus to a drain resistance to ground, called a brake resistor, to reduce voltage. In the case of overvoltage, undervoltage, overcurrent, or incorrect commutation edges within a certain number of commutations, a base block circuit may be utilized to block the base of each power transistor and enforce an off state. A fault LED may also be displayed at the controller [].

Most externally mounted incremental encoders measure the rotational motion of the axle shaft. The angles are measured at constant time intervals, giving the measurements needed to compute speed, delta angle, and delta time. Speed is computed by dividing the delta angle by the delta time. With an integer sine and arcsine lookup table stored in memory and integer computations used for speed, computed trigonometric relationships can determine the magnitude of the PWM duty cycle to approximate a sinusoidal waveform during commutation. Position information from the encoder is used to synthesize two sinusoids, one being 120 degrees phase shifted from the other and these signals are then multiplied by a throttle or torque command so that the amplitudes of the sine waves are proportional to the desired output. Unlike Hall sensors, high-resolution incremental optical encoders only measure changes in relative position. Direction, velocity and acceleration may be derived in software, but it is not possible to determine the absolute position of a stationary axle, at rest with respect to the encoder [].

Successful motor commutation requires some initial phase-finding technique when the location of the field vector unknown at power-up. The Stepper phase finding technique consists of energizing the stator winding and tracking the initial rotor movement to the nearest magnetic pole. To avoid incorrect phase finding at a null position of zero torque condition, the technique implements a small open-loop commutation move and allows the motor to cog and stutter slightly before encoder acquisition occurs, and the closed loop commutation sequence can begin. Even with a high resolution position encoder where precise velocity information may be available, at slow speeds the signal is likely to be contaminated with quantization noise []. The encoder will also increases system cost and complexity. Encoder cables and mounting considerations add critical points of failure to the system.

While a high-resolution encoder driven sinusoidal commutation is an improvement over Hall-effect driven trapezoidal commutation, it is still not the optimal control sequence for the PMSM. The motor is an inductive system with complex field linkages changing with respect to time. Any load induces temporary slip in the rotor flux linkage with respect to the rotor. Instantaneous voltage is the applied control variable, however it is only the instantaneous component of the current waveform that lags the rotor flux at 90 degrees as transferred to the rotor via the rotor-stator field linkage that contributes to motor torque output. All other voltages and currents are converted to system inefficiencies, such as heat, and cogging torque. In sinusoidal commutation, commutation is performed first and is followed by some control of the resulting sinusoidal current command signals. The controllers in a sinusoidal system are therefore exposed to the time variant currents and voltages of the motor, and motor performance is limited by bandwidth and phase shift of the controllers [].

MRDT PMSM VFD

Flux field orientation seeks to optimize the torque production of a PMSM by utilizing real time matrix transformations and filters to precisely separate the torque producing flux linkage components from the flux linkage components that only contribute to cogging, heating, hysteresis, speed, and other forms of parasitic power absorption []. The transform operations are well known, and in order to efficiently execute the mathematical calculations and abstract the application from the details of the implementation, these operations have traditionally been implemented in custom OEM ASIC's, or a third party license agreements for FPGA IP cores, to be installed in industrial drives alongside microcontrollers, typically in stationary installations, and typically at standard industrial voltages of 120, 220, and 408 VAC.

The cost of digital signal processing cores capable of hardware accelerated mathematical operations that model motor flux in real time has fallen significantly over the last decade, primarily due to Moore's law, which describes the exponential growth in silicon processing and computing density that has occurred over the past half century. Recent digital motor control DSP hardware solutions now often come preloaded with many parameterized application programmer interface definitions, allowing access to Read Only Memory (ROM) function calls for object oriented instantiation of common field orientation modules related to flux tracking operations.

The 2017 MRDT PMSM VFD leverages the vendor specified algorithms available in ROM on the Texas Instruments C28x DSP core, utilizing hardware accelerated implementations of rotor flux position and motor parameter estimating sliding mode observers, tightly coupled to Clarke and Park transforms feeding current mode PI filters. With only simple and direct PWM triggered ADC measurements internal to the drive itself, optimal dynamic torque response is provided without external sensor readings, thereby eliminating the added cost, complexity, and fragility of the various motor shaft sensor implementations, while still providing an improved motor performance when compared against traditional low voltage synchronous commutation schemes. By acquiring the position of rotor flux, stator current vector, and their linkage components at orthogonal angles, the VFD may derive the optimal applied signals necessary for optimal torque production at the motor.

The flux oriented coordinate transformations are collectively referred to as the DQ transform, and this technique is able to realize the optimal control sequence for the PMSM by leveraging the ability to align a complex plane consisting of a direct (D) axis and quadrature (Q) axis oriented with respect to the flux vector. The result is transformation from that of a complicated flux coupled AC motor magnetic field model, into a simple linear system of a two axis. A direct flux axis represents a single scalar motor flux variable responsible synchronous speed, and the second axis represents the single scalar motor flux variable responsible for flux linkage induced torque. Speed and Torque producing flux may therefore be directly monitored and controlled independently. As part of the coordinate transformation, an estimation of the motor's speed and the position of the rotor can also be derived from the observed stator currents, effectively removing the need for any encoder from the design, and reducing the system cost and fragility [4].

The zero-direct-quadrature transform, also called the alpha-beta transform, is often used in the context of electrical engineering with three-phase circuits to rotate the reference frames of AC waveforms such that they become DC signals. In reality, any three phase values of the PMSM are oscillating in such a way that the net vector is spinning about the Z axis, in a balanced system. From Trigonometry, in order to specify a vector's magnitude and angle, a coordinate system with only two axis is needed, removing the redundant phase. It is possible to rotate the reference frame, and therefore model the machine as a single sine axis coil and a single cosine axis coil, such that changes due to net vector spinning are canceled out. Simplified calculations can then be carried out on these DC quantities before performing the inverse transform to recover the actual three-phase AC results [4].

The Clarke transformation is a motor specific implementation on the zero-direct-quadrature transformation, modeled as a tensor that rotates the reference frame of a three-by-three element matrix in an effort to simplify analysis. Typically, motor terminal currents are measured by triggering ADC channels at the peak of the PWM cycle, at the instance where all upper power switches are on, to avoid any PWM transients. This requires synchronous trigger channels from the enhanced PWM modules to the ADC modules. The two shunt currents for an isolated stator are scalar values readings are taken for each phase, added together, and the sum is negated, in order to derive the third phase. This is then reflected about the magnetic axis to get the third vector. The three current vectors exist in the stationary coordinate frame where a, b and c are the phase axes, with α and β being a fixed Cartesian coordinate frame aligned to phase a. The forward Clarke transform operation uses three multiplications and a single subtraction to convert the three balanced abc phase currents into two phase-balanced $\alpha\beta$ currents. Despite the effectiveness of this transformation, many complications still arise. The initial dynamics of the system are highly nonlinear, and the rotor flux is not usually measurable. Quantities such as the rotor resistance value vary considerably with conditions, and have a significant impact on the system. When tracking a variable speed signal with respect to a stationary signal, an inherent sample and tracking skew is created, inducing phase delay to the system. In the $\alpha\beta$ frame, the expression of torque still depends on the position of the rotor flux and still prevents any easy solution of the electrical differential equation [4].

The inputs to the flux estimator are the motor output phase currents and voltages expressed in α-β coordinate frame, and it is possible to fully reconstruct the system state from these output measurements, using a filter based estimation technique to track a series of measurements observed over time.

Tracking Filters, or Alpha/Beta filters, are second order low pass Infinite Impulse Response filters in which feedforward compensation is added to minimize the phase delay. When a simple second order low pass Infinite Impulse Response filters is implemented as two cascaded integrators, the output of each stage can be monitored and act as the derivatives of the tracking variable. For example, in a cascaded two pole IIR filter tracking rotor position, the rotor velocity may be taken of the first integrator feedback section, and rotor acceleration may be taken off the second integrator feedback section, effectively estimating two variables with respect to the filtered measurement variable [4].

Sharp-cut-off notch filters can help eliminate narrow-band mechanical resonance, removing energy that would otherwise excite resonant modes and possibly make the system unstable. In a second order filter, a zero may be added at the same location as the unstable pole to cancel the pole by pole-zero cancellation, effectively converting the system to a more stable first order response [4].

In the standard series two pole amplifying integrator realization of the PSMS tracking filter, complex poles may be avoided, by setting the amplitude of the inner gain to the same quantity as the R over L of the motor under operation. R over L refers to the ratio of resistance and inductance values that would be measured over the total inductive path of the stator winding, and is effectively the inverse of the system's time constant. Once this operating point is realized, the amplitude and phase of the filter may be set independently. Any gain change in the tracking system shifts the amplitude ratio up or down, but does not affect the phase angle. Any change in the time delay affects the phase angle, but not the amplitude ratio [5].

The Park transformation is next adopted in order to completely eliminate the effect of time-varying inductances, by combining all stator and rotor quantities into single rotating reference frame. The Park operation transforms the two-axis orthogonal stationary reference frame quantities into rotating reference frame quantities. Similar to jumping onto a merry go round before taking photos of a subject, the transform jumps the $\alpha\beta$ stator field reference on the rotating frame of the rotor field, resulting in a two phase DQ system [5].

In a two-phase DQ system that rotates at the electrical speed of the rotor, the D axis is directly aligned with the electrical position of the rotor flux and produces no torque. In this frame, the electrical expression of the torque becomes identical to the phase alignment of the Q axis, and is independent from the rotor position. In practice, the Forward Park transform takes the angle of the rotor flux and calculates trigonometric expansions by utilizing lookup tables pre calculated for the expected operating ranges [5].

Finally, discontinuous control may also replace the equivalent continuous measurement information about the unmeasured flux states. By switching estimate states asymptotically closer to true unmeasured values, the Sliding mode brings the estimated state error to zero in finite time. When employing the flux estimator to replace a shaft encoder, the location of the field vector is still unknown at power-up. Consequently, the field vector location must be determined using some initial phase-locking technique. The Dither technique locates the field position through high frequency (10k+) PWM injection. By setting a known stator vector orientation with a very high frequency at very low duty cycle, the filter can wait several microseconds to observe the initial direction of the field acceleration, as calculated from changing current response. Based on the initial acceleration, the angular extent of the region containing the armature field vector can be continually reduced in the sliding mode, by repeated process of successive approximation. The Dither technique is only applied to a static rotor at rest, and once the system has been initialized to a tracking state, the Field Orientation routine takes over, and the Dither ceases. The goal is to have smallest possible time and position change during the sequence of dithering. When the Dither is initialized, the sliding mode quickly converges the system by application of a discontinuous control signal, in order to force the system to slide along the normal continuous behavior, in discrete steps. The estimation error switches rapidly from one continuous structure to another with some inherent zig-zag like overshoot, before converging to a localized window of stable observation. This nonlinear high gain observer method has attractive noise resilience properties and provides quite similar to the common Kalman filter. Despite the statistical noise and other inaccuracies, good acquisition of a field orientation can be made by the model-driven filter to produce estimates of unknown variables that tend to be even more accurate than those based on a single physical shaft measurement alone [5].

For motors with unknown parameters, equivalent measurement information about the unknown resistance and inductance values of the motor can also be estimated, using identification mode of the controller. Successive series of high frequency injection vectors are applied, first in the voltage mode, and next in the current mode, to the motor terminals at very low average voltages. The small deltas in the current readings drawn during the injection are utilized to perform run time parameter identification routines, commonly referred to as a motor Autotune routine, or Motor Identification routine. In the same manner that the Dither technique leverages the inductive time response of the motor and the convergent tracking capability of the sliding mode, the Motor Identification routine repurposes these same filters to calculate system parameters, by monitoring the system response of the predetermined motor terminal signal injection test sequences. Parameters needed by the real-time responsive control mode of the drive are extrapolated by the Autotune routine, and saved in non-volatile memory. Autotuning a motor is typically recommended upon first time connection to any drive capable of executing the routine [5].

CONCLUSION

In the 2015 URC, the MRDT fielded a Rover utilizing six commercially available trapezoidal digital motor controllers, marketed for electric scooter applications, and utilizing a Hall-effect commutated controller without adequate overcurrent sense and protection. Hall-effect sensor failure was a major source of a multitude of system fault conditions. Cable runs from the sensor were plagued by intermittent connections leading to electrical jitter and motor stutter. Errors in Hall-effect readings led to improper timing sequences and cogging torque. Misaligned commutations increased as discrete segment switching motor cogging became more pronounced with increases in quantization noise. With inadequate overcurrent sensing, fault conditions led to back emf build up across the DC Link Bus. With no motor telemetry available from the drives, the operators continued to drive the vehicle beyond specified overcurrent limits and through periods of high jitter commutation misfires. These effects were especially pronounced at low or zero speed, even for light loads. Without adequate protection diodes to freewheel and drain off reverse current, the power transistor stages become back-biased, the capacitors vented, and the integrated circuits smoked from thermal runaway, catastrophically destroying a motor drive during the competition's terrain traversal, and leaving the Rover unable to finish the task.

In 2016, MRDT fielded a more expensive commercially available Hall-effect commutated trapezoidal digital motor controller with added overcurrent sense and protection. Hall-effect cabling issues still led to overcurrent conditions and backfires in the commutation and lack of telemetry still led to operation of the vehicle beyond specified overcurrent limits. The added overcurrent protection functionality did, however, register the fault condition, shutting down and protecting a motor drive from failure during the 2016 competition. The Rover was only able to recover by remotely power cycling the drive's individually regulated bus via a separate power controller IoT endpoint subsystem, thereby clearing the run-time fault condition and allowing the Rover to complete the task.

The most recent 2017 URC, Gryphon was the only Rover fielded to operate a model-driven observer tracking the quadrature axis flux linkage of each individual motor. As a result, Gryphon showed a dynamic torque response on the Mars-like terrain. Gryphon also presently has a variety of embedded endpoint IoT device subsystems that may benefit in the future from the data acquired by model-driven observers at each motor. As new iterations are developed, captured motor data from competition tasks may be analyzed to better target operating conditions for specific motors and specific terrain. An autonomy controller on Gryphon presently has access to a GPS, IMU data, and a preloaded satellite acquired map of the competition terrain. In the future, by subscribing to motor data locally, topological terrain where high torque output or instantaneous power applied is measured may be tagged by autonomic software routines, in order to facilitate terrain learning classifications. Motor capabilities may be directly incorporated into strategic navigation of the terrain, with the rover itself cataloging areas to approach indirectly, to approach at speed, or to avoid all together. The modular drivetrain system provides a framework for which to developed added value for a distributed network of electric vehicle capabilities.

REFERENCES

[1] The Mars Society, "Requirements & Guidelines," [Online]. Available: http://urc.marssociety.org/home/requirements-guidelines.

[2] Missouri S&T Mars Rover Design Team Official GitHub [Online]. Available: https://github.com/MST-MRDT/.

[3] Rik De Doncker Duco W.J. Pulle André Veltman Advanced Electrical Drives Analysis, Modeling, Control

[4] Werner Leonhard , "Control of Electrical Drives", Springer-Verlag Berlin Heidelberg New York 2001, Accessed 2017, June 1.

[5] Teaching Old Motors New Tricks, Part 1 thru 5
Dave Wilson, Texas Instruments, Published Feb 19, 2014, Accessed 2017, June 1
https://training.ti.com/teaching-old-motors-new-tricks-c2000-5-part-series

[6] Teaching Your PI Controller to Behave, Part 1 thru 10
Dave Wilson, Texas Instruments
https://e2e.ti.com/blogs_/b/motordrivecontrol/archive/2015/07/20/teaching-your-pi-controller-to-behave-part-i

TELEMETRY AND DATA LOGGING IN A FORMULA SAE RACE CAR

Aaron Schultz
Dr. Michael Marcellin (Advisor)
The University of Arizona

ABSTRACT

The problem with designing and simulating a race car entirely through CAD and other computer simulations, is that the real world behavior of the car will differ from the results outputted from CFD and FEA analysis. One way to learn more about how the car actually handles, is through telemetry and data logging of many different sensors on the car while it is running at racing speeds. This data can help the engineering team build new components, and tune the many different systems on the car in order to get the fastest time around a track as possible.

KEY WORDS

Formula SAE, Data Logging, Data Acquisition, Telemetry

INTRODUCTION

Formula SAE is an engineering competition where students are given the task of designing, building, and testing an open wheel, autocross racecar. Teams must work within a strict set of rules to optimize their car for a low cost, high performance setup. Each year, a new car is built entirely from scratch, using knowledge and data gathered from previous testing.

The goal for this system is to accurately log data to an excel spreadsheet, and then display that data to a simple GUI so that an engineer can see what is currently happening with the car. This sensor data is critical, because computer simulations are not always completely accurate. By finding discrepancies between these computer simulations and real-world data, an engineer can change the setup of the car until they reach peak performance.

There are a few different parts to the data acquisition (DAQ) system. First, the different sensors around the car measure parameters such as acceleration, wheel speed, steering angle, etc., and output different voltages depending on the measured value. Second, the Engine Control Unit (ECU) measures these sensors, and controls outputs such as fuel injection and shifting. The ECU also sends data to the Data Logger through a Controller Area Network (CAN) Bus. The third part of the system is the Data Logger, which records sensor and ECU data to help engineers tune other parts of the car's handling. Lastly, a laptop is needed to be able to view recorded values from the Data Logger, and manipulate outputs in the ECU.

INITIAL SYSTEM DESIGN

The DAQ system had to be designed with a few constraints in mind. First, the system had to be reasonably priced, with a maximum cost of $500. Second, the system had to be lightweight, as any extra weight in a racing car causes slower lap times, even if only by a few tenths of a second. Third, every part of the system had to give a decent performance gain, because having more sensors does not justify slower lap times. Finally, the system could not be overly complex, because of the limited time for research and production. The decision matrix below was used to justify which sensors would be helpful. The scoring ranged from 1 to 5, where 1 is the least ideal, and 5 is the most ideal. Only sensors with a score higher than 12 were used.

Sensor Description	Cost	Weight	Performance	Complexity	Total
Steering Angle	5	5	3	5	18
Suspension Travel	2	4	4	4	14
Accelerometer	4	4	5	4	17
Gyroscope	4	4	5	4	17
Thermal Tire Cameras	1	1	4	1	7

Tire Camber Measurement	2	3	3	3	11
Wheel Speed Sensors	3	4	2	4	13

The second challenge with our system was sending data from the Data Logger, and storing it on an excel spreadsheet on a computer. Our team found two ways to tackle this problem. The first way to transmit data would include transmitting through a WiFi antenna on the car, and into a WiFi antenna directly connected to a computer. The second method included setting up two arduino uno modules that could send data over bluetooth radio. We scored the decision matrix below from 1 to 20, with 1 being the most ideal and 20 being the least ideal solution.

System Description	**Parts**	**Cost**	**Complexity**	**Total**
WiFi antenna		9	10	22
	Mobile Antenna	5	4	
	Base Station Antenna	8	5	
Arduino Radio		8	12	20
	Mobile Arduino	4	6	
	Base Station Arduino Uno	4	6	

CONCLUSION

Since our telemetry system is still in progress, we can’t test our data transmission feature yet. So far, we have done extensive testing on reading sensor data accurately and logging the results. This feature has already proven to be helpful, as we have been able to tune our car to compensate for discrepancies between our simulation data and our real world data. Our team is working to quickly develop the wireless transmission feature, because this will help our engineers fix various problems much faster.

OPTIMIZING CODED 16-APSK FOR AERONAUTICAL MOBILE TELEMETRY

Michael Rice, Chad Josephson
Department of Electrical & Computer Engineering
Brigham Young University
Provo, Utah, USA
mdr@byu.edu, chadcjosephson@gmail.com

Erik Perrins
Electrical Engineering & Computer Science Department
University of Kansas
Lawrence, Kansas, USA
esp@ku.edu

ABSTRACT

This paper investigates the application of 16-APSK modulation to aeronautical mobile telemetry. The peak-to-average power ratio vs. code rate tradeoff is mapped to an optimization problem involving spectral efficiency and the constellation parameters. The optimization results produce a theoretically optimum solution that is 3.95 times more spectrally efficient as uncoded SOQPSK-TG. When implementation losses and the available IRIG 106 LDPC code rates are factored in, the advantage drops to 3.20 times the spectral efficiency of SOQPSK-TG.

INTRODUCTION

In the wake of recent spectral reallocations, such as the AWS-3 auctions that reallocated 1755 – 1780 MHz and 2155–2180 MHz from US Government use to commercial wireless services [1], spectral efficiency in aeronautical mobile telemetry (AMT) has become an even more important topic. One of the efforts to increase spectral efficiency has been an investigation of the application of non-binary linear modulations in the AMT environment. But linear modulations achieve improved spectral efficiency (relative to SOQPSK-TG) at the expense of amplitude variations, quantified by the peak-to-average power ratio (PAPR). Whereas SOQPSK has a PAPR of 0 dB, band-limited linear modulations have a PAPR greater than 0 dB. This necessitates lowering the

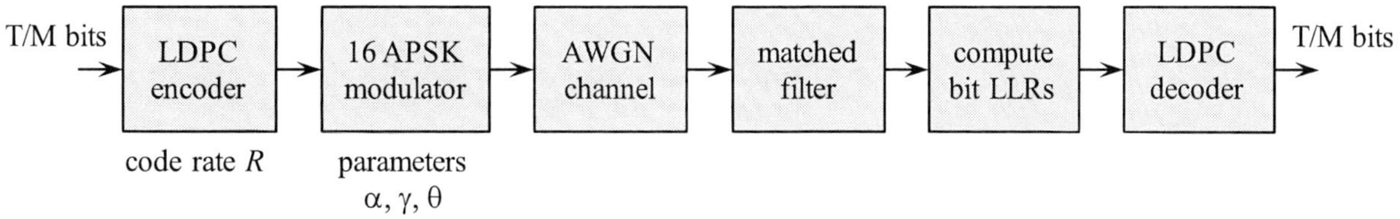

Figure 1: A block diagram an LDPC-coded 16-APSK system for aeronautical telemetry.

operating point of the RF power amplifier to accommodate the amplitude variations without distortion. Lowering the operating point reduces the available transmit power which, in turn, reduces the link margin or transmitter-to-receiver distance. This loss can be recovered using an error correcting code. But error correcting codes increase the bandwidth and therefor reduce the spectral efficiency gains. It might be that the code rate required to compensate for reduced RF power produces a system whose spectral efficiency is not appreciably better than that of uncoded SOQPSK-TG!

Motivated by the good properties of the APSK family of modulations adopted by the digital video broadcast, system 2 (DVB-S2) satellite standard [2], Shaw and Rice [3] explored the use of 16- and 32-APSK with turbo codes in the AMT. The results showed that 16-APSK with rate-$4/5$ turbo codes is able to achieve spectral efficiency approximately 3 times that of uncoded SOQPSK-TG.

Here the performance of 16-APSK with the LDPC codes defined in Appendix R of IRIG 106-15 [4] is explored. The impact of pulse shape and constellation parameters on PAPR is quantified and used to describe the PAPR vs. code rate trade-off. This trade-off is recast to an optimization problem involving spectral efficiency and the constellation parameters. The optimization is used to identify the 16-APSK parameters that maximize the theoretically optimum code. The solution is a set of modulation and code parameters that is 3.95 times more spectrally efficient than uncoded SOQPSK-TG. When implementation losses and the limited IRIG 106 LDPC code rates are factored in, the advantage drops to 3.20 times the spectral efficiency of SOQPSK-TG.

PROBLEM FORMULATION

The block diagram of the system under consideration is shown in Figure 1. Telemetry bits are encoded using an LDPC encoder with rate $0 < R < 1$. The coded bits are used to modulate a carrier using 16-APSK. The complex-valued lowpass equivalent of the 16-APSK modulated carrier is

$$s(t) = \sum_k a_k p(t - kT_s) \tag{1}$$

where $a_k \in \mathcal{C}$ is the k-th symbol which is one of the constellation points in the set $\mathcal{C}$ shown in Figure 2, $p(t)$ is a unit-energy version of the square-root raised-cosine pulse shape [5] with excess bandwidth $0 < \alpha \leq 1$, and T_s is the symbol time.

The received signal is $r(t) = s(t) + w(t)$ where $w(t)$ is the additive thermal noise modeled as

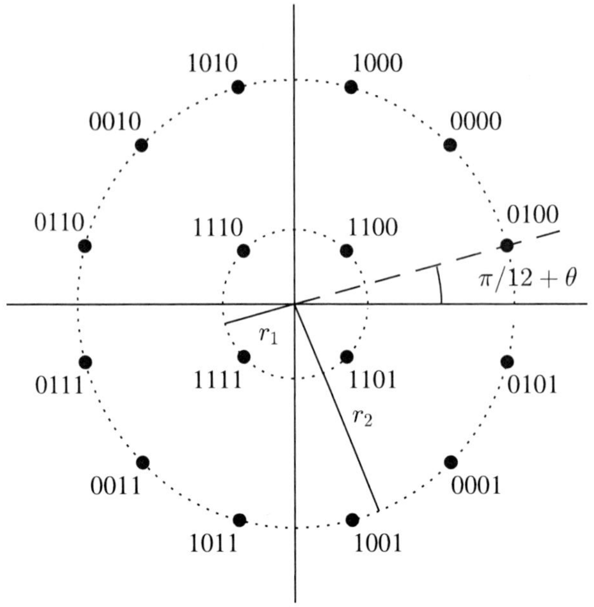

Figure 2: The 16-APSK constellation with the DVB-S2 bit-to-symbol mappings.

a complex-valued normal random process whose in-phase and quadrature components are real-valued wide-sense stationary normal random processes with zero mean and power spectral density N_0 W/Hz [6]. The received signal is matched filtered and sampled at the optimum sampling instants (perfect synchronization is assumed). The log-likelihood ratios (LLRs) are computed for each of the four bits corresponding to each symbol. The log-likelihood ratios form the inputs to the LDPC decoder. The LDPC decoder outputs are estimates of the input bits. The bit error rate is measured by the difference between the bits at the LDPC encoder input and the bits at the LDPC decoder output.

The block diagram shows the four parameters to be optimized in this paper.

MAJOR FACTORS INVOLVED IN THE OPTIMZATION

A. *CONSTELLATION PROPERTIES*

The 16-APSK constellation shown in Figure 2 comprises four points on a circle of radius r_1 and twelve points on a circle of radius r_2, with $r_2 > r_1$. The points on the outer circle are offset by the angle θ as shown. The constellation is parameterized by $\gamma = r_2/r_1$ and θ. For example, the DVB-S2 standard defines $\gamma = 3.15, 2.85, 2.75, 2.70, 2.60$, and 2.57, all with $\theta = 0$, for $R = {}^2/_3$, ${}^3/_4$, ${}^4/_5$, ${}^5/_6$, ${}^8/_9$, and ${}^9/_{10}$, respectively, for 16-APSK [2].

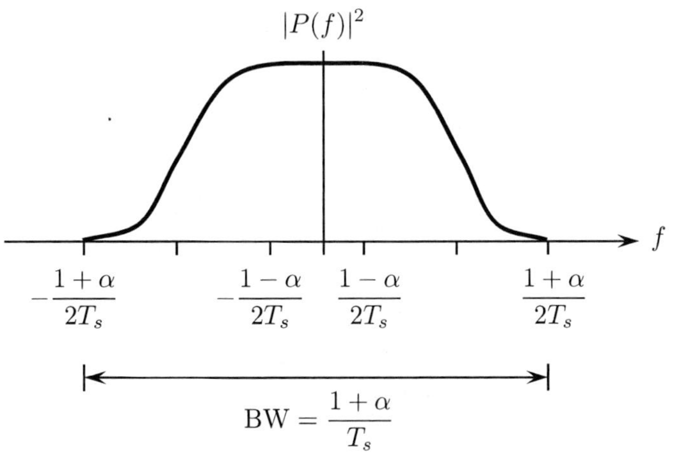

Figure 3: The frequency-domain representation of the SRRC pulse shape.

The average energy of the constellation is

$$E_{\text{avg}} = \frac{1}{16}\sum_{m=0}^{15} |c_m|^2 = \frac{r_1^2 + 3r_2^2}{4} = \frac{1+3\gamma^2}{4} r_1^2, \tag{2}$$

where $c_0, \ldots, c_{15}$ are the 16 complex-valued points in the constellation $\mathcal{C}$.

B. PULSE SHAPE PROPERTIES

The square-root raised-cosine (SRRC) pulse shape ideally provides absolute bandlimiting without producing intersymbol interference [5]. The spectrum of the SRRC pulse shape is illustrated in Figure 3. Observe that the RF bandwidth is defined by the two-sided bandwidth of $|P(f)|^2$ and is parameterized by α, called the "rolloff factor" or "excess bandwidth." The RF bandwidth may be expressed in terms of code rate and the input (information) bit rate:

$$\text{BW} = \frac{1+\alpha}{4R} \times \text{ bit rate.} \tag{3}$$

C. PEAK-TO-AVERAGE POWER RATIO

The power in $s(t)$ given by (1) is proportional to $|s(t)|^2$. The peak power proportional to $\max_t |s(t)|^2$ and the mean (or average) power is $\text{E}\{|s(t)|^2\}$ where the expectation is over both the symbol sequence and time. The peak-to-average power ratio (PAPR) is the ratio of the two:

$$\text{PAPR} = \frac{\max_t |s(t)|^2}{\text{E}\{|s(t)|^2\}}. \tag{4}$$

PAPR is important for RF power amplifier design. RF power amplifier efficiency (measured by ratio of RF output power to DC input power) is maximized when the RF power amplifier oper-

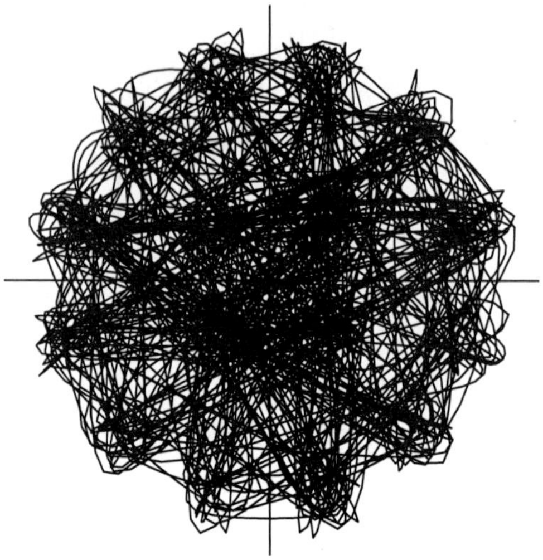

Figure 4: The phase trajectory (quadrature component vs. in-phase component) for 16-APSK ($\gamma = 2.75$, $\theta = 0$) and the SRRC pulse shape with $\alpha = 0.5$.

ates in full saturation. Operating in full saturation imposes amplitude distortion on the transmitted waveform. This distortion causes unwanted sidelobes (called "spectral regrowth") in the spectrum of modulated carriers with amplitude variations. For this reason, waveforms with no amplitude variations, or PAPR = 1 (0 dB), such as PCM/FM, SOQPSK-TG, and ARTM CPM [4] have been preferred in aeronautical telemetry. When a waveform with PAPR > 1 is used, the RF power amplifier must operate in its "linear region" which necessarily reduces both the peak and the average output power. The reduction in average output power is usually called output back-off (OBO). Here, we assume the required OBO is the PAPR. Under this assumption, PAPR is a direct measure of the required OBO.

There are two contributors to PAPR [7]. The first contributor is the arrangement of constellation points. If all the constellation points are on a circle (such as BPSK and QPSK) then the constellation's contribution to the PAPR is zero. Otherwise, the arrangement of points contributes to PAPR. The second contributor is the waveform variation due to the pulse shape. This is illustrated by the phase trajectory (quadrature component vs. in-phase component) of 16-APSK with the SRRC pulse shape with $\alpha = 0.5$ illustrated in Figure 4. The distance from the origin to any point on the phase trajectory represents $|s(t)|$. If $|s(t)|$ were a constant, the phase trajectory would be a circle. Clearly, this is not the case. Note that the peak values of $|s(t)|$ occur in the "overshoot" associated with the constellation points on the outer circle.

The PAPR is a complicated function of the pulse shape parameter α and the constellation parameters γ and θ. We first explore PAPR as a function of α. Figure 5 shows a plot of PAPR for five different 16-ary constellations: the traditional square 16-QAM constellation, the unconstrained optimum (minimum probability of error) constellation, the constrained optimum (minimum probability of error) constellation where the points are constrained to a regular grid, the 16-APSK constellation with $\gamma = 2.75$ and $\theta = 0$, and 16-PSK [5]. To compute the PAPR, a discrete-time version of (1) was generated at a sample rate equivalent to 10 samples/symbol. Because the PAPR

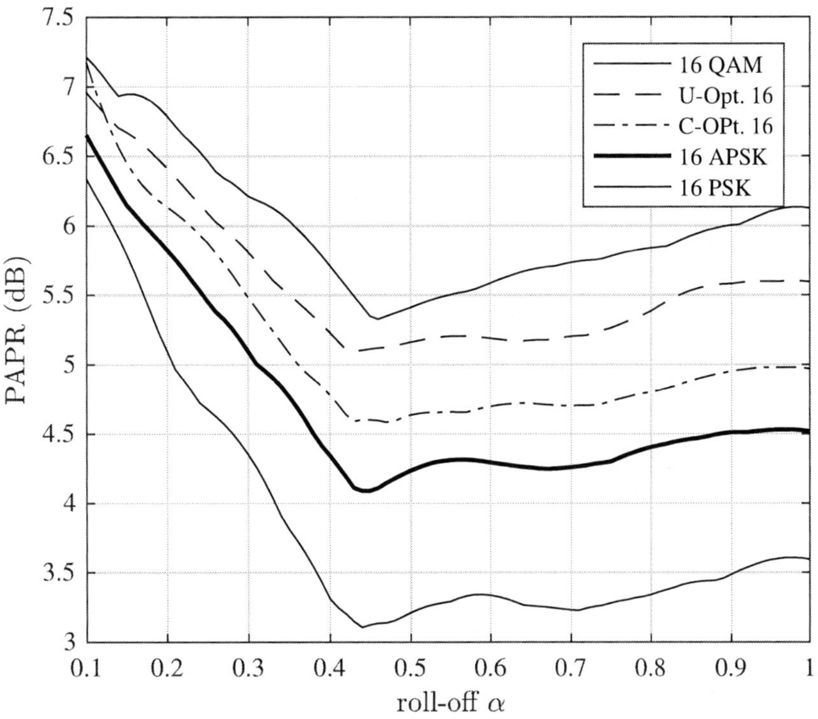

Figure 5: PAPR (dB) vs. SRRC roll-off α for a number of different 16-ary constellations.

depends on the symbol sequence, the symbol sequence corresponding to four repetitions of the length-$32{,}767$ PN sequence (sometimes called the "PN15" sequence because $2^{15} = 32{,}768$) was used as a bit sequence representative of an encrypted telemetry data bit sequence. The results show that the traditional square 16-QAM constellation has the highest PAPR, followed by the unconstrained and constrained optimum constellations. The PAPR of 16-APSK is about 1.5 dB less than that of 16-QAM. Because the 16-PSK constellations positions all the constellation points on a circle, 16-PSK has the lowest PAPR, but it also has the worst bit error probability and therefore requires a lower code rate than the other options.

For all constellations, PAPR is quite high for $\alpha < 0.4$. This demonstrates that even though small α decreases the occupied bandwidth [see Equation (3)], the bandwidth reduction is achieved at the cost of increased PAPR. The optimum (in the sense of minimum PAPR) α is $0.4 < \alpha < 0.5$.

For $\alpha = 0.5$ PAPR for 16-APSK as a function of γ and θ may now be explored. The results are shown in Figure 6. The plot shows that PAPR is a strong function of γ but only weakly dependent on θ.

D. CODE RATE

As explained in the Problem Formulation, a rate-R code, for $0 < R < 1$, produces $1/R$ coded bits for each input (information) bit. If the input (information) bit interval is T_b seconds, then the coded bit interval is $T_c = RT_b$. Because 16-APSK uses four bits per symbol (see the bit-to-symbol labels

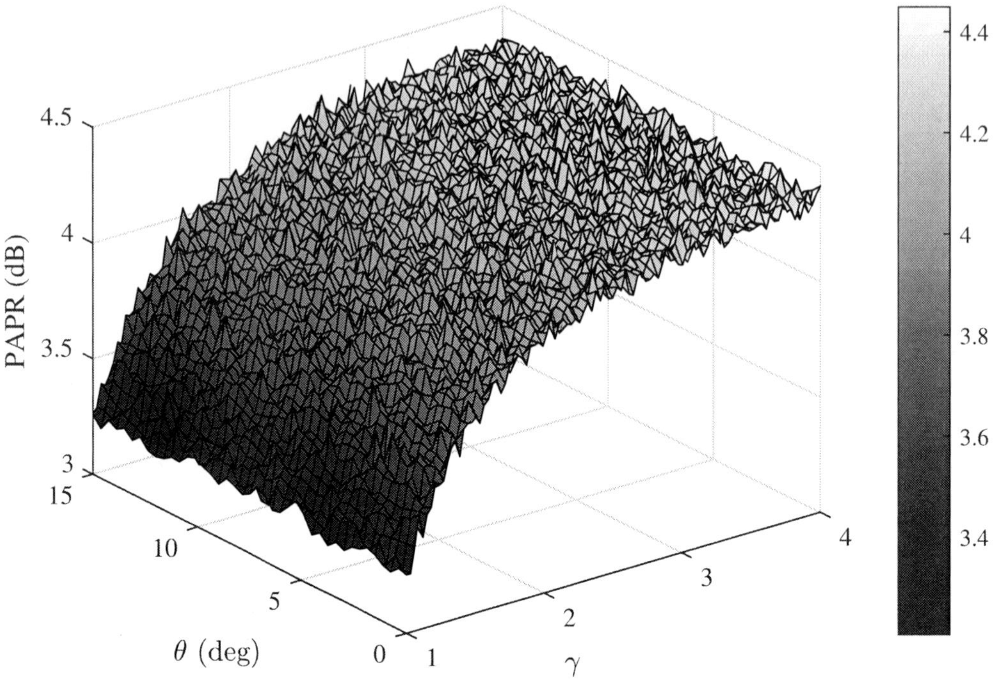

Figure 6: Peak-to-average power ratio (PAPR) for 16-APSK as a function of γ and θ.

in Figure 2), the symbol time in Equation (1) is $T_s = 4T_c$. The resulting bandwidth expansion is inversely proportional to the code rate as indicated in Equation (3).

When attempting to optimize a coded system, the performance of 16-APSK for an arbitrary code rate is of interest. One might envision designing a large number of LDPC codes, each with a different code rate, and simulating the performance of each one for all interesting (γ, θ)-variants of 16-APSK. Unfortunately, the bit error rate performance of coded systems is notoriously difficult to simulate. In light of this difficulty, we resort to concepts from information theory to bound the performance of coded 16-APSK.

Let $x \in \mathcal{C}$ be a constellation point and let $y = x + n$ be the corresponding matched filter output. Given the assumptions for $w(t)$ outlined in the Problem Formulation, n is a complex-valued circularly symmetric normal random variable with zero mean and variance $2\sigma^2$ [i.e., the real part of n (in-phase component) and the imaginary part of n (quadrature component) have the same variance σ^2]. Consequently, y is conditionally normal with density function

$$f(y|x) = \frac{1}{2\pi\sigma^2} \exp\left\{ -\frac{1}{2\sigma^2} |y - x|^2 \right\}. \tag{5}$$

For notational purposes below, we write $y = y_R + jy_I$ and $x = a + jb$, and express the conditional density function (5) as the joint density

$$f(y_R, y_I|x) = \frac{1}{2\pi\sigma^2} \exp\left\{ -\frac{1}{2\sigma^2} \left[(y_R - a)^2 + (y_I - b)^2 \right] \right\}. \tag{6}$$

The symmetric information rate between x and y is the mutual information [8] between x and y

assuming a uniform distribution on x (i.e., the constellation points are equally likely):

$$I(x;y) = \frac{1}{16}\sum_{m=0}^{15}\int_{-\infty}^{\infty}\int_{-\infty}^{\infty} f(y_R, y_I|c_m)\log_2\left\{\frac{f(y_R, y_I|c_m)}{1/16\sum_{m'=0}^{15} f(y_R, y_I|c_{m'})}\right\} dy_R dy_I. \quad (7)$$

The units for $I(x;y)$ are bits/symbol. For example, because 16-APSK uses four bits per symbol, using a rate-$1/2$ code with 16-APSK results in 2 bits/symbol, a rate-$3/4$ code results in 3 bits/symbol, and so on.

The symmetric information rate is an *upper bound* on the number bits/symbol for which it is possible to achieve reliable communications [8]. The corresponding code rate is $I(x;y)/4$. Thus, the interpretation of these results is that is possible to achieve reliable (error-free) bit error rate performance with coded 16-APSK as long as the code rate $R < I(x;y)/4$.

An example illustrates the concept. The symmetric information rate (7) for 16-APSK with $\gamma = 2.75$ and $\theta = 0$ is plotted[1] in Figure 7. $I(x;y) = 1$ bit/symbol at $E_b/N_0 = 0$ dB. This means that reliable communication can be achieved at $E_b/N_0 = 0$ dB using a code with $R < 1/4$. Similarly, $I(x;y) = 2$ bits/symbol at $E_b/N_0 = 2$ dB which means reliable communications can be achieved at $E_b/N_0 = 2$ dB using a code with $R < 1/2$; and $I(x;y) = 3$ bits/symbol at $E_b/N_0 = 4.55$ dB which means reliable communications can be achieved at $E_b/N_0 = 4.55$ dB using a code with $R < 3/4$. Note that as E_b/N_0 increases, the symmetric information rate approaches 4 bits/symbol, which means at these high values of E_b/N_0, a high rate code ($R \to 1$) may be used.

A closely related concept is the maximum code rate for a non-zero probability of error. The relationship between the decoder output bit error probability p, the code rate R and the signal-to-noise ratio for 16-APSK is given by [9]

$$R = \frac{1/4\, I(x;y)}{1 - H_2(p)}, \quad (9)$$

where $H_2(p)$ is the entropy of a binary random variable and is given by [8]

$$H_2(p) = -p\log_2(p) - (1-p)\log_2(1-p). \quad (10)$$

For fixed code rate, (9) relates the decoder output probability of error to the signal-to-noise ratio E_b/N_0 for a given constellation. This relationship is plotted in Figure 8 for 16-APSK with $\gamma = 2.75$

[1]Here we have followed the convention of expressing the symmetric information rate in terms of the equivalent uncoded signal-to-noise ratio E_b/N_0. (E_b is the uncoded bit energy and N_0 is the power spectral density of the inphase and quadrature components of the noise.) Because the modulation is 16-APSK, we have $E_{\text{avg}} = 4E_c$, where E_c is the coded-bit energy, and $E_c = RE_b$. Consequently, E_{avg}/N_0 is the over-the-air signal-to-noise ratio at which the synchronizers must operate and E_b/N_0 is the signal-to-noise ratio for the uncoded bits (and without the corresponding bandwidth expansion). The relationship between σ^2 in (7) and the equivalent uncoded signal-to-noise ratio E_b/N_0 is

$$\sigma^2 = \frac{E_{\text{avg}}}{8R\left(\dfrac{E_b}{N_0}\right)}, \quad (8)$$

where E_{avg} is given by (2) and $R = I(x;y)/4$.

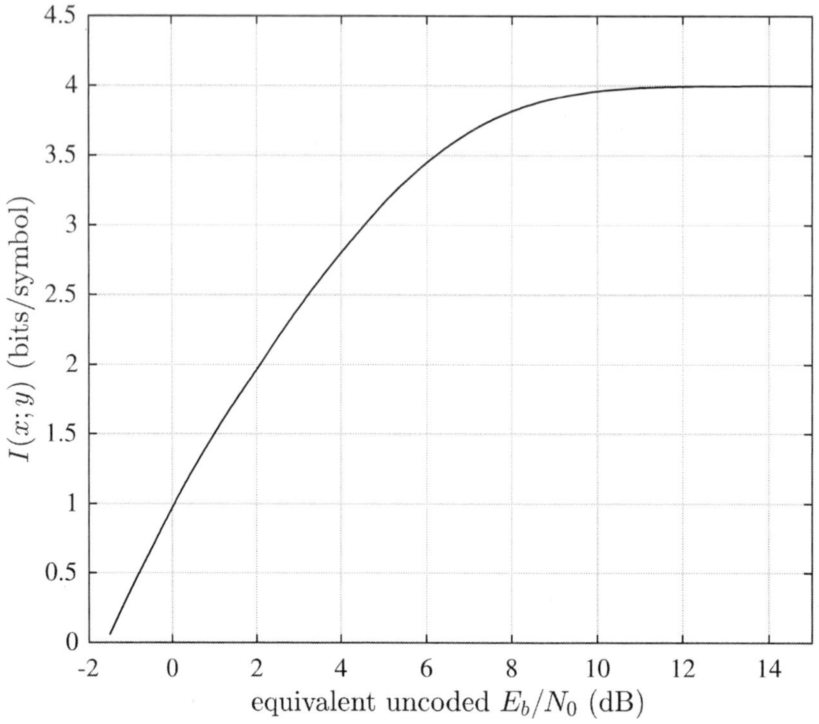

Figure 7: The symmetric information rate (7) vs. E_b/N_0 for 16-APSK with $\gamma = 2.75$ and $\theta = 0$.

and $\theta = 0$ for code rates $1/2$, $2/3$, $3/4$, and $4/5$. (These rates were selected because Appendix R of IRIG 106-15 [4] defines codes of rates $1/2$, $2/3$, and $4/5$.) The bit error probability vs. E_b/N_0 curve for uncoded SOQPSK-TG is included for reference.

The interpretation is that the code curves define a theoretical boundary. A realizable system must operate to the right of its corresponding curve. For example, when using this version of 16-APSK with a rate-$4/5$ code, the equivalent E_b/N_0 must be greater than 4.55 dB to achieve $p < 10^{-6}$. If $E_b/N_0 > 1.55$ dB cannot be achieved, then either a higher p must be tolerated, or a lower code rate must be used. The bit error probably curve for uncoded SOQPSK-TG, calculated from the analytical expression [10, Equation (15)], is also included for reference. The number of deciBels separating the uncoded SOQPSK-TG curve and one of the code curves upper bounds the coding gain associated with the code and corresponding constellation. It should be pointed out that codes used in practice have coding gains less than those predicted here.

The curves in Figure 8 are consistent with the symmetric information rate of Figure 7. For example, the $R = 1/2$ curve in Figure 8 asymptotically approaches $p = 0$ slightly above $E_b/N_0 = 2$ dB. Because $R = 1/2$ corresponds to $I(x;y) = 2$ bits/symbol, the curve in Figure 7 crosses 2 bits/symbol just above $E_b/N_0 = 2$ dB as well. Similarly, the $R = 3/4$ curve in Figure 8 asymptotically approaches $p = 0$ at $E_b/N_0 = 4.55$ dB and the curve in Figure 7 crosses 3 bits/symbol at the same value of E_b/N_0.

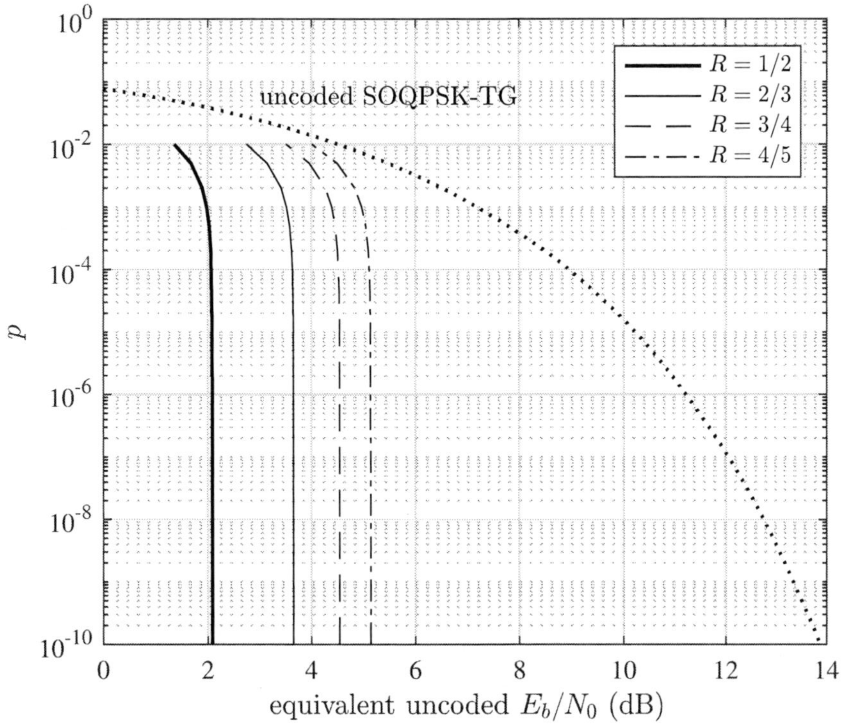

Figure 8: A graphical interpretation of (9): decoder output probability of error vs. signal-to-noise ratio for code rates 1/2, 2/3, 3/4, and 4/5 for 16-APSK with $\gamma = 2.75$ and $\theta = 0$. The bit error probability curve for SOQPSK-TG is included for reference.

E. SPECTRAL EFFICIENCY

The spectral efficiency of coded 16-APSK is the ratio of (information) bit rate to RF bandwidth and has units bits/s/Hz. The spectral efficiency is [see Equation (3)]

$$\eta = \frac{4R}{1+\alpha} \qquad \text{bits/s/Hz.} \tag{11}$$

The best spectral efficiency is achieved when R is as close to 1 as possible and α is as close to 0 as possible. But these are competing demands. As α decreases, the PAPR increases, which increases the OBO, which, in turn, demands a lower code rate.

OPTIMIZATION

With all the pieces of the previous section now in place, we are in a position to optimize the constellation parameters. The optimization criterion is spectral efficiency (11) which, for fixed α, is proportional to the code rate R. Because the code rate R is proportional to the symmetric information rate $I(x; y)$, maximizing spectral efficiency is the same as maximizing code rate.

The optimization is outlined as follows:

1. Select a reference operating point, $(E_b/N_0)^*$. The operating point is defined by the desired probability of bit error p. The goal is "error-free" performance, which is open to some interpretation. In the example that follows, we choose $p^* = 10^{-10}$ as the definition of "error free" ($p = 10^{-10}$ corresponds to 1 error every 17 minutes for a 10 Mbit/s link or 1 error every 83 minutes for a 2 Mbit/s link). For uncoded SOQPSK-TG, $p = 10^{-10}$ requires (see Figure 8)

$$\left[\left(\frac{E_b}{N_0}\right)^*\right]_{\text{dB}} = 13.9 \text{ dB}. \tag{12}$$

2. The 16-APSK operating point is determined by the OBO, which we equate to the PAPR. Thus, for a given (γ, θ) pair, the corresponding PAPR(γ, θ) defines the 16-APSK operating point as

$$\left[\frac{E_b}{N_0}(\gamma, \theta)\right]_{\text{dB}} = \left[\left(\frac{E_b}{N_0}\right)^*\right]_{\text{dB}} - \Big[\text{PAPR}(\gamma, \theta)\Big]_{\text{dB}} \tag{13}$$

3. The symmetric information rate corresponding to the 16-APSK constellation defined by the (γ, θ) pair at E_b/N_0 given by (13) is calculated from (7) using a plot similar to Figure 7.

4. The (γ, θ) pair that maximizes $I(x; y)$ defines the optimum constellation.

We demonstrate the optimization using $\alpha = 0.5$. The symmetric information rate for (γ, θ) pairs of Figure 6 are shown in Figures 9 and 10. Figure 9 shows the three-dimensional plot of $I(x; y)$ vs. (γ, θ). The plot displays "horizontal" peaks suggesting that like the PAPR, $I(x; y)$ is more strongly dependent on γ than θ. This motivates exploring slices of $I(x; y)$ along lines of constant θ. The results are shown in Figure 10. The slice corresponding to $\theta = 0$ (the black line in Figure 10) is the maximum symmetric information rate for all values of γ. The maximum along the $\theta = 0$ curve is at $\gamma_{\text{opt}} = 2.3477$, $I_{\text{opt}}(x; y) = 3.9512$ bits/symbol, which corresponds to a code rate of $R_{\text{opt}} = 0.9878$.

The optimization procedure may be modified to account for implementation loss. The two implementation losses considered here are those for SOQPSK-TG detection and decoding for the AR4JA LDPC codes.

1. The implementation loss for SOQPSK-TG may be estimated by comparing the BER curve on page 11 of the data sheet [11] with the analytical curve for SOQPSK-TG, see for example Equation (15) of [10]. This comparison yields the SOQPSK-TG implementation loss $L_{\text{SOQPSK}} = 1$ dB.

2. The implementation loss for the length-4096 AR4JA LDPC codes [4, Appendix R] is obtained by comparing the theoretical optima of Figure 8 with the simulation results presented in Figure 7 of [12]. The curves for the length-4096 AR4JA LDPC codes in Figure 7 of [12] are about 1.5 dB to the right of theoretical bounds in Figure 8 for the $R = 1/2$, $2/3$, $4/5$. Assuming the same difference for the $R = 3/4$ code, we can set the implementation loss for the LDPC decoder at $L_{\text{LDPC}} = 1.5$ dB.

L_{SOQPSK} moves the operating point to the right and L_{LDPC} moves the operating point to the left. Incorporating these losses into the optimization procedure alters the operating point in step 2 as follows:

$$\left[\frac{E_b}{N_0}(\gamma,\theta)\right]_{\text{dB}} = \left[\left(\frac{E_b}{N_0}\right)^*\right]_{\text{dB}} - \left[\text{PAPR}(\gamma,\theta)\right]_{\text{dB}} + L_{\text{SOQPSK}} - L_{\text{LDPC}}. \quad (14)$$

The results corresponding to using (14) in place of (13) are shown in Figures 11 and 12. The behavior is identical to that for the ideal case with no implementation loss, except the rates are slightly lower. Figure 12 shows the slices Figure 11 along planes of constant θ. The slice corresponding to $\theta = 0$ (shown in black) produces the optimum symmetric information rate. The maximum, shown by the star, occurs at $\gamma = 2.3477$ (the same as for the unconstrained case) and is $I_{\text{opt}}(x; y) = 3.9253$ bits/symbol (or a code rate of $R_{\text{opt}} = 0.9813$).

The optimization results are summarized by the spectral efficiency plot in Figure 13. (Figure 14 is a zoomed-in view.) The spectral efficiencies for the unconstrained optimization with and without implementation losses are 2.6342 and 2.6169, respectively. The optimum points correspond to code rates of 0.9878 (no implementation loss) and 0.9813 (with implementation loss).

Also included for reference are the results corresponding to the seven DVB-S2 constellation variants. For the parameters used here, the optimum constellation and code pair identified by the optimization routine is slightly better than the best DVB-S2 option.

Because these rates do not correspond to one the AR4JA LDPC code rates in IRIG 106-15, the highest rate not greater than the unconstrained rate must be used. The spectral efficiency corresponding to these points is also plotted in the figures. But when mapped to the available AR4JA LDPC code rates, all options achieve the same spectral efficiency, but the optimum code/constellation pair has a slight link margin advantage.

Finally, the operating point for SOQPSK-TG is shown for reference. The -50 dBc bandwidth for SOQPSK-TG was used to compute its spectral efficiency. The results show that LCPC-coded 16-APSK is capable of achieving spectral efficiencies a factor of 3.95 times (unconstrained with no implementation loss), 3.93 times (unconstrained with implemented loss), and 3.20 times (AR4JA rate constrained with implementation loss) the spectral efficiency of uncoded SOQPSK-TG.

CONCLUSIONS

The 16-APSK modulation together LDPC code rates presents the system designer with a peak-to-average power ratio vs. code rate trade-off. This trade-off was recast to an optimization problem involving spectral efficiency and the constellation parameters. The optimum 16-APSK, parameterized by the (γ, θ) pair, was identified using the optimization developed in this paper. The unconstrained optimization result is 3.95 times more spectrally efficient than uncoded SOQPSK-TG. When implementation losses and the limited number of IRIG 106 LDPC code rates are factored in, the advantage drops to 3.20 times the spectral efficiency of SOQPSK-TG.

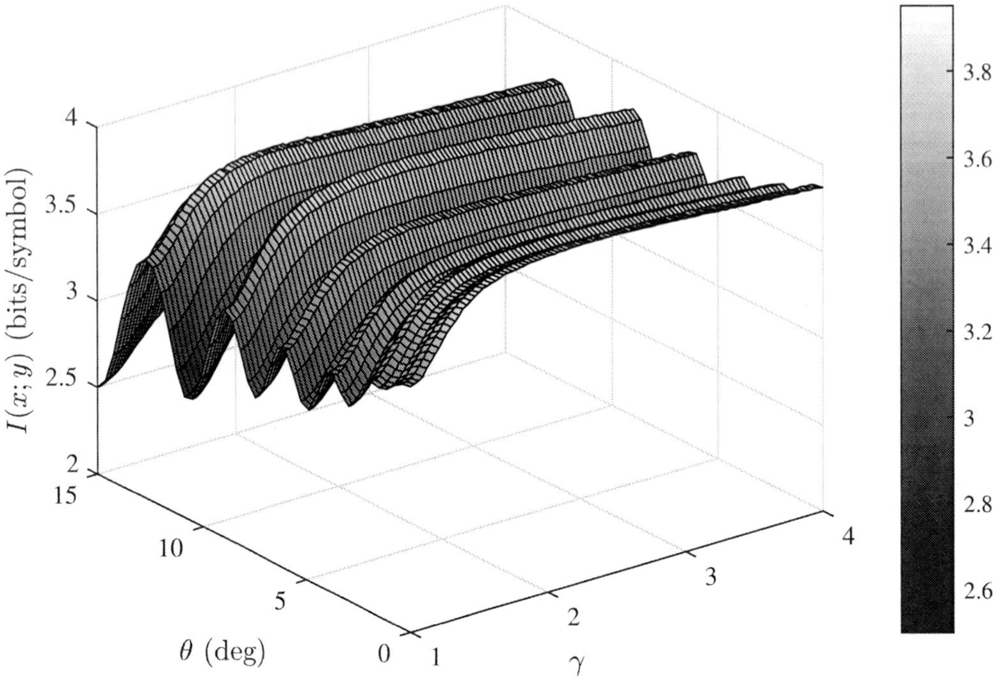

Figure 9: The symmetric information rates $I(x; y)$ corresponding to the γ, θ pairs of Figure 6.

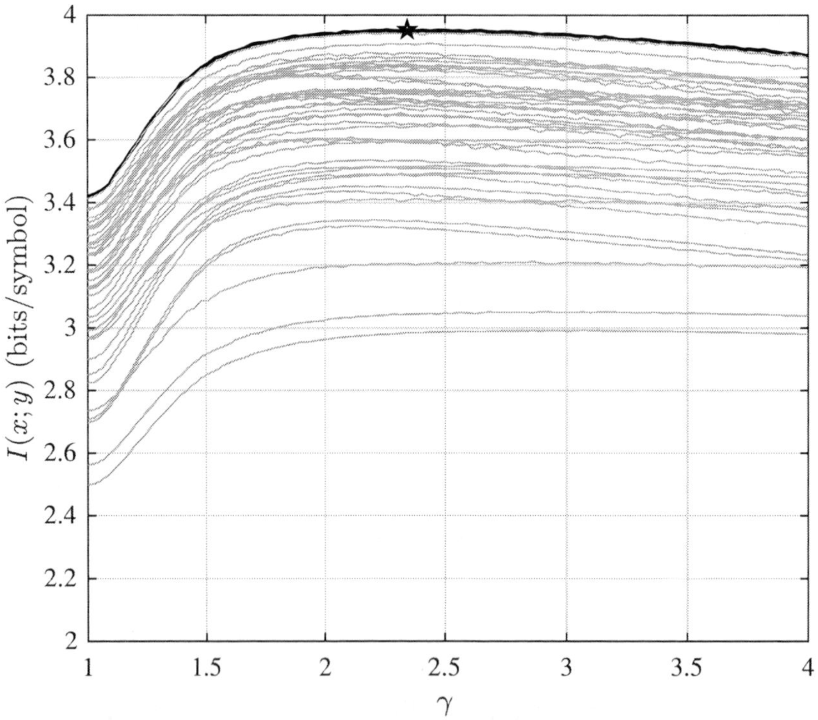

Figure 10: Slices of the three-dimensional plot of Figure 9 through constant-θ planes. The black line corresponds to $\theta = 0$. The star is placed at the maximum symmetric information rate.

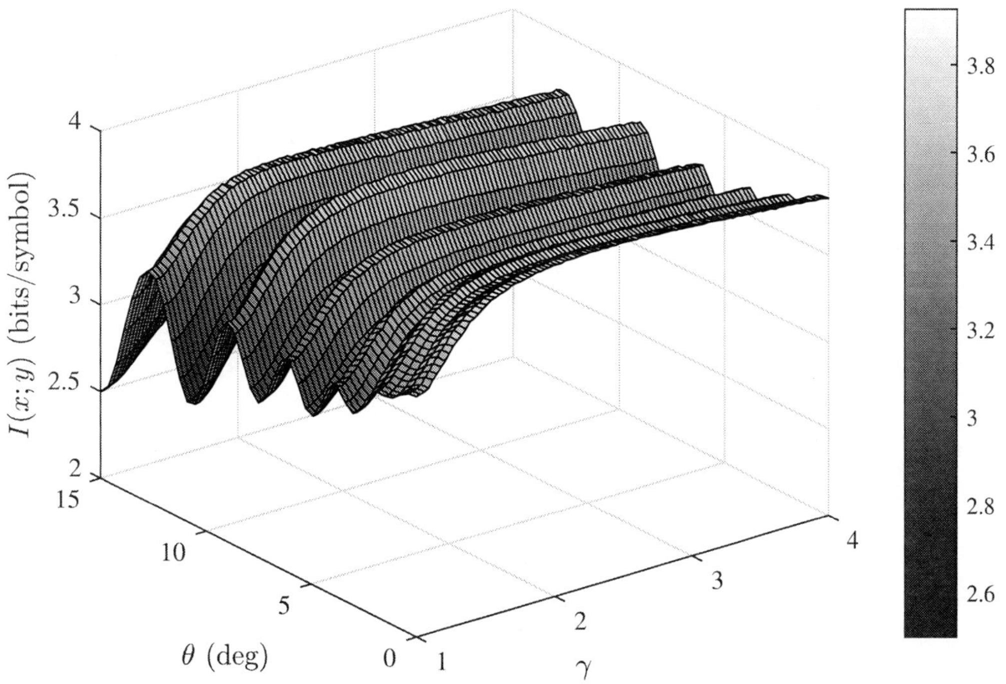

Figure 11: The symmetric information rates $I(x;y)$ corresponding to the γ, θ pairs of Figure 6 and including implementation losses $L_{\text{SOQPSK}} = 1$ dB and $L_{\text{LDPC}} = 1.5$ dB (cf. Figure 9).

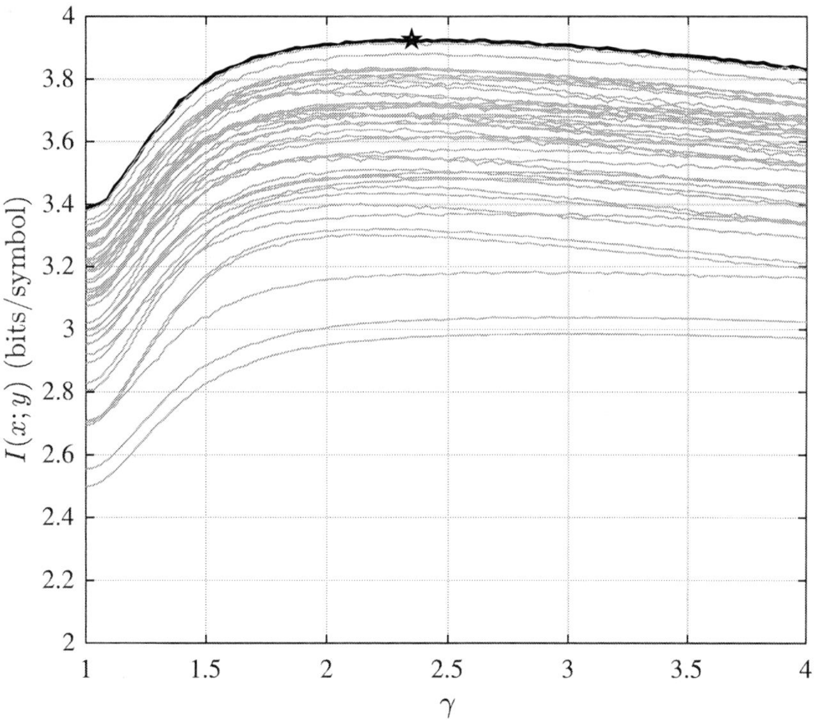

Figure 12: Slices of the three-dimensional plot of Figure 11 through constant-θ planes and including implementation losses $L_{\text{SOQPSK}} = -1$ dB and $L_{\text{LDPC}} = -1.5$ dB (cf. Figure 10). The black line corresponds to $\theta = 0$.

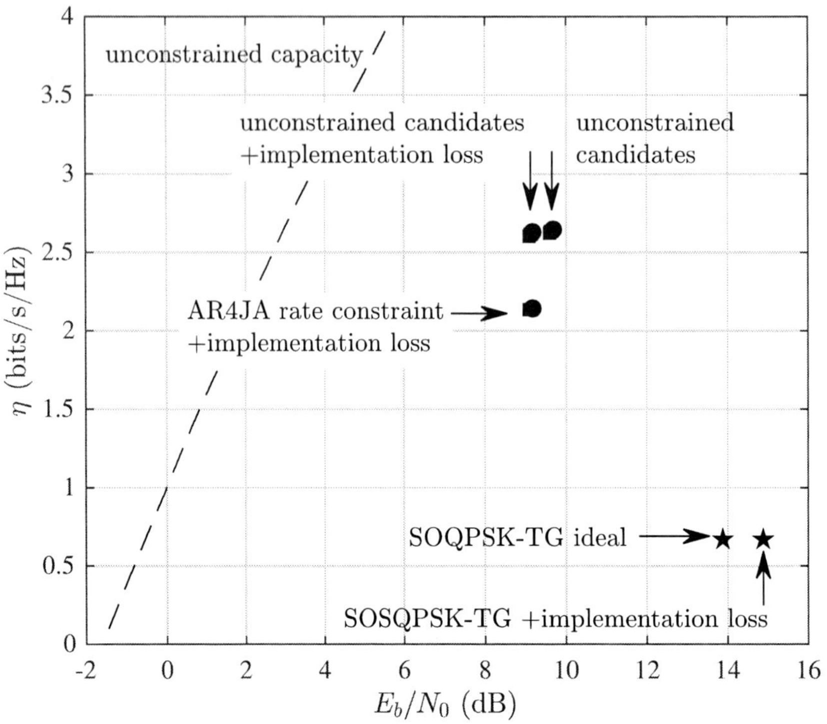

Figure 13: Spectral efficiency η SOQPSK-TG, optimum 16-APSK, and the DVB-S2 options corresponding to a target bit error probability of $P_b = 10^{-10}$ (cf. Figure 14).

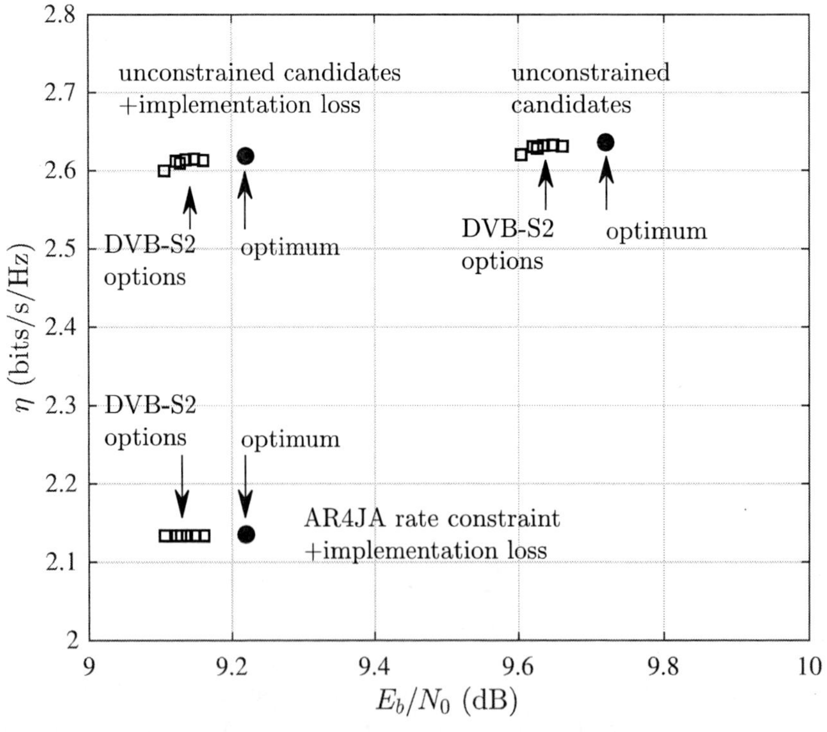

Figure 14: A close-up view of the spectral efficiency plot of Figure 13.

ACKNOWLEDGEMENTS

This project is managed by the Test Resource Management Center (TRMC) and funded through Spectrum Access R&D Program via Picatinny Arsenal under Contract No. W15QKN-15-9-1004. The Executing Agent and Program Manager work out of the AFTC.

REFERENCES

[1] Federal Communications Commission, "Auction 97: Advanced Wireless Services (AWS-3)." Available at http://wireless.fcc.gov/auctions/default.htm?job=auction_summary&id=97.

[2] ETSI EN 302 207 v1.1.2, "Digital video broadcasting (DVB): Second generation framing structure, channel coding and modulation systems for broadcasting, interactive services, new gathering and other broadband satellite applications," June 2006.

[3] C. Shaw and M. Rice, "Turbo-coded APSK for aeronautical telemetry," *IEEE Aerospace and Electronic Systems Magazine*, pp. 37–43, April 2010.

[4] Secretariat, Range Commanders Council, White Sands Missile Range, New Mexico, *IRIG Standard 106-15: Telemetry Standards*, 2015. (Available on-line at http://www.irig106.org/docs/106-15).

[5] M. Rice, *Digital Communications: A Discrete-Time Approach.* Upper Saddle River, NJ: Pearson Prentice-Hall, 2009.

[6] J. Proakis and M. Salehi, *Digital Communications.* New York, NY: McGraw-Hill, fifth ed., 2008.

[7] F. Mahmood, E. Perrins, and L. Liu, "Comprehensive energy analysis and modeling of wireless handset transceiver systems," 2017. Under review *IEEE Transactions on Green Communications and Networking*.

[8] T. Cover and J. Thomas, *Elements of Information Theory.* Hoboken, NJ: John Wiley & Sons, second ed., 2006.

[9] T. Moon, *Error Correction Coding: Mathematical Methods and Algorithms.* Hoboken, NJ: John Wiley & Sons, 2005.

[10] E. Perrins, "FEC systems for aeronautical telemetry," *IEEE Transactions on Aerospace and Electronic Systems*, vol. 49, pp. 2340–2352, October 2013.

[11] Quasonix, "3rd Generation Rack-Mount RDMS™ Telemetry Receiver." Data sheet. Available at http://www.quasonix.com/sites/default/files/qsx_rackmount_thirdgen_rdms_datasheet_0.pdf.

[12] J. Hamkins, "Performance of low-density parity-check coded modulation," in *Proceedings of the IEEE Aerospace Conference*, (Big Sky, MT), 6–13 March 2010.

PERFORMANCE RESULTS USING DATA QUALITY ENCAPSULATION (DQE) AND BEST SOURCE SELECTION (BSS) IN AERONAUTICAL TELEMETRY ENVIRONMENTS

Mark Geoghegan
Robert Schumacher
Quasonix, West Chester, OH

ABSTRACT

Flight test telemetry environments can be particularly challenging due to RF shadowing, interference, multipath propagation, antenna pattern variations, and large operating ranges. In cases where the link quality is unacceptable, applying multiple receiving assets to a single test article can significantly improve the overall link reliability. The process of combining multiple received streams into a single consolidated stream is called Best Source Selection (BSS). Recent developments in BSS technology include a description of the maximum likelihood detection approach for combining multiple bit sources, and an efficient protocol for providing the real-time data quality metrics necessary for optimal BSS performance. This approach is being standardized and will be included in Appendix 2G of IRIG-106-17. This paper describes the application of this technology and presents performance results obtained during flight testing.

KEY WORDS

Best Source Selection (BSS), Data Quality Metric (DQM), Data Quality Encapsulation (DQE), Aeronautical Flight Testing

INTRODUCTION

Aircraft testing uses real-time RF telemetry for safety-of-flight and performance monitoring. This data allows the Test Conductor and engineering team to monitor the status of various aircraft systems and analyze critical performance metrics during flight. Decisions on whether or not to proceed with testing, modify or skip particular tests, or to change mission goals during a flight rely heavily on having accurate, up-to-date downlink telemetry data. Further, for "single-shot" tests where the test article is lost or destroyed, repeating the test is costly, or impossible. Therefore, it is mandatory that the telemetry system provides a reliable communications link with the aircraft under test. Unpredictable outages in connectivity or having to overly restrict the operating range can disrupt testing, require test point repetition, or force testing to be cancelled. This wastes valuable personnel and equipment resources costing a company time and money.

In cases where the link quality is unacceptable, a common solution is to use receive diversity techniques to obtain multiple recovered data streams with different (hopefully independent) error characteristics. These individual outputs can be processed to produce a single 'best source' result. Significant performance improvements can be obtained relative to the individual signals. Examples include polarization diversity, frequency diversity, and spatial diversity.

This paper presents recent technology developments and discusses the new IRIG-106-17 Appendix 2G standard [1] that describes a BSS system based on maximum likelihood 'bit detection' and an efficient encapsulation protocol for providing the real-time data quality metrics necessary for optimal BSS performance. This paper also describes the application of this technology and presents performance results obtained during flight testing.

BACKGROUND

There is a long history of successfully improving the performance of a communication link through the use of multiple receive channels. By time aligning and comparing received copies of each transmitted bit, a composite stream can be produced whose likelihood of error is statistically as good or better than that of the individual channels. A block diagram of a simple BSS system is shown in Figure 1. The simplest example is a majority vote scheme where three received data streams are compared and the 'majority' decision is used to produce the combined output. One drawback to this approach is that majority voting is only possible with an odd number of observations. Furthermore, two weak 'ones' may actually be inferior to a single strong 'zero' for certain types of channels. Although the concept of combining multiple noisy copies of a transmission to produce a better output is straightforward, achieving the full potential of this type of system presents several challenges.

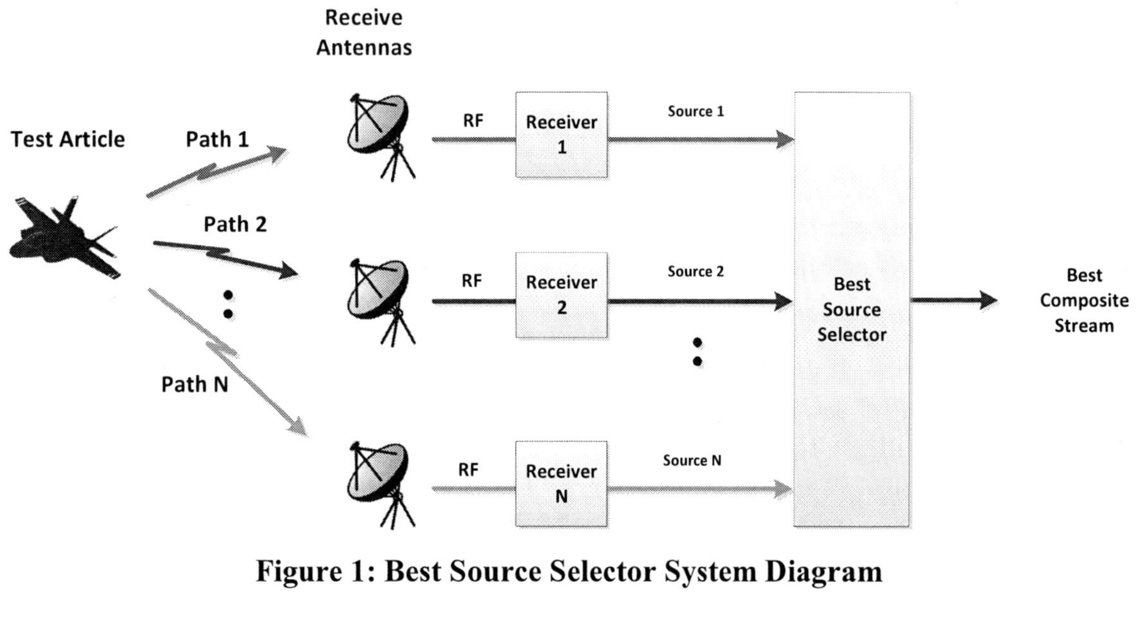

Figure 1: Best Source Selector System Diagram

The majority vote example illustrates the need for some type of confidence information to be supplied along with the binary data decisions. Developing a suitable confidence metric is challenging in itself as there are many possible indicators of bit quality including signal-to-noise ratio (SNR), error vector magnitude (EVM), measured errors in known segments of the transmission, and the number of errors corrected by forward-error-correction (FEC) schemes.

Unfortunately, it is unclear which metric, or combination of these metrics, provides the best performance, in all range environments. For example, SNR provides an excellent estimate of error rate in Additive White Gaussian Noise (AWGN), but performs poorly in multipath environments. Even worse is the fact that many of these measurements are typically unavailable due to the absence of coding, known framing patterns, or encryption. Even if a good confidence metric is generated, there is still the problem of transporting it independently across the test range along with the data, or encapsulating it into the data stream itself. Although proprietary schemes and equipment have tried to address these issues, the lack of performance standards and vendor interoperability have hindered the adoption of this powerful capability.

Multiple Bit Source Detection and a Standardized Encapsulation Protocol

The first question of the 'optimum' confidence metric was answered by Rice and Perrins [2], who concluded that the bit decision ($\mathcal{N}_0$, $\mathcal{N}_1$), along with its bit error probability (p_n), is sufficient to achieve maximum-likelihood bit detection (MLBD) performance. The calculation compares the sums of the log-likelihood ratios for channels with a 'zero' versus a 'one' to make its decision as shown in Figure 2. The detector behaves like a majority voter when all channels have the same probability of error, and a best source selector when there is a dominant channel.

$$\sum_{n\in\mathcal{N}_0} \log\left(\frac{1-p_n}{p_n}\right) > \sum_{n\in\mathcal{N}_1} \log\left(\frac{1-p_n}{p_n}\right) \rightarrow 0$$

$$\sum_{n\in\mathcal{N}_1} \log\left(\frac{1-p_n}{p_n}\right) > \sum_{n\in\mathcal{N}_0} \log\left(\frac{1-p_n}{p_n}\right) \rightarrow 1$$

MLBD Output

Figure 2: Maximum-Likelihood Bit Detection calculation from [2]

To illustrate the potential enhancement in link performance, a plot for the special case of equal channel probabilities is reproduced from [2] which shows dramatic improvements in link quality even with a relatively small number of low quality channels. Note that this plot assumes a binary symmetric channel and that the channels are statistically independent. Based on this mathematical result, the optimal approach is to accurately calculate and provide the probability of bit error along with each bit.

The second issue involves transportation of the confidence information. Many installations use receivers in remote locations and have limited connectivity (simple clock and data streams) to a central processing site, where multiple data streams converge and are processed by a BSS. Therefore, a natural solution is to encapsulate the additional BEP information directly into the data stream on the wired side of the range infrastructure. This approach doesn't require any changes or increase to the over-the-air transmission rate, as it adds the confidence information on the ground side only to provide the MLBD with the means to achieve optimal performance. An encapsulation protocol for efficiently realizing this capability is described in Appendix 2G, 'Standards for Data Quality Metrics and Data Quality Encapsulation', of IRIG 106-17.

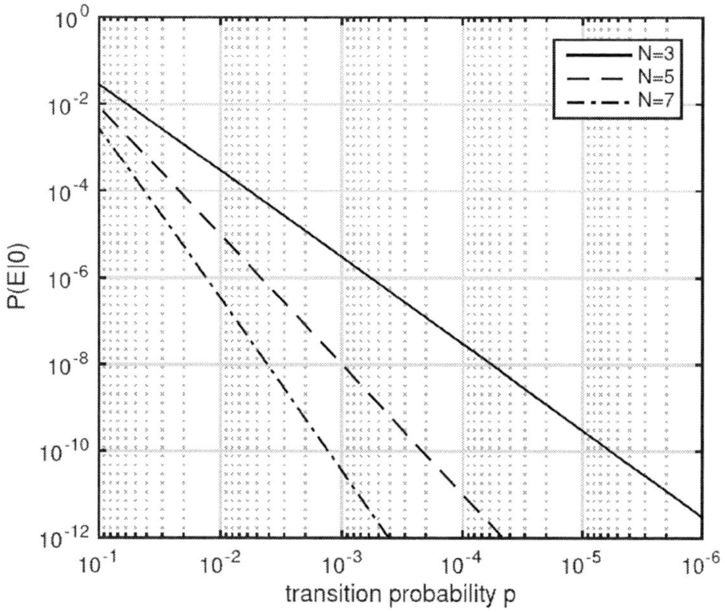

Figure 3: MLBD performance with equal channels ($p_n = p$) from [2]

The appendix consists of two separate parts, a standard for Data Quality Metric (DQM) and a standard for Data Quality Encapsulation (DQE). The DQM standard describes how to map the estimated Bit Error Probability (BEP) into a 16 bit word, but it does not define how the telemetry receiver performs its BEP estimation. Note that this approach does not depend on any particular modulation scheme, coding selection, transmission rate, or a priori knowledge of the data being transmitted. It is up to the receiver to accurately estimate the real-time BEP under the current channel conditions. Given the candidate decisions and their associated BEP, the BSS calculation is straightforward.

The DQE standard describes how to format the received telemetry data with the associated DQM for transport. A summary of the DQM and DQE protocol is illustrated in Figure 4 and Figure 5. The result is the definition and format of an optimal metric and an encapsulation protocol for a high-performance, interoperable, BSS system. Note that the BSS overhead on the ground side is at most (48+1024)/1024 = 4.7% and as low as (48 + 16384)/16384 = 0.3%. Again, there is no change or expansion of the over-the-air transmission format or rate.

The estimated BEP is first converted to a Likelihood Ratio (LR) between 0 and 1, converted to the log domain and quantized to 16 bits. The table below shows how the BEP, LR, and DQM are related. The DQE protocol consists of 48 bits that are periodically inserted into the data stream to convey the estimated quality for use by the BSS. For detailed information consult the standard [1].

k is the exponent for lowest estimated bit error probability
n number of bits in the DQM field

LR is the Likelihood Ratio

For this standard ***k*=12, *n*=16**.

$$DQM = \frac{-log_{10}(LR)}{k} * 2^n$$

$$LR = \frac{BEP}{(1 - BEP)}$$

BEP	LR	DQM
0.5	1.00	0
1e-01	1.11111e-01	5211
1e-02	1.01010e-02	10899
1e-03	1.00100e-03	16382
1e-04	1.00010e-04	21845
1e-05	1.00001e-05	27307
1e-06	1.00000e-06	32768
1e-07	1.00000e-07	38229
1e-08	1.00000e-08	43691
1e-09	1.00000e-09	49152
1e-10	1.00000e-10	54613
1e-11	1.00000e-11	60075
1e-12	1.00000e-12	65535

Figure 4: Summary of the IRIG 106-17 Data Quality Metric (DQM)

16 Bits	12 Bits	4 Bits	16 Bits	1024 – 16384 Bits
SW	RSV	VER	DQM	PAYLOAD

a. SW = Sync Word (16 bits). The Sync word is a fixed value of 0xFAC4.
b. RSV = Reserved (12 bits). Reserved for future use. These bits shall be set to zero (0) until used.
c. VER = IRIG 106 Version number (4 bits). Version number shall start with Version 0 (0000) for IRIG 106-17.
d. DQM = Data Quality Metric (16 bits). This field will contain the DQM value as defined in Paragraph 3.0.
e. PAYLOAD = Telemetry data payload to which the DQM value applies. The DQM and the Data Payload are contained in the same block. The minimum payload size shall be 1024 bits and the maximum size shall be 16384 bits. Payload size can be any multiple of 32 bits between the minimum size and maximum size.

Figure 5: Summary of the IRIG 106-17 Data Quality Encapsulation (DQE)

FLIGHT TEST RESULTS

In order to determine the potential performance improvement achievable in an actual flight test environment, a BSS system was tested in a military setting. The equipment consisted of three Quasonix RDMS receivers and a NetAcquire Advanced Correlating Source Selector (A-CSS). A standard PN test pattern was used as the transmission source and the recovered data and estimated BEP were encapsulated into the DQE stream and provided to the BSS. The resulting 'best' output was monitored by a BERT and the error statistics were logged. In addition to the BSS result, the embedded BERT capability within each receiver was used to log the error statistics of each receiver channel. This allows direct comparison of the performance of each site and the improvement achieved by processing all sites. The error statistics were converted to a Link Availability score which indicates the percentage of time the link conditions were better than the 10^{-5} BER threshold.

A simplified diagram of the BSS setup is depicted below in Figure 6. Although not shown in the diagram, the receivers are typically located at remote antenna sites with their clock and data signals being transported to a central location via range infrastructure. Several combinations of modulation format and rate were used over a variety of channel conditions. These included runway testing, up and away, and flight paths through challenging areas of the range. The focus of this test is on the improvement achievable with the BSS under real-world operating conditions.

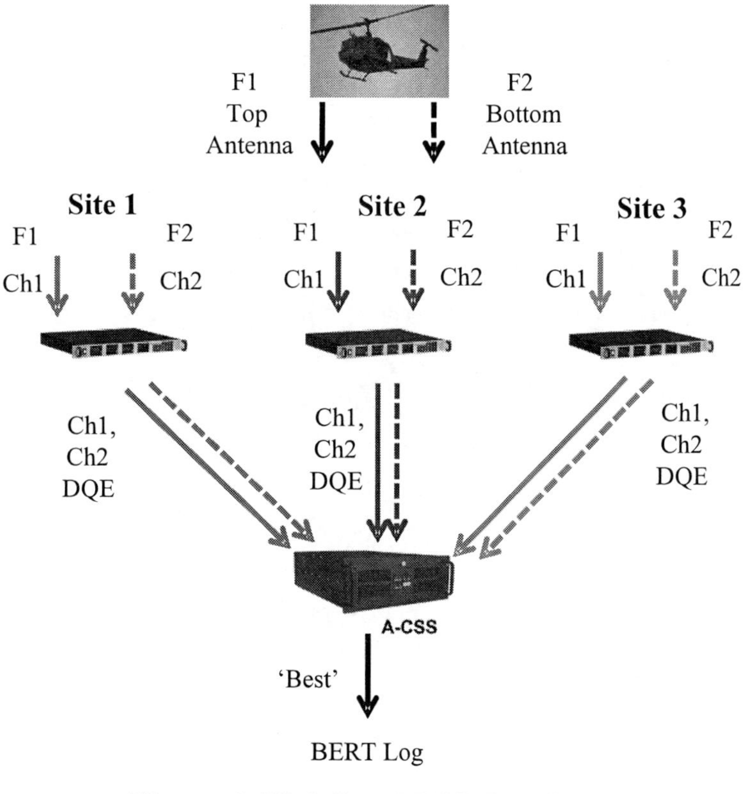

Figure 6: High Level BSS Test Setup

YUMA Proving Grounds (BSS with Multiple Receive Sites)

Flight testing at YPG used a flight profile that covered a variety of ranges and altitudes over diverse terrain, including open desert, mountains, and take-off and landing areas. Figure 7 shows the flight profile and relative positions of the receiving sites which are reproduced from [3]. During the four flights, several configurations of modulation, data rate, and coding were tested. Figure 8 shows the overall link availablility from the top (blue) and bottom (green) antennas for each of the individual sites as well the BSS (black) as measured by a known PN23 test pattern. Also included for comparison is the best performance achievable from a single site (brown). For this particular setup the bottom antenna outperformed the top, and using both antennas provided a substantial additional improvement. Finally, processing all the sites through the BSS yielded significantly better results and culminated with 34 minutes of error-free flight (0 bit errors over 1.02E10 bits at 5 Mbps) over this very challenging test course. This spectacular result illustrates the potential of the BSS system and associated technology to solve difficult telemetry links.

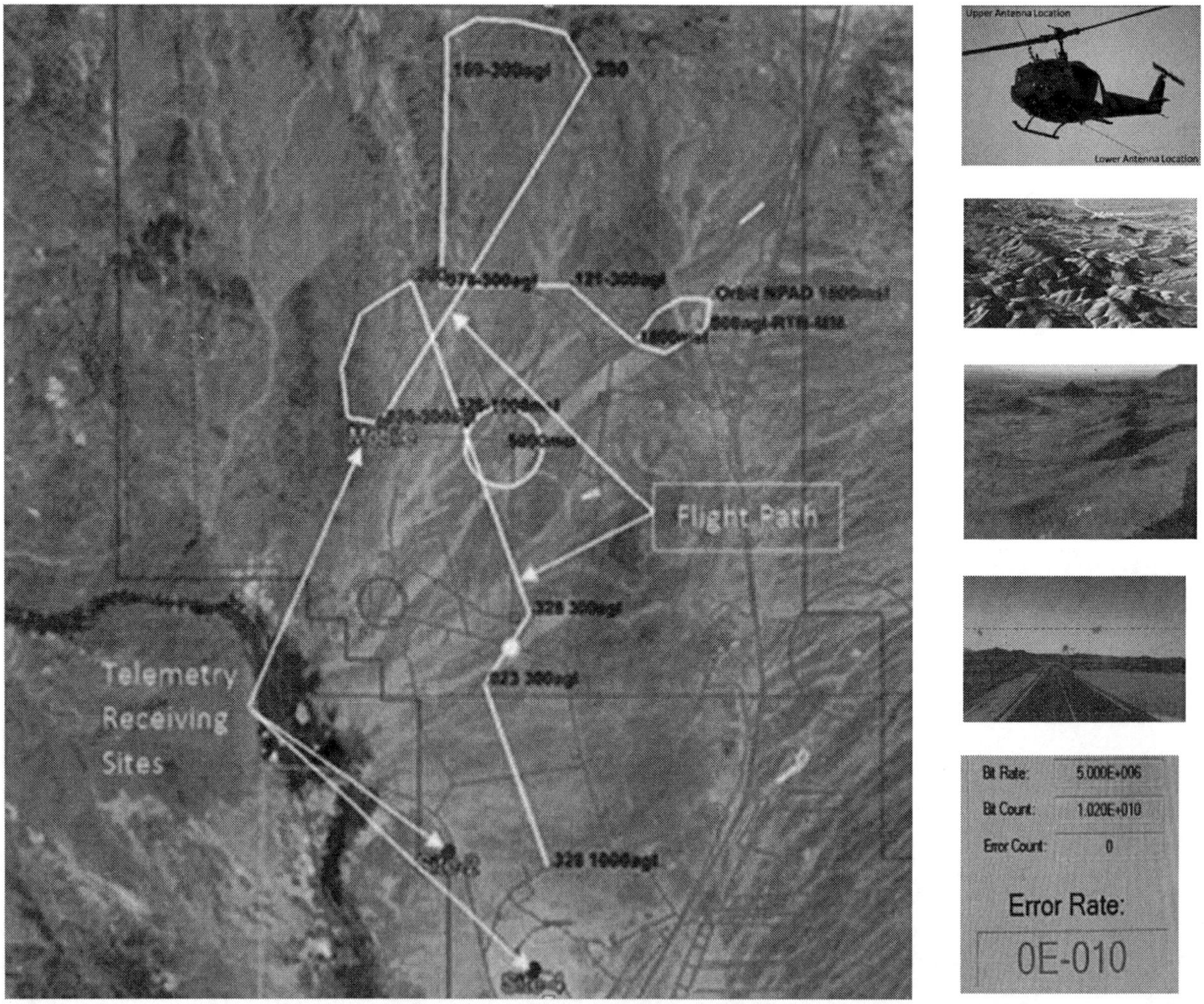

Figure 7: Flight profile from YPG flight testing [3]

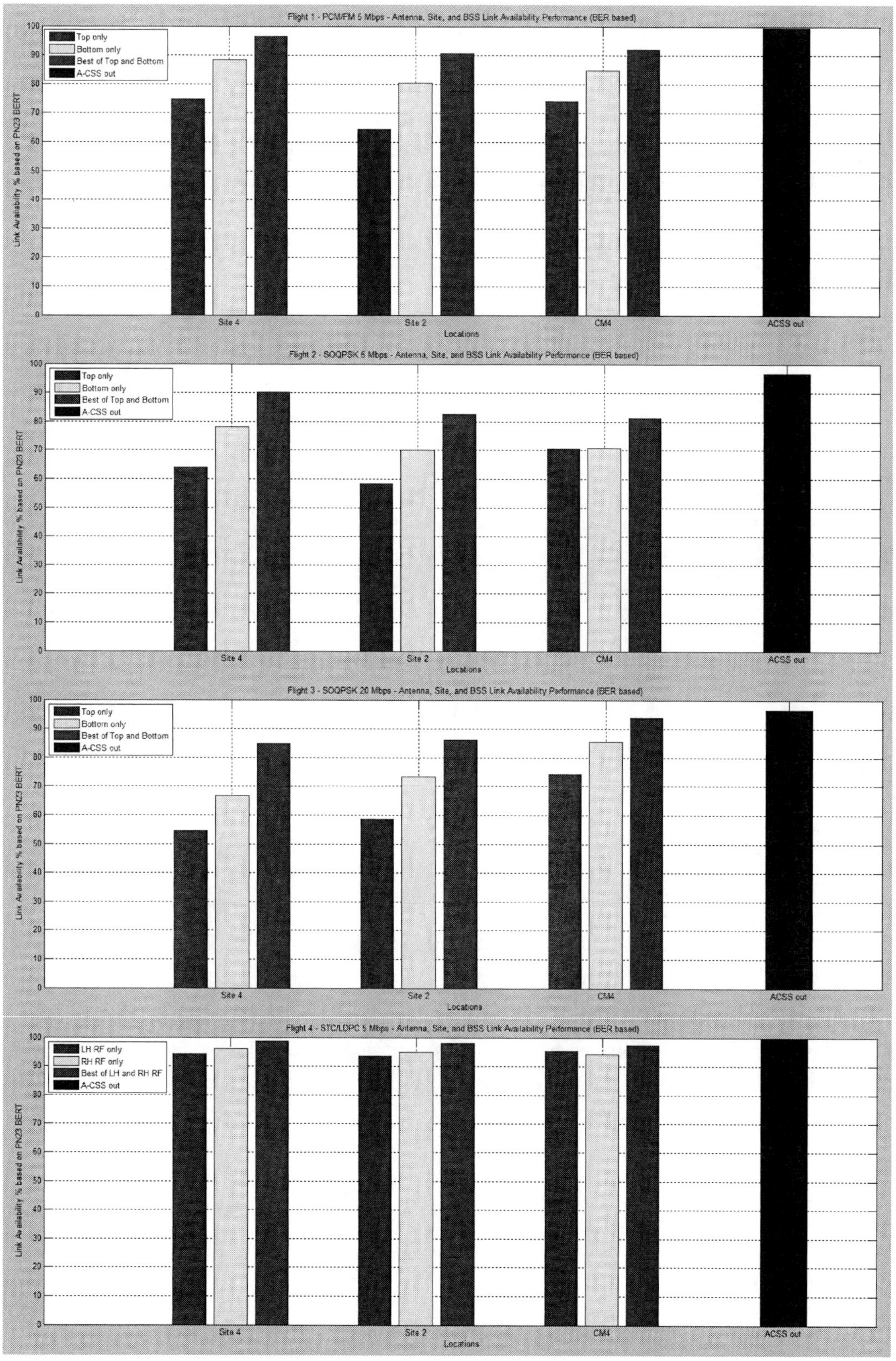

Figure 8: Link Availability of Individual Receiving Sites and BSS

BSS with a Single Receive Site

In addition to the testing performed at YPG, this BSS technology has also been field tested at a commercial avionics facility with similar results. While the most impressive performance gains were seen when multiple receiving sites were available, it was also shown that the performance with just a single site can be significantly improved if multiple received streams are available with different error characteristics. A common example includes frequency and polarization diversity as shown in Figure 9. Furthermore, specialty processing modes such as signal combining or adaptive equalization can also be included to yield additional improvement under certain types of RF channel conditions.

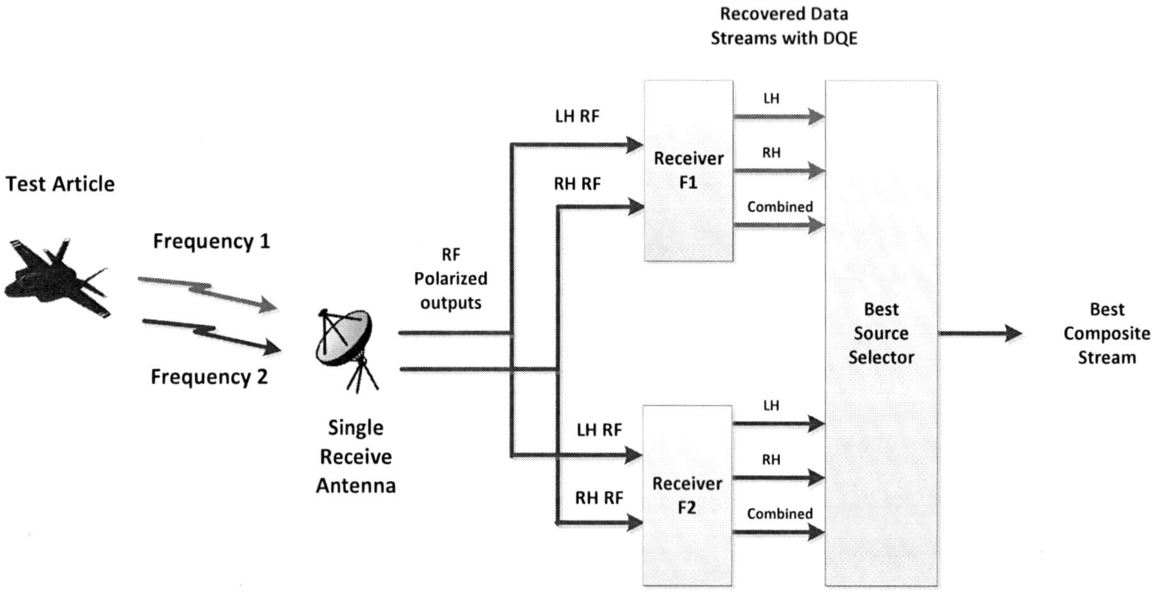

Figure 9: Single Site BSS example using both Frequency and Polarization Diversity

Another application of the BSS is to add simple fixed antennas around problem areas such as the runway or hangar areas. This is a cost-effective approach that can often solve stubborn connectivity issues. The BSS makes it easy to add in extra receive streams that occasionally outperform the primary TM equipment.

CONCLUSION

This paper has presented a BSS system based on the protocol described in 'Standards for Data Quality Metrics and Data Quality Encapsulation' in Appendix 2G of IRIG-106-17, along with mathematical and operational flight test results. The process involves recovering the data from RF signal, estimating the bit error probability, mapping it into a 16 bit DQM word, encapsulating it into a DQE telemetry data stream, and processing two or more streams with an optimal multiple bit detection algorithm. The resulting 'best' composite stream was shown to significantly outperform the individual streams, particularly when multiple receive sites were available. Configurations for a single site application were also discussed using the BSS to process streams from different polarizations, frequencies, or even simple fixed antennas to provide coverage in challenging locations.

During flight testing, one of the runs was completely error-free for 34 minutes over a variety of challenging link conditions. This level of performance illustrates the potential improvement attainable with this technology. In summary, the combination of the maximum likelihood bit detector algorithm and the development and standardization of the data quality metric and encapsulation protocol have led to an efficient BSS implementation that offers significant opportunities to improve wireless telemetry links.

REFERENCES

[1] Telemetry Group, Range Commanders Council, IRIG Standard 106-17 Appendix 2G, Telemetry Standards, Secretariat, Range Commanders Council, U.S. Army White Sands Missile Range, NM, 2017.

[2] M. Rice and E. Perrins, "Maximum Likelihood Detection from Multiple Bit Sources," in Proceedings of the International Telemetering Conference, Las Vegas, NV, October 2015.

[3] M. Diehl, J. Swain, and T. Wilcox, "Rotary-Wing Flight Tests to Determine the Benefits of Frequency and Spatial Diversity at the Yuma Proving Ground," in Proceedings of the International Telemetering Conference, Las Vegas, NV, October 2016.

ON IF-TO-BASEBAND TRANSLATION AND RESAMPLING IN SAMPLED-DATA RECEIVERS

Michael Rice, Chris Hogstrom, Chris Nash
Department of Electrical & Computer Engineering
Brigham Young University
Provo, Utah, USA
mdr@byu.edu, chris.hogstrom@gmail.com, christopherlarrynash@gmail.com

ABSTRACT

This paper summarizes the design of a discrete-time quadrature downconversion and resampling processor that operates on samples of a 70 MHz IF signal. The unique properties of discrete-time processing—aliasing due to resampling bandpass signals and polyphase filter decompositions—are applied to create a low-complexity approach that does not require any arithmetic operations at the IF sample rate. The required tasks are performed in two stages: a downsample-by-2 operation followed by a more traditional resampler based on a polyphase filterbank.

INTRODUCTION

A fundamental operation in receiver/demodulator front ends is frequency translation from the transmitted radio frequency (RF) to I/Q baseband. This is usually accomplished in two (or more) steps. The first step(s) involve translation from RF to an intermediate frequency (IF). Most receivers used in aeronautical telemetry use the industry-standard 70 MHz as the IF frequency. The last step involves a quadrature mixer and a pair of lowpass filters to produce the inphase component (I) and the quadrature component (Q) in parallel. For digital communications, synchronization, detection filtering, and bit decisions are all performed at I/Q baseband.

The architecture used to accomplish this has evolved as digital technology has matured. The evolution is illustrated in Figure 1. First generation demodulators perform all operations using continuous-time processing as illustrated in Figure 1 (a). The carrier phase PLL is usually closed at the quadrature mixers as shown so as to perform coherent quadrature downconversion. Second-generation demodulators, shown in Figure 1 (b), sample the inphase and quadrature components and perform all the synchronization and detection tasks using discrete-time processing. For this reason, the I/Q downconversion is performed using a free-running VCO tuned to the IF frequency. Third-generation demodulators move the A/D converter to the IF signal and sample the IF signal directly as illustrated in Figure 1 (c). The normal situation is to use bandpass sampling techniques such as those described in [1, Section 2.7].

In third-generation demodulators, the first task performed after sampling is the quadrature downconversion as shown Figure 1 (c). The structure shown here mimics the first-generation continuous-time counterpart. This task can be replaced by more efficient multi-rate processing techniques. The application of multirate discrete-time processing to perform this task is the subject of this paper.

Before proceeding, two items must be kept in mind.

1. The first follows from Euler's formula
$$e^{-j\Omega_0 n} = \cos(\Omega_0 n) - j\sin(\Omega_0 n). \tag{1}$$
Euler's formula allows for a more compact representation of the quadrature downconversion operation as illustrated in Figure 2. In the figure, single lines indicate real-valued signals whereas double lines indicate complex-valued signals.

2. The second item is the requirement for *resampling*. This is because synchronization and detection require a sample rate equivalent to an integer multiple (usually 2 or 4) of the bit rate. Because it is rare that the IF sample rate ($1/T$) is a multiple of the bit rate, a resampling operation must be performed. This is illustrated in Figure 3 (a). The resample factor is explained in the next paragraph.

Recent work on pilot-assisted equalization in aeronautical telemetry [2]–[13] is the motivation for the problem considered in this paper. Here, a continuous-time telemetry receiver translates a 10.3125 Mbit/s SOQPSK-TG signal from RF (L or C band) to a 70 MHz IF. The IF signal is sampled in prepration for the application of discrete-time equalization techniques. The desire is produce I/Q samples at a sample rate equivalent to 2 samples/bit, or 20.625 Msamples/s.

The challenge with implementing the block diagram in Figure 3 (a) is the complex heterodyne operation following the ADC. First, the samples of the sinusoids must be produced (or stored and accessed) at the IF sample rate. Second, the multiplications required for the complex heterodyne operation must be performed at the IF sample rate.

For this design, quadrature downconversion and resampling are performed in two steps. The first step involves quadrature downconversion and downsampling by an integer factor D. This is because proper selection of D and the IF sample rate removes the requirement to perform the multiplications by the complex exponential at the high sample rate. The second step is devoted purely to resampling and is based on a polyphase filterbank approach. The end result is the system of Figure 3 (b).

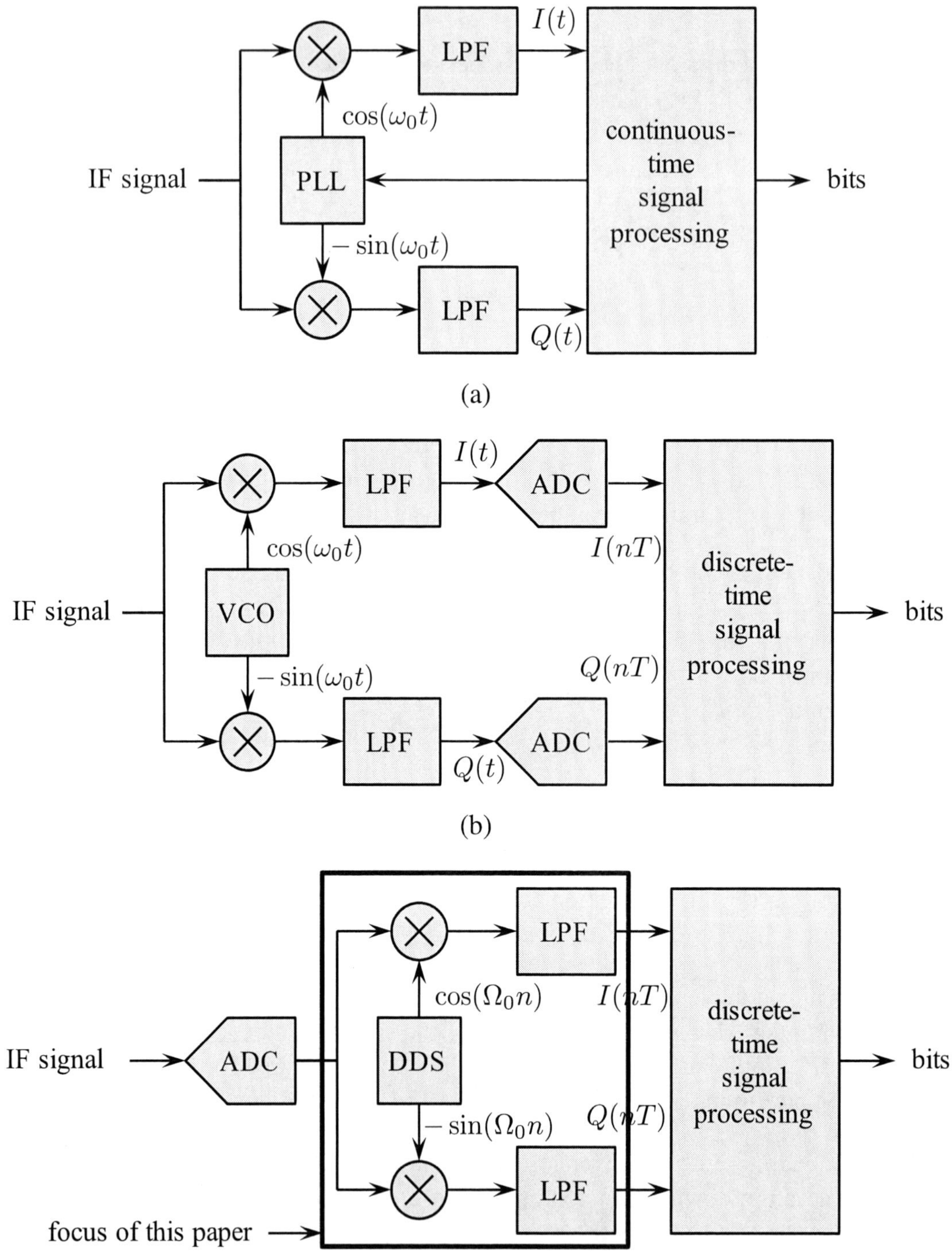

Figure 1: Generational evolution of receiver/demodulator architectures for digital communications: (a) first generation; (b) second generation; (c) third generation.

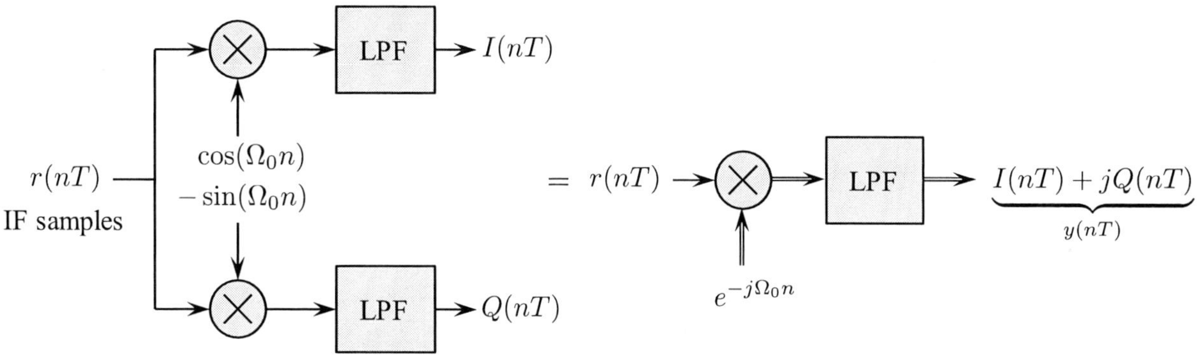

Figure 2: An illustration of the complex-valued equivalent processing based on Euler's formula (1).

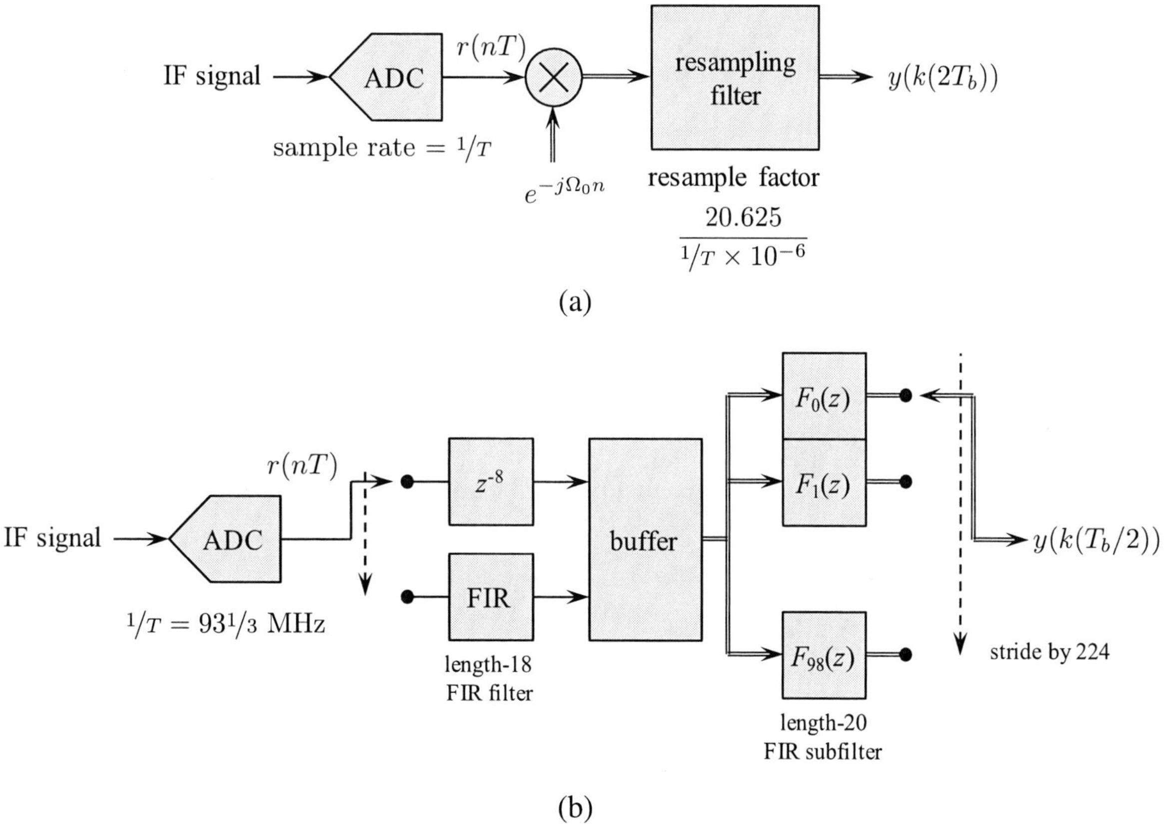

Figure 3: The required resampling operation: (a) the general problem formulation; (b) the solution using two polyphase filterbanks described in this paper.

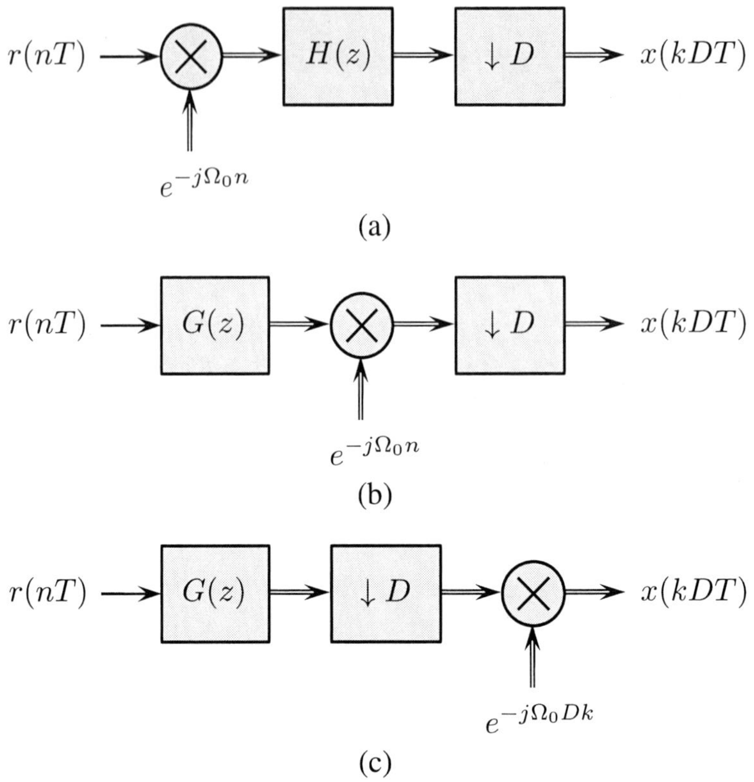

Figure 4: Three equivalent versions of downconvert/filter/downsample: (a) downconversion → low-pass filter → downsample; (b) bandpass filter → downconversion → downsample; (c) bandpass filter → downsample → downconversion.

FIRST STAGE DESIGN

Bandpass Processing in Sampled Data Systems. The first stage performs quadrature downconversion and downsampling by an integer factor D. The situation is illustrated in Figure 4 (a). The output of the low-pass filter is

$$x(nT) = \sum_{\ell=0}^{L-1} h(\ell)x(n-\ell)e^{-j\Omega_0(n-\ell)} \tag{2}$$

which may be re-expresssed as

$$= e^{-j\Omega_0 n}\sum_{\ell=0}^{L-1} \underbrace{h(\ell)e^{j\Omega_0 \ell}}_{g(\ell)} x(n-\ell) = e^{-j\Omega_0 n}\sum_{\ell=0}^{L-1} g(\ell)x(n-\ell). \tag{3}$$

In Equation (3), the filter coefficients $g(n)$ are result of heterodyning the low-pass filter coefficients $h(n)$. Consequently, $g(n)$ is the impulse response of a band-pass filter centered at Ω_0 rads/sample. Equation (3) defines the system shown in Figure 4 (b).

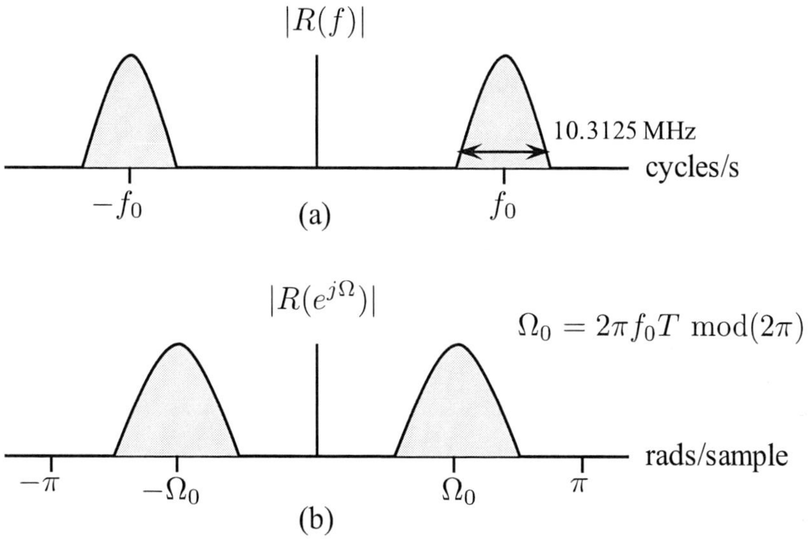

Figure 5: The relationship between spectrum of the continuous time IF signal (a) and the sampled IF signal (b).

The connection between the relationship (2) [illustrated in Figure 4 (a)] and the relationship (3) [illustrated in Figure 4 (b)] in the frequency domain is shown in Figures 5 and 6. The relationship (2) describes a frequency translation of the input signal spectrum to the left by Ω_0 rads/sample; this is illustrated by the black arrows in Figure 6 (a). This operation moves the desired spectrum (the positive-frequency component in this case) to baseband where it is isolated by the low-pass filter $H(z)$ as shown. In contrast, the relationship (3) creates a band-pass filter centered at Ω_0 rads/sample as shown in Figure 6 (b). The positive-frequency component of the input signal centered at Ω_0 is isolated at the band-pass filter output. The band-pass filter output is then translated to baseband as shown. The point is that both system are equivalent: they both produce the same result.

Polyphase Filterbank. Returning to Figure 4, Figure 4 (c) moves the downsample operation in front of the post-filter heterodyne. This move creates the familiar downsample-filter structure ideally suited to a polyphase filterbank. The polyphase decomposition of $G(z)$ may be derived by ordering the z-transform as follows. Let L be the length of $F(z)$. Assuming the downsample factor D divides the filter length L (i.e., $L = qD$ for some postive integer q), the z-transform may be expressed as

$$\begin{array}{rlllll} G(z) & = g(0) & + \; g(D)z^{-D} & + \; \cdots & + \; g((q-1)D)z^{-(q-1)D} \\ & + \; g(1)z^{-1} & + \; g(D+1)z^{-(D+1)} & + \; \cdots & + \; g((q-1)D+1)z^{-((q-1)D+1)} \\ & \vdots & & & \\ & + \; g(D-1)z^{-(D-1)} & + \; g(2D-1)z^{-(2D-1)} & + \; \cdots & + \; g(qD-1)z^{(qD-1)} \end{array} \tag{4}$$

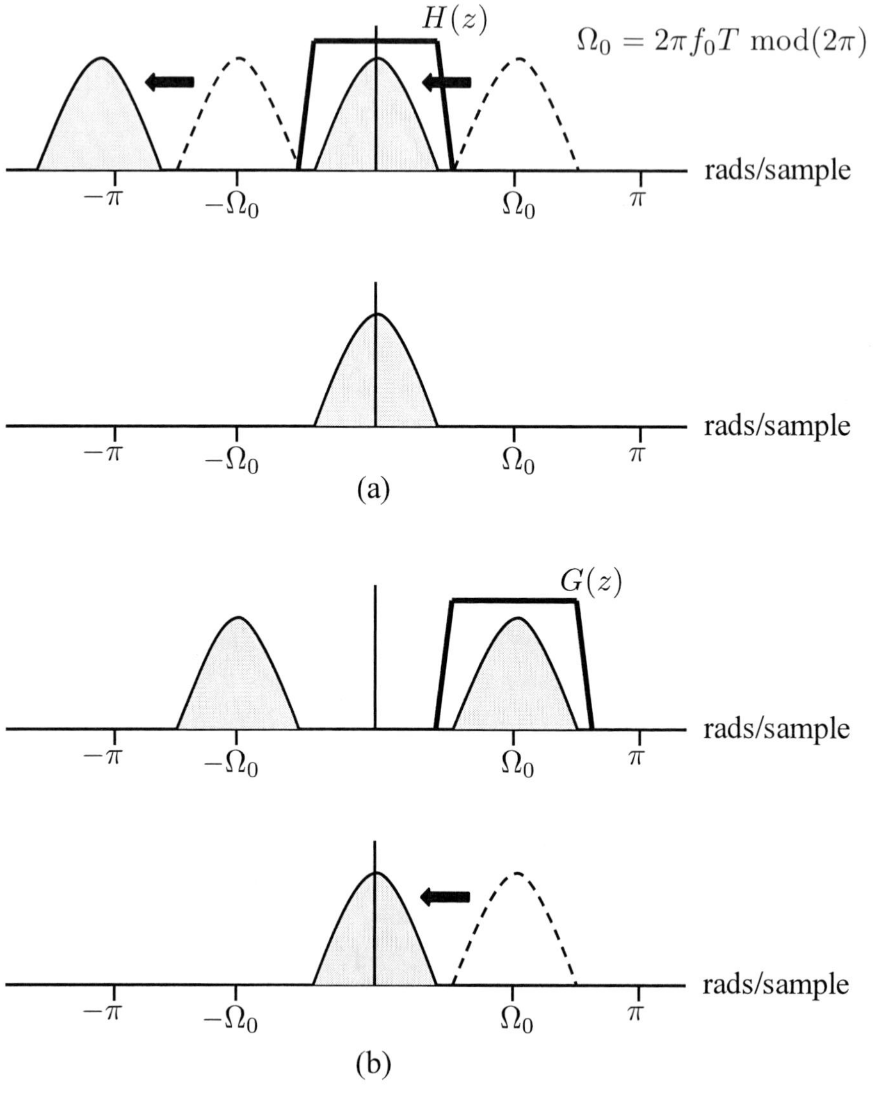

Figure 6: Frequency-domain interpretations: (a) frequency-domain interpretation of the relationship (2) [illustrated in Figure 4 (a)]; (b) frequency-domain interpretation of the relationship (3) [illustrated in Figure 4 (b)].

$$
\begin{aligned}
G(z) \;=\;& g(0) &+\;& g(D)z^{-D} &+\;& \cdots &+\;& g((q-1)D)z^{-(q-1)D} \\
+\;& z^{-1}\left[g(1)\right. &+\;& g(D+1)z^{-D} &+\;& \cdots &+\;& \left. g((q-1)D+1)z^{-(q-1)D}\right] \\
\vdots \\
+\;& z^{(D-1)}\left[g(D-1)\right. &+\;& g(2D-1)z^{-D} &+\;& \cdots &+\;& \left. g(qD-1)z^{(q-1)D}\right].
\end{aligned}
\tag{5}
$$

The polynomials inside the square brackets in each row define a z-transform in z^D. Consequently, the z-transform of $G(z)$ may be expressed as a sum of D z-transforms in z^D:

$$
G(z) = G_0(z^D) + z^{-1}G_1(z^D) + \cdots z^{-(D-1)}G_{D-1}(z^D), \tag{6}
$$

where

$$
G_m(z^D) = g(m) + g(m+D)z^{-D} + g(m+2D)z^{-2D} + \cdots + g(m+(q-1)D)z^{-(q-1)D}. \tag{7}
$$

The application of (6) to the system of Figure 4 (c) is illustrated in Figure 7 (a). Sliding the downsample operation from the filter output to the filter input produces the system shown in Figure 7 (b). Here the commutator operating on the input samples performs the downsample by D.

Using the substitution $g(\ell) = h(\ell)e^{j\Omega_0 \ell}$, $G_m(z)$ may be expressed as

$$
\begin{aligned}
G_m(z) &= h(m)e^{j\Omega_0 m} + h(m+D)e^{j\Omega_0(m+D)}z^{-1} + h(m+2D)e^{j\Omega_0(m+2D)}z^{-2} + \cdots \\
&\qquad + h(m+(q-1)D)e^{j\Omega_0(m+(q-1)D)}z^{-(q-1)} \\
&= e^{j\Omega_0 m}\left[h(m) + h(m+D)e^{j\Omega_0 D}z^{-1} + h(m+2D)e^{j\Omega_0 2D}z^{-2} + \cdots \right. \\
&\qquad \left. + h(m+(q-1)D)e^{j\Omega_0(q-1)D)}z^{-(q-1)}\right] \\
&= e^{j\Omega_0 m}H_m\left(e^{-j\Omega_0 D}z\right).
\end{aligned}
\tag{8}
$$

Applying the relationship (8) to the system in Figure 7 produces the system shown in Figure 7 (c).

Equipped with the system in Figure 7 (c), we are now in a position to make design decisions to simplify the mathematical operations.

1. To eliminate the complex exponential operations at the filterbank inputs, the IF sample rate must be selected so that $\Omega_0 = 0, \pi/2$, or π. $\Omega_0 = 0$ is not possible because this requires colocating the positive and negative frequency components of the continuous-time IF spectrum in Figure 5 (a) at $\Omega_0 = 0$. This cancels the quadrature component of the IF signal. A similar problem occurs with $\Omega_0 = \pi$. The only workable solution is $\Omega_0 = \pi/2$. Here each input sample is "multiplied" 1, j, -1, or $-j$. "Multiplication" by 1 requires no resources, and "multiplication" by -1 is a simple sign switch. Similarly, multiplication by $\pm j$ simply assigns the real-valued input sample to the imaginary component of the filter input.

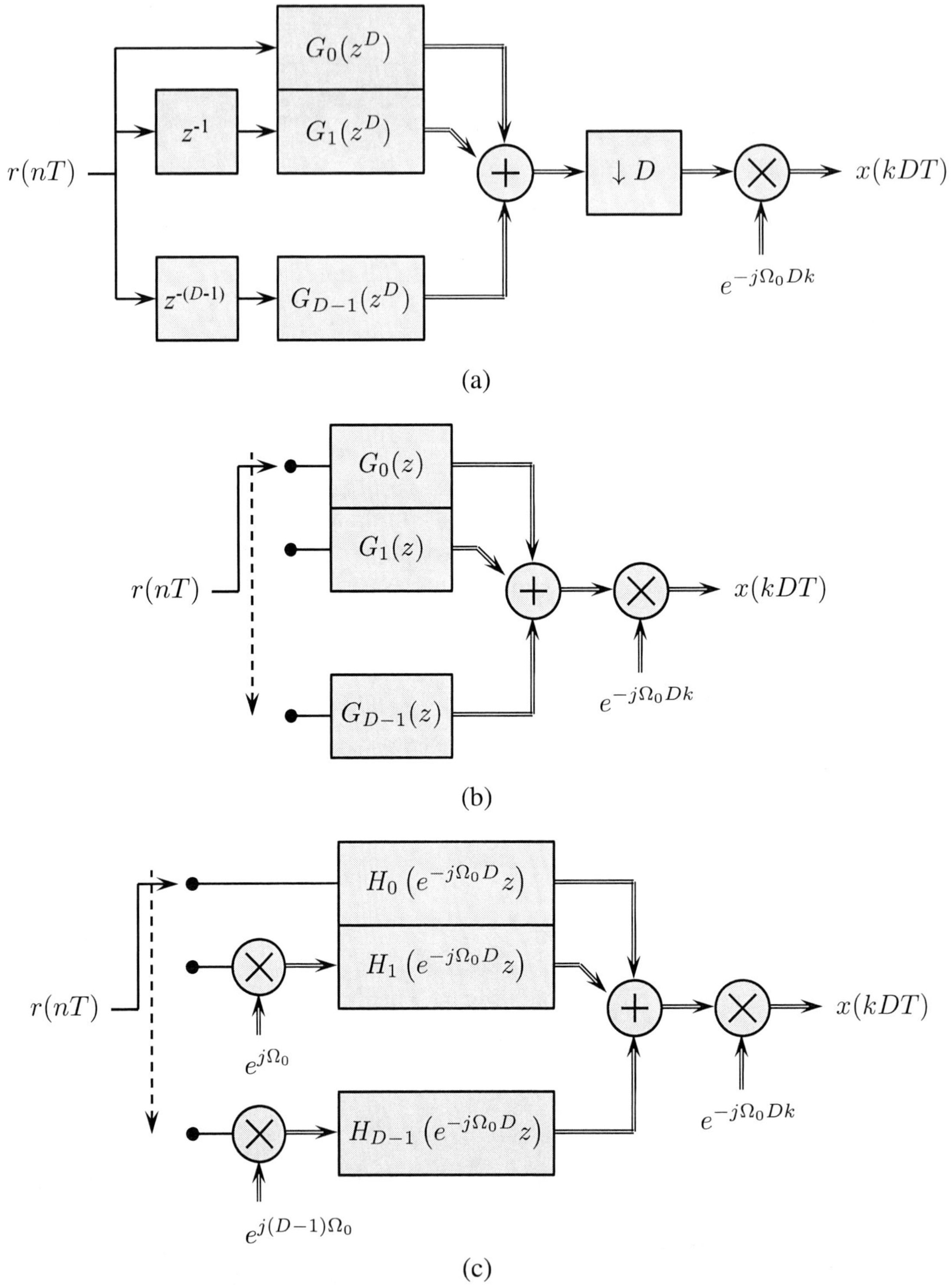

Figure 7: A polyphase decomposition of the system of Figure 4 (c): (a) the system of Figure 4 (c) with the polyphase decomposition of $G(z)$ used in place of $G(z)$; (b) the system of part (a) with downsample operation translated to the input; (c) the polyphase bandpass filter $G(z)$ replace by the equivalent version of the low-pass filter $H(z)$.

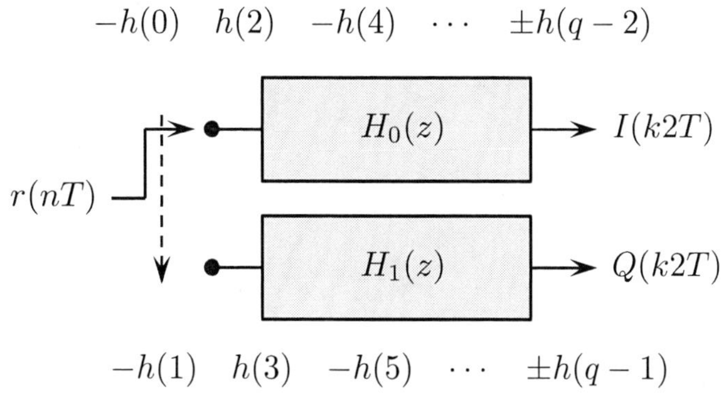

Figure 8: The special case of the system in Figure 7 (c) for $\Omega_0 = \pi/2$ and $D = 2$.

2. The complexity of the polyphase filter bank can be reduced by a factor of 2 if all of the filter coefficients are real. Equation (8) shows that this occurs when when

$$\Omega_0 D = \pi \bmod(2\pi) \quad \text{or} \quad \Omega_0 D = 0 \bmod(2\pi). \tag{9}$$

In the first case, $g_m(n) = \pm h_m(n)$. In the second case, $g_m(n) = h_m(n)$. Using $\Omega_0 = \pi/2$, as argued previously,

$$\begin{aligned} \Omega_0 D &= \frac{\pi}{2} D = \pi \implies D = 2 \\ \Omega_0 D &= \frac{\pi}{2} D = 2\pi \implies D = 4. \end{aligned} \tag{10}$$

Because using $D = 2$ simplifies the resampling filter in the second stage, $D = 2$ is assumed in the remainder of this paper.

3. The post filter heterodyne operation involves the complex exponential

$$e^{-j\Omega_0 Dk} = e^{-j\pi k} = (-1)^k. \tag{11}$$

Thus, the final heterodyne operation reduces to sign alternations applied to the filter output. It is easy to show that the sign alternations may be moved to filter coefficients. (This sign alternation is in addition to the sign alternations in item 2.)

The special case of the system in Figure 7 (c) corresponding to $\Omega_0 = \pi/2$ and $D = 2$ is illustrated in Figure 8. The multiplier on the input to $H_1(z)$ reduces to $e^{j\Omega_0(D-1)} = e^{j\pi/2} = j$. This is not really a multiplication, but rather assigns all odd-indexed IF samples to the imaginary portion of the output. Because the imaginary portion is the quadrature component (see Figure 2), the lower branch produces the samples of the quadrature component at half the IF sample rate. The upper branch produces samples of the inphase component at half the IF sample rate.

With $\Omega_0 = \pi/2$ and $D = 2$, the remaining issues are IF sample rate and prototype filter design.

IF Sample Rate. The IF sample rate F_s that places the spectrum of the IF samples at $\Omega_0 = \pi/2$ rads/sample satisfies the relationship [1]

$$kF_s \pm \frac{1}{4}F_s = f_0 \tag{12}$$

where f_0 is the IF center frequency. Using $f_0 = 70$ MHz, the values of F_s produced by (12) that accommodate a bandwidth greater than 10 MHz are given in Table 1. For our problem, any of these would work [see Figure 5 (a)]. By far, the most common IF sample rate is the second entry $93^1/3$ Msamples/s. Consequently, we use this value in our design.

Prototype Filter Design. Moving to the prototype filter design. The spectral requirements for this filter are summarized in Figure 9. Because the downsample factor is $D = 2$, the ideal lowpass filter is one with a cut-off frequency of $\pi/2$. A practical filter requires a transition band and a specification for the minimum allowable stop-band attenuation. Here we chose 60 dB for the stop-band attenuation (a common value in sampled-data systems) and, because the IF signal is oversampled, we use $0.4\pi < \Omega < 0.6\pi$ as the transition band. The filter length is the shortest length that achieves the desired stop band attenuation. The result is the length $L = 35$ filter illustrated in Figure 10. The top plot shows the impulse response $h(n)$ for $-17 \leq n \leq 17$. The lower plot in Figure 10 plots the DTFT of the prototype filter. The filter easily meets the stop-band attenuation requirement.

Observe that the $h(n) = 0$ for n non-zero and even and that $h(0) = 1$. This is characteristic of *half-band filters* [1]. Consequently, the top subfilter of Figure 8 is simply $H_0(z) = z^{-8}$. The subfilter $H_1(z)$ is

$$H_1(z) = h(-17) - h(-15)z^{-1} + h(-13)z^{-2} - \cdots - h(17)z^{-17}. \tag{13}$$

Note the sign reversals here. The sign reversals are required to compensate for the fact that using $F_s = 93\,^1/3$ Msamples/s with $f_0 = 70$ MHz produces a spectral reversal.[1]

The polyphase filterbank of Figure 8 forms the first stage of the final solution shown in Figure 3 (b).

SECOND STAGE DESIGN

The output of the first stage produces a complex-valued sample sequence, where the real part corresponds to the inphase component of the IF signal and the imaginary part corresponds to the quadrature component of the IF signal, all sampled at half the IF sample rate (the sample rate here is

[1]Spectrum reversal here occurs because the negative-frequency component of the continous-time signal aliases to the positive frequency component of the sampled sequence. Consequently, the negative frequency component of the sample sequence is the desired sequence. This is change can be made by changing the sign on the complex exponential mixer input in Figure 4 (a) and tracking this change through the subsequent development.

Table 1: A partial list of IF sample rates that alias the IF spectrum to the quarter sample rate.

F_s (Msamples/s)	Usable Bandwidth (MHz)	Spectrum Swap
280	140	No
$93\,^1/_3$	$46\,^2/_3$	Yes
56	28	No
40	20	Yes
$31\,^1/_9$	$15\,^5/_9$	No
$25\,^5/_{11}$	$12\,^8/_{11}$	Yes
$21\,^7/_{13}$	$10\,^{10}/_{13}$	No

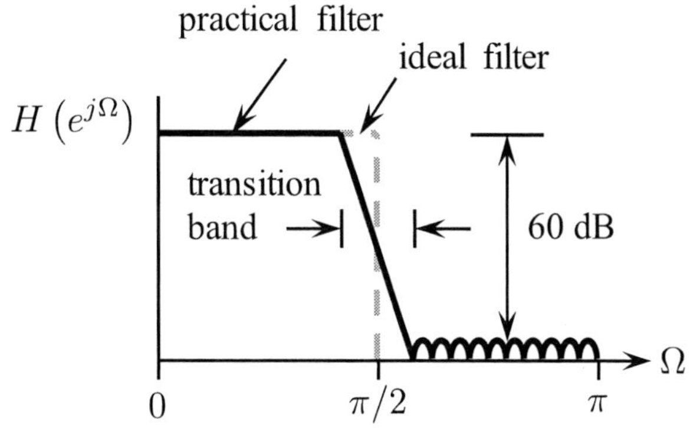

Figure 9: Spectral requirements for the prototype low-pass filter $H(z)$.

$46\,^2/_3$ Msamples/s). The desired sample rate is 2 samples/bit or 20.625 Msamples/s. Consequently the requirement for the second stage is to resample the output of the first stage by the factor

$$R = \frac{20.625}{46\,^2/_3} = \frac{99}{224}. \tag{14}$$

The brute force method for accomplishing this is to upsample the first stage output by 99 then downsample the result by 224. The upsample operation uses the familiar "upsample $\rightarrow$ filter" structure and the downsample operation uses the familiar "filter $\rightarrow$ downsample" structure. The approach is illustrated in Figure 11, where the (ideal) filter requirements are also shown.

The brute force approach of Figure 11 may be simplified in two steps.

1. The first step is to recognize that the cascade of a low-pass filter with cut-off frequency $^\pi/_{99}$ rads/sample with another low-pass filter with cut-off frequency $^\pi/_{224}$ rads/sample is simply a low-pass filter with cut-off frequency $^\pi/_{224}$. This is shown in the top portion of Figure 12.

2. Even with the replacement of the cascade of the two low-pass filters by a single low-pass filter, the system shown in the top portion of Figure 12 is still not practical. This is because the

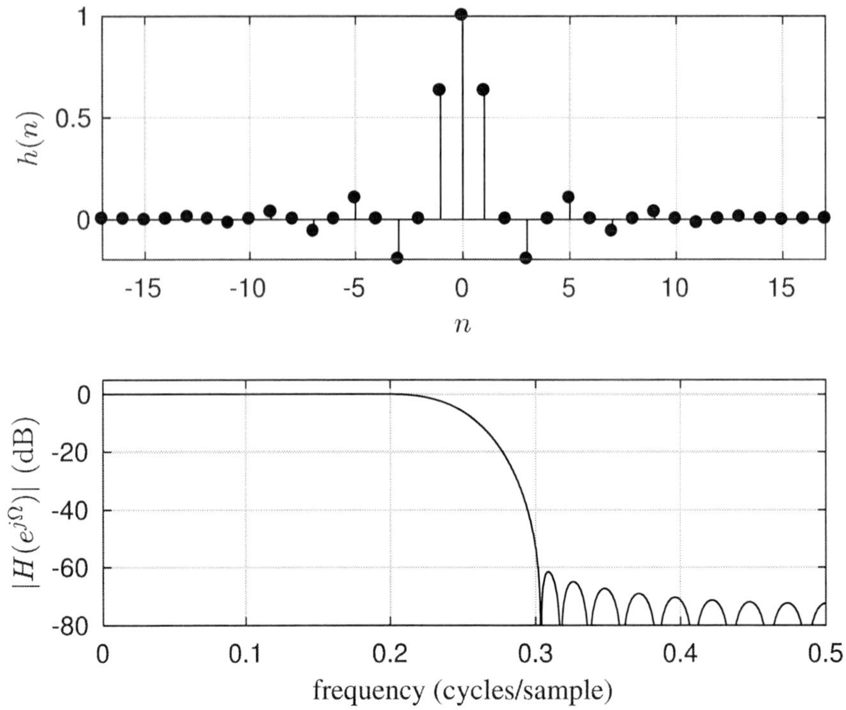

Figure 10: The prototype lowpass filter design.

filter must operate at the intermediate sample rate 99 times the input sample rate! Assuming the length of $F(z)$ is a multiple of 99, a polyphase decomposition may be used to perform the upsample operation. This is shown in the lower portion of Figure 12. The downsample operation simply strides through the polyphase filterbank outputs as shown.

Prototype Filter Design. The spectral requirements for the prototype filter $F(z)$ are summarized in Figure 13. The ideal filter has a cut-off frequency at $\pi/224$ rads/sample. The practical filter must have a transition band. Given the signal properties, a narrow transition band must be used. The filter was designed using $\pi/224 < \Omega < 2\pi/224$ rads/sample as the transition band. As before a stop

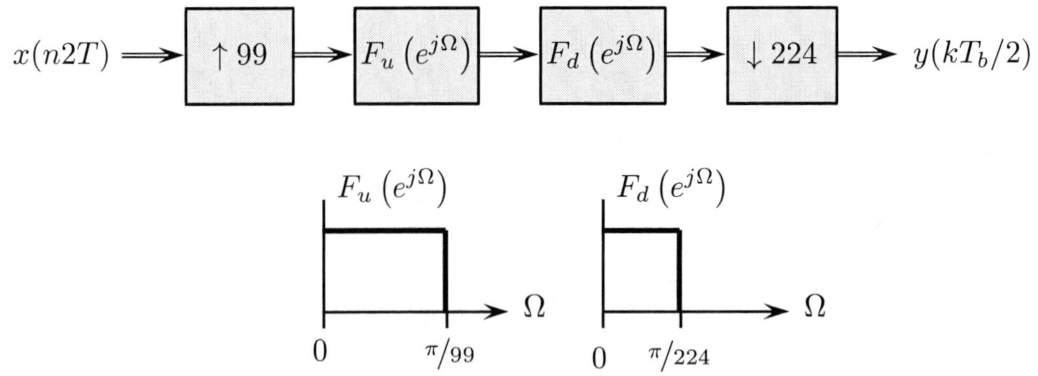

Figure 11: A brute force approach to resampling by $99/224$.

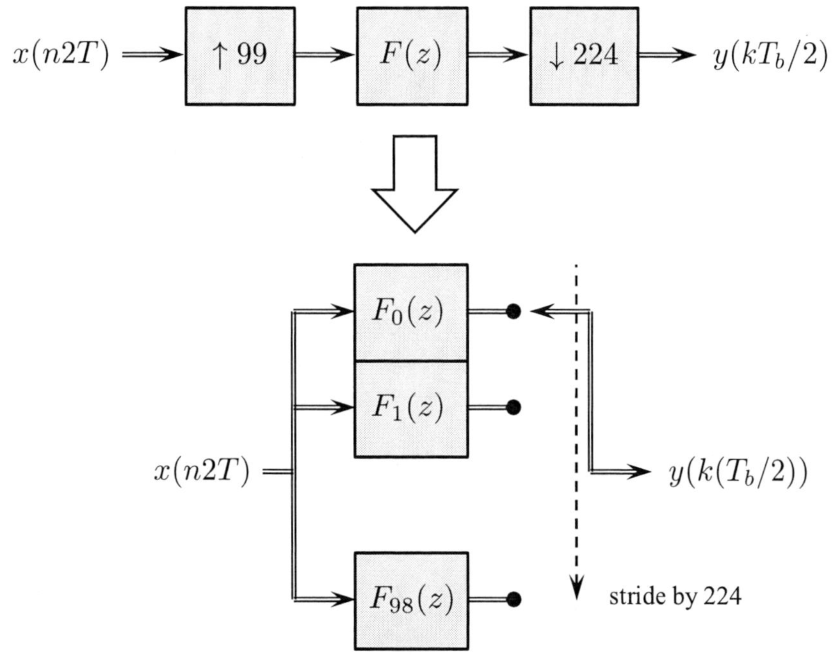

Figure 12: A block diagram of the resampling operation: (top) a conceptual block diagram; (bottom) the polyphase filterbank implementation.

band attenuation of 60 dB is also required. The filter length—defined as the smallest multiple of 99 that meets the stop-band attenuation requirements—is $20 \times 99 = 1{,}980$. The filter and its DTFT are shown in Figure 14.

Operation. The system shown in the lower portion of Figure 12 operates at half the IF sample rate on the input and at 2 times the bit rate on the output. Conceptually, each input sample produces 99 possible outputs, of which only 1 in 224 is required. The first few occurrences in the pattern are illustrated listed in Table 2. The pattern alternates between computing one output for every two inputs and computing one output for every three inputs in a structured way. The table ends with the 224-th input sample that produces the 99-th output sample (i.e., the resampling by $^{99}/_{224}$ means 99 output samples are produced for every 224 input samples). The pattern of subfilter indexes repeats at this point.

In general, the k-th output $y(k)$ is computed from the m-th subfilter using samples $x(n), x(n-1), \ldots, x(n-19)$:

$$y(k) = \sum_{\ell=0}^{19} f_m(\ell) x(n-\ell). \tag{15}$$

The block diagram in the lower portion of Figure 12 suggests that the digital hardware implementation comprises 99 length-20 FIR filters. But because the data in each filter is the same, the reality is that only a single length-20 FIR be implemented in hardware. When the proper samples have been clocked into the filter register, the coefficients corresponding to the desired subfilter are used to compute the output.

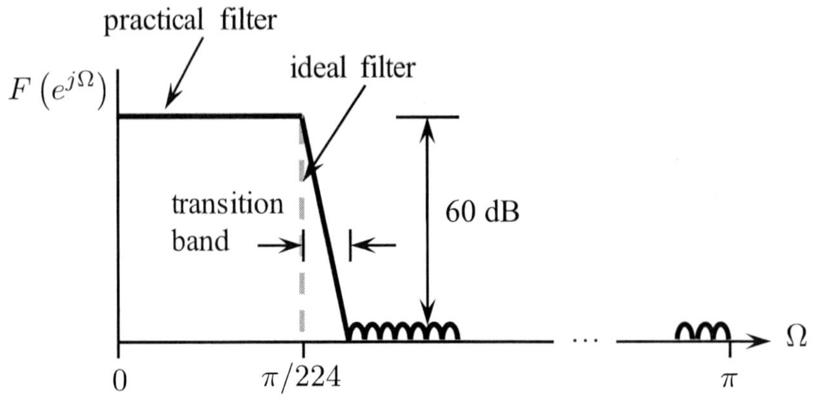

Figure 13: Spectral requirements for the prototype low-pass filter $F(z)$.

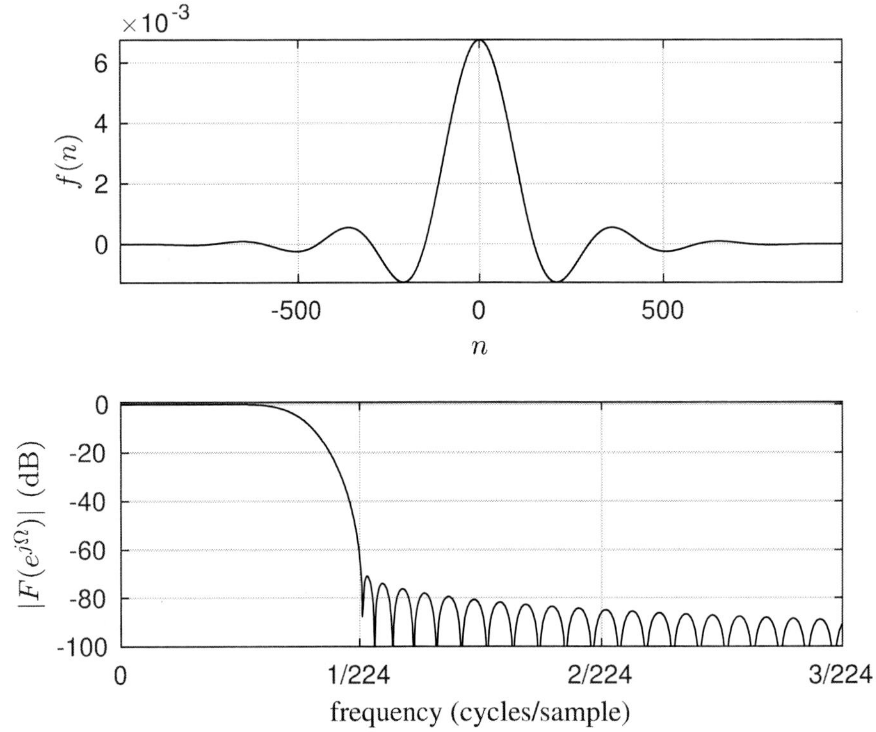

Figure 14: The prototype lowpass filter design for the resampling filter.

The convolution (15) may also be interpreted as an inner product:

$$y(k) = \begin{bmatrix} f_m(0) & \cdots & f_m(19) \end{bmatrix} \begin{bmatrix} x(n) \\ \vdots \\ x(n-19) \end{bmatrix}. \tag{16}$$

Because many digital hardware platforms are equipped with high-performance built-in inner product engines, the interpretation (16) can be leveraged to advantage. As the index pointer slides through the first stage outputs $x(n)$, each desired output is computed by forming the inner product of the length-20 first stage outputs and the appropriate vector of filter coefficients.

Table 2: The relationship between the input sample index, the polyphase filterbank index, and the output sample index.

input sample index	polyphase filter	output sample index
0	F_0	0
1	–	–
2	F_{26}	1
3	–	–
4	F_{52}	2
5	–	–
6	F_{78}	3
7	–	–
8	–	–
9	F_5	4
10	–	–
11	F_{31}	5
⋮		
221	F_{73}	98
222	–	–
223	–	–
224	F_0	99

CONCLUSIONS

This paper summarized the design of a discrete-time quadrature downconversion and resampling processor that operates on samples of 70 MHz IF signal. The unique properties of discrete-time processing—aliasing due to resampling bandpass signals and polyphase filter decompositions—were applied to create a low-complexity approach that does not require any arithmetic operations at the IF sample rate. The required tasks were performed in two stages. The basic principles were highlighted along the way to provide a roadmap for designs that meet different requirements.

ACKNOWLEDGEMENTS

This project is funded by the Test Resource Management Center (TRMC) Test and Evaluation, Science & Technology (T&E/S&T) Program through the U.S. Army Program Executive Office for Simulation, Training, and Instrumentation (PEO STRI) under Contract No. W900KK-13-C-0026

(PAQ). The Executing Agent and Program Manager work out of the AFTC.

REFERENCES

[1] M. Rice, *Digital Communications: A Discrete-Time Approach.* Upper Saddle River, NJ: Pearson Prentice-Hall, 2009.

[2] M. Rice, A. McMurdie, and E. Perrins, "A low-complexity preamble detector for iNET-formatted SOQPSK," in *Proceedings of the IEEE Military Communications Conference*, (Baltimore, MD), pp. 718–723, 6–8 October 2014.

[3] M. Rice and E. Perrins, "On frequency offset estimation using the iNET preamble in frequency selective fading channels," in *Proceedings of the IEEE Military Communications Conference*, (Baltimore, MD), pp. 706–711, 6–8 October 2014.

[4] M. Rice, M. Saquib, M. S. Afran, A. Cole-Rhodes, and F. Moazzami, "On the performance of equalization techniques for aeronautical telemetry," in *Proceedings of the IEEE Military Communications Conference*, (Baltimore, MD), pp. 456–461, 6–8 October 2014.

[5] A. Cole-Rhodes, H. Umuolo, F. Moazzami, and M. Rice, "A block processing approach to CMA equalization of SOQPSK for aeronautical telemetry," in *Proceedings of the IEEE Military Communications Conference*, (Baltimore, MD), pp. 456–461, 6–8 October 2014.

[6] M. Rice, C. Hogstrom, C. Nash, J. Ravert, A. Cole-Rhodes, F. Moazzami, M. Saquib, M. S. Afran, E. Perrins, and K. Temple, "Some initial results for data-aided equalizer experiments at Edwards AFB," in *Proceedings of the International Telemetering Conference*, (Glendale, AZ), 8–10 November 2016.

[7] C. Hogstrom, C. Nash, and M. Rice, "Locating and removing preamble sequence in aeronautical telemetry," in *Proceedings of the International Telemetering Conference*, (Glendale, AZ), 8–10 November 2016.

[8] J. Ravert and M. Rice, "On frequency offset compensation for equalized SOQPSK," in *Proceedings of the International Telemetering Conference*, (Glendale, AZ), 8–10 November 2016.

[9] M. Rice, M. S. Afran, and M. Saquib, "Performance limits on joint estimation of frequency offset, channel, and noise variance in aeronautical telemetry," in *Proceedings of the International Telemetering Conference*, (Las Vegas, NV), 26–29 October 2015.

[10] M. Rice, M. Saquib, and M. S. Afran, "MMSE equalization for aeronautical telemetry channels," in *Proceedings of the International Telemetering Conference*, (San Diego, CA), 20–23 October 2014.

[11] M. Rice, M. Saquib, M. S. Afran, A. Cole-Rhodes, and F. Moazzami, "A comparison of three equalization techniques for iNET-formatted SOQPSK-TG," in *Proceedings of the International Telemetering Conference*, (San Diego, CA), 20–23 October 2014.

[12] M. Rice, M. Saquib, and E. Perrins, "Estimators for iNET-formatted SOQPSK," in *Proceedings of the International Telemetering Conference*, (San Diego, CA), 20–23 October 2014.

[13] A. Cole-Rhodes, H. Umuolo, F. Moazzami, and M. Rice, "Real-time CMA equalization for SOQPSK for aeronautical telemetry," in *Proceedings of the International Telemetering Conference*, (San Diego, CA), 20–23 October 2014.

PERFORMANCE OF DIRECTIONAL MODULATION SYSTEMS IN FADING CHANNELS

Elam Curry and Deva K. Borah
Klipsch School of Electrical & Computer Engineering
New Mexico State University
Las Cruces, NM 88003, USA
email: mrelam1@nmsu.edu, dborah@nmsu.edu

Faculty Advisor:
Deva K. Borah

ABSTRACT

Physical layer security techniques have received increasing interests in recent years due to their ability to provide additional security and their ability to be combined with conventional higher layer security measures. One such physical layer security technique is the directional modulation (DM), where the goal is to provide unhindered communication in an intended direction while hindering the communication in unintended directions. In this paper, we study the effect of Rician fading on a system employing a DM technique that minimizes the intersymbol distance in unintended directions. The performance of the system is investigated for varying ratios of the line-of-sight and scattered signal power, or Rician K-factors. The effect of Rician fading on the bit error rate (BER) performance in intended and unintended directions is studied.

INTRODUCTION

As the data transmitted using a wireless communication system is vulnerable to interception, security is an important aspect of any wireless communication system. Historically, wireless communication security efforts have been taken care of using effective encryption of the data [1]-[2]. Recently, physical layer security measures have received increasing attention, as they can be combined with existing encryption techniques, allowing for further increased security. There are many methods that can provide physical layer security [3]-[5], and they include the beamforming and directional modulation (DM) methods [6]-[8]. The beamforming approach reduces the signal-to-noise ratio (SNR) in unintended transmission directions while maintaining sufficient SNR in an intended direction. In contrast, the goal of the DM approach is to distort the received constellation along the directions of the eavesdroppers, while maintaining an undistorted received constellation to the intended receiver.

In [6], a DM transmit vector design method is proposed where the minimum inter-symbol distance is minimized in unintended directions and kept unaffected in an intended direction. This is accomplished by iterating between a quadratic minimization stage and a linear transportation problem stage. The proposed DM algorithm is shown to provide increased security in comparison to a traditional beamforming array. However, [6] does not consider the presence of fading. Thus, the results given in [6] correspond to the line of sight (LOS) channel component only.

In this paper, we investigate the effect of Rician fading on a DM system employing the DM algorithm proposed in [6]. We formulate an analytical expression for the average bit error rate (BER) of a DM system in the presence of Rician fading. As finding a closed-form expression for the average BER is difficult, we adopt a Monte Carlo integration approach. We study the BER performance of the DM system for various ratios of LOS and scattered signal power. The effect of increasing the number of transmit antennas is studied with regard to diminishing the BER degradation effect of fading as well as increasing security.

Notations: We use bold lower case letters for column vectors, and bold upper case letters for matrices. For a set of vectors, $\mathbf{b}_i$ denotes the i-th vector, and $b_{i,m}$ is the m-th element of vector $\mathbf{b}_i$. For a matrix $\mathbf{A}$, $A_{m,n}$ denotes the (m,n)-th element. $\mathrm{Re}(\cdot)$ and $\mathrm{Im}(\cdot)$ are the real and imaginary parts of a complex number. The notation $[\cdot]^T$ denotes transpose, $[\cdot]^H$ is the Hermitian operator, $||\cdot||_l$ is the l-norm, $\binom{n}{k}$ denotes $n!/(n-k)!k!$, $\mathbf{0}_{m,n}$ is a zero matrix of size $m \times n$, $\mathcal{CN}(m,\sigma^2)$ denotes the complex Gaussian distribution with mean m and variance σ^2, and $E[\cdot]$ denotes expectation.

SYSTEM MODEL

We consider an N-element uniform linear array (ULA) with inter-antenna spacing d_a located at a distance d from an intended receiver as shown in Figure 1. DM provides an undistorted M-ary

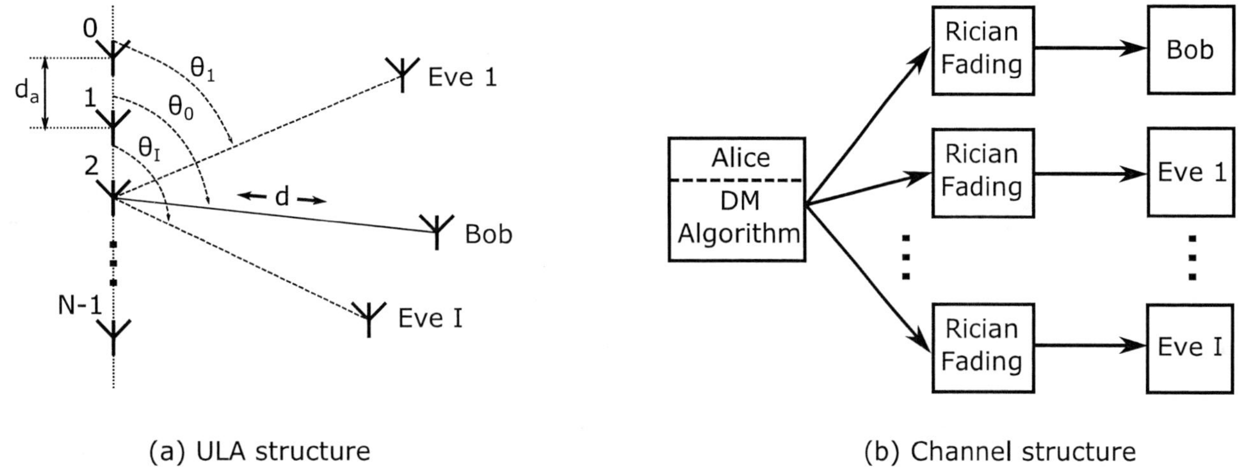

Figure 1: A ULA that transmits through a Rician fading channel to an intended receiver and I eavesdroppers.

constellation $\mathcal{U}$ in an intended direction θ_0, while distorting that constellation in multiple unintended directions θ_i, $i = 1, 2, \ldots, I$. The complex $N \times 1$ overall Rician channel vector is expressed as [9]

$$\mathbf{h}(\theta) = \sqrt{\frac{K_r}{1+K_r}}\mathbf{g}(\theta) + \sqrt{\frac{1}{1+K_r}}\mathbf{c}, \tag{1}$$

where K_r is the Rician K-factor, $\mathbf{c}$ is the scattered channel component with elements distributed as $\mathcal{CN}(0,1)$, and $\mathbf{g}(\theta)$ is the LOS channel component. The LOS channel component for a receiver located in the direction θ is calculated as

$$\mathbf{g}(\theta) = [1, \exp(j\phi_1(\theta)), \dots, \exp(j\phi_{N-1}(\theta))]^T, \tag{2}$$

where $\phi_n(\theta) = (2\pi/\lambda)nd\cos\theta$, $n = 1, \dots, N-1$, and λ is the wavelength. The transmission of symbol $u_m \in \mathcal{U}$ corresponds to an array excitement vector $\mathbf{x}_m = [x_{m,1}, x_{m,2}, \dots, x_{m,N}]^T$. The noiseless received symbol in the direction θ due to u_m is

$$s_m(\theta) = \sqrt{P_r}\mathbf{h}^H(\theta)\mathbf{x}_m = \sqrt{\frac{K_r P_r}{1+K_r}}\mathbf{g}^H(\theta)\mathbf{x}_m + \sqrt{\frac{P_r}{1+K_r}}\mathbf{c}^H\mathbf{x}_m, \tag{3}$$

where $P_r = P_t d^{-\alpha}$ is the received power, P_t is the transmitted power, and α is the path loss exponent. For a given transmitted symbol $\mathbf{x}_m$ and transmission direction θ,

$$E[s_m(\theta)] = \sqrt{\frac{K_r P_r}{1+K_r}}\mathbf{g}^H(\theta)\mathbf{x}_m \quad \text{and} \quad \text{var}(s_m(\theta)) = \frac{P_r}{1+K_r}||\mathbf{x}_m||^2.$$

There are a total of $K = \binom{M}{2}$ symbol pairs. The symbol pair (m, p) is also referred to as the k-th symbol pair with intersymbol distance $D(k, \theta) = |s_m(\theta) - s_p(\theta)|$. The DM signal optimization is accomplished via minimization of $D_l(k, \theta_i) = |\mathbf{g}^H(\theta_i)\mathbf{x}_m - \mathbf{g}^H(\theta_i)\mathbf{x}_p|^2$ along the undesired directions θ_i, $i = 1, \dots, I$. This requires: (1) the assignment of specific symbols pairs to given unintended directions, and (2) the design of array excitement vectors $\mathbf{x}_m$, $m = 1, \dots M$ for the specific symbol pair assignments along the unintended directions.

THE DM ALGORITHM

Since the DM algorithm is based on the transmit directions θ_i, we consider only the LOS channel component while performing the DM design. The DM transmit vector design method proposed in [6] is used. The DM algorithm consists of three important parts: (1) Symbol pair assignment (SPA), (2) Transmit vector optimization (TVO), and (3) the iterations between the SPA and TVO stages. The SPA stage is formulated as a transportation problem, where the K symbol pairs are assigned to the I unintended directions and a dummy direction θ_{I+1} is introduced to balance the supply and demand. The $K \times (I+1)$ decision matrix $\mathbf{Y}$ has elements $y_{k,i} \in \{0, 1\}$, where $y_{k,i} = 1$, $i = 1, 2, \dots, I$, indicates the selection of symbol pair k for direction θ_i. Column $(I+1)$ of $\mathbf{Y}$ has elements $y_{k,I+1} \in \{0, 1, \dots, S_{max}\}$ that absorb excess supply, where S_{max} is the maximum number of times a symbol pair can be assigned. The corresponding $K \times (I+1)$ cost matrix is $\tilde{\mathbf{C}}$ with elements $\tilde{C}_{k,i} = D_l(k, \theta_i)$, $i = 1, 2, \dots, I$, and $\tilde{C}_{k,i} = 0$ for $i = I+1$. The i-th column of $\mathbf{Y}$ is denoted as $\mathbf{y}_i$, and the i-th column of $\tilde{\mathbf{C}}$ is denoted as $\tilde{\mathbf{c}}_i$. The decision vector, which gives symbol pair assignments, and cost vector are then $\mathbf{y} = [\mathbf{y}_1^T, \mathbf{y}_2^T, \dots, \mathbf{y}_I^T]^T$ and $\tilde{\mathbf{c}} = [\tilde{\mathbf{c}}_1^T, \tilde{\mathbf{c}}_1^T, \dots, \tilde{\mathbf{c}}_I^T]^T$ respectively. The final transportation problem is expressed as

$$\min_{i,j} \ \mathbf{y}^T\tilde{\mathbf{c}} \tag{4}$$

$$\text{s.t.} \sum_{k=1}^{K} y_{k,i} = 1, i = 1, \dots, I, \quad \sum_{k=1}^{K} y_{k,I+1} = KS_{max} - I, \quad \text{and} \quad \sum_{i=1}^{I+1} y_{k,i} = S_{max}, \ k = 1, \dots, K.$$

The TVO stage seeks to minimize [6]

$$|D_l(k,\theta_i)|^2 = |\mathbf{g}^H(\theta_i)\mathbf{x}_m - \mathbf{g}^H(\theta_i)\mathbf{x}_p|^2 = (\mathbf{x}_m - \mathbf{x}_p)^H \mathbf{g}(\theta)\mathbf{g}^H(\theta)(\mathbf{x}_m - \mathbf{x}_p), \tag{5}$$

for $i = 1, \ldots, I$ given symbol pair assignments. Let **x** be $2NM \times 1$ vector defined as

$$\mathbf{x} = [\text{Re}(\mathbf{x}_1^T),\ \text{Im}(\mathbf{x}_1^T),\ \ldots, \text{Re}(\mathbf{x}_M^T),\ \text{Im}(\mathbf{x}_M^T)]^T. \tag{6}$$

Next, let **B** be the $2IN \times 2IN$ block diagonal matrix whose i-th block corresponds to unintended direction θ_i and is found as

$$\tilde{\mathbf{B}}(\theta_i) = \begin{bmatrix} \text{Re}(\mathbf{g}(\theta_i)\mathbf{g}^H(\theta_i)), & -\text{Im}(\mathbf{g}(\theta_i)\mathbf{g}^H(\theta_i)) \\ \text{Im}(\mathbf{g}(\theta_i)\mathbf{g}^H(\theta_i)), & \text{Re}(\mathbf{g}(\theta_i)\mathbf{g}^H(\theta_i)) \end{bmatrix}.$$

In order to control BER performance in the intended direction, a new constraint is introduced, and we denote η as the amplitude reduction in the intended direction θ_0 due to the DM algorithm. Define the $2M \times 2MN$ block diagonal matrix **H** whose i-th block is

$$\tilde{\mathbf{H}}_0 = \begin{bmatrix} \text{Re}(\mathbf{g}^T(\theta_0)), & \text{Im}(\mathbf{g}^T(\theta_0)) \\ -\text{Im}(\mathbf{g}^T(\theta_0)), & \text{Re}(\mathbf{g}^T(\theta_0)) \end{bmatrix}.$$

Lastly, define the vector

$$\mathbf{u} = [\text{Re}(u_1),\ \text{Im}(u_1),\ \ldots,\ \text{Re}(u_M),\ \text{Im}(u_M)]^T \tag{7}$$

Defining a matrix $\mathbf{C} = \mathbf{A}^T\mathbf{B}\mathbf{A}$, where the matrix **A** consists of 1, 0, and -1 as defined in [6], the TVO problem is expressed as

$$\begin{aligned} & \min_{\mathbf{x}} \frac{1}{2}\mathbf{x}^T(\mathbf{C}^T + \mathbf{C})\mathbf{x} \\ \text{s.t.} \quad & \mathbf{H}\mathbf{u} - N\eta\mathbf{u} = \mathbf{0}_{2M,1} \\ & -1 \le \text{Re}(x_{m,n}) \le 1 \\ & -1 \le \text{Im}(x_{m,n}) \le 1 \\ \text{for } & m = 1,\ \ldots, M, \ \text{ and } \ n = 1,\ \ldots, N \end{aligned} \tag{8}$$

The DM algorithm is initialized with a random symbol pair-direction assignment. The TVO stage is performed and the solution is used to produce the cost matrix $\tilde{\mathbf{C}}$ for the SPA stage. The new assignments from the SPA stage are then passed to the TVO stage. This process continues until the symbol pair assignment does not change. Because the solution of the DM algorithm is not guranteed to be globally optimal, the DM algorithm is performed with multiple symbol pair assignments, and the solution that offers the minimum solution to (8) is finally selected.

AVERAGE BIT ERROR PROBABILITY

We consider received symbols in the presence of complex Gaussian noise with zero mean and variance N_o, where N_o is the noise spectral density. We then define the SNR as $\gamma = E_b/N_o$, where

$$E_b = \frac{1}{M\log_2 M}\sum_{m=1}^{M} E[|s_m(\theta_0)|^2], \tag{9}$$

and $E[|s_m(\theta_0)|^2]$ is calculated as

$$E[|s_m(\theta_0)|^2] = \frac{K_r P_r}{1+K_r}|\mathbf{g}^H(\theta)\mathbf{x}_m|^2 + \frac{P_r}{1+K_r}||\mathbf{x}_m||_2^2. \tag{10}$$

Due to the presence of fading, it is necessary to determine the average BER, which can expressed for Gray coded systems as

$$\bar{P}_b \approx \frac{1}{\log_2 M}\int_{\mathbb{C}} P_e(\mathbf{g},\mathbf{x})p_{\mathbf{c}}(\mathbf{x})d\mathbf{x}, \tag{11}$$

where $P_e(\mathbf{g},\mathbf{c})$ is the symbol error rate (SER) for a particular channel realization $\mathbf{h}$, and $p_{\mathbf{c}}(\mathbf{x})$ is the probability density function (PDF) of $\mathbf{c}$. Using the nearest neighbor approximation, $P_e(\mathbf{g},\mathbf{c})$ can be expressed as [10], [11]

$$P_e(\mathbf{g},\mathbf{c}) \approx \frac{1}{M}\sum_{m=1}^{M} N_m Q\left(\frac{d_{min}^{(m)}}{\sqrt{2N_0}}\right), \quad \text{with } d_{min}^{(m)} = \min_{p=1,\dots,M,p\neq m} |s_m(\theta) - s_p(\theta)|, \tag{12}$$

where N_m is the number of nearest neighbors for $s_m(\theta)$, and $Q(x) = 1/\sqrt{2\pi}\int_x^\infty \exp(-y^2/2)dy$. Using (12) in (11), along with the complex Gaussian PDF of $\mathbf{c}$, allows for an analytical evaluation for $\bar{P}_b$. However, obtaining a closed-form expression for the average BER in the presence of Rician fading is difficult. Therefore, we compute (11) numerically. We adopt a Monte Carlo integration approach. We randomly generate T channel vectors $\mathbf{c}_i$ according to (1) and find the SER $P_e(\mathbf{g},\mathbf{c}_i)$ using (12) for each channel vector. The average BER is then

$$\bar{P}_b = \frac{1}{T\log_2 M}\sum_{i=1}^{T} P_e(\mathbf{g},\mathbf{c}_i) \tag{13}$$

In order for (13) to give an accurate estimate of the average BER, it is necessary to choose T to be large. Increasing T decreases the variance of $\bar{P}_b$ but simultaneously increases the computational complexity of using (13).

RESULTS AND DISCUSSION

Our results consider quadrature phase shift keying (QPSK) modulation with $M = 4$ symbols and $K = 6$ symbol pairs. We fix $\eta = 0.5$ in (8), $S_{max} = 8$, and SNR $\gamma = 10$ dB. The considered ULA has $d_a = \lambda/2$ spacing, $\alpha = 2.2$, and is located at a distance of $d = 1000$ m from the intended receiver. In the DM algorithm, we consider $I = 20$ unintended directions with the θ_i's distributed evenly on either side of θ_0.

Figure 2 shows the received QPSK constellation for the DM system in the intended direction $\theta_0 = 90^o$ and unintended direction $\theta_i = 65^o$ with and without the presence of $K_r = 10$ Rician fading. The DM system maintains the QPSK constellation in the intended direction $\theta_0 = 90^o$. In the unintended direction $\theta_i = 65^o$, the distance between symbols is reduced significantly by the DM algorithm. In the presence of Rician fading, the QPSK constellation is randomly distorted for both the intended and unintended directions. We use a single realization of $\mathbf{h}$ with $\mathbf{c} = [-0.5594 + j0.1600,\ -1.0487 + j0.1396,\ -0.3061 + j1.2109,\ 0.1718 - j0.0177,\ 1.0188 + j1.0130]$. The

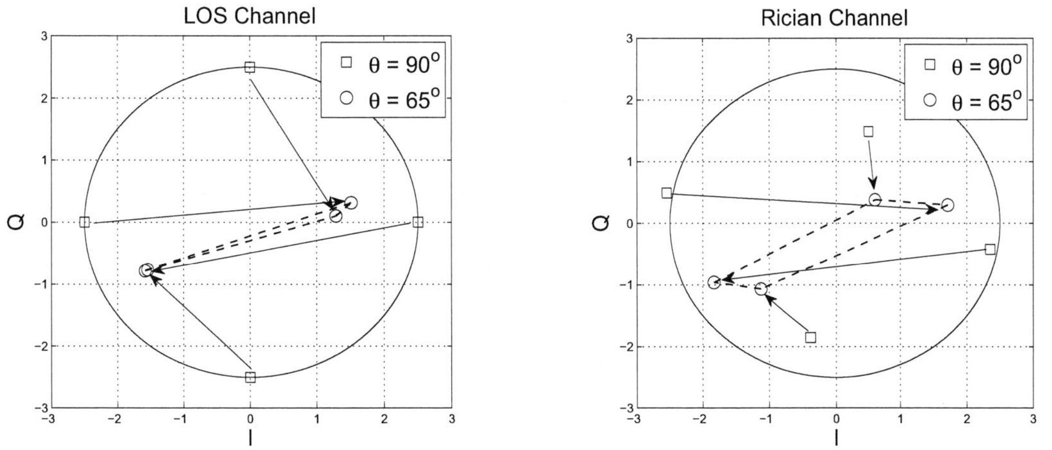

Figure 2: The received QPSK constellation for the intended direction $\theta_0 = 90^o$ and unintended direction $\theta_i = 65^o$. We use $K_r = 10$.

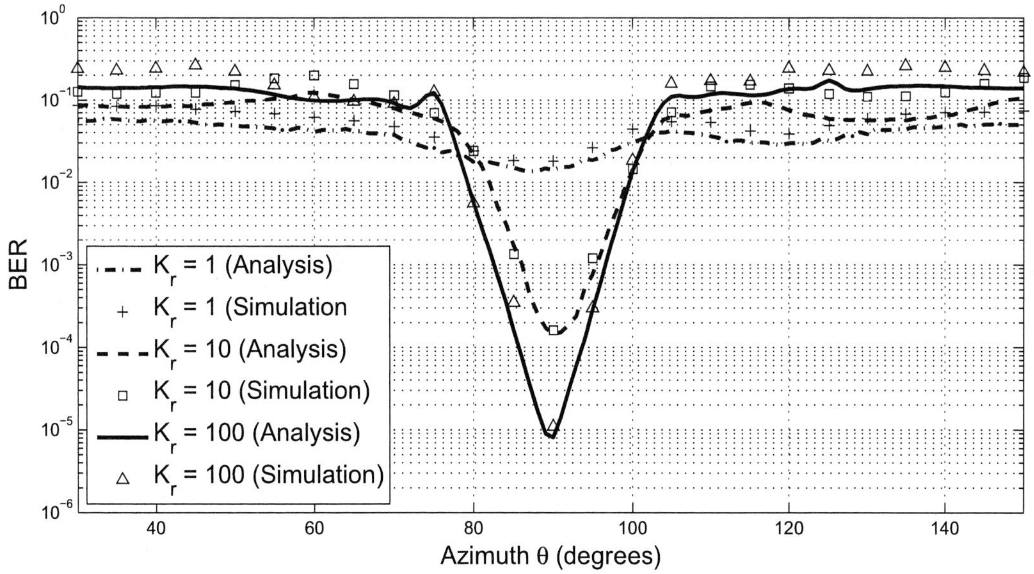

Figure 3: BER vs Azimuth angle for various K_r values with $N = 5$.

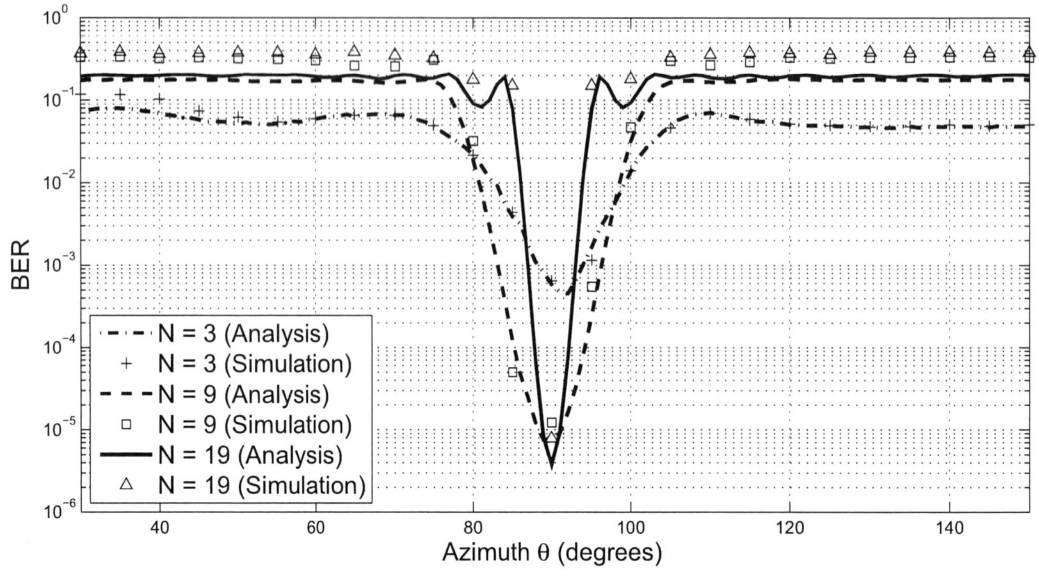

Figure 4: BER vs Azimuth angle for various N values with $K_r = 10$.

Rician channel results in the minimum distance among symbols in the intended direction being decreased, while also spreading symbols apart in the unintended direction, improving the BER performance for eavesdroppers for this specific realization of the channel.

In Figure 3, the BER performance of the DM system is shown for various K_r values with intended direction $\theta_0 = 90^o$. The Monte Carlo integration approach for obtaining $\bar{P}_b$ is shown to closely match with the simulation results. For low K_r values, the BER performance is degraded in the intended direction. Additionally, the BER is lower in unintended transmission directions because of symbol spreading due to fading. As one would expect, the results shown for $K_r = 100$ are similar to those presented in [6] as the LOS channel component is highly dominant.

Figure 4 presents the BER performance of the DM system for various values of N with $K_r = 10$ and intended direction $\theta_0 = 90^o$. The results indicate that there are dual benefits of increasing N. First, increasing N results in a narrow, low BER region with poor, angle-flat BER performance in unintended directions. Secondly, due to increased transmitter diversity, increasing N helps reduce the BER degradation due to fading. The figure suggests that a ULA with high N will result in high secrecy in unintended directions and low BER in the intended direction even in the presence of fading.

Figure 5 shows BER performance for various N values with $K_r = 5$ and intended direction $\theta_0 = 60^0$. For this figure only, we use 3 unintended directions for $\theta < \theta_0$ and 17 unintended directions for $\theta > \theta_0$. We first note that because K_r is lower than in Figure 4, the BER performance is worse in the intended direction. Secondly, the main lobe for the $N = 19$ case in Figure 4 has $\bar{P}_b \leq 10^{-1}$ for an angle range of about 10.5^o, whereas the main lobe for the $N = 19$ case in Figure 5 has $\bar{P}_b \leq 10^{-1}$ for an angle range of about 12°.

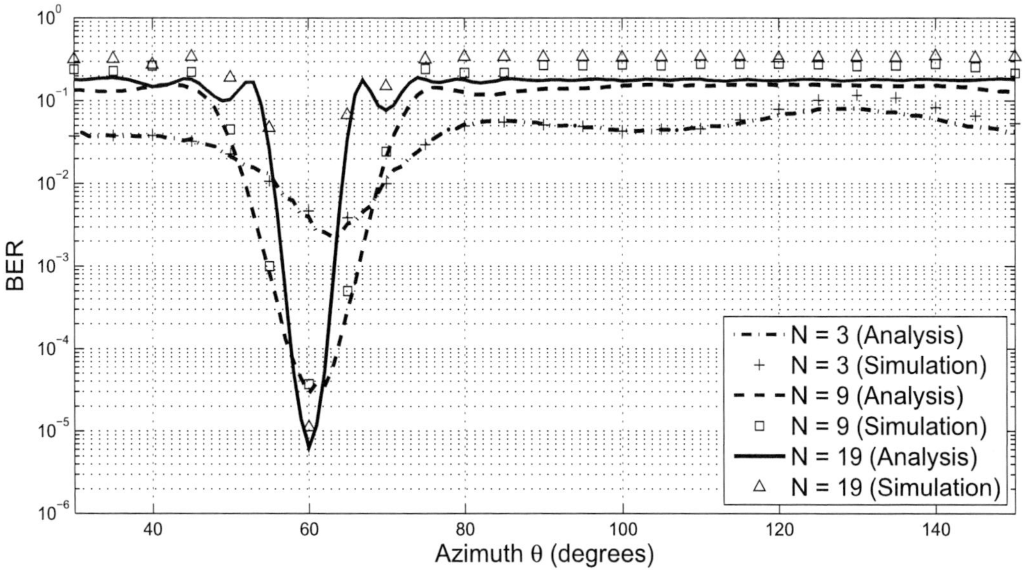

Figure 5: BER vs Azimuth angle for various N values with $K_r = 5$.

CONCLUSION

We consider a DM communication system as in [6] in the presence of Rician fading. An analytical average BER evaluation method is proposed using a Monte Carlo integration approach. Discussions and results are provided for a number of physical scenarios as well as Rician K-factors. It is shown that fading can degrade the BER in the intended direction while simultaneously improving BER in the unintended directions. The results presented suggest that a higher number of transmit antennas can reduce the effect of BER degradation as well as increase security in the presence of fading.

REFERENCES

[1] M. Jalili and H. Sarbazi-Azad, "Endurance-aware security enhancement in non-volatile memories using compression and selective encryption," *IEEE Transactions on Computers*, vol. 66, pp. 1132–1144, Jul. 2017.

[2] W. Ma, Q. Wu, Y. Tan, C. Wang, Q. Wu, and H. Hu, "Using all-or-nothing encryption to enhance the security of searchable encryption," in *Proc. IEEE International Conference on Computational Science and Engineering*, pp. 104–108, Sydney, Australia, Dec. 2013.

[3] D. Wang, B. Bai, W. Chen, and Z. Han, "Achieving high energy efficiency and physical-layer security in AF relaying," *IEEE Transactions on Wireless Communications*, vol. 15, pp. 740–752, Jan. 2016.

[4] Y. Zheng, M. Schulz, W. Lou, Y. T. Hou, and M. Hollick, "Highly efficient known-plaintext attacks against orthogonal blinding based physical layer security," *IEEE Wireless Communications Letters*, vol. 4, pp. 34–37, Feb. 2015.

[5] S. R. Aghdam and T. M. Duman, "Physical layer security for space shift keying transmission with precoding," *IEEE Wireless Communications Letters*, vol. 5, pp. 180–183, Apr. 2016.

[6] R. M. Christopher and D. K. Borah, "Optimization of symbol distance metric in directional modulation systems," in *Proc. International Telemetering Conference*, Glendale, AZ, Nov. 2016.

[7] M. P. Daly and J. T. Bernhard, "Directional modulation and coding in arrays," in *Proc. IEEE International Symposium on Antennas and Propagation (APSURSI)*, pp. 1984–1987, Spokane, WA, Jul. 2011.

[8] F. Sohrabi and W. Yu, "Hybrid analog and digital beamforming for mmWave OFDM large-scale antenna arrays," *IEEE Journal on Selected Areas in Communications*, vol. 35, pp. 1432–1443, Apr. 2017.

[9] C. Liu and R. Malaney, "Location-based beamforming and physical layer security in Rician wiretap channels," *IEEE Transactions on Wireless Communications*, vol. 15, pp. 7847–7857, Nov. 2016.

[10] S. Benedetto and E. Biglieri, *Principles of Digital Transmission with Wireles Applications*. Kluwer Academic/Plenum Publishers, New York, 1999.

[11] Y. Sun, D. K. Borah, and E. Curry, "Optimal symbol set selection in GSSK visible light wireless communication systems," *IEEE Photonics Technology Letters*, vol. 28, pp. 303–306, Feb. 2016.

LINK QUALITY METRICS FOR ADAPTIVE CODING AND MODULATION WITH SOQPSK AND OFDM

Enkuang D Wang, Timothy J Brothers, and Richard T Causey
Georgia Tech Research Institute
Atlanta, Georgia 30318
{enkuang.wang, timothy.brothers, richard.causey}@gtri.gatech.edu

ABSTRACT

In previous work, we presented a link quality metric for adaptive modulation and coding of two standard telemetry waveforms, orthogonal frequency-division multiplexing (OFDM) and shaped-offset quadrature phase shift keying (SOQPSK). That metric unified error vector magnitude (EVM) and Godard dispersion, for OFDM and SOQPSK, respectively, in the contexts of additive white Gaussian noise (AWGN) channels. In this paper, we present an alternative metric based on low density parity check (LDPC) decoding iterations. We show this new metric to be an equally effective substitute for AWGN channels but also to be applicable to a wider variety of channels, including those with multipath and interference. Furthermore, we show the metric to be robust within sub-optimal, lower complexity receiver architectures.

INTRODUCTION

The migration toward the integrated network enhanced telemetry (iNET) system introduces a new challenge of defining adaptive modulation and coding algorithms for the telemetry community. Adaptive modulation and coding schemes have been widely used in several Long Term Evolution (LTE) and Wi-Fi standards [1-2]. Similar adaptive algorithms have not yet been applied to aeronautical telemetry. In order to accomplish this task, the adaptive algorithm must employ two distinct modulation waveforms defined in iNET: 1) telemetry group SOQPSK (SOQPSK-TG) and 2) OFDM. Solving this task involves a multi-scheme adaptation, in which both SOQPSK and OFDM are used under the same decision rule. Georgia Tech Research Institute (GTRI) has investigated adaptive modulation and coding scheme for aeronautical telemetry, developed an algorithm that employs both modulation types, validated its performance via simulations, and shown operation in a laboratory environment. The performance of an adaptive scheme has shown to have 30% to 340% goodput performance improvement over a baseline scheme in various static telemetry channel models [1].

To further raise the technology readiness level (TRL) of using adaptive schemes in aeronautical telemetry, GTRI continued this effort by implementing the design onto a software defined radio (SDR) platform for test range demonstration. The transmitter hardware is composed of a National Instruments (NI) PXIe-7975R FlexRIO FPGA module, an NI 5791 RF transceiver module, an NI 8135 controller. The receiver hardware is composed of a Linux server with a Xilinx VC707 development card, a 4DSP FMC150 analog to digital converter card, and a custom built RF downconverter.

This paper focuses adaptation selection criteria implemented in the VC707 card. The field programmable gate array (FPGA) implementation is processing constrained and as such the efficiency of the adaptation rule must be mathematically simplified for the FPGA. The first part of the paper will discuss the hardware system and the adaptation algorithm developed in previous efforts. This will be followed by a description of a new link quality metric that exploits LDPC decoder information and an update on the system architecture requirement to enable this feature. The conclusion shows that utilizing the LDPC decoder iteration count can lead to a simpler adapter architecture that can be implemented in an FPGA.

BACKGROUND

In our previous work [2], we implemented a transceiver that employs both iNET waveforms: SOQPSK-TG and OFDM. In addition, LDPC [3] was used as the forward error correction (FEC) code. The top-level architecture of this system is shown in Figure 1. The objective of developing such a system is to demonstrate an adaptive algorithm that can adjust the waveform and FEC code based on an approximation of the link quality. One note of interest pertaining to Figure 1 is that Link Dependent Adaptive Radio's (LDAR) adapter is driven by the signal-to-noise ratio (SNR) estimator, which is approximated using EVM for OFDM and dispersion for SOQPSK.

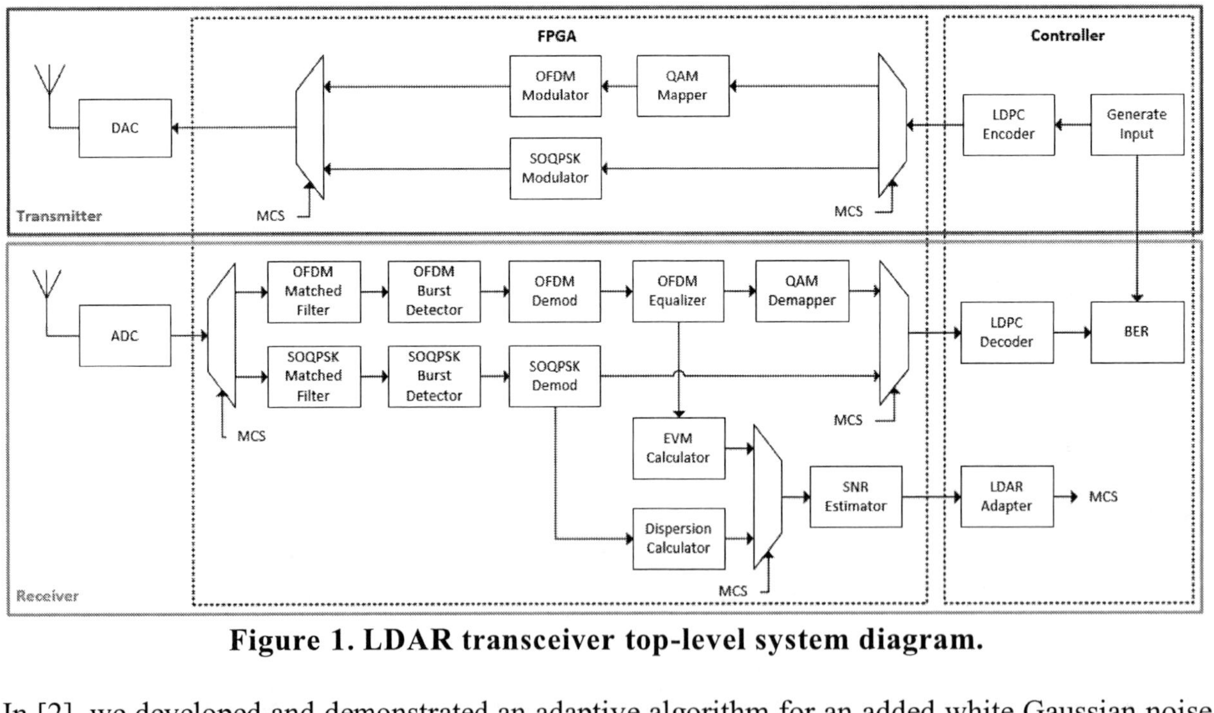

Figure 1. LDAR transceiver top-level system diagram.

In [2], we developed and demonstrated an adaptive algorithm for an added white Gaussian noise (AWGN) channel. This is done by using an approximation between dispersion and SNR. The problem arises when the channel is no longer AWGN. The dispersion equation for non-AWGN channel, derived in [4], is shown in (1)

$$Dispersion = 2\sum_{l=0}^{L-1}\sum_{m=0}^{M-1}|h_l|^2|h_m|^2 + \sum_{l=0}^{L-1}|h_l|^4 + 2\left[\left(\frac{EVM}{\frac{1}{N}\sum_{jj=1}^{N}\left|H_{jj}^{-1}\right|^2}\right)^2 + \frac{EVM}{\frac{1}{N}\sum_{jj=1}^{N}\left|H_{jj}^{-1}\right|^2}\right] - 1 \quad (1)$$

As can be seen, the complexity of (1) is much greater than the assumptions made in [2]. GTRI investigated the assumptions made in [4] to understand if (1) could be used in a realistic channel. The investigation motivated the work presented in this paper.

LINK QUALITY METRIC

Previously, we had used an approximation of SNR to determine the adaptation algorithm. The waveform and FEC code would be adjusted depending on the approximated SNR. To expand this to a generic form, a link quality metric is defined. When we refer to the "link," we are speaking about the received signal strength versus all physical impairment (e.g. multipath propagation, interference, and noise) of the environment over which the communications take place. The "link" also encompasses the choice of modulation and coding as well as the transceiver implementation. From the perspective of the system designer, the link quality metric should take all of these elements into account: mode of operation, transceiver complexity, and the physical communications link.

The link metric can be seen as a measure of link quality from the viewpoint of the receiver. Ideally, it would incorporate all the physical channel impairments as well as any non-ideal implementations in other parts of the transceiver. If we focus specifically on the LDPC decoder for a quantization of the link quality metric, this would return a consistent number regardless of waveform or complexity of the system. Specifically, we are going to examine the number of iterations required for the LDPC to converge.

To design an adaptive algorithm that can be efficiently implemented onto an FPGA, GTRI investigated the potential of using LDPC decoder iteration count as a driver for adaptive modulation and coding algorithms. The LDPC decoder used here is a soft decision decoder developed by NASA's JPL [3], which makes use of the log-likelihood ratio (LLR) information that contains some intrinsic knowledge of the channel properties. The decoder, then, gains some extrinsic knowledge of the code constraints during one of its decoding iterations. This information is used as the a priori information for the subsequent decoding iterations until the valid codeword is found or the maximum number of iterations has passed. An in-depth explanation of a soft decision LDPC decoder can be found in Chapter 2 of [5].

Using the soft-decision decoder, the bit error rate (BER) and LDPC decoder iteration plots for all available LDAR modulation and coding schemes (MCS) for an AWGN channel is shown in Figure 2. In this figure, the black lines represent no LDPC code, and the blue, red, and magenta lines represent 4/5, 2/3, and 1/2 rates LDPC for all modulations, respectively. The maximum decoding iterations for all three LDPC code rates are set to 10, which means the decoder has a maximum of 10 tries to perform the syndrome check to determine if the codeword is valid. It can be clearly seen in Figure 2 that there is a direct correlation between the BER curves and the number of LDPC iterations. As the SNR increases for a particular MCS, the BER and decoder iterations both decrease.

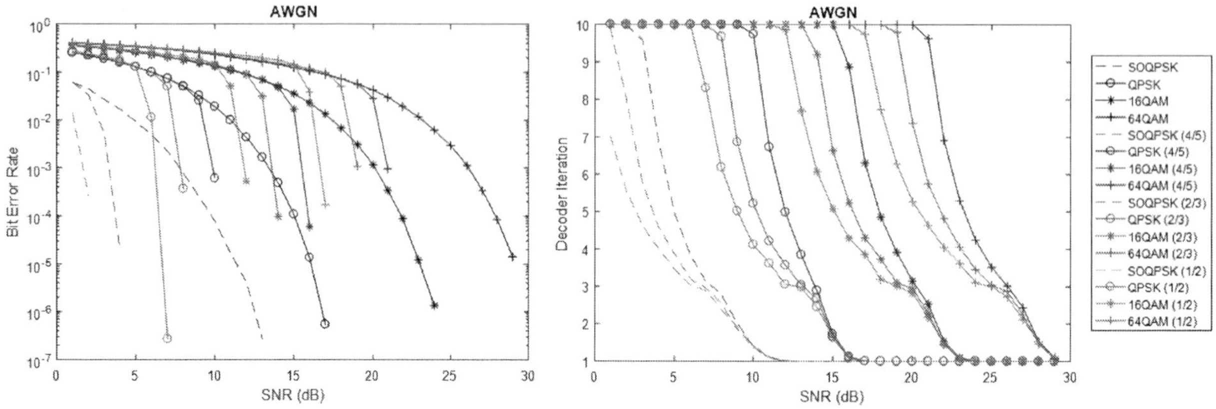

Figure 2. BER and decoder iteration counts for all LDAR MCS selections.

Another important note relating to Figure 2 is that at certain SNRs, the performance of all three code rates start to converge. Using SOQPSK (dash lines) as an example, at approximately 8 dB, the number of iterations for all three code rates slowly converge as the SNR increases. This behavior indicates that as the SNR increases the effects of the LDPC codes slowly diminishes and converges with the uncoded BER curves. This information gives the system designer an advantage of knowing when to switch to a MCS with higher data throughput or back-off to a more robust MCS to overcome channel impairments.

SYSTEM DESIGN

The BER and decoding iterations performance shown in Figure 2 highly correlates with how the communications system is designed. This section highlights some of the important blocks that are required to produce the results shown throughout this paper. One of the most important components is the LDPC encoder and decoder pair since the channel metric we are evaluating relies heavily on the number of LDPC decoding iterations. Another important component pertaining to the performance of the system is the equalizer. The following sub-sections highlight the equalizers used in both SOQPSK and OFDM demodulators.

A. SOQPSK

Due to the system requirement for soft decision inputs, a symbol-by-symbol (SxS) OQPSK detector is employed over a Viterbi detector. The soft decision of an SxS OQPSK detector is defined by the real and imaginary parts of the symbol when the sample number is even and odd, respectively [6]. The block diagram for SOQPSK simulation is shown in Figure 3, where the bits are encoded and modulated. The modulated signals are then filtered by one of the multipath channels shown in Figure 5. To overcome the inter-symbol interferences (ISI) resulting from the multipath propagation, an equalizer is used for the SOQPSK signal. Various equalization techniques for SOQPSK have been studied in [7]. The results have shown the performance difference between a 4-state trellis detector and a symbol-by-symbol detector with equalizer is around 0 to 2 dB. The minimum mean-squared error (MMSE) equalizer used here is shown in (2),

$$c(n) = \boldsymbol{R}_{yy}^{-1} R_{xy} \tag{2}$$

where $\boldsymbol{R}_{yy}$ is a convolution matrix derived from the autocorrelation of the received preambles, and R_{xy} is the cross-correlation of the transmitted and received preambles. The derived equalizer coefficients, $c(n)$, are used to attempt to *undo* the effects of the multipath channel.

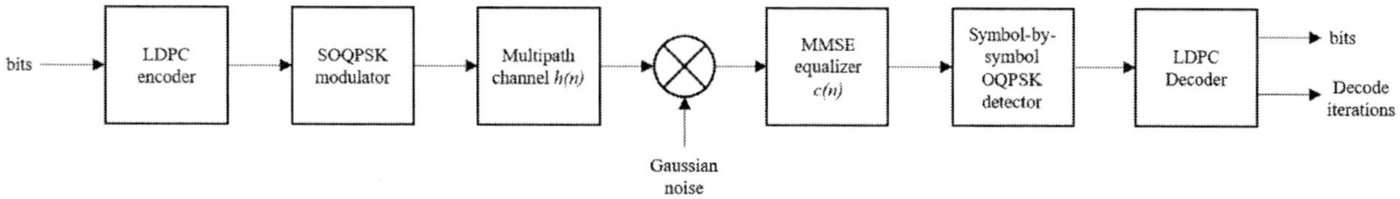

Figure 3. Block diagram for the simulation procedure.

B. OFDM

For OFDM, the block diagram is shown in Figure 4. The soft decisions are computed in the symbol demapper. The difference between SOQPSK and OFDM is that the receiver signal processing is mainly done in the frequency domain for OFDM. A major advantage of OFDM is its ability to easily adapt to frequency-selective channels without complex time-domain equalization. In the frequency domain, the equalizer becomes a one tap multiplication per data subcarrier, and allows for a simpler and more efficient receiver design. A more in-depth description for each of these blocks in Figure 4 can be found in [2].

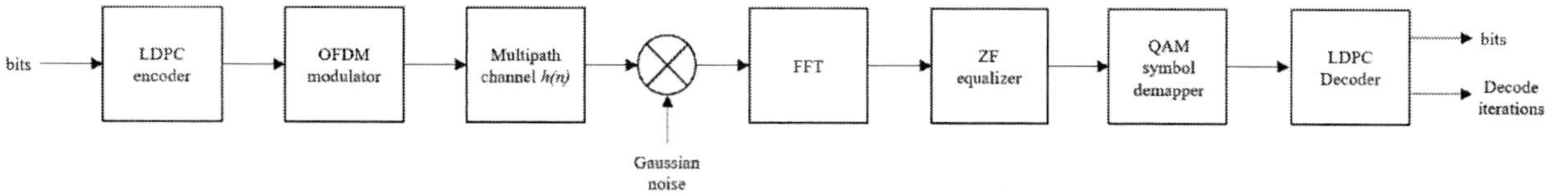

Figure 4. Block diagram for OFDM simulation.

PERFORMANCE RESULTS

To truly validate the usefulness of this technique, some multipath channels are used. The simulations were performed over 10 representative channels derived from one of the channel sounding measurements conducted at Edwards Air Force Base [8]. The corresponding frequency response plots are shown in Figure 5.

The simulated results for these channels are shown in Figures 6 and 7. In both figures, each subplot represents a mode of operation (or MCS), the y-axis of each subplot shows the channel number, and the x-axis shows the SNR in dB. The BER threshold plots of all LDAR modes of operation in all 10 channels are shown in Figure 6. In this figure, the MCS of each subplot is identified by its title, and the BER threshold is defined as 1×10^{-5}, meaning that the line is no longer displayed when the BER is greater than the defined threshold. If the BER is lower than the threshold for a given

SNR, the lines show a green square. The top row of the Figure 6 does not have any FEC code applied and the resulting BER threshold is reached sooner. The second row is 4/5 code rate, third row is 2/3 code rate, and fourth row is 1/2 code rate.

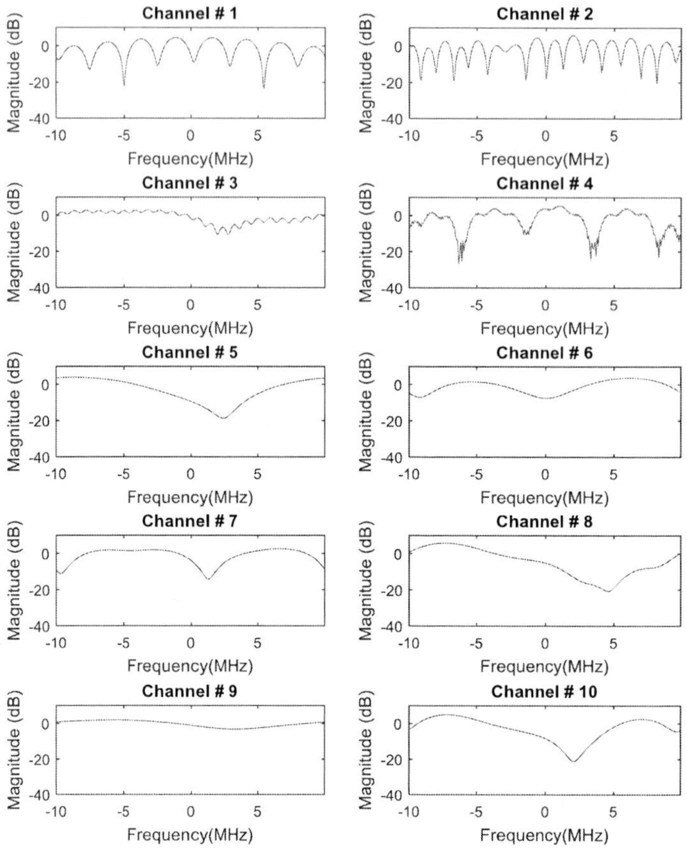

Figure 5. Frequency response plots of the representative channels at Edwards AFB.

In Figures 6, it can be seen that the system becomes more resilient to noise as the code rate decreases. This shows a relation between the link quality metric and the SNR/BER for these channels. The previous LDAR system utilizes SNR approximations to determine the adaptation selection. However, as can be seen in Figure 6, there is a high degree of uncertainty based upon the channel of operation. Switching the MCS based off of an SNR approximation alone would not be enough to take into account the BER caused by the individual channels. For instance, channel 6 can be seen to have a higher bit error rate for the same SNR value as compared to channel 9 for all MCS selections. The previous LDAR system attempted to predict the most robust MCS based on an assumption as to the channel of operation. Using a single, unified link quality metric can account for all of these uncertainties, the LDAR adaptive algorithm can be simplified. Based on the current definition of "link," the adapter would view the mode of operation, the channel properties, as well as the equalizer effects as part of the link quality metric.

The number of LDPC decoder iterations for each MCS is shown in Figure 7. The maximum number of decoding iterations is set to 10. In this figure, the color scale on the right reflects the number of iterations for convergence. In the case where the decoder does not converge, the color is shown in white. When Figures 6 and 7 are compared, it can be seen that there is a direct correlation between the convergence rate and the BER for all modes of operations in all ten channels. Utilizing the LDPC decoder iteration count, a link quality metric can be obtained. This metric captures a complete characterization of the transceiver complexity, channel, and the transmission mode. This allows for even more accurate control of the adaptability than the approximation of SNR since it includes an added layer of link quality information.

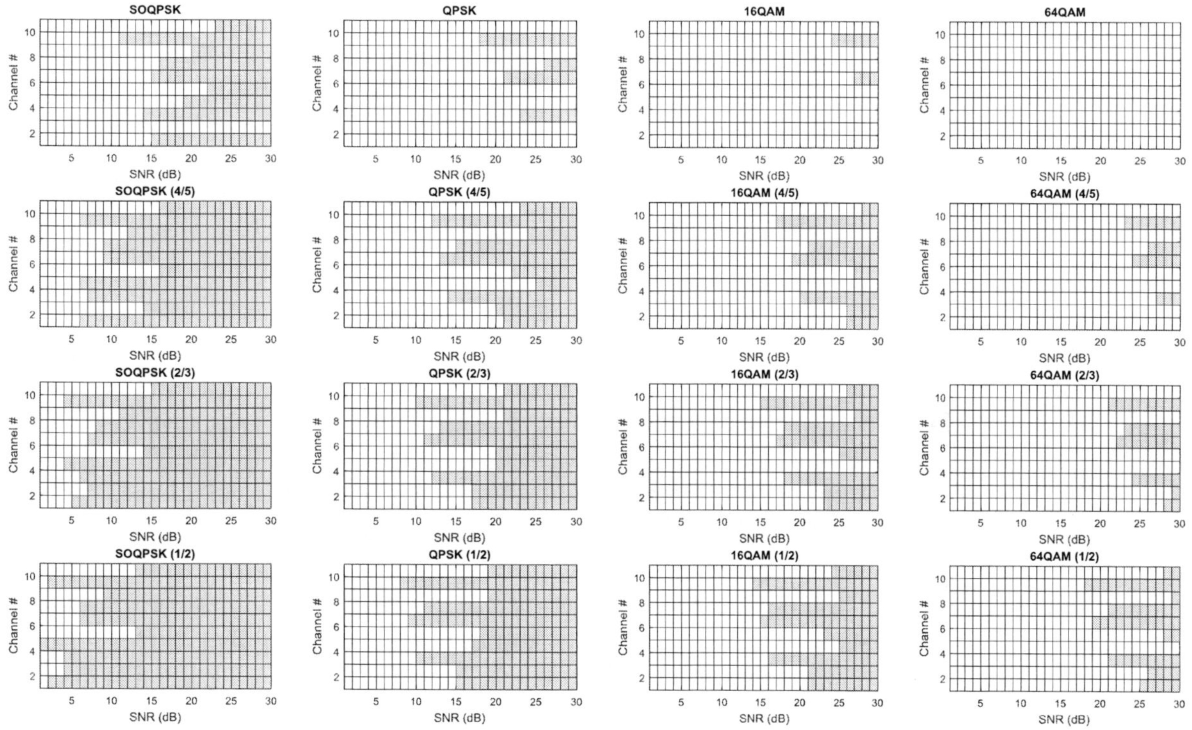

Figure 6: LDAR MCS BER thresholds in all ten Edwards AFB channels.

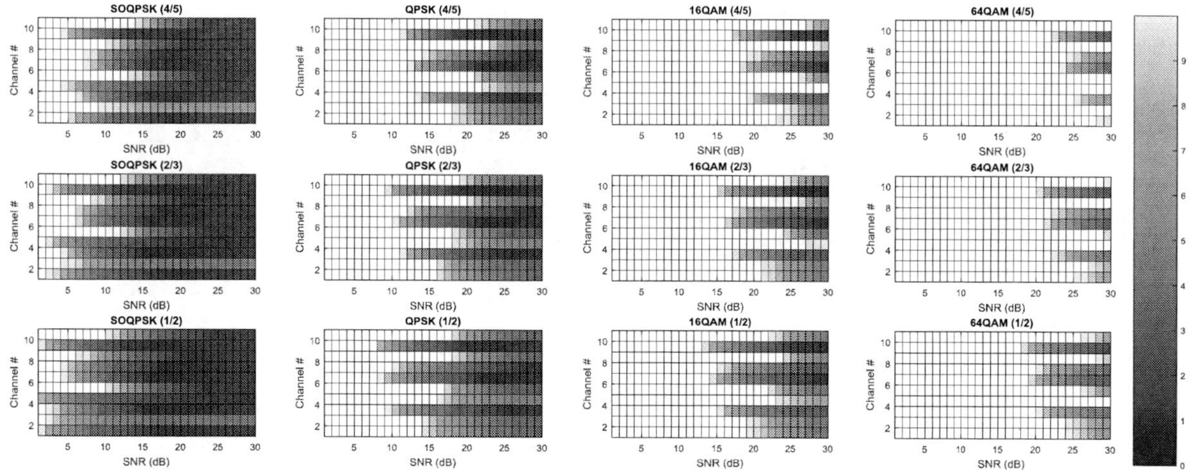

Figure 7: LDPC iteration counts for all LDAR MCS in all ten Edwards AFB channels.

Figure 8 illustrates the top-level block diagram utilizing the LDPC decoder iteration count to adapt the transmission mode of operations. This not only reduces computational complexity but also provides a more accurate characterization over the link quality.

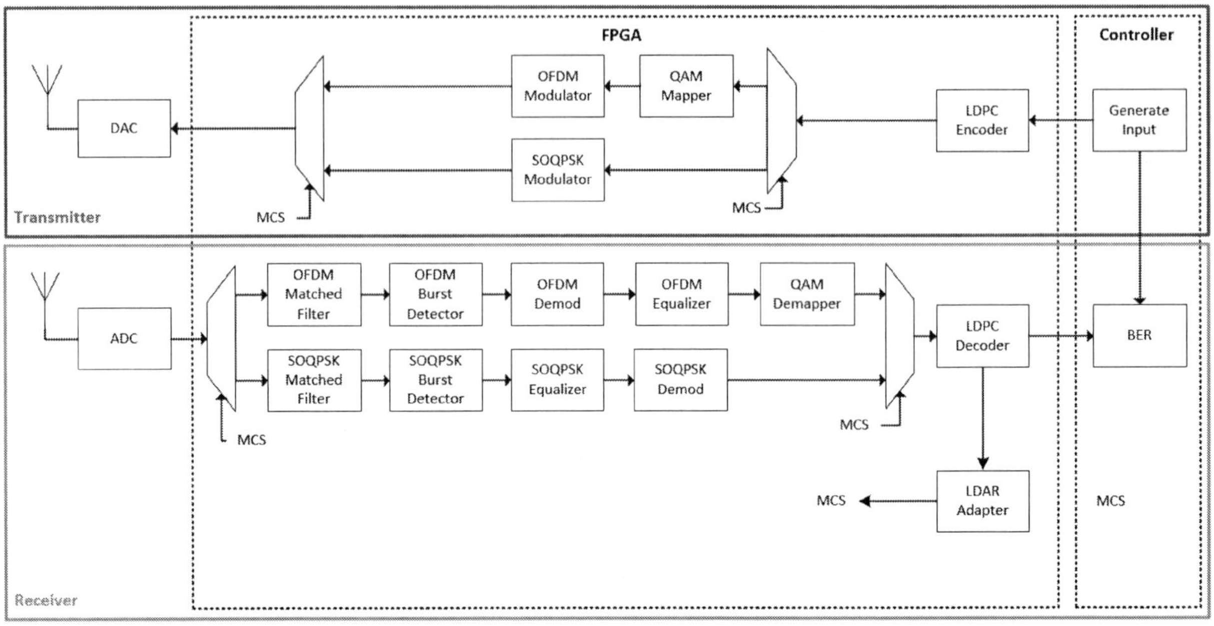

Figure 8: Top-level architecture for next generation LDAR transceiver.

The adaptation scheme now simplifies to incrementing or decrementing the MCS based on a low and a high threshold applied to the LDPC iteration count. These thresholds could be adjusted based on the acceptable BER for the given test. Given the case of a rapidly deteriorating channel (such as an aircraft executing a fast maneuver), the number of decoding iterations for a particular modulation would suddenly increase over a short amount of time, prompting the adapter to "back-off" and jump to an MCS that is more robust. Conversely, if the LDPC iteration counts decrease more slowly over time, a more spectrally efficient waveform such as OFDM can be selected. Table 1 shows the possible combinations of modulations and code rates along with the MCS selection and the throughput. Please note that the current LDAR transceiver have not implemented all of the LDPC codes and the MMSE equalizer for SOQPSK-TG.

Table 1: Adaptation Selection

MCS	Modulation	Code Rate	Throughput (Mbps)
1	SOQPSK	1/2	3.75
2	SOQPSK	2/3	5
3	SOQPSK	4/5	7.5
4	QPSK (OFDM)	1/2	9.6
5	QPSK (OFDM)	2/3	12.8
6	QPSK (OFDM)	4/5	15.36
7	16 QAM (OFDM)	1/2	19.2
8	16 QAM (OFDM)	2/3	25.6
9	64 QAM (OFDM)	1/2	28.8
10	16 QAM (OFDM)	4/5	30.72
11	64 QAM (OFDM)	2/3	38.4
12	64 QAM (OFDM)	4/5	46.08

CONCLUSION AND FUTURE WORK

This paper has defined a link quality metric for adaptability that includes the transceiver design complexity and the channel properties. We have shown that the LDPC iteration count is a valid metric to utilize for the link quality estimation. Finally, a transceiver architecture with a simplified adapter has been realized in hardware to exploit the LDPC decoder iteration count to adjust the modulation and coding scheme (MCS) selection based on the link quality.

The future work includes completing system integration, hardware in the loop tests, and a field test of the LDAR system.

ACKNOWLEDGEMENT

This project is funded by the Test Resource Management Center (TRMC) Test and Evaluation/Science & Technology (T&E/S&T) Program through the U.S. Army Program Executive Office for Simulation, Training and Instrumentation (PEO STRI) under Contract No. W900KK-13-C-024. The public release number is 412TW-PA-17322.

REFERENCES

[1] E. Wang, B. T. Walkenhorst and J. Han, "A Simulation Testbed for Adaptive Modulation and Coding in Airborne Telemetry," in *Proceedings of the International Telemetering Conference*, San Diego, CA, Oct. 2014.

[2] E. Wang and T. Brothers, "OFDM and SOQPSK transceiver hardware implementation with preliminary results," in *Proceedings of the International Telemetering Conference*, Glendale, AZ, 2016.

[3] Consultive Committee for Space Data System (CCSDS), "Low density parity check codes for use in near-Earth and deep space applications (131.1-O-2 Orange Book)," Sep. 2007.

[4] J. Han, B. Walkenhorst and E. Wang, "Adaptive modulation schemes for OFDM and SOQPSK using error vector magnitude (EVM) and Godard dispersion," in *Proceedings of the International Telemetering Conference*, San Diego, CA, October 2014.

[5] S. Johnson, Iterative error correction: turbo, low-density parity-check and repeat-accumulate, Cambridge University Press, 2010.

[6] E. Perrins, "FEC systems for aeronautical telemetry," *IEEE Trans on Aerospace and Electronic Systems,* vol. 49, no. 4, pp. 2340-2352, 2013.

[7] M. Rice, M. S. Afran, M. Saquib, A. Cole-Rhodes and F. Moazzami, "A comparison of three equalization techniques for iNET-formatted SOQPSK-TG," in *Proceedings of the International Telemetering Conference*, San Diego, CA, October 2014.

[8] M. Rice and M. Jensen, "A comparison of L-band and C-band multipath propagation at Edwards AFB," in *Proceedings of the International Telemetering Conference*, Las Vegas, NV, October 2011.

[9] integrated Network Enhanced Telemetry (iNET), "Radio Access Network (RAN) Standard," 2013.

[10] E. Hosseini and E. Perrins, "Timing, Carrier, and Frame Synchronization of Burst-Mode CPM," *IEEE Transactions on Communications,* vol. 61, no. 12, pp. 5125-5138, 2013.

[11] J. Chen, A. Dholakia, E. Eleftheriou, M. Fossorier and X.-Y. Hu, "Reduced-complexity decoding of LDPC codes," *IEEE Transaction on Communications,* vol. 53, no. 7, pp. 1288-1299, 2005.

[12] D. N. Godard, "Self-recovering equalization and carrier tracking in two-dimensional data communication systems," *IEEE Trans. Acoust., Speech, Signal Process.,* vol. 31, no. 2, pp. 459-472, Apr. 1983.

[13] 3GPP TS 36.521-1 V9.3.0, "User equipment (UE) conformance specification, radio transmission and reception, part 1: conformance testing".

[14] IEEE 802.11 working group, "IEEE P802.11n/D0.02," Feb 2006.

MULTI-BIT BIT-FLIPPING ALGORITHM FOR COLUMN WEIGHT-4 LDPC CODES

Xin Xiao[†], Bane Vasić[†], Shu Lin[‡]
[†]Dept. of Electrical and Computer Engineering
[†] University of Arizona
Tucson, 85719
[‡] University of California, Davis
[†]{7xinxiao7, vasic}@email.arizona.edu
[‡]shulin@ucdavis.edu

ABSTRACT

Low-density parity-check (LDPC) codes with column weight-4 are widely used in many communication and storage systems. However, traditional hard decision decoding algorithms such as the bit-flipping (BF) algorithm suffer from error floor due to trapping sets in LDPC codes. In this paper, to lower error floor of the BF algorithm over the Binary Symmetric Channel (BSC), we design a set of decoding rules incorporated within the BF algorithm for column weight-4 LDPC codes. Given a column weight-4 LDPC code, the dominate error patterns of the BF algorithm are first specified, and according to the designed rules, additional bits at both variable nodes (VN) and check nodes (CN) provide more information for the BF algorithm to identify the dominate error patterns, so that the BF algorithm could deliberately flip some bits to break them. Simulation results show that the modified BF algorithm eliminates all 4-error patterns and lowers the *Bit Error Rate* (BER) for at least two orders of magnitude with a trivial increment of complexity.

INTRODUCTION

LDPC codes are one of most promising candidates for error control in high speed applications such as data storage and deep space communications systems. *Belief-Propagation* (BP) based iterative decoding algorithms to decode LDPC codes have been shown to approach Shannon capacity[1, 2] in various channels. Soft-decision BP algorithms, such as *Sum Product Algorithm* (SPA) [3], show excellent error performance especially for LDPC codes with variable degree no greater than 4. However, the computational complexity of soft-decision BP algorithms is high and the decoding speed is limited due to real number calculations and real number storage. Hard-decision BP algorithms hence become popular especially in data storage systems because of its low computational complexity and high speed. The major drawback of hard-decision BP algorithms is that they suffer from high error floor, which degrades the error performance.

Among different hard-decision BP algorithms, the Bit-Flipping (BF) algorithm has the highest speed and least decoding complexity, making itself a good candidate in fault-tolerant memories[4,

5]. Furthermore, the decoding speed neither relies on the column weight nor on the row weight of LDPC codes. As a typical hard-decision BP algorithm, the BF algorithm still suffers from high error floor over the BSC for finite length codes.

In recent decade, many modifications on the BF algorithms have been proposed to improve the error performance. In [6, 7, 8], soft information from a channel is incorporated into the BF Algorithm, which cannot be used over the BSC. Furthermore, people also consider using randomness to add more information and slow down the decoding process. Miladinovic *et al.* [9] introduce a probability $p < 1$ in the BF algorithm such that once a variable node (VN) is suggested to be flipped by the number of its unsatisfied neighbors(CNs), it will be flipped with probability p. Ivanis *et al.* [10] incorporate a randomness in gradient descent bit flipping algorithm[8] to help the algorithm correct trapping set errors. They then extend the Probability-GDBF algorithm in [11] based on the principle of multiple decoding attempts and random re-initializations, which combines a new threshold adaptation method to further improve the immunity to the noisy registers and logic gates. Instead of applying randomness, Nguyen *et al.* [12] introduce one additional bit for each VN to indicate the strength and one additional bit for each CN to represent the reliability. This approach also adds more information and slows down the decoding process in a deterministic way by reducing the strength of a VN before flipping it.

Different from Nguyen's work which is designed for column weight-3 LDPC codes, in this paper, we develop a set of decoding rules incorporated with the BF algorithm for finite length column weight-4 LDPC codes over the BSC. Inspired by [12], given a column weight-4 LDPC code, we first analyze the dominate error patterns of the BF algorithm. To specify these dominate error patterns whenever they happen, we add one more bit for each VN and multiple additional bits for each CN to provide more information. The additional bit at each VN still represents the strength, but the multiple additional bits for each CN play a role not only as the reliability as in [12], but also as a flag to indicate some stable dominate error patterns. The "flag" feature is another difference from [12], since the stable dominate error patterns are not encountered in [12]. As long as the dominate error patterns are identified in decoding process, some VNs(bits) will be deliberately flipped according to the designed rules to break these error patterns. Simulation results show that the modified BF algorithm could eliminate all 4-error patterns and lower at least 2 order of magnitude in terms of BER level. As the increment of decoding complexity is trivial, the proposed multi-bit BF algorithm runs at fast speed, which is suitable for applications that require high throughput.

The rest of the paper is organized as follows. Section II introduces the preliminaries. Section III shows the proposed multi-bit BF algorithm. Section IV gives the numerical results and analysis and Section V concludes this paper.

PRELIMINARIES

Let $\mathcal{C}$ be an (n, k) binary LDPC code given by the null space of a $m \times n$ parity check matrix $\mathbf{H}$. G is the Tanner graph of $\mathcal{C}$, a bipartite graph with two sets of nodes and a set of edges. The two sets of nodes are: n VNs (bits) $V(G) = \{1, 2, ...n\}$ and m CNs $C(G) = \{1, 2, ...m\}$. $E(G)$ denotes the set of edges. A vector $\mathbf{z} = (z_1, z_2, ...z_n)$ is a codeword in $\mathcal{C}$ iff $\mathbf{zH}^T = \mathbf{0}$, where $\mathbf{H}^T$ is the transpose of $\mathbf{H}$. A d_v-left-regular (d_c-right-regular) LDPC code has a Tanner graph G in which all VNs (CNs) have degree d_v (d_c). A (d_v, d_c)-regular LDPC code is d_v-left-regular and d_c-right-regular. In this paper, we only consider $(4, d_c)$-regular LDPC codes, where the column weight is 4 and row weight is d_c. A subgraph of a bipartite graph G is a bipartite graph U such that $V(U) \subseteq V(G), C(U) \subseteq C(G)$ and $E(U) \subseteq E(G)$. An induced subgraph on a set of variable nodes $V_d \subseteq V(G)$ is a bipartite graph U with $V(U) = V_d, C(U) = \{c \in C(G) : \exists v \in V_d, (v, c) \in E(G)\}$ and $E(U) = \{(v, c) \in E(G) : v \in V_d\}$. The length of the shortest cycle in the Tanner graph G is called the girth g of G.

Assume the transmission of the all-zero codeword over the BSC. Supposed that $\mathbf{r} = (r_1, r_1, ...r_n)$ is the channel output vector, where $r_i \in \{0, 1\}$. Let $\widehat{\mathbf{x}}^\ell = (\widehat{x}_1^\ell, \widehat{x}_2^\ell, ...\widehat{x}_n^\ell)$ be the decision vector and $\mathbf{s}^\ell = (s_1^\ell, s_2^\ell, ...s_m^\ell)$ be the syndrome vector of $\widehat{\mathbf{x}}^\ell$ after the ℓ-th iteration, where ℓ is a positive integer. Then, $\mathbf{s}^\ell = \widehat{\mathbf{x}}^\ell \mathbf{H}^T$. A check node c is said to be satisfied at the beginning of the ℓ-th iteration if $s_c^{\ell-1} = 0$, otherwise it is unsatisfied.

For any VN v in a Tanner graph G, let $N_s^\ell(v)$ and $N_u^\ell(v)$ be the number of satisfied CNs and unsatisfied CNs adjacent to v at the beginning of the ℓ-th iteration, respectively. The standard BF algorithm operates as follows:

- in each iteration, the BF algorithm in parallel flips the VNs v if $N_u^\ell(v) > d_v/2$ and set the VNs v to be initial value if $N_u^\ell(v) = d_v/2$;
- the BF algorithm operates until either $\mathbf{s}^\ell = \mathbf{0}$ or the preset maximum number of iterations I_{max} is reached.

MULTI-BIT BIT-FLIPPING ALGORITHM

In this section, we introduce the multi-bit BF algorithm for column weight-4 LDPC codes. First we specify some harmful error patterns of column weight 4 LDPC codes encountered in the BF decoding.

A. *Two-Bit for Variable Node to eliminant dominant 3-error pattern*

Consider a $(4, d_c)$-regular LDPC code $\mathcal{C}$ with girth 8. Without loss of generality, supposed that a subset of VNs: $V_d = \{v_1, v_2, v_3, v_4, v_5\}$ forms three 8-cycles in the Tanner graph G of $\mathcal{C}$, and the induced subgraph on V_d is shown in Fig.1(a)(b). Consider the BF decoding with the initial condition that only v_1, v_3 and v_5 are in error. Fig.1(a) and (b) show the first and second iteration, respectively, where ○ represents the correct VN, ● denotes the erroneous VN, □ indicates the satisfied CN, and ■ implies the unsatisfied CN at the end of $\ell - 1$ iteration. It can be observed that by the standard BF decoding, all 5 VNs are flipped in every iteration, and the set of erroneous VNs is oscillating between $\{v_1, v_3, v_5\}$ and $\{v_2, v_4\}$. Therefore, for this initial 3-error pattern, the standard BF decoding will eventually end in failure. A possible idea is to find a way to only flip one set of $\{v_1, v_3, v_5\}$ and $\{v_2, v_4\}$. We use a similar approach in [12] to slow down the flipping process.

Specifically, as $N_u^\ell(v) \in \{0, 1, 2, 3, 4\}$, we assign 2 bits to create four states for each VN as: {Strong Zero, Weak Zero, Weak One, Strong One}, and denote them by $\{0_s, 0_w, 1_w, 1_s\}$. The assignment is reasonable in the sense that the states are "quantitative" soft information. Let $Q = \{0_s, 0_w, 1_w, 1_s\}$, and $\mathbf{q}^\ell = (q_1^\ell, q_2^\ell, ...q_n^\ell)$ be a state vector where each $q_j^\ell \in Q$ represents the state

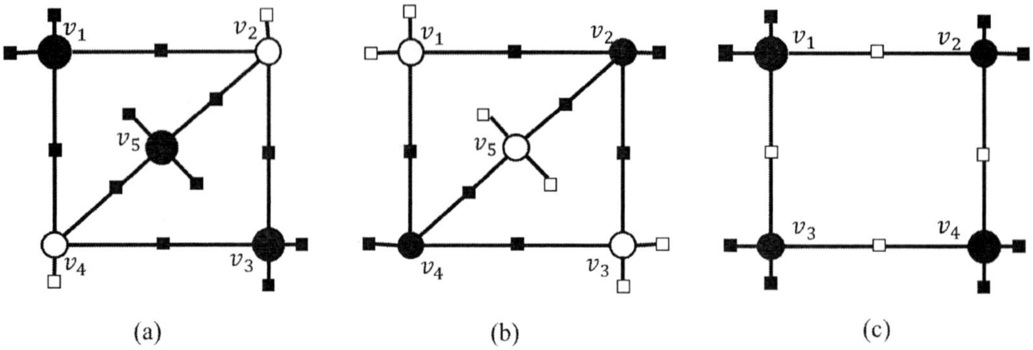

Figure 1: Dominant error patterns of $d_v = 4$ LDPC codes. (a) 3-error pattern in the first iteration of the BF decoding. (b) 3-error pattern in the second iteration of the BF decoding. (c) Stable 4-error pattern in the 2-bit joint BF decoding.

Table 1: **State Transition Function** $f_1(q_v^{\ell-1}, N_u^{\ell}(v))$

		$N_u^{\ell}(v)$							$N_u^{\ell}(v)$				
		0	1	2	3	4			0	1	2	3	4
$q_v^{\ell-1}$	0_s	0_s	0_s	0_s	0_w	1_s	$q_v^{\ell-1}$	1_w	1_s	0_w	0_w	0_s	0_s
	0_w	0_s	1_w	1_w	1_s	1_s		1_s	1_s	1_s	1_s	1_w	0_s

of VN j in ℓ-th iteration. Then the decision vector $\widehat{\mathbf{x}}^{\ell}$ is determined by $\mathbf{q}^{\ell}$ as follows: $\widehat{x}_j^{\ell} = 0$ if $q_j^{\ell} = 0_s$ or 0_w; $\widehat{x}_j^{\ell} = 1$ if $q_j^{\ell} = 1_s$ or 1_w. $\mathbf{q}^{\ell}$ is updated as a finite state machine based on $\mathbf{q}^{\ell-1}$ and $N_u^{\ell}(v)$, as illustrated in Table 1. Each VN v is initialized to $q_v^0 = 0_s$ if $r_v = 0$, otherwise $q_v^0 = 1_s$. With state vector and state transition function $q_v^{\ell} = f_1(q_v^{\ell-1}, N_u^{\ell}(v))$ given in Table 1, the state of initial 3-error pattern in Fig.1(a) changes from $(q_1^1, q_2^1, q_3^1, q_4^1, q_5^1) = (1_s, 0_s, 1_s, 0_s, 1_s)$ to $(q_1^2, q_2^2, q_3^2, q_4^2, q_5^2) = (0_s, 0_w, 0_s, 0_w, 0_s)$, resulting in a decoding success. Similarly, other initial 3-error patterns can also be eliminated by this modified BF algorithm. Therefore, the 2-bit joint BF algorithm guarantees to correct up to 3 errors for column weight-4 LDPC codes.

B. Multi-Bit for Check Node to eliminate dominant 4-error patterns

B..1 Stable 4-error pattern

Introducing one additional bit for each VN causes new dominate error patterns of size 4. To further improve the error performance, we explore the possibility of removing dominate 4-error patterns. Without loss of generality, assuming that a subset of VNs: $V_d = \{v_1, v_2, v_3, v_4\}$ forms a 8-cycle in G, and the induced subgraph on V_d is shown in Fig.1(c). Consider the initial condition where only $\{v_1, v_2, v_3, v_4\}$ are erroneous. Supposed that the rest VNs not in Fig.1(c) remain correct during decoding process and are not to be discussed. With the decoding rules given in Table1, the state $(q_1^{\ell}, q_2^{\ell}, q_3^{\ell}, q_4^{\ell})$ will be stuck at $(1_s, 1_s, 1_s, 1_s)$, and all CNs stay in the same state: either satisfied or unsatisfied. It is reasonable to argue that if some CN stays in unsatisfied state for several consecutive iterations, it has a probability to be in a stable 4-error pattern isomorphic to Fig.1(c). Furthermore, a possible approach to break the stable situation is to reduce the VN's strength, which again slows down the decoding process. Since the unsatisfied CNs provide us important information, we give the following definition.

Definition 1 *For each check node $c \in C(G)$, the flag of c, denoted by $Fl(c)$, is defined as the number of consecutive iterations during decoding process where c is stuck at unsatisfied state. A check node c is called Flagged-check node if $Fl(c) \geq \theta$. For each variable node $v \in V(G)$, the energy of v, denoted by $E(v)$, is defined as the number of Flagged-check nodes adjacent to v.*

In this paper, $Fl(\cdot)$ is limited to no greater than 3 and θ is predefined to 2, thus two more bits are required for each flag. With Flagged-CNs, we now modify the updating rules for each VN v symmetrically as follows:

$$q_v^\ell = f_2(q_v^{\ell-1}, N_u^\ell(v), E(v)) = \begin{cases} 0_w, \text{if } q_v^{\ell-1} = 0_s, \text{and } E(v) \geq 2 \\ 1_w, \text{if } q_v^{\ell-1} = 1_s, \text{and } E(v) \geq 2 \\ f_1(q_v^{\ell-1}, N_u^\ell(v)), \text{otherwise} \end{cases} \tag{1}$$

With Eq.1, in each iteration, the state of initial stable 4-error pattern in Fig.1(c) becomes:

$$(q_1^0, q_2^0, q_3^0, q_4^0) = (q_1^1, q_2^1, q_3^1, q_4^1) = (q_1^2, q_2^2, q_3^2, q_4^2) = (1_s, 1_s, 1_s, 1_s)$$
$$(q_1^3, q_2^3, q_3^3, q_4^3) = (1_w, 1_w, 1_w, 1_w), (q_1^4, q_2^4, q_3^4, q_4^4) = (0_w, 0_w, 0_w, 0_w)$$

thus the decoding ends in a valid codeword in $\mathcal{C}$.

B..2 Oscillating 4-error pattern

The flag feature may fail if some oscillating error configurations happen durning decoding process. Without loss of generality, supposed that a subset of VNs: $V_d = \{v_1, v_2, v_3, v_4, v_5, v_6\} \subseteq V(G)$ has an induced subgraph shown in Fig.2(a)(b). If in some iteration ℓ, the state vector becomes $(q_1^\ell, q_2^\ell, q_3^\ell, q_4^\ell, q_5^\ell, q_6^\ell) = (1_w, 1_w, 1_s, 0_w, 0_w, 1_s)$, then the state of V_d will oscillate between $(1_w, 1_w, 1_s, 0_w, 0_w, 1_s)$ and $(0_w, 0_w, 1_s, 1_w, 1_w, 1_s)$, as shown in Fig.2(a) and (b). We call the oscillating phenomenon in Fig.2(a)(b) Type-1 oscillating error configuration. To briefly analyze Type-1, only v_3 and v_6 are strong VNs and stable, which are easier to be identified than other weak VNs. Furthermore, the adjacent CNs of v_3 and v_6 share the same state stability: one stable satisfied CN, one stable unsatisfied CN, and two unstable CNs that change state (satisfied or unsatisfied) alternately. Another similarly oscillating case is given in Fig.2(c)(d), where the state alternates between $(1_w, 1_s, 0_w, 1_s, 1_w, 0_w)$ and $(0_w, 1_s, 1_w, 1_s, 0_w, 1_w)$. We call the second case Type-2 oscillating error configuration. In Type-2, only $\{v_2, v_4\}$ are strong and stable. All their neighbors are oscillating, and in every iteration, both v_2 and v_4 are connected to two unsatisfied CNs. Such shared state information can be used to identify the strong VNs and reduce their strength. We follow the same way in [12] by introducing one more bit for each CN to describe its reliability.

Specifically, in each iteration, every check node has one out of four states: {previously-satisfied (PS), newly-satisfied (NS), newly-unsatisfied (NU), previously-unsatisfied (PU)}. Denote the state of a check node c in ℓ-th iteration by $\mathcal{T}_\ell(c)$. A satisfied (unsatisfied) check node has state PS (PU) if it was satisfied (unsatisfied) in the previous decoding iteration, otherwise it has state NS (NU), i.e.,

$$\mathcal{T}_\ell(c) = (1 - s_c^\ell)(1 - s_c^{\ell-1})\mathbf{PS} + s_c^\ell(1 - s_c^{\ell-1})\mathbf{NU} + (1 - s_c^\ell)s_c^{\ell-1}\mathbf{NS} + s_c^\ell s_c^{\ell-1}\mathbf{PU} \tag{2}$$

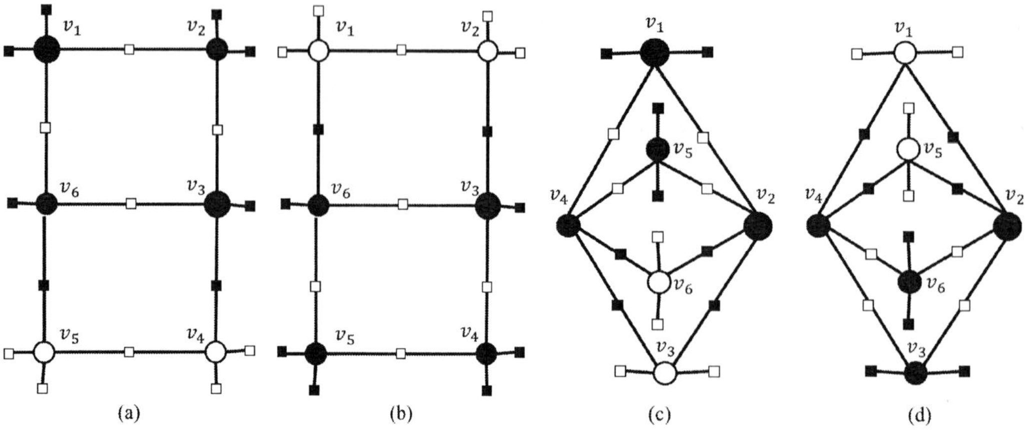

Figure 2: Oscillating 4-error patterns: (a)Odd iterations of Type-1 oscillating error configuration. (b)Even iterations of Type-1 oscillating error configuration. (c)Odd iterations of Type-2 oscillating error configuration. (d)Even iterations of Type-2 oscillating error configuration.

Then for any variable node v, the reliability of its neighbors can be described as a 4-tuple: $\mathcal{Y}(v) = (N_{ps}(v), N_{pu}(v), N_{ns}(v), N_{nu}(v))$, where $N_X(v)$ represents the number of CNs adjacent to v that have state $X, X \in \{\text{PS,NS,NU,PU}\}$. It is straightforwards to identify the strong VNs in Type-1 and Type-2 by $\mathcal{Y}(v) = (1,1,1,1)$ and $(0,0,2,2)$, respectively. The algorithm will reduce the strength once it detects such strong VNs.

The two oscillating cases in Fig.2 now can be successfully decoded:

- Type-1: $(1_w, 1_w, 1_s, 0_w, 0_w, 1_s) \rightarrow (0_w, 0_w, 1_w, 1_w, 1_w, 1_w) \rightarrow (1_w, 1_w, 0_w, 0_w, 0_w, 0_w) \rightarrow (0_s, 0_s, 1_w, 0_s, 0_s, 1_w) \rightarrow (0_s, 0_s, 0_s, 0_s, 0_s, 0_s)$;
- Type-2: $(1_w, 1_s, 0_w, 1_s, 1_w, 0_w) \rightarrow (0_w, 1_w, 1_w, 1_w, 0_w, 1_w) \rightarrow (1_w, 0_w, 0_w, 0_w, 1_w, 0_w) \rightarrow (0_s, 1_w, 0_s, 0_s, 1_w, 0_s) \rightarrow (0_s, 0_s, 0_s, 0_s, 0_s, 0_s)$.

Combining the stable and oscillating error patterns, a strong VN will be reduced strength if the following condition holds:

$$E(v) \geq 2 \text{ or } \mathcal{Y}(v) \in \{(1,1,1,1), (0,0,2,2)\} \tag{3}$$

Eq. 1 is then modified into following formula:

$$q_v^\ell = f_3(q_v^{\ell-1}, N_u^\ell(v), E(v), \mathcal{Y}(v)) = \begin{cases} 0_w, \text{if } q_v^{\ell-1} = 0_s, \text{and Eq. 3 holds} \\ 1_w, \text{if } q_v^{\ell-1} = 1_s, \text{and Eq. 3 holds} \\ f_1(q_v^{\ell-1}, N_u^\ell(v)), \text{otherwise} \end{cases} \tag{4}$$

Other 4-error patterns can be removed in a similar way. We summarize the proposed multi-bit bit-flipping algorithm below.

NUMERICAL RESULTS

In this section, we simulate the multi-bit bit-flipping algorithm on a column weight-4 LDPC code. The code length is 1296, and code rate is 0.5. Fig.3 shows the simulation results of BER performance over the BSC. The blue curve stands for the standard BF decoding, and the red curve

Algorithm 1 Multi-Bit Bit-Flipping Algorithm

Initialization: $\forall v \in V(G), q_v^0 \leftarrow 0_s$ if $r_v = 0$; $q_v^0 \leftarrow 1_s$ if $r_v = 1$. $l \leftarrow 1$
while $\mathrm{s}^\ell \neq \mathbf{0}$ and $l < I_{max}$ **do**
    $\forall v \in V(G)$, update q_v^ℓ according to Eq. 4;
    $\forall c \in C(G)$, update $\mathcal{T}_\ell(c)$ according to Eq. 2;
end while

represents multi-bit BF decoding. The crossover probability is α ranging from 0 to 0.1. The smaller α is, the better the channel is. It can be observed that the multi-bit BF decoding outperforms the standard BF decoding when $\alpha < 0.045$. At $\alpha = 0.01$, it lowers 2 order of magnitude in terms of BER level. Simulation shows that no 4-error pattern is found when $\alpha < 0.01$. Noted that the proposed decoding may work worse than the standard BF decoding when $\alpha \geq 0.045$. The reason is that there are so many small dominate error patterns when the channel is noisy, and the correlation among them cannot be isolated from multi-bit BF algorithm. Since only four more bits are added in proposed decoding, the computational complexity is trivial.

CONCLUSIONS

In this paper, we introduced a multi-bit bit-flipping algorithm for column weight-4 LDPC codes. We specified the dominate error patterns and designed a set of rules to slow down the decoding process. Simulation results show that proposed algorithm can remove all 4-error patterns and lower the BER for at least two orders of magnitude over the BSC at the price of a trivial increment of complexity.

REFERENCES

[1] D. J. MacKay and R. M. Neal, "Near shannon limit performance of low density parity check codes," *Electronics letters*, vol. 32, no. 18, p. 1645, 1996.

[2] T. J. Richardson and R. L. Urbanke, "The capacity of low-density parity-check codes under message-passing decoding," *IEEE Transactions on information theory*, vol. 47, no. 2, pp. 599–618, 2001.

[3] F. R. Kschischang, B. J. Frey, and H.-A. Loeliger, "Factor graphs and the sum-product algorithm," *IEEE Transactions on information theory*, vol. 47, no. 2, pp. 498–519, 2001.

[4] S. Brkic, P. Ivanis, G. Djordjevic, and B. Vasić, "Taylor-kuznetsov fault-tolerant memories: a survey and results under correlated gate failures," in *Proc. 10th Int. Conf. on Telecommun. in Modern Satellite, Cable and Broadcasting Services (TELSIKS 2013)*, (Nis, Serbia), pp. 455–461, Oct. 16–19 2013.

[5] B. Vasić and S. K. Chilappagari, "An information theoretical framework for analysis and design of nanoscale fault-tolerant memories based on low-density parity-check codes," *IEEE Transactions on Circuits and Systems I: Regular Papers*, vol. 54, pp. 2438–2446, Nov. 2007.

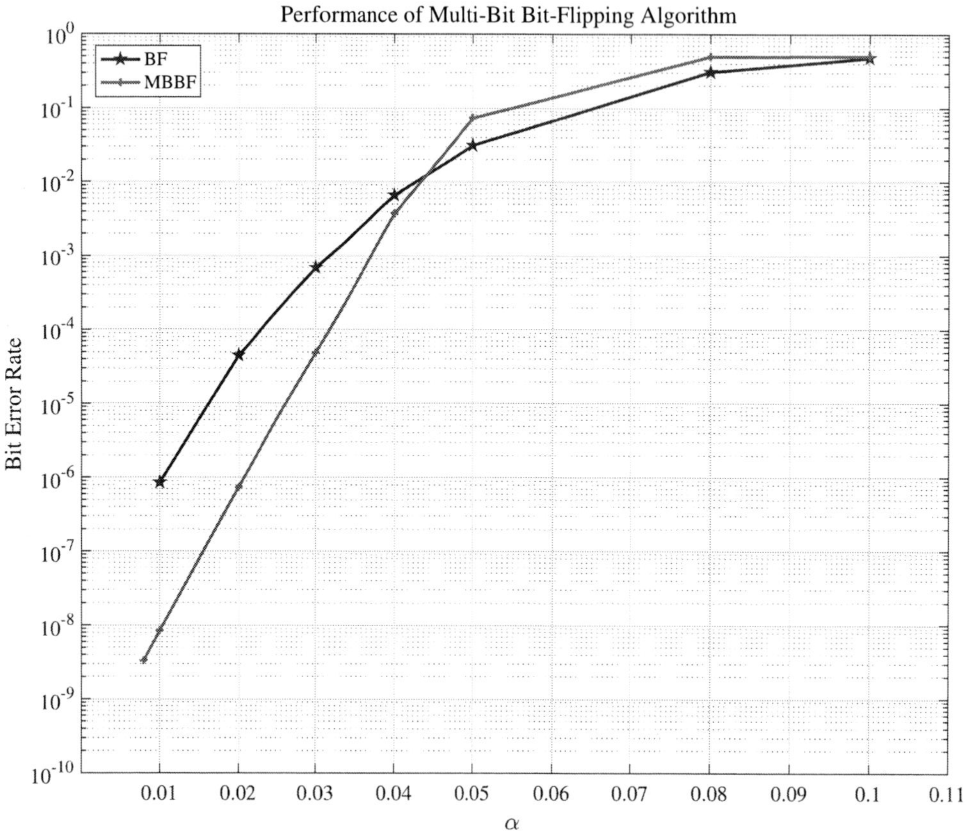

Figure 3: BER Performance of Multi-Bit Bit-Flipping Algorithm over the BSC

[6] T. M. N. Ngatched, F. Takawira, and M. Bossert, "A modified bit-flipping decoding algorithm for low-density parity-check codes," in *IEEE International Conference on Communications, 2007. (ICC'07).*, pp. 653–658, 2007.

[7] X. Wu, C. Ling, M. Jiang, E. Xu, C. Zhao, and X. You, "New insights into weighted bit-flipping decoding," *IEEE Transactions on Communications*, vol. 57, Aug. 2009.

[8] T. Wadayama, K. Nakamura, M. Yagita, Y. Funahashi, S. Usami, and I. Takumi, "Gradient descent bit flipping algorithms for decoding ldpc codes," in *International Symposium on Information Theory and Its Applications, 2008. (ISITA 2008).*, pp. 1–6, 2008.

[9] N. Miladinovic and M. P. Fossorier, "Improved bit-flipping decoding of low-density parity-check codes," *IEEE Transactions on Information Theory*, vol. 51, pp. 1594–1606, April 2005.

[10] O. Al Rasheed, P. Ivaniš, and B. Vasić, "Fault-tolerant probabilistic gradient-descent bit flipping decoder," *IEEE Communications Letters*, vol. 18, pp. 1487–1490, Sep. 2014.

[11] P. Ivanis, O. Al Rasheed, and B. Vasić, "Mudri: A fault-tolerant decoding algorithm," in *IEEE International Conference on Communications, 2015 (ICC2015)*, pp. 4291–4296, 2015.

[12] D. V. Nguyen and B. Vasić, "Two-bit bit flipping algorithms for LDPC codes and collective error correction," *IEEE Transactions on Communications*, vol. 62, pp. 1153–1163, April 2014.

LDPC CODED APSK FOR AERONAUTICAL TELEMETRY

Sumant Pathak
Electrical Engineering and Computer Science
University of Kansas
Lawrence, KS, 60045
sumant.pathak@ku.edu

Faculty Advisor: Erik Perrins

ABSTRACT

This paper presents the performance of capacity-approaching low-density parity check (LDPC) coded amplitude phase shift keying (APSK) over aeronautical telemetry channels. We show the bit-error rate results for code-rates of 1/2, 2/3, 3/4, and 4/5 with 16 and 32 point constellations. Results are presented and compared between an optimal and sub-optimal reduced-complexity demodulating system. We also compare the results with SOQPSK-TG under similar channel conditions and provide an estimate of backoff needed for implemention with power amplifiers.

INTRODUCTION

Forward-error correction (FEC) codes have become nearly ubiquitous in wireless communication systems and provide excellent power efficiency in terms of coding gain [1]. However, this efficiency comes at the cost of reduced spectral efficiency and increased implementation complexity. The decreased spectral efficiency can be mitigated by combining the FEC code with a more spectrally efficient modulation format [2]. Implementation complexity can be reduced by using sub-optimal complexity reduction techniques by sacrificing part of the gained power efficiency.

The FEC codes used in this paper are low-density parity check (LDPC) codes. LDPC were proposed by Gallager in 1960 [3]. The codes were ignored until the work of Mackay [4] brought them back into consideration. LDPC codes have the advantage of high coding gains along with a parallelizable decoding algorithm. The LDPC codes used in this paper belong to the Accumulate, Repeat-by-4, and Jagged Accumulate (AR4JA) family of codes developed at the Jet Propulsion Laboratory (JPL) for deep space applications [5]. These codes were later adopted in the IRIG 106 standard [6].

Amplitude-phase shift keying is a linear modulation that is well suited for telemetry systems. APSK consists of constellation points arranged in concentric circles. This provides better resilience to distortion from non-linear power amplifiers when compared to other linear modulations with similar spectral efficiencies such as QAM [7]. The APSK constellations used in this paper have been standardized in the Digital Video Broadcasting - Satellite - Second generation (DVB-S2) [8] standard for applications in digital television broadcasting.

Combinations of FEC codes and different modulations (coded modulations) is not new [9]. Coded modulations have also been previously studied for aeronautical telemetry applications. SOQPSK-TG [10] has been combined with serially concatenated convolutional codes (SCCC), parralel concatenated convolutional codes (PCCC), and LDPC codes. APSK has also been considered with PCCC (also known as Turbo Codes) [11]. Turbo-coded APSK was also studied with power amplifier models used in telemetry systems and it was found that Turbo-coded APSK with backoff can match the performance of SOQPSK-TG with improved spectral efficiency [12]. The AR4JA LDPC codes have also been coupled with different linear modulation formats including 16 and 32 APSK in [13].

This paper describes a system with LDPC codes paired with 16 and 32 point APSK constellations. We simulate the performance of the system over an Additive White Gaussian Noise (AWGN) channel. An overview of the system model is shown in Figure 1. The flow of this paper will follow this figure. Initial sections will describe each block in the system in detail, followed by a presentation of our simulation results and conclusions.

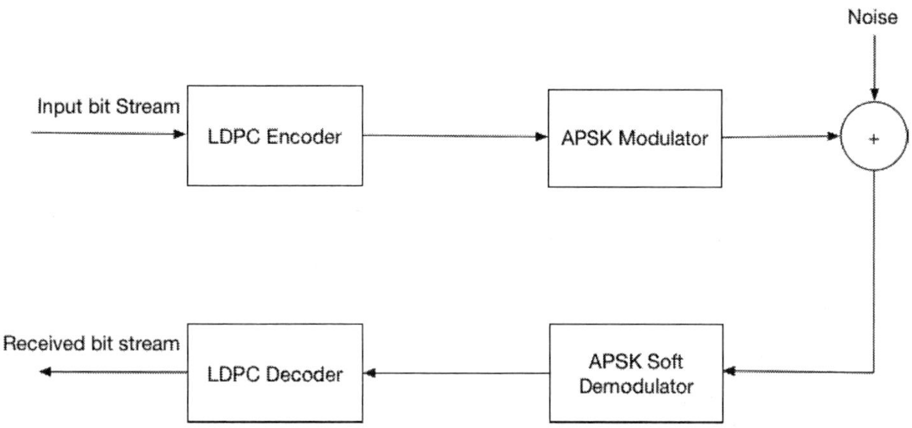

Figure 1: System overview

LDPC ENCODER

Our proposed system uses the AR4JA LDPC codes developed at JPL [5] and adopted into the IRIG 106 standard for aeronautical telemetry [6]. These codes are linear and binary and a codeword can be generated by multiplication of the message vector with a $k \times n$ generator matrix over GF(2). These codes are also systematic, which means each n length codeword vector conatins the k length message vector followed by $n - k$ redundant bits. This means the encoder need only store the last $n - k$ columns of the generator matrix, since the intitial k columns will be an identity matrix.

Also, the AR4JA codes are quasi-cyclic, which means the generator matrix is constructed of circulant sub-matrices. The encoder need only store the first column of each submatrix, and can subsequently generate the entire matrix by appropriately shifting these vectors. To generate a codeword, the encoder instead of brute-force matrix multiplication with the entire generator matrix need only generate the redundant parity bits and concatenate them with the message vector. An efficient algorithm specfied in [14] describes an encoding process to generate the parity bits by exploiting the systematic and quasi-cyclic properties of the code.

APSK MODULATOR

APSK is a linear modulation format in which constellation points are arranged in concentric circles. This enables APSK to have a lower peak-to-average power ratio (PAPR) than other linear modulation formats with similar spectral efficiencies. This makes APSK ideal for telemetry systems since the lower PAPR reduces distortions by non-linear power amplifiers operating in the saturation region [15]. The modulator can use either a 16 or 32 point constellation. The 16 point option will have a better PAPR at the cost of spectral efficiency. We use the constellations specified in the DVB-S2 standard for digital broadcast television which specifies the phase offsets and bit mapping. The ratio of radii of the rings of the constellation depends on the rate of the LDPC code used and is also specficied in [15] and [13]. It has been shown that these constellations provide a good compromise between minimizing PAPR and maximizing mutual information [7].

The modulator in our system receives the codeword vector from the LDPC encoder and divides it into blocks of 4 (for 16-APSK) or 5 bits (for 32-APSK). These blocks are then mapped to a symbol from the selected constellation. The modulator then applies a transmit pulse shape and upconverts the resulting complex baseband signal to a passband signal and sends it over the AWGN channel. Unlike Continuous Phase Modulation (CPM) systems, this system will not be able to drive the power amplifier at full saturation since the APSK constellation is not immune to non-linear distortions. The system would then need to incorporate backoff to ensure the signal stays in the linear region of the power amplifier. We discuss the backoff the system can operate under in the results section.

SOFT DEMODULATOR

The soft demodulator is the first block in the receiver subsystem. The term ‘soft’ implies that the demodulator does not make a ‘hard’ decision on the received noisy signals. Instead, the receiver returns the likelihood of each bit in the block of bits the symbol represents. The likelihoods are represented by log-likelihood ratios (LLRs). For a received signal r, the LLR for the ith bit is

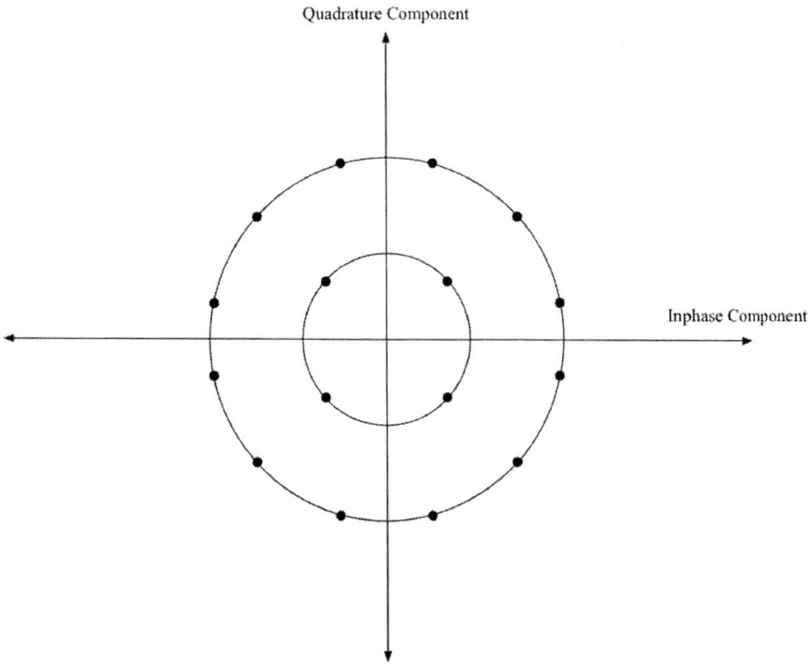

Figure 2: 16-APSK constellation described in [8]

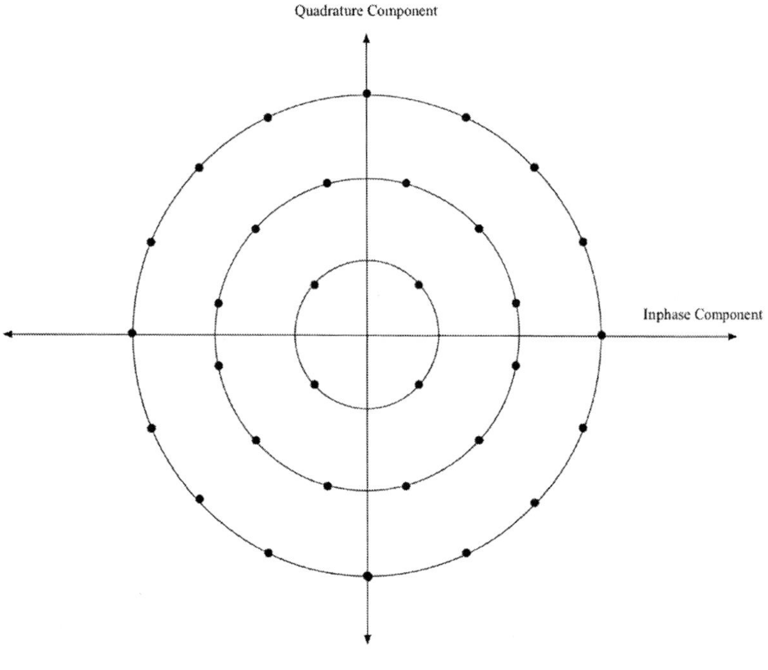

Figure 3: 32-APSK constellation described in [8]

$$\lambda_i = \ln\left(\frac{Pr(b_i = 0|r)}{Pr(b_i = 1|r)}\right). \quad (1)$$

We assume the bits are *a priori* equiprobable, thus λ_i becomes

$$\lambda_i = \ln\left(\frac{Pr(r|b_i = 0)}{Pr(r|b_i = 1)}\right) \quad (2)$$

when Baye's rule is applied. For an AWGN channel, the exact LLR is shown to be [13]

$$\lambda_i = \ln\left[\frac{\sum_{\mathbf{s}:b_i=0} \exp\left(\frac{< r, \mathbf{s} >}{2\sigma^2}\right)}{\sum_{\mathbf{s}:b_i=1} \exp\left(\frac{< r, \mathbf{s} >}{2\sigma^2}\right)}\right] \quad (3)$$

where $\mathbf{s}$ is the APSK symbol, σ^2 is the noise variance, and $< r, \mathbf{s} >$ represents the inner product between r and $\mathbf{s}$.

The above method calculates the exact LLR at the expense of finding multiple inner products, a division, and a logarithm. A more computationally efficient technique [13] considers only the inner product with the constellation point with the least euclidean distance from the received noisy symbol. This reduces Eq. 3 to

$$\lambda_i = \frac{1}{2\sigma^2}\left(2 < r, \mathbf{s}^0_{\text{min}} - \mathbf{s}^1_{\text{min}} > + ||\mathbf{s}^0_{\text{min}}||^2 - ||\mathbf{s}^1_{\text{min}}||^2\right) \quad (4)$$

where $\mathbf{s}^j_{\text{min}} = \arg\min_{\mathbf{s}:b_i}(||r - \mathbf{s}||^2)$. This closest point can be found in the 16-APSK constellation without any multiplications by dividing it into Voronoi regions [13] and finding the closest point through a combination of comparisons with $\text{Re}\{r\}, \text{Im}\{r\}, \text{Re}\{r\} \pm \text{Im}\{r\}$ and $\arg(r)$. The closest point in the 32-APSK constellation can be found by finding the closest three points on each ring and calculating the Euclidean distance. The LLRs for all bits in a codeword are sent to the LDPC decoder.

LDPC DECODER

Each LDPC code is associated with a (n, m) parity check matrix H. This parity check matrix can be represented graphically by a Tanner graph which is a bipartite graph [1]. The graph contains two types of nodes; bit nodes and check nodes. The decoding algorithm iteratively updates the check nodes with its associated bit nodes, and then updates the bit nodes with its associated check nodes. The optimal algorithm we use is described in [1]. The number of iterations can be varied, and the variation in coding gain with number of iterations for APSK have been presented in [13]. The complexity in this algorithm comes from the check node update metric which is given by

$$\eta_{m,n}^{[l]} = -2\tanh^{-1}\left(\prod_{i\in N_{m,n}} \tanh\left(-\frac{\lambda_i^{[l-1]} - \eta_{m,i}^{[l-1]}}{2}\right)\right) \tag{5}$$

where $N_{m,n}$ represents the bits associated with the mth check except bit n. It is obvious that the compution of the hyperbolic and inverse hyperbolic tangents is computationally expesive.

We also present results using a more computationally efficient [16] but sub-optimal "scaled-min" algorithm. This algorithm is identical to the optimal algorithm but with the check node update metric given by

$$\eta_{m,n}^{[l]} = -K\left(\prod_{i\in N_{m,n}} \text{sign}(-\lambda_i^{[l-1]} + \eta_{m,i}^{[l-1]}) \times \min_{i\in N_{m,n}} \{|-\lambda_i^{[l-1]} + \eta_{m,i}^{[l-1]}|\}\right) \tag{6}$$

where K is a constant. For our simulations we use $K = 0.75$. This greatly reduces complexity since the hyperbolic tangent functions have been replaced with multiplications, additions, and comparisons. This comes at a cost of coding gain which we show in our results varies from 0.1 to 0.5 dB.

RESULTS

In our simulations we considered code rates of 1/2, 2/3, 3/4, and 4/5 with information bit block sizes of 1024, 4096 (for rates 1/2, 2/3, and 4/5), 768, and 1536 (for rate 3/4). The BER performance of rates 1/2, 2/3, and 4/5 with APSK have been presented in [13]. Here we show the loss in coding gain with the reduced complexity algorithm. Table 1 shows the loss in performance for each combination of code rate, block size, and constellation. Figures 4 and 5 show the BER performance for rate 3/4 codes with the optimal and the reduced complexity algorithm.

We also compare the performance of our system with SOQPSK-TG. This comparison shows the capability of the system with backoff while still performing at least on par with an SOQPSK-TG system. For this comparison we use the sub-optimal decoder with 50 iterations. Also, we make this comparison at a bit error rate of 10^{-10}. To do this we extrapolate our results to 10^{-10} BER and compare it to the theoretical probability of error bound for SOQPSK-TG [17]. The results for rate 3/4 codes are shown in Figure 6. The backoff capability for all code rates, block-lengths, and constellations is summarized in Table 2

The backoff capability in Table 2 represents the power amplifier backoff the system can bear before the system performs worse than an SOQPSK-TG system. It can be seen that a lower code rate and higher block size can provide a higher backoff capability, however this comes at a cost of lower spectral efficiency and higher latency times.

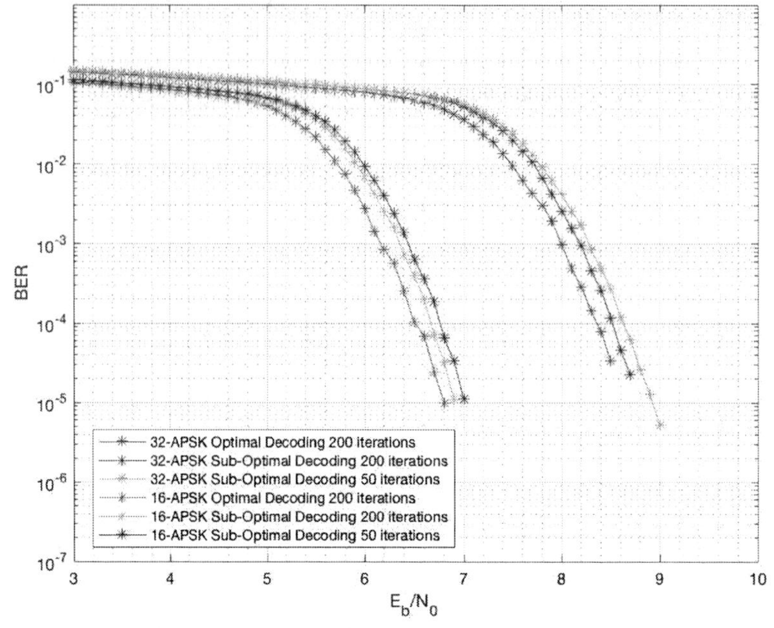

Figure 4: Bit Error Rate of Rate 3/4 codes with block size of 768 bits.

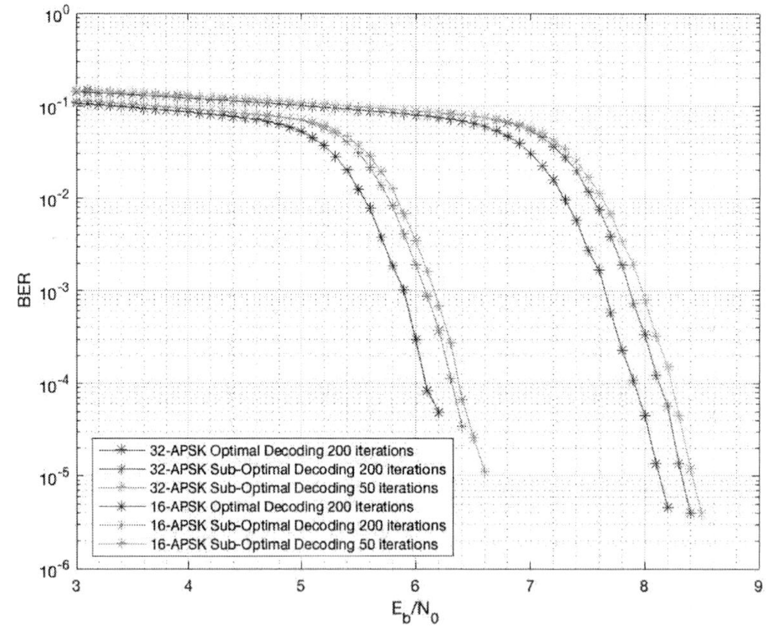

Figure 5: Bit Error Rate of Rate 3/4 codes with block size of 1536 bits.

Constellation	Block Size	Code Rate	Loss in Coding Gain
16-APSK	768 bits	3/4	0.25 dB
	1024 bits	1/2	0.38 dB
		2/3	0.37 dB
		4/5	0.20 dB
	1536 bits	3/4	0.29 dB
	4096 bits	1/2	0.50 dB
		2/3	0.40 dB
		4/5	0.10 dB
32-APSK	768 bits	3/4	0.30 dB
	1024 bits	1/2	0.48 dB
		2/3	0.40 dB
		4/5	0.20 dB
	1536 bits	3/4	0.25 dB
	4096 bits	1/2	0.11 dB
		2/3	0.20 dB
		4/5	0.22 dB

Table 1: Loss in coding gain for different code rate, blocksize, and constellation combinations.

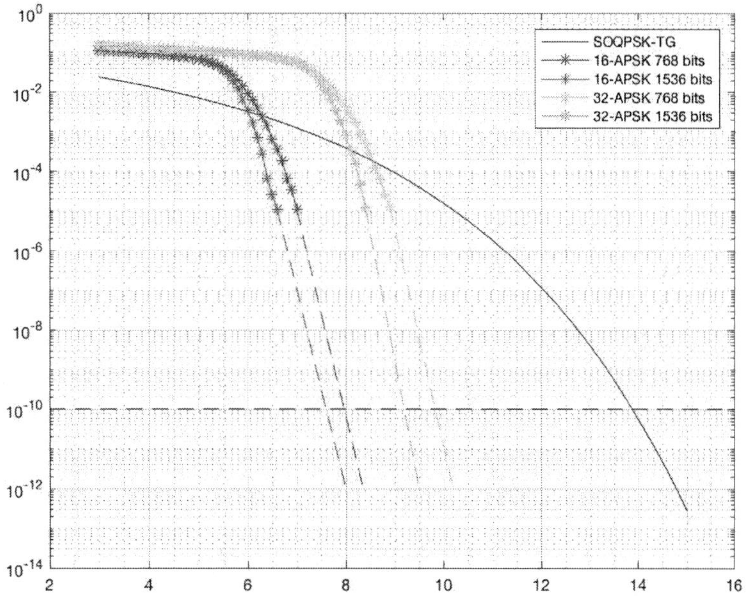

Figure 6: Comparison of Rate 3/4 codes with SOQPSK. The dashed lines represents extrapolation.

Constellation	Block Size	Code Rate	Backoff Capability
16-APSK	768 bits	3/4	5.14 dB
	1024 bits	1/2	5.51 dB
		2/3	6.09 dB
		4/5	5.56 dB
	1536 bits	3/4	6.28 dB
	4096 bits	1/2	9.38 dB
		2/3	8.17 dB
		4/5	6.60 dB
32-APSK	768 bits	3/4	4.00 dB
	1024 bits	1/2	6.50 dB
		2/3	5.16 dB
		4/5	3.60 dB
	1536 bits	3/4	4.69 dB
	4096 bits	1/2	7.47 dB
		2/3	6.17 dB
		4/5	5.06 dB

Table 2: Backoff capibility for different code rate, blocksize, and constellation combinations.

CONCLUSION

Coded-APSK systems provide an alternative to traditional CPM based telemetry systems with improved spectral efficiency. APSK, unlike CPM, is suseptible to non-linear distortions when used with power amplifiers. We have shown an LDPC coded APSK system which can be used with backoff to mitigate these distortions. The loss in performance due to complexity reduction techniques is documented and the backoff capbility of the system is shown. Future work includes analyzing the performance of this system when used with different power amplifier models and backoff required to match or better existing SOQPSK-TG systems while increasing spectral efficency.

REFERENCES

[1] T. K. Moon, *Error correction coding Mathematical Methods and Algorithms.* John Wiley and Sons, 2005.

[2] J. G. Proakis, *Digital communications.* McGraw-Hill, New York, 1995.

[3] R. Gallager, "Low-density parity-check codes," *IRE Transactions on Information Theory*, vol. 8, no. 1, pp. 21–28, 1962.

[4] D. J. MacKay, "Good error-correcting codes based on very sparse matrices," *IEEE Transactions on Information Theory*, vol. 45, no. 2, pp. 399–431, 1999.

[5] K. S. Andrews, D. Divsalar, S. Dolinar, J. Hamkins, C. R. Jones, and F. Pollara, "The development of turbo and LDPC codes for deep-space applications," *Proceedings of the IEEE*, vol. 95, no. 11, pp. 2142–2156, 2007.

[6] Range Commanders Council Telemetry Group, Range Commanders Council, White Sands Missile Range, New Mexico, *IRIG Standard 106-15: Telemetry Standards*, 2015,. (Available on-line at http://www.irig106.org).

[7] R. De Gaudenzi, A. Guillén i Fàbregas, and A. Martinez, "Turbo-coded APSK modulations design for satellite broadband communications," *International Journal of Satellite Communications and Networking*, vol. 24, no. 4, pp. 261–281, 2006.

[8] European Telecommunications Standards Institute (ETSI), *Digital Video Broadcasting (DVB), Second generation framing structure, channel coding and modulation systems for broadcasting, interactive services, news gathering and other broadband satellite applications (DVB-S2)*, 2006.

[9] G. Ungerboeck, "Channel coding with multilevel/phase signals," *IEEE Transactions on Information Theory*, vol. 28, no. 1, pp. 55–67, 1982.

[10] E. Perrins, "FEC systems for aeronautical telemetry," *IEEE Transactions on Aerospace and Electronic Systems*, vol. 49, no. 4, pp. 2340–2352, 2013.

[11] C. Shaw and M. Rice, "Turbo-coded APSK for telemetry," in *International Telemetering Conference Proceedings*, International Foundation for Telemetering, 2007.

[12] C. Shaw and M. Rice, "Turbo-coded APSK for aeronautical telemetry," *IEEE Aerospace and Electronic Systems Magazine*, vol. 25, no. 4, pp. 37–43, 2010.

[13] J. Hamkins, "Performance of low-density parity-check coded modulation," in *Aerospace Conference, 2010 IEEE*, pp. 1–14, IEEE, 2010.

[14] Consultative Committee for Space Data Systems (CCSDS, *Low density parity check codes for use in near-Earth and deep space applications*. 131.1-O-2 Orange Book, Sept. 2007.

[15] R. De Gaudenzi, A. G. i Fabregas, and A. Martinez, "Performance analysis of turbo-coded APSK modulations over nonlinear satellite channels," *IEEE Transactions on Wireless Communications*, vol. 5, no. 9, pp. 2396–2407, 2006.

[16] J. Chen, A. Dholakia, E. Eleftheriou, M. P. Fossorier, and X.-Y. Hu, "Reduced-complexity decoding of LDPC codes," *IEEE Transactions on Communications*, vol. 53, no. 8, pp. 1288–1299, 2005.

[17] E. Perrins and M. Rice, "Reduced-complexity approach to iterative detection of coded SO-QPSK," *IEEE Transactions on Communications*, vol. 55, no. 7, pp. 1354–1362, 2007.

PREDICTIVE TELEMETRY MODELS FOR AEROSPACE FLIGHT MISSIONS

Amir Liaghati, Nick Chang, Mahsa Liaghati, John Blumer
The Boeing Company
1100 Redstone Gateway Dr
Huntsville, AL 35824
amir.l.liaghati@boeing.com
chen.j.chang@boeing.com
mahsa.l.liaghati@boeing.com
john.h.blumer@boeing.com

ABSTRACT

Designing an efficient telemetry system is one of the most challenging tasks in the aerospace industry due to the limited telemetry bandwidth and the restricted radio frequency spectrum allocation. In a traditional telemetry system design, a large percentage of the allotted bandwidth is consumed by the overhead information to follow various standards and protocols which results in inefficiency of output data. A dynamic simulation model is created by using the Matlab/Simulink to analyze the telemetry system performance, and it is also used to predict the overall downlink data, including sensor instrumentation, flight critical information, video, and the Consultative Committee for Space Data System (CCSDS) overhead over the mission event timeline.

INTRODUCTION

A typical telemetry system consists of multiple data acquisition avionic boxes for acquiring instrumentation data; a telemetry processing box for collecting instrumentation provided by multiple data acquisition boxes, which repackages the received data to end user's required data format. It then sends the reformatted data to the downlink transmitter. Figure 1 illustrates a typical telemetry system design.

However, a large portion of the allocated downlink telemetry bandwidth is consumed by the CCSDS overhead [1, 2, 3]. One of the biggest disadvantages for using the CCSDS protocol is the overhead size in one downlink telemetry stream is between 20% and 60%, depending on the type of Forward Error Correction (FEC), Channel Access Data Unit (CADU) frame size, and the telemetry processing efficiency.

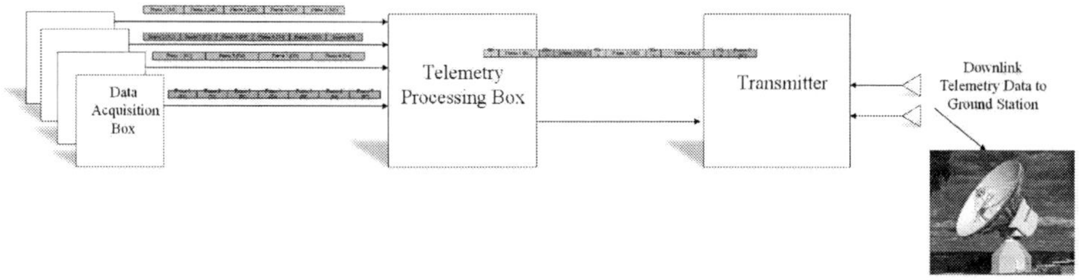

Figure 1: Typical Telemetry System Architecture.

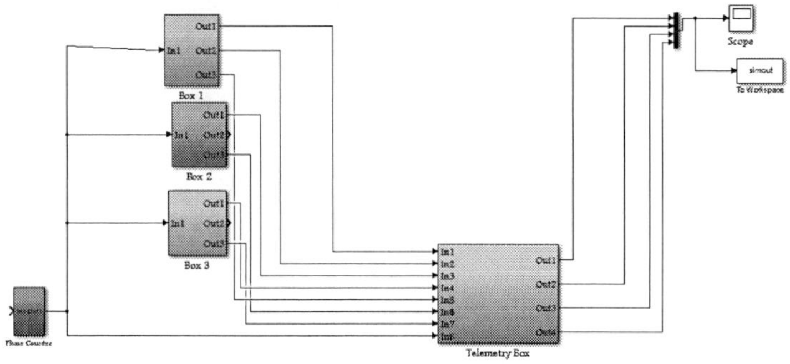

Figure 2: Telemetry Model in Simulink

PREDICTIVE TELEMETRY MODEL

A predictive Matlab/Simulink model for benchmarking the telemetry system performance is developed, analyzed, and validated in a large space program. The main objective of this model is to predict the overall data, overhead, and filled packets correlating to a particular space mission profile. This model consists of several avionics boxes that transmit data to the telemetry processing box with different data rate as inputs. The output of this model is coming from the telemetry processing box, with the reformatted data with different downlink rates during different phases of a mission. Figure 2 shows the telemetry model and figure 3 shows the telemetry box model in Simulink.

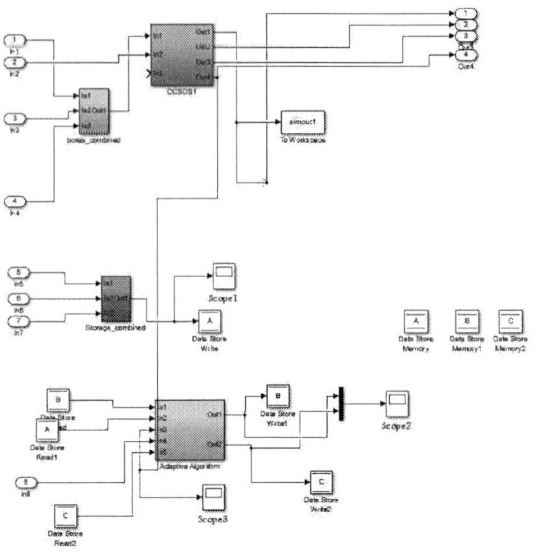

Figure 3: Telemetry Box in Simulink

VARIABLE PROCESSING TIME TO MINIMIZE OVERHEAD

[h] In a conventional telemetry box design, the telemetry box has a fixed processing time τ. Oc-

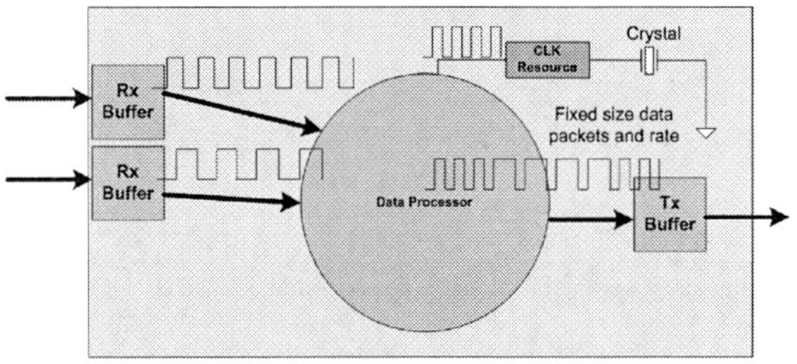

Figure 4: Variable Output Data Rate Block Diagram.

casionally, the total input data rate received by the telemetry processing box is not large enough to fill the CADU, an Idle Packet is used which reduces system performance. In the dynamic modeling approach, the box dynamically changes its processing time to $\tau + \alpha$, α is the time that box waits for more data to arrive until the optimum threshold is reached. This efficient approach can reduce the overhead and overall data for transmission. Figure 4 is a simplified functional diagram of the dynamic telemetry processing model with variable output rate. Figure 5 shows the result of variable output data rate. It is shown that when τ has a smaller value the overall data and overhead will be larger for the same data rate from the avionics boxes. However, when τ increases, the overall data and overhead decreases. This implementation could be used in missions where output rate is significantly smaller, for instance, transmission to satellites with a smaller transmission rate (e.g., 256kbits/s). Also, the filled data are reduced when the processing time is increased. In our previous work [4], we have shown a detailed study on the transition between packets. When the data rate is very small, at least one packet per processing time needs to be transmitted. However, since there are not much data to fill the packet, Idle pattern data, also known as filled data will be added in the packet. As a result, most of the packet is filled with non-essential filled data. To avoid that, in this work we suggest varying the processing time, so more data could arrive to fill up the packet. As a result, most of the packets will be filled with actual data for transmission. Figure 6 shows an example of correlation between data,, overhead, and fill data as data rate is increasing. It can be seen that as data is increasing, filled data decreases until the transition between packets occur.

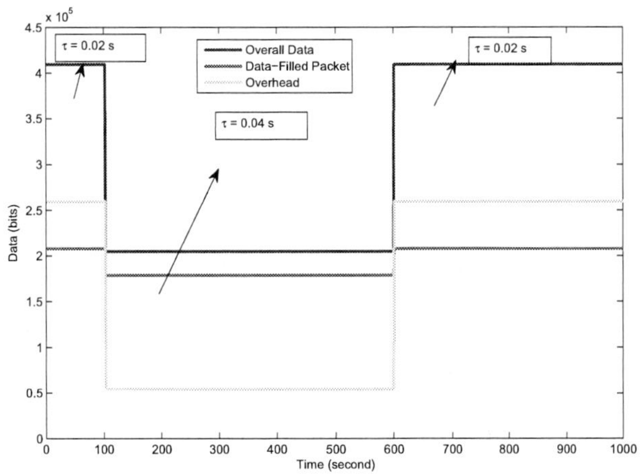

Figure 5: Result for Variable Output Data Rate.

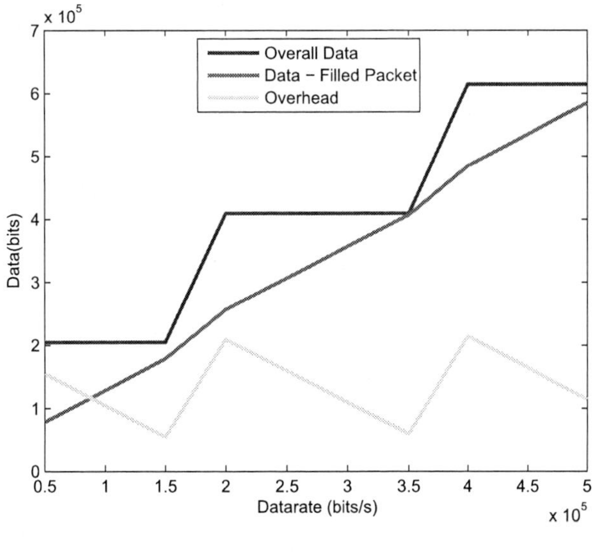

Figure 6: Correlation between data, overhead, and filled data for different data rates.

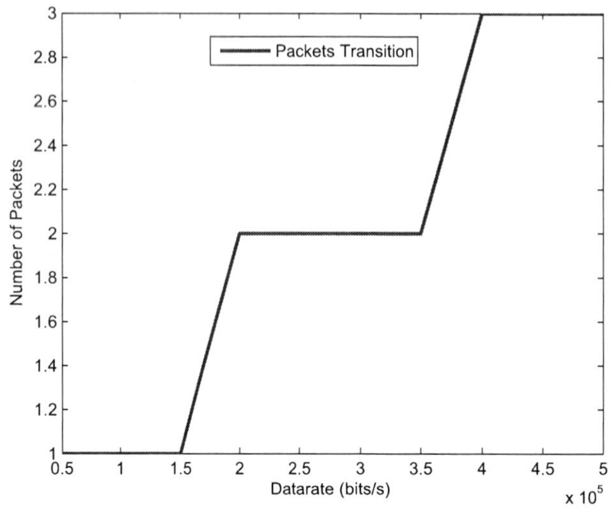

Figure 7: Correlation between number of packets for different data rates.

When the transition happens, that is where the filled data are minimized and most of the packet is filled with actual data which implies that this is the most optimum transmission. As mentioned before, our main goal is to maximize the data transmission and minimize the filled data. Figure 7 shows the number of packets transmitted versus data rate. As data rate is increasing, the number of packets transmitted stays the same regardless of filled data or actual data. When data rate is small enough that only one packet per processing time is transmitted, the majority of that packet might be filled with filled data, and by waiting longer, it allows more data to arrive for more optimization of the data transmission.

STORAGE/ REAL-TIME TRANSMISSION

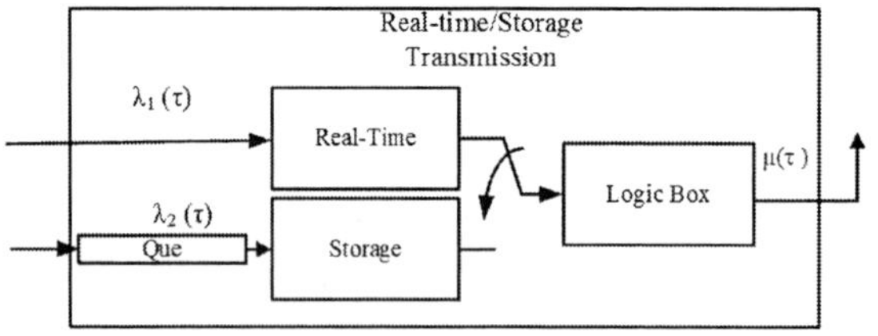

Figure 8: Avionics Box Real-Time/Storage Transmission

In this section, a new model is designed for storage and real-time transmission. This capability allows users to transmit important information real-time, and to store some portion of the data with lesser priorities. During some phase of mission, there is more room to transmit the storage data. As a result, this model transmits the stored data during those phases. Figure 8 shows the real-time/storage model diagram. λ_1 is the real-time rate, λ_2 is the storage input rate, and μ is the output rate. In this model, we consider three main cases.

A. *Case 1: Real-time on, Storage off*

In case 1, real-time transmission is on, and since real-time is a priority, it will be transmitted with the same rate. Note, $\mu(\tau)$ can be greater than $\lambda_1(\tau)$. For this case, we assumed that they will have the same rate. Since storage is off, the storage will contain the same data as previously without any changes. The following equation is used when case one occurs.

$$\lambda_1(\tau) = \mu(\tau) \tag{1}$$

$$storage(n) = storage(n-1); \quad \text{for } n > 1 \tag{2}$$

where $\lambda_1(\tau)$ is the real-time input rate and $\mu(\tau)$ is the output rate.

B. *Case 2: Real-time on, Storage on*

In case 2, real-time and storage are both on. Since real-time is a priority, it will be transmitted in real time. However, since storage is on, storage data will be stored in the buffer to be transmitted when the input rate real-time is off in case 3. In other words, the amount of storage is increasing in the buffer. The following equation is used when case two occurs.

$$\lambda_1(\tau) = \mu(\tau) \tag{3}$$

$$storage(n) = \lambda_2(\tau).\phi + storage(n-1) \quad \text{for } n > 1 \tag{4}$$

where ϕ) is the processing time.

C. *Case 3: Real-time off, Storage on*

In case 3, input rate real-time is off, and storage transmission is on. As a result, the output rate will transmit data from storage. It depends on the output rate μ, storage in the buffer could decrease or increase. If $\mu > \lambda_2$ The storage buffer is decreasing, and if $\mu < \lambda_2$, the storage buffer is increasing similar to case 2. The following equation is used when case three occurs.

$$\lambda_1(\tau) = 0 \tag{5}$$

$$storage(n) = \lambda_2(\tau).\phi + storage(n-1) - \mu(\tau).\phi \quad \text{for } n > 1 \tag{6}$$

RESULTS

Figure 9 and 10 show storage results of a random mission example. It can be seen that in some duration storage is increasing, and in some phases of mission storage is decreasing. These results can help the user when designing a mission profile. One of the main objectives of real-time/storage transmission is to downlink the storage prior to mission completion. This model can help users study different mission profile and design the most optimum data rates for both real-time and storage.

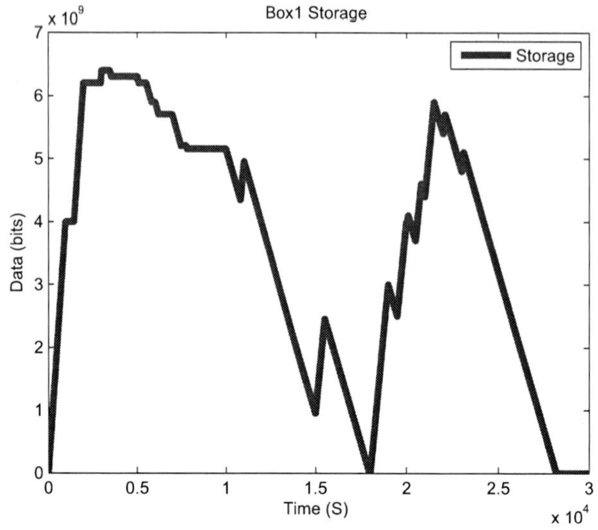

Figure 9: Telemetry Model in Simulink

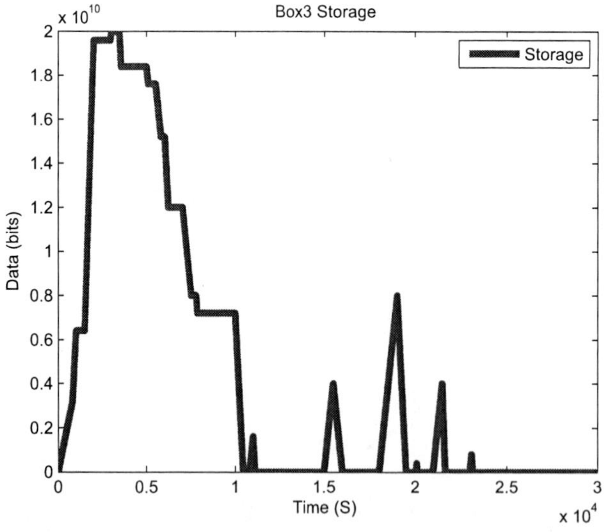

Figure 10: Telemetry Box in Simulink

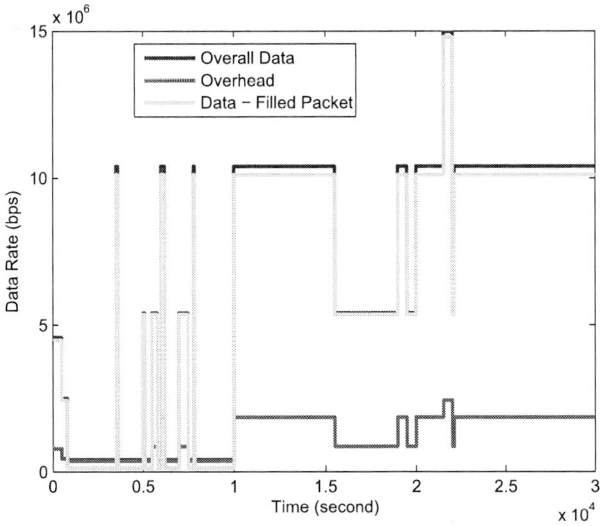

Figure 11: Telemetry Model Result

Figure 11 shows the result of the entire mission. Note, this mission profile is a random mission with different input/output rates for demonstration purposes only. The overall data is the overall data after adding the CCSDS overhead plus the filled data. The overhead is the overall CCSDS overhead plus the filled data. Note, in designing a good mission profile model, users need to make sure that the overall data is always below the output rate allowable by the RF transmitter.

CONCLUSIONS

In this work, we modeled a telemetry system in Simulink which can predict the overall data, CCSDS overhead, and filled data for the entire mission profile. We propose to use dynamic processing time to optimize data transmission by minimizing filled data/overhead. In addition, a model is used for a real-time/storage transmission for the avionics boxes which allows user to transmit more important data real-time and store the ones with less priority to transmit later during some phases of the mission. This model predicts when the all of the storage data is transmitted. The goal is to transmit the storage for each box prior to mission completion. Overall, this model helps predict and study different mission profiles and to minimize overhead/filled data.

REFERENCES

[1] Consultive Committee for Space Data Systems (CCSDS), "Encapsulation service (131.1-B-2 Blue Book)," Oct. 2009.

[2] Consultive Committee for Space Data Systems (CCSDS), "Aos space data link protocol (732.0-B-2 Blue Book)," July 2006.

[3] Consultive Committee for Space Data Systems (CCSDS), "Tm synchronization and channel coding (131.0-B-2 Blue Book)," August 2011.

[4] A. Liaghati, N. Chang, M. Liaghati, and A. Maffei, "A novel scheme for telemetry system data rate optimization," in *2016 IEEE Aerospace Conference*, pp. 1–7, March 2016.

RAPIDLY RECONFIGURABLE SYSTEM MANAGEMENT

Patrick J. Noonan Jr., Austin J. Whittington, Hakima Ibaroudene, Myron L. Moodie
Southwest Research Institute®
San Antonio, Texas
patrick.noonan@swri.org, austin.whittington@swri.org, hakima.ibaroudene@swri.org, myron.moodie@swri.org

ABSTRACT

The growth of network and distributed technologies in flight test instrumentation (FTI) has provided the benefits of flexibility, scalability, and compatibility with prevalent computing capabilities. However, to achieve these capabilities, the complexity of each piece of FTI and the overall system has increased dramatically. Even with systems composed of equipment from a single vendor, it is important to have management systems that provide the flexibility to adapt quickly to various system configurations and present unified information to the flight test users. The growth of network technologies and then standardized approaches such as iNET standards becoming accepted IRIG 106 standards is leading to the growth of multi-vendor systems. These multi-vendor systems further increase the need for rapidly reconfigurable management systems. This paper describes a constraints engine we have developed to enable flexible system management systems and reflects on how these techniques have been used successfully in the iNET System Manager.

KEYWORDS

Flight Test Instrumentation, Constraints, Rete, MDL, XML

INTRODUCTION

The growth of network and distributed technologies in flight test instrumentation (FTI) has provided the benefits of flexibility, scalability, and compatibility with prevalent computing capabilities. However, to achieve these capabilities, the complexity of each piece of FTI and the overall system has increased dramatically. Even with systems composed of equipment from a single vendor, it is important to have management systems that provide the flexibility to adapt quickly to various system configurations and present unified information to the flight test users.

Network-based telemetry systems have unprecedented amounts of flexibility due to the ability to manipulate configuration during a test. This flexibility, along with ever increasing data demands, imposes requirements not only on the devices acquiring, recording, and processing the data, but also on the underlying network transporting that data. Furthermore, the amount of spectrum available for tests is ever decreasing, so that optimal utilization of spectrum is required. This leads to complexities involved in managing and configuring the devices and the underlying network, which need to be carefully handled. In addition, the need to conduct multiple concurrent tests, remotely manage the FTI on a Test Article (TA), reconfigure devices in the middle of a test, reconfigure the test system for a different test, and dynamically modify certain parameters of the test system to maximize test efficiency are all now requirements of system management.

The flexibility and complexity of dynamic status and control is increased further when using a bidirectional communication link with the TA. This increase in flexibility creates challenges for test operations. For instance, a user might need to re-route data on the TA for a given test. To ensure proper configuration and performance, a network topology and loading prediction may be needed in order to forecast the potential impacts of the re-configuration, not just within the TA, but across the systems that transport data to the ground stations. The growth of network technologies and then standardized approaches such as Integrated Network Enhanced Telemetry (iNET) standards becoming accepted Inter-Range Instrumentation Group (IRIG) 106 standards has made these scenarios a reality. This flexibility and complexity, along with the growth of multi-vendor systems, has only increased the need for rapidly reconfigurable management systems.

Highly specialized applications, or management systems, are needed to fully manage network-based telemetry systems and are central in maintaining the system working seamlessly, reliably, and effectively. A key aspect of this is the speed and ease with which FTI and system level configurations can be built and deployed. This paper describes a constraints engine we have developed to enable flexible system management systems and reflects on how these techniques have been used successfully in the iNET System Manager.

CONFIGURATION VALIDATION

Constraints

Constraints are rules that state or control what something can or is allowed to do. These can range from how much weight someone can pick up without hurting themselves, to the tensile strength of a material, to the top speed of a car. In FTI, constraints generally describe the capabilities and limitations of devices and networks. A data acquisition device may have an upper limit on its sample rate, or a network switch can only handle 100 megabits per second from any individual data source.

Correct-by-Construction

In the practical case, these constraints are used as validation rules. A system integrates the vendor constraints that describe the devices, system constraints that describe the capabilities of all devices that participate in the system, and the user constraints that standardize or simplify device configuration [1]. User input is checked as they create a configuration, or at the very least in a verification action that can be triggered, and inputs are evaluated against the set of interwoven constraints, which point out any problems with the current set of values. As the user develops a test, the constraints ensure that the configuration will be valid, both for the individual devices and within the larger system.

Constraints Languages

Many of the existing flight test telemetry configuration languages use an XML format, including MDL, XidML, and TMATS. As such, XPath, the W3C query language for selecting nodes from XML documents, was a natural choice to describe the constraints on those files. This language, in conjunction with XForms, another XML-based technology made for gathering and processing XML data, was used by the Boeing Company's Modular Instrumentation Setup Tool (MIST), an implementation which uses constraints in the manner described above to aid its users in creating configuration using the correct-by-construction techniques [2].

As we have continued to explore the application of constraints to FTI, a more powerful constraints language, W3C's Rule Interchange Format (RIF) [3], has been identified as a logical successor to XPath. RIF was developed for expressing rules which computers can execute, which indeed matches this use case for constraints. RIF's main advantage over XPath is its lack of immediate mapping to XML. While XPath's direct mapping makes it easy to create constraints, it does require deep knowledge of the XML schema being constrained, and is then targeted to a specific version of that schema. Depending on the magnitude of changes between versions, this may be a minor issue, or it could require whole rewrites of the constraints. RIF can be indirect, with an individual constraint referring to nouns defined elsewhere. An XPath constraint has to define a specific instance, such as the path to a measurement in MDL:

```
MDLRoot/MeasurementDomains/MeasurementDomain/Measurements/Measurement
```

A RIF path could simply refer to a Measurement and have the particulars of what that means defined in a separate file matching the schema version currently in use. This means that only the mappings from these nouns to their specific instances would need to be updated between versions (or even sufficiently compatible schemas), instead of the XPath statements, which could contain complex logic that is hard to translate while maintaining certainty that the intent of the constraint is being preserved.

This also means that RIF constraints can refer to concepts and objects more abstractly, or be as specific as necessary. A RIF rule can reference a Transducer, which is covered by the broader term Device in MDL, and again a second rule can define what that term resolves to. Those developers or users creating new rules, given a sufficient set of prebuilt nouns, can use terms which are more familiar to them to describe the constraints, which are then independent of any particular schema but for the mapping rules.

However, a switch to RIF is not incompatible with the XPath-based approach for a particular XML-based solution. Because the references in a RIF statement will eventually reduce down to XML entities, given that it is describing the constraints on XML configuration files, it is possible to create an XPath expression (or expressions, in cases where a rule may be too complex to capture with only one) that describe the same constraint for a particular schema. Thus, a set of RIF constraints can be used with an existing XPath system, after this conversion process takes place.

Research Areas

There are several active research areas seeking to improve these technologies. Constraint satisfaction technologies can be used to automatically build configurations, given a set of goals and the constraints that a configuration must be valid within [4]. Users could give the system a desired measurement list, and such a satisfaction system would come back with a set of possibilities, each of which acquires all of the specified measurements with varying combinations of devices or parameters. The user could then refine a search, eliminating devices they don't want to consider or optimizing for other factors such as weight and space, knowing that any generated configuration will be valid within the system's bounds.

There is also the area of meta constraint validation, validating the constraints themselves to make sure their interrelation does not lead to a validation paradox [5]. For example, the case where a vendor specifies a finite impulse response (FIR) filter as the only option for a particular acquisition technique, and a user constraint gives the rule that all filters must be infinite impulse response (IIR). In this case, if a user set up a measurement using that technique, there are no valid filters usable, which is a failure that could be detected by this higher-level validation.

CONSTRAINTS ENGINE

In previous systems, XForms has been used for the validation technology, evaluating the constraints and updating its object model as the user inputs new data. However, XForms provides a lot of presentation layer functionality that is not necessary in a tool that already has its own database and user interface (UI) frontend. Thus, these types of applications need a more specialized tool, a constraints engine that can work with existing objects and simply provide running validation of configurations to the rest of the system, rather than doing all the processing, storing, and rendering that a full XForms system uses.

This gives the blueprint of what the constraints engine does. It monitors a set of objects that reflect the current state of the database, and on a change, it evaluates the new information and determines whether any constraints have failed. In the case that they did, the constraints engine sends a report on each object that failed and the error message to show, and the UI decides how to display the message. By dropping the requirements for any rendering or other presentation layer, the focus becomes performance and efficiency.

The main processing idea behind the constraints engine is the usage of the Rete algorithm [6], a standard pattern matching algorithm for implementing production rule systems, through an existing rules engine. A high-level illustration of this algorithm is shown in Figure 1 [7]. In general, the Rete algorithm matches facts against production rules. The production rules are "if/then" statements, such that if an initial condition is matched, the "then" action is fired. The Rete algorithm is generally used for business rules, such as "when a customer hits 10,000 airline miles, award them special status", but we will show how this algorithm can be used for our constraint evaluation use case.

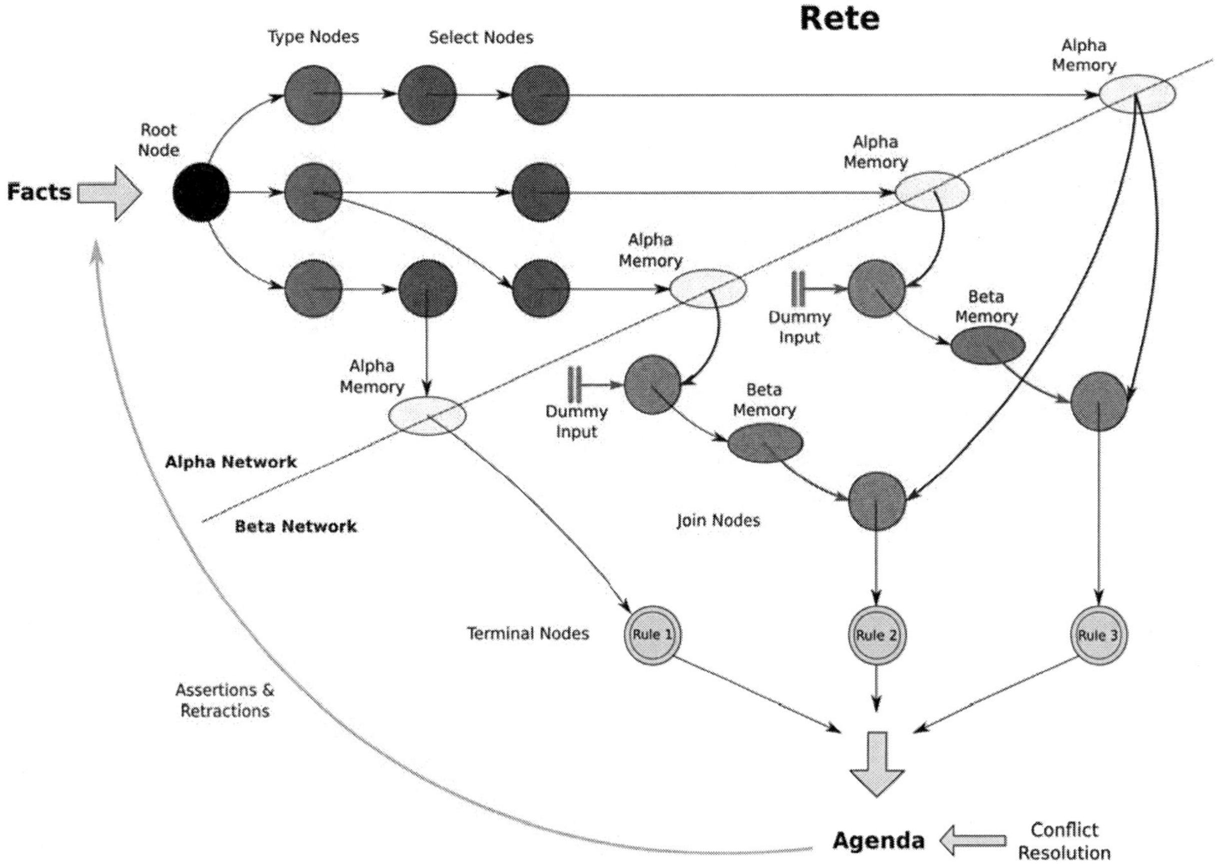

Figure 1 – Rete Algorithm

Before any useful evaluation can take place, we first convert our RIF constraints into Rete production rules, where the antecedent (the "if") is the failure case of the constraint, and the consequent (the "then") is the resulting error message and the triggering element. In this way, any rule which matches the conditions of failure will return the information necessary for the UI to display what went wrong to the user.

As the user builds up a configuration, the new facts (stored in the objects) are added to the Rete network. The Alpha Network generally categorizes facts based on simple conditional tests, and files facts into alpha memory as conditional chains are satisfied. For example, in FTI, such a chain could be "is this a piece of hardware", then "is this network-connected", and "is it made by vendor A", eventually categorizing as a Vendor A DAU. When a new fact is added, or an existing fact is modified, other facts do not have to be reevaluated because they are already stored in the network. The Beta Network joins between different memory elements. In our case, this applies after nodes have been categorized, and will be where many rules are evaluated, as it is very common that a constraint depends on more than one part of a system. For example, such a rule could be "on a Vendor A DAU, each measurement has an upper sample rate bound of 10000 Hz." Then, within that beta node, a list of measurements could be joined with the list of DAUs to determine which measurements need to satisfy this check. Finally, the rules are evaluated, and if such a check was to fail, the user would receive the error information. Note that the particular facts and commonality between alpha and beta nodes will be done programmatically by the rules engine implementing the Rete network, and so will not have to be explicitly captured by hand.

In this application, we do not currently need to make use of any conflict resolution, assertion, or retraction functionality. The Rete algorithm is only being used to validate the user's input and return errors. Thus, the constraints only provide errors, and do not cause other facts to be asserted or retracted, and because no facts are changing, conflict resolution is unnecessary, with no ordering dependency on which an error message is revealed, as the UI will display them all regardless.

SYSTEM MANAGEMENT

Rapidly reconfigurable system management is being realized by the System Manager application development for the iNET program, which provides an initial user application for describing, configuring, monitoring, controlling, coordinating, and visualizing the operations of a Telemetry Network Standards (TmNS)-based FTI system [8]. A key aspect of System Manager is the ability to describe flight test configurations and generate MDL configuration files used to configure FTI. The flexibility provided by MDL allows almost everything in a flight test system to be described, from low-level sensors to complex RF networks. However, with this flexibility comes complexity, and that is where the use of constraints can greatly simplify the user's choices.

The Southwest Research Institute (SwRI®) developed system management design is based around the principles of extensibility, modularity, and separation of concerns. Because of these principles, system management can occur with various levels of constraints or even none at all. The modular constraints engine described above can be seamlessly integrated with existing objects backed by a database, and the separation of the RIF constraints from code allows for user experience updates at any level of the constraints hierarchy (FTI, system, user) without the need for rebuilding the application. By taking advantage of these techniques, System Manager is itself rapidly reconfigurable, that is it can be quickly adapted to new or modified FTI, system, and/or user level requirements.

The result of rapidly reconfigurable system management is the ability to use constraints evaluation to alert the user of errors in the configuration and provide meaningful error messages so that the errors can be quickly addressed. An example of this is shown in Figure 2, where the user is in the process of configuring a Link Manager that is managing 10 concurrent test missions with each test mission consisting of two RF links (an uplink and a downlink).

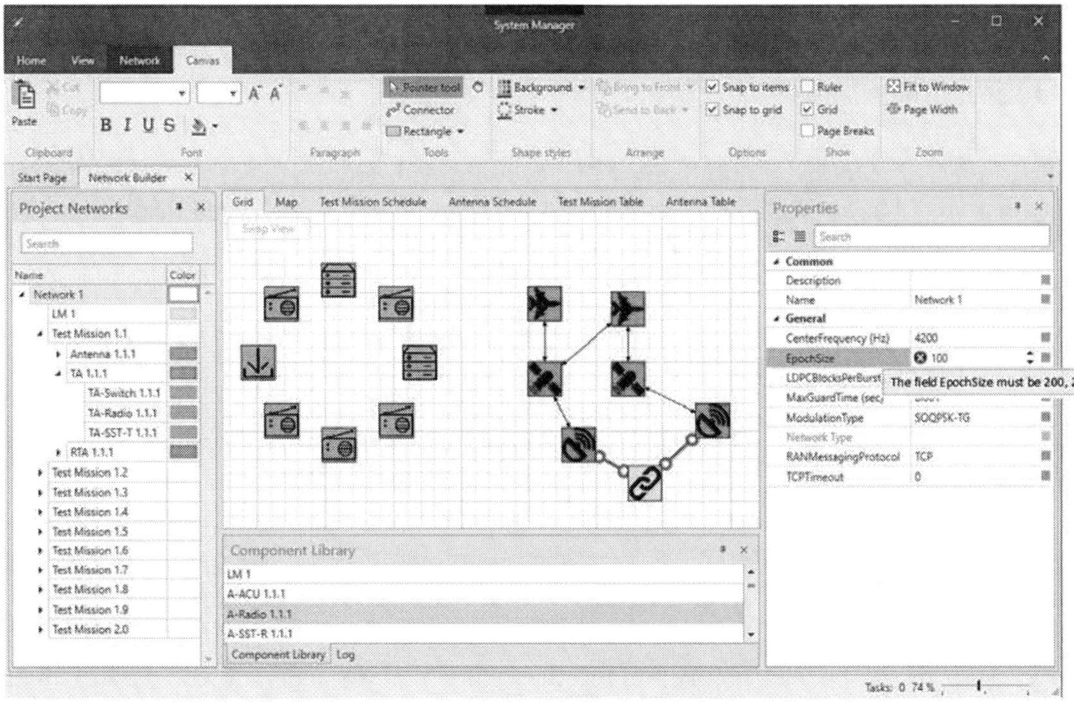

Figure 2 – Constraints Evaluation in System Manager

As shown in the example, the user has selected an epoch size of 100 ms which generates not only an error, but also an informative message stating that the epoch size must be either 200, 250, 500, or 1000 ms. This occurs because the constraints engine has evaluated a RIF rule stating that the Link Manager cannot operate correctly at an epoch size less than or equal to 100 ms if the guard bands are 1 ms and the burst size is 16 blocks per burst. Using this information, the user can quickly correct the error and continue describing the configuration. This small example shows how combining a familiar, user-friendly UI with the powerful constraints engine can lead the user to a configuration that is on its way to being correct-by-construction.

CONCLUSION

The rise of network technology in FTI and standardized approaches to system management have opened new capabilities and flexibility. Maximizing the benefit of these network systems requires equally flexible and rapidly reconfigurable management systems. Through our experience with a variety of commercial and military flight test programs we have developed a number of techniques that can be leveraged to build system management systems for the specific needs of a variety of flight test requirements. One of these applications, the iNET System

Manager, uses an extensible and modular design to incorporate a constraints engine that allows for rapid reconfiguration when FTI, system, and/or user level requirements change.

REFERENCES

[1] Noonan, P.J., H. Ibaroudene, A.J. Whittington, M.L. Moodie. "Simplifying Flight Test Configuration with Constraints." Proceedings of the International Telemetering Conference, Glendale, Arizona, November 2016

[2] Neumann, M., J. Moore, S. Pantham, M.L. Moodie, P.J. Noonan, A.J. Whittington. "An Adaptable Constraints-based Metadata Description Language (MDL) System for Flight Test Instrumentation Configuration." Proceedings of the European Telemetry and Test Conference, Nuremberg, Germany, May 2016

[3] https://www.w3.org/TR/rif-overview/

[4] Apt, Krzysztof. *Principles of constraint programming*. Cambridge University Press, 2003.

[5] Moskal, J., M. Kokar, J. Morgan. "Semantic Validation of T&E XML Data." Proceedings of the International Telemetering Conference, Las Vegas, Nevada, October 2015

[6] Forgy, C.L. "Rete: A fast algorithm for the many pattern/many object pattern match problem." *Artificial intelligence* 19.1 (1982): 17-37.

[7] https://commons.wikimedia.org/w/index.php?curid=15901661

[8] Noonan, P.J., T.A. Newton, G.C. Willden, T.B. Grace, W.A Malatesta. "iNET System Manager." Proceedings of the International Telemetering Conference, San Diego, California, October 2014

EFFICIENT DATA STORAGE, SAMPLING, AND RETRIEVAL BY LEVERAGING OPEN SOURCE TECHNOLOGIES

Micah Ferrill
Avionics Test & Analysis Corporation
4000 Eagle Point Corporate Drive
Birmingham, AL 35242
ferrillm@avtest.com

ABSTRACT

This paper demonstrates the use of open-source software tools to manage large data sets.

Advances in technology have greatly reduced the cost of data storage and processing systems. The ability to handle large amounts of data efficiently while retaining fine-grain control of the data retrieval process becomes a challenge. In particular, traditional data processing applications are inadequate to handle the large data sets typically encountered in IRIG-106 Chapter 10[1] data files.

We answer this challenge by using readily available, open-source tools that efficiently store and retrieve IRIG-106 Chapter 10 data to/from a file-based database.

We will demonstrate a method that facilitates a separation between the parsing of raw input data and the display of desired information at a user-defined sample rate. This open-source based solution provides a low-cost, reliable, and efficient means for handling large amounts of data at a high rate of speed.

INTRODUCTION

The problem we have to solve is displaying data from large collections of IRIG 106 Chapter 10 data potentially spanning multiple files. The issue that arises, of course, is that parsing multiple large, binary data files simultaneously can be a very demanding process and potentially limits us from real-time playback. Eventually, distributed data storage and processing may hold an answer, but for some uses this is much too heavy of a solution, and a system capable of running on a laptop would be more applicable.

Our intent is to showcase how we solve this problem while leveraging free and open-source software. By reusing previous work done in this space and incorporating various mature and well-vetted open source tools, we have built a solution that will facilitate real-time playback on common hardware while also architecting to facilitate the future leveraging of modern distributed systems design.

EXISTING WORK

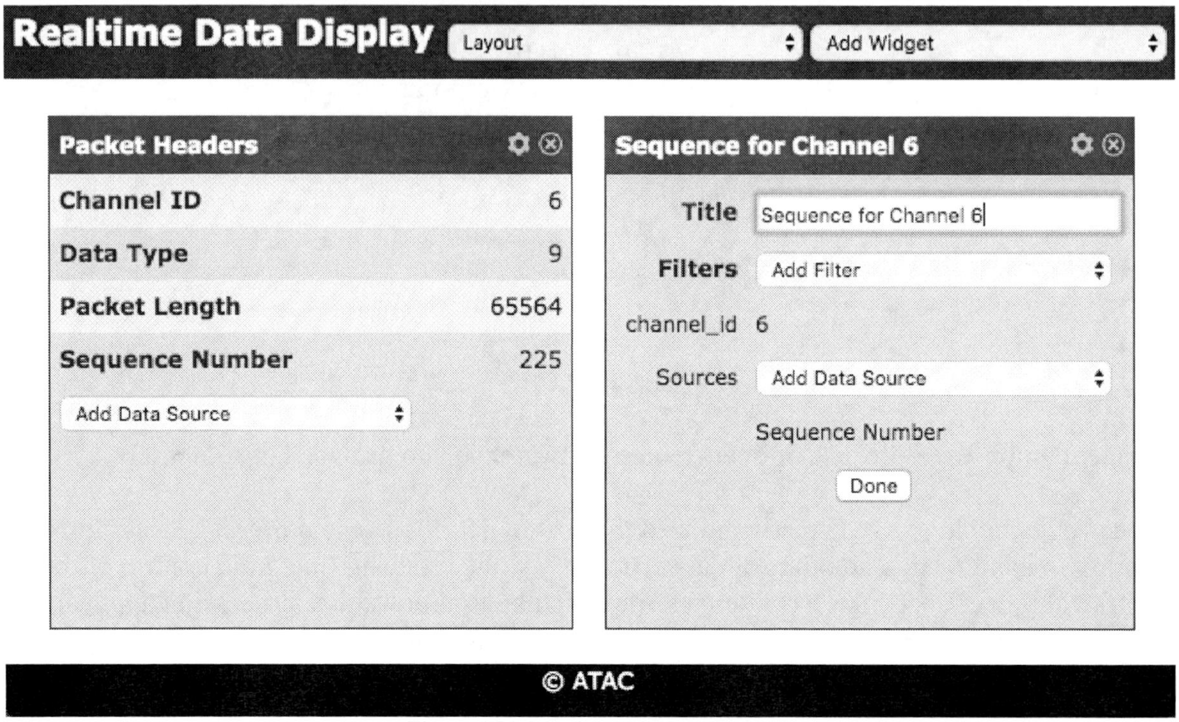

Figure 1 Lightweight Display Tool Demo

In previous years at ITC, we have demonstrated various iterations of a lightweight data display tool using a browser-based user interface. Much of the backend data processing and storage mechanism could be easily repurposed to jumpstart our new effort. One of our existing software libraries used for the data processing is a Chapter 10 parsing library written in Python[2], which was also reused for this project.

OPEN SOURCE

The Python programming language has been a favorite of ours for some time, not least because of the speed with which tools can be developed. The language also can be extended using C/C++ to enhance performance while maintaining the easy to use language characteristics of Python.

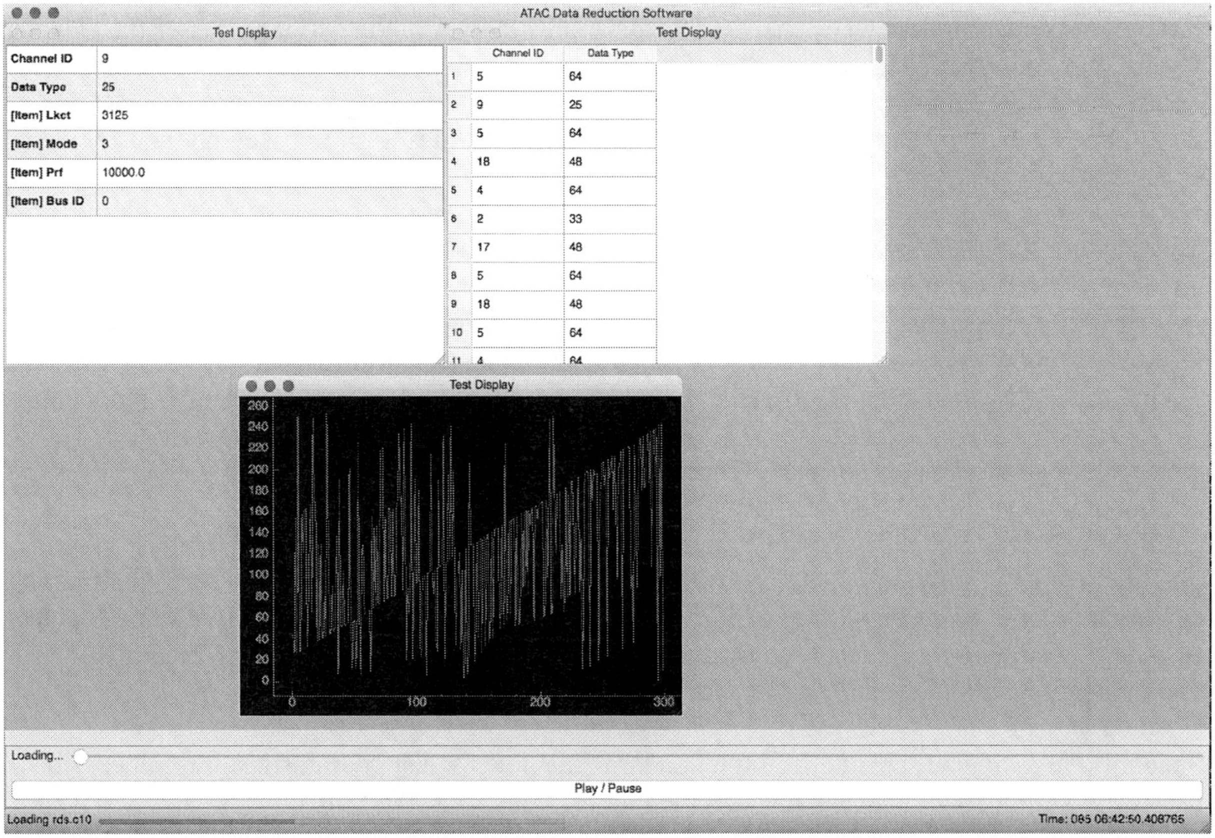

Figure 2 Draft Qt Native Interface (macOS)

We also built a cross platform OS-native user interface for the lightweight display tool by utilizing the Qt windowing toolkit[3] which has seen use in a wide array of successful projects. Qt is particularly useful to us in that it's cross-platform and written in C++, but it also has a number of Python wrappers. This allows us to prototype rapidly in Python while still having the flexibility to rewrite elements using C++ if faster performance is needed.

The SQLite database engine[4] has been the go-to embedded database option for quite some time. It is ubiquitous across all of the major browsers, smartphones, and desktop operating systems. SQLite can be used for temporary in-memory databases, or file-based data stores. With many similarities to other database engines, SQLite is also an excellent choice for prototyping with its simplicity of implementation. It also comes embedded into Python by default, making it ideal for the initial stages of this project.

THE SOLUTION

Our initial solution consists of 3 parts:

1. The loader module
2. Data displays and configuration (within the new Qt-based UI)
3. Self-adjusting update rate

Firstly, the loader module takes chapter 10 data from file[s] or a UDP stream, parses the Chapter 10 structure and stores packets in a SQLite database along with absolute time (computed based on RTC and time packets), channel ID, and data type. This allows us to reference information within the database by time as well as filter at the packet level if desired based on measurements of interest.

The database interface structure, while simplistic in appearance, has a number of interesting benefits. For one thing, the interface (as the loader or display see it) has been abstracted away from the specific database back-end. By having a module that provides a common interface to the various components we can centralize any database optimizations to that module as well as having the ability to change database back-ends without having to rewrite the entire application.

This technique has already seen a lot of use in Python and web development in the ORM (object-relational mapper) world. In some cases, not only the database interface, but the schema definition is abstracted away from the database engine into code. Using this technique one can quickly develop a solution on a lightweight database such as SQLite, and then deploy the same code to a production service using a more complex database system such as MySQL or Postgres by simply changing the database configuration. Part of the reason we borrowed from this approach is to facilitate future support for distributed data storage and computing.

Secondly, in a module separate from the loader is the new Qt display interface. From here, the user may configure and manage measurement definitions consisting of:

1. Measurement name – for human-readable reference
2. Packet description - channel ID and/or data type if appropriate
3. Message description – Mil-Std-1553 command word, ARINC 429 label, etc.
4. Size and offset of measurement data within specified message

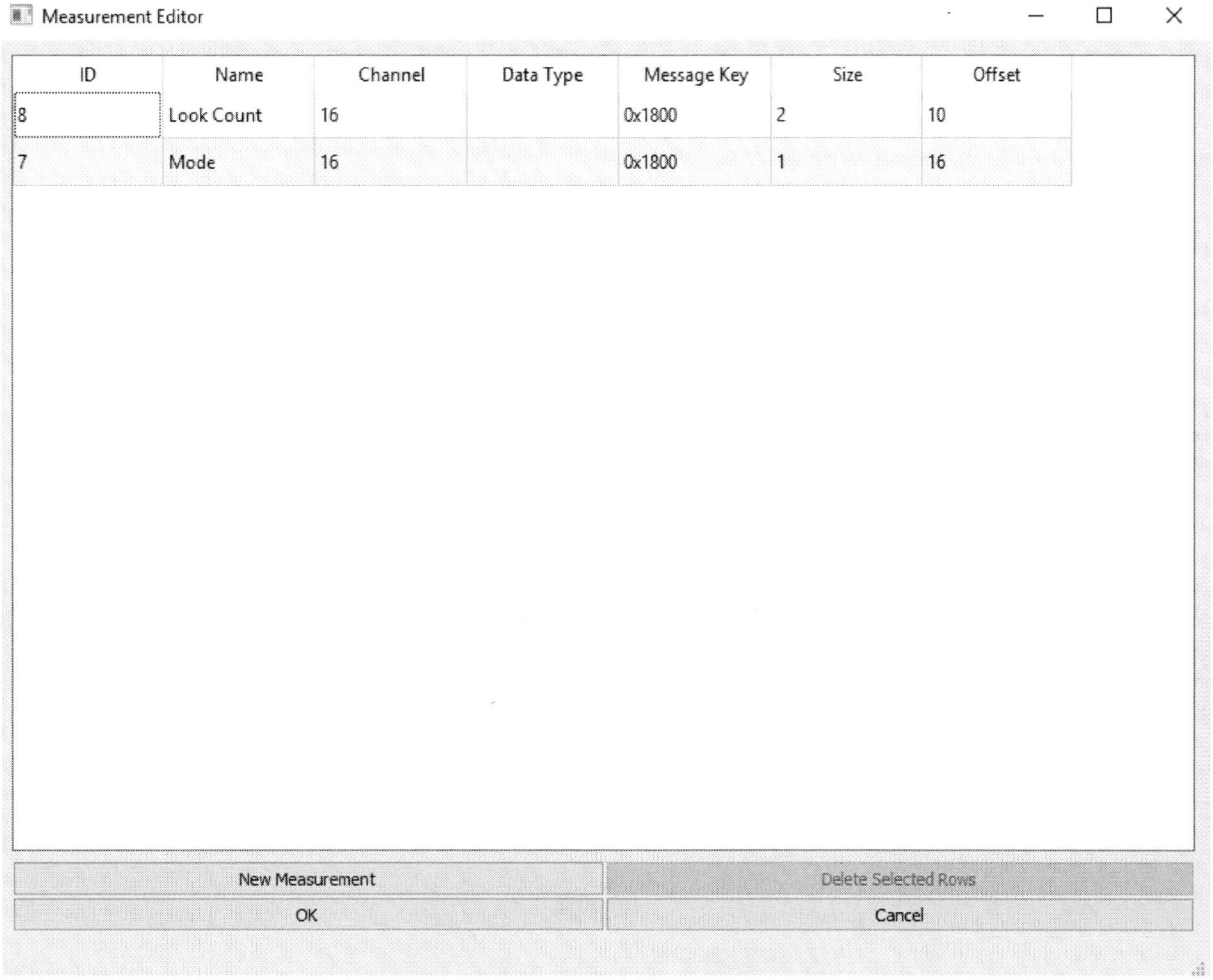

Figure 3 Measurement Editor (Windows 10)

Finally, the last piece of the puzzle is an auto-adjusting sample rate for display updates. When playback starts, the first packet is loaded and its absolute time tracked. The next update asks for the first packet later than that timestamp. To ensure real-time performance of the overall process we then take the duration of an update, and add that to the time offset for the next update. Thus, when an update is of a duration that would put us out of sync we "skip" a few packets to remain synced to the clock. By utilizing this method, we can then focus on optimizing individual displays without having to worry about the overall application falling out of sync.

FUTURE DEVELOPMENT PLANS

Eventually, the measurement editor interface will be expanded with import functions for TMATS or CSV. Other formats may be considered in the future. File-level filtering may also be a consideration to handle conflicting channel IDs.

Derived parameters are also an important consideration, and we plan to solve that problem in two ways: firstly, by making common transforms readily available through the measurement editor, and secondly by providing a scripting engine for complex derived measurements.

As referred to above, we are also exploring distributed computing and storage options for the future. By architecting a certain level of abstraction away from the specific database engine used, and having a distinction between the loading and displaying processes we have striven for a future-leaning design.

We hope to use these techniques and technologies well into the future to provide efficient, low-cost solutions with rapid development turn-around.

REFERENCES

[1] IRIG 106 Chapter 10 – irig106.org
[2] Python - python.org
[3] Qt - qt.io
[4] SQLite – sqlite.org

NEAR-FIELD HOMING AND GUIDANCE PLANNING FOR AUTONOMOUS NAVIGATION SYSTEMS

Vincent R Radzicki [1], Abhejit Rhajagopal [1], and Hua Lee [1]
[1]Dept. of Electrical and Computer Engineering
University of California, Santa Barbara
Santa Barbara, CA 93106
vradzicki@umail.ucsb.edu

Faculty Advisor:
Hua Lee

ABSTRACT

Advances in planning and controls algorithms for Unmanned Autonomous Vehicles (UAVs) have led to a substantial increase in a wide variety of applications. An important task for UAVs is automated high-precision homing-and-docking. This requires the UAV to autonomously estimate its relative bearing to the home docking station and plan its optimal approach accordingly.

This paper presents the design of homing and navigation system for UAVs that can operate in near-field scenarios. The system incorporates a dual-transmitter/receiver design and through a modified angle of arrival and motion estimation routine, the UAV can determine its relative bearing to the homing station while simultaneously planning the optimal approach. The approach planning algorithm will be described, along with theoretical analysis and simulated results documenting its performance in comparison to other techniques.

INTRODUCTION

Remote navigation and tracking remains a critically important task for vehicle automation and control [1]. As robotics and control systems have improved in recent years, more complicated tasks of Unmanned Autonomous Vehicles (UAVs) have been completely automated without the need for manual guidance or intervention. UAV systems can now navigate over far distances along a set of way-points fully autonomously and adapt to obstacles along the way. One task that remains difficult to automate, however, is precision homing and docking of UAVs with its respective base station.

During the homing and docking operation, the UAV must track its spatial position with respect to the base station with high precision as it makes its approach [2]. Tracking telemetry systems for both aeronautical and nautical vehicles developed currently, allow for navigation on large scales on the order of kilometers. The tracking sensors in these systems, such as radar and sonar systems, are designed to operate in far-field scenarios where targets are very far from the homing beacon, and typically their perofrmance will degrade significantly as the target enters the near-field of the transceivers [3]. Therefore upon successful guidance to a homing station, the automated guidance and tracking system will relieve control to manual pilotting for the more difficulty landing or docking maneuvers. Current geo-spatial tracking techniques such as the Global Positing System (GPS) can achieve sufficient precision for automated docking given adequate calibration, but its accuracy can vary greatly depending on many different environmental factors and is also susceptible to jamming and spoofing attack. A remote tracking system for UAVs with accurate performance in near-field scenarios to aid in automated homing and docking would therefore be highly desirable. These systems should also have high power and computationally efficient, so as to not over-exceed the vehicles payload.

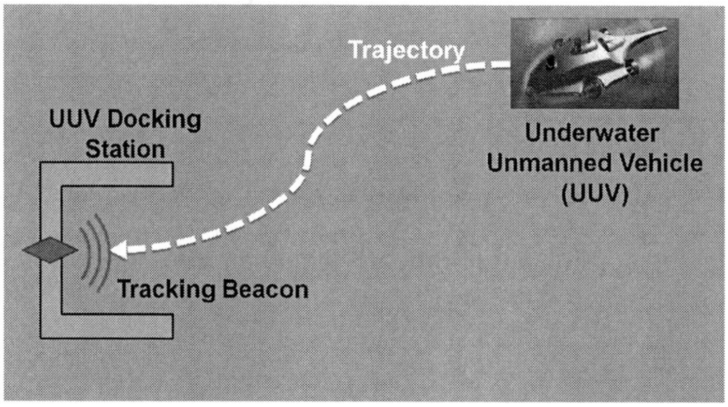

Figure 1: UUV homing and docking conceptual diagram

This paper presents a remote sensing and tracking solution for homing and docking exercises of UAVs based upon a novel bearing angle estimation routine that can operate in near-field sensing scenarios. The original motivations for this work stemmed from past efforts on designing automated homing-and-docking procedure for Unmanned Underwater Vehicles (UUVs) [4, 5]. The UUVs operating mode consisted of autonomously navigating through a surveillance mission, and upon returning to its home base station it required manual piloting to complete the more complicated docking procedure. To conserve pilot time and communication bandwidth, the preferred solution would have been to completely automate the approach and docking of the UUV with the home base. A conceptual diagram of this scenario can be seen in Fig. 1. The underwater environment imposed very low-power and size constraints on any tracking system to be integrated with the UUV, as well as restricted the use of optical and GPS tracking systems. To satisfy these requirements, we proposed a sonar tracking system that could be integrated into the communications systems of the UUV. In the next section we first provide a brief description of classical bearing-angle estimation in sonar systems, along with a description of their limitations in near-field scenarios. Our proposed solution to improve the accuracy of these techniques is then presented along with analysis of its performance and implementation details. Theoretical and numerical results are given in support of the improvement claims, as well as a brief discussion on the implications for future work.

BEARING ANGLE ESTIMATION FOR HOMING AND DOCKING NAVIGATION

A. *Far-Field Bearing Angle Estimation*

Bearing angle estimation was an early tool used in radio and sonar navigation for aircraft and sea-craft. Knowledge of the bearing and heading of the craft were critical pieces of information for pilots who faced limited visibility. The tracking systems consisted of a homing beacon that transmited a radar or sonar pulse with reasonable enough power to be distinguished from background noise, and two or more receivers on the vehicle that measure the transmitted pulse. Assuming that the target was in the far-field of the transmitter, at least hundreds of wavelengths in distance, then the incoming wave-front would be approximately planar. This technique is commonly referred to as mono-pulse bearing-angle estimation due to a single repetitive pulse being transmitted by the homing beacon, and a conceptual diagram of this system can be seen in Fig. 2.

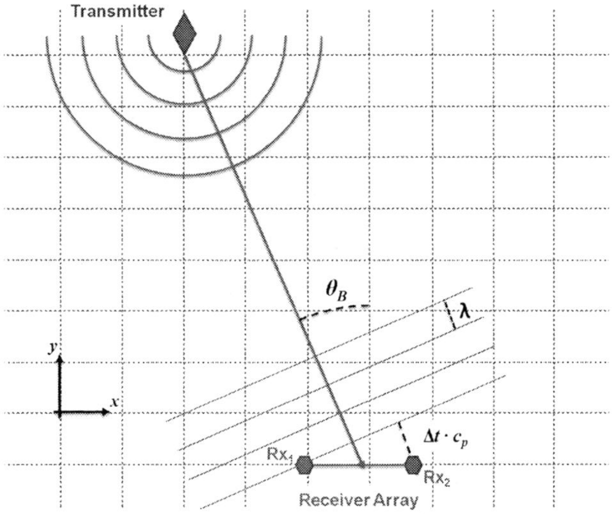

Figure 2: Mono-pulse bearing angle estimation system.

Through trigonometric relationships, it can be shown how to relate the bearing angle θ_B to propagation delay of the pulse Δt between the two receivers and the speed of propagation of the wave c_p.

$$\sin(\theta_B) \approx \frac{\Delta r \cdot c_p}{D} \tag{1}$$

Calculating the time-delay between the pulse measured at the first and second receiver, along with knowledge of the distance between the receivers D, provides enough information to estimate the bearing angle θ_B [6].

As can be seen in Fig. 2 the wave-front becomes more spherical as the receiver array approaches the homing beacon. This degrades the estimation accuracy of the bearing angle and limits the systems ability to navigate autonomously. Furthermore, in this configuration the time-delay Δt and thus the bearing angle are dependent on the tilt of the receiver array relative to the transmitter array. These drawbacks illustrate the need for improvements in near-field estimation scenarios.

B. *Near-field Bearing Angle Estimation using Dual-Transceiver Arrays*

To increase the near-field accuracy of the bearing angle estimation, we propose a slight modification of the mono-pulse estimation routine. Including an additional transmitter at the homing beacon can significantly improve the bearing angle estimation accuracy and allow for automated homing and docking. This dual-transceiver approach with two transmitters acting as homing beacons and two receivers integrated on the UUV is illustrated in Fig. 3.

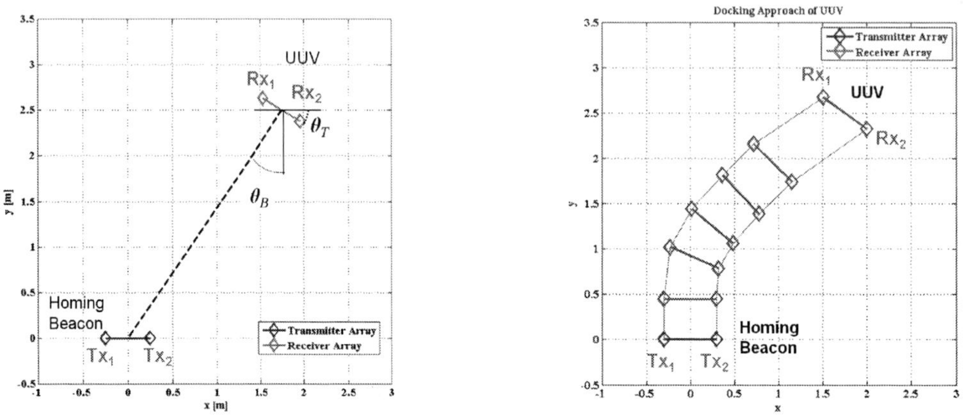

(a) Dual transceiver UUV and homing beacon (b) UUV approach guided by the homing beacon

Figure 3: Dual Transceiver Approach

The goal of this method is to aid in autonomous trajectory planning of the UUV, so that it can approach and maneuver into its home dock along a smooth trajectory as seen in Fig. 3b. This requires an estimate of the bearing angle of the UUV θ_B as well as its relative tilt angle with respect to the transmitter array θ_T. If the UUV can measure both of these remotely, they can be periodically adjusted during the approach by the system controller to ensure a smooth docking procedure. An optimal approach path will be one that ensures both the tilt angle θ_T and the bearing angle θ_B approach zero at approximately the same time and at a reasonable speed.

For the mono-pulse bearing angle-estimation routine, the difference in propagation distance between the transmitter and both receivers was needed to determine θ_B. For the Dual-transceiver approach, there are four total propagation paths. Let r_{ij} be defined as the propagation path between the ith transmitter and the jth receiver. The new geometry and propagation paths of each signal can be seen in Fig. 4. The time delay corresponding to each propagation path will be

$$t_{ij} = \frac{r_{ij}}{c_p} \tag{2}$$

so that if the transmitter sends the signal $s_0(t)$, the signal measured at receiver j from transmitter i will be

$$s_{ij}(t) = s_0(t - t_{ij}) \tag{3}$$

Given that the transmitter and receiver do not share an absolute time base as in a GPS system, only the difference in the time-delays can be measured. In the far-field scenario, the time delay

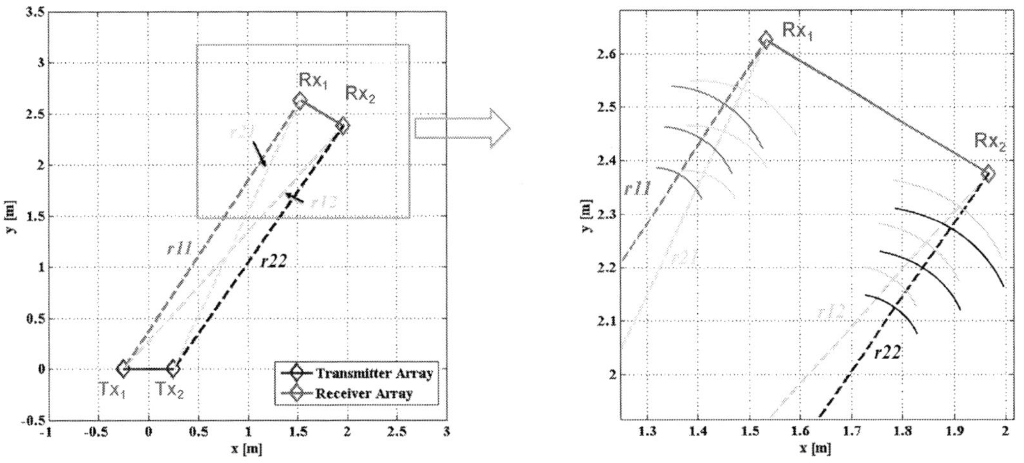

(a) Dual Transceiver Propagation Distances (b) Dual Transceiver Incoming Wave-fronts

Figure 4: Dual Transceiver Geometry

differences between the transmitters $t_{11} - t_{12}$ or $t_{22} - t_{21}$ could be used to estimate the bearing angle θ_B as in Equ. (1). However, as discussed a large error term distorts the estimate as the receiver approaches the transmitter beacon. However, if the difference of both these time-delays is calculated the main error contributors tend to cancel preserving the accuracy of the original bearing angle estimate. A new estimation parameter Δ_1 can then be defined as the difference between the two delays multiplied by the propagation speed c_p

$$\begin{aligned} \Delta_1 &= \frac{c_p}{2}\left((t_{11} - r_{21}) - (t_{22} - t_{12})\right) \\ &= \frac{1}{2}\left((r_{11} - r_{21}) - (r_{22} - r_{12})\right) \end{aligned} \tag{4}$$

where the factor of $1/2$ is meant to apply an averaging effect. The bearing angle estimation can then be calculated using the same trigonometric relationship

$$\sin(\theta_T) \approx \frac{\Delta_1}{D} \tag{5}$$

The other advantage of the dual-transceiver approach, is the propagation delays can also be used to estimate the tilt angle θ_T. Referring again to the geometry in Fig. 4, it can be seen that θ_T will be approximately proportional to the difference $r_{11} - r_{22}$. This quantity can be computed again by the time-delay difference between t_{11} and t_{22} as

$$\begin{aligned} \Delta_2 &= c_p \cdot (t_{11} - t_{21}) \\ &= (r_{11} - r_{21}) \end{aligned} \tag{6}$$

The second estimation parameter Δ_2 can then be used to estimate the tilt angle through

$$\sin(\theta_T) \approx \frac{\Delta_2}{D} \tag{7}$$

The dual-transceiver method effectively parameterizes the bearing and tilt estimation problem into two quantities defined as the difference in time-delays between the transmit signals, all of which can be calculated using classical signal processing techniques with only measurements made at the receiver array without the need for a common time-base. Implementing the signal processing procedures requires careful transmit signal and array design that is discussed in the following section.

C. Signaling Techniques for Dual-Transceiver Arrays

For the mono-pulse estimation routine, the choice of transmit waveform is relatively unconstrained. For the dual-transceiver approach described, designing the transmit signals so as to avoid interference and aliasing at the receiver array is an important consideration. It is also important to not impose a signaling scheme that imposes a high-level of complexity on both the transceiver hardware and estimation algorithm.

A Continuous Wave (CW) transmit signal consisting of a single frequency component is a low cost transmit signal that imposes a relatively low computational burden on the receiver to estimate time and phase delays, which comprise the required parameters discussed earlier. Let us define the base transmit waveform as

$$s_0(t) = e^{j\omega_0 t} \tag{8}$$

The signals radiated from transmitter one and two need to be differentiated at the receiver so the respective time-delays t_{ij} from each transmitter can be calculated. The signals will therefore need to be orthogonal along some basis. A conventional approach to achieve this, is to modulate the CW signal with two orthogonal time-codes before transmission. A low-cost method is to apply on-off keying modulation to the transmit signal so that transmitters are never transmitting simultaneously. A transceiver architecture that implements this approach with two orthogonal modulated transmit signals is illustrated in Fig. 5.

With the orthogonal coding of the CW waveform, each transmit signal can be disambiguated at both receivers. The delayed CW signal measured at the jth receiver from the ith transmitter will be

$$s_{ij}(t) = e^{j\omega_0(t-t_{ij})} \tag{9}$$

If the receiver moves only a small distance over the pulse-width of the transmitted signal, then the relative phase-difference between each signal can be used to estimate the relative propagation difference between each waveform. Using a quadrature receiver, the phase component of the $s_{ij}(t)$ can be computed as

$$\begin{aligned}\phi_{ij} &= \angle\left(s_{ij}(t) \cdot e^{-j\omega_0 t}\right) \\ &= \angle e^{-j\omega_0 t_{ij}} = \angle e^{-j2\pi\frac{r_{ij}}{\lambda}}\end{aligned} \tag{10}$$

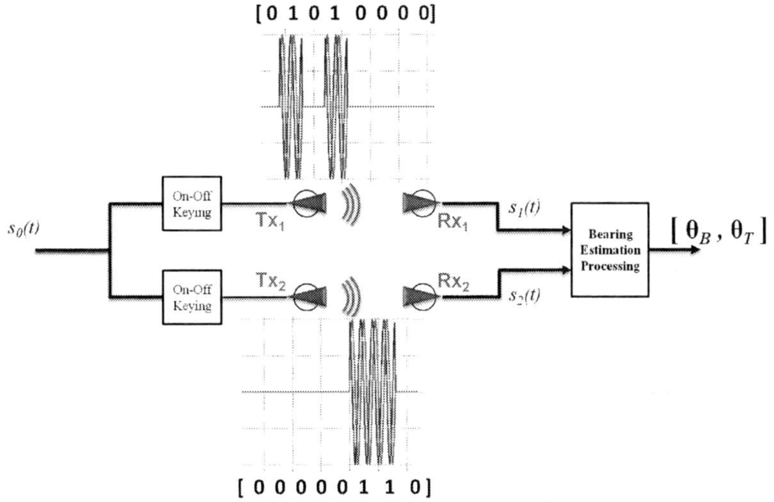

Figure 5: Transceiver Architecture

The first estimation parameter to calculate the bearing angle can be calculated as

$$\begin{aligned}\Delta_1 &= \frac{\lambda_0}{2\pi}\left(\frac{\phi_{11}-\phi_{21}}{2}-\frac{\phi_{22}-\phi_{12}}{2}\right)\\ &= \frac{r_{11}-r_{21}}{2}-\frac{r_{22}-r_{12}}{2}\end{aligned} \tag{11}$$

The second estimation parameter to calculate the tilt angle can also be calculated as

$$\begin{aligned}\Delta_2 &= \frac{\lambda_0}{2\pi}(\phi_{11}-\phi_{22})\\ &= (r_{11}-r_{22})\end{aligned} \tag{12}$$

The second equality in both Equations (11) and (12). will be valid if the propagation differences do not exceed $\frac{\lambda}{2}$, which can be enforced by requiring the distance between the transmitters and receivers D to be

$$D \leq \frac{\lambda_0}{2} \tag{13}$$

To determine the accuracy of the dual-transceiver approach a numerical simulation was run that implemented the solution proposed here. For this simulation the transmit frequency was set to $5kHz$ and in underwater environments the speed of propagation of sonar signals is approximately $c_p = 1484m/s$ so that the wavelength is $\lambda \approx 30cm$. The array separation was then set to $D = 15cm$. The homing trajectory was pre-computed to fit a fourth order parametric equation to optimize the acceleration profile similar to minimum snap trajectory planning utilized in popular UAV planning algorithms [7]. The approach path along with the angular estimation results can be seen in Fig. 6.

As can be seen in the Fig. 6b, the bearing angle estimation θ_B is within less than a degree of the ground-truth known to the simulation model. The tilt angle estimate θ_T also closely follows

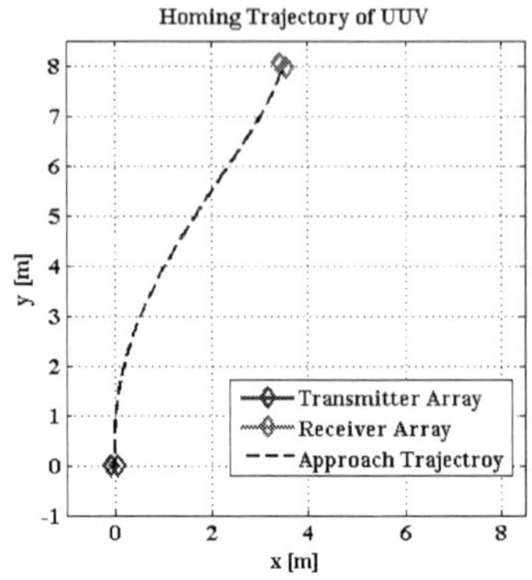

(a) UUV Homing Approach

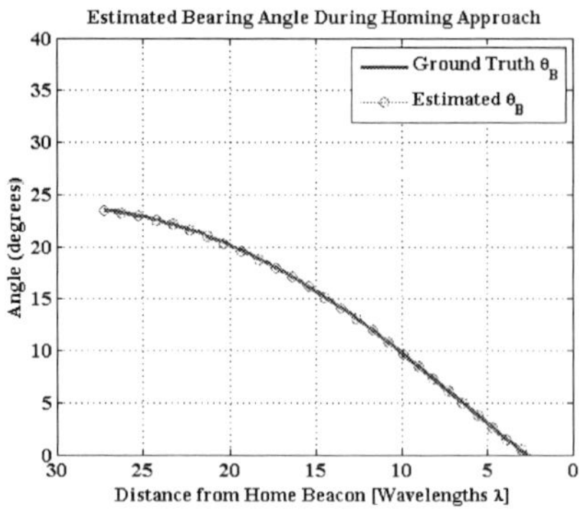

(b) Bearing Angle Estimation Results

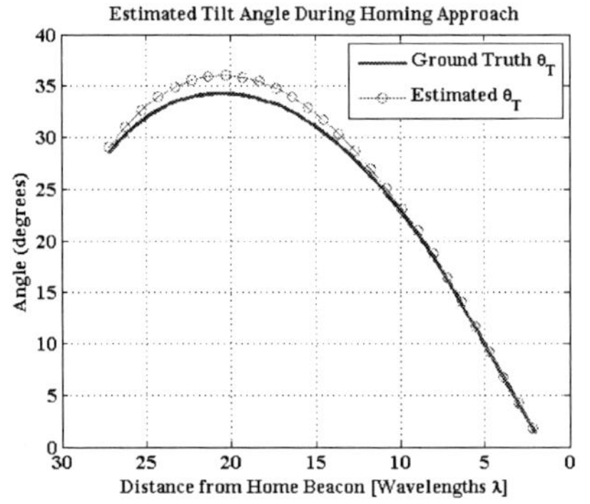

(c) Tilt Angle Estimation Results

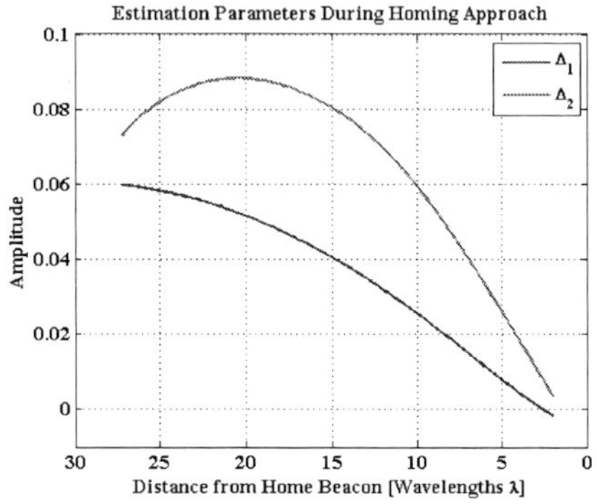

(d) Parameter Estimation Results

Figure 6: Numerical simulation results of double-transceiver angle estimation routine

the ground truth with a maximum error less than 3 degrees. The results also are not dependent on distance from the homing beacon and increase in accuracy as the UUV approaches its home dock. With the addition of a second transmitter and imposing orthogonality conditions using on-off modulation keying, this method allows for accurate and low-cost estimation of the UUV angular position. The estimation performances improves as the UUV is close to homing beacons, making it an ideal sensing technique to aid in automated homing-and-docking of UAVs.

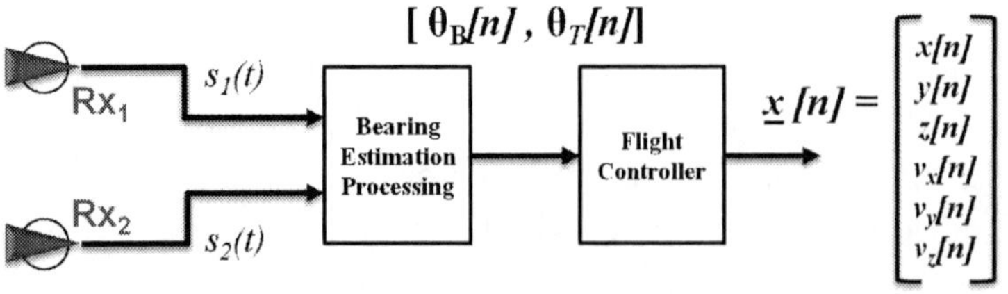

(a) Integrated Sensor and Controller

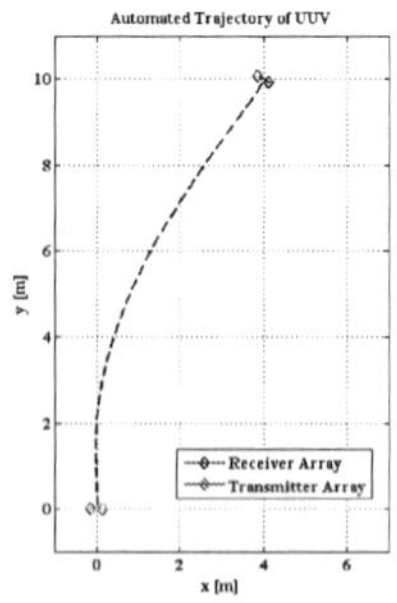

(b) Automated Homing Approach

Figure 7: Integrated sensor and controller used for automated homing and docking of UAV systems

DISCUSSION

The estimation performance improvements of the proposed methodology have been outlined in the previous section, however, other advantages also exist for UUV homing and docking. The estimates can be calculated in real-time due to the relatively low computational complexity required, and the algorithm can also be estimated in either analog or digital systems. At time n, both estimation parameters $\Delta_1[n]$ and $\Delta_2[n]$ can be calculated quickly using only time-delay estimation and summing/difference circuits. It can be easily shown that as the estimation parameters approach zero, so to will the bearing angle estimate

$$As\ \mathbf{\Delta_1}[\mathbf{n}] \rightarrow 0\ so\ will\ \theta_{\mathbf{B}}[\mathbf{n}] \rightarrow 0$$

This same condition also applies to the tilt angle.

$$As\ \mathbf{\Delta_2}[\mathbf{n}] \rightarrow 0\ so\ will\ \theta_{\mathbf{T}}[\mathbf{n}] \rightarrow 0$$

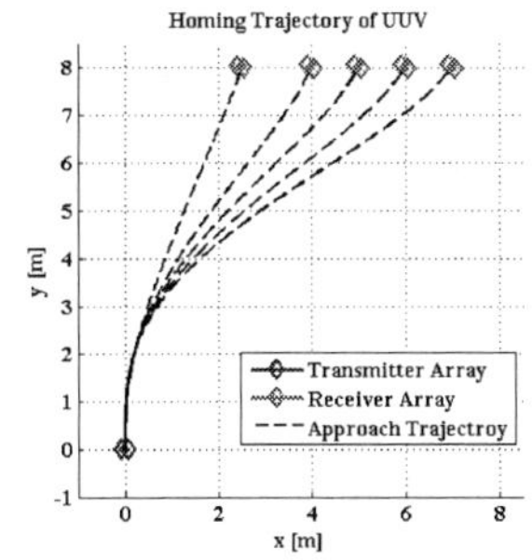

(a) 5 Homing Approaches of UUV

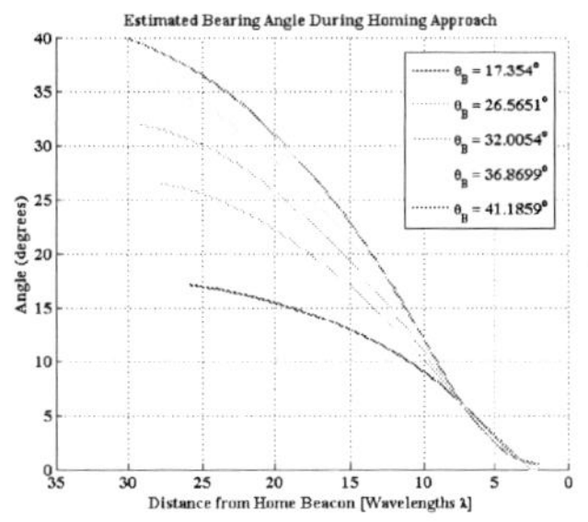

(b) Bearing estimation results

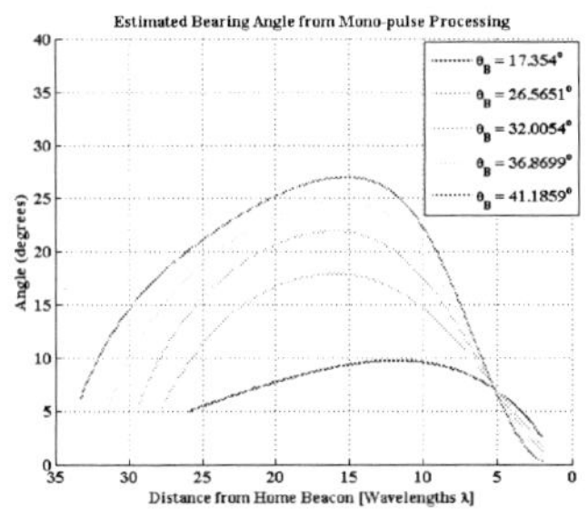

(c) Mono-pulse results

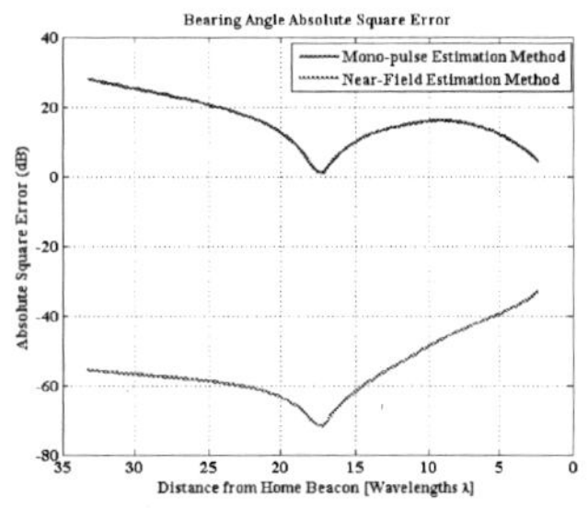

(d) Average error comparison

Figure 8: Performance comparison of proposed dual-transceiver estimation routine and mono-pulse estimation routine

These two conditions imply that a controller of any UAV system, can use these parameters as feedback to determine the optimal approach to the home dock. The convergence rate at which the parameters are minimized can be tuned to ensure a smooth approach, similar to gradient descent optimization methods used in advance controls systems. This general approach has been illustrated in Fig. 7. To compare the performance of the proposed methodology to classical mono-pulse processing a simulation model was again computed for a sequence of various in-bound bearing angles. Utilizing the same settings as the simulations run in Fig. 6 the bearing angle estimation results were computed and plotted in Fig. 8b and 8c. In all 5 cases the double-transceiver approach out-performs the mono-pulse method by an average error of 60dB shown in Fig. 8d. The changing position of the receiver array greatly affects the bearing angle estimation results as the UUV homes in on the docking station for the mono-pulse technique, but the double-transceiver routine accurately computes the bearing angle for the entire approach trajectory.

CONCLUSION

A novel angle-of arrival method has been presented for automated homing and docking of UUV systems. Although this work was motivated and constrained by the requirements of UUV systems, much of the theory and results also apply to other UAV systems operating in a variety of scenarios. UAVs that are required to operate with low-SWAP payloads as well as in GPS-denied environments could benefit greatly from methods [8]. The proposed tracking system utilizes current communications and sensing technology in a novel way to achieve improve results. Future work would focus on integration of the feedback control system with the estimation process to leverage computation gains, and hardware implementation of these methods.

REFERENCES

[1] R. Rout and B. Subudhi, "Narmax self-tuning controller for line-of-sight-based waypoint tracking for an autonomous underwater vehicle," *IEEE Transactions on Control Systems Technology*, vol. 25, pp. 1529–1536, July 2017.

[2] P. Jantapremjit and P. A. Wilson, "Optimal control and guidance for homing and docking tasks using an autonomous underwater vehicle," in *2007 International Conference on Mechatronics and Automation*, pp. 243–248, Aug 2007.

[3] D. K. Barton, "Modern radar system analysis," *Norwood, MA, Artech House, 1988, 607 p.*, 1988.

[4] H. Lee, "Signal processing agorithms for unmanned undersea vehicles navigation and collision avoidance.," *The Journal of the Acoustical Society of America*, vol. 125, no. 4, pp. 2537–2537, 2009.

[5] H. Lee, "Optimization of computation efficiency in underwater acoustic navigation system," *The Journal of the Acoustical Society of America*, vol. 139, no. 4, pp. 1909–1913, 2016.

[6] W. S. Burdic, *Underwater acoustic system analysis.* Prentice Hall, 1991.

[7] D. Mellinger and V. Kumar, "Minimum snap trajectory generation and control for quadrotors," in *Robotics and Automation (ICRA), 2011 IEEE International Conference on*, pp. 2520–2525, IEEE, 2011.

[8] G. Balamurugan, J. Valarmathi, and V. P. S. Naidu, "Survey on uav navigation in gps denied environments," in *2016 International Conference on Signal Processing, Communication, Power and Embedded System (SCOPES)*, pp. 198–204, Oct 2016.

INVESTIGATING UNKNOWN IRIG CHAPTER 4, CLASS I OR II FORMATS

Wayne Rettig
Principle Field Support Engineer
Lumistar, Inc.
2270 Camino Vida Roble Suite L Carlsbad, CA 92011
wrettig@lumistar.net

ABSTRACT

This describes one approach to investigating an unknown IRIG 106, Chapter, 4 Class I or Class II Pulse Code Modulation (PCM) format. The assumption is that you are supplied with decrypted data and clock signals for the unknown PCM stream. This technique is optimized for 16, 12, or 8-bit Word Minor Frames. Standard IRIG 106, Appendix C, Table C-1, Frame Synchronization (Frame Sync) values of 32, 24, 16-bit patterns are simpler to investigate. Other IRIG Frame Sync patterns can also be investigated, but are not dealt with in this document. This document will utilize a sample 24-bit Frame Sync pattern, because it will produce Endian issues in the recorded data.

INTRODUCTION

The system used to investigate the unknown PCM Frame is Windows based. The Decommutator utilizes a mode that records the raw data from the Decom Current Value Table (CVT) memory onto the Hard Drive as 16-bit Words that are Little Endian swapped. The 16-bit Word 0x1234 in CVT memory is stored onto the HD as 0x3412. Using a combination of real-time Raw Data Buffer, Frame Buffer and recorded data inspected with a Hexadecimal Viewer will give insight into the unknown Frame. Using this information to update the Decom settings and repeating this inspection process, eventually a successfully lock on the unknown PCM Frame format will be achieved.

PROCEDURE

Setup the Decommutator (Decom) for PCM Frame investigation as shown in Figure 1.

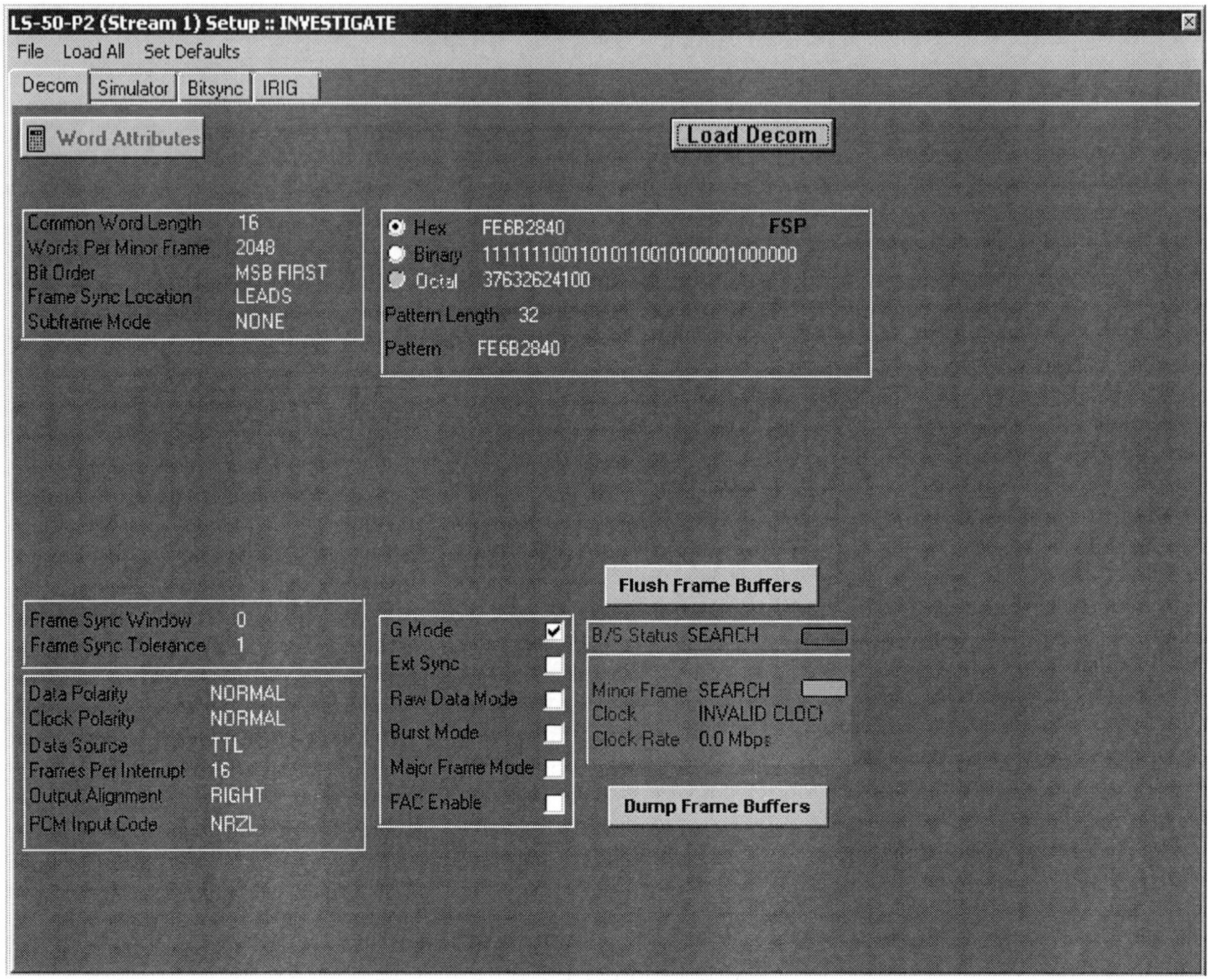

Figure 1: Initial Decom Setup

This configuration is optimal for 16, 12, 8, bit Words. **Note:** the G Mode selected above allows locking onto the Frame Synchronization (Frame Sync) pattern if found. However, even if no Frame Sync pattern is found, the Decom will collect buffers of data and then record them to the hard drive when placed in the archiving mode.

Since the format is unknown, we will try 32, 24 and 16-bit Frame Sync patterns until the pattern is found. This is accomplished by trying each Frame Sync pattern and observing the Raw Data Buffer (indicated by the term **FPI** in the banner). Starting with the 32-bit Frame Sync pattern 0xFE6B2840, the Raw Data Buffer is inspected for this Frame Sync pattern in Figure 2.

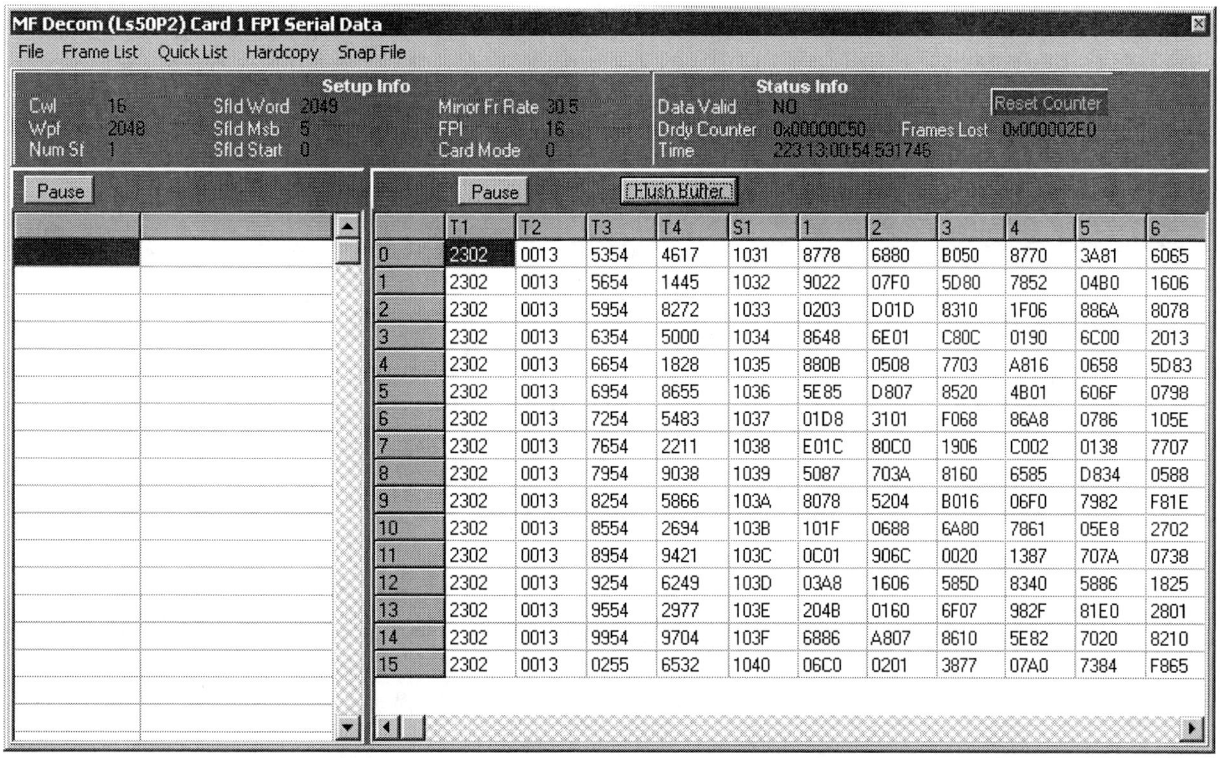

Figure 2: Raw Data Buffer, no 0xFE6B2840 pattern found

Reconfigure the Decom to look for 24-bit Frame Sync pattern 0xFAF320 then inspect the Raw Data Buffer as in Figure 3.

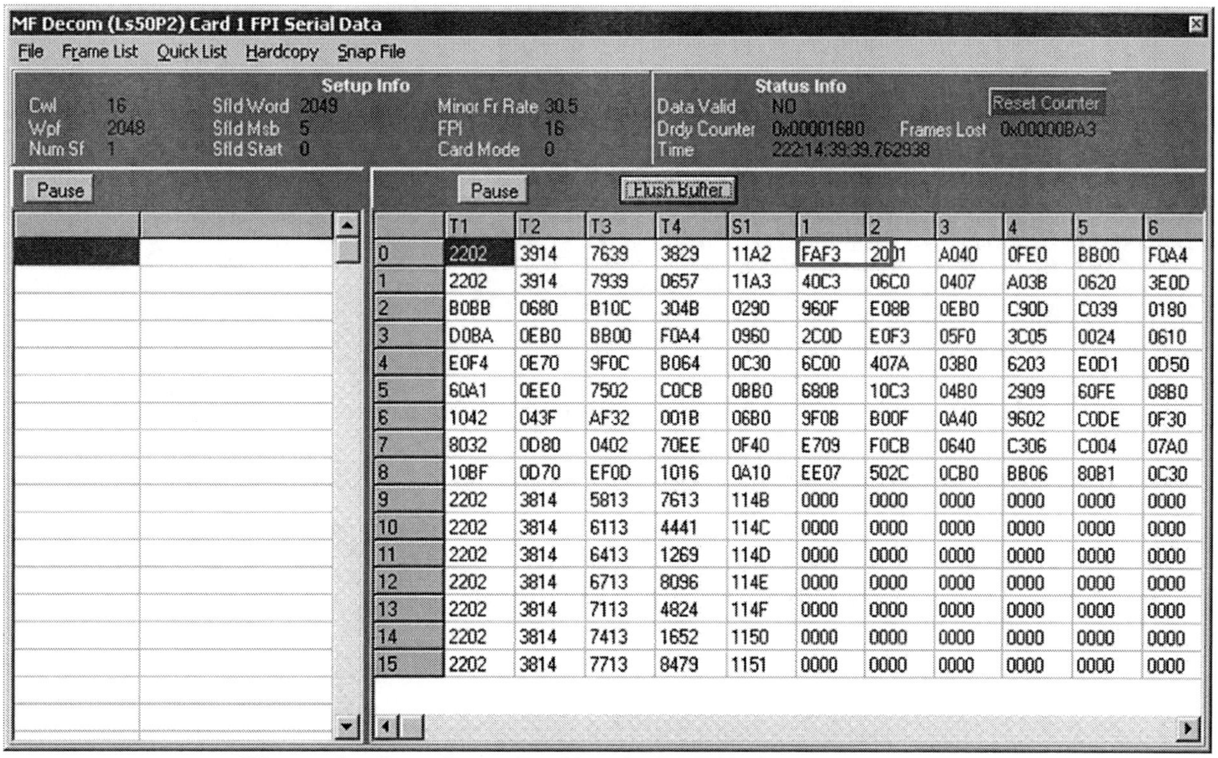

Figure 2: Raw Data Buffer, 0xFAF320 pattern found

The unknown PCM Frame uses a 24-bit Frame Sync. Record Raw Data and review this recording using a Hex Viewer looking for the Frame Sync. Since the recorded data is Little Endian swapped, look for the pattern 0xF3FA as in Figure 3.

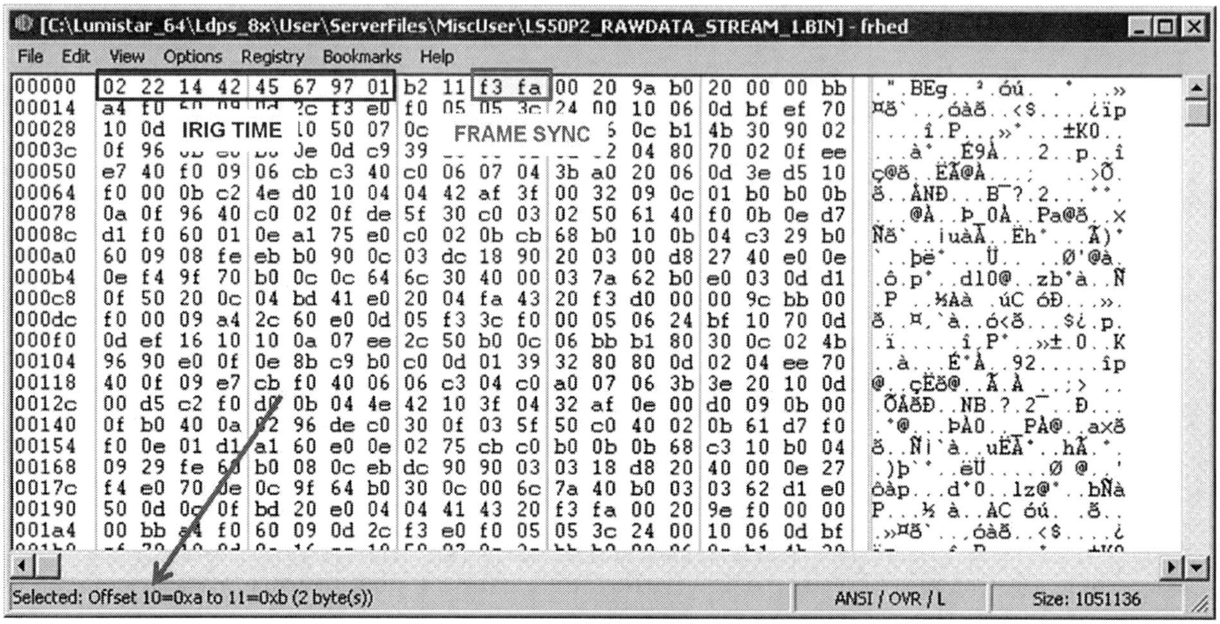

Figure 3: First Frame Sync location using a Hex Viewer

The first Frame Sync found has a 10 byte offset value. **Note:** this Decom will prepend 10 bytes of data in front of the Frame Sync. This is comprised of an IRIG Time Stamp and various status flags. The next Frame Sync occurrence is shown in Figure 4.

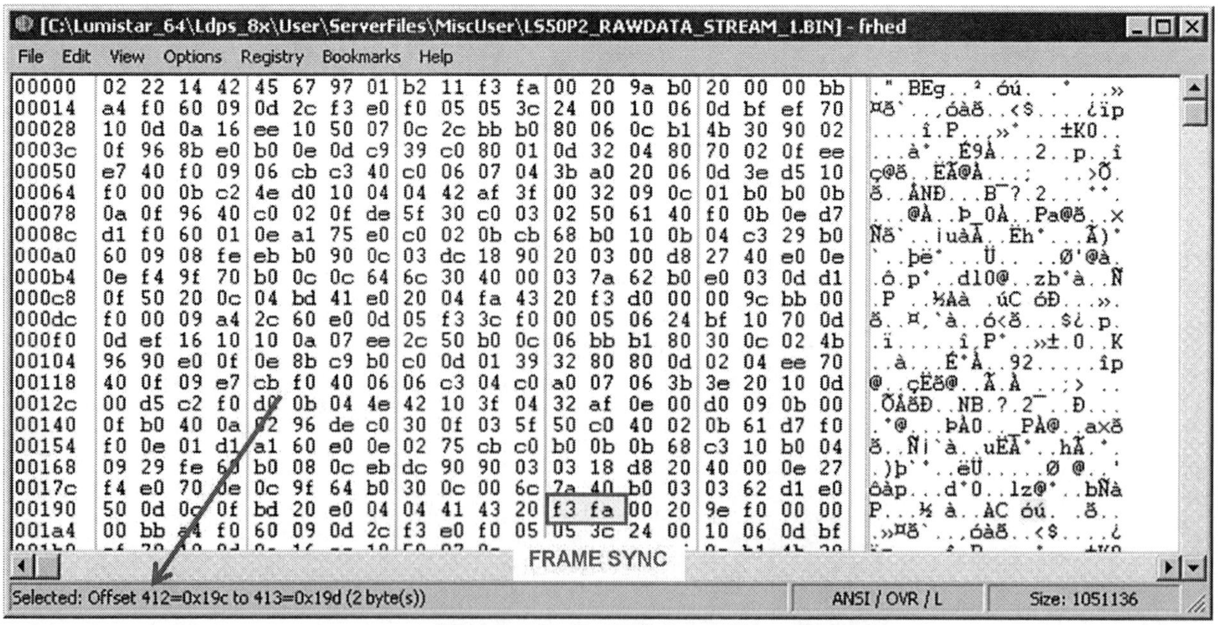

Figure 4: 2nd Frame Sync location

This Frame Sync has a 412 byte offset value.

Looking 10 bytes to the left of the Frame Sync, shows there is no IRIG timestamp and Status bytes..Therefore Minor Frame length (the number of bytes between Frame Syncs) can be calculated.

(**412** bytes - **10** bytes) = **402** bytes or **201** 16-bit Words

Since the Decom is set for 16 bit Words, reconfigure the Decom to look for **201** 16-bit word Minor Frames. The incoming data now indicates a Frame Lock in Figure 5.

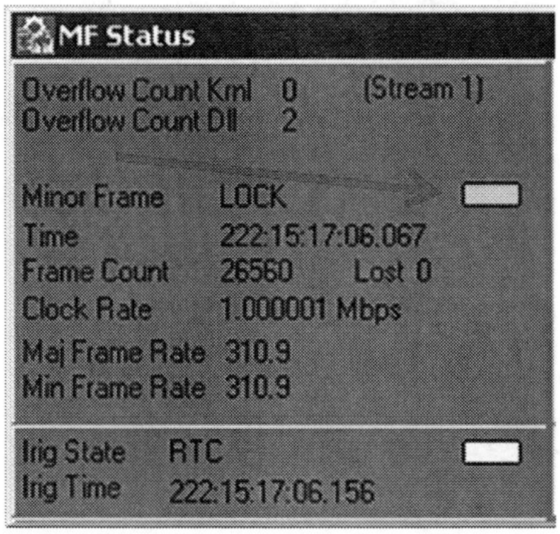

Figure 5: Frame Lock Indicator

The Frame Lock affirms the Minor Frame length or it is a multiple of the Minor Frame length. To determine which is the case, look in the middle of the Raw Data Buffer in Figure 6.

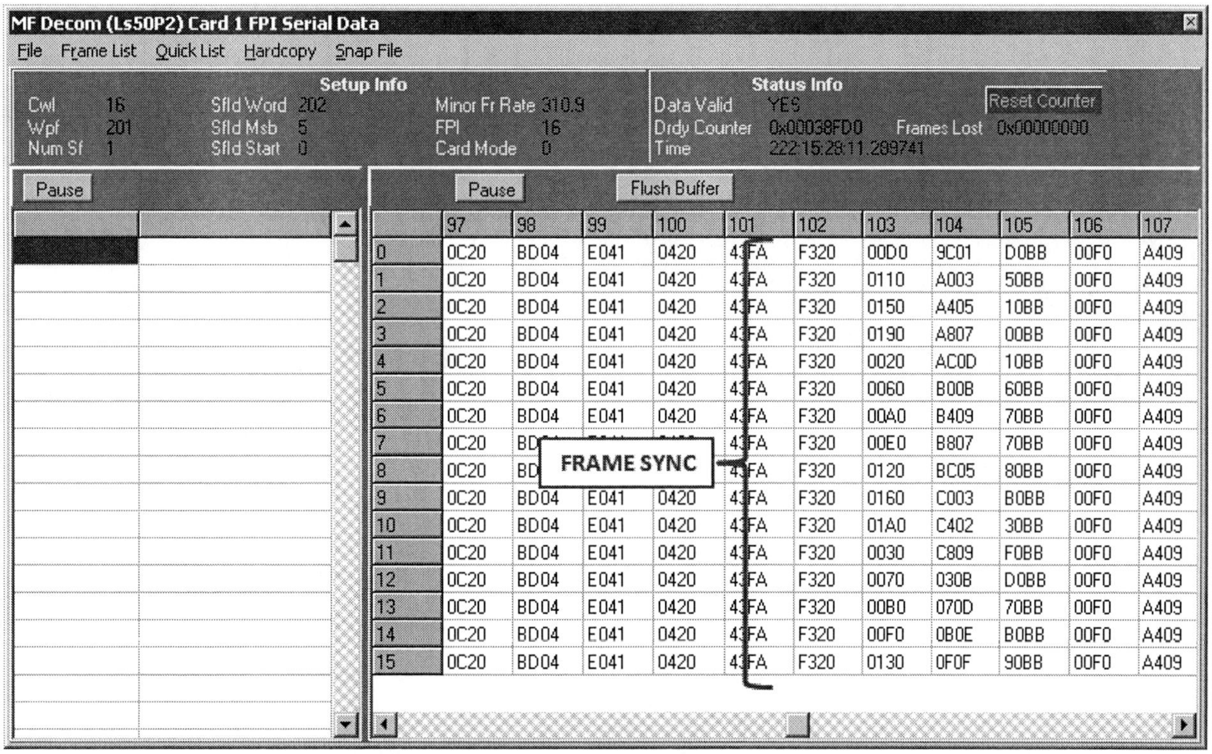

Figure 6: Raw Data Buffer Mid Frame Inspection

This Frame Sync is offset by one byte midway down the current Minor Frame. Recalculating to reduce the Minor Frame length.

(**201** 16-bit words / **2**) = **100.5** 16-bit Words

Set the new Decom settings to **101** 16-bit Words with Word **101** set to **8** bits in length. Again look for a Frame Sync mid Minor Frame in the Raw Data Buffer, Figure 7.

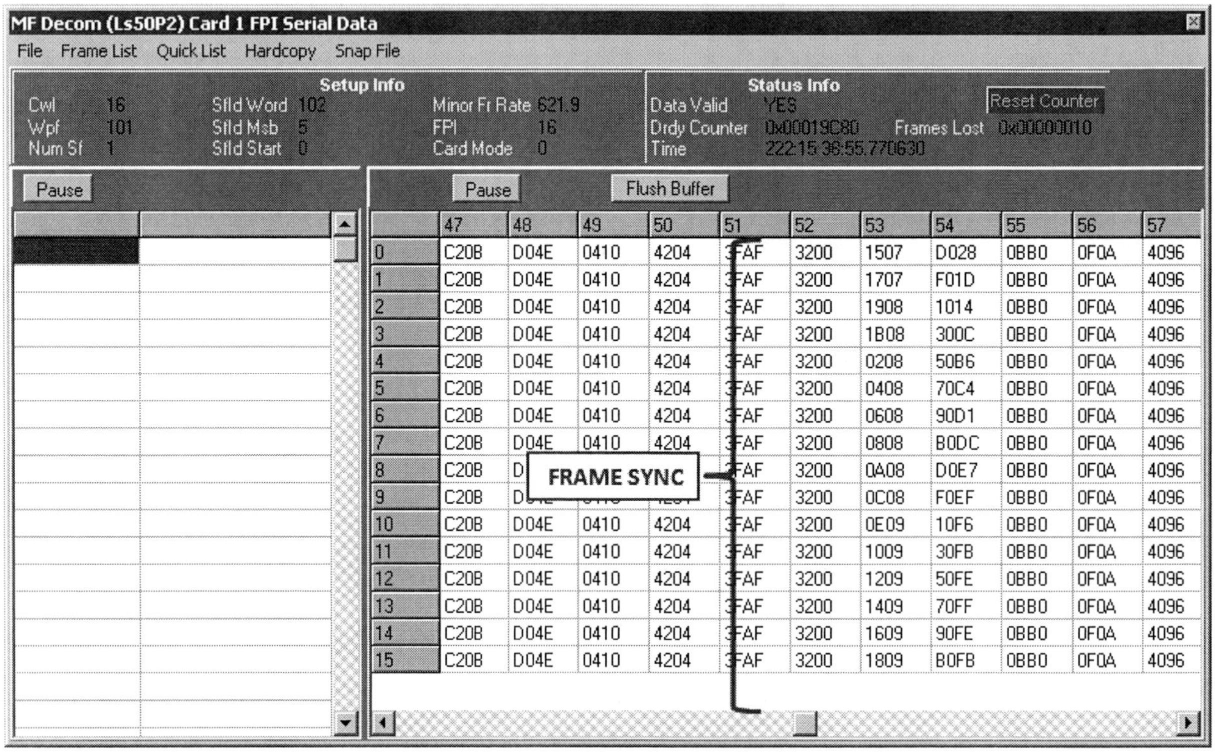

Figure 7: Raw Data Buffer, 2ndrMid Frame Inspection

This indicates the Minor Frame length is still twice the required size, recalculating.

(**100.5** 16-bit words / **2**) = **50.25** 16-bit Words

Change the Decom settings to **51** 16-bit Words with Word **51** set to **4** bits in length and look at the Raw Data Buffer for a Frame Sync mid Minor Frame - none was found. The Minor Frame length has been determined. Take the Decom out of the Raw Data mode to allow it to lock appropriately on the Minor Frame with no SFID.

Since the 16-bit Minor Frame Word count requires a variable word, calculate the Minor Frame length in 12 bit words.

(**50.25** Word * **16** bits/Word) / **12** bits/Word = **67** 12-bit Words

After setting these new values into the Decom, look at the Frame Buffer (indicated by NO **FPI** term in the banner), there are no dropped Frames seen in Figure 8.

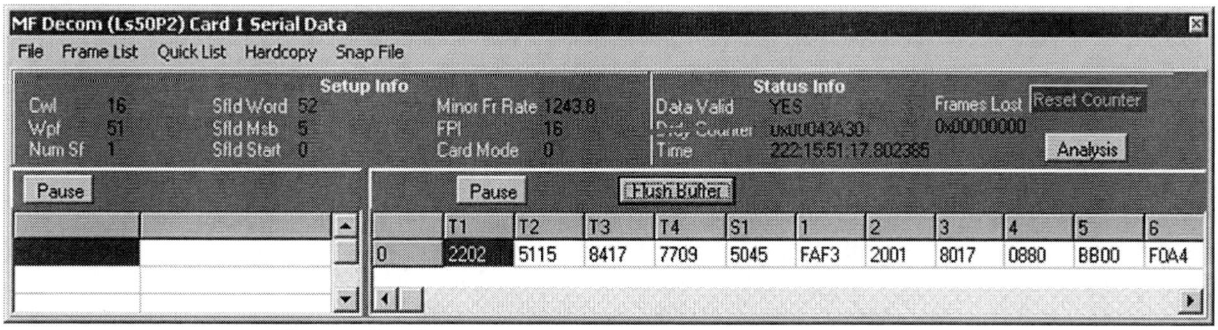

Figure 8: Frame Buffer Showing No Lost Frames

Change the raw data buffer size to maximum (255 with this hardware). This buffer will be Frame aligned for easier inspection as seen in Figure 9.

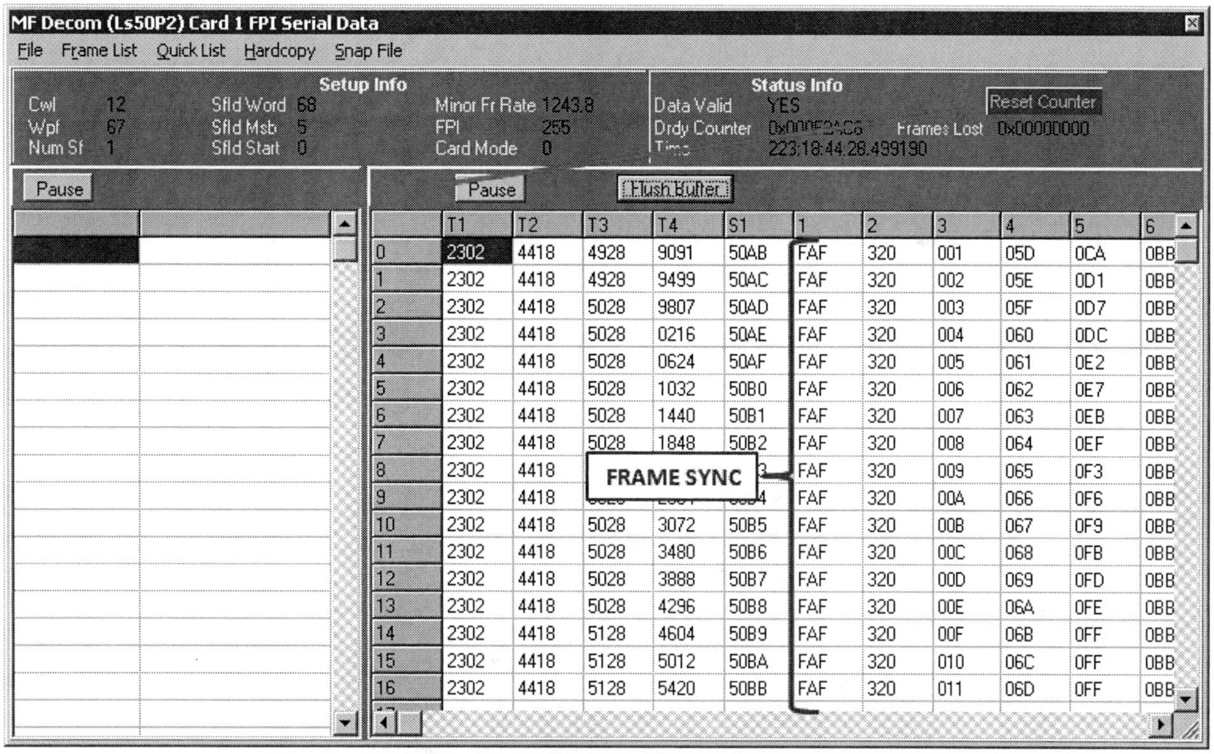

Figure 9: Raw Data Buffer Shows No Lost Frames

Now use the Decom's Snap File feature which records a snapshot of the entire Raw Data Buffer into a text file. Using Excel to import this text file and organize the data into columns. Look for possible Subframe ID (SFID) data in Figure 10.

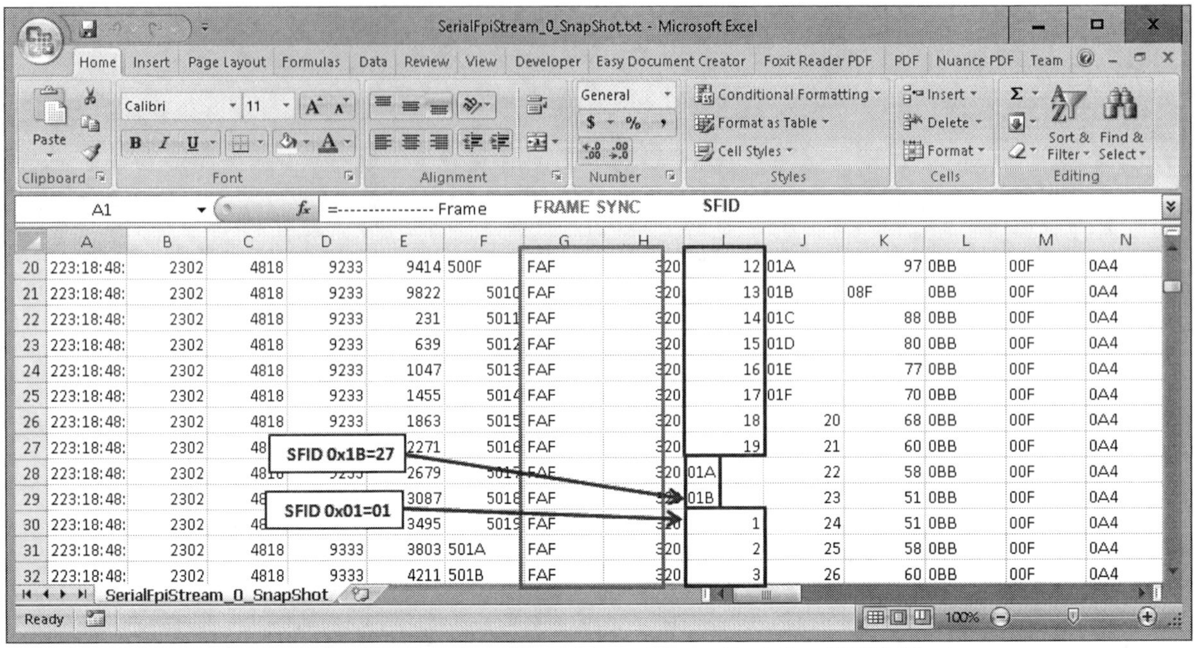

Figure 10: Reviewing Raw Data Snapshot for SFID Data

The SFID is seen to rollover at 27 and starts with 1, indicating a total of 27 Minor Frames. Using this new information and making note of the SFID bit alignment, make final Decom settings for this Frame as shown in Figure 11.

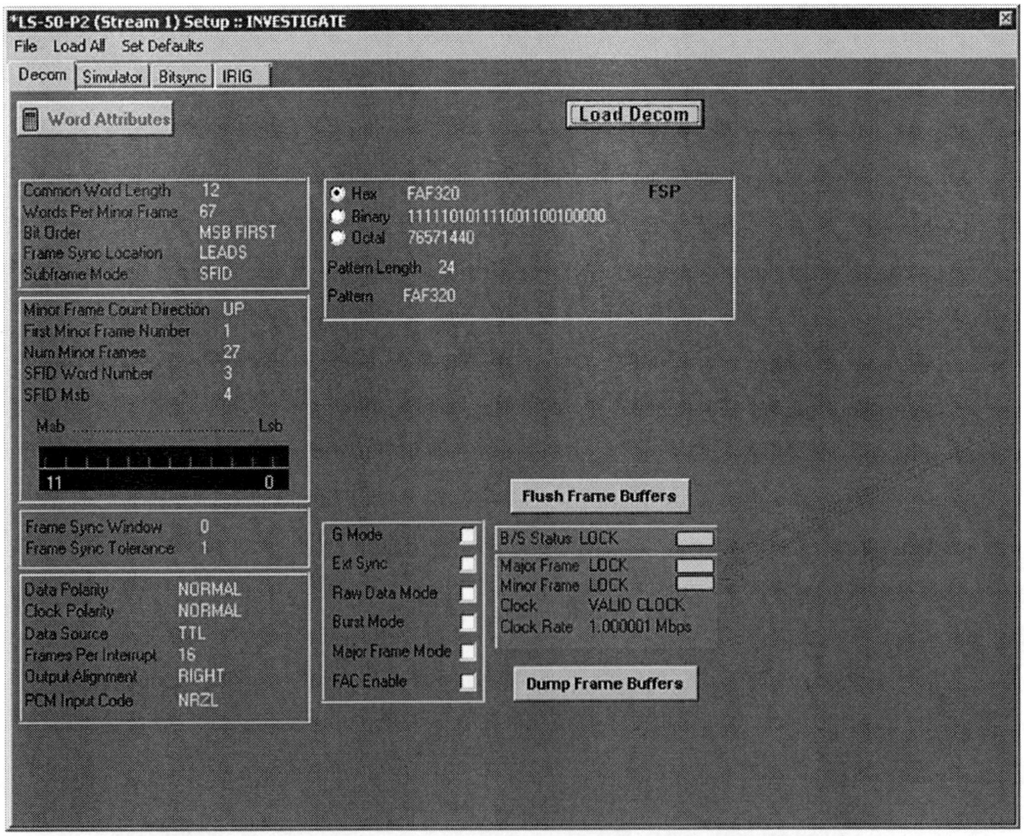

Figure 11: Final Decom Configuration with Solid Locks

Finally, Looking at the Frame Buffer, no lost Frames are seen in Figure 12.

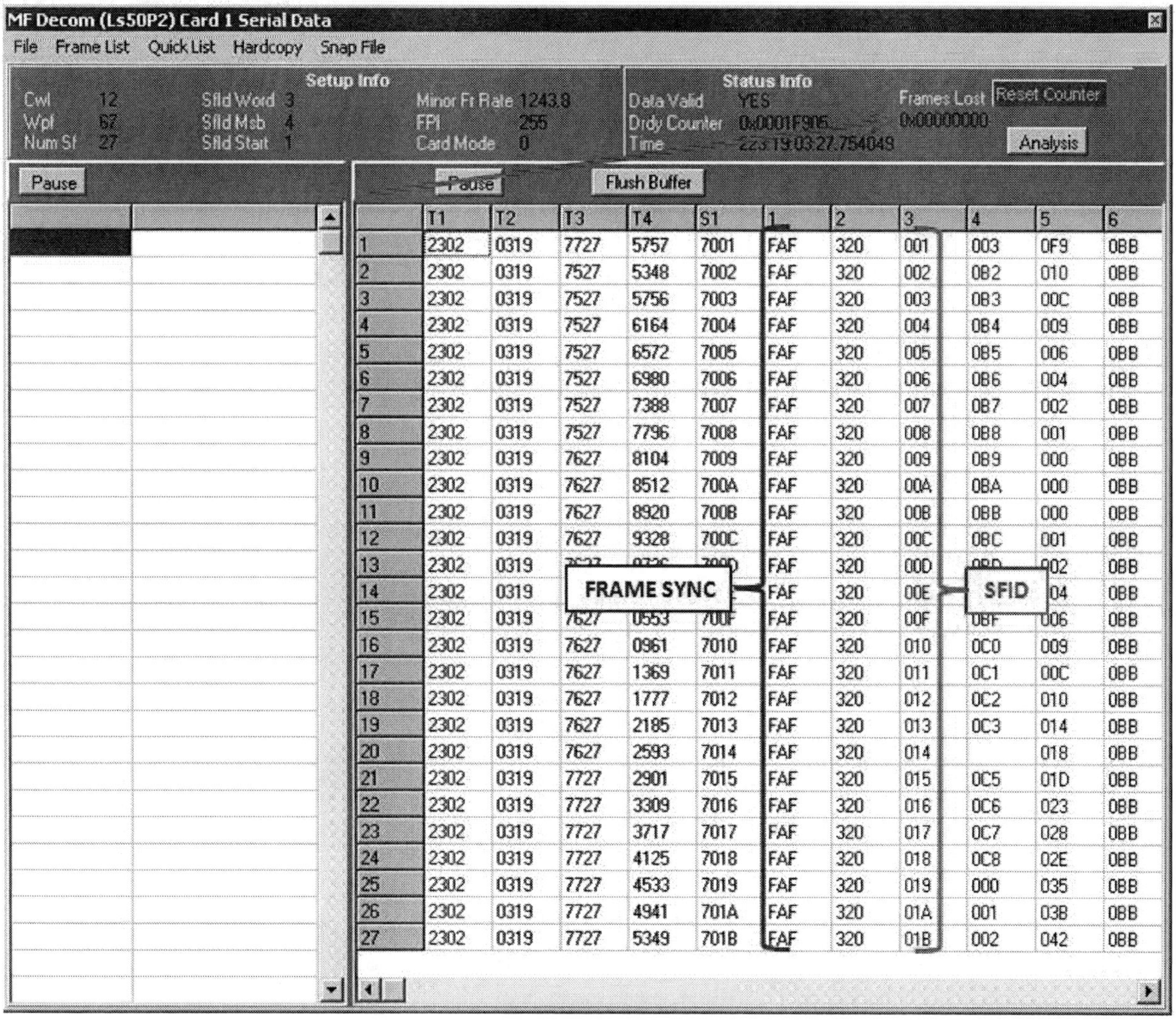

The IRIG Chapter 4 Frame Format has been determined and can be recorded for later data reduction by Analysts.

CONCLUSIONS

This example was based on a 12-bit word, 24 bit Frame Sync Encoder. If this had been based on an encoder using 8-bit words the SFID would be bit shifted. This shift would still be identified with this technique. To further investigate the SFID alignment, there is a binary Data Radix display mode for the Raw Data and Frame Buffers. The data can be paused and the displayed data can be reviewed to more easily identify the SFID rollover pattern in this binary form.

This investigative approach has been successfully used for 32 and 16-bit Frames Syncs also. Other word length data and Frame Syncs will require changing the bits per word after finding the proper Frame Sync pattern to align bit boundaries. Resulting fill data will need to be taken into account when determining the final Decom configuration.

REFERENCES

[1] Range Commanders Council Telemetry Group, Range Commanders Council, White Sands Missile Range, New Mexico, *IRIG Standard 106-15: Telemetry Standards*, 2015

[2] Lumistar Inc., "P2" Platform PCM Decommutator LS-50-P2 (R5) Technical Manual, Document U500501, August 2008

ASSPA. ASSISTANT FOR FLIGHT TEST EXECUTION

Pedro Rubio, Rodrigo Lopez
Airbus D&S - Flight Test

1. ABSTRACT

The objective of this paper is to describe the tools that have been developed to ease the Flight Test Engineers (FTE) the execution of Flight Test and ensure the highest quality of the final product. This first product is mainly devoted to Production Flight Test (PATM), where the profile and the tests to be performed are almost the same for all flights.

As during the PATM there are many checks to do and many data to gather, the idea was to execute in an automatic way as many test points as could be possible. To cope with this requirement we have to develop an Automatic Flight Phase Detection in order to start some applications in the FTE Computer without any human intervention.

So, we need a smart assistant to predict the following test points that will be executed.

In addition of the previous point, it was also needed to send the results of the data gathering and calculation to a "Digital Flight Test Card" to avoid the use of paper.

With both aids, Automatic Flight Phase Detection and Digital Flight Test Card, the efficiency of the execution of tests has been highly improved.

2. INTRODUCTION

During the delivery of an aircraft, Flight Test organization has to perform one or more flights to demonstrate that the aircraft is ready to be delivered to the customer. This demonstration takes place covering the requirements written in a document call PATM (**P**roduction **A**cceptance **T**est **M**anual). This document contains all the test points, operations procedures and maneuvers needed to complete the tests.

In a general case, it is necessary to make complex calculations in order to identify the flight phase, an aircraft condition or calculate a repetitive computation and generate evidences of the correct aircraft operation.

3. TRADITIONAL APPROACH

Years ago, this PATM was a paper document that the Flight Test Engineers (FTE onwards) use to follow the flight profile and take notes of the results. Afterwards, on ground, this document must be introduced in a digital document, database…

The only source of information was the aircraft itself, which means that FTE have to look at the aircrafts instruments and write down the results in the paper document. Obviously this was very inefficient, therefore next approach to ease the job of the FTE and improve efficiency was to introduce a Flight Test Instrumentation (FTI onwards).

Using this traditional FTI, many information displays were available for the FTE, but these displays only show FTI parameters or simple calculations with them. Even though, they were able to gather much more information from the aircraft to fulfill the PATM, however the workload during the flight test was very high and also it was needed a lot of effort after the flight to complete the document and to introduce the data in some digital format.

Some improvement in the process was a must; some aids must be put in place.

4. AMS (ADVANCED MONITORING SYSTEM)

The AMS is mainly composed by the so called SMART OBJECTS.
This suite was the first attempt to help the FTE collect all the data needed to complete the PATM.

A SMART object usually has several phases during its cycle of life:

- **Initial conditions**. Check that some parameters or even calculated parameters are inside boundaries.
- **Run**. During this phase, SMART object is acquiring data and doing some calculations until some conditions have been reached.
- **Results**. SMART object does some calculations and gives numeric results and the final status of the test.

One example of complex calculations, shown in **Figure 1: Smart Object for Stabilizations** is when some maneuvers needs a previous stabilization. The criteria of validation are based on the vertical difference between the values of the linear regression (least squares) in the first and in the last sample of the time slice for the following parameters:

- MACH (Speed)
- SAT (Ambient Temperature)
- ALT (Altitude)
- EGT (Engine Temperature)
- Others

A manual detection delegates the responsibility of the quality of the stabilization to the intuition and expertise of the FTE. By the aid of a SMART object there is an automatic detection based in the rule of calculating the criteria of validation every 10 seconds using the data from the previous 3 minutes.

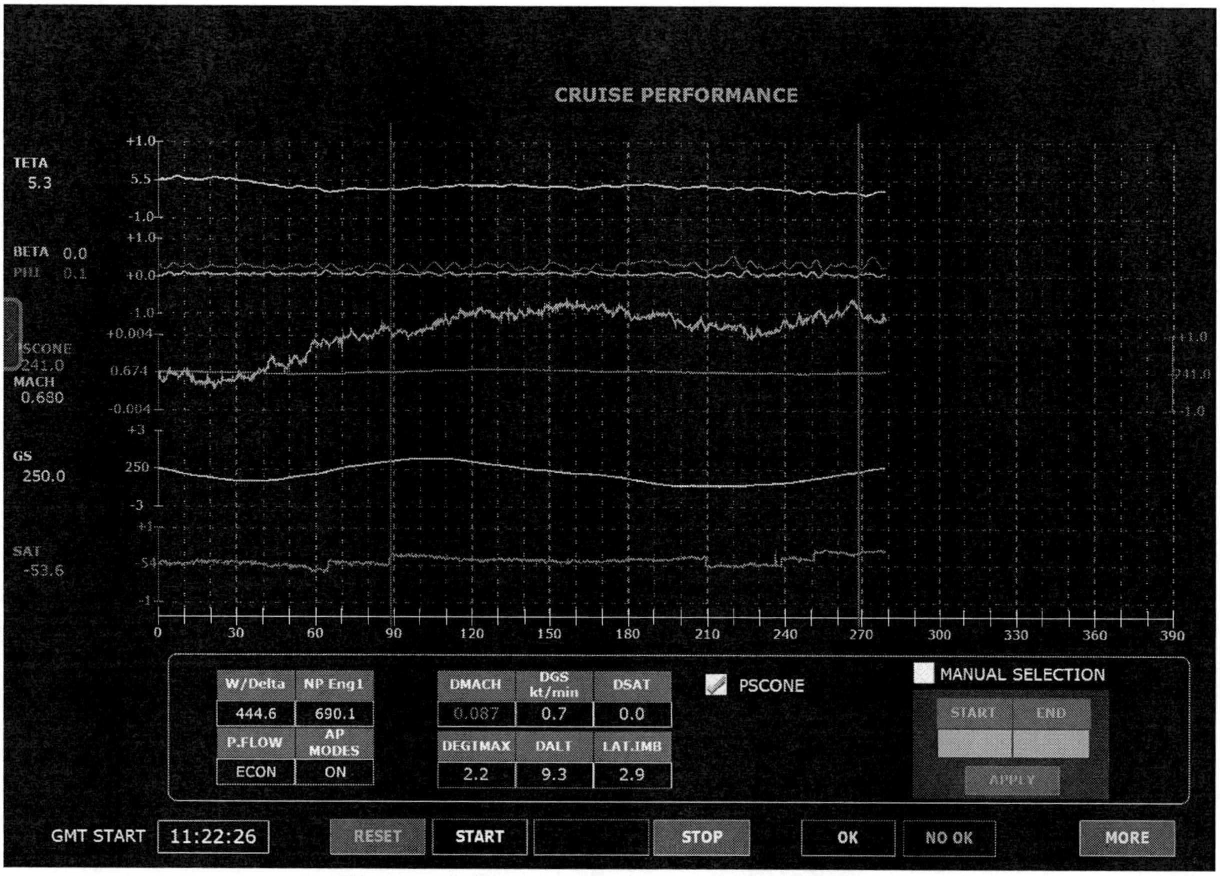

Figure 1: Smart Object for Stabilizations

Every single object contains a set of buttons, status and warnings that manage the smart object operations in real time.

- **START** Automatic tasks are started.
- **STOP** Automatic tasks are stopped and data are not deleted.
- **RESET** Reset the component to initial status, deleting all data.

Every Smart Object indicates its status

- **RUNNING** Indicates that the component is activated and running automatic tasks.
- **FINISHED** Indicates that the component is finished and suggests a result.
- **STOPPED** Indicates that the component has been stopped and all tasks are stopped.

When a component is activated (RUNNING) it starts to perform the programmed computation, and when all the computations are finished, it suggests to the user **OK** or **NO OK** depending on the results

- **OK** indicates that all tasks have been performed and all checks are OK.

- **NO OK** indicates that all tasks have been performed and some of them have some problem.

However, even if the suggestion has been NO OK, the FTE has the capability to override the final output (ie. Minimal discrepancies in some parameters…)
Another feature implemented in the SMART Object architecture is the capability of generates result files. This facilitates the FTE to generate evidences and automatic reports even on board. In **Figure 2: Smart Object Output File,** a sample of this output file is shown


```
<PATM VERSION="" PROGRAM="A400M" PROTOTYPE="0007" TEST="F0002" DATE_TEST="13/03/2013">
  <OUTPUTS>
    <TEST_POINT ID="07" NAME="PATM_TAKE_OFF" RESULT="OK" TIME_TP="58946">
      <OUTPUTS>
        <OUTPUT ID="07_01_OAT" VALUE="7.937769" />
        <OUTPUT ID="07_01_QNH" VALUE="0" />
        <OUTPUT ID="07_01_PALT" VALUE="179" />
        <OUTPUT ID="07_01_DPCB" VALUE="1" />
        <OUTPUT ID="07_01_OVB" VALUE="1" />
        <OUTPUT ID="07_01_PACKS" VALUE="1" />
        <OUTPUT ID="07_01_NAI" VALUE="0" />
        <OUTPUT ID="07_01_WAI" VALUE="0" />
        <OUTPUT ID="07_01_OBOGS" VALUE="0" />
        <OUTPUT ID="07_01_OBIGGS" VALUE="0" />
        <OUTPUT ID="07_01_UTC" VALUE="08:53:21" />
        <OUTPUT ID="07_01_PWRTABLE" VALUE="95.52448" />
        <OUTPUT ID="07_01_PWRTRG" VALUE="90.469" />
        <OUTPUT ID="07_01_PWRTABTRGDIF" VALUE="5.055473" />
        <OUTPUT ID="07_01_PWRTRGCB" VALUE="0" />
        <OUTPUT ID="07_01_CAS" VALUE="81.16449" />
        <OUTPUT ID="07_02_UTC" VALUE="03:53:29" />
        <OUTPUT ID="07_02_PWRTABLE" VALUE="100.6119" />
        <OUTPUT ID="07_02_PWRTRG" VALUE="100.75" />
        <OUTPUT ID="07_02_PWRTABTPGDIF" VALUE="-0.1381073" />
        <OUTPUT ID="07_01_PWR1" VALUE="100.7272" />
        <OUTPUT ID="07_01_PWR2" VALUE="100.7429" />
        <OUTPUT ID="07_01_PWR3" VALUE="100.7429" />
        <OUTPUT ID="07_01_PWR4" VALUE="100.5772" />
        <OUTPUT ID="07_01_NP1" VALUE="841.3197" />
        <OUTPUT ID="07_01_NP2" VALUE="843.3735" />
        <OUTPUT ID="07_01_NP3" VALUE="843.6405" />
        <OUTPUT ID="07_01_NP4" VALUE="843.9855" />
        <OUTPUT ID="07_01_EGT1" VALUE="778.8824" />
        <OUTPUT ID="07_01_EGT2" VALUE="753.0271" />
        <OUTPUT ID="07_01_EGT3" VALUE="749.3499" />
        <OUTPUT ID="07_01_EGT4" VALUE="771.0232" />
        <OUTPUT ID="07_01_NPCB" VALUE="1" />
        <OUTPUT ID="07_02_PWRTRGCB" VALUE="1" />
        <OUTPUT ID="07_01_VIBE1" VALUE="1.13812" />
```


Figure 2: Smart Object Output File

With this concept of AMS, the workload of the FTE has been reduced but it is still high in some phases of the test flight.

In many cases, the FTE is late to start the display manually, and we could have five or six smart objects in the same flight phase.

Next step is to automatized as much as possible the execution of the flight test points, for this purpose ASSPA integrate AMS with automated Flight Phase detection to launch different displays when triggered condition are reached.

The final idea is to develop an intelligent enough assistant who manages the entire process of accepting an aircraft, in the most autonomous way possible.

5. ASSPA (AUTOMATIC ASSISTANT PRODUCTION AIRCRAFT)

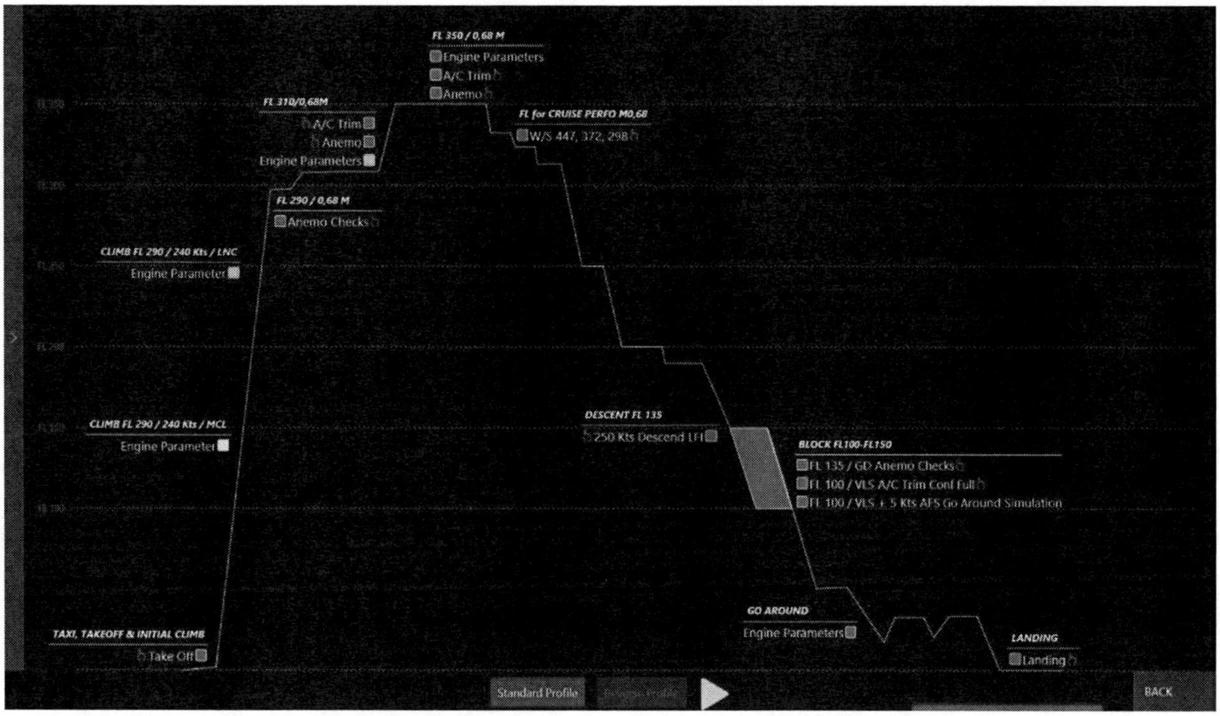

Figure 3: ASSPA Flight Profile

The first step in the process is the selection by the FTE of all test point to be executed.
We will have several default profiles, which can be modified.

In **Figure 3: ASSPA Flight Profile**, there is an standard Flight Profile used during a PATM flight, the idea behind is that most of the checks to be done may run autonomously, triggered by some conditions that predict the beginning of the screen, so we need a:

- Automatic Flight Phase detection.
  What part of the profile are we at?
  Here, we can identify which screens are necessary to activate.
- Automatic Execution of the Displays.
  What conditions must be fulfilled?
  Here, we can identify when the displays begin to calculate.

Once the Flight Phase has been detected and the triggered conditions match some display, the display runs automatically.
We can see in the figure 4, blue displays activated and yellow display running.

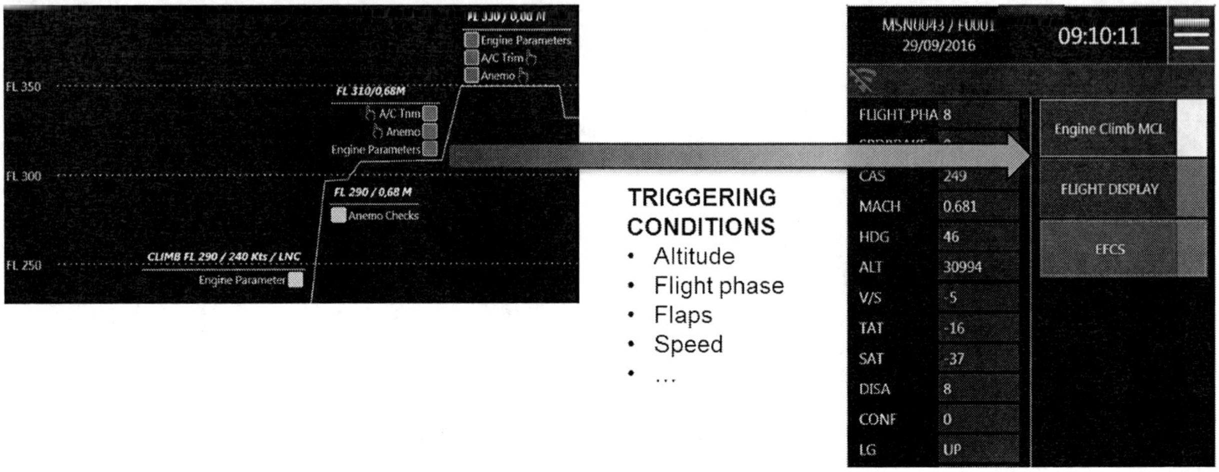

Figure 4: ASSPA Automatic Display

So, we can have the following status:

- Display waiting for Flight Phase.
- Display launched and waiting for initial conditions (stabilization, minimum time…)
- Display running.
- Display finished successfully and data gathered.
- Display finished with some errors, data gathered.

Additionally, the status information of the display is shown in the main screen and in the Flight Profile screen.

Any of these states could be blinking, that means some human input is needed, normally the confirmation of a successful display.

Additionally, we have developed an architecture for gathering results and store in a Database. From this database is possible to generate a PATM document fulfilled using the DOMA Application Framework.

6. DOMA

Document **O**riented **M**onitoring and **A**nalysis is a Framework developed for digitalized a Tagged Word document. The idea is to avoid the use of paper on board. That implies that the documents used for taking notes with the aircraft parameter results must be written by an automatic system.
Starting from a Word document, DOMA is able to read the document that has been tagged using a strict labeling method which aid to identify the kind of parameter that must be associated with

each token. For example, check-box, boxes for values and tables are associated to parameters from FTI or results of the SMART Objects calculations.

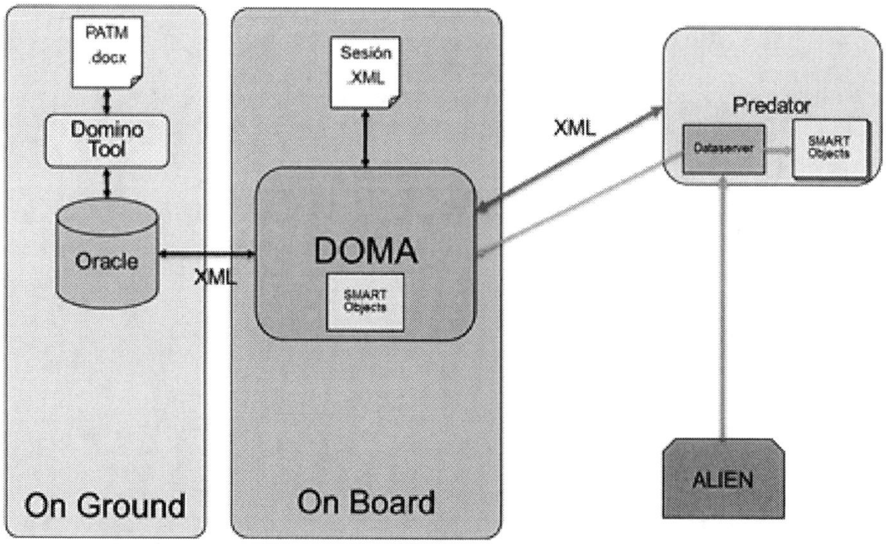

Figure 5: DOMA Document management process

In **Figure 5: DOMA Document management process** the overall process is shown; it consists on the following steps:

- A Word Document is generated according to a Tagging system rules in order to know the parameter associated to each element of the document and also the type of element.
- DOMA on ground tool (DOMINO) parsers the document and store the information in a Database.
- Before the test, DOMA download the information in a Tablet. That allows managing the document in a stand-alone mode. The document is stored in XML format
- During the test, FTI data is acquired with ALIEN (compact data acquisition system) and sent to a Server PC (PREDATOR)
- During the test, data is gathered and stored in the tabled using XML file.
- Once the test has finished, in ground, data is uploaded to the database.
- DOMINO can regenerate the Word Document with the information of the tests needed to fill the document.

Figure 6 and 7 show a piece of the PATM word document and how is displayed in DOMA.

AIRBUS MILITARY ME-A4-DT-110206 Edic./ Issue B PAG. 274

8 INITIAL SYSTEM CHECK AFTER LIFT OFF

Note: Consider to use ACMS/ALIEN report.

After lift off when stabilized on climb path, record engine VIB indications .

	ENG 1	ENG 2	ENG 3	ENG 4
N2 (%)				
N3 (%)				
FF (KG/H)				
OIL QT (QT)				
ENG OIL PRESS (PSI)				
ENG OIL TEMP (°C)				
PROP OIL PRESS (PSI)				
PROP OIL TEMP (°C)				
VIB ENG				
VIB PROP				
TRQ (%)				

- The ENG VIB levels should not exceed 3 C.U. ☐
- The PROP VIB levels should not exceed 3 C.U. ☐

In SD BLEED page:

- Check neither abnormal indications nor abnormal behavior in the bleed system ☐

Figure 6: MSWord PATM document extraction

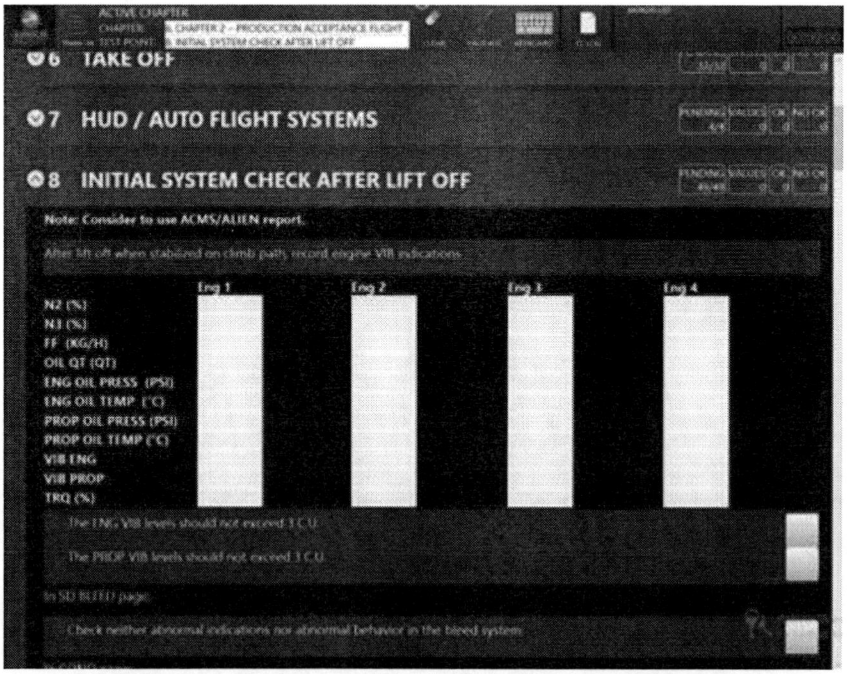

Figure 7: Same as Fig. 6 digitalized in DOMA

DOMA application runs on a MS Windows 10 based tablet PC, taking the advantage of touching screen, multi-touch capabilities…

Inputs in the digitalized PATM document can be one of the following:

- Direct human input. Only human being can answer
    - A/C is clear for start engines, clear safety area…

- Direct parameters gathered from FTI
    - Height, rpm, speed…

- Simple calculations made inside DOMA
    - Average, minimum, maximum…

- Complex computations coming from ASSPA

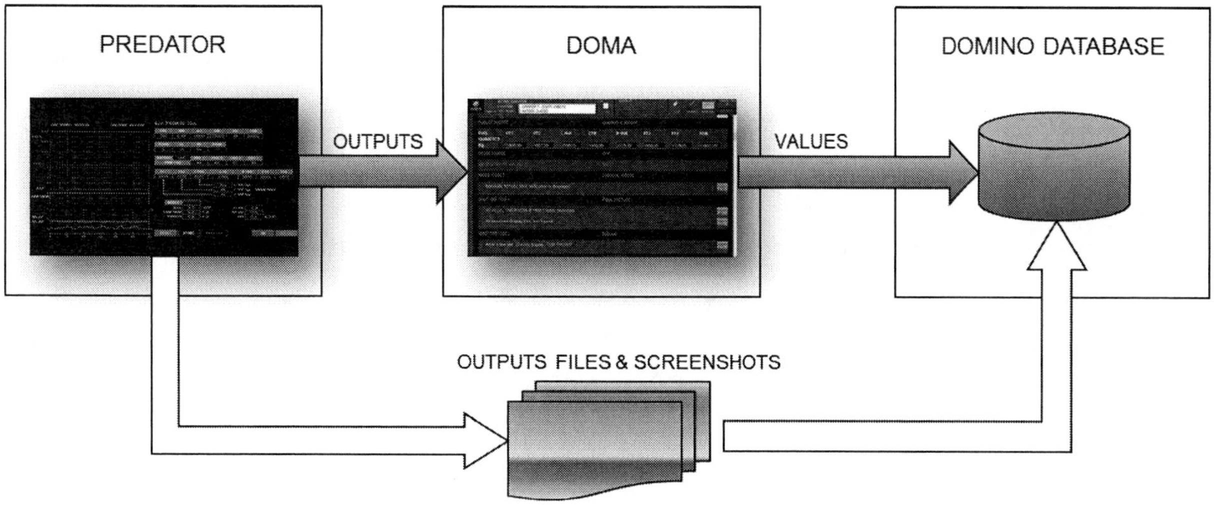

Figure 8: DOMA Communication

7. CONCLUSIONS

Currently, with ASSPA and DOMA we are getting a lot of accuracy and quality in the data gathered and reliability when passing each test point.
Greatly facilitating the work of the FTE and that he is focused on the test being performed correctly and do not have to be aware of the data gathering, wasting time on it.

The analysis of each test point is performed in depth in real time, which represents a significant saving to avoid subsequent flights to a ground analysis.
If some point has not been done correctly, the FTE may require repeating it immediately.

With ASSPA, the FTE has a powerful assistant that helps them at all times with the way to go.

USE OF LTE FOR CELLULAR RANGE TELEMETRY (CRTM): A SIMULATION STUDY

Arvind Krishnamoorthy
Nokia Networks
1455 W. Shure Drive, Arlington Heights, IL 60004
arvind.krishnamoorthy@nokia.com

ABSTRACT

The 3GPP LTE standard was primarily designed for terrestrial cellular usage. The potential use of LTE for cellular range telemetry produces some unique technical challenges for LTE technology. In this paper we examine these challenges in detail, and through the use of system level simulations identify the sensitivity of the performance of LTE-based cellular range telemetry system performance to key factors such as the ground station antenna pattern and orientation, the propagation model assumed, the number of transmit and receive antenna at the ground station, altitude of the airborne test article, etc.

INTRODUCTION

The 3GPP LTE standard is well known throughout the world as the current standard in terrestrial cellular service. Recognizing the persistent and explosive increase in data usage, the LTE standard was developed with a strong emphasis on spectral efficiency because radio frequency bandwidth is a precious natural resource. Building on previous cellular technologies, low latency and mobility in terrestrial deployments were other key considerations for LTE standard design. Although recent 3GPP standardization efforts consider providing LTE service to aerial vehicles and the interference posed by these, these efforts are still in preliminary stages. The CRTM project presents unique design challenges that are different from terrestrial cellular systems. Key among these difference are

1. **Traffic pattern**: Traditional cellular networks demonstrate asymmetric traffic patterns with downlink (network infrastructure to user equipment) dominating by a factor of 10. Hence terrestrial network design concentrates on the downlink. However in the case of telemetry, the user equipment (planes) is the data source and the telemetry application needs a strong uplink (from user equipment to network) which is quite the opposite of what is seen in terrestrial networks. To avoid confusion, we will refer to the user equipment to network link as the Air-to-Ground (A2G) link.

2. RF propagation and geometry that is quite different from traditional cellular networks.

    (a) **Antenna design**: In traditional cellular networks, the coverage is along the azimuth and antennae positioned and antenna patterns designed to provide coverage for a certain azimuth angle range. As a result, terrestrial design is focused on the horizontal sectorization. For the telemetry application with planes flying anywhere from 500 ft above ground to 50000 ft, the concept of horizontal sectorization might be less relevant than vertical sectorization. So antenna design must be rethought for the telemetry application.

(b) **High speed environment**: While the 3GPP LTE standard was designed with certain high speed scenarios in mind, the telemetry application will certainly stretch these to the limit considering the planes might be flying up to twice the speed of sound. Special consideration must be given to designing for these situations while maintaining a reliable A2G link.

(c) **Radio channel conditions**: Terrestrial designs usually design for rich scattering environments which provide a lot of multipath, frequency selectivity and spatial multiplexing opportunities. On the other hand, telemetry application is likely to be in line-of-sight (LOS) conditions and so the design has to be appropriate.

To understand the challenges posed by the telemetry application, we take a two step approach.

1. As a first step, a Monte-Carlo simulation was conducted to understand the extent of the Doppler shift experienced by airborne UEs. This helps pose requirements on the Doppler shift detection/compensation scheme at the UE. Any residual unmitigated Doppler shift would manifest in the reverse link (A2G link) resulting in a performance degradation compared to ideal Doppler shift mitigation. To model this performance degradation, a link simulator could be used to quantify this performance degradation as an equivalent SINR shift. This would need to be done in many different SINR/Doppler shift conditions.

2. The SINR shift obtained by the link level simulator is proposed to be used in the system level simulator. The system level simulator could then be used as a design/verification tool to compare various design choices at a system level. Since Nokia has productized system level simulators that reflect faithfully the algorithms in Nokia's LTE eNB, these simulators represent the best and most reliable choices of modeling and estimating the performance of the Nokia Flexi product in the unique telemetry application.

This paper is organized as follows - first, a set of channel models that are used in both the link level characterization study and system level simulations are described. Following the channel model description, the Doppler shift characteristics of the telemetry application are described that enable requirement setting for the airborne UE. Following this, a brief description of the link level characterization of an airborne terminal and results from a first round of system level simulations of the telemetry application are presented.

RISING TO THE CHALLENGE

A. Channel Models

The channel models used in this report are listed in Table 1. In determining these channel models, we drew inspiration from [1, 2, 3, 4, 5, 6].

Table 1: Channel Models

Frequency band	Channel Model 1	Channel Model 2 High altitude ($>$ 3300 m)	Channel Model 3 Low altitude ($\leq$ 3300 m)
S band (2.0 GHz)	LoS only	Ricean + 1 tap channel $\{$(0 dB, 0 ns, Ricean K=16), (-16 dB, 200 ns, Rayleigh tap)$\}$	Ricean + 2 tap channel $\{$(0 dB, 0 ns, Ricean K=8), (-9 dB, 200 ns, Rayleigh tap), (-15 dB, 600 ns, Rayleigh tap)$\}$
C band	LoS only	Ricean + 1 tap channel $\{$(0 dB, 0 ns, Ricean K=25), (-25 dB, 200 ns, Rayleigh tap)$\}$	Ricean + 2 tap channel $\{$(0 dB, 0 ns, Ricean K=16), (-17 dB, 200 ns, Rayleigh tap), (-23 dB, 600 ns, Rayleigh tap)$\}$

The K-factors assumed in Table 1 for the S-band and C-band are conservative compared to those recorded in [3]. Understanding that a lower K-factor Ricean channel will negatively influence the performance of Doppler compensation schemes creates a tradeoff situation: high carrier frequency implied by the C-band causes higher absolute Doppler shift and higher operating pathloss than S-band, while lower K-factor in S-band can adversely impact performance in S-band compared to C-band.

B. Doppler Characterization

The Doppler shift in a communication system can be calculated using the equation

$$f_d = \frac{\vec{v}.\vec{r}_{1,u} f_c}{c}. \tag{1}$$

According to (1), the Doppler observed at the airborne terminal depends on both the speed and the direction in which the airplane is traveling relative to the ground station. A Monte-Carlo simulation was carried out in which airborne UEs were randomly placed at different altitudes from 150 m to 15 km. These UEs were also assigned random velocity vectors with random direction and a speed that is uniformly distributed between 400 km/hr and 3000 km/hr. Equation (1) was used to calculate the actual Doppler shift observed for each airborne UE. As shown in Figure 1, the expected distribution of Doppler as a function of location and speed differs quite significantly from Doppler calculated from just the speed. For example, in Figure 1 the Doppler experienced at 200 m altitude at 4.7 GHz (C-band) frequency is less than 3 kHz 95% of the time. Of course at low altitudes, the maximum air speed of the airplane has been assumed to be 800 km/hr. At high altitudes (15 km), allowing a max air speed of 3000 km/hr (MACH 2.5), the 90 %-ile Doppler is 8 kHz. In contrast, the subcarrier spacing in an LTE system is around 15 kHz.

At lower carrier frequency of 2.0 GHz (S-band), the behavior of the Doppler distribution is similar to the C-band Doppler distribution but proportionally reduced in magnitude as indicated by (1). These observations allows a relaxation of the Doppler correction requirements at the airborne UE.

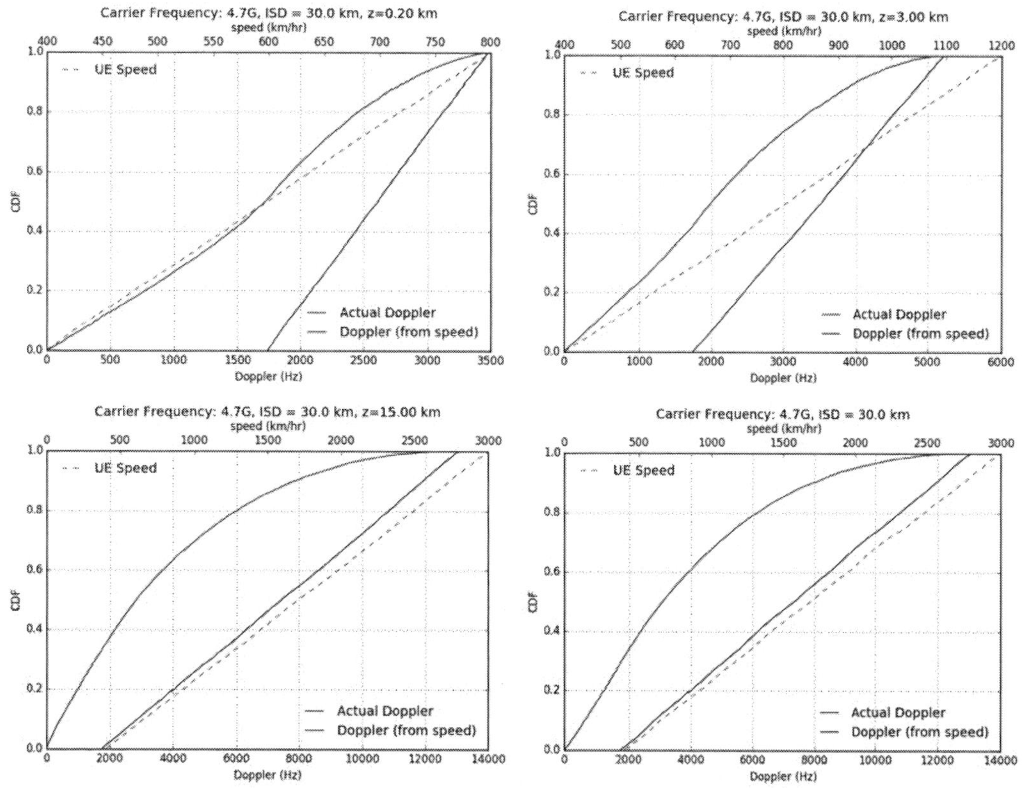

Figure 1: Doppler distribution for 4.7 GHz at different altitudes (200m, 3 km, 15 km, random).

C. *Antenna Pattern*

An antenna pattern representative of that used for in-flight coverage of commercial aircraft was used for this study and is shown in Figure 2. The vertical antenna pattern in Figure 2 is quite different from regular terrestrial antenna patterns in that it is wider and therefore more suited to covering a large swath of altitudes making it suitable for the telemetry application.

D. *System Simulation Results*

The system level characterization for the CRTM project utilized MoRSE. MoRSE stands for Mobile Radio Simulation Environment a Nokia system level simulator that considers the interaction of many users in many cells; works dynamically and models time-dependent effects on TTI granularity; allows simulation of hexagonal or real network layouts; and is closely aligned to the Nokia product specific algorithms and parameters. MoRSE focuses on the modeling of the air-interface, user plane aspects (related to the transmission of user data), and layer 1 and 2 algorithms.

E. *General configuration for System Level Simulations*

System level simulations make a large number of assumptions regarding the layout of the cells, the number of users simulated, the traffic pattern assumed, the channel models, etc.. These are summarized in the Table 2.

Table 2: Key System Simulation Assumptions

Parameter	Value Description
Layout	Hexagonal layout 12cells (4 sites), regular wrapped
ISD	30, 40, 50 km
LTE mode & bandwidth	FDD, 15 MHz
Transmission modes	2x2 TM4 in DL 1x4 SIMO in UL (IRC receiver at eNodeB)
Operational band	C band (4.7 GHz) S band (2.0 GHz)
Output power	eNodeB: 49 dBm total (40 W/antenna) UE: 37 dBm (5 W)
Penetration loss	0 dB
Antenna height	eNodeB: 33m (100 ft) UE: variable per simulation scenario
Modulation	QPSK/16QAM - UL QPSK/16QAM/64QAM - DL
Scheduler	Nokia Scheduler Channel unaware for UL & DL
Shadowing	Log-normal $\sigma = 7$ dB
Fast fading	Channel models from Section A.
Traffic model	Full buffer or Constant bitrate (DL/UL : 2/8 Mbps)
Load (# UEs in the simulation)	12 cells, 8 UEs total, maximum 2 UE per cell
PDCCH	Dynamic PCFICH (Nokia algorithm)
Power control	Open Loop, $P_{0,PUSCH} = -80$ dBm, $\alpha = 0.8$
UL MU-MIMO	disabled
PUCCH	2 PRBs
eNodeB antenna gain and pattern	13.0 dBi, pattern in Figure 2
UE antenna gain and pattern	-3.0 dBi, omni pattern
Noise Figure	eNodeB: 3 dB UE: 7 dB
Doppler Compensation	Ideal

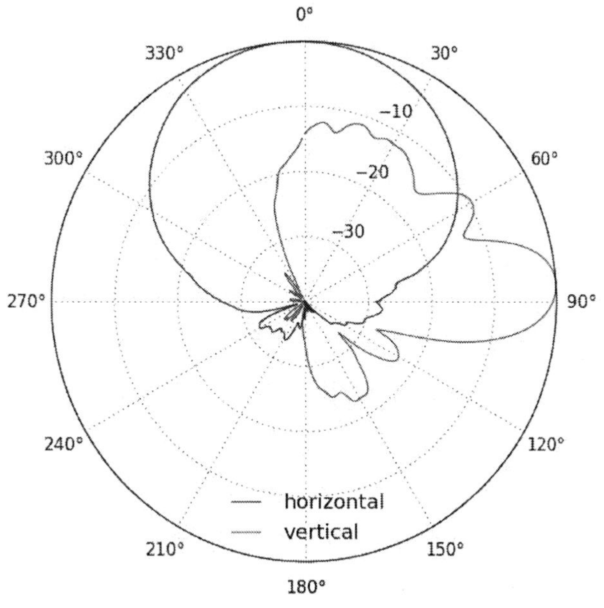

Figure 2: Representative antenna pattern used for in-flight coverage of commercial aircraft

F. *Simulation Scenarios*

1. Static altitude simulations - UE is assumed to be at a fixed altitude for the entire simulation, while movement in the xy plane is permitted with speed falling in a specified range.

   - C band: high altitude CFF, random speed between 450 km/hr to 3000 km/hr, altitude: {5000, 10000, 15000, 20000m}, ISD: {30, 40, 50 km}, eNodeB antenna uptilt: {5, 10, 15 degree}, check the UL/DL UE TP - {Total 36 cases}
   - C band: low altitude CFF, random speed between 450 km/hr to 1200 km/hr, altitude: {150m, 1500m, 3000m}, ISD: {30, 40, 50 km}, eNodeB antenna uptilt: {5, 10, 15 degree}, check the UL/DL UE TP - {Total 27 cases}
   - C band: static channel, random speed between 450 km/hr to 3000 km/hr, altitude: {10000m, 20000m}, ISD: {30, 40, 50 km}, eNodeB antenna uptilt: {5, 10, 15 degree}, check the UL/DL UE TP - {Total 18 cases}
   - S band: high altitude CFF, up to 3000 km/hr, altitude: {5000, 10000, 15000, 20000m}, ISD: {30, 40, 50 km}, eNodeB antenna uptilt: {5, 10, 15 degree}, check the UL/DL UE TP {Total 36 cases}
   - S band: low altitude CFF, up to 1200 km/hr, altitude: {150m, 1500m, 3000m}, ISD: {30, 40, 50 km}, eNodeB antenna uptilt: {5, 10, 15 degree}, check the UL/DL UE TP {Total 27 cases}
   - S band: static channel, up to 3000 km/hr, altitude: {10000m, 20000m}, {30, 40, 50 km}, eNodeB antenna uptilt: {5, 10, 15 degree}, check the UL/DL UE TP {Total 18 cases}

2. Dynamic altitude simulations - UE in Dynamic movement (at start time, UEs can be at any altitude in the range provided, (x, y, z) motion is allowed, no limit on how fast movement along z-axis can be)

    - C band: static channel, 450km/hr $\leq$ speed $\leq$ 3000 km/hr, 4000m $\leq$ altitude $\leq$ 20000m, ISD: $\{30, 40, 50$ km$\}$, eNodeB antenna uptilt: $\{5, 10, 15$ degree$\}$, check the UL/DL UE TP - $\{$Total 9 cases$\}$
    - S band, static channel, 450km/hr $\leq$ speed $\leq$ 3000 km/hr, 4000m $\leq$ altitude $\leq$ 20000m, ISD: $\{30, 40, 50$ km$\}$, eNodeB antenna uptilt: $\{5, 10, 15$ degree$\}$, check the UL/DL UE TP - $\{$Total 9 cases$\}$
    - C band: high altitude CFF, 450km/hr $\leq$ speed $\leq$ 3000 km/hr, 4000m $\leq$ altitude $\leq$ 20000m, ISD: $\{30, 40, 50$ km$\}$, eNodeB antenna uptilt: $\{5, 10, 15$ degree$\}$, check the UL/DL UE TP - $\{$Total 9 cases$\}$
    - C band: low altitude CFF, 450km/hr $\leq$ speed $\leq$ 1200 km/hr, 150m $\leq$ altitude $\leq$ 4000m, ISD: $\{30, 40, 50$ km$\}$, eNodeB antenna uptilt: $\{5, 10, 15$ degree$\}$, check the UL/DL UE TP - $\{$Total 9 cases$\}$
    - S band, high altitude CFF, 450km/hr $\leq$ speed $\leq$ 3000 km/hr, 4000m $\leq$ altitude $\leq$ 20000m, ISD: $\{30, 40, 50$ km$\}$, eNodeB antenna uptilt: $\{5, 10, 15$ degree$\}$, check the UL/DL UE TP - $\{$Total 9 cases$\}$
    - S band, low altitude CFF, 450km/hr $\leq$ speed $\leq$ 1200 km/hr, 150m $\leq$ altitude $\leq$ 4000m, ISD: $\{30, 40, 50$ km$\}$, eNodeB antenna uptilt: $\{5, 10, 15$ degree$\}$, check the UL/DL UE TP TP - $\{$Total 9 cases$\}$

The simulator is configured to operate with up to eight airborne terminals. The starting (x, y, z) location and the velocity vector of the airborne terminal were randomly selected as per the simulation case. The average session life time of an airborne terminal was set to 5 seconds and at the end of the session, the statistics of the terminal are recorded and the terminal is resurrected at a different (x, y, z) location with a different velocity vector. This process repeats until the simulation ends after 1200 seconds of operation. This procedure of 'birthing' and 'resurrecting' the airborne terminal multiple times allows adequate sampling of the entire three-dimensional space around the cluster of sites.
To keep things simple, this simulation study assumed ideal Doppler compensation, i.e., the Doppler shift could be compensated and the degradation in throughput due to either high Doppler shift or low K-factor would not be assimilated into the system simulation results.

G. *Simulation Results*

In this section, a brief description of the results of system level simulations are provided. In Figure 3, the average UL (air-ground) throughput is plotted as a function of inter-site distance (ISD), altitude, and antenna uptilt for high altitude scenarios. It is observed that there is improvement in UL throughput with higher ISD and higher uptilt. The dependence on uptilt is quite clear because increased uptilt provides higher antenna gain for reception of air-ground transmissions. There appears to be less dependence on inter-site distance than on uptilt.

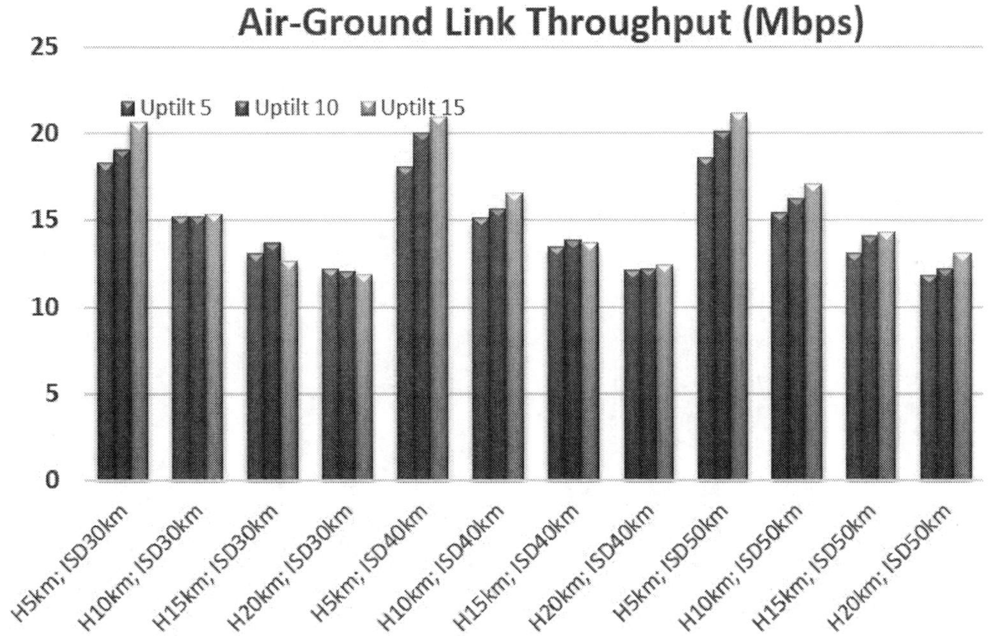

Figure 3: Avg. air-ground link throughput (at C-band) for high alt. channel model 2 vs. ISD and Uptilt

A similar plot of UL throughput for low altitude flight patterns in shown in Figure 4, where it is observed that there is a slight preference for smaller ISD and lower uptilt. This contradicts the need for performance in the high altitude cases. Figure 5 shows a composite view of both Figure 40 and Figure 41, where these observations are more easily made. These results imply that the optimum value of ISD and uptilt vary as a function of altitude for the fixed antenna design assumed in the study. Performance requirements such as desired air-ground link throughput across full range of altitudes used at a test range such as Edwards Air Force Base, will determine alternate antenna strategies such as broader ground station antenna patterns, vertical sectorization, or a combination of both. Other practical considerations such as site location and backhaul availability are also critical in determining ISD.

ACKNOWLEDGEMENTS

The results in this paper would not have been possible without the help of my Nokia colleagues - Yj (James) Deng, Lily Wang, Han Chao, Yang Bin, Xiao-Pai Bai, and Joseph Shannon. I owe them my gratitude and recognize that this paper is as much their creation as it is mine.

CONCLUSIONS

Based on the analysis and simulation above results, the following are concluded:

- The Doppler distribution modeled in Section A. indicates that the Doppler calculated from geometry is much lower than a pure speed based calculation over most of the aircraft speeds

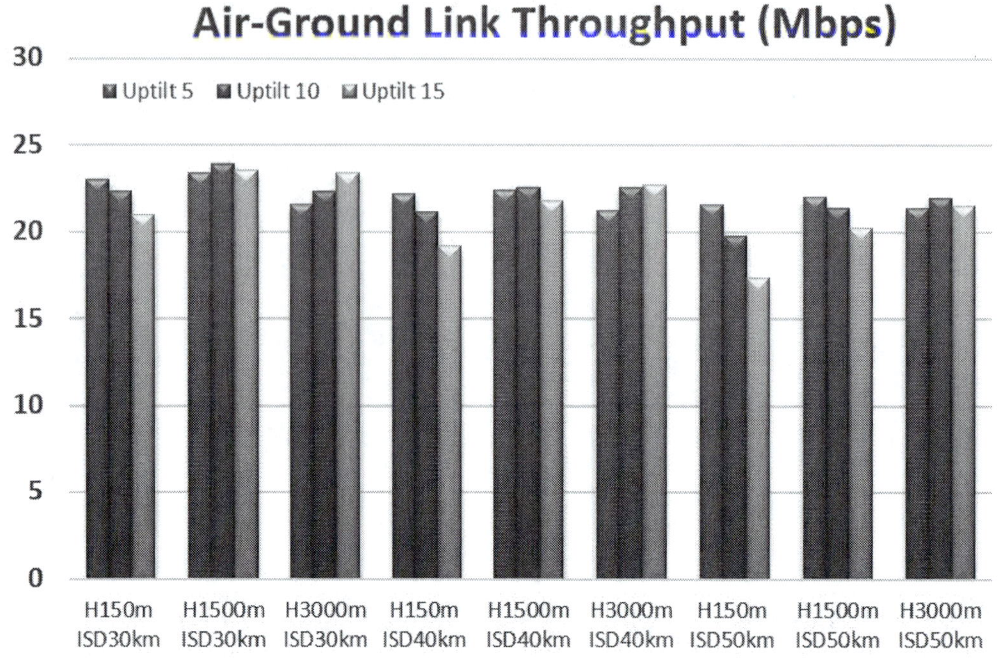

Figure 4: Avg. air-ground link throughput (at C-band) for low alt. channel model 3 vs. ISD and Uptilt

and locations anticipated on the test range. This reduces the Doppler compensation requirement on the airborne UE.

- From the C-band simulation results in Section G., the key observation is that the optimum ISD and antenna uptilt are a function of the throughput desired at a certain altitude for the existing antenna patterns.

The system simulations and analysis in this report were completed for the C-band (4.7 GHz) and S-band (2.0 GHz) static altitude and dynamic altitude scenarios in Section F. The dynamic altitude simulation results are omitted for the sake of brevity.

REFERENCES

[1] D. W. Matolak and R. Sun, "Unmanned aircraft systems," *IEEE Vehicular Technology Magazine*, vol. 1, pp. 79–85, Jun. 2015.

[2] D. W. Matolak, "Air-ground channels and models: Comprehensive review and considerations for unmanned aircraft systems," *Proc. of IEEE Aerospace Conference*, pp. 1–17, Mar. 2012.

[3] D. W. Matolak and R. Sun, "Air-ground channels for UAS: Summary of measurements and models for l- and c-bands," *Proc. Integrated Communications Navigation and Surveillance (ICNS)*, pp. 8B2–1–8B2–11, 2016.

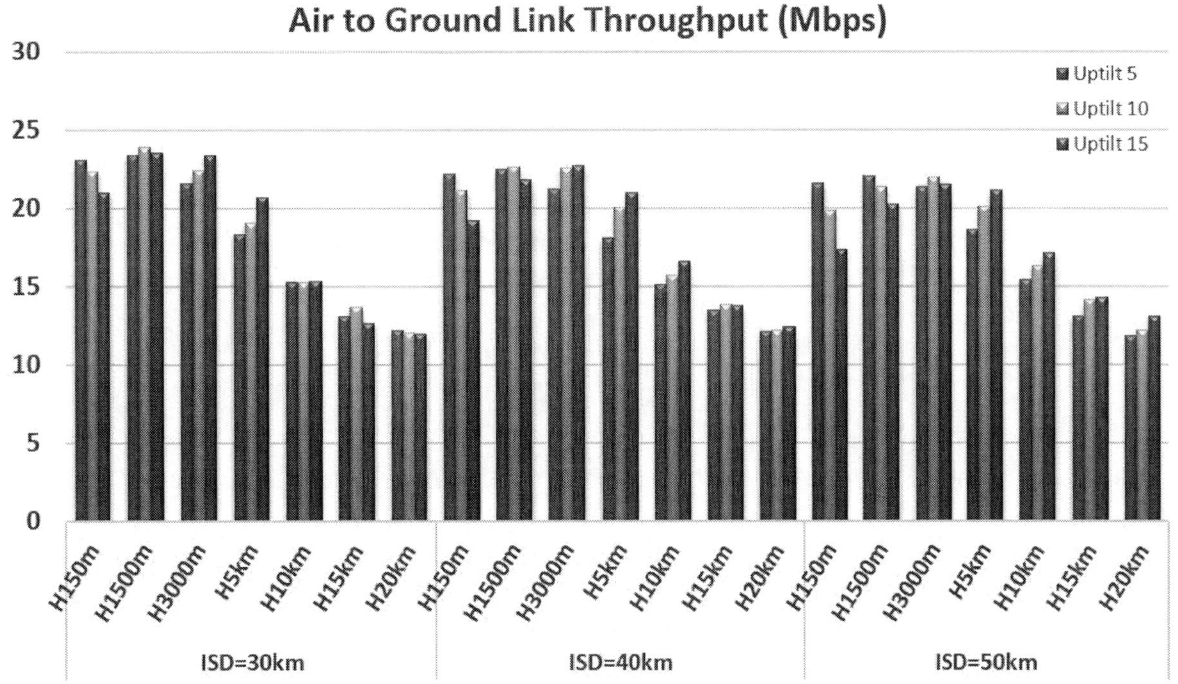

Figure 5: Composite view of average air-ground link throughput (at C-band) vs. ISD and Uptilt

[4] R. Sun and D. W. Matolak, "Air-ground channel characterization for unmanned aircraft systems: the mountainous environment," *IEEE/AIAA 34th Digital Avionics Systems Conference (DASC)*, pp. 5C2–1–5C2–9, 2015.

[5] T. S. Rappaport, *Wireless Communications: Principles and Practice*. Upper Saddle River, NJ: Prentice Hall, 2nd ed., 2002.

[6] M. Rice and A. Davis, "A multipath channel model for wideband aeronautical telemetry," in *Proc. MILCOM*, vol. 1, pp. 622–626, 2002.

SEAMLESS HANDOVER AT HIGH SPEEDS FOR CELLULAR RANGE TELEMETR

Vinayak Hegde
Nokia Networks
1455 W. Shure Dr, Arlington Heights, IL 60004
vinayak.hegde@nokia.com

ABSTRACT

Cellular based Range Telemetry provides a lot of benefits over legacy Aeronautical Mobile Telemetry, such as:

- Higher capacity, allowing multiple test articles (i.e. telemetry radio links) which have active telemetry data measurements exchanged in near real time.
- Bi-directional radio links, that can allow new signaling with the avionics equipment on the test articles.
- Higher performance provided by spectrally efficient OFDMA waveform used in LTE and the dynamic link adaptation that is built into the LTE waveform.

With the CRTM cellular approach, the overall range coverage area is broken into multiple coverage areas, each served by a single cell. To maintain seamless service throughout the test flight, CRTM must support seamless mobility as the aircraft moves between the coverage areas of these cells. This is achieved using the connected mode handover procedures supported by LTE technology.
Traditional LTE networks rely on signal strength measurements made by the User Equipment (UE) to make handover decisions. The measurement system and handover procedures have been designed to support mobility at speeds under 500 km/hr. To support mobility at higher speeds, the following need to be taken into account:

- **LTE User Equipment (UE) RF Signal Measurements**: Accuracy of measurements depends on UE's capability to measure the signal strength of a neighbor cell while synchronized to its serving cell. The higher Doppler shifts expected in an AMT environment due to the higher test article speeds and the higher frequency of operation along with the difference in Doppler shifts between the serving cell and neighbor cells complicates cell measurements.

- **Handover Processing Time**: The serving eNodeB needs to process the measurements from the UE and initiate handover in a timely manner. Delays in handover processing can result in radio link failures. The higher speeds of test articles in the AMT environment reduces the time available to the eNodeB for handover processing.

This paper explains the handover procedure in 4G LTE, with the focus on:

- Measurement reports that support handover decisions

- Decisions impacting data service delays during handover

impacts to the handover processing when the UE (i.e. test article in CRTM) is moving at very high speed.

TERMINOLOG

- AMT: Aeronautical Mobile Telemetry
- CRTM: Cellular Range Telemetry
- eNodeB: Ground Station of the LTE network
- EPC: Evolved Packet Core
- HSS: Home Subscriber Server
- LTE: Long Term Evolution
- MME: Mobility Management Entity
- PCRF: Policy and Control Function
- PGW: Packet Gateway
- RAN: Radio Access Network
- SGW: Serving Gateway
- UE: User Equipment (Airborne node in CRTM)
- VoLTE: Voice over LTE

INTRODUCTION

Legacy telemetry systems are facing technical headwinds in meeting demands for throughput, spectrum availability and test capability. To address these concerns, one of the proposed solutions is to migrate the existing radio communication system used at test ranges to a cellular approach, gaining the wireless communication technology advances available in commercial systems. The figure below shows the migration from a traditional AMT system to Cellular based AMT system.

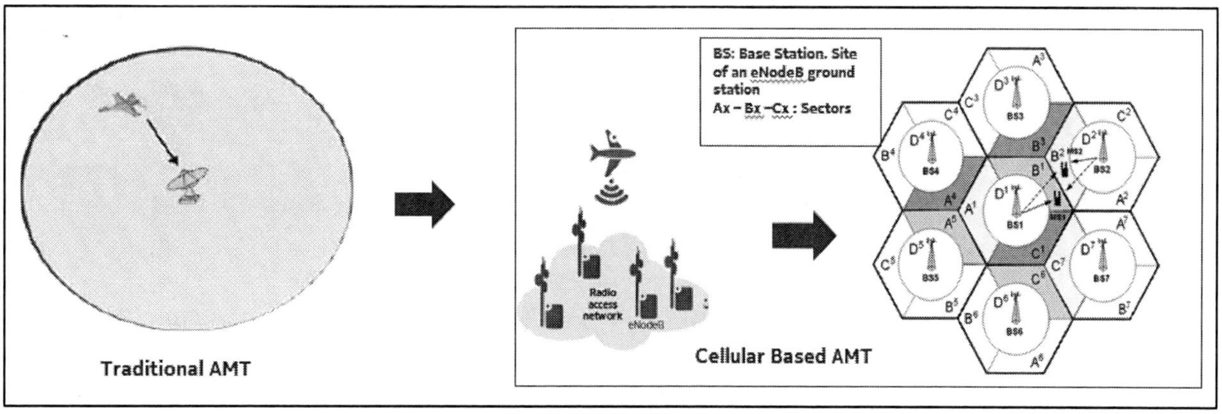

Figure 1: Migration to Cellular Range Telemetry

4G LTE Cellular technology provides a high capacity and high performance scalable architecture for the next generation telemetry systems. However, commercial LTE cellular technology has been designed for terrestrial application and needs enhancements to meet the goals of a telemetry application. The key technical challenge areas are:

- Doppler Shift due to high speeds

- Ground Station Antenna Patterns for vertical and horizontal coverage
- Air to Ground Throughput
- Seamless Handover at the high speeds

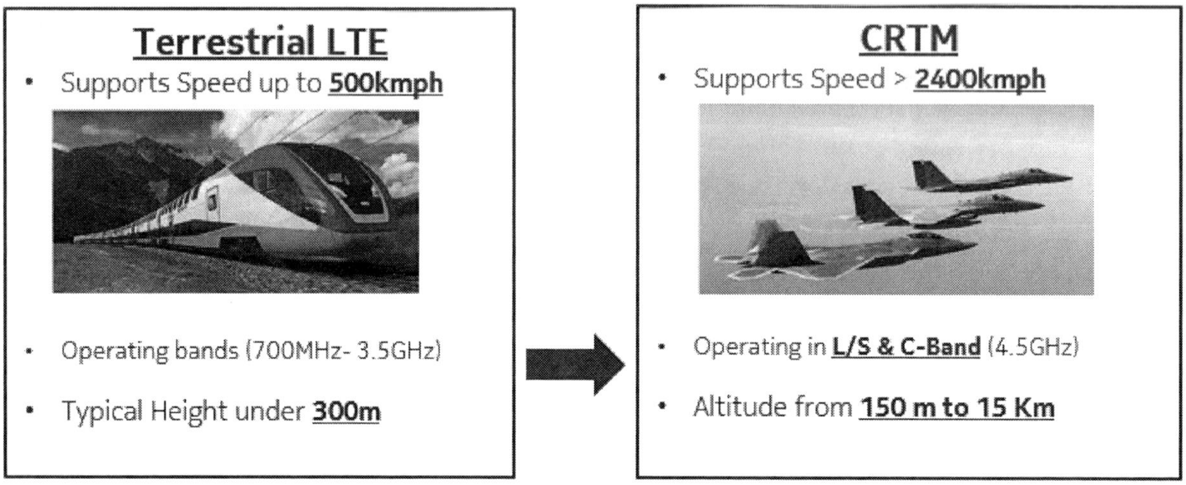

Figure 2: LTE Technology evolution to Cellular Range Telemetry

This paper explains the concepts of handover in an LTE network and covers the design challenges for performing a seamless handover at the extremely high aircraft speeds in a telemetry application. It is assumed that the reader has a basic understanding of the 3GPP LTE architecture and network elements.

3GPP LTE Handover Overview

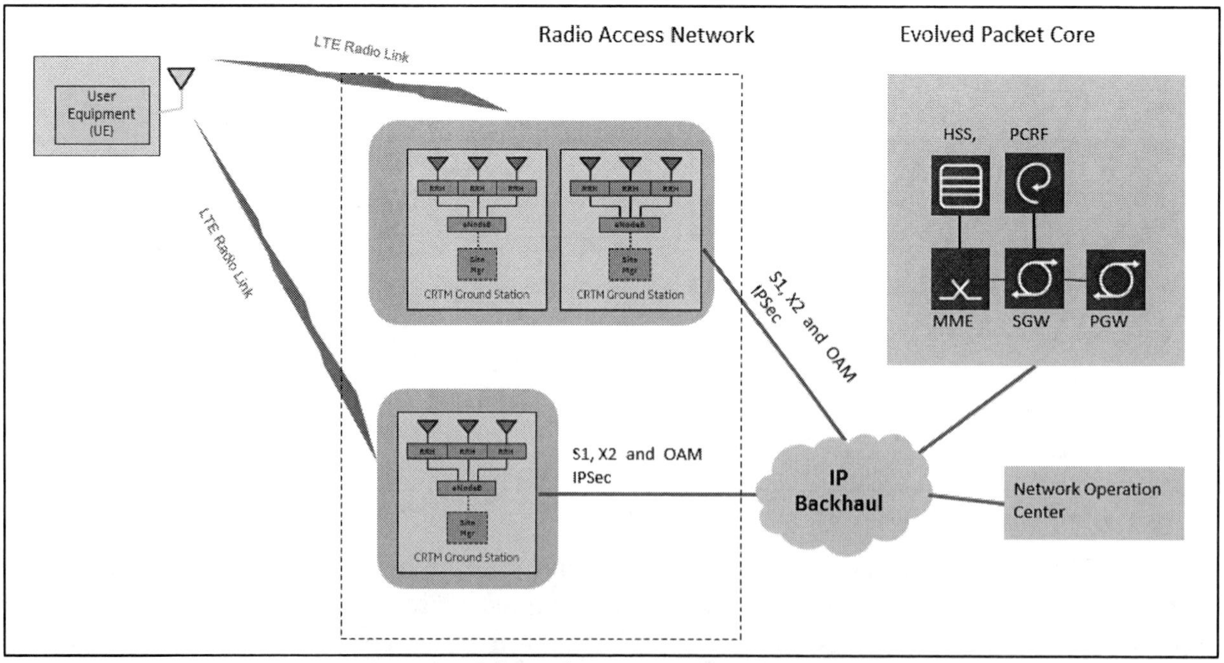

Figure 3: High Level LTE Network Architecture

Shown in figure above, a typical LTE network is comprised of:

- Radio Access Network (RAN): The RAN provides the radio link data and management functionality of the cellular network. It consists of Ground Stations, known as eNodeB in LTE terminology, that provide the RF coverage for a geographical area (Cell). The eNodeBs are connected to the LTE Evolved Packet Core Network (S1 & OAM interface) and to other eNodeBs in the RAN (X2 interface).
- Evolved Packet Core (EPC): The EPC provides the management and control of the radio link and the data path between the UE and the network operation center.
- User Equipment (UE): The UE is at the other end of the radio link in a cellular network. It provides the radio data link and IP network layer services to end user applications. For CRTM, the airborne modem in the test article represents the UE entity of a LTE network.

Handover is a term used in the 3GPP standards to define the procedure used by the LTE network to provide continuous and seamless data service as the UE moves through the different cells of the network. As the UE moves from the coverage of one cell to another cell, the handover procedure is initiated to maintain seamless mobility. The basic objectives of handover procedures are:

- Quality of Service (QoS) should be maintained all the time, both during and after handover.
- UE should able to continue its normal services before and after handover. For example, a voice call before handover should be maintained after handover as well.
- Data integrity should be maintained through handover.

In a LTE network, the handover decisions are made by the network (eNodeB) using measurement information provided UE. The UE receives a measurement configuration on establishing and radio connection with a cell (serving cell) and based on this configuration makes periodic measurement of the Received Signal Strength and/or Quality of the serving cell's and neighboring cell's radio signals. There are different types of measurement configurations that define the radio conditions that will trigger a measurement report from the UE. For e.g. the configuration could be set to report if:

- The serving cell's signal is lower than an operator configured threshold
- The serving cell's signal is lower than a threshold and a neighbor cell's signal is better than a threshold
- The neighbor cell's signal is better than the serving cell's signal by a configured threshold.

The eNodeB based on the measurement reports from the UE then makes handover decisions:

- whether to initiate a handover to a neighbor cell (target cell) and
- to which target cell should the handover be made if there are more than one good candidate target cells.

The selection of the measurement configuration type and thresholds are based on the radio network planning and desired performance of the operator. Poor selection of the measurement configuration can result in too early or too late handoffs. The end outcome would be dropped connections or longer delays in establishing service.

There are different types of mobility scenarios supported within a LTE network, mainly:

- Intra-eNodeB Handover
- Inter-eNodeB Handover

A) Intra-eNodeB Handover

When the UE moves between the coverage areas of two cells that are served by the same eNodeB, an intra-eNodeB handover is performed. For such handovers, there is little signaling between the eNodeB and the core network (MME, Serving Gateway and Packet Gateway). Shown below is the message sequence flow for a CRTM scenario where the test article (UE) moves across cells that are served by the same ground station. Refer to [1] & [2] for further procedure and message details.

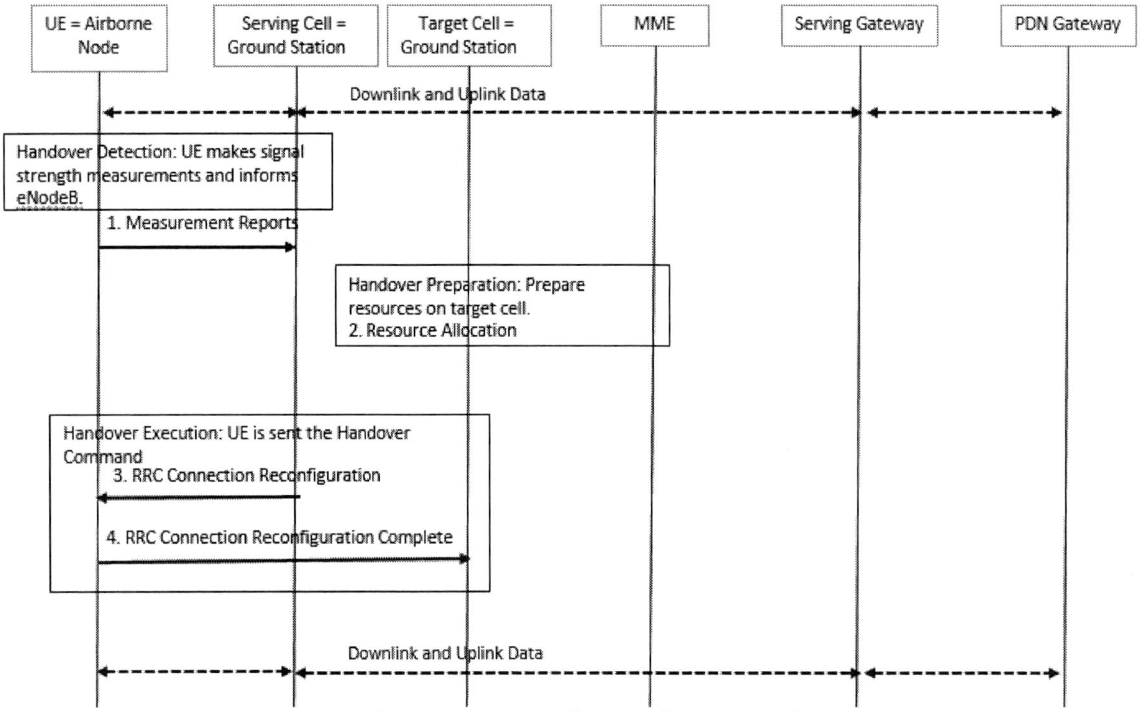

Figure 4: Intra eNodeB Handover Procedure

1) UE makes serving and target cells signal strength measurement and the measurement report is sent.
2) The serving cell eNodeB determines the best handover target cell and in this scenario, it is another cell served by the same eNodeB. It prepares the resources on the target cell.
3) The serving cell sends to the UE the handover command. The handover command points the UE to the target cell and the radio configuration it is assigned on the target. From this point the data exchange with UE is paused.
4) The UE connects to the target cell and acknowledges the handover.
5) Data exchange is resumed.

Data exchange between UE and eNodeB is paused between steps 3 & 4. Data gets buffered at the eNodeB and UE for this period and the transmission of the buffered and new data begins once the service is resumed at step 5. The data service interruption duration is around 30 to 50ms.

B) Inter-eNodeB Handover

When the UE moves between the coverage areas of two cells that are served by the different eNodeB, an inter-eNodeB handover is performed. The LTE core network needs to be involved for such handovers and there can be multiple variants depending on the Operator's core network deployment and which entities within the core network are involved. For CRTM, the LTE core network does not have a significant impact. Shown below is the message sequence flow for an inter-eNodeB handover scenario where the test article (UE) moves across cells that are served by the different ground stations with no need of switching data paths in Serving Gateways. Refer to [1], [2], [3] & [4] for further procedure and message details.

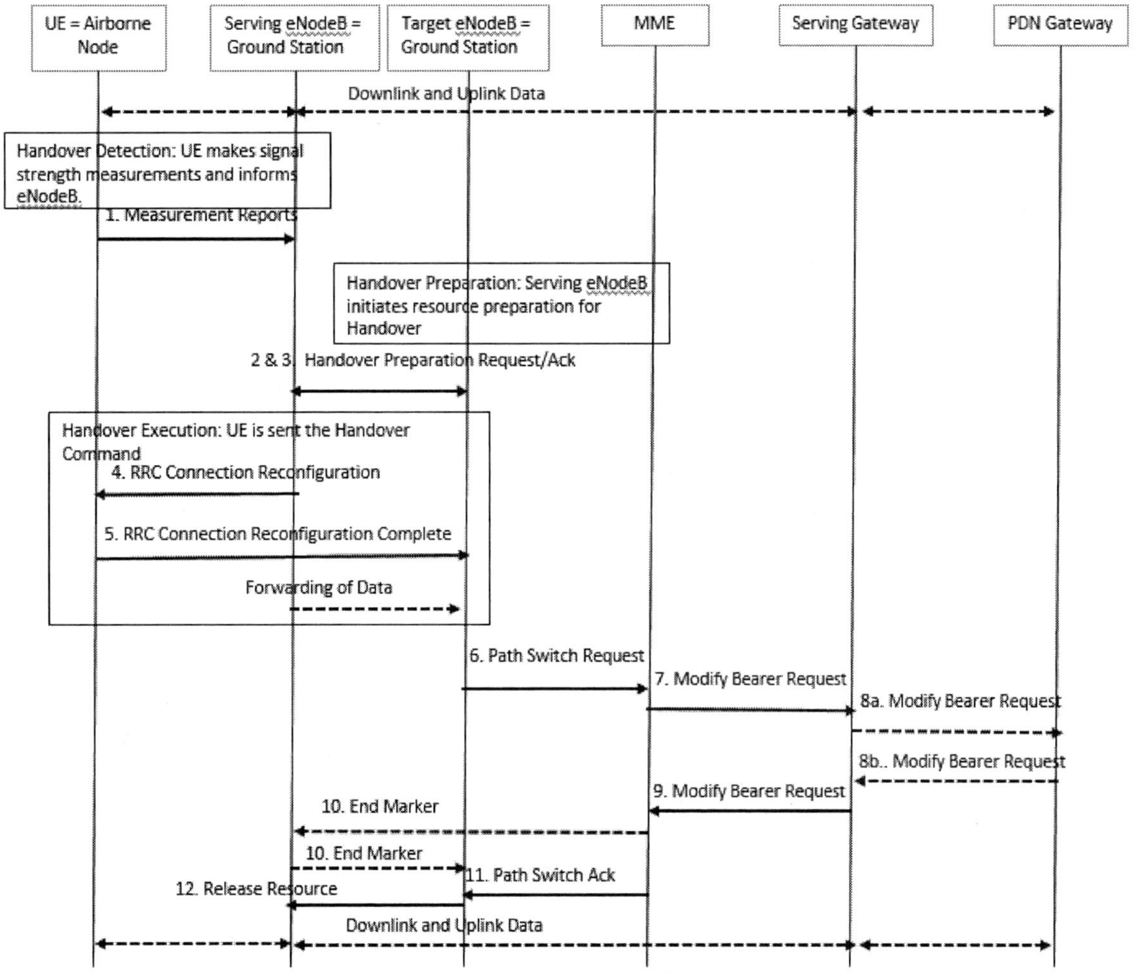

Figure 5: Inter eNodeB Handover Procedure

1) UE makes serving and target cells signal strength measurement and the measurement report is sent.
2) The serving cell eNodeB determines the best handover target cell and in this scenario, it is another cell on a neighbor eNodeB. The source eNodeB initiates a Handover Request.
3) The target eNodeB prepares the resources on the target cell and sends an acknowledge to the source eNodeB. The acknowledge message contains the handover command.

4) The serving cell sends to the UE the handover command. The handover command points the UE to the target cell and the radio configuration it is assigned on the target. From this point the data exchange with UE is paused. Any unsent data on the source cell and any data for the UE received from the serving gateway is forwarded to the target eNodeB.
5) The UE connects to the target cell and acknowledges the handover.
6) The target eNodeB informs the MME to switch the downlink Path from the serving eNodeB to the target.
7) The MME sends a Modify Bearer Request to the Serving Gateway the path modification information.
8) The steps a) & b) are conditional. It depends on whether the Serving Gateway has received the User Location Information IE and it needs to inform the PDN Gateway
9) The Serving Gateway starts sending downlink packets to the target eNodeB using the newly received path information. A Modify Bearer Response message is sent back to the MME.
10) In order to assist the reordering function in the target eNodeB, the Serving GW shall send one or more "end marker" packets on the old path immediately after switching the path.
11) The MME confirms the Path Switch Request message with the Path Switch Request Ack message. At this point the Handover is considered completed.
12) By sending Release Resource the target eNodeB informs success of the handover to source eNodeB and triggers the release of resources.

Data exchange between UE and eNodeB is paused between steps 4 & 11. Any received data by the source eNodeB is forwarded to the target eNodeB. Data gets buffered at the target eNodeB and UE for this period and the transmission of the buffered and new data begins once the service is resumed at step 11. The data service interruption duration is around 50 to 70 ms.

Challenges for Handover at High Speed

Commercial LTE is built around applications for terrestrial use, UEs move at speeds up to 500 km/hr. UEs typically have sufficient time and capability to perform handover measurements at these speeds and complete a handover. Also, commercial applications (VoLTE, FTP, Web browsing etc.) are more tolerant to brief interruptions in data connectivity. However, as the data is streaming from the test article continuously in a telemetry application and is sensitive to latency, longer pauses during handover or due to handover failures are of a bigger concern. The paragraphs below address the key challenges for performing successful handovers and minimizing the interruptions to data service.

A) UE Neighbor Cell Measurements

- Accuracy of neighbor cell measurements depends on the UE's capability to sync to candidate neighbor cells.
- In a conventional LTE system, a neighboring cell may be a few hundred Hz offset from the serving cell due to Doppler shift from UE motion. Since this is a small fraction of the 15 kHz sub-carrier spacing, the impact of this frequency misalignment on neighbor cell

measurement quality (accuracy) is generally acceptable and the same frequency correction used for the serving cell is general used for the neighbor cells.

- At very high speeds and at higher frequencies, the signals will experience a Doppler shift significantly higher than terrestrial LTE. For e.g. a typical Doppler shift would be +/- 250 Hz for a terrestrial UE, but for CRTM operating at C-Band frequencies and at speeds of up to 2400 km/hr the Doppler shift can be +/- 8 kHz. This means that the UE, which is tracking the Doppler shift relative to the serving cell, will see a significant Doppler shift on signals received from each neighbor cell. This shift can represent a significant fraction of the 15 kHz sub-carrier spacing. Unless independent Doppler correction is provided for each cell, the resulting inter-subcarrier interference degrades the cell measurement accuracy.

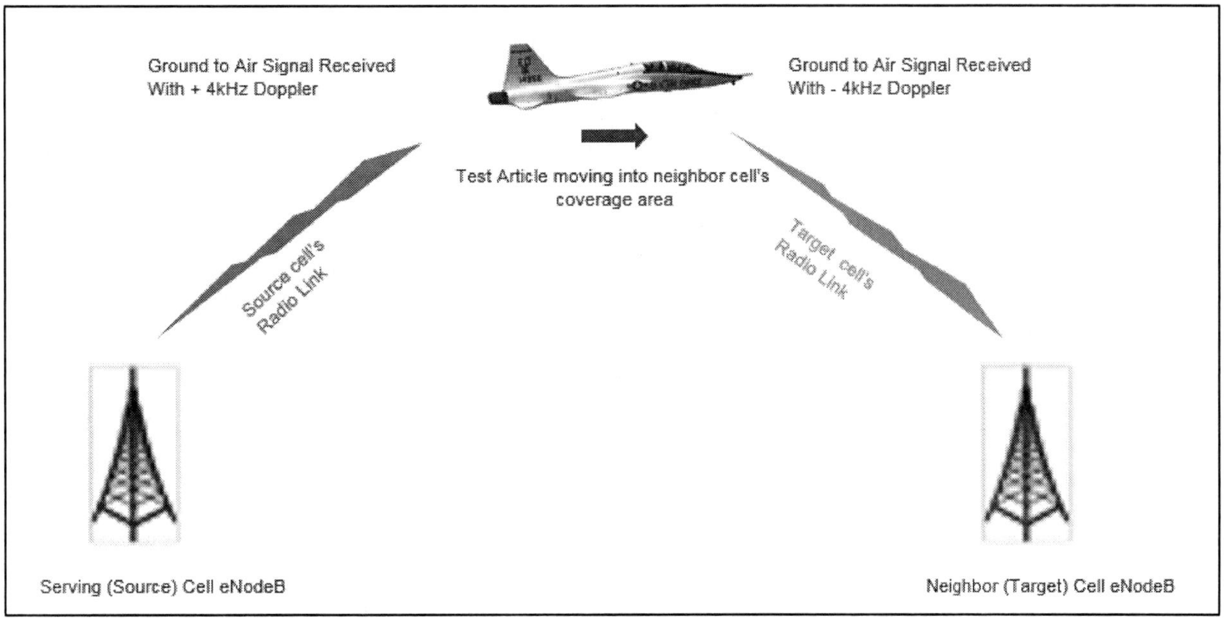

Figure 6: Doppler Shifts between Serving and Neighbor cell signals

- Inaccurate measurements can either delay the trigger for Handover or lead to selection of a wrong target cell for the handover. This will result in loss of connectivity and thus cause longer pauses in data exchange.
- In order to mitigate the high Doppler offset difference between the serving and measured neighbor cell, UE design changes are necessary for the neighbor cell measurements.

B) Timing Advance

- Once the UE is given the command to leave the serving cell and connect to the target cell (Step # 3 of Intra- and Step # 4 of Inter-eNodeB handover), the UE attempts to make a connection with the target cell. To do this, it selects an initial timing advance value when sending its first transmission (Random Access Connection Request). Due to high speeds and the possibility of having cells larger than 100 km in radius, the initial timing advance computation is a challenge in a CRTM deployment.
- Incorrect Timing Advance will result in the target eNodeB being unable to detect the Random Access Connection Request, which will eventually lead to a handover failure.
- Design changes will be needed at the UE to select an initial timing offset that it can use to adjust the timing of the Random Access Request transmission.

C) Handover Processing Time

- Handover processing for commercial LTE systems is designed for supporting speeds up to 500 km/hr. The time between the UE detecting that a measurement threshold has been met to the point where the serving cell informs the UE to connect to the specified target cell is typically in the order of 100-150ms.
- For CRTM, the test article is moving at very high speeds, and this limits the amount of time available for a system to do a handover. A very fast UE can move out of the coverage area of the serving cell and into the coverage area of the neighbor cell very quickly. Taking too long to initiate a handover can result in a handover failure. This in turn would result in call drops and a longer interruption in data service.

CONCLUSIONS

This paper shows the major challenges for extending 3GPP LTE seamless handover in terrestrial applications to the high speeds of a telemetry application. Enhancements are needed to the eNodeB and UE design to handle the doppler shifts caused by high speeds and higher operating frequencies and to the timing advance and handover processing algorithms on the eNodeB and UE.

Nokia is building upon its commercial Air to Ground LTE solution to address these challenges for Cellular Range Telemetry. ([5][6][7])

REFERENCES

[1] 3GPP TR 24.301: "Non-Access-Stratum (NAS) protocol for Evolved Packet System (EPS)".

[2] 3GPP TS 36.331: "Evolved Universal Terrestrial Radio Access (E-UTRA); Radio Resource Control (RRC) protocol specification".

[3] 3GPP TS 36.413: "Evolved Universal Terrestrial Radio Access Network (E-UTRAN); S1 Application Protocol (S1AP)".

[4] 3GPP TS 36.423: "Evolved Universal Terrestrial Radio Access Network (E-UTRAN); X2 Application Protocol (X2AP)".

[5] Apparatus, method and computer program for determining a frequency offset, Nokia Patent European Patent # EP2490389A1

[6] Access to a wireless communications network by a transceiver equipment adjusting the timing advance value used for sending a Random Access Channel preamble to a base station broadcasting cell radius information, Nokia European Patent #EP2427018A1

[7] Method for random access to a wireless or mobile communication network, and corresponding transceiver equipment, Nokia European Patent #EP2408353A1

PERFORMANCE OF THE EXTENDED TIME DIVERSITY TECHNIQUE IN REAL FLIGHT TEST CONDITIONS

Grégory Blanc, Alexandre Skrzypczak, and Jean-Guy Pierozak

ZODIAC AEROSPACE - ZODIAC DATA SYSTEMS,
5 esplanade Anton Philips - F - 14460 Colombelles - France

{alexandre.skrzypczak, gregory.blanc, jeanguy.pierozak}@zodiacaerospace.com
http://www.zodiacaerospace.com

ABSTRACT

The emission of the telemetry signal is required over minimum two different antennas to keep the telemetry link available during a maneuver of a flying object. If nothing is made at the transmitter side, the telemetry link can be fully lost as both signals may have an opposite phase. This is the well-known "2-antennas problem", also known as the "porcupine effect". In 2016, Zodiac Data Systems (ZDS) presented a pragmatic and cost effective concept named eXtended Time Diversity (XTD) which is dedicated to combat the porcupine effect. The efficiency of this concept was demonstrated through lab tests, as well as its robustness in presence of multipath. The goal of this paper is to present the performance of the Extended Time Diversity in real conditions, after a series of flight tests.

Keywords: eXtended Time Diversity, "porcupine effect", real flight test

INTRODUCTION

This article describes all tests performed at ZDS Aquitaine facility on a telemetry link during a helicopter flight to evaluate the performance of eXtended Time Diversity in real conditions for both ARTM Tier 1 (SOQPSK) and ARTM Tier 2 (MH-CPM) by involving side-by-side comparisons of traditional two-antenna transmissions and, eXtended Time Diversity (XTD) transmissions. The objective is to finalize the Proof-of-Concept of XTD to mitigate porcupine effect.

Besides, all transmissions during the real flight have been recorded, so they can be replayed to reproduce the real dynamics of the missions while being in a laboratory. Currently, not all results have been analyzed. This article will focus on the description of the mission setup. Final results will be given during the presentation.

TEST CONFIGURATION

To eliminate aircraft location and antenna placement as a variable in the comparison/performance evaluation, the airborne system should be able to produce simultaneous XTD and traditional two-antenna transmissions, in two different frequencies.
The airborne platform is an "Alouette" model helicopter.
Transmitters are two XTD prototypes from Emhiser Research Inc which is a S-band only model that comes with a limited power per output which has been calibrated (10dBm). On the transmitters, the change of parameters like Differential Encoding and LDPC code, or Modulation Scheme (SOQPSK or ARTM CPM) can be performed by a simple ASCII command on a serial link.
They are a single–data, dual-output transmitters able to delay one channel compared to the other, thus preventing co-channel interference (two-antenna problem).
Both channels are synchronized (apart from the delay between them). It is crucial that there is not jitter between them in order to facilitate the ground processing relying on an adaptive equalizer.
This prototype is a "2" x "3" transmitter board in a metallized plastic box housing. It is battery-powered, capable of holding a charge for about three and half hours, and is able to generate an internal PN15 Pseudo-Random binary Sequence. The battery can be removed, and the transmitter may be powered with an external 28V power supply with banana plugs. The battery capacity is more than 3 hours for the transmitter consumption.

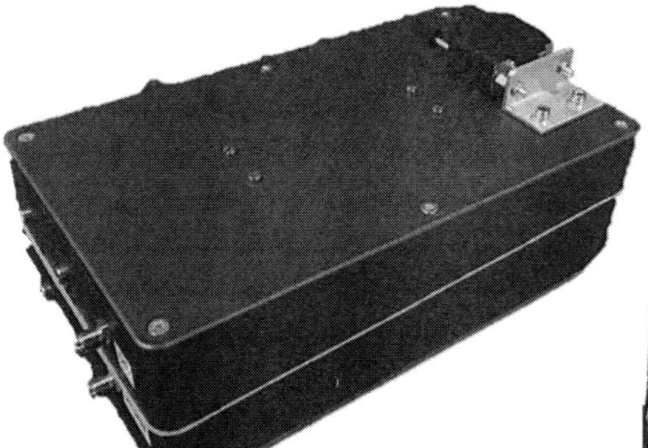

XTD Prototype Transmitter housing

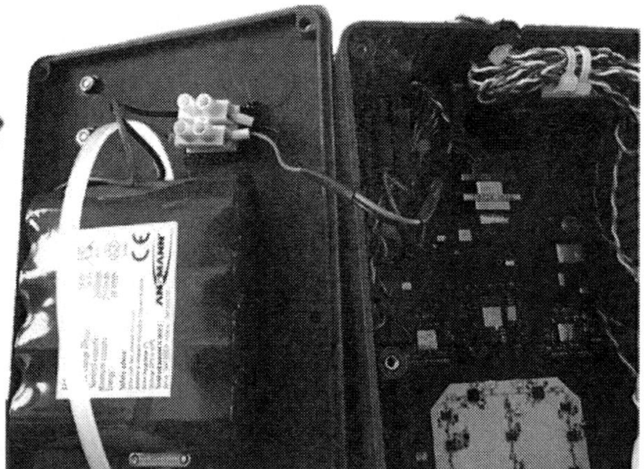

Inside

The antenna ground system is a Sparte 1.8m antenna located on ZDS Arcachon premises.
The receivers are ZDS RTR NeXT and RTR-eX receivers.

1. Airborne System Configuration

The airborne system has been implemented with antennas at 90° (as it is mostly used) as follows:

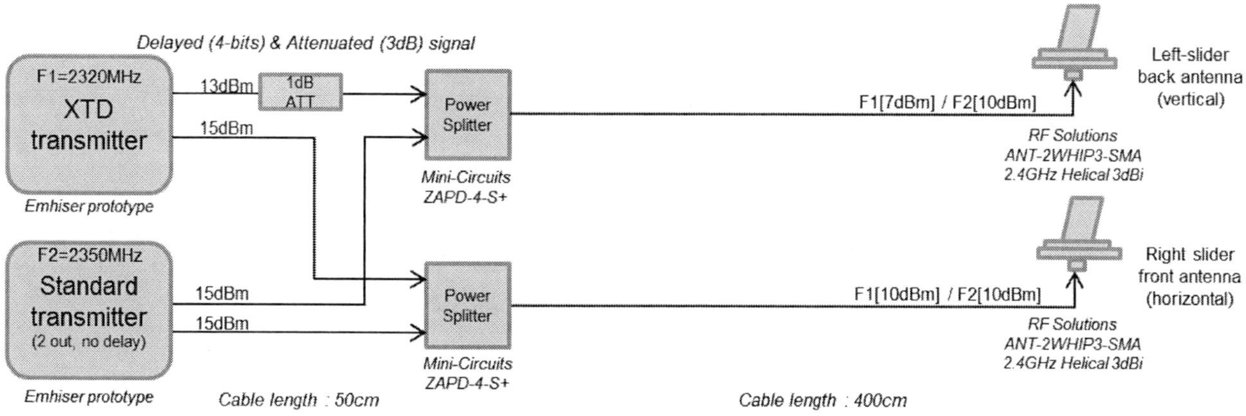

Figure 1: A block diagram of the Airborne System

Real configuration :

Figure 2 : Aircraft

Figure 3: Transmitters inside the aircraft

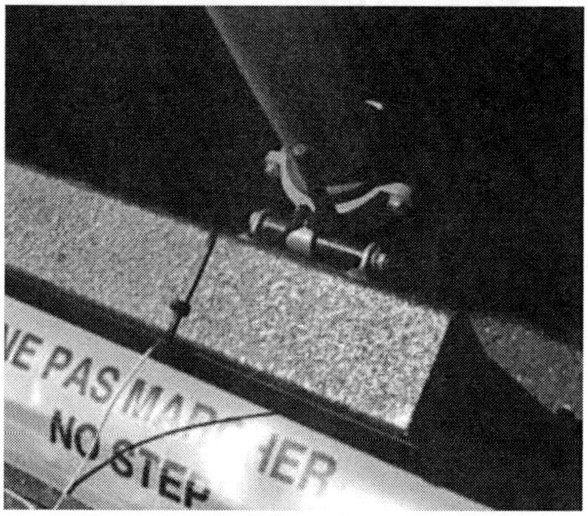

Figure 4: Left slider antenna, vertical

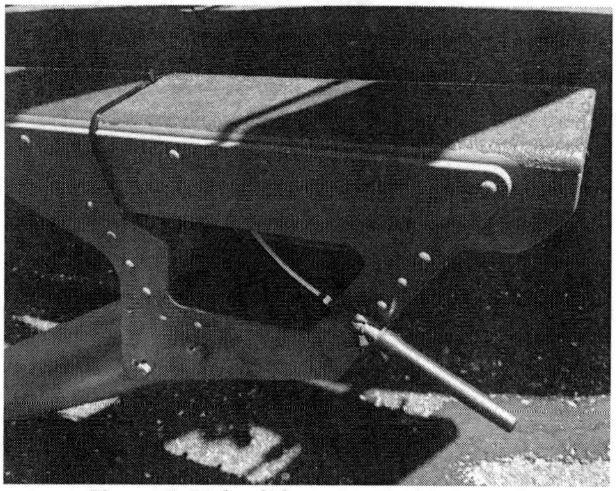

Figure 5: Right slider antenna, horizontal

2. Ground System Configuration

The ground station configuration is implemented as follows:

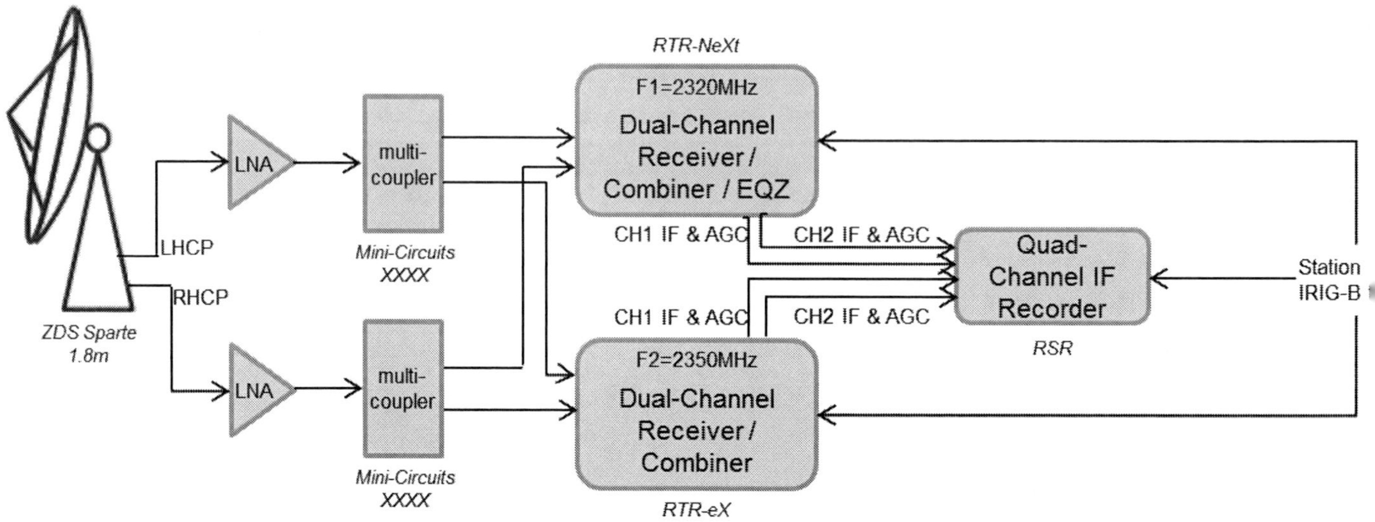

Figure 6: A block diagram of the Ground System

Real configuration :

Figure 7: Shelter and Antenna

Figure 7: Sparte 1.8m

Figure 8: Antenna Rack => multicouplers

Figure 9: Top/RTR with EQZ - Bottom/Reference RTR

Figure 10: RSR Quad Channel IF recorder

3. Missions configurations

Three missions were conducted: SOQPSK-TG, ARTM CPM and SOQPSK-TG + LDPC 4k 2/3.

3.1. Set 1: SOQPSK-TG

Parameter	**XTD Link**	**Reference Link**
Data	PN15 (4095-bit length)	PN15 (4095-bit length)
Data Rate	5 Mbits/s	5 Mbits/s
Delay	4-bit	
Attenuation of delayed signal	3 dB	
Modulation	SOQPSK-TG	SOQPSK-TG
Differential encoding	Enabled	Enabled
LDPC	Disabled	Disabled
Carrier Frequency Band	2320MHz	2350MHz
Power per transmit antenna	7dBm / 10dBm	10dBm / 10dBm

3.2. Set 2: ARTM CPM

Parameter	**XTD Link**	**Reference Link**
Data	PN15 (4095-bit length)	PN15 (4095-bit length)
Data Rate	5 Mbits/s	5 Mbits/s
Delay	4-bit	
Attenuation of delayed signal	3 dB	
Modulation	ARTM CPM	ARTM CPM
Differential encoding	N/A	N/A
LDPC	N/A	N/A
Carrier Frequency Band	2320MHz	2350MHz
Power per transmit antenna	7dBm / 10dBm	10dBm / 10dBm

3.3. Set 3: SOQPSK-TG + LDPC 4K 2/3

Parameter	**XTD Link**	**Reference Link**
Data	PN15 (4095-bit length)	PN15 (4095-bit length)
Data Rate	5 Mbits/s	5 Mbits/s
Delay	4-bit	
Attenuation of delayed signal	3 dB	
Modulation	SOQPSK-TG	SOQPSK-TG
Differential encoding	Disabled	Disabled
LDPC	Enabled (4k 2/3)	Enabled (4k 2/3)
Carrier Frequency Band	2320MHz	2350MHz
Power per transmit antenna	7dBm / 10dBm	10dBm / 10dBm

The attenuation of the delayed signal was tuned to 2dB on the transmitter, but an additional 1dB attenuator has been added at the output of the delayed signal to improve the XTD behavior.

4. Flight profile

For each mission, the same flight profile has been used (cf. Figure 12 Planned Flight Profile & Figure 13 Planned Flight Profile)

Sequence	Evolution	Aircraft speed			Aircraft altitude		Distance % antenna	Target	Target scrolling	Target	Flight duration
		(km/h)	(m/s)	(knot)	(meters)	(feet)	(meters)	(degres)	(°/s)	(°/s2)	(mn)
PR#0	Transmitter configuration										
Tango / 70	Triangle	70	19		500	1 500	1 000	27	1,0		
Tango / 150	Triangle	150	42		500	1 500	1 000	27	2,1		
Sierra / 70	S shape trajectory	70	19		500	1 500	1 000	27	1,0		
Sierra / 150	S shape trajectory	150	42		500	1 500	1 000	27	2,1		
Romeo / PR#3	Spinning	0	0		500	1 500	1 000	27	-		
Charlie / 300 ft	Small circle over buildings	70	19		100	300	400	14	2,7		
Delta Victor / PR#1	Take-off / landing vertical				500	1 500					
Delta Alpha / PR#1	Take-off / landing like an airplane				500	1 500					
PR#0	Transmitter re-configuration										
TOTAL											-

Figure 11: Planned flight profile description

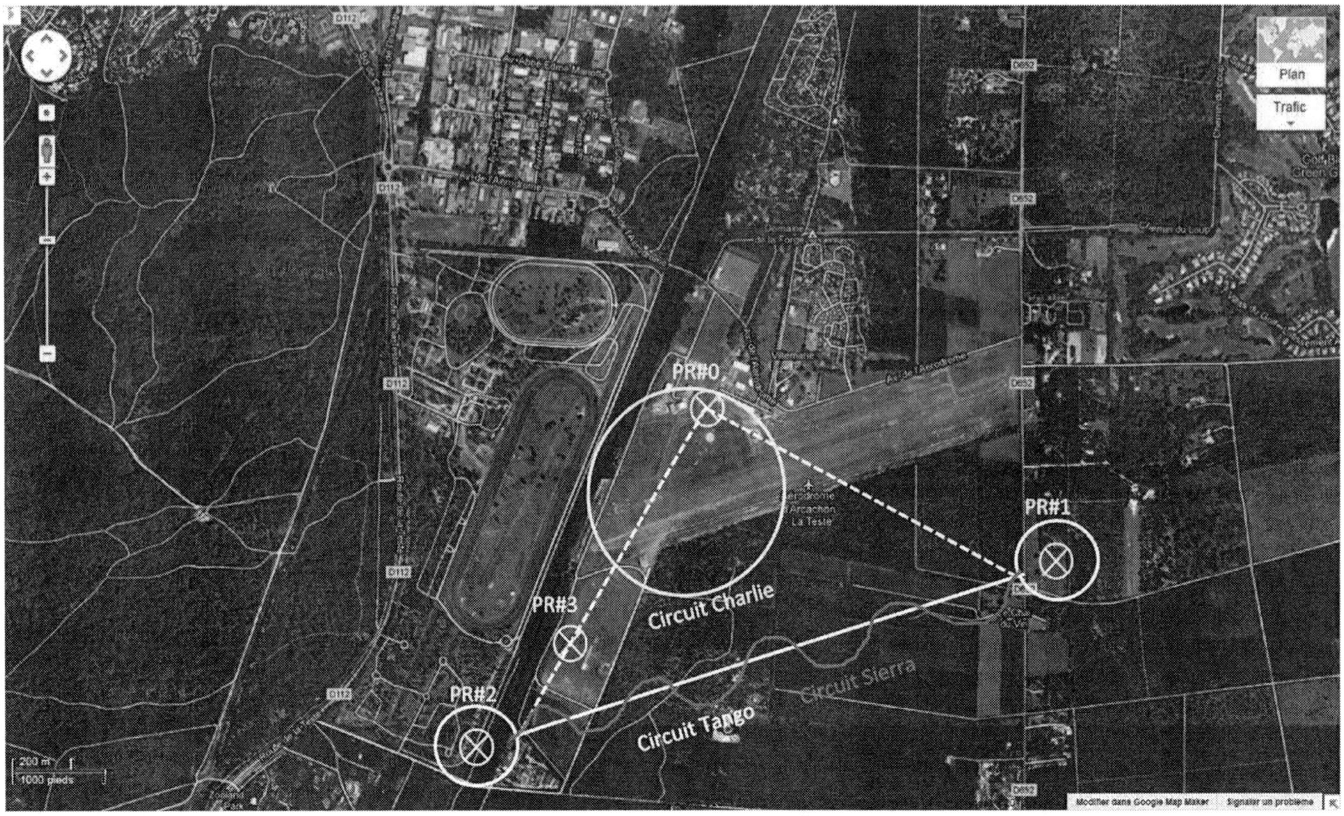

Figure 13 : Planned flight profile

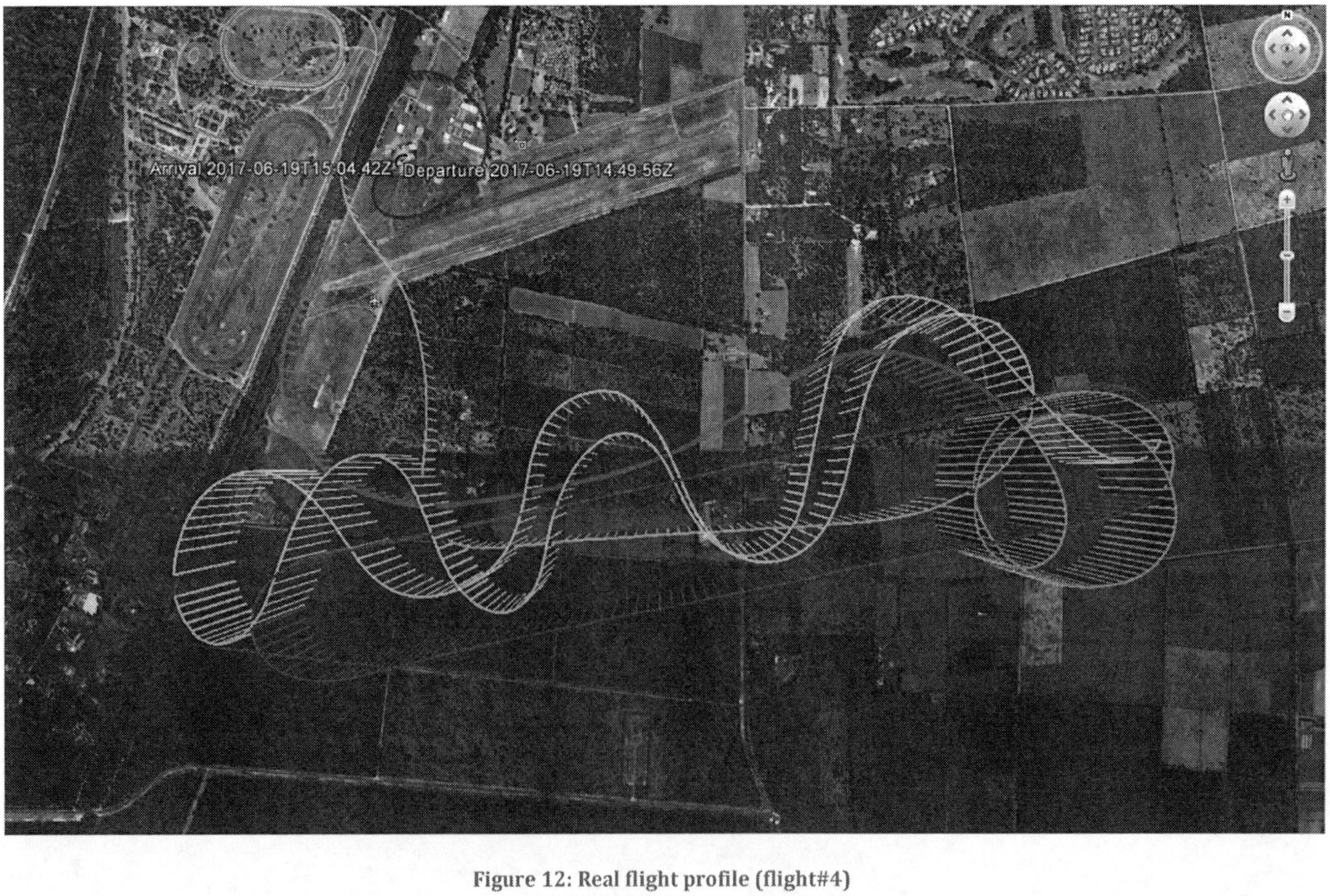

Figure 12: Real flight profile (flight#4)

4.1. *Sequences descriptions*

The various sequences of the flight profiles will give different conditions for both porcupine effect and multipath:

- PR#0 : landing for transmitter configuration – strong signal - multipath effect, enhanced by the blades
- Tango sequence : straight lines at two different speeds – different antenna look-angles - no ground multipath effect (high altitude)
- Sierra sequence : S-shapes lines at two different speeds – alternate visibility between transmitting antenna – no ground multipath (high altitude)
- PR#1 and PR#2 : strong U-turn – high dynamic change - no ground multipath
- PR#3 : spinning : alternate visibility between transmitting antenna – no ground multipath (high altitude)
- Delta Victor and Delta Alpha : different profiles of take-offs and landings – strong ground multipath

4.2. *Flights*

Three flights were made:

- Flight #1: set 1 (SOQPSK-TG) – 3dB attenuation on the XTD delayed channel
  An 1dB attenuator has been inserted at the delayed output of the XTD transmitter.
- Flight#2: set 2 (ARTM-CPM) - 3dB attenuation on the XTD delayed channel
  Frequencies are back to Set 1
- Flight#3: set 3 (SOQPSK-TG+LDPC) - 3dB attenuation on the XTD delayed channel
  It has to be noticed that the reference receiver had no LDPC option, so it was not possible to have the real time BER for the reference transmission.

4.3. *Measurements*

Main objective was to record the different flights, and for each receiver, display real time AGC for each RF channel and display BER (on PN15) on the combined channel.

Flight number	RSR file	Start record	Stop record	Duration
#1: set 1 (SOQPSK-TG) – 3dB	XTD-ARC_19062017_135825_142529	170-13:58:26	170-14:25:29	27'03"
#2: set 2 (ARTM-CPM) 3dB	XTD-ARC_19062017_144820_150416	170-14:48:20	170-15:04:16	15'56"
#3: set 3 (SOQPSK-TG+LDPC - 3dB	XTD-ARC_19062017_151011_152317	170-15:10:11	170-15:23:17	13'06"

RESULTS

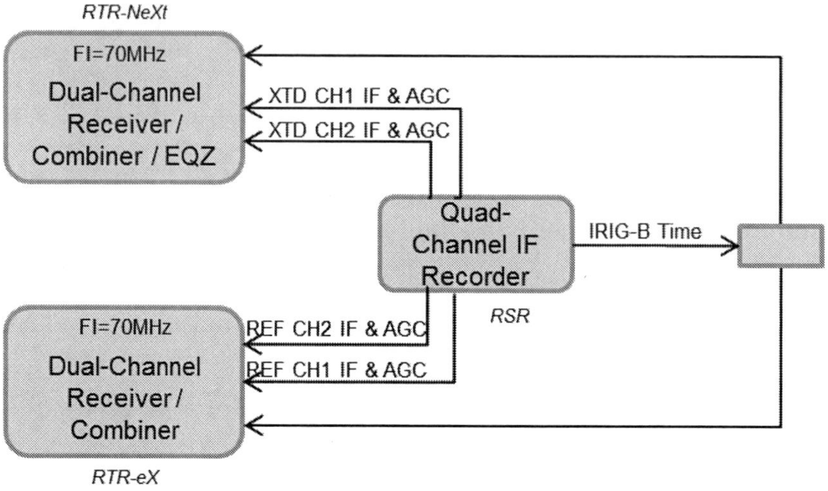

The replay setting allows reproducing of the recorded IF signal, with the benefit of AGC information for the Pre-D combiner, and IRIG-B Time for synchronizing the receivers.

As mentioned in introduction, the results are not yet completely analyzed, and will be given at the presentation. The results will focus on the comparison between the reference setup and the XTD setup on the following items:

- AGC level: the variation of the AGC level over time gives an indication of the porcupine effect. A system designed to mitigate the porcupine effect is expected make the AGC level more stable than the reference system.
- BER: a system designed to mitigate the porcupine effect shall not degrade the bit-error rate compared to the reference system.

ENERGY EFFICIENT CELLULAR NETWORK USING ADAPTIVE USER CLUSTERING ALGORITHM FOR SPARSELY POPULATED AREA

Hailu B Kassa

Doctorate of Engineering Candidate, Department of Electrical and Computer Engineering, Morgan State University, Baltimore, USA, hailu.kassa@morgan.edu

Faculty Advisor: Farzad Moazzami, Yacob Astatke

ABSTRACT

In this paper, an algorithm for distance aware energy efficient Base Stations (BSs) is proposed, which exploits the knowledge of the distance between the Mobile User (MUs) and the BS. The proposed algorithm changes the cell radius depending on user distribution in the cell. In this case adaptive (dynamically changing) concentric circles are virtually drawn so that the UEs in the same concentric circle can get constant power from the BS. It means that the mobile users(MUs) are clustered based on the distance from the BS. The energy consumption has been evaluated without sacrificing significant offered Quality of Service (QoS) on the cellular networks. The proposed scheme aids to achieve energy saving as a result of reduced transmit power based on reducing the radius of the concentric. As the radius reduces the transmit power is also reduced logarithmically. The system uses Omni directional antenna which covers all 360 degree of the cell at once. The simulation result shows that as the distance from the BS decreases, the transmission power decreases and the energy also decreases. The distance or the radius variation dynamically is based on the number of users in that track comparing to the minimum required number of threshold users. The result shows that an average of 0.1762 dB or 1.04 Watt which is 9.45% of the maximum transmit power can be saved by dynamically varying the base station the radius at the area of low traffic load for a single scan.

INTRODUCTION

Recent analysis by manufacturers and network operators has shown that current wireless networks are not very energy efficient, particularly the base stations by which terminals access services from the network. Green radio technology describes one of the most promising research directions in reducing the energy consumption as well as the carbon emissions of future base stations. Given the worldwide growth in the number of mobile subscribers, the move to higher-data-rate mobile broadband, and the increasing contribution of information technology to the overall energy consumption of the world, there is a need on environmental grounds to reduce the energy requirements of radio access networks [3]. The Green Radio program sets the aspiration of achieving a hundredfold reduction in power consumption over current designs for wireless communication networks. This challenge is rendered nontrivial by the requirement to achieve this reduction without significantly compromising the quality of service (QoS) experienced by the network's users.

Traditional traffic engineering in cellular networks was focused on traffic measurement, characterization and control, aiming to carry the largest amount of traffic load while satisfying the required quality of service (QoS) using limited radio resources, e.g., wireless channels [4]. The new dimension of energy consumption can be similarly added into this problem, since energy is also an important type of radio resource, except that the objective now becomes to minimize energy consumption per traffic bit, for energy efficiency maximization under given QoS constraints [5]. Typically, cellular network operators allocate their radio resources based on the worst case principle, which is to allocate available resources based on usage requirements at peak traffic load [6]. This leads to significant waste of resources, including energy consumption, during periods with low traffic load [7]. To save energy, a shutdown or switch-off under-utilized BSs scheme was proposed at low traffic load conditions [8]. A method to accommodate peak traffic load with secondary (micro-, pico- or femto-) BSs that have smaller coverages was presented in [9]. Furthermore, combining these two methods was explored by, for example, employing the shut-down strategy to smaller secondary BSs in [10]. In [12], the authors explored the tradeoff between the operating power and the embodied power contained in the manufacturing process of infrastructure equipment from a life-cycle perspective. However, most of the above traffic studies in wireless networks neglect realistic transmission of a single base station based on the traffic density of virtual area clustering to divide the transmit power accordingly. To the best of our knowledge, there is no research work has been conducted using this technique to bring adaptive transmit power adjustment on a single base station based on traffic density.

The rest of the paper is organized as follows. Section II introduces the system model and in section III the power model is described in detail. In Section IV simulation results are presented based on numerical values considered. Finally, Section V concludes this paper.

SYSTEM MODEL

In this paper, energy efficiency is computed for the downlink cellular system considering the density of the MUs which is found by counting and adding down those users and dividing by their respective area to find the specific density. Counting users is performed considering channel state information as signal to noise ratio (SNR). If the specific density is less than the average density, which is the threshold density (found from the two variables: total coverage and total number of MUs), then the cell radius shrinks to a new lower radius position and the transmit power is adjusted based on this radius. The proposed system model consists of one serving base station and multiple mobile users which create point to multipoint communication. The BS transmits signal with Omni directional antenna. The total transmit power is shared by all users in the range of five kilometers. The cell is divided by a number of virtual concentric circles in a given cell using an adaptive user clustering technique. The users in the cell are randomly distributed. The system power model estimates the total power supply where our focus is on the downlink communication, i.e., from BS to MU.

The proposed algorithm, which can perform the combination of users clustering and adaptive transmit power allocation is specified below. The parameters used here are basically number of users and the cell radius. Assume R_x, R_i, Uo, Ux are radius of the concentric circle, minimum radius, number of users to be marginalized initially and the number of users between the consecutive concentric circles respectively.

1: Initialize; $R=R_{max}$ (max radius), $P=P_{max}$,Uo=0
2: Find the no. of users, Ux between Rx and R_{x-1}
3: Uo=Uo+Ux
4: Compare Uo and Th
5: **if** Uo > Th then
6: Assign power based on R_x
7: T time interval
8: Go to step 1
9: Else
10: $R_x=R_{x-1}$
11: Compare R_x and R_i
12: **if** $R_x = R_i$
13: Go to Step 6
14: **Else**
15: Go to Step 2

POWER MODEL

The problem formulation for the overall system goes through the following procedures:

- Count and add the number of users until it is greater than or equal to the threshold value (Th) from outer to inner tracks. N=N1+N2+N3+…+Nx,
- If N=0 at the beginning, then the base station sleeps (uses minimum transmit power) and no need of clustering.
- The base station wakes up immediately when the users enter to the active zone (radius)
- Find the number of users N_x at each track
- If the number of users of a track N_x at the far distance from the base station is instantly falls below the threshold value, then R decreases to Rm_{-1}(to the radius of the next concentric circle) and the new path loss will be calculated using Okumura Hatta models based on R_{m-1} such that $P_L=247.41+35.22*log(R_{m-1})$ and the new transmit power, P_T will change since it depends on the path loss effect.

$$P_T=P_R*L \quad (1)$$

Where P_T, is transmitted power, P_R, is minimum received power, and the path loss, L, from reduced Okumura Hatta propagation model can also be written as:

$$L=247.41+35.22*log(R_m), \quad (2)$$

Where $R_{m,}$, is the mobile user's distance from the BS in the mth concentric circle.

NUMERICAL RESULTS

In this part, simulation results for power minimization techniques of a base station with clustered users in the cell have been shown. All the simulations are carried out using MATLAB software. The analysis for each result is discussed in detail after different system parameter considerations. Table1 provides system parameters that are used in the simulation and their corresponding values. The values for the different simulation parameters are selected in such a way that the whole system meets some standard regulations and have practical applicability.

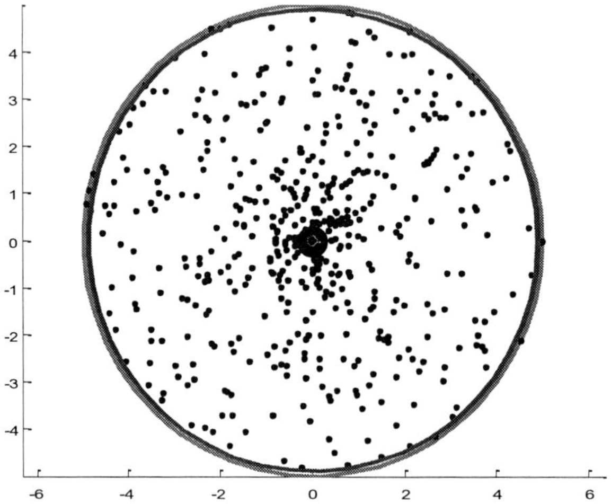

Figure.1: BS with Full Coverage

TABLE 1: SIMULATION PARAMETERS AND THEIR CORRESPONDING VALUES

Parameters	Symbol	Value (if any)
Maximum Distance	R	2km,3km,5 km
frequency	f	900 Mhz
Base station Antenna Height	h_{te}	200 meter
Mobil station Antenna Height	h_{re}	30 meter
Wave length	λ	Depends on frequency
Antenna gain at the transmitter	G_{te}	-
Speed of light	C	$3*10^8$ m/s
Antenna gain at the receiver	G_{Area}	1
Environment gain	G_{AREA}	1
Number of users	n	Adaptive
Received power	p_r	10mw
User threshold	thr	-
Minimum DSP power for each iteration	P_{dsp}	0.001
Omni directional Antenna	-	-

Figure.1 shows the random distribution of users over the cell's maximum coverage. In this case the cell radius is 5kms and the transmitted power reaches maximum. But in Figure.2, the cell is clustered in three regions depending on the number of users and minimum radius. By smoothly moving down to the center, the MEs are counted and the density is compared with the threshold users' density. The blue colored line is the position where the density of the users counted is less or equal to the threshold density and as a result the radius shrinks to 4.3 km from 5 km. Therefore, by doing this the BS saves 0.1762 dB or 1.04 Watt for the radius reduction. From Figure.1 and Figure.2, the Omnidirectional antenna transmits power for the whole of five kilometers radius. If the number of users at the last track is less than the threshold value (density), then the radius shrinks to the second outer most track as shown on Figure.2 and the transmit power is adjusted based on that reduced distance.

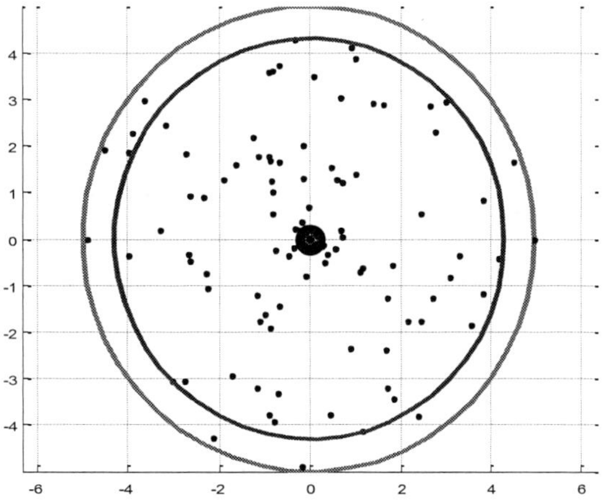

Fig.2: BS with Reduced Distance

We also compared the propagation models in Figure.3 based on the distance of randomly distributed MUs throughout the cell for the transmit power calculation as stated in equation (1). From this we have found that the Okumura Hatta model is better than the Cost 231 and Free Space for its lower power and practical feasibility. Using the Okumura Hatta Model we show the relationship between the transmit power and the distance between of MUs and the BS.

As it is shown on Figure.4, the transmit power logarithmically increases as the distance increases. Figure.5 depicts the transmit power pattern of different scan times. As the number of users and the distribution varies from time to time, the power pattern also varies with those parameters. At iteration three there are no any users between 0.5 and 2.4 kilometers and thus there is no power dissipation at this point however the middle area power is not the main concern.

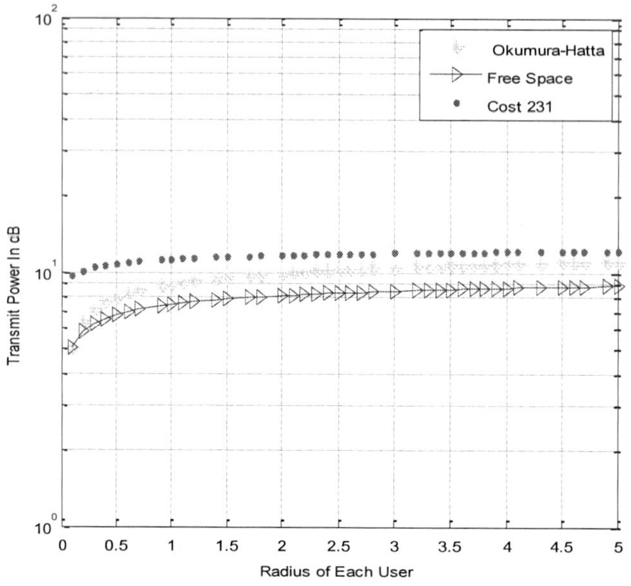

Figure.3: Transmit Power at different Propagation Models

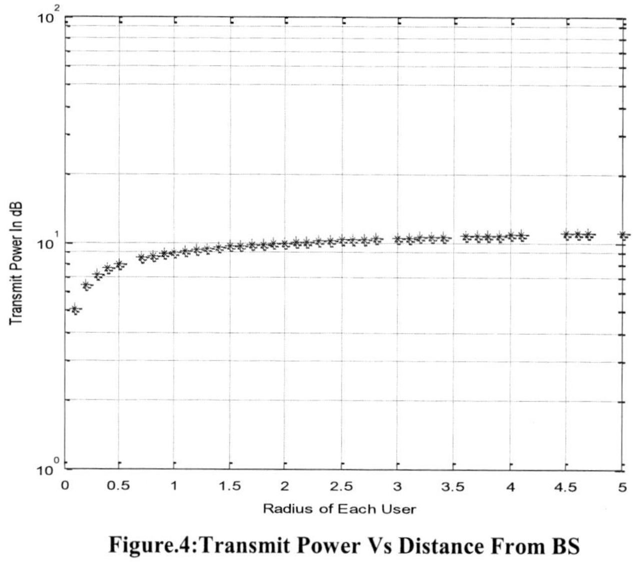

Figure.4:Transmit Power Vs Distance From BS

Here the main point is that as the minimum and maximum distances of each iteration vary the saved power also varies. From Figure 6 the bars represent, the maximum transmit power at maximum coverage, the power with respect to the reduced distance and the gain which is obtained from the difference of the two respectively. This result is of only one iteration or scan time and as the BS performs a number of scans within a second, the saved power will ultimately increase with time. Figure 6 is the graphical representation of Figure 2.

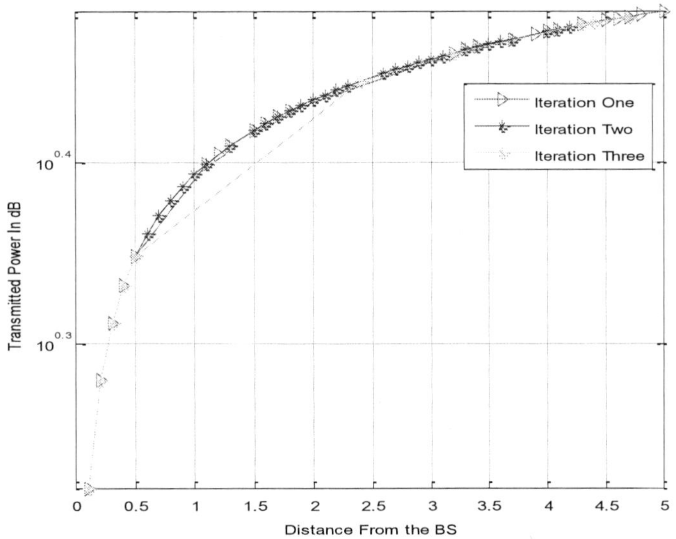

Figure.5: Transmit Power Pattern at Three iterations

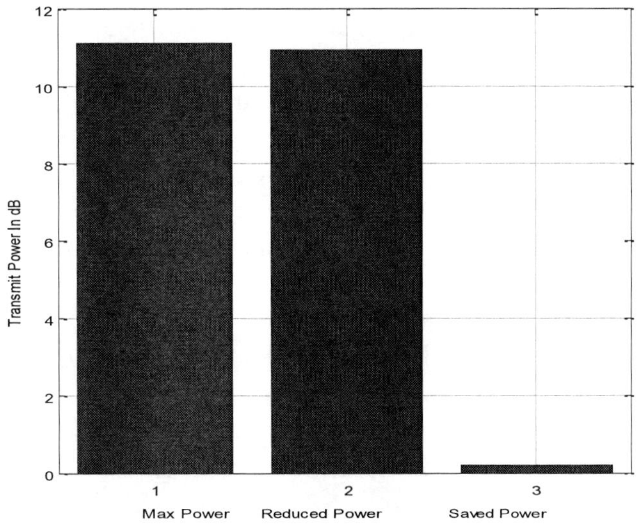

Figure.6: Saved Power for one Scan Time

CONCLUSIONS

The simulation results discussed in section IV have shown that we have studied different approaces in order to reduce the transmit power. We have shown that the energy efficiency can be achieved at the BS based on several factors, including traffic load. Dynamically adjusting the cell radius based on traffic statistics in the specific area has brought greater energy savings. Our simulations show that from a single scan the BS has saved 0.1762 dB or 1.04 Watt which is 9.45% of the maximum transmit power. As a future work, it can be extended to the technique to which the power adjustment is sensitive to only the last user's distance.

REFERENCES

[1] A. Fehske, G. Fettweis, J. Malmodin, and G. Biczok, "The global footprint of mobile communications: the ecological and economic perspective," *IEEE Commun. Mag.*, vol. 49, no.8, pp. 55–62, Aug. 2011

[2] C. Han, T. Harrold, S. Armour, I. Krikidis, *et al.*, "Green radio: radio techniques to enable energy efficient wireless networks," *IEEE Commun. Mag.*, vol. 49, no. 6, pp. 46–54, June 2011

[3] D. E. Everitt, "Traffic engineering of the radio interface for cellular mobile networks," *Proc. IEEE*, vol. 82, no.9, pp. 1371–1382, Sep. 1994

[4] H. Bogucka and A. Conti, "Degrees of freedom for energy savings in practical adaptive wireless systems," *IEEE Commun. Mag.*, vol. 49, no. 6, pp. 38–45, June 2011

[5] P. Trimintzios, G. Pavlou, P. Flegkas, P. Georgatsos, A. Asgari, and E. Mykoniati, "Service-driven traffic engineering for intra-domain quality of service management," *IEEE Netw.*, vol. 13, no. 3, pp. 29–36, May 2003

[6] E. Hwang, K. J. Kim, J. J. Son, and B. D. Choi, "The power-saving mechanism with periodic traffic indications in the IEEE 802.16e/m," *IEEE Trans. Veh. Technol.*, vol. 59, no. 1, pp. 319–334, Jan. 2010

[7] E. Oh, B. Krishnamachari, X. Liu, and Z. Niu, "Toward dynamic energy efficient operation of cellular network infrastructure," *IEEE Commun. Mag.*, vol. 49, no. 6, pp. 56–61, June 2011

[8] W. Wei and G. Shen, "Energy efficiency of heterogeneous cellular network," in *Proc. 2010 IEEE VTC – Fall*, pp. 1–5

[9] I. Ashraf, F. Boccardi, and L. Ho, "SLEEP mode techniques for small cell deployments," *IEEE Commun. Mag.*, vol. 49, no. 8, pp. 72–79, Aug. 2011

[10] R. Wang, J. Tsai, C. Maciocco, T. Y. Tai, and J. Wu, "Reducing power consumption for mobile platform via adaptive traffic coalescing," *IEEE J. Sel. Areas Commun.*, vol. 29, no. 8, pp. 1618–1629, Sep. 2011

[11] I. Humar, X. Ge, L. Xiang, J. Ho, and M. Chen, "Rethinking energy efficiency models of cellular networks with embodied energy," *IEEE Netw.*, vol. 25, no. 2, pp. 40–49, Mar.–Apr. 2011

[12] L. Xiang, F. Pantisano, R. Verdone, X. Ge, and M. Chen, "Adaptive traffic load-balancing in green cellular networks," in *Proc. 2011 IEEE PIMRC*, pp. 41–45

SPECTRUM RELOCATION FUND TRANSITION AGILITY CHALLENGE

Josef Von Niederhausern, Terry Wade, Michael Hagg,
Matthew Morgan, Greg Uhland, Alfredo Berard, Lorin Klein
96th Test Wing
Eglin AFB, FL 32542

ABSTRACT

Due to Auction 97 frequency sell off and anticipated higher usage of C-Band, space time code (STC) and low-density parity-check (LDPC) code were evaluated for latency, characterization, and performance during flight. The benefit of an STC path can be observed by measuring the channel power of a dual antenna and comparing it to the contributions from each antenna independently. The STC provides a benefit only when both antennas are visible to a receiver and when the combined result of a dual antenna system would destructively add. The Eglin Spectrum Reallocation Fund (SRF) Project transitions the United States Air Force (USAF) aeronautical mobile telemetry (AMT) from operations in frequencies auctioned by the Department of Commerce (DOC) as part of Auction 97. This paper describes the AMT test methods used and upgrades accomplished by the 96th Test Wing in order to meet requirements of Auction 97.

INTRODUCTION

The SRF requirement was two-fold, provide for a frequency agile and spectral efficient telemetry system; secondly assess the benefits to mission effectiveness of space-time code (STC) and low-density parity check (LDPC). In order to assess the impact of integrating STC and LDPC systems on Eglin, AFB, the latency overhead of the transmitter and receiver were evaluated. Historically, Eglin AFB has been primarily an L/S band range. To meet requirements of Auction 97, various tri-band antennas were surveyed for suitability. To assess the utility of STC on Eglin AFB an aircraft was modified with both an STC/LDPC capable transmitter and a baseline unencoded transmitter. Channel power measurements were taken of the aircraft in J-PRIMES using either a signal generator to provide dBi or the transmitter to provide effective isotropic radiated power (EIRP). Traditional telemetry systems are configured prior to flight. A requirement of SRF is to allow for the frequencies and/or modulation codes to be dynamically changed based on airborne real-time requirements. Finally a real-time software tool was developed to assess the utility of STC. This paper describes the test methods and resultant upgrades accomplished by the 96th Test Wing (96 TW) in order to meet requirements of Auction 97.

STC AND LDPC MEASUREMENTS AND TEST METHOD

Critical to the modernization of telemetry systems is the evaluation of latency overhead due to new encoding schemes (STC and LDPC). The STC and LDPC are not new coding technologies, the 96 TW had to come up with a standard method of evaluating transmitters from different manufacturers. A major constraint of the 96 TW was to stay within a millisecond latency. The use of STC and LDPC require overhead consisting of framing and parity information which introduce latency; the user data rate is accelerated to allow additional information, Figure 1 - Overhead Bits. Measuring latency was accomplished by generating a pattern of zeros or ones then inserting a short burst of a different pattern, see Figure 2 - Measuring Latency.

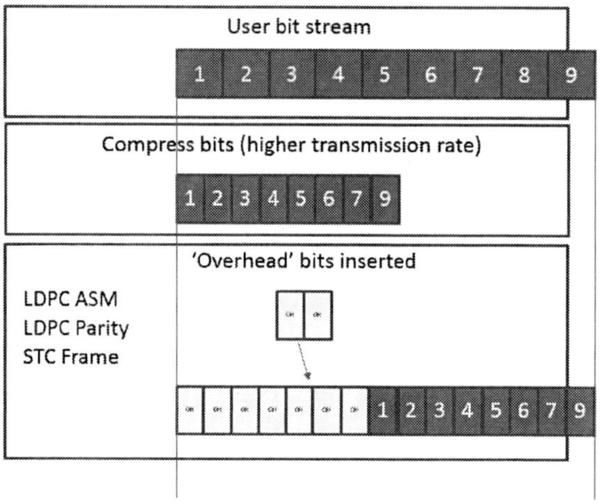

Figure 1 - Overhead Bits

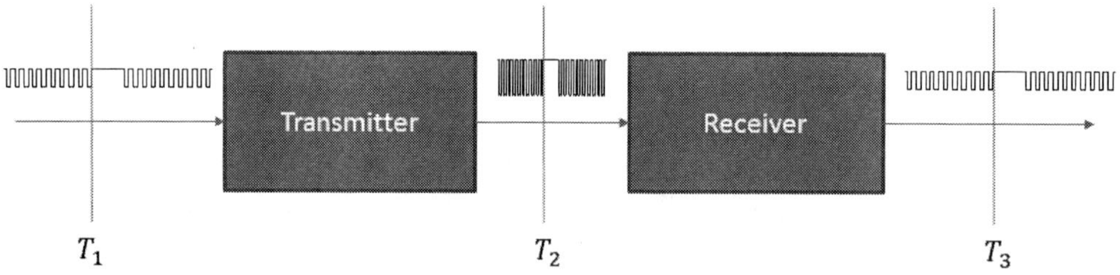

Figure 2 - Measuring Latency

System Latency L_s is given by: $L_S = T_3 - T_1$

Transmitter Latency L_t is given by: $L_T = T_2 - T_1$

A simulator was used to create a bit stream and a discrete trigger coincident with a burst of different data. A Mixed Domain Oscilloscope (MDO) was used to measure transmitter coding latencies shown below in Figure 3.

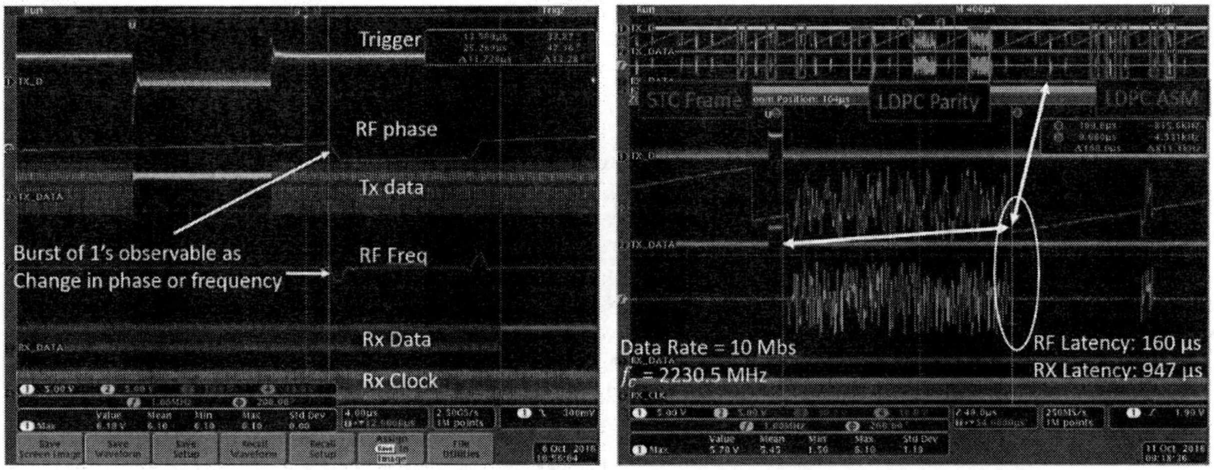

Figure 3 – Mixed Domain Oscilloscope RF Latency Measurements

The transmitter latency varied depending on where in the frame (STC or LDPC) the burst occurred. For a data rate of 10 megabits per second (Mbps), transmitter latencies at the start of the frame were 14 microseconds (µs) while at the end of the frame they approached 160 µs. From a system perspective latency the contribution of LDPC and STC coding is fixed as the receiver compensates for the transmitter packet 'jitter'. Contributing factors include data rate, coding formats, block size, and vendor implementation.

JOINT-PREFLIGHT INTEGRATION OF MUNITIONS AND ELECTRONIC SYSTEMS (J-PRIMES) TESTING

As part of the aircraft modification tri-band antennas were assessed for suitability. Antenna patterns were taken on a 3-foot diameter radial ground plane and are shown in Figure 4. Testing reflects that prior to Military Aircraft Temporary Class 2 Modification (T2), antennas should be measured to verify vendor antenna patterns across all bands.

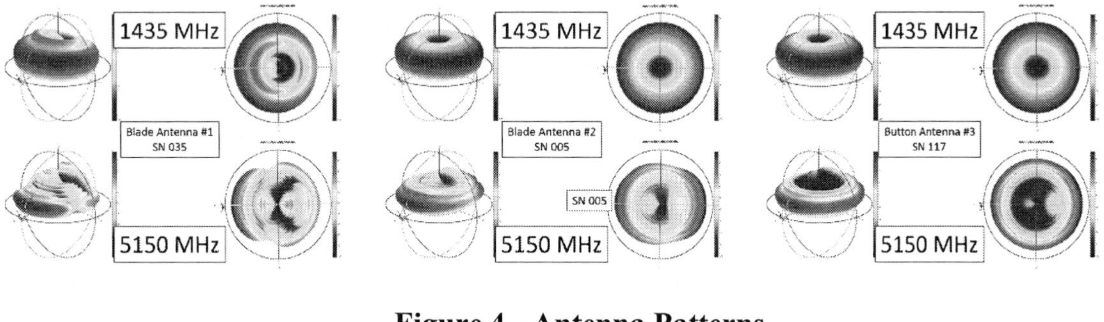

Figure 4 - Antenna Patterns

The test aircraft was placed in the J-PRIMES anechoic chamber (Figure 5) and mapped EIRP. Signal strength, polar, and azimuth readings were recorded creating vector (r, θ, φ) from the center of the rotating antenna. Aircraft measurements were taken with signal generators as well as transmitters. Figure 6 Left shows the diversity combined benefit of an STC signal; it should be noted that the power measurement for the non-STC signal is lower than anticipated by ~6 dB due to test setup. Since the two signals are orthogonal, the non-combined benefit of STC can be observed with a signal generator by first mapping the dual antenna configuration and then the upper and lower antenna independently as shown in Figure 6 – Diversity Combined STC benefit (left) and Non-Combined benefit (right).

Figure 5 - Test Aircraft in J-PRIMES

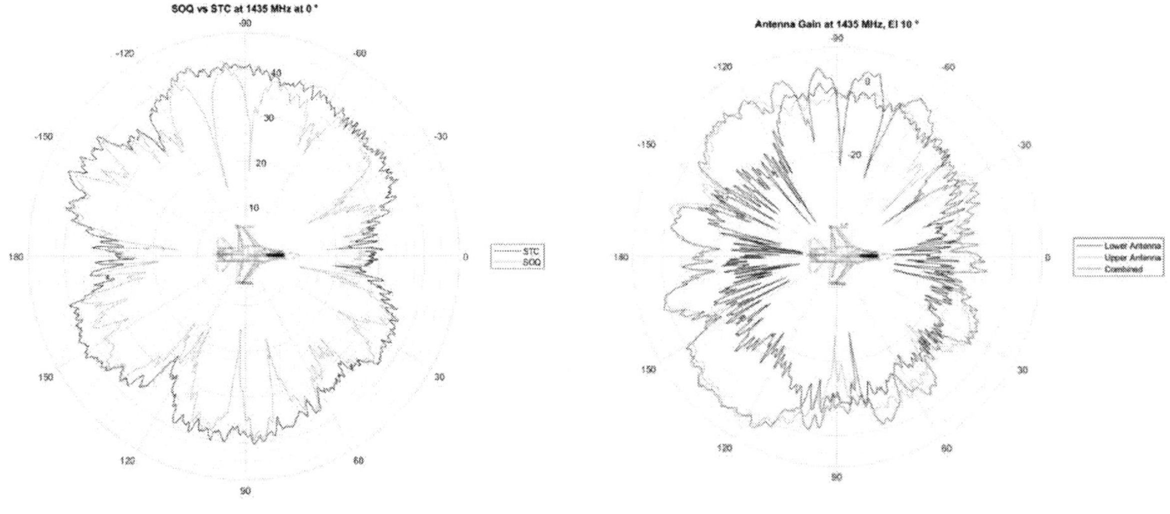

Figure 6 – Diversity Combined STC benefit (left) and Non-Combined benefit (right)

T2 MODIFICATION

The 96 TW addressed challenges to provide for spectral agility and mission effectiveness: allow the transmitter frequency to be changed in flight, resolve the "dual antenna problem" [1, 2], and increase the link margin of the system without increasing power. The airborne configuration consisted of the installation of new Cockpit Control Display Unit (CCDU), Ethernet Serial Instrumentation Remote (eSIR), low-loss cables, tri-band antennas, and instrumentation tray. The SRF instrumentation tray consisted of two transmitters: a Baseline, and an STC/LDPC Spectrally Efficient Airborne Transmitter (SEAT) as shown in Figure 7. The Baseline and SEAT transmitter signals were combined so the performance could be evaluated from the same set of antennas [3]. A third transmitter, labeled Legacy, was used to provide for aircraft time-space position information (TSPI) and command and control status. Power measurements were lower for the baseline transmitter than expected due to the absence of isolators between the baseline splitter and combiners.

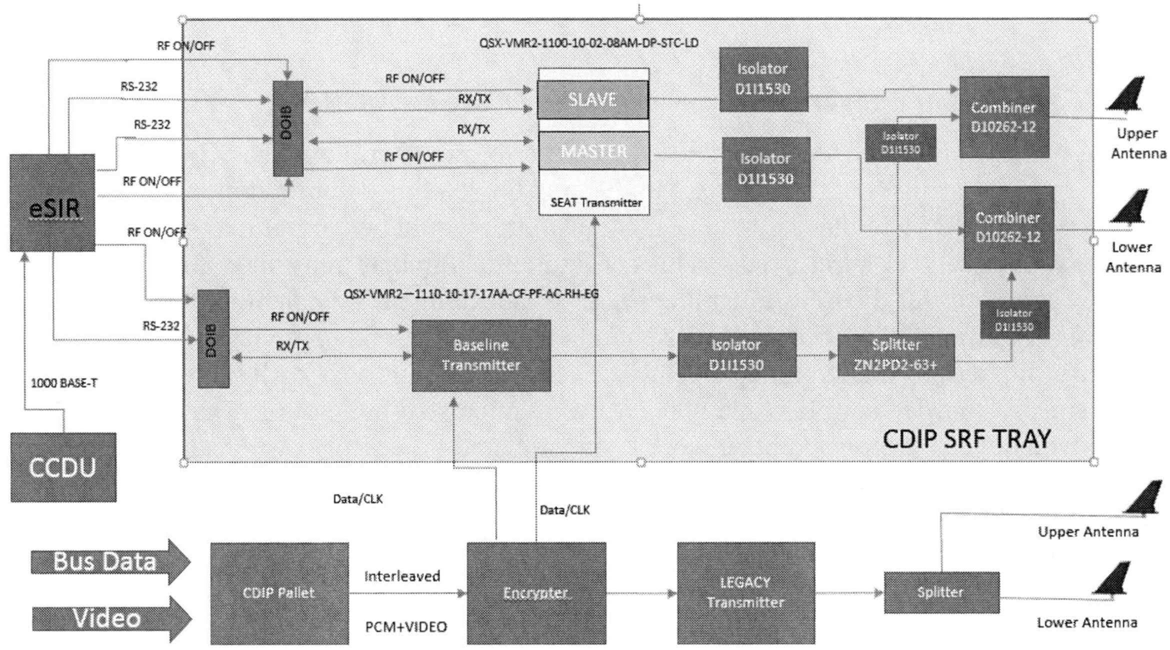

Figure 7 - SRF Instrumentation Tray

The CCDU was used to control the transmitters using a combination of serial commands and discrete inputs from the eSIR. The eSIR maintained an internal resource state for each transmitter, using periodic serial polling to guarantee that its state was accurate. The Cockpit CCDU interacted with the resource state in the eSIR via Constrained Application Protocol (CoAP) packets. Changes made to the internal resource state of the eSIR were translated into the appropriate serial commands and discrete inputs to the transmitters. To maintain simplicity of implementation and troubleshooting, CoAP payloads were limited to text strings. Resource definitions which returned multiple parameters packaged the data using JavaScript object notation (JSON). In this manner, a single CoAP GET request would return all the state data in one packet, rather than using multiple packets to query each individual resource. The master and slave transmitters were logically presented as a single transmitter in the eSIR. For the flight test, a transmitter's state was represented by the parameters in Figure 8:

Resource	Example
Name	"Legacy"
RF	"ON" or "OFF"
Frequency	"2450.5"
Bit_Rate	"10"
Modulation	"SOQPSK"
STC	"ON" or "OFF"
LDPC	"ON" or "OFF"
Differential_Encoding	"ON" or "OFF"
Source	"Internal" or "External"
Power_Amp	"10"
Temp	"45.4"
Fault	"FAULT"

Figure 8 – Transmitter State Representation

A simple "Config" resource was also defined to allow for predetermining configuration profiles for different combinations of transmitter parameters. By sending a configuration number to the eSIR via a CoAP POST to that resource, each transmitter would be configured with all the parameters required for a specific test. This configuration number, transmit enable (RF), and the frequency selections for the transmitters represented the parameters available for command from the CCDU. Multiple CCDUs could be used simultaneously with no complication, as each would query the eSIR for information independently. The eSIR, acting as a CoAP server, did not need to maintain any knowledge of how many clients were interested in its state. The nature of CoAP as a lightweight UDP protocol kept network bandwidth and processing overhead to a minimum, with no noticeable degradation in performance as multiple CCDUs were added to the network switch. Using this control system, the transmitters were configured to generate internal pseudorandom (PRN)-15 data, as well as transmit aircraft data. When transmitting PRN-15 data, a direct single bit comparison of STC and STC/LDPC could be accomplished between the legacy and SRF transmitter. Both the transmitters under test were provided with variable power and max power of 20 Watts. Bit rates of 10, 20, 30, and 40 Mbps were tested.

FLIGHT TEST AND REAL TIME DATA

A bit error analysis software application was developed for use with both real time and post mission data reduction. Transmitter performance as a bit error rate over one second was blended with TSPI. Flight testing included the use of one receiving site on the Eglin AFB land range, one on the flightline and one 100 nautical miles to the east. Each receiving site simultaneously provided two PRN-15 pattern signals, one from the Baseline transmitter and from the SEAT. The real-time TSPI was provided from either instrumentation on the jet, from the legacy stream, or and Advance Range Data System II (ARDS II) pod. As shown in Figure 9, all six telemetry streams throughout a 4g right-hand turn, each ribbons indicates the bit error rate (BER) over a second. White vectors are shown indicating range and bearing to each telemetry site. Coloring of the ribbon trails were configured for a yellow, at a BER of 10^{-7}, and red, for a BER of 10^{-6}.

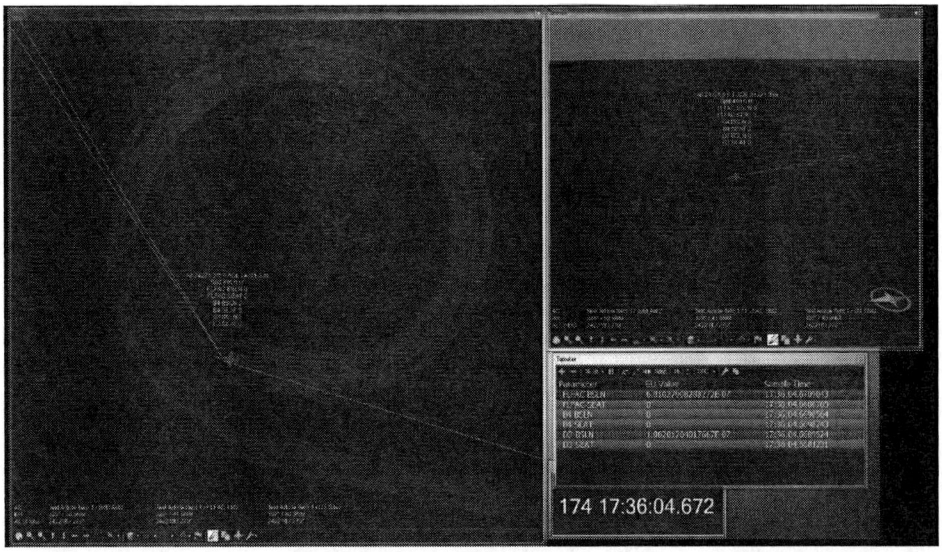

Figure 9 - NetView Bit Error Rate (BER) versus time-space position information (TSPI)

Each receiver site was instrumented with an IRIG 106-17 Chapter 10 [4] recorder capturing combiner outputs from the baseline, SEAT, and TSPI on the legacy channel. Spectrum analyzers and a camera were installed to capture the channel power measurements and front panel display from each receiver. To demonstrate the utility of STC, the power was significantly reduced by 11 dB from operational levels so that the uuencoded SOQPSK-TG would exhibit errors. Conceptually this can be viewed as a circle that approaches the EIRP as link margin is consumed in free-space path loss as shown in Figure 10. Additionally, testing was performed with both systems, Baseline and SEAT, at full power. The flight plan included left and right orbits at 60 and 100 NM with aileron rolls in sets of three to accommodate both test points.

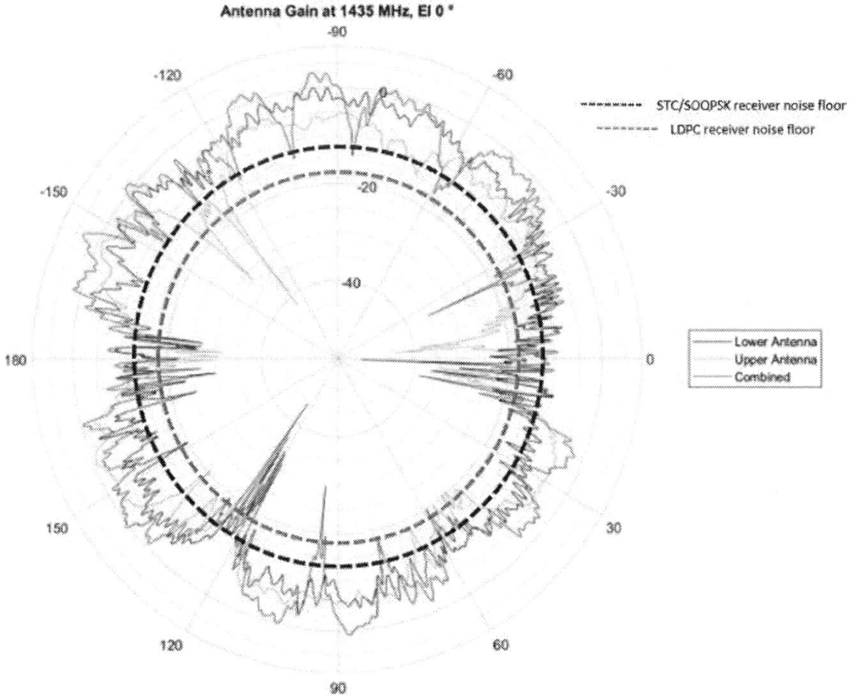

Figure 10 - Arbitrary Rx sensitivity floors of STC and LDPC

FLIGHT TEST RESULTS

In general, improvements were seen while using STC with occasional unexpected anomalies. For the purposes of this paper, we will discuss performance of the SEAT configured for STC and Baseline configured for SOQPSK-TG at 10 Mbps configured for low power and short range. Data for the left hand orbit taken at 55 NM is shown in Figure 11. Numerical and graphical results are shown in Table 1, Figure 12 - Cumulative Errors, and Figure 13 – Bit Error Rate. Anomalous combiner performance was periodically observed on the STC receiver which degraded the overall performance in every segment of flight as shown in Figure 14 - Left Orbit 55 NM segment zoom. At the time of this paper, the cause of the anomalies was not known. However, the reduction or elimination of the anomalies would greatly improve performance. Some of these anomalies are more apparent and impactful than others.

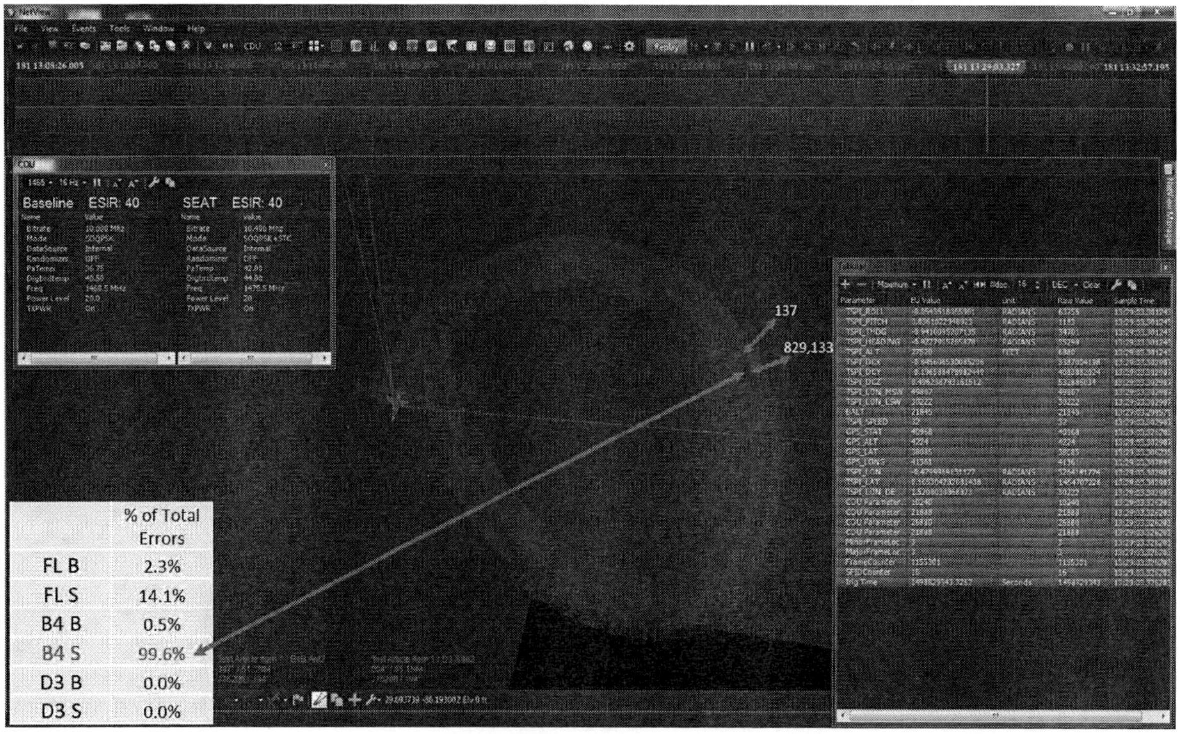

Figure 11 - Left Orbit 55 NM

Link AvailabilityL_A is given by: $L_A = \frac{T_{bits} - T_{err}}{T_{bits}}$

Where $B_t$the total are bits over an interval and B_e are the bits in error over the same interval.

The percentage of baseline errors $\%B_{err}$ given by: $\%B_{err} = \frac{Seat\ T_{err}}{Baseline\ T_{Err}}$

Table 1 - Left Hand Orbit 55 NM

	T_{Errors}	T_{Bits}	T_{BER}	$\%B_{err}$	L_A
FL Baseline	*55,412,497*	*5,908,842,880*	*9.38E0-3*		*99.0622%*
FL Seat	*8,706,875*	*5,908,833,152*	*1.47E-03*	*15.7128%*	*99.8526%*
B4 Baseline	*58,918*	*5,908,833,152*	*9.9 E-06*		*99.9990%*
B4 Seat	*831,951*	*5,908,833,152*	*1.41E-04*	*1412.05%*	*99.9859%*
D3 Baseline	*2,299,874*	*5,908,408,416*	*3.89E-04*		*99.9611%*
D3 Seat	*85,023*	*5,908,833,152*	*1.44E-04*	*3.70%*	*99.9986%*

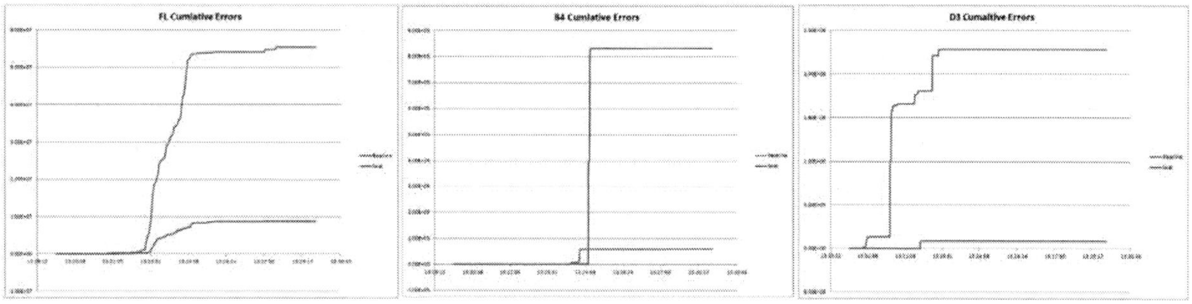

Figure 12 - Cumulative Errors

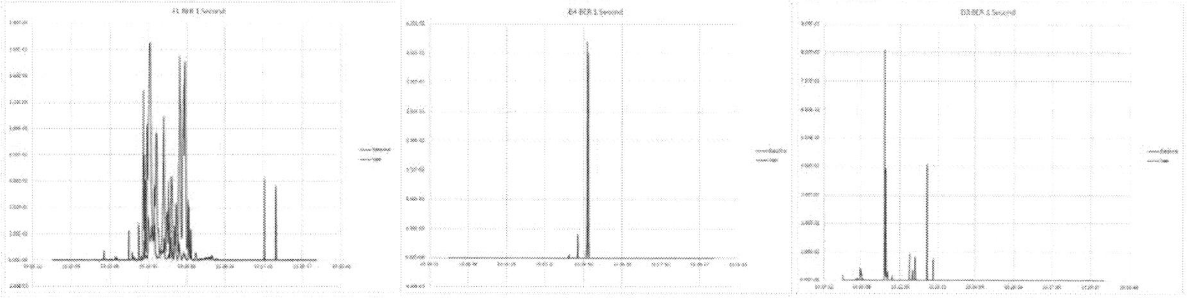

Figure 13 – Bit Error Rate

Figure 14 - Left Orbit 55 NM segment zoom

A test comparison between STC and LDPC was performed which clearly demonstrated the benefit of LDPC; see Figure 8 Arbitrary Rx sensitivity floors of STC and LDPC. Dedicated testing of LDPC was not accomplished due to other higher priority testing requirements supported by the test aircraft. Additional testing is planned as resources become available.

CONCLUSION

In this paper, we identified methods to measure the latency of a STC/LDPC coded system, accounting for both transmit and receive components. Demonstrated the measurement of an STC channel power providing comparison with a dual antenna system. Developed a TSPI/BER tool for which ranges can use free of charge to evaluate dynamic link performance in real-time and post mission. Validated that benefit of STC is heavily dependent on the airframe and antenna placement. Revealed from J-PRIMES plots that it was evident that the 'dual antenna' problem may be addressed with power offsets between the two antennas. The LDPC showed positive results; additional testing is required to assess the degree of benefit in range operations.

REFERENCES

[1] M. Jensen, M. Rice, and A. Anderson. Aeronautical telemetry using multiple-antenna transmitters. *IEEE Transactions on Aerospace and Electronic Systems*, 43(1):262–272, Jan 2007.

[2] M. Rice, "Space-Time Coding for Aeronautical Telemetry: Part I – System Description," in *Proceedings of the International Telemetering Conference*, Las Vegas, NV, October 2011.

[3] M. Rice and K. Temple. Space-time coding for aeronautical telemetry: Part II — Experimental Results in *Proceedings of the ITC*, Las Vegas, NV, October 2011.

[4] Range Commanders Council Telemetry Group, Range Commanders Council, White Sands Missile Range, New Mexico, IRIG Standard 106-17: Telemetry Standards, 2017. (Available on-line at http://www.wsmr.army.mil/RCCSite/Pages/default.aspx).

Telemetry Range Support Aircraft (TRSA) Program: Providing the Navy with Next Generation TM and Range Support

J. Kyle Roudebush, Jose Hernandez, Scott Kujiraoka and Kenneth Sanchez
NAVAIR-Point Mugu

ABSTRACT

The aircraft currently being used to support the Navy's mission of Telemetry Reception, Range Safety as well as Range Surveillance/Clearance are reaching the end of their useful life. As a result, there are ongoing efforts to procure a new aircraft and integrate these mission systems in order to continue the support of critical Naval Test Range operations. This paper will detail the current efforts being undertaken to upgrade a Gulfstream 550 to perform Range Support missions for the Multi-service Government Test Ranges.

KEY WORDS

Airborne Telemetry, Range Support Aircraft, Airborne Command Transmitters

BACKGROUND

NAVAIR relies on different kinds of Research, Development, Test and Evaluation (RDT&E) aircraft including Range Support Aircraft (RSA) for all users of NAVAIR ranges. At the Naval Air Warfare Center Weapons Division (NAWCWD), the RSA are considered key pieces of NAVAIRs RDT&E infrastructure for test items on the vast open-air/sea ranges at China Lake and Point Mugu, California.

Weapons and systems under test at these ranges often travel beyond the line-of-sight of ground-based instrumentation, therefore there is a need for airborne platforms such as the RSA to provide over-the-horizon range safety and clearance, telemetry tracking, relay and recording, missile destruct (if necessary), photometric tracking and recording, and other range management and support functions while far away from land-based systems. As a result, these aircraft have been an integral part of the ranges since their establishment in the early 1940s.

From the 1950s until today, there have been numerous types, models and series of RSA that have been developed, evolved, modified, and then stricken when the airframe's useful life was expended. In the early 1980s NAVAIR developed eight

NP-3D aircraft for use in the RSA role. These aircraft were re-worked versions of the original P-3A airframe. They were extensively modified for the Sea Range support mission. Three of these had sophisticated "billboard-type" telemetry antennas bolted onto the tail of the aircraft. In addition to telemetry systems, the aircraft included the ability to monitor communications, clear and monitor range safety hazards, control range boundaries, and coordinate all of the entities necessary for a live flight test event.

The current NP-3D billboard antenna systems operate only in the S-band frequency range for telemetry, are more than 30-years old, and both the antenna and the controlling computer system are degrading to the point where they will be inoperable in the near future. The single remaining billboard NP-3D, after many depot cycles, will soon reach the end of its airframe service life and is becoming critically unsupportable.

This unique airborne RDT &E system is the only long-range airborne system capable of fully supporting Navy, Air Force, Army and other Department of Defense (DoD) and non-DoD programs in the gathering and analysis of long-range telemetry data. Furthermore, frequency consolidation by the Federal Communications Commission (FCC) and the increased complexity and scope in testing for major programs of record require a significant growth in telemetry capability over the current limited S-band system.

Retaining the ability to collect decision-quality test data over long periods of time and distances, and ensuring the safety of aircraft systems, operators and the range is a critical piece in sustaining the range infrastructure. DoD and Navy strategic and tactical weapons systems continue to require test points beyond the line-of-sight of range control rooms – especially on vast sea ranges and over the open ocean. The ability to conduct range safety functions and hazard pattern control while airborne and away from DoD ranges (such as hypersonic weapons) is also a requirement.

NEXT GENERATION RSA

A collaborative effort has been ongoing since 2006 to ensure all requirements for a replacement RSA are established. The acquisition team consists of AIR-5.0 Test & Evaluation leadership, AIR-2.0 Contracts, AIR-5.2 Ranges and AIR-5.1 test squadrons for mission equipment and airframe expertise, AIR-6.0 Logistics for sustainment, and Tactical Airlift Program Office (PMA-207) as the designated program manager for Commercially Derived Aircraft (CDA).–This project to develop a new RSA is divided into two phases: Phase I is for the procurement of a replacement RSA CDA airframe with common avionics and related range safety systems under PMA-207, and Phase II will provide the RSA CDA aircraft as government furnished property to a company chosen through a full and open competition. This company will develop a tri-band (L, S and C) telemetry system, and procure and integrate the telemetry system along with additional mission equipment into the RSA CDA airframe. The telemetry effort will be paid for with Central Test and Evaluation Investment Program (CTEIP) funding via the Commercial Based aircraft Instrumentation and Telemetry System (CBITS) project.

CBITS originated as the Next Generation Range Support Aircraft (NGRSA) in FY2005. NGRSA transitioned into a T&E/S&T-managed risk reduction effort consisting of two projects, Airborne Telemetry Phased Array (AirPA) and Universal Beamforming Technology (UBT), both successfully completed in FY2015. Based on the successful progress of both T&E/S&T projects during FY2014, the Navy finalized the acquisition strategy for procurement of a Gulfstream G550 as the Telemetry Range Support Aircraft (TRSA) platform and CTEIP initiated the CBITS project into Phase 0.
CBITS will provide L-, S- and C-band multi-target telemetry tracking capability for a FY18 build and a FY21 Initial Operational Capability (IOC). The G550 AEW airframe was selected based on its size, weight, power, and cooling properties. It has the capacity to host robust telemetry systems and antennas – up to a total area of 125 square feet. This new, modern, phased-array telemetry system will have the capability to support major programs in complex, robust and dynamic test environments for the next 25 years.

CURRENT STATUS

In FY2016, PMA 207 awarded the contract to procure the G550 Airborne Early Warning (AEW) aircraft with delivery of a “green” aircraft planned in the 4th Quarter FY2018. Navy PMA 207 also approved the acquisition strategy for the Phase II CBITS development and integration of all other aircraft mission systems with subsequent CTEIP CBITS milestone I approval to enter design. In October 2016 the CBITS Phase II Request for Proposal was released to industry.

RSA FEATURES

In FY2017, PMA 207 and the CBITS team completed Phase II source selection and the contract award is planned in the August 2017 timeframe. TRSA Initial Operational Capability (IOC) is scheduled for the end of FY2021.

The value of range data for RDT&E customers lies in the fact that it provides critical technical and performance data for systems under test – information such as velocities, accelerations, engine temperatures, fuel status, engine turbine speeds, and outer temperature – and enables real-time analysis and decision making during complex test events. In addition, quality telemetry data supports rapid post-test failure analysis and critical milestone decisions. In order for the program to move forward and make the best time/money/performance-based decisions, it's hard to beat valuable decision-quality data from airborne telemetry.

Figure 1: Typical Test Scenario

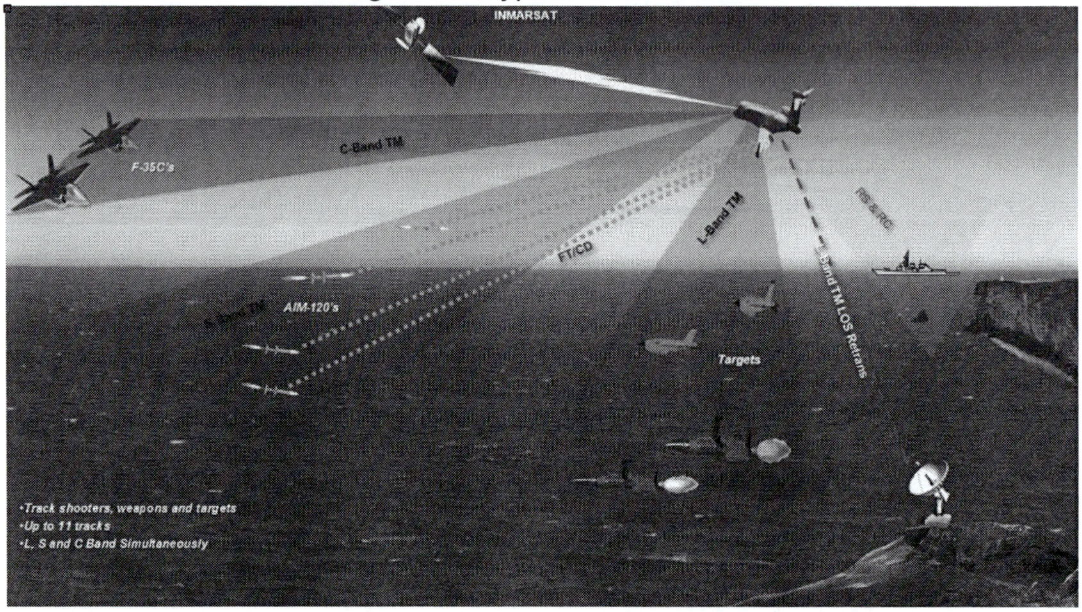

As shown in Figure 1, test scenarios typically require simultaneous tracking of multiple assets in L-, S and C-bands to support complex and robust test scenarios. Telemetry now requires augmenting the existing L- and S-band spectrum with C-band for launch platforms and weapons in order to test long-range and supersonic/hypersonic threats beyond the line-of-sight. Figure 2 is an artistic rendition of what the TRSA will look like when completed.

Figure 2: Artistic Rendition of TRSA

SUMMARY

The TRSA will eventually become the much needed replacement for the existing NP-3D aircraft at NAWCWD. This paper provided the background and status for the establishment of the CBITS and TRSA Programs.

AUTHOR INDEX

AUTHOR INDEX